Advances in Natural and Technological Hazards Research

Volume 52

The book series entitled *Advances in Natural and Technological Hazards* is dedicated to serving the growing community of scholars, practitioners and policy makers concerned with the different scientific, socio-economic and political aspects of natural and technological hazards.

The series aims to provide rapid, refereed publications of topical contributions about recent advances in natural and technological hazards research. Each volume is a thorough treatment of a specific topic of importance for proper management and mitigation practices and will shed light on the fundamental and applied aspects of natural and technological hazards.

Comments or suggestions for future volumes are welcomed.

Gopal Krishna Panda · Rajib Shaw ·
Subodh Chandra Pal · Uday Chatterjee · Asish Saha
Editors

Landslide: Susceptibility, Risk Assessment and Sustainability

Application of Geostatistical and Geospatial Modeling

Springer

Editors
Gopal Krishna Panda
Department of Geography
KISS Deemed University
School of Tribal Resource Development
Bhubaneswar, India

Subodh Chandra Pal
Department of Geography
The University of Burdwan
Bardhaman, India

Asish Saha
Department of Geography
The University of Burdwan
Bardhaman, India

Rajib Shaw
Graduate School of Media and Governance
Keio University Shonan Fujisawa
Fujisawa, Japan

Uday Chatterjee
Department of Geography
Bhatter College
Vidyasagar University
Kharagpur, Paschim Medinipore, India

ISSN 1878-9897 ISSN 2213-6959 (electronic)
Advances in Natural and Technological Hazards Research
ISBN 978-3-031-56590-8 ISBN 978-3-031-56591-5 (eBook)
https://doi.org/10.1007/978-3-031-56591-5

This Springer imprint is published by the registered company Springer Nature Switzerland AG
The registered company address is: Gewerbestrasse 11, 6330 Cham, Switzerland

Paper in this product is recyclable.

Dedicated to

Young Scholars in the Field of Environmental Science, Hazard and Disaster Management, Geospatial Science, Geography, Social Science and Sustainability Science and Policy Makers

Foreword

Landslide is one of the major natural disasters in the world and it causes extensive damages to life, infrastructure, properties and environment. In the last two decades, nearly 4.8 million people affected by landslides and more than 18000 people died all over the world. The frequency of landslide occurrences is significant in the tropical and sub-tropical countries, especially due to heavy rainfall and flooding. The anthropogenic activities are the major causes of landslides in the mountainous terrain. Proper planning and preventive measures need to be taken before implementing any developmental activities in the hilly terrain. In the near future, the landslide occurrences are expected to increase due to climate change and associated incidents.

In this background, the geo-statistical and geo-spatial techniques play a vital role in mapping landslide susceptibility zone, assessment of risk, vulnerability and management of landslides. It gives me an immense pleasure to write a foreword note to an edited book entitled "Landslide: Susceptibility, Risk Assessment and Sustainability-Application of Geostatistical and Geospatial Modeling." The book is brought out by the editors Dr. Gopal Krishna Panda, Dr. Rajib Shaw, Dr. Subodh Chandra Pal, Dr. Uday Chatterjee and Mr. Asish Saha through Springer publication.

The editors are well experienced and expertise in their respective field, made significant contributions to the natural disaster risk assessment and management.

The book is well conceived and incorporated 27 chapters under two broad headings namely Landslide Hazard Susceptibility and Sustainability. The editors have given much importance to include articles in diversified components such as landslide risk, vulnerability, assessment procedures like Frequency ratio, Artificial Intelligence, AHP and impact of climate change on landslide hazard. The book covers major case studies from various countries including highly vulnerable parts of India, Japan, Indonesia, Bangladesh and Iran. I extend my appreciation to the editors for their teamwork and brought a comprehensive book for landslide. This book is very useful to the Postgraduate students, Research scholars, Planers, Policies makers, NGOs, Experts and Geoscientists who are all involved in Natural disaster studies.

Prof. Dr. S. Anbazhagan
Senior Professor and Director
Centre for Geoinformatics
and Planetary Studies
Professor and former Head
Department of Geology
Periyar University
Salem, India

Former Vice Chancellor
Dr. J. Jayalalithaa University
Nagapattinam, India

Preface

Landslides are the major natural hazards of the mountainous regions of the world which often turn into disasters resulting in colossal loss of life and property. Their devastating effects are felt on the economy of the people, their infrastructure and the environment as well. Recent phenomena of climate change are adding a new dimension to the landslide hazards with increasing magnitude, frequency and geographic spread over the areas where these hazards had been unnoticed. Besides the natural causes of landslides, human occupancy over the mountain slopes and valleys, development of infrastructure and communication networks in the hilly terrains are often adding new disaster risks over the existing situations. Social impacts of the landslides are also visible in the form of displacement of people, disruption in the health care system, psychological stress and access to education. It has also been observed that the adverse impacts and global toll from the landslides are the heaviest for the developing countries. Studies on the database of the global distribution of landslides reveal that most of these hazards are found in the South and South East Asia, the Himalayan region and in the South and Central Americas. The present book is an attempt toward understanding these hazards, their complexities of origin, adverse impacts and their management using the advanced tools of science and technology and more specifically the methods and approaches of geo-statistical and geo-spatial analysis and modeling.

This book has been an attempt to capture the nature and characteristics of different type of landslides from diverse geographical setups, their susceptibility, varied causes of origin and effects on the human habitat as well as the challenges to develop effective modalities and strategies for prevention, mitigation and disaster risk reduction for sustainable development. The studies incorporated in the book also highlight the need to develop cooperation and convergence among the different stakeholders and work together toward developing an early warning system, improving land use planning, proactive governance, community participation and policy intervention to reduce the risk of landslides. The book has come out unique in demonstrating the application of advanced tools of geo-statistical and geo-spatial modeling, computer-based simulations, machine learning and artificial intelligence in assessing and managing the landslide hazards. The case studies presented in the book indicate the complex nature of the problem and the significance of the landslide inventory maps prepared on real-time basis using remote sensing data and geo-informatics in risk assessment and evaluation of landslide susceptibility zones.

This book comprises 27 chapters divided into two sections, i.e., (I) Landslide Hazard Susceptibility and (II) Sustainability. The book gives an extensive coverage of the landslide problems and its management prevailing across the Himalayas, Indonesian hills, Japanese Archipelago and the central hills of Iran. Each chapter presents a case study using the advanced tools of stochastic simulation, geo-spatial applications, statistical analysis and policy interventions in landslide management and disaster risk reduction. The papers highlight various approaches of how to undertake study and research on landslide hazard assessment and susceptibility and their triggering factors at the local and regional level. Some of the chapters in the book have attempted the application of different methods of physical and social vulnerability analysis and risk assessment, factors triggering the propensity to the landslide hazards and policy interventions aiming toward stabilization and community resilience in a holistic manner. Several chapters have indicated the use and application of slope safety and community involvement, adoption of geotechnical standards in slope stabilization, early warning system and emergency response, database management of landslide events and risk mitigation measures using advanced tools of geophysical and social engineering to strengthen the knowledge base and application skills of the practitioners, policymakers, researchers and academicians.

We feel that the book has a tremendous promise for use in teaching novices and disaster managers and a source of reference for the researchers about a variety of scientific approaches, social principles and policy guidelines on landslide susceptibility and risk reduction for sustainability. The book as a whole is expected to not only give a contemporary documentation of a rapidly increasing and important field of research in natural hazard and disaster management, but also present many exciting strategies to help in establishing a productive framework for future practices in landslide risk management. It is specifically designed to serve the community of undergraduates, postgraduates, researchers, policymakers, environmental scientists, disasters managers, NGOs, social scientist as well as public and private sector enterprises and academicians from the field of Environmental Sciences, Geography, Hydrology, Natural Hazards, Remote Sensing and GIS, Forestry, Soil Science, Geology, Social Sciences, etc. in the study of landslide susceptibility and sustainability.

Bhubaneswar, India — Gopal Krishna Panda
Fujisawa, Japan — Rajib Shaw
Bardhaman, India — Subodh Chandra Pal
Kharagpur, Paschim Medinipore, India — Uday Chatterjee
Bardhaman, India — Asish Saha

Acknowledgments

We extend our profound gratitude to the contributors, colleagues, publishers, researchers and readers who played a vital role in the development and publication of "Landslide: Susceptibility, Risk Assessment and Sustainability: Application of Geostatistical and Geospatial Modeling." Their expertise, support, guidance and unwavering dedication significantly influenced the creation of this comprehensive resource addressing the deeper understanding of the challenges posed by landslides and the imperative for sustainable solutions. Special recognition is due to our families for their unwavering support, understanding and encouragement during the demanding process of producing this resource. The dedication of readers to sustainability aligns with our vision, and we acknowledge and value their commitment to positive change. We also express our gratitude to the publishing team at Springer for their exceptional guidance, expertise and commitment to excellence. Their meticulous editing, design and production efforts have transformed the book into a thorough and easily accessible resource for readers.

Bhubaneswar, India — Gopal Krishna Panda
Fujisawa, Japan — Rajib Shaw
Bardhaman, India — Subodh Chandra Pal
Kharagpur, Paschim Medinipore, India — Uday Chatterjee
Bardhaman, India — Asish Saha

Acknowledgement

[illegible]

Contents

Editors and Contributors

About the Editors

Dr. Gopal Krishna Panda is Emeritus Professor at the Kalinga Institute of Social Sciences (Deemed University) and former Professor and Head of the Department of Geography at Utkal University, Bhubaneswar, Odisha, India. He has been involved in teaching and research in the area of Coastal Geomorphology, natural hazards and disaster management, climate change and environment focusing on program evaluation, capacity building and value addition. He was awarded Emeritus Fellowship of University Grants Commission in 2016 and Senior Research Fellowship of Indian Council of Social Science Research, New Delhi in 2019 for his significant contribution to the field of geography. He was the Gold Medalist of his batch in post-graduation from Utkal University in the year 1997. He was the visiting fellow of the Liverpool School of Tropical Medicine, Liverpool University availing British Council Fellowship in 1995. He was the Principal Investigation for a major research project sponsored by the UNDP and OSDMA to prepare the "Odisha State Natural Hazards and Disaster Vulnerability Analysis and Risk Assessment Report" for the state of Odisha. He was one of the advisory board members of the World Bank Supported ICZM project of the Government of Odisha. He has written and edited a number of books of which important ones are "Indigenous Knowledge and Disaster Risk Reduction: Insight Towards Perception, Response, Adaptation and Sustainability" of Springer

Nature, "Indigenous People and Nature: Insights for Social, Ecological, and Technological Sustainability" of Elsevier, Eastern Ghats of Odisha: Environment, Resources and Development of Odisha Environmental Society, Natural Hazards and Disaster Management-Concept and Approaches by Artech Publishers and Recent Advances in Application of Geospatial Technology in Natural Resource Management of Rawat Publications, New Delhi, He has worked for many natural and international agencies like WHO, UNDP, DFID and public sector organizations like OSPCB, OSDMA of the state.

Dr. Rajib Shaw is a professor at the Graduate School of Media and Governance at Keio University, Japan. He is also the Senior Fellow of Institute of Global Environmental Strategies (IGES) Japan and the Chairperson of SEEDS Asia and CWS Japan, two Japanese NGOs. Earlier, he was the Executive Director of the Integrated Research on Disaster Risk (IRDR) and was a Professor at Kyoto University. His expertise includes disaster governance, community-based disaster risk management, climate change adaptation, urban risk management, and disaster and environmental education. Professor Shaw is the Chair of the United Nations Science Technology Advisory Group (STAG) for disaster risk reduction; and also, the Co-chair of the Asia Science Technology Academic Advisory Group (ASTAAG). He is also the CLA (Coordinating Lead Author) for Asia chapter of IPCC's 6th Assessment Report. He is the editor-in-chief of the Elsevier's journal "Progress in Disaster Science," and series editor of a Springer book series on disaster risk reduction. Prof. Shaw has published more than 45 books and over 300 academic papers and book chapters.

Dr. Subodh Chandra Pal Assistant Professor has been working in the Postgraduate Department of Geography, The University of Burdwan, Purba Bardhaman, West Bengal, India. He received his Ph.D. in Geography from Visva-Bharati (Central University in India) in 2015. His main fields of expertise are fluvial geomorphology, pedo-geomorphology, land degradation, soil erosion, gully erosion, landslide vulnerability, climate change, water resources monitoring and management and published nearly two hundred international and national research articles in various renowned journals (Springer, Elsevier, Taylor & Francis group, Wiley, MDPI and Nature Publishing groups.). He published one author book from Springer Nature publishing house on Climate Change Impact on Soil Erosion in Subtropical Environment: Application of Empirical and Semi-empirical Models. He is the Editor of four running Edited books be published by Elsevier and Springer Publishing house. His research work has been funded by the University Grants Commission (UGC), Government of India and Department of Science and Technology (DST), Government of India. Dr. Pal has served as a reviewer for many International reputed journals. He is an Editorial board member of MDPI Water journal and Frontiers of Earth Science journal. Since 2022, Stanford University has recognized Dr. Pal as one of the world's top 2% scientists.

Dr. Uday Chatterjee is an Assistant Professor at the Department of Geography, Bhatter College, Dantan, Paschim Medinipur, West Bengal, India and an Applied Geographer with a Doctoral Degree in Applied Geography at Ravenshaw University, Cuttack, Odisha, India. His areas of research interest cover Urban Planning, Social and Human geography, Applied Geomorphology, Hazards and Disasters, Environmental Issues, disaster governance, community-based disaster risk management, climate change adaptation, urban risk management and disaster. He has delivered 10 invited lectures in University Grants Commissions (UGC) sponsored national seminars and various academic departments of different colleges in India. In addition, he presented 20 papers in national and international seminars/conferences held in India as well as chaired and co-chaired more than 5 technical sessions.

He has successfully guided project dissertations to undergraduate students. He has also conducted (Convener) one Faculty Development Programme on "Modern methods of teaching and advanced research methods" sponsored by the Indian Council of Social Science Research (ICSSR), Government of India. Currently, Dr. Uday Chatterjee has completed the Special Issue (S.I) of Urbanism, Smart Cities and Modelling, Geojournal, Springer as a Lead Editor, book series editor Development in Environmental Science, Elsevier (https://www.elsevier.com/books-and-journals/book-series/developments-in-environmental-science). His research work has been funded by the West Bengal Pollution Control Board (WBPCB) Government of West Bengal, India. He has served as a reviewer for many International journals. Currently, Dr. Chatterjee doing the international project, in collaboration with Indonesia, Malaysia and Japan funded by APN (Asia Pacific-Global Change Research) and he is the guest editor of the special issue Social Ecology, Human-Well-being and Sustainability, Global Social Welfare journal, Springer. He has published 30 research papers, 10 edited books, 2 author books, 15 book chapters and 2 conference proceedings.

Asish Saha is currently engaging as a Ph.D. Research Scholar in the Department of Geography, The University of Burdwan, Bardhaman, West Bengal, India. He received his Master's degree in Geography from The University of Burdwan and Post Graduate Diploma in Geoinformatics from the Department of Science and Technology, Government of West Bengal, in collaboration with Maulana Abul Kalam Azad University of Technology, Kolkata. His fields of expertise are Climate Change, Climate Variability, Regional Climate Modeling, Hydro-climatic extremes, Water Resources Management, Water Quality, Hazard and Risk Assessment and Statistical Modeling. He has published more than eighty international and national research articles in various renowned journals like Springer, Elsevier, Willey and Taylor & Francis group. His research work has been funded by the University Grants Commission (UGC), India.

Contributors

Atif Ahamad Department of Geology, Aligarh Muslim University, Aligarh, India

Wali Akhtar Department of Geology, Aligarh Muslim University, Aligarh, India

Md. Shakib Al Fahad Department of Geography and Environment, University of Dhaka, Dhaka, Bangladesh

Abdullah Al Rakib Department of Urban & Regional Planning, Rajshahi University of Engineering & Technology (RUET), Rajshahi, Bangladesh

Sazzad Al Toufiq Shuvo Department of Urban & Regional Planning, Rajshahi University of Engineering & Technology (RUET), Rajshahi, Bangladesh

Ershad Ali Department of Geography, Ananda Chandra College, Jalpaiguri, West Bengal, India

Anjuman Ara Department of Geography and Environmental Studies, Faculty of Biological Sciences, University of Chittagong, Chittagong, Bangladesh; Department of the Built Environment, Faculty of Engineering and Science, Aalborg University, Aalborg East, Denmark

Putu Aryastana Department of Civil Engineering, Warmadewa University, Denpasar, Indonesia

Mahdi Yousefi Nejad Attari Department of Industrial Engineering, Bonab Branch, Islamic Azad University, Bonab, Iran

Vibhore Bakshi Department of Urban and Regional Planning, School of Planning and Architecture, Bhopal, India

Rowdra Dip Chackroborty Department of Urban & Regional Planning, Rajshahi University of Engineering & Technology (RUET), Rajshahi, Bangladesh

Shovan Lal Chattoraj Geosciences Department, Indian Institute of Remote Sensing, Dehradun, India

Ishika Chowdhury Department of Earth Sciences and Remote Sensing, JIS University, Agarpara, West Bengal, India

P Danuta Mohan Disaster Management Studies Department, Indian Institute of Remote Sensing, Dehradun, India

Payel Das Department of Geography, Rabindra Bharati University, Kolkata, India

Fahime Lotfian Delouyi Department of Mechanical Engineering, Faculty of Engineering, University of Zabol, Zabol, Iran

Listya Dewi Master Program of Infrastructure and Environmental Engineering, Postgraduate Program, Warmadewa University, Denpasar, Indonesia

Nayan Dey Department of Geography, Pandit Ravishankar Shukla University, Raipur, India

Himisha Dixit School of Life Sciences, Central University of Himachal Pradesh, Dharamshala, HP, India

Bhagyasree Dolui Department of Geography, Queen Mary's College, Chennai, Tamil Nadu, India

Babak Ejlaly Department of Industrial Engineering, Malek Ashtar University of Technology, Tehran, Iran

Poulomi Ganguli Agricultural and Food Engineering Department, Indian Institute of Technology Kharagpur, Kharagpur, West Bengal, India

Asutosh Goswami Department of Earth Sciences and Remote Sensing, JIS University, Agarpara, West Bengal, India

MD. Toufiq Hossain Department of Urban & Regional Planning, Rajshahi University of Engineering & Technology (RUET), Rajshahi, Bangladesh

Md. Sakhawat Hossen Department of Geography and Environment, University of Dhaka, Dhaka, Bangladesh

Md. Humayain Kabir Institute of Forestry and Environmental Sciences, University of Chittagong, Chittagong, Bangladesh;
Wegener Center for Climate and Global Change, University of Graz, Graz, Austria;
Department for Knowledge and Communication Management, University for Continuing Education Krems, Krems an der Donau, Austria

Labib Intisar Department of Urban & Regional Planning, Rajshahi University of Engineering & Technology (RUET), Rajshahi, Bangladesh

Abdulla-Al Kafy Department of Urban & Regional Planning, Rajshahi University of Engineering & Technology (RUET), Rajshahi, Bangladesh

Mitthan Lal Kansal Water Resources Development and Management, Indian Institute of Technology, Roorkee, Uttarakhand, India

Taskin Kavzoglu Departments of Geomatics Engineering, Faculty of Engineering, Gebze Technical University, Gebze, Kocaeli, Turkey

Rohan Kumar Department of Chemical and Petroleum Engineering, Lovely Professional University, Punjab, India

Priyanka Majumder Department of Earth Sciences and Remote Sensing, JIS University, Agarpara, West Bengal, India

Anusha Mondal Department of Earth Sciences and Remote Sensing, JIS University, Agarpara, West Bengal, India

Danish Monga Agricultural and Food Engineering Department, Indian Institute of Technology Kharagpur, Kharagpur, West Bengal, India

Sayani Mukherjee Department of Earth Sciences and Remote Sensing, JIS University, Agarpara, West Bengal, India

K. M. Nafee Department of Geography and Environment, University of Dhaka, Dhaka, Bangladesh

Nandakishore Amity Institute of Geoinformatics and Remote Sensing, Amity University, Noida, India

Biswajit Nath Department of Geography and Environmental Studies, Faculty of Biological Sciences, University of Chittagong, Chittagong, Bangladesh

Benita Nathania World Research Institute Indonesia, South Jakarta, Jakarta, Indonesia

Baishali Ojha Department of Geography, Rabindra Bharati University, Kolkata, India

Apurv Bhogibhai Patel Department of Urban and Regional Planning, School of Planning and Architecture, Bhopal, India

Subrata Kundu Paul Department of Geography, Ananda Chandra College, Jalpaiguri, West Bengal, India

Mohadese Hoseingoli Poorasl Department of Industrial Engineering, Bonab Branch, Islamic Azad University, Bonab, Iran

Jason Pajimola Punay Department of Physics, College of Science, Bicol University, Legazpi City, Philippines

Julia Quaicoe Department of Geography and Regional Planning, University of Cape Coast, Cape Coast, Ghana

Saswati Roy Assistant Professor, Department of Geography, Sarsuna College, Kolkata, West Bengal, India

Bipul Chandra Sarkar Department of Geography, Ananda Chandra College, Jalpaiguri, West Bengal, India

Suhel Sen Department of Geography, Lovely Professional University, Phagwara, Punjab, India;
Department of Geography, Bhairab Ganguly College, Belgharia, West Bengal, India

Hari Harjanto Setiawan National Research and Innovation Agency (BRIN), Jakarta, Indonesia

Martiwi Diah Setiawati Institute for the Advanced Study of Sustainability (UNU-IAS), United Nations University, Tokyo, Japan;
Research Center for Oceanography, National Research and Innovation Agency (BRIN), Jakarta, Indonesia

Swati Sharma Amity Institute of Geoinformatics and Remote Sensing, Amity University, Noida, India

Tariq Siddique Department of Geology, Aligarh Muslim University, Aligarh, India

I Nengah Sinarta Master Program of Infrastructure and Environmental Engineering, Postgraduate Program, Warmadewa University, Denpasar, Indonesia

Sachchidanand Singh Water Resources Development and Management, Indian Institute of Technology, Roorkee, Uttarakhand, India;
Western Himalayan Regional Center, National Institute of Hydrology Jammu, Jammu & Kashmir, India

J. S. Sudharsan School of Energy and Environment, NICMAR University, Pune, India

Evangelin Ramani Sujatha Centre for Advanced Research in Environment, School of Civil Engineering, SASTRA Deemed to be University, Thanjavur, India

Mahdi Ibrahim Tanjung Ministry of Public Works and Housing (PUPR), Bandung, Indonesia

Alihan Teke Departments of Geomatics Engineering, Faculty of Engineering, Gebze Technical University, Gebze, Kocaeli, Turkey

Md. Khayrul Islam Tuhin Department of Geography and Environment, Dhaka College, Dhaka, Bangladesh

Md. Sofi Ullah Department of Geography and Environment, University of Dhaka, Dhaka, Bangladesh

Naima Umar Assistant Professor, Department of Geography, School of Liberal Arts, Noida International University, Greater Noida, Uttar Pradesh, India

Vipin Upadhyay School of Life Sciences, Central University of Himachal Pradesh, Dharamshala, HP, India

Harsh Varshney Department of Geology, Aligarh Muslim University, Aligarh, India

Putu Ika Wahyuni Master Program of Infrastructure and Environmental Engineering, Postgraduate Program, Warmadewa University, Denpasar, Indonesia

Jackson Chang Hian Wui Preparatory Center for Science and Technology, University Malaysia Sabah (UMS), Sabah, Malaysia

R. M. Yuvaraj Department of Geography, University of Madras, Chennai, Tamil Nadu, India

Part I
Landslide Hazard Susceptibility

Chapter 1
Landslide Risk and Vulnerability; Real Issues, Thoughts and Perspectives

Julia Quaicoe

Abstract Landslides are among the most dangerous geological hazards of the world as they involve the movement of earth materials under the influence of gravity. Landslides can occur as a result of many different disturbances, but they are also strongly influenced by certain environmental factors such geology, precipitation, land cover changes and population. This Chapter sought to explore the use geostatistical and geospatial models to assess the contributions of the environmental factors to landslide risks and vulnerabilities in Ghana. Through the use of the weighted overlay method, it was found that even though precipitation was assigned the highest weight of 35%, areas which were identified as high landslide risk areas rather experienced the lowest levels of precipitation in the country. Thus, the high landslide risk was increased due to the geological setting which made the location much prone after any little amount of precipitation. The level of vulnerabilities is also on the high side due to increasing population and the resulting improper land use planning. Based on these, it was concluded that even though landslides could exert a significant multidimensional impact on a region's life, livelihoods and geomorphology whiles their occurrence could not be predicted, the use of geostatistical and geospatial modelling technologies could help integrate available datasets to be able to ascertain the individual contribution of such factors to the level of risks and vulnerabilities.

Keywords Landslide risk · Precipitation · Slope · Population · GIS

1.1 Introduction

Landslides, quite and often overlooked geological phenomena, have slowly but persistently had major environmental, economic and social impacts worldwide (Froude and Petley 2018). When soil, rock, and organic materials slide down a hill

J. Quaicoe (✉)
Department of Geography and Regional Planning, University of Cape Coast, Cape Coast, Ghana
e-mail: juliana.appiah@ucc.edu.gh

G. K. Panda et al. (eds.), *Landslide: Susceptibility, Risk Assessment and Sustainability*, Advances in Natural and Technological Hazards Research 52,
https://doi.org/10.1007/978-3-031-56591-5_1

due to gravity, it is referred to as a *landslide* (Highland and Bobrowsky 2008). Landslides can occur as a result of many different disturbances, but they are also strongly influenced by certain environmental factors such intense rains, human activity, and litho-structural propensity (Hungr et al. 2014). The form of landslide occurrence may depend on some factors including the triggering factor, angle of slope, type of material involved, velocity and its' impact (Froude and Petley 2018). Thus, some may occur suddenly while others may occur more gradually over a period of time when exposed to some of the afore mentioned conditions.

Landslides have a significant and complex history in Ghana, reflecting the country's geological diversity, climate patterns, and human activities. Ghana, as a topical country is characterized by varying geologic, topographic and climatic factors across it's sixteen regions thereby facing notable risks and varying levels of vulnerabilities that call for a closer attention (Anderson et al. 2011). Records of landslides in Ghana, though may not extend as far back as ancient civilizations, their occurrence has been noted over the years, with some key historical milestones and trends. These geological events occurred periodically in regions with hilly terrain and steep slopes, especially in the southwestern, eastern and central parts of the country (Nsengiyumva et al. 2019; Angnuureng et al. 2022). According to Campion and Venzke (2013), these early landslides were often localized and had limited impact on the population due to low levels of urbanization and infrastructure development (Campion and Venzke 2013).

During this era, Ghana was primarily agrarian, with a significant portion of the population engaged in subsistence farming (Boateng 2017). Communities were largely rural and dispersed across the landscape, often located in proximity to fertile valleys and riverbanks. According to Tacoli (2017), the low population density and limited urbanization meant that landslides, when they did occur, typically affected only localized areas. The primary impacts of these early landslides were felt by individual farming communities or isolated settlements, where they might lead to crop damage, the displacement of a few families, or temporary disruption of local trade routes. However, due to the limited extent of infrastructure development, such as roads and buildings, the societal impact of these landslides was relatively minimal. The population of Ghana expanded considerably after independence and as people moved from nearby rural areas in quest of economic possibilities, urban centres including Accra, Kumasi, Cape Coast, Koforidua, Sunyani, and Sekondi-Takoradi grew. This urbanization trend led to increased construction and land development on hillsides and slopes, which were once areas that were less densely populated. Because of this, there are now more people and infrastructure at risk of landslides (Abd Majid and Rainis 2019).

As Ghana transitioned into a less agrarian country, the interplay of population growth, urbanization, deforestation, and mining activities significantly altered the environmental dynamics. Mining activities, including illegal mining known as *galamsey* has surged, while rapid population growth and urbanization have also increased leading to environmental degradation and heightened landslide potential in mining areas such as Tarkwa Nsuaem, Amenfi East and Prestea Huni-Valley Districts (Tschakert and Singha 2007; Afriyie et al. 2016). Owusu-Ansah et al. (2019) reveal that areas around the Weija dam near Accra, the capital of Ghana where encroachers continue to explode sites with dynamites to blast rocks, weakened the base of hills putting residents at risk. These changes have significantly altered the landscape and intensified landslide risks. In Table 1.1 is a record of landslide events and the resulting effects they have had various aspects of the environment.

Landslides in Ghana have escalated both in frequency and severity, marking a worrisome trend that has brought significant challenges and impacts to various regions of the country. The findings from a study conducted by Alam (2020) argue that the anthropogenic factors contributing to the climate change-induced rise in landslide occurrences are unplanned development activities, overpopulation, habitation along hill slopes, and inadequate disaster risk mitigation measures. Once Landslides are localized and less prominent, they became increasingly critical concern, highlighting the need for proactive measures, research, and disaster management strategies to mitigate the growing risks Alaneme and Okotete (2018) which demonstrates the necessity of study, preventive actions, and disaster management plans to reduce the escalating dangers.

Whiles proactive measures such as the development of early warning systems, effective communication and community preparedness have been used to reduce the effects of such natural phenomenon in much developed countries, such efforts are still evolving in most developing countries including Ghana. Among the strategies to enhance the development of such proactive measures on a national scale as Ghana is the application of Geostatistical and Geospatial modelling techniques. According to McKinley et al. (2011), Geostatistical modelling techniques employs techniques such as kriging, simulation, and Bayesian inference to predict, model uncertainty, and optimize spatial sampling design for geomorphic properties that vary in both space and

Table 1.1 A record of Landslide in recent times in dates, location, causes and significant effects (*Source* Modified after Daily Graphic, 19th June, 2015)

Date of landslide	Location of event	Cause	Effect
1933, June 15th	Akuapim-Togo Ranges (Volta region)	Heavy downpour	No major disaster recorded
1968	Jamasi-Mampong (Ashanti region)	Heavy downpour	Road block
1972	Afram Plains (Eastern region)	Heavy downpour	Not recorded
2005, March 5th	Afram Plains (Eastern region)	Heavy downpour	Community cut-off
2010, June 20	Pantang-Mamfe (Eastern region)	Heavy downpour	Road block
2010, October 6th	Adukrom-Yelensi (Eastern region)	Heavy downpour	Three (3) lives, many property
2012, October 25th	Wassa Assikuma/Nkran (Prestea-Huni Valley, Western region)	Heavy downpour	About 20 acres of Cocoa farms, livelihoods

time. In simpler terms, in order to be able to develop strategies for managing landslide risks and vulnerabilities, a range of datasets including geologic, topographic, climatic and remote sensing data are required for integration, mapping, prediction and other multivariate analysis. And these are very well provided by Geostatistical modelling techniques and patterns can be provided. Geospatial techniques on the other hand provide the platform, techniques as well as the processes to extract and visualise the patterns and trends resulting from the use of the Geostatistical techniques.

The employment of these methods can result in management and mitigation strategies based on which effective communication, early warnings as well as education to ensure community preparedness can be centred. However, available literature on landslide research in Ghana hardly portrays works done based on such strategies. For instance, Glade (2003) and Zheng (2006) emphasise the effect of deforestation and land clearing as a major contributor to landslide susceptibility. Buah-Kwofie et al. (2019) examined how a slope unit and landslide susceptibility analysis were affected by the Digital Elevation Model (DEM) data parameters utilizing a Geographic Information System (GIS). Also, Adu-Boahen (2020) assessed residents' knowledge of mass wasting in the Weija catchment of Ghana whiles Amponsah et al. (2022) reviewed natural hazards governance in Ghana. In order to make some contributions to control landslides, Dankyi et al. (2023) identified the variables affecting rock fall and used the Modified Colorado Rock Fall Hazard Rating System to assess the slope face (MCRHRS) identified the variables affecting rock fall and used the Modified Colorado Rock Fall Hazard Rating System to assess the slope face (MCRHRS) whiles Min et al. (2023) analysed the combination of soil stability and slope analysis methods to design a retaining wall for a mudslide prevention in Accra, Ghana.

It is evident from the preceding discussions that landslide risk analysis in Ghana, apart from being localized to few places, has also not considered the possible effects and strength of the contributions of the other geomorphic factors including climate, landcover, geology and population. However, in order to be able to develop robust mitigation and management strategies, these factors should be considered in addition to robust risk and vulnerability assessment methodologies including the use of geospatial and geostatistical modelling techniques. The objectives of this Chapter therefore are: to assess the contributions of various geomorphic factors, including climate, landcover, geology, and population on landslide risk and vulnerability in Ghana; to explore the effectiveness of geostatistical and geospatial modelling in analysing landslide risk and vulnerabilities; and to formulate mitigation and management strategies based on a holistic assessment that includes the influences of climate, landcover, geology, and population.

1.2 Rational of the Study

Landslides are among the geomorphic phenomena that shape and evolve landscapes over geologic timescales. Even though these processes are essential for the rejuvenation of natural landscapes, they however pose significant risks to people and infrastructure within the environment because of their characteristics such as their uncertainty and sudden nature of occurrence as well as their high frequency of occurrence (Shinohara and Kume 2022). In developing countries such as Ghana, limited methodologies to understanding these hazardous events has meant a limited understanding of its operation within the country.

Thus, the rational of this study is to contribute to the understanding of the interactions and effects of various landslide risks and vulnerability factors through the use of Geospatial and Geostatistical modelling techniques. This can contribute to the development of evidence-based policies and public awareness creation in a season of climate change and increasing population growth.

1.3 Materials and Methods

Using the spatial analytic tools of ArcGIS 10.5 software, a spatial database for the causative elements was initially created in order to conduct landslide susceptibility analysis in the study area. The database comprises datasets pertaining to landslide inventories as well as the causative factors of landslides, including slope, aspect, land use, precipitation, and geology. The weights of these factors were later determined by assessing their association with the occurrence of landslides. The accuracy of these results was subsequently confirmed. There is a lack of accepted standards or recommendations pertaining to the selection of triggering components in various statistical methodologies employed for the purpose of landslide susceptibility mapping. According to Ayalew and Yamagishi (2005), it is important to select parameters that are both operational and measurable, taking into account the specific characteristics of a given area (Ayalew and Yamagishi 2005). The presence of a single parameter may serve as a significant determinant for the incidence of landslides in a certain region. However, it is commonly observed that the simultaneous influence of two or more causative variables contributes to the occurrence of landslides, in conjunction with the triggering factor.

Generally, the types of landslide-causing factors that were chosen were based on the study area and the data that was available. A total of five factors were chosen for this purpose: slope, aspect, geology, land use, and rainfall. The coordinate system used was WGS 1984 UTM zone 30N, and each pixel size was 30 m $\times$ 30 m. All of the maps that showed causal factors were then transformed into raster maps. This was accomplished by creating a rating system wherein numerical values are assigned to the factors and their corresponding classifications. The rasterized dataset and all the maps of the causal factors were put on top of each other, and the spatial analyst tool in ArcGIS was used to get the information. A grading system was created using survey data on related causative elements for landslides as well as expert information about landslide causes from earlier research. This was accomplished by creating a rating system wherein numerical values are assigned to the factors and their corresponding classifications. The rasterized dataset and all the maps of the causal factors were put on top of each other, and the spatial analyst tool in ArcGIS was used to get the information. A grading system was created using survey data on related causative elements for landslides as well as expert information about landslide causes from earlier research.

The factors were given a numerical ranking in this chapter according to their importance on a scale of 1–9. On a 0–9 ordinal scale, weights were also assigned to the classes of the components; a larger weight indicates a greater influence for the occurrence of landslides (Fig. 1.1).

Topographic parameters, such as slope, were obtained from the Advanced Spaceborne Thermal Emission and Reflection Radiometer (ASTER) elevation data (https://search.earthdata.nasa.gov/search), which was used to build the Digital Elevation Model (DEM), which has a cell size of 30 m by 30 m. The Modis website (https://ladsweb.modaps.eosdis.nasa.gov/) provided land use maps for download. Using rainfall data from the Ghana Meteorological Agency and the IDW interpolation technique of the spatial analyst tool in ArcGIS, the precipitation map was created from twenty-two rain gage stations inside the boundaries of the study area. The International Modelling and Mapping Centre (IMMC), UCC, provided the geology data.

According to Sadr et al. (2014), Shit et al. (2016), Wang et al. (2021), the weighted overlay method is a straightforward, sufficient, and uncomplicated approach for assessing possible landslide areas. The assessment of landslide susceptibility

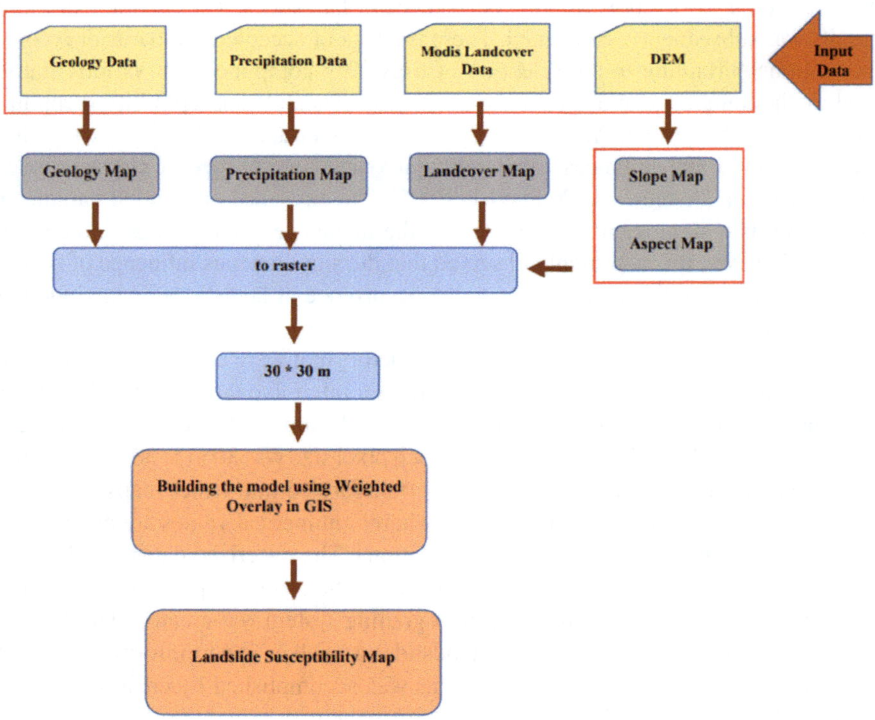

Fig. 1.1 Flowchart for landslide suscepitibility modelling

mapping in this work was done using five environmental factor maps. To run the analysis, all the cells from the map layers need to be classed into a similar preference scale, like 1–9, where 9 is the most desirable. Using the Weighted Overlay Model, all theme maps were combined.

$$s = \frac{\sum w_i s_{ij}}{\sum w_i}$$

where, S is the spatial unit value in the output map, W_i is the weight of the ith factor map, and S_{ij} is the ith spatial class weight of the ith factor map and susceptibility landslide map was created by integrating these weighted component maps into a GIS platform and applying WOM analysis.

1.4 Results and Discussion

1.4.1 Geological Characteristics of Ghana

The topography of Ghana is a defining characteristic, marked by rolling hills and rugged mountains. Ghana's geological diversity plays a role in the country's susceptibility to landslides. These geological formations possess distinct properties that can influence landslide risk at locations where they are found (Will et al. 2020). The Accraian formation, predominantly composed of sandstones and shale, characterizes areas in the southern part of Ghana, covers an area of about 11.7 km^2 in the vicinity of Accra and unconformably overlies the Dahomeyan basement complex. The series consists of quartz-grits, gentle folded sandstone, shale and mudstones. The total thickness of the series is unknown due to the extensive faulting in the area which has obscured the relationship of the various rock units. The properties of this formation, which include relatively porous and unconsolidated sandstones, make the region susceptible to landslides during heavy rainfall. The sandstone's porous nature facilitates water infiltration, leading to soil saturation and reduced cohesion leading to landslide, especially in areas with steep topography.

From Fig. 1.2, the Basal Sandstone formation is found in the Northern part of Ghana. This formation's properties include sandstones and shale. The presence of shale can promote the development of preferential sliding planes during soil saturation. Moreover, the porous nature of sandstone, combined with intense rainfall in these regions, elevates landslide risk. Birimian sediments characterize much of Ghana's geology, particularly in some parts of the Northern, the Middle and Southern parts of the country. The Birimian volcanics however, appear in patches across the country. These rocks, known for their variable properties, interact with factors like rainfall and topography. For instance, volcanic rocks can be porous, allowing for water infiltration, while the varied mineral content within these formations may lead

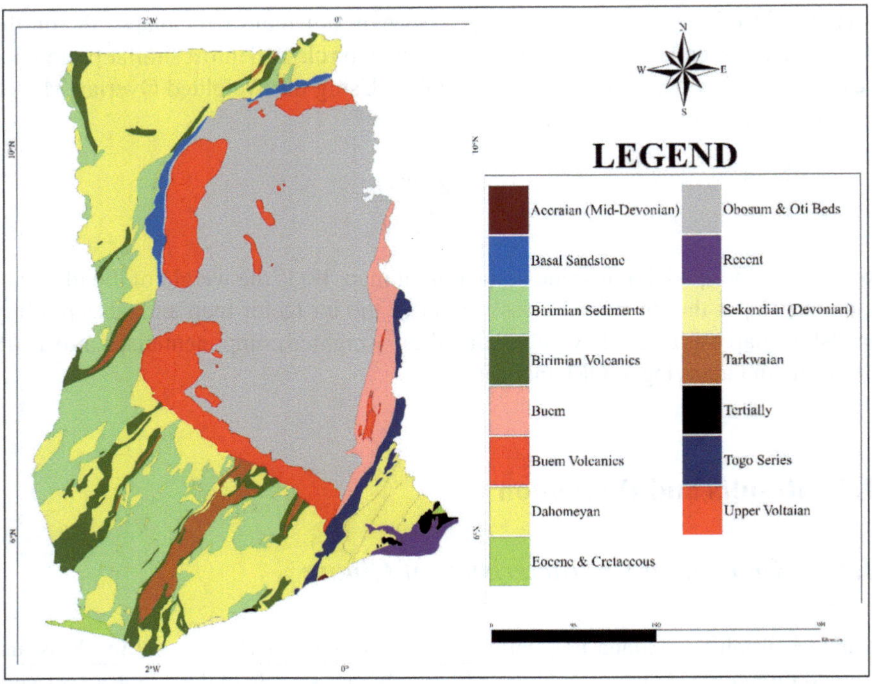

Fig. 1.2 Geological map of Ghana (*Source* Ghana Geological Service Department)

to differential weathering and erosion. These factors increase landslide vulnerability, especially in areas where topography accentuates the impact of geological properties.

The Buem formation is significant in the Volta Region, particularly in the northern and northwestern part of the Togo Series. It is composed of basaltic, andesitic, and trachytic lavas, agglomerates, tuffs, and jaspars, as well as calcareous, argillaceous, sandy, and ferruginous shales, sandstones, arkose, greywacke, and conglomerate. The presence of shale can make the region susceptible to landslides when saturation occurs. Additionally, Buem Volcanoes, composed of various volcanic rocks, have properties that can exacerbate landslide risks, especially in hilly terrain, where soil instability is more pronounced. The Dahomeyan formation lies in the southeastern part of the country from Accra to intersect the Ghana-Togo boundary near Agome in Togo. The system underlies the Accra Plains and has a total area of approximately 7,000 km^2, along with the Eocene and Cretaceous rocks. The Dahomeyan consist of hornblende and biotite gneisses, migmatites, granulites, schists, some of which are rich in garnet and a little marble. Intruded in the Dahomeyan are granites, nepheline, syenite and dykes of porphyry, aplite and dolerite.

The Obosum and Oti Beds, which contain sandstones, shale, and limestone, are found in the central and northern parts of Ghana. The presence of porous sandstones and erodible shale can enhance soil instability, while the vulnerability of limestone to dissolution during heavy rainfall elevates landslide potential. The Sekondian formation in the western region of Ghana and also found on series of disconnected outcrop between Cape Coast and Dixcove granites, and is characterized by shale and sandstone. Shale's capacity to promote sliding planes and the porous nature of sandstone contribute to the region's vulnerability to landslides during intense rainfall.

The Togo Series of rocks form a range of hills and mountains that trend northeast from the Ghana-Togo border near Palime to the Volta River between Kpong and Anum. The range extends as far west of Accra to the southwest of the Volta. Outliers of the Togo rocks, the Ho-Abutia hills and a few other tiny hills to the east of the main range are likewise made of rocks of the same age. With a few exceptions where undisturbed shales and sandstones are still visible, the alternating arenaceous and argillaceous strata that once made up the Togo Series have now given way to phyllites, schists, and quartzites. Although hornstones, jaspar sand, and hematite quartz schist are also found, quartzite, quartz-sericite, sericite-quartz schist, and phyllites are the most common types of rocks. Shale and sandstones can be found in the Upper northern Voltaian formation.

1.4.2 Precipitation Characteristics of Ghana

Precipitation plays a pivotal role in influencing landslide risk across various regions in Ghana as indicated in the history of landslides in Ghana and emphasised by Fan et al. (2020). The recorded data from the different weather stations in the Northern, Middle, and Southern belts of the country shows rainfall patterns that can either increase or decrease the potential for landslides. Figure 1.3 explores the average annual precipitation variation over the entire land area of Ghana with some areas falling within a lower range of about 756.315–1,037.172 mm, and others falling within a relatively higher range of about 1,513.782–1,841.449 mm.

In regions characterized by relatively low precipitation levels, such as Navrongo, Tamale, Bole, Yendi in the northern belt stations, the soil often tends to retain some degree of moisture and cohesion. However, during seasonal or periodic heavy rainfall events, these areas can experience increased soil saturation, which may compromise stability and elevate the risk of landslides. In the middle belt, including stations like Wenchi and Keta Karachi, the lower precipitation levels still make these areas susceptible to landslides when significant rainfall occurs. The relatively stable ground can become prone to landslides during exceptional weather events, especially if the soil is already compromised by factors like geological instability, deforestation, or human activities (Rahman et al. 2014; Shibasaki et al. 2016).

Conversely, areas with high precipitation levels, such as Kumasi, Sehwi Bekwai, Akim Oda and Axim, experience substantial rainfall as depicted on Fig. 1.3. Ng and Shi (1998) affirms that in regions where rainfall is high, the soil often becomes saturated during the rainy season, reducing its stability. The combination of frequent heavy rains and specific geological and topographical features can result in a higher susceptibility to landslides (Fan et al. 2017). Comparatively, areas with higher rainfall

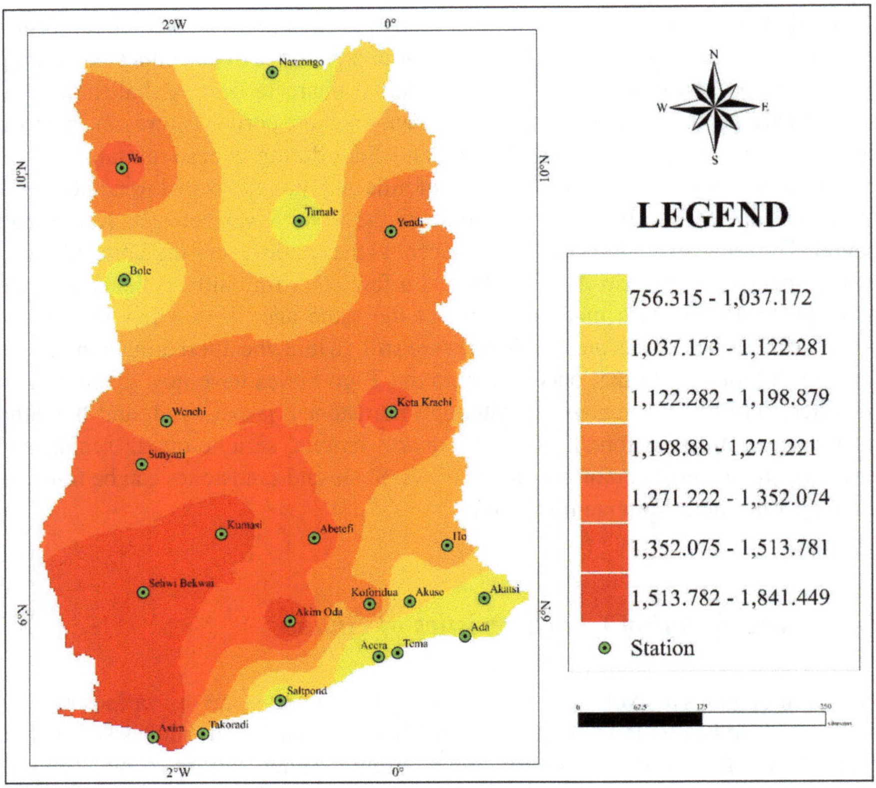

Fig. 1.3 Interpolated rainfall characteristics in Ghana (*Source* Ghana Meteorological Service Department)

levels are more likely to witness landslides, especially during prolonged or intense rainfall periods. According to Jakob (2022), the continuous influx of moisture can weaken the soil structure, making it more prone to sliding and the elevated risk is particularly evident in regions with hilly terrain, as the water runoff and erosion further contribute to instability.

1.4.3 Land Cover and Land Use Characteristics of Ghana

In Ghana, land use patterns have a significant influence on the frequency and severity of landslides. Figure 1.4 shows about nine diverse land cover classes in Ghana including Tree cover which are predominantly in the Southern western and a few in the South Eastern portion of the country. Trees cover vast regions in the country, with dense forests primarily in the Western, Western north and Central regions and Ashanti Regions.

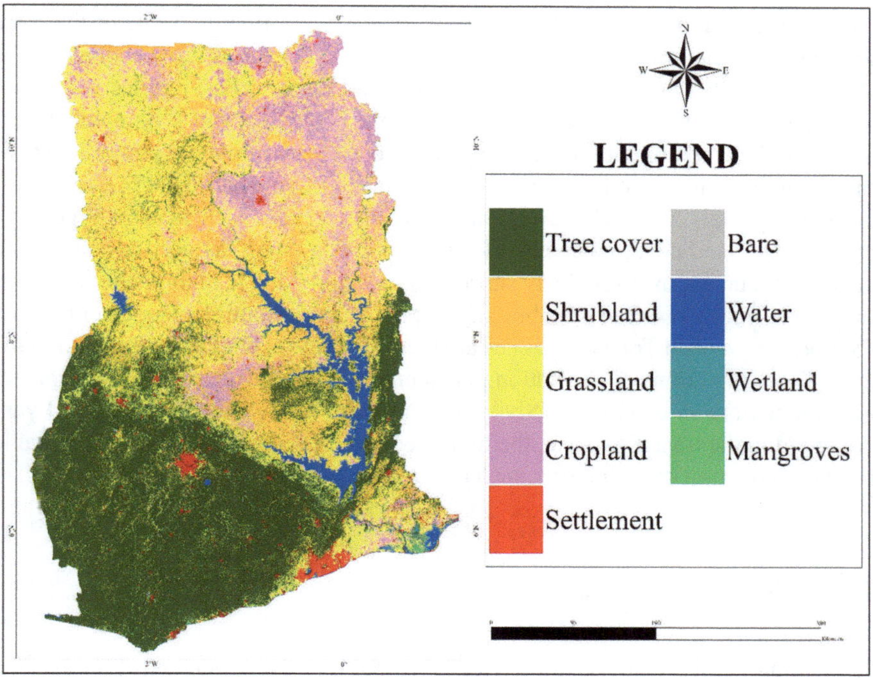

Fig. 1.4 Land cover classes of Ghana (*Source* Global land survey)

The Shrublands and Grasslands make up the Savanna region of the country. They are quite extensive extending from the Middle belt which is also known as the transitional belt of the country to the Northern regions of the country and are characterised depending on the proportions of tree to shrub to herbs (Bagamsah 2005). The transitional belt comprises the upper part of the Ashanti and Eastern regions, through the Bono regions to the main Savanna region up the Northern regions. Within these area, moist semi-deciduous forest transitions into the savanna forest and then into the guinea savanna vegetation type (Agyare et al. 2016). Thus, Shrublands have a lot of open wooded trees relative to the herbs while Grasslands have a lot of the herbs relative to the wooded tree. The Northern regions are mainly the arid part of the country mostly known as the Savannah regions, thus, these distinct land cover. The croplands are the farmlands or farming areas and the vast majority of the Croplands are also located in the North-East regions and some few others across the other sections of the Savannah, and North-western sections of the country. However, some few croplands also show up in the Ashanti and South-eastern part of the country. Settlements are scattered across Ghana, with urban centres like Accra and Kumasi representing the highly populated areas. Bare areas are uncommon but show up in small patches in the Northern Region. There are many lakes around the nation, but one of the most famous is Lake Volta, located in the Volta Region on the Volta River. It is the biggest artificial lake in the entire globe. There are wetlands all along the coast, particularly close to the estuary of large rivers. In the Greater Accra Region, some of these are located along the Volta Delta and the Densu Delta. Mangroves grow well along the shore, especially in Ghana's Western, Central, and Eastern coast regions.

These diverse land cover classes showcase the ecological and geographical variability of the land area of Ghana. They are also are very important determinants of the behaviour, distribution and frequency of landslides in a particular region (Glade 2003; Garg et al. 2015). However, unlike the other factors such as such as geology and climate, land cover characteristics have been described as very dynamic with changes occurring over relatively short periods mainly due to anthropogenic activities and normally result in destabilizing slopes. The wet Semi-deciduous Forest and the Forest Savannah Transition zones are the two main vegetation types found in the area. These two forms of vegetation play a major role in the degree of development and fluctuations in economic activity. For instance, while cattle, cashew, and yam are found in the transition zone, the moist semi-deciduous forest zone is favourable for the production of cash crops, such as cocoa.

1.5 Risk Assessment

The comprehensive analysis of precipitation, geology, land cover, and elevation data of Ghana has yielded valuable insights into the distribution of landslide risk in Ghana. The results, distinctly highlight three risk categories: High Landslide Risk, Medium Landslide Risk, and Low Landslide Risk areas. As indicated earlier, the weighted overlay method was employed by assigning weights to the factors which were considered. Thus, since historic landslide events in Ghana have mostly been triggered by precipitation, it was assigned the highest weight, and then slope, geology and land use. The result is presented in Fig. 1.5.

1.5.1 Influence of Physical Factors on Landslide Risk in Ghana

Notably, the areas of High Landslide Risk are concentrated predominantly along the coastal regions of Ghana stretching from some parts of the Western region, through the Central to the South-eastern sections of the countries including the Volta regions as well as the extreme North Eastern regions. These regions, lie within the relatively low average annual rainfall regions of the country which receive about 750–1000 mm amount of precipitation. Thus, even though in the weighted overlay analysis, precipitation was assigned the highest weight, areas within the country like the Western, Ashanti and the Brong Ahafo regions which receive the highest levels of precipitation of about 1200–1800 mm of annual average rainfall rather fall within the medium landslide risk areas whiles areas like the upper parts of the transitional belts

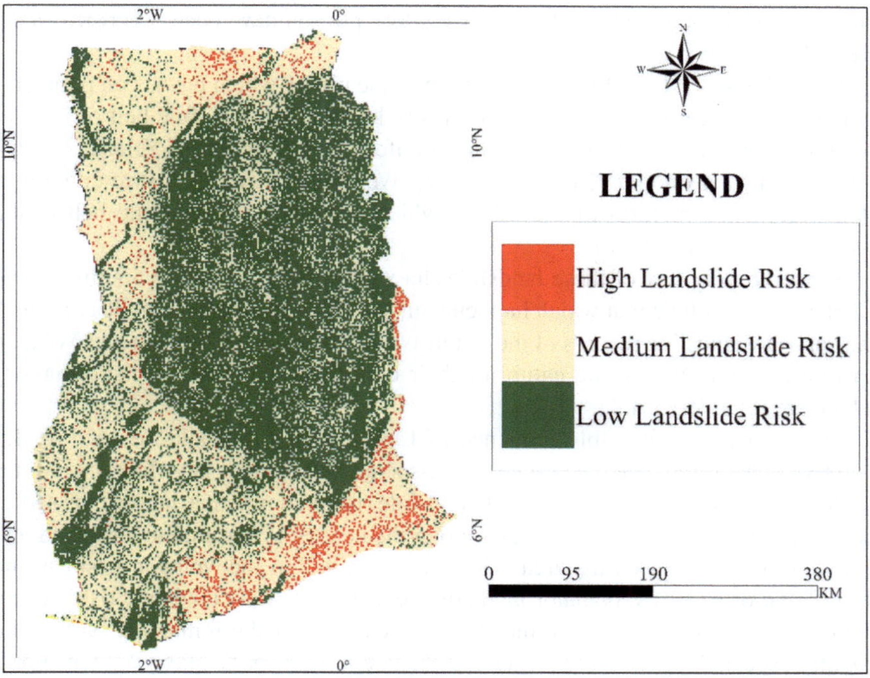

Fig. 1.5 Distribution of landslide risk in Ghana (*Source* Weighted overlay analysis)

to some sections of the Northern (Voltaian Basin) which receive relatively higher average annual precipitation of about 1100–1300 mm of precipitation lie within the Low landslide risk areas.

The Geologic factor was assigned the 3rd highest weight and this was mainly done to understand the effect that geologic features such as basement rocks, distribution of formations as well as the geological units within the area have on its landslide risk (Avanzi et al. 2004; Bartelletti et al, 2017). Within the areas categorized as high landslide risk areas, geological units identified include the recent, tertiary, and the outcrops of Dahomeyan which are found to the South Eastern part of the country from Greater Accra region to the Eastern and Volta regions while the main geologic units identified in the South Western direction west of Greater Accra including the Central and Western regions include the Brimian volcanic, Birimian sediments and the Sekondi Sandstone. According to Darko et al. (2019), the South Eastern parts of the country lies on the Dahomeyan basin complex which falls within the Dahomeyan belt resulting from the Pan-African orgogeny which occurred about 600 million years ago. Thus, areas within this section of the country lie within an active tectonic belt of the country. However, the basement rocks in the South Western part of the country as well as those of the Voltaian basin have been designated as part of the Eburnean complex as well and relative to the basement of the south east which have been described as the mobile belt, the South Western portion have been described as the stable belt.

It is quite understandable now that most of the places which have been popularly in the news concerning landslides are mostly located in the mobile belt of Ghana-Afram Plains (1972); 2005); Pantang-Mamfe (2010); Adukrom-Yelensi (2010); Lower Manya Krobo (2015); Aburi (2019); Weija-Kasoa (2022) just to cite a few. Below are some pictures of Landslides which occurred in the mobile belt of the country (Figs. 1.6 and 1.7).

However, very few of these landslides have been recorded in the South-West of Ghana. Some include that which happened in Wassa Assikuma/Nkran (2012). Thus, the very few landslide records of the South Western section, the middle and Voltaian basin are mainly due to the nature of their underlying basement rock formations which are relatively stable.

In reviewing the possible influence of Land cover to landslide risks, Fig. 1.4 portrays that various landcover types are associated with High landslide risk. Major landcover types within which the high landslide risk predominates include the settlement, grassland, tree cover and water along the coast line and mainly cropland and grassland in the Upper East area. The enter area within the Voltaian basin of the Volta Lake or water as its major land cover experiences Low landslide risks, while the medium landslide risks are mostly experienced over the remaining part of the country especially in the Brong-Ahafo, Ashanti and some parts of the Western-north regions of the country.

Fig. 1.6 Landslide scene at Aburi, located on 5.8512° N, 0.1729° W (Ghana). *Source* Esther, L. (9th November 2019). Aburi Landslide: Geoscientists warn of a looming danger. Ghana web. https://www.ghanaweb.com/GhanaHomePage/NewsArchive/Aburi-Landslide-Geoscientists-warn-of-a-looming-danger-797779

Fig. 1.7 Landslide scene at Weija-Kasoa, located on 5.5666° N, 0.3335° W (Ghana). Source: Esther, L. (24th May 2022). Landslide scare hits residents around Weija-Kasoa ridge. *Myjoyonline.* https://www.myjoyonline.com/landslide-scare-hits-residents-around-weija-kasoa-ridge/

1.6 Vulnerability Assessment

The vulnerability of Ghana has been assessed in terms of total number of people living in the various landslide risk zones based on the Population and housing census data (2021). The result is presented in Fig. 1.8.

The result shows that, out of the three highly populated regions of the country (Greater Accra, Ashanti and Eastern), two of the regions are within the High Landslide risk areas, while one (Ashanti), is within the medium landslide risk zone. And quite remarkably, in plotting the locations where most of Ghana's landslides have occurred, they coincided with the two locations where landslide risks are highest and population are also very high. Nonetheless, very few land-slide recorded events coincided with the areas marked as prone to medium landslide risks sones marked medium and apparently no record for areas marked are low landslide risk zones in the Ghana. Again, from Fig. 1.8, it can be seen that the total number of people living in the low landslide risk zones are relatively lower than those found in the medium to high land-slide risk zones of the Ghana.

In effect from Fig. 1.8, that southern portions including the Greater Accra and Eastern Regions are highly populated and are also highly vulnerable to landslide in Ghana due to their location in high-risk areas. Ashanti Region even though highly populated but depicts low to medium level of landslide vulnerability. Again, some regions in the southeastern border extending upward in the Volta region, even though

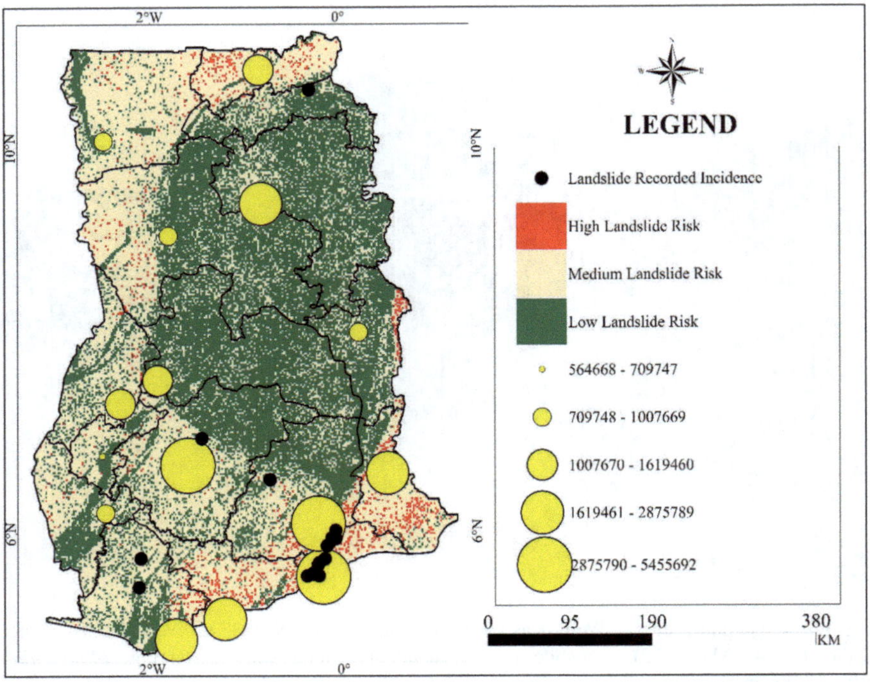

Fig. 1.8 Landslide risk and vulnerability map of Ghana

are moderately populated exhibit very high risk to landslides making residents of those areas potentially vulnerable to landslide risk as in the case of the Greater Accra and some areas in the Eastern Region of Ghana.

Conversely, the Northern Region, characterized by a very moderately low population, showcases a low level of vulnerability to landslides, making it a safer living environment. In the western portions of the country, including Bono and the Bono East regions, we observe low to medium landslide risk. The North Eastern portion is prone to high landslide risk, with areas in the North Eastern region demonstrating more vulnerability than their counterparts in the western part of the country, taking into account population.

1.7 Limitations of Study

Some limitations to this current study include the following: the study was done on a national scale using relatively coarse data from purely secondary sources and so may have some level of effect of the accuracy of results. Also, the study assumed all factors as static including climate and landcover which is very different from reality thereby having the potential of also affecting the results as well; and lastly data weighting was subjectively done without any sensitivity analysis to varying them to examine the effect it can produce.

1.8 Recommendation

To address the varying vulnerabilities across the country, we recommend a multi-pronged approach.

i. The Government and individual stakeholders should Launch extensive public awareness campaigns in high-risk regions, particularly in the southern areas, to educate the population about the dangers of landslides and promote safety land use measures.
ii. There should be an implementation of stringent regulations and building codes in high-risk regions to ensure that construction practices are resilient to landslides. Retrofitting and reinforcing existing infrastructure should also be prioritized.
iii. Disaster Management body and researchers should help develop and enhance early warning systems that can provide timely alerts to residents in landslide-prone areas, especially in the southern regions.
iv. Residents should be encouraged to plan rural developments in low- to medium-risk regions to reduce the pressure on densely populated high-risk areas.
v. There should be huge investments in ongoing geological and hydrological research to continuously update landslide risk maps. Implement regular monitoring of rainfall patterns, geological conditions, and land cover changes to improve risk assessment accuracy.

By heeding these recommendations, Ghana can work towards a more resilient and secured future, where the vulnerability to landslides is reduced, particularly in the highly vulnerable areas.

1.9 Conclusion

The nature of the Ghanaian landscape is a very major contributor to the nation's medium landslide risk susceptibility as also emphasised by the Global Facility for (Disaster Reduction and Recovery Annual Report (2020)). This has been confirmed by this research as most of the high landslide risk areas are along the South Central to Eastern sections including some parts of the Volta Region which also exhibit very high vulnerability as well as the Upper East region within the Northern region of the country. The most important factor within the Ghanaian landscape contributing to this according to the weighted overlay analysis is the geological make-up and structure of the country rather than the nature of precipitation. Thus, the middle sections to the southern Western sections of the country which experience medium to very high precipitation levels are exposed to medium landslide risks whiles the

South Eastern regions region which is experiences relatively low rainfall amounts but is the country's mobile belt in terms of geological characteristic is exposed to high landslide risk. However, varying land cover types also can affect the magnitude of risk when they do occur. One of the most significant contributors to landslides in Ghana is deforestation. In many regions, especially in the hilly and forested areas, communities engage in extensive deforestation for various purposes, including agriculture and logging. A recent example of this is the Atewa Range in the Eastern Region. The relentless clearing of forests for cocoa farming and logging has not only altered the natural landscape but has also left the soil exposed and susceptible to erosion during heavy rains (Asase et al. 2009). The removal of tree cover disrupts the soil's ability to retain moisture, increasing the risk of landslides. The degradation of these forested areas has both ecological and socio-economic consequences, and landslides have become a recurrent threat to communities in these regions.

In conclusion, Landslides can exert a significant multidimensional impact on a region's life, livelihoods and geomorphology (Crozier 2010; Nowicki Jessee et al. 2020). However, the interplay of variables such as the geological composition, climate variabilities, elevation and land cover intensify these effects (LePage et al. 2012). Though their occurrence mostly cannot be predicted, the use of geostatistics and geospatial modelling technologies can help integrate these datasets to be able to ascertain the individual contribution of such factors to the level of risks and vulnerabilities. In the case of Ghana, the research has shown that mostly the various land cover within the mobile belt of the country increases the risk and vulnerability of the people within the location. With such information, certain proactive measures can be developed in terms of land use planning in these very risky areas to mitigate the effects of the risks when they do occur especially in developing countries. Thus, such underlying factors in addition to the increasing population and to some extent changes in the precipitation patterns are important factors to be considered in Landslide risk and vulnerability assessment. Furthermore, the use of geostatistics and geospatial technologies can help in the manipulation and evaluation of these datasets to assess the effect of these key factors regionally.

The following areas could also be considered for future studies: Dynamic modelling of the impact of climate change, land-use changes, and population on landslide risk and susceptibility; a comparison in some geostatistical and geospatial modelling techniques in studying landslide risks and vulnerabilities.

Acknowledgements I would like to acknowledge the support of my National Service Personnel (Mr. Nathaniel Appiah) and one of the graduate students in the Department of Geography and Regional Planning (Mr. Samuel Kofi Asiamah Nyave) for their support during the period of putting together this Chapter.

References

Abd Majid N, Rainis R (2019) Aplikasi Sistem Maklumat Geografi (GIS) dan Analisis Diskriminan dalam Pemodelan Kejadian Kegagalan Cerun di Pulau Pinang, Malaysia. Sains Malays 48:1367–1381. https://doi.org/10.17576/jsm-2019-4807-06

Adu-Boahen K (2020) Evaluation of sustainable management practices of water resources in Ghana: perspectives from the Fosu and Essei Lagoons. Afr Soc Sci Human J 1(1):82–100

Afriyie K, Ganle JK, Adomako JAA (2016) The good in evil: a discourse analysis of the *galamsey* industry in Ghana. Oxf Dev Stud 44:493–508. https://doi.org/10.1080/13600818.2016.1217984

Agyare C, Boakye YD, Bekoe EO et al (2016) Review: African medicinal plants with wound healing properties. J Ethnopharmacol 177:85–100. https://doi.org/10.1016/j.jep.2015.11.008

Alam E (2020) Landslide hazard knowledge, risk perception and preparedness in Southeast Bangladesh. Sustainability 12:6305. https://doi.org/10.3390/su12166305

Alaneme KK, Okotete EA (2018) Critical evaluation of seismic activities in Africa and curtailment policies—a review. Geoenviron Disasters 5:24. https://doi.org/10.1186/s40677-018-0116-2

Amponsah TY, Danuor SK, Wemegah DD, Forson ED (2022) Groundwater potential characterisation over the Voltaian basin using geophysical, geological, hydrological and topographical datasets. J Afr Earth Sc 192:104558. https://doi.org/10.1016/j.jafrearsci.2022.104558

Anderson MG, Holcombe E, Blake JR et al (2011) Reducing landslide risk in communities: evidence from the Eastern Caribbean. Appl Geogr 31:590–599. https://doi.org/10.1016/j.apgeog.2010.11.001

Angnuureng DB, Brempong KE, Jayson-Quashigah PN et al (2022) Satellite, drone and video camera multi-platform monitoring of coastal erosion at an engineered pocket beach: a showcase for coastal management at Elmina Bay, Ghana (West Africa). Reg Stud Mar Sci 53:102437. https://doi.org/10.1016/j.rsma.2022.102437

Asase A, Ofori-Frimpong K, Ekpe PK (2009) Impact of cocoa farming on vegetation in an agricultural landscape in Ghana. Afr J Ecol 48:338–346. https://doi.org/10.1111/j.1365-2028.2009.01112.x

Avanzi GDA, Giannecchini R, Puccinelli A (2004) The influence of the geological and geomorphological settings on shallow landslides. An example in a temperate climate environment: the June 19, 1996 event in northwestern Tuscany (Italy). Eng Geol 73(3–4):215–228

Ayalew L, Yamagishi H (2005) The application of GIS-based logistic regression for landslide susceptibility mapping in the Kakuda-Yahiko Mountains, Central Japan. Geomorphology 65:15–31. https://doi.org/10.1016/j.geomorph.2004.06.010

Bagamsah TT (2005) The impact of bushfire on carbon and nutrient stocks as well as albedo in the savanna of northern Ghana, vol 25. Cuvillier Verlag

Bartelletti C, Giannecchini R, D'Amato Avanzi G, Galanti Y, Mazzali A (2017) The influence of geological–morphological and land use settings on shallow landslides in the Pogliaschina T. basin (northern Apennines, Italy). J Maps 13(2):142–152

Boateng P (2017) Land access, agricultural land use changes and narratives about land degradation in the savannahs of northeast Ghana during the pre-colonial and colonial periods. Soc Sci 6:35. https://doi.org/10.3390/socsci6010035

Buah-Kwofie A, Humphries MS, Pillay L (2019) Dietary exposure and risk assessment of organochlorine pesticide residues in rural communities living within catchment areas of iSimangaliso World Heritage Site, South Africa. Environ Sci Pollut Res 26:17774–17786. https://doi.org/10.1007/s11356-019-05046-9

Campion BB, Venzke J-F (2013) Rainfall variability, floods and adaptations of the urban poor to flooding in Kumasi, Ghana. Nat Hazards 65:1895–1911. https://doi.org/10.1007/s11069-012-0452-6

Crozier MJ (2010) Landslide geomorphology: An argument for recognition, with examples from New Zealand. Geomorphology 120:3–15. https://doi.org/10.1016/j.geomorph.2009.09.010

Dankyi PO, Dei LA, Nyarko BK (2023) Rockfall hazards along Aburi-Accra Road, Ghana. Malays J Trop Geogr 49(1):38–51

Darko JAA, Ahenkorah I, Aakyiir MN (2019) Petrogenesis, geochemistry and structural evidence of the neoproterozoic Pan-African orogenic event in Ghana, West Africa. SN Appl Sci 1:293. https://doi.org/10.1007/s42452-019-0319-5

Fan W, Wei X, Cao Y, Zheng B (2017) Landslide susceptibility assessment using the certainty factor and analytic hierarchy process. J Mt Sci 14:906–925. https://doi.org/10.1007/s11629-016-4068-2

Fan L, Lehmann P, Zheng C, Or D (2020) Rainfall intensity temporal patterns affect shallow landslide triggering and hazard evolution. Geophys Res Lett 47. https://doi.org/10.1029/2019GL085994

Froude MJ, Petley DN (2018) Global fatal landslide occurrence from 2004 to 2016. Nat Hazard 18:2161–2181. https://doi.org/10.5194/nhess-18-2161-2018

Garg A, Leung AK, Ng CWW (2015) Comparisons of soil suction induced by evapotranspiration and transpiration of S. heptaphylla. Can Geotech J 52:2149–2155. https://doi.org/10.1139/cgj-2014-0425

Glade T (2003) Landslide occurrence as a response to land use change: a review of evidence from New Zealand. Catena (Amst) 51:297–314. https://doi.org/10.1016/S0341-8162(02)00170-4

Highland LM, Bobrowsky P (2008) The landslide handbook—a guide to understanding landslides: Reston, Virginia, US. Geological Survey Circular 129

Hungr O, Leroueil S, Picarelli L (2014) The Varnes classification of landslide types, an update. Landslides 11:167–194

Jakob M (2022) Landslides in a changing climate. In: Landslide hazards, risks, and disasters. Elsevier, pp 505–579

LePage BA, Jacobs BF, Williams CJ (2012) Insights from paleohistory illuminate future climate change effects on wetlands. Global change and the function and distribution of wetlands. Springer, Netherlands, Dordrecht, pp 3–59

McKinley JM, Atkinson PM, Lloyd CD et al (2011) How porosity and permeability vary spatially with grain size, sorting, cement volume, and mineral dissolution in fluvial Triassic sandstones: the value of geostatistics and local regression. J Sediment Res 81:844–858. https://doi.org/10.2110/jsr.2011.71

Min SJ, Mayuka RN, Joseph AJ, Cosmos BC, Samuel AS, Qiaomin ZQ (2023) Analysis of the combination of the soil stability and the slope analysis methods to design a retaining wall on the Akoasa Mountain Mudslide (Accra, Ghana). Int J Comput Eng 4(1):17–32

Ng CWW, Shi Q (1998) A numerical investigation of the stability of unsaturated soil slopes subjected to transient seepage. Comput Geotech 22:1–28. https://doi.org/10.1016/S0266-352X(97)00036-0

Nowicki Jessee MA, Hamburger MW, Ferrara MR et al (2020) A global dataset and model of earthquake-induced landslide fatalities. Landslides 17:1363–1376. https://doi.org/10.1007/s10346-020-01356-z

Nsengiyumva JB, Luo G, Amanambu AC et al (2019) Comparing probabilistic and statistical methods in landslide susceptibility modeling in Rwanda/Centre-Eastern Africa. Sci Total Environ 659:1457–1472. https://doi.org/10.1016/j.scitotenv.2018.12.248

Owusu-Ansah JK, Dery JM, Amoako C (2019) Flood vulnerability and coping mechanisms around the Weija Dam near Accra, Ghana. GeoJournal 84:1597–1615. https://doi.org/10.1007/s10708-018-9939-3

Rahman A, Khan AN, Collins AE (2014) Analysis of landslide causes and associated damages in the Kashmir Himalayas of Pakistan. Nat Hazards 71:803–821. https://doi.org/10.1007/s11069-013-0918-1

Sadr S, Moosavian SAA, Zarafshan P (2014) Dynamics modeling and control of a quadrotor with swing load. J Robot 2014:1–12. https://doi.org/10.1155/2014/265897

Shibasaki T, Matsuura S, Okamoto T (2016) Experimental evidence for shallow, slow-moving landslides activated by a decrease in ground temperature. Geophys Res Lett 43:6975–6984. https://doi.org/10.1002/2016GL069604

Shinohara Y, Kume T (2022) Changes in the factors contributing to the reduction of landslide fatalities between 1945 and 2019 in Japan. Sci Total Environ 827:154392. https://doi.org/10.1016/j.scitotenv.2022.154392

Shit PK, Bhunia GS, Maiti R (2016) Potential landslide susceptibility mapping using weighted overlay model (WOM). Model Earth Syst Environ 2:21. https://doi.org/10.1007/s40808-016-0078-x

Tacoli C (2017) Food (in)security in rapidly urbanising, low-income contexts. Int J Environ Res Public Health 14:1554. https://doi.org/10.3390/ijerph14121554

Tschakert P, Singha K (2007) Contaminated identities: Mercury and marginalization in Ghana's artisanal mining sector. Geoforum 38:1304–1321. https://doi.org/10.1016/j.geoforum.2007.05.002

Wang G-L, Wang Z-Y, Duan L-J et al (2021) Susceptibility of circulating SARS-CoV-2 variants to neutralization. N Engl J Med 384:2354–2356. https://doi.org/10.1056/NEJMc2103022

Will TM, Höhn S, Frimmel HE et al (2020) Petrological, geochemical and isotopic data of Neoproterozoic rock units from Uruguay and South Africa: correlation of basement terranes across the South Atlantic. Gondwana Res 80:12–32. https://doi.org/10.1016/j.gr.2019.10.012

Zheng F-L (2006) Effect of vegetation changes on soil erosion on the Loess Plateau. Pedosphere 16:420–427. https://doi.org/10.1016/S1002-0160(06)60071-4

Chapter 2
Landslide Susceptibility Mapping by Using Geospatial Technique: Reference from Hofu City, Yamaguchi Prefecture, Japan

Benita Nathania and Martiwi Diah Setiawati

Abstract Hofu City in the southern part of Honshu Island, Japan, was hit by a severe landslide on July 21, 2009. This chapter aims to determine the sites where the landslides are most likely to happen by creating a susceptibility map within the case study sites. We used remote sensing and geographic information systems (GIS) datasets that include ALOS AVNIR-2 satellite imagery, the digital elevation models (DEM), geology records, the local rain gauge data, and geographical representation of the history of landslides. Seven parameters, including land cover, elevation, slope, aspect, geology, and boundary extraction, were integrated using logistic regression with an isohyet map (i.e., from the rainfall dataset) to model a landslide susceptibility map (LSM). Following the research's findings, landslides were more likely to occur at elevations between 50 and 350 m, slope angles between 5 and 50 degrees, slope directions northeast and north, all of the land cover types and lithological types of granodiorite, fan deposits, and middle terrace. Among those statics' parameters, elevation, land cover, and slope were the most significant in determining the landslide susceptibility model. Further, almost half of Hofu City was categorized as high and very high susceptible areas. Moreover, out of the 928 inventory landslide dataset, 916 were placed at the very high susceptibility areas of the LSM.

Keywords Landslides · Susceptibility map · Hofu City · Yamaguchi prefecture · Logistic regression

B. Nathania
World Research Institute Indonesia, South Jakarta, Jakarta, Indonesia
e-mail: benita.nathania@wri.org

M. D. Setiawati (✉)
Institute for the Advanced Study of Sustainability (UNU-IAS), United Nations University, Shibuya-Ku, Tokyo, Japan
e-mail: martiwi1802@gmail.com

Research Center for Oceanography, National Research and Innovation Agency (BRIN), Ancol Timur, Jakarta, Indonesia

G. K. Panda et al. (eds.), *Landslide: Susceptibility, Risk Assessment and Sustainability*, Advances in Natural and Technological Hazards Research 52,
https://doi.org/10.1007/978-3-031-56591-5_2

2.1 Introduction

Landslides are unpredictably destructive catastrophes that cause severe risks to construction, people's lives, economic growth, and business. These are common catastrophic disasters, resulting in extensive damage and putting individuals and societies at risk (Zhao et al. 2021; Chang et al. 2023). Also, landslides make up roughly twelve percent of any natural catastrophe, causing more than eleven thousand mortality yearly, with a financial loss of $4.5 billion (Froude and Petley 2018; WHO 2022). Japan is particularly susceptible to landslides because it has exceptional topographical and geological circumstances, which include unstable ground and granite structures, high rains, as well as tectonic activity. Arguably, the most prevalent landslide-prone region in Japan is the southern region, which is located near the Pacific Ocean. Also, there are several landslides and flash floods there every year.

The Chugoku Mountain in Western Japan is well-known for its hilly environment and sloping topography. These locations have been recognized as particularly vulnerable to landslides and have been categorized as high-risk areas of these types of natural disasters by many studies on landslide modeling. In addition, the global incidence of landslides is now much affected by changing climates. The susceptibility and occurrence of landslides are significantly impacted by climate change-induced components, like intense precipitation, warmer temperatures, and rising sea levels (Pei et al. 2023; Qiu et al. 2019). To offer reliable and trustworthy estimates, it is crucial to integrate climate variables when developing Landslide Susceptibility Mapping (LSM) models.

According to previous studies, landslides are susceptible to occurring in many geographic regions. Using several landslide model predictions could lessen the harm resulting from landslides (Pradhan and Lee 2010). The LSM has been simpler to create in the last 20 years due to significant advancements in computational capacity, remote sensing (RS), and geographic information systems (GIS) (Mirus et al. 2020; Moragues et al. 2023). Such developments are transforming the field of landslide research.

The disastrous landslides in Hofu City in 2009 that claimed over 14 lives underscored the need to analyze landslide susceptibility in Western Japan precisely (ICHARM 2020). This analysis is essential for reducing the destruction of structures and fatalities. As a result, several academics have used mapping landslide susceptibility as an essential initial phase in the strategies, assessment, and mitigation of landslides. This technique has been used to create landslide vulnerability maps in several publications, including (Yu et al. 2021; Cheng et al. 2022; Kohno and Higuchi 2023). Nevertheless, the landslide susceptibility assessments developed in Japan have mainly concentrated on significant transportation lines and particular sites in extremely susceptible provinces. In order to address landslide threats effectively, establishing and creating the LSM at various spatial resolutions is crucial (Persichillo et al. 2016). This modeling could help with the deployment of early warning and surveillance systems, strategic planning for risk reduction, and the development of feasible and effective land development strategies for policymakers (Roccati et al. 2021).

In general, the LSM was generated into two main approaches: knowledge-driven and data-driven. The knowledge drive was heavily dependent on expert judgment (e.g., the Analytical Hierarchy Process (AHP)), whereas this kind of approach is useful when the inventory data are limited. The AHP, for example, is a widely used knowledge-driven approach for the LSM worldwide (e.g., Kohno and Higuchi 2023; Zangmene et al. 2023; Panchal and Srivastava 2022). Given the rapid advancement of computational and GIS approaches, various complexities of data-driven models for performing LSM have recently been proposed. There are two kinds of data-driven models that exist: machine learning and statistical models (Yu et al. 2021; Cheng et al. 2022). Recently, machine learning has been used so vast and generally has high predictive accuracy (Cheng et al. 2022). However, their fundamental principles are complex and hard to describe directly. As a result, these methods are unsuitable for studying the interaction among landslides and variables (Cheng et al. 2021). Therefore, in this chapter, we proposed a statistical model approach that could overcome those challenges.

Numerous GIS and RS approaches have been used to evaluate the LSM over the past decade. Quantitative and qualitative techniques were employed extensively with such approaches (Cheng et al. 2021; Kohno and Higuchi 2023). The LSM using logistic regression (LR) represents a primarily quantitative approach that provides a practical and useful means of mapping landslide susceptibility groups through numerous proxies. Investigators worldwide have frequently used the LR approach to examine the link between historical landslide disasters and various categories of possibly significant elements. The LR model is especially well-suited for evaluating landslide vulnerability in hilly settings, claim Ali Mohammadi et al. (2014). Even though a severe landslide hit Hofu City in 2009, a lack of research assesses the LSM. This assessment is crucial for the study area's local strategic disaster risk reduction planning. Therefore, this chapter aims to use an LR model to develop the LSM and use GIS and RS approaches to identify the physical elements that influence the likelihood of an incident using the pilot study of Hofu City, Japan.

2.2 Database and Methodology

2.2.1 Study Area

This assessment was carried out in Hofu City, Yamaguchi Prefecture, Chugoku region, the westernmost region of Honshu, Japan's largest island. The entire area is 188.59 km^2 and has an elevation ranges from 0 to 630 m. Hofu is south of Yamaguchi and faces the Sea of Suo in the Seto Inland Sea. From the rocky slopes of Migita in the north to Mount Ohira in the east and an island mountain on the southern seaside, the city is nearly encircled by mountains and rice fields. The Saba River flows from Tokuji town (legally part of Yamaguchi City) to the Seto Sea. Hofu City faced severe landslides due to extreme rainfall in July 2009. Moreover, this area also faced depopulation from 2010 to 2020 by 2.3%. Figure 2.1 depicts the research site.

Furthermore, Hofu City has an annual average rainfall of 1652 mm, with the highest precipitation in summer (μ = 222 mm/ month) and the lowest during

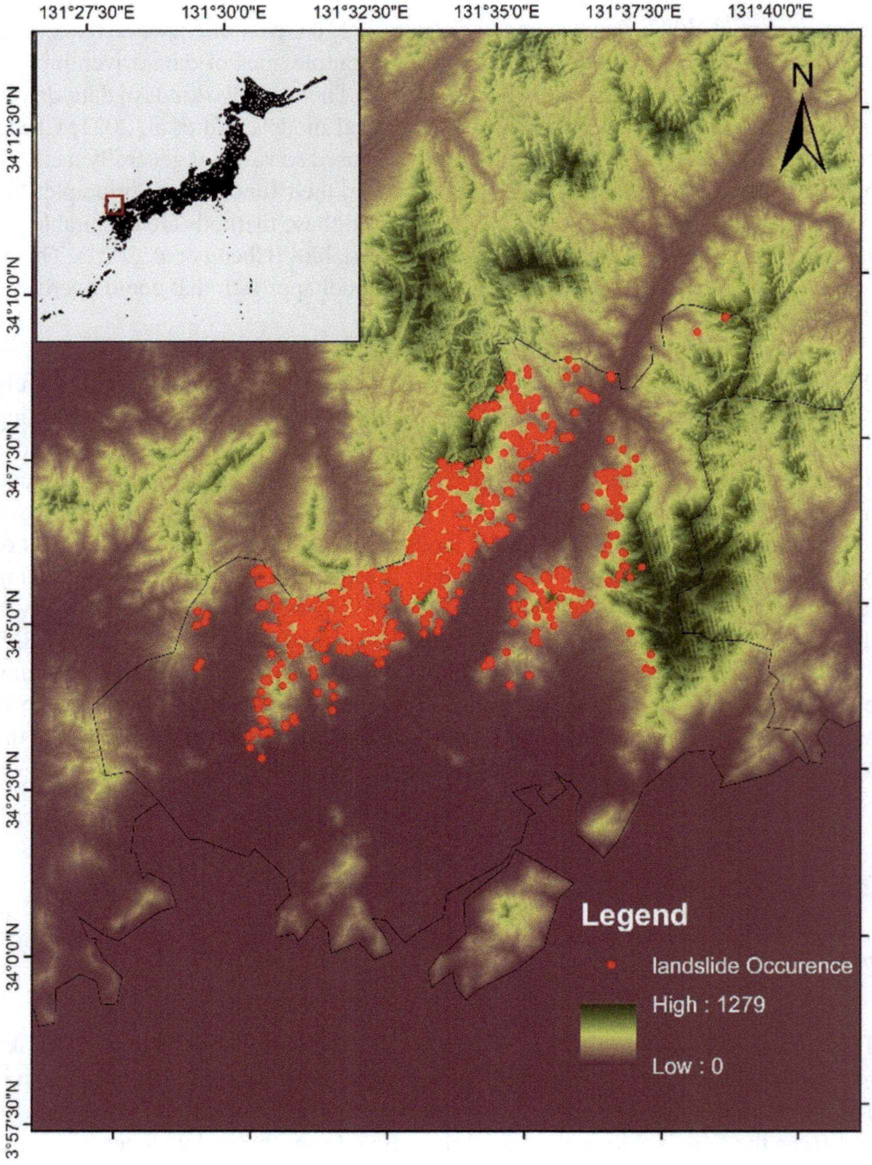

Fig. 2.1 Landslide inventory map in Hofu City, Japan

the winter season (59 mm/month) (JMA, 2022). Throughout 1990–2022, July was recorded as the highest monthly average rainfall (i.e., 300mm/ month), while December was the lowest (52 mm/month) (JMA, 2022). Moreover, the highest monthly rainfall in Hofu City was 812 mm/month in July 2009; therefore, a severe landslide occurred during this time (JMA, 2022). Also, the study area has a dominant granitic rock, followed by marine and marine sediment and pelitic schist (GSI 2014), whereas the inventory landslide dataset is mainly located along the granitic rock. Moreover, according to a previous study by Chigira et al. 2011 and Yamashita et al. 2017, the incidents in the study area mainly occurred in weathered granite, whereas this type of rock is prone to sliding due to extreme rainfall. Yamashita et al. 2017 also revealed that the granite areas had thinner soil and more considerable shear strength, which could cause a slip plane.

2.2.2 *Dataset*

2.2.2.1 Map of Landslide Inventory

This dataset was obtained from the Ministry of Land, Infrastructure, Transport and Tourism (MLIT), Chugoku District, Japan. The data consists of landslide locations and areas. This inventory map recorded landslide events on 21 July 2009 (Fig. 2.1). We utilized this dataset for validation of the susceptibility map.

2.2.2.2 Remote Sensing Data

In this chapter, we used the ALOS Advanced Visible and Near Infrared Radiometer type 2 (AVNIR-2) satellite dataset to get the land cover information within the study area. This collection contains data from a visible and near-infrared (NIR) radiometer applied to observe terrestrial and coastal regions., which has 10m spatial resolution images. Table 2.1 explains the attribute details of the ALOS AVNIR-2 images.

In this chapter, ALOS AVNIR-2 was utilized to build a land cover map within the pilot sites. The image of ALOS AVNIR-2 acquired on July 14, 2009, was processed using ArcGIS 10.8.2.

2.2.2.3 Rainfall Data

Rainfall data used in this chapter were obtained from the Japanese rain-gauged dataset called the Automated Meteorological Data Acquisition System (AMEDAS). This dataset is freely available and could be downloaded at http://www.jma.go.jp/jma/index.html. Rainfall data on a specific date (21 July 2009), both daily and hourly, were collected from rain gauges in Hofu City and surrounding cities located near

Table 2.1 AVNIR-2 details (JAXA 2008)

Number of Bands	Four
Wavelength	Band 1: 0.42 to 0.50 μm Band 2: 0.52 to 0.60 μm Band 3: 0.61 to 0.69 μm Band 4: 0.76 to 0.89 μm
Spatial resolution	10 m
Swath width	70 km
Signal-to-noise ratio	Higher than 200
MTF	Band 1,2,3: higher than 0.25 Band 4: higher than 0.20
Number of detectors	7000/band
Pointing angle	−44 to + 44 degree
Bit length	8 bits

the study area. The rainfall data in each station was downloaded and interpolated to create an isohyet map.

2.2.2.4 Geology and Elevation Data

Geology and elevation dataset was collected from the Geospatial Information Authority of Japan (GSI), and they were downloaded from the GSI homepage at www.gsi.go.jp. Geology data were downloaded in raster data format, while the elevation data were downloaded in 10 mesh of point data. Ten mesh has a spatial resolution of 0.4 arc-second (1 arc-second = 30.86 m). The elevation data was interpolated to build a Digital Elevation Model (DEM) and then derived from the slope map and aspect map.

2.2.3 Method

This chapter applied three main data sources: satellite remote sensing data from ALOS AVNIR-2, rainfall data from AMEDAS, and elevation and geology data from GSI. These data were derived to generate a landslide parameter map using the Weight of Evidence (WOE) approach to demonstrate the relevance of environmental factors or variable groups and the LR modeling to determine the likelihood of incidence or absence of landslides. Figure 2.2 describes the overview of this chapter.

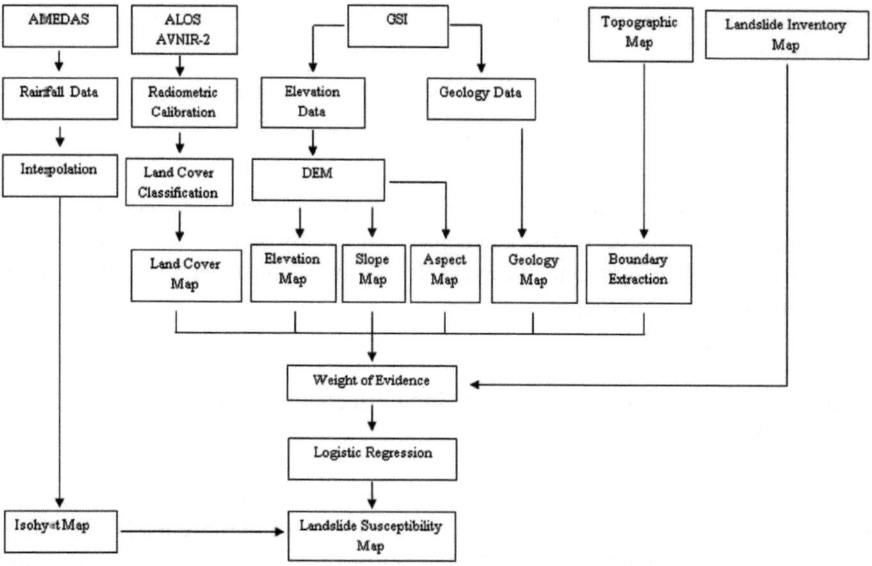

Fig. 2.2 Flowchart of the landslide susceptibility map

2.2.3.1 The Pre-Processing of ALOS AVNIR-2

This satellite image has eight-bit values denoting brightness metrics, generally known as the Digital Numbers (DN). The DN value is varied according to the byte type of the satellite RS dataset; thus, converting it into common features such as radiance value is necessary. The procedure for converting DN to radiance was described by Eq. 1 (Bouvet et al. 2007).

$$L_\lambda = GAIN_{rescale} \times QCAL + BIAS_{rescale} \tag{2.1}$$

where L_λ is the spectrum radiance (W/m2/sr/m), $GAIN_{rescale}$ indicates the rescaled gain, QCAL represents the DN, and $BIAS_{rescale}$ is referred to as the rescaled bias. Table 2.2 summarizes the rescaling gains and biases utilized for the DN-to-radiance conversion.

Normalization for solar irradiance could reduce between-scene variability in comparatively "clear" images by translating the spectral radiance into a reflectance.

Table 2.2 GAIN and BIAS rescaling for conversion of DN to radiance in ALOS AVNIR-2

Band	$GAIN_{rescale}$	$BIAS_{rescale}$
1	0.5888	0
2	0.5730	0
3	0.5020	0
4	0.8350	0

The algorithm for converting radiation to reflectance is described in Eq. (2.2) (Sah et al. 2012).

$$\rho_\lambda = \pi \times L_\lambda \times \frac{d^2}{ESUN_\lambda} \times \cos\theta_s \tag{2.2}$$

The ρ_λ defines as the reflectance (no unit), L_λ represents the spectral radiance, d^2 represents the Earth-Sun distance in units of astronomy according to a navigational guidebook, ESUN_λ represents the average solar irradiances, while the θ_s defined as the solar zenith angle (degrees).

2.2.3.2 Land Cover Classification Using Object-Based Image Analysis OBIA

This is a technique of image analysis that derives information from pictures by using objects in the context of an image pixel-by-pixel (Burnett and Blaschke 2003). This is a two-step technique that begins with the segmentation of images and ends with image categorization. Throughout the segmentation stages, the image is initially separated into homogeneous and neighboring regions that consider region context, spectral data, texture, and shape into account. After, the image is grouped by employing the K Nearest Neighbor classification approach, a process that considers the object's Euclidean distance in the n-dimensional area to the parts of the training sample, in which n is set by the number of attributes of the object applied during the categorization. The length of the distance of two planar locations having coordinates (x, y) and (a, b) is obtained using the Euclidean distance method as shown below:

$$dist((x, y), (a, b)) = \sqrt{(x - a)^2 + (y - b)^2} \tag{2.3}$$

2.2.3.3 Land Cover Accuracy Assessment

This chapter employed error matrices and Cohen's kappa (k) as an accuracy assessment. This accuracy assessment can be beneficial in developing algorithms that anticipate different groupings and classify images. Moreover, the assessment aids in determining how precise and beneficial the model is. In order to generate an error matrix, a database of locations by which the true categorization can be determined is required. Since the location(s) were examined on the ground, the columns include reference data or established categorization. Rows represent projected categories. Kappa can be applied to provide a metric for assessing the degree of coherence of simulation results and actual (Congalton 1991) or as a proxy for whether the numbers in an error matrix reflect an outcome that is considerably more accurate (Chen et al. 2010). Equation 4 shows the Kappa formula.

$$k = \frac{N \sum_{i=1}^{r} x_{ii} - \sum_{i=1}^{r} (x_{i+} \times x_{+i})}{N^2 - \sum_{i=1}^{r} (x_{i+} \times x_{+i})} \tag{2.4}$$

2.2.3.4 Rainfall Isohyet Map

We defined isohyet as a line derived from a weather map to connect dots that get equivalent rainfall quantities throughout a specific period. This map illustrates rainfall statistics. The contoured lines link regions that receive equivalent precipitation, while a color scale is often applied to make the distinction among them. Further, isohyet maps were generated by interpolating the precipitation data from the in-situ rain gauge dataset (Chow et al. 1988).

The arithmetic-mean approach is the most basic way of calculating the average amount of rainfall in a particular area. This entails averaging the rainfall volumes measured at various gauges. This strategy works well if the rain gauge is evenly distributed across the region and each gauge measurement stays within the average value.

2.2.3.5 Digital Elevation Model (DEM)

The initial GSI elevation points were transformed into a DEM of the raster dataset using an interpolation method of Inverse Distance Weighted (IDW). This interpolation approach used a linearly weighted ensemble of a series of features from the sample to calculate the value of each cell. Moreover, the weight was determined by the inverse distance.

Aspect and slope were generated from DEMs and determined by comparing the point's elevation to its neighbors. The slope was expressed as "a horizontal tangent to a topographic surface" as an equation of a DEM for a given position (Burrough 1986). Moreover, slope and aspect were calculated by comparing the elevations of points in a 3 × 3 neighborhood. Further, the slope and aspect at one point were calculated using an elevation dataset in that spot location and eight other points. Equation (2.5) describes the initial procedure employed to compute the slope.

$$\text{Sloperadians} = \text{ATAN}\left(\sqrt{\left(\left[\frac{dz}{dx}\right]^2 + \left[\frac{dz}{dy}\right]^2\right)}\right) \tag{2.5}$$

The slope is typically expressed in degrees, where employs the equation as follows (6):

$$\text{Slopedegree} = \text{ATAN}(\sqrt{\left[\frac{dz}{dx}\right]^2 + \left[\frac{dz}{dy}\right]^2}) * 57.29578 \tag{2.6}$$

The aspect indicates the downslope direction of the greatest extent of the variation in magnitude of one grid to the next. In other words, the slope direction could be referred to as the aspect. Also, the compass-derived vector of the aspect will be represented by the values in the finalized raster. The aspect of a raster surface is the direction of the highest alteration rate in the z-value from each cell. Aspect is stated as a positive number with values from 0 to 359.9, identified clockwise from north. A typical aspect of computation is shown in Eq. 2.7:

$$\tan(slope) = \frac{\text{dz/dx}}{\text{dz/dy}} \tag{2.7}$$

where (dz/dx) denotes the horizontal surface while (dz/dy) is the vertical surface. The arithmetic-mean method is the most accessible approach to measuring the average rainfall in a specific area. It entails averaging the rainfall depths measured at various gauges. This approach works well if the gauges are spread evenly across the region and each gauge measurement stays within the average value.

2.2.3.6 WOE and LR Model

WOE is a statistical analysis according to the Bayesian probability model in which the distribution is defined based on evidence (Bonham-Carter 1994). The relevance of variable classes is expressed by W + (positive weight) and W- (negative weight), correspondingly to the formulas (2.8) and (2.9). When a parameter class has a positive W + , landslides have a greater likelihood of taking place. Meanwhile, the landslide susceptibility decreases when the negative W-value indicates that the class is absent. Likewise, a negative W + suggests that the category is not conducive to landslides occurring, yet a positive W − suggests that the variable category's absence increases landslide susceptibility. Weights nearly "0" show no association between the parameter class and the frequency of landslides. Table 2.3 (Bonham-Carter 1994) indicated that the weight computation was according to a cross-tabulation of areas of the parameters class and the landslide area.

Table 2.3 Area cross-tabulation (*Source* Bonham-Carter 1994)

	Parameter class		
Landslide area	Yes	No	Total
Yes	T_{11}	T_{12}	T_1
No	T_{21}	T_{22}	T_2
Total	$T_{.1}$	$T_{.2}$	$T.$

$$W^{+} = \ln\left[\frac{\frac{T_{11}}{T_{21}}}{\frac{T_{1}}{T_{2}}}\right] = \ln[\frac{T_{11}T_{2}}{T_{21}T_{1}}] \quad (2.8)$$

$$W^{-} = \ln\left[\frac{\frac{T_{12}}{T_{22}}}{\frac{T_{1}}{T_{2}}}\right] = \ln[\frac{T_{12}T_{2}}{T_{22}T_{1}}] \quad (2.9)$$

The multivariate statistical method known as logistic regression (LR) can determine the likelihood of a categorical regressand, including the likelihood (i.e., Yes) and "No" of landslides, according to the association among the response and predictors variables, such as land cover, lithology, slope gradient, isohyet, etc. Then, the probabilistic values are between zero to one, representing the susceptibility of landslides. Mining resource estimations (Agterberg et al. 1993) and the likelihood of landslide analysis (Feby et al. 2020) have both benefited from the use of LR. The benefits of LR include the fact that it can be employed when variables have conditional dependency and when variables have several classes or are continuous. The logistic equations f(z) underpins LR was described as follows:

$$f(z) = \frac{1}{1+e^{-z}} \quad (2.10)$$

where,

$$z = \alpha_0 + \beta_1 X_1 + \beta_2 X_2 + \cdots + \beta_x X_x \quad (2.11)$$

LR involves calculating the coefficients β1, β2,...βk, as well as the constant or intercept, α0. The parameters of X1, X2, and Xk represent the predictor factors that are associated with landslides.

2.2.3.7 Process of WOE and LR Model

First, all the predictor factors (i.e., elevation, slope, aspect, geology) are converted into the same grid size of 10 × 10 m and overlaid with the evidence point. Figure 2.3 explains the procedure of the weight calculation. The figure describes 6 × 6 pixels for each environmental factor, representing the presence and absence of the incident likelihood in Hofu City.

In order to determine the weight of all parameter classes, the cross-tabulation table was created and computed by pixel-based analysis, as shown in Table 2.4. For example, the elevation parameter for 150 m is shown in black (8 pixels), 200 m is shown in green (28 pixels), and the landslide evidence point is shown in a white dot (Fig. 2.3a). The W + and W- values are calculated to identify where the landslide is likely to occur.

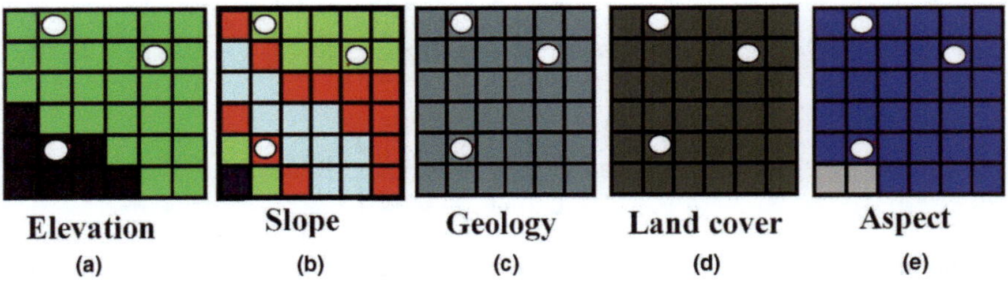

Fig. 2.3 The pixel sampling of each environmental factor

Table 2.4 Area cross-tabulation of elevation

	Elevation 150 m (black pixel)				Elevation 200 m (green pixel)		
Landslide (Σpixel)	Present	Absence	Total	Landslide (Σpixel)	Present	Absence	Total
Present	1	2	3	Present	2	1	3
Absence	7	26	33	Absence	26	7	33
Total	8	28	36	Total	28	8	36

Elevation 150 m Elevation 200 m

$$W^{+} = \ln\left[\frac{T_{11}T_2}{T_{21}T_1}\right] = \ln\left[\frac{1x33}{7x3}\right] = 0.452$$

$$W^{+} = \ln\left[\frac{T_{11}T_2}{T_{21}T_1}\right] = \ln\left[\frac{2x33}{26x3}\right] = 0.053$$

$$W^{-} = \ln\left[\frac{T_{12}T_2}{T_{22}T_1}\right] = \ln\left[\frac{2x33}{26x3}\right] = 0.017$$

$$W^{-} = \ln\left[\frac{T_{12}T_2}{T_{22}T_1}\right] = \ln\left[\frac{1x33}{7x3}\right] = 0.165$$

The weight calculation process for other environmental factors was also applied in the same steps. The weight raster of each environmental factor was generated from the weight calculation. The weight raster of elevation, slope, geology, land cover, and aspect are described in Fig. 2.4.

The process of the logistic regression model performed in GIS was done by overlaying all environmental factors weight raster (Fig. 2.5a) to the landslide evidence point and boundary of the pilot sites (Fig. 2.5b) for calculating the LSM value.

The value for landslide susceptibility ranges from 0 to 1. The number of 0 indicates that there is no chance of a landslide likelihood in this region. A higher value means a higher likelihood of landslides. As per the calculation in this particular sampling area, the landslide susceptibility value for all pixels is 0.008 (Fig. 2.6).

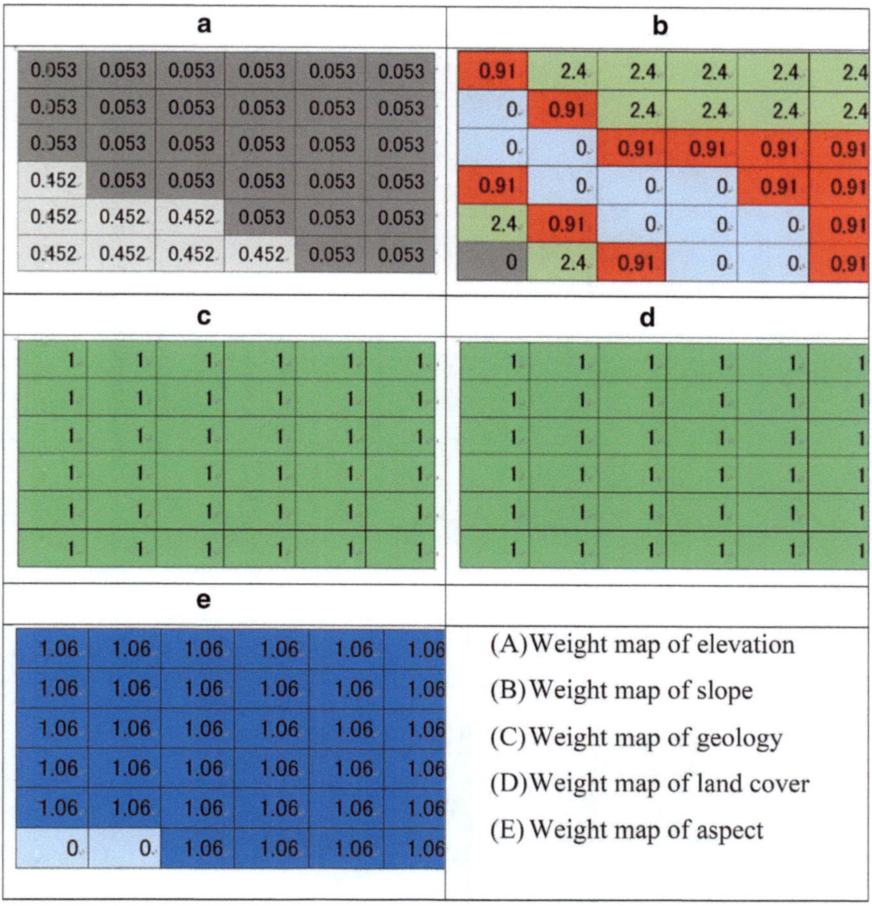

Fig. 2.4 The weight raster of elevation, slope, geology, land cover, and aspect

2.3 Result

2.3.1 Dem

The original dataset of elevation points obtained from GSI has a spatial resolution of 0.4 arc-second. The elevation point was used in creating the DEM map, slope map, and aspect map, which will be used as the environmental variables to estimate the LSM. Data processing started from the interpolation of elevation points through the IDW method. The interpolation result was raster data in a spatial resolution of 10 m. The spatial reference of the data has been converted into World Geodetic System (WGS) 1984 with the coordinate system of Universal Transverse Mercator (UTM) zone 52 north.

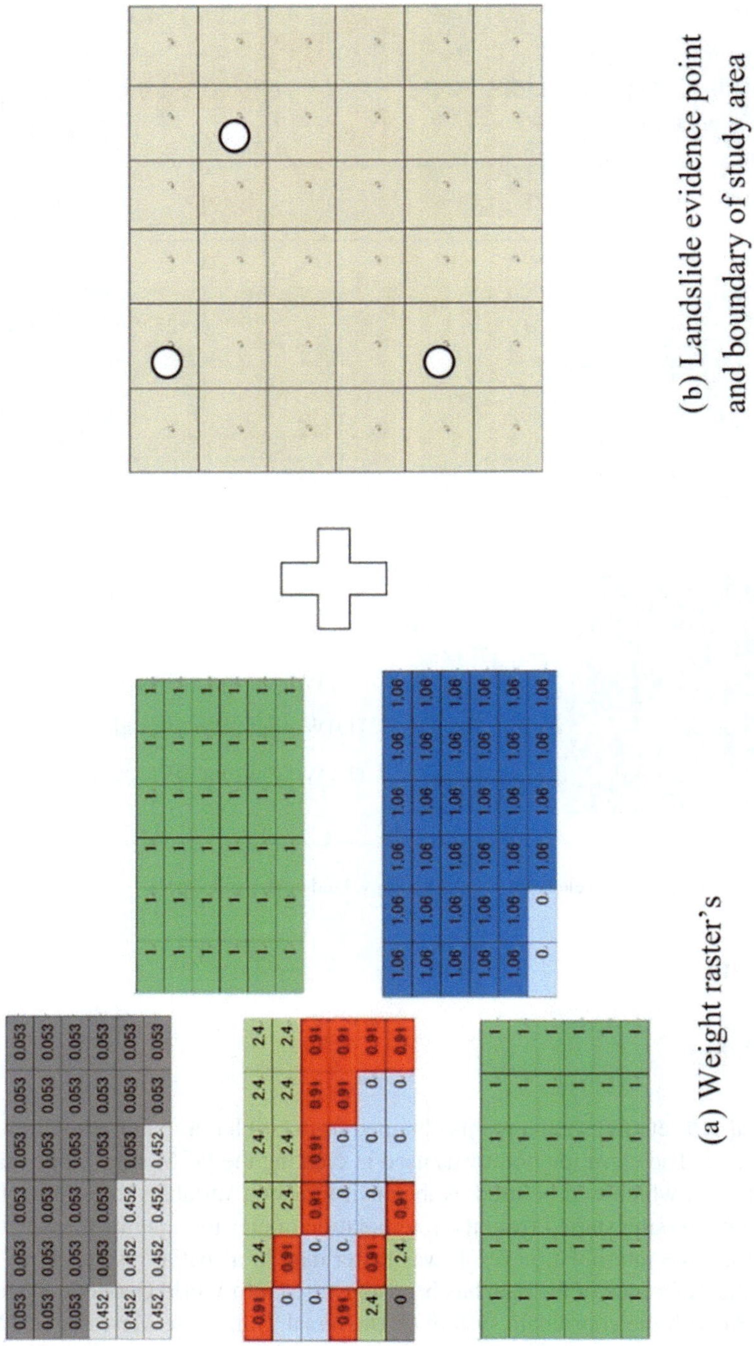

Fig. 2.5 Logistic regression process

Fig. 2.6 The simulation of landslide susceptibility value 3

0.008	0.008	0.008	0.008	0.008	0.008
0.008	0.008	0.008	0.008	0.008	0.008
0.008	0.008	0.008	0.008	0.008	0.008
0.008	0.008	0.008	0.008	0.008	0.008
0.008	0.008	0.008	0.008	0.008	0.008
0.008	0.008	0.008	0.008	0.008	0.008

The elevation, slope, as well as aspect map that was produced from the GSI's elevation point, is displayed in Fig. 2.7. The slope varied from 0 to 64 degrees with an average of 15.7 degrees, while the altitude ranged from 0 to 630 m with an average of 141 m. Furthermore, most of the lowland area was surrounded by elevations greater than 300 m, with nearly half (48.7%) of the research area having an elevation of less than 100 m. Additionally, approximately 41.6% of the Hofu City has a slope of less than 10 degrees, while around 17% and 41.4% have moderate to high slopes. Also, the aspect was concentrated in the south and southeast, where the slope direction might be conceptualized as this physical component.

2.3.2 *Geological Map*

Geological data were downloaded from GSI websites in Tiff format. Similar to the elevation data, geological data were also converted into the spatial reference of the WGS 984 with the UTM zone 52 north coordinate system. Japan's 1:200,000 scale geological map acted as the source of the lithology unit keys included in the geological map. Lithology was classified into 14 classes: fan deposits, granite, granodiorite, lower terrace, non-alkaline, marine and non-marine sediments, middle terrace, non-alkaline mafic volcanic rocks, dune deposits, politic gneiss, politic schist, reclaimed land, mafic schist, and water. Figure 2.8 depicts the spatial variation of the lithology map within Hofu City.

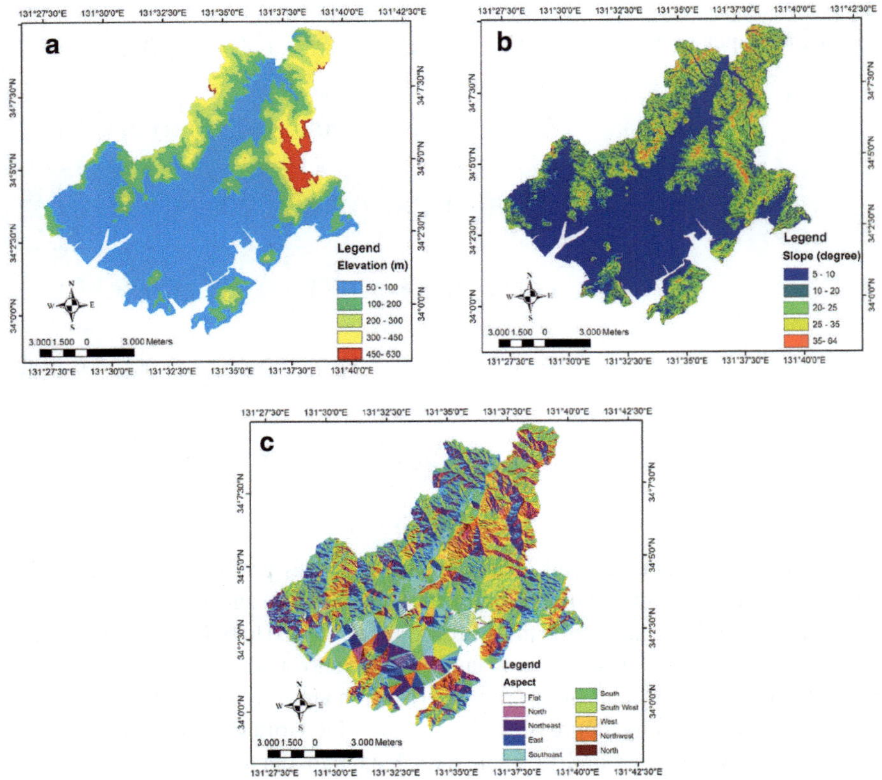

Fig. 2.7 The spatial distribution of **a** Elevation, **b** Slope, and **c** Aspect within the study area

2.3.3 The Spatial Distribution of Land Cover

The spatial distribution of land cover has been generated using an ALOS AVNIR-2 image acquired on July 14, 2009. Image processing started with converting the DN image into absolute radiance value, followed by radiance to reflectance conversion. Moreover, we used the OBIA approach to cluster the land cover into several categories within the region. The land cover map is shown in Fig. 2.9.

As shown in the figure, forest land accounts for 57.3% of the pilot sites, followed by urban area (26.4%), agricultural land (12.8%), bare soil (2.2%), and water (1.4%).

An accuracy assessment is needed to verify the result of the classification. Typically, it was determined by comparing it to reference data, which is deemed to represent its actual condition accurately. Moreover, a confusion matrix is typically used to describe the findings of an accuracy assessment. The Kappa statistic was applied as a consensus on predicted models and the actual condition. As the reference, the high-resolution image provided by Google Maps (Copyright of Google MapIT, Zenrin) with the time acquisition of 3 February 2006 was used. The accuracy assessment result is shown in Fig. 2.10. From this analysis, the overall accuracy was 90.83%,

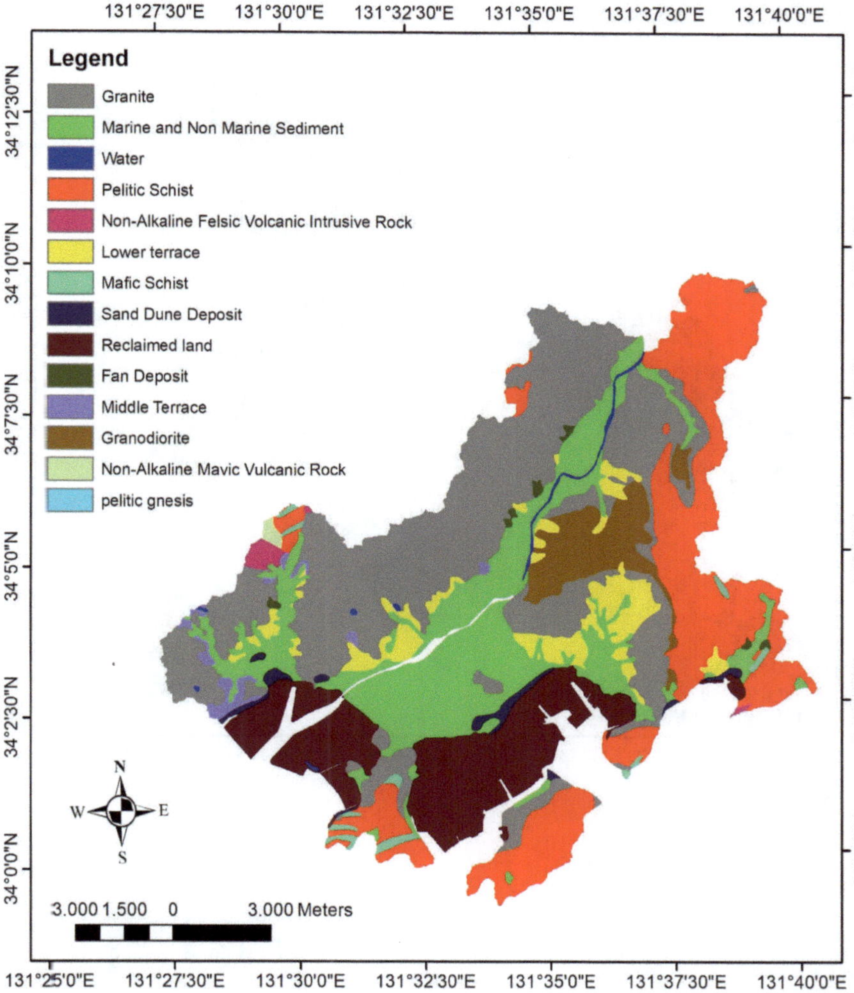

Fig. 2.8 Lithology map

with a Kappa coefficient of 0.89. Solid/ good classification performance was indicated by Kappa values greater than 0.80. Meanwhile, classification performance was considered medium once the Kappa values fell between 0.40 and 0.80, while it is called low if below 0.40 (Jensen 2005).

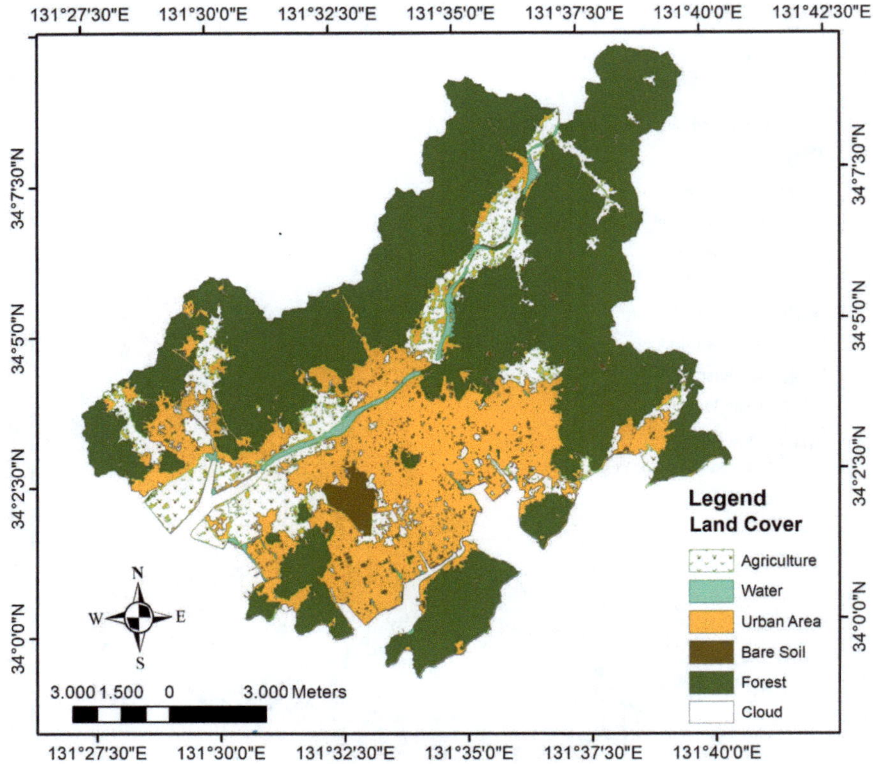

Fig. 2.9 Land cover map

Overall Accuracy = (109/120) 90.8333%
Kappa Coefficient = 0.8900

Ground Truth (Percent)

Class	EVF: Layer: a	EVF: Layer: b	EVF: Layer: c	EVF: Layer: f	EVF: Layer: s
Unclassified	0.00	0.00	0.00	0.00	0.00
Agriculture F	90.00	0.00	10.00	0.00	5.00
Bare Soil	0.00	95.00	0.00	0.00	0.00
Cloud	0.00	0.00	90.00	0.00	0.00
Forest	5.00	0.00	0.00	100.00	0.00
Settlement	5.00	5.00	0.00	0.00	95.00
water	0.00	0.00	0.00	0.00	0.00
Total	100.00	100.00	100.00	100.00	100.00

Ground Truth (Percent)

Class	EVF: Layer: w	Total
Unclassified	0.00	0.00
Agriculture F	5.00	18.33
Bare Soil	0.00	15.83
Cloud	0.00	15.00
Forest	15.00	20.00
Settlement	5.00	18.33
water	75.00	12.50
Total	100.00	100.00

Fig. 2.10 Land cover accuracy assessment

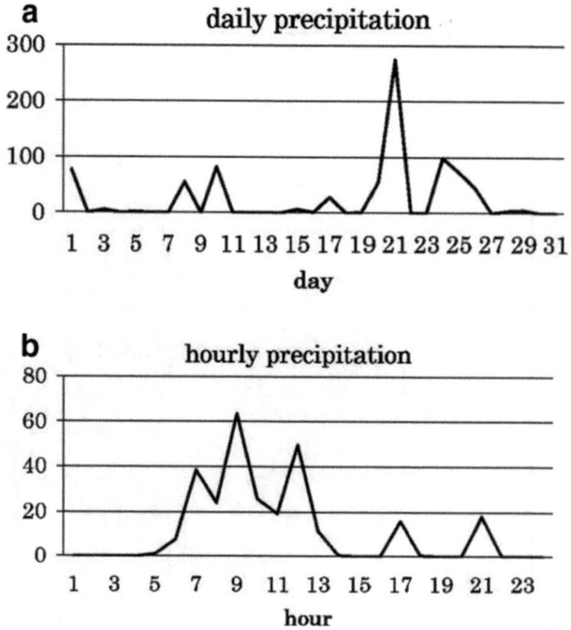

Fig. 2.11 Daily **a** and hourly **b** precipitation (mm) in Hofu City during the landslide event in July 2009

2.3.4 *Rainfall Map*

In late July 2009, a stationary rain front (Baiu Front) brought significant precipitation across Chugoku and northern Kyushu districts. An intense rainstorm was recorded in the pilot study on July 21, 2009. An enormous amount of precipitation was reported on that particular day, more than 270 mm, and the city's most significant hourly rainfall was 63.5 mm/hour. These rains caused numerous small landslides on the mountain slopes around Hofu. Figure 2.11a shows the daily rainfall rate in the landslide event, and Fig. 2.11b depicts the hourly rate in the landslide event on July 21, 2009.

The precipitation dataset was obtained from rain gauges at the pilot study and surrounding city. Moreover, the precipitation dataset was taken on July 21, 2009, which triggered many shallow landslides. The data was stored in point-type data, which was interpolated through the IDW method to create the rainfall map. Figure 2.12 shows the interpolation results in the raster data set, which describes rainfall distribution. During that day, the highest precipitation occurred in the lowland area, which is illustrated by the purple color.

2.3.5 *The LSM*

The LSM was conducted after we defined the landslide's physical factors. All factors were then reclassified based on the WOE approach using the Arc-Spatial Data

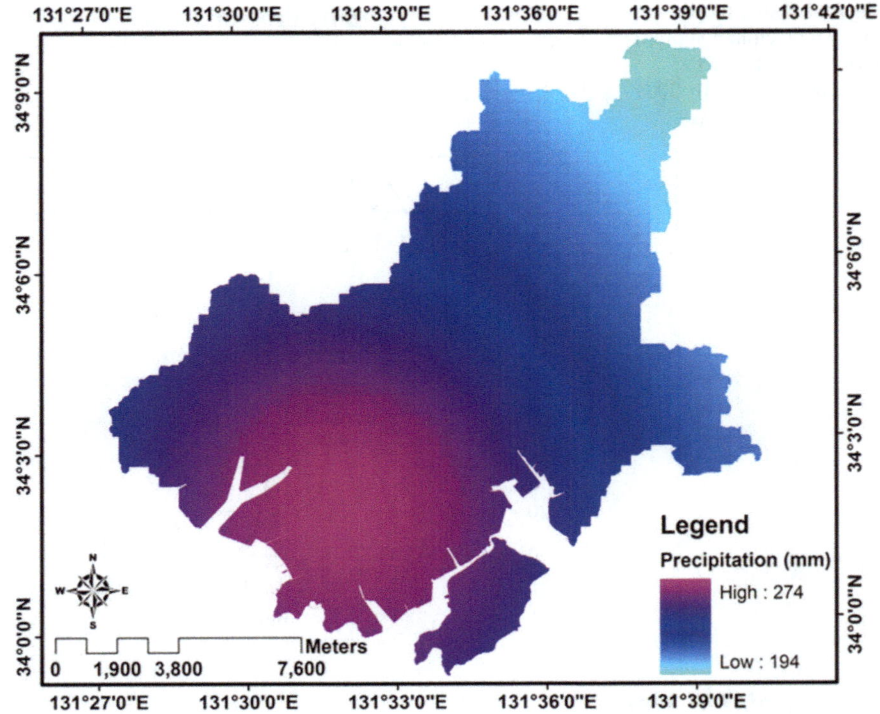

Fig. 2.12 Spatial distribution of rainfall in the study area on July 21, 2009

Modeller (Arc-SDM) tool in ArcGIS 10.8.2 to determine the relationship and the influence of each environmental variable during the incident. The weights of each environmental factor were calculated through the WOE tool in the Spatial Data Modeller. Weights and contrast indicate the relationship and the effect of landslide occurrence. WOE of all environmental factors is described in Table 2.5.

The reclassification of elevation is shown in Table 2.5A. The weights and contrast show that classes with elevations ranging from 50 to 350 m (positive W +) were more likely to have landslides than the other classes. Table 2.5B describes slope reclassification and WOE. It shows W + was 0 for slope gradient classes < 5° and > 50°, which means that the possibility of disaster likelihood within the slope gradient was zero. Meanwhile, the W + was positive for slope range > 5° and < 50°; this implies that gradients of slopes higher than 5° and lower than 45° indicate that this range of slope gradient was vulnerable to the occurrences of landslide. Moreover, the aspect was classified by numbers 1 to 9, respectively, referring to flat, north, northeast, east, southeast, south, southwest, west, and northwest (Table 2.5C). The value and contrast of W + revealed that slope directions of north and northeast were suitable for this type of natural disaster. Table 2.5D defines the reclassification and WOE for the lithology parameter. The result stated that the lithology class numbers 10, 11, and 12 (respectively referring to granodiorite, middle terrace, and fan deposits)

Table 2.5 WOE of all environmental factors: (A) Elevation, (B) Slope, (C) Aspect, (D) Lithology, (E) Land Cover

(A)				(B)			
Class	W +	W−	Contrast	Class	W +	W−	Contrast
550–630	0	0	0	55–64	0	0	0
500–550	−2.611	0.0137	−2.6247	50–55	0	0	0
450–500	−3.1688	0.0249	−3.1937	45–50	0.5223	−0.0013	0.5236
400–450	−2.9889	0.0416	−3.0305	40–45	0.277	−0.0032	0.2801
350–400	−0.7235	0.0361	−0.7596	35–40	0.4889	−0.0236	0.5125
300–350	0.2454	−0.0324	0.2778	30–35	0.6277	−0.094	0.7218
250–300	0.6665	−0.1873	0.8538	25–30	0.7953	−0.3298	1.1251
200–250	0.8239	−0.4317	1.2556	20–25	0.7964	−0.7599	1.5564
150–200	0.9054	−0.9251	1.8305	15–20	0.7612	−1.6617	2.4229
100–150	0.8308	−1.7449	2.5757	10–15	0.6508	−2.9454	3.5963
50–100	0.6633	−4.5054	5.1688	5–10	0.5342	−4.8568	5.391
0–50	0	7.6021	−7.6021	0–5	0	7.6021	−7.6021
(C)				(D)			
Class	W +	W−	Contrast	Class	W +	W−	Contrast
9	−0.0351	0.005	−0.0401	14	0	0	0
8	−0.1192	0.0368	−0.156	13	0	0	0
7	−0.1715	0.1016	−0.2731	12	0.4753	−0.0346	0.5099
6	−0.1085	0.1212	−0.2297	11	0.3108	−0.0246	0.3354
5	−0.1037	0.2277	−0.3314	10	0.2694	−0.022	0.2914
4	−0.0246	0.1167	−0.1413	9	−0.8276	0.1314	−0.959
3	0.0149	−0.235	0.2499	8	−0.8791	0.1447	−1.0238
2	0.3798	−0.0171	0.3969	7	−0.9131	0.154	−1.0671
1	0	7.6021	−7.6021	6	−1.0856	0.2203	−1.3059
				5	−1.0584	0.2227	−1.2811
				4	−1.5159	0.5358	−2.0517
				3	−1.5253	0.5444	−2.0697
				2	−1.8333	0.9687	−2.802
				1	0	7.5975	−7.5975
(E)				Note			
Class	W +	W−	Contrast				
6	0	0	0	(A) WOE of Elevation			
5	0.7619	−4.2649	5.0268	(B) WOE of Slope			
4	0.7249	−4.385	5.1099	(C) WOE of Aspect			
3	0.3601	−4.9488	5.3089	(D) WOE of Lithology			
2	0.1069	−3.8698	3.9767	(E) WOE of Land Cover			
1	0	7.8166	−7.8167				

Table 2.6 LR Coefficient

Environmental factor	Coefficient
Constant value	−4.8630
Elevation	−3.9264
Aspect	−0.1820
Slope	−1.9248
Geology	0.2117
Land cover	−2.0987

were conducive to the likelihood of landslides. Moreover, land cover was classified by numbers 1 to 6 (Table 2.5E), referring to water, agriculture, urban area, bare soil, forest, and cloud. A positive contrast suggests that the category is favorable for the threat of landslides, whereas a negative contrast indicates that the class is not. Based on the WOE calculation in Table 2.5E, agriculture, urban area, bare soil, and forest were suitable conditions for landslide incidents. In this assessment, the WOE was calculated for all environmental factors except for rainfall.

The logistic regression tool from SDM combined the point landslide inventory data, evidence maps, and weight of evidence table. The tool generates tables with the environmental factor coefficients and computes the posterior probability. The coefficients represent the degree of significance of the different environmental component types. A positive coefficient suggests that the category increases susceptibility level, whereas a negative lessens the likelihood of landslide occurrence. The results of LR coefficients are listed in Table 2.6.

The LR calculation shows that the susceptibility value ranged from 0 to 0.00895. A value of 0 indicates very low susceptibility or low likelihood that a landslide will hit in this area; the higher the value, the higher the likelihood of the disaster. Figure 2.13 describes the classification of susceptibility distribution of landslide occurrences. We classified the susceptibility map into five classes, where red indicates the highest susceptibility and blue indicates no susceptibility.

2.4 Discussion

In this chapter, the LSM was created utilizing LR techniques and five physical parameters comprising land cover, slope, elevation, aspect, lithology, and excessive precipitation as the trigger factors. These physical factors gave a good overview and effectiveness in the landslide mapping (Hong 2023). In terms of DEM preparation, the elevation point provided by GSI Japan made the availability of a resolution elevation map possible. Moreover, the application of the IDW method in elevation point interpolation performed good results in creating a raster image of DEM.

Hofu City has a variety of elevations, from 0 to 630 m (Table 2.5A), with slope gradients varying between zero to sixty-four degrees (Table 2.5B). According to the

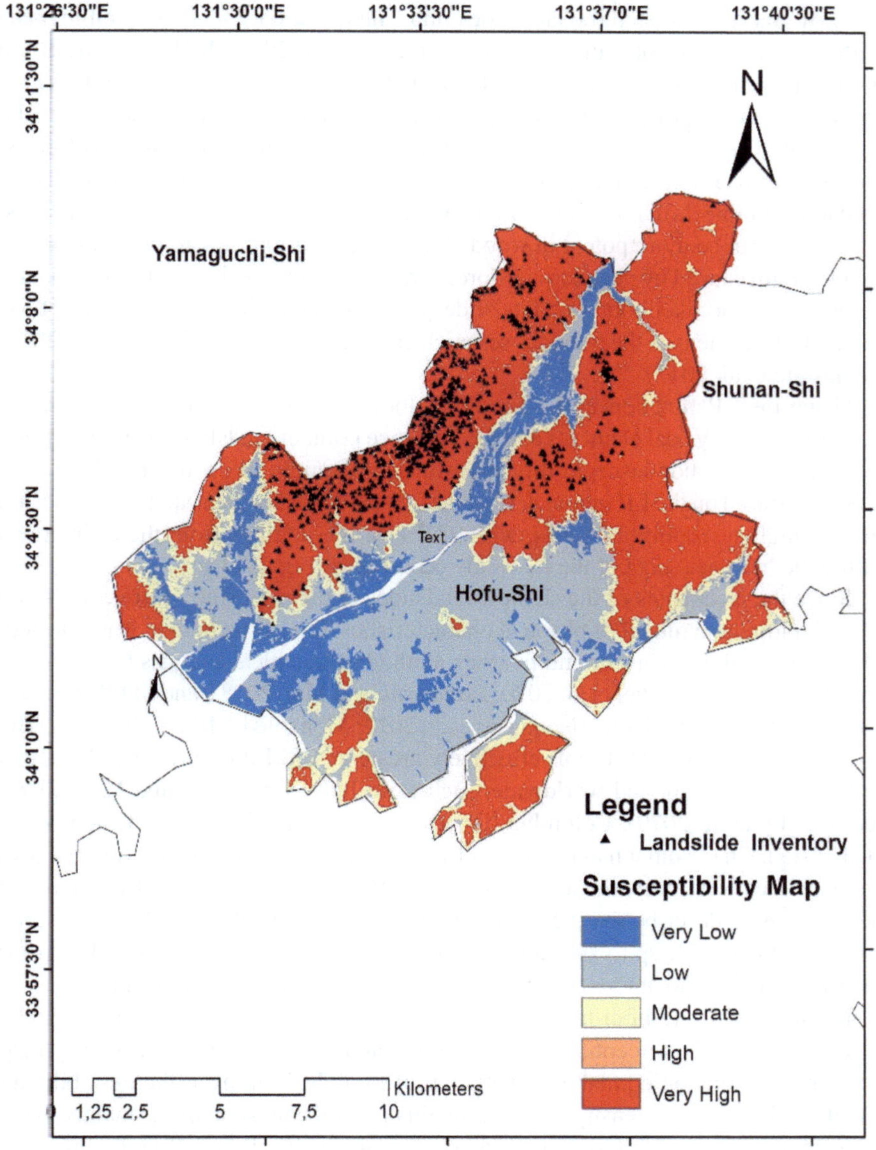

Fig. 2.13 Landslide susceptibility map

inventory map, the disaster arises at elevations and slopes ranging from 50 to 550 m and 5° to 50°, with the most significant frequent incidence occurring at 200 m and 25° with the most likely land cover type of forest land (Figs. 2.7, 2.9, 2.13). Meanwhile, based on the weight computation, landslides tend to get increasingly prevalent between 50 to 350 m of elevation and in the slope angle zone of 5° to 50°(Table 2.5 A, B). Further research revealed that soil and colluvium is an additional element that leads to shallow landslides; thus, landslides are less common in areas with slope angles greater than 35 degrees (Chen et al. 2015, 2018). As a result, a highly steep slope is the most vulnerable zone for landslide incidents all of the time. In addition, the slope directions north, northeast, and northwest were conducive to the incidence of landslides (Fig. 2.7c and Table 2.5C). Also, territories such as forests, bare soil, urban areas, and agricultural land were conducive to the emergence of landslides (Table 2.5E). Even having identical slope angles and lithology qualities, a landslide can be more potent in wooded highland locations due to the reduction in root strength caused by vegetation. Moreover, the granite was the most lithologically favorable for the likelihood of a landslide (Fig. 2.8 and Table 2.5D). This result has the same agreement as the previous research conducted in Japan (Chigira et al. 2011; Yamashita et al. 2017).

Using the LR function in the Arc-SDM tool, the weights and rasters (evidence rasters) of the physical elements and the evidence point of landslides were integrated to assess the landslide susceptibility. In Hofu, the susceptibility score ranged from 0 to 0.00895. Based on the LR output, 916 of the 928 landslide incident spots were located in very high-susceptibility zones, with 98.7% of the findings from the estimations fitting the historical data of the landslide occurrence (Fig. 2.13).

Since the mechanics of the shallow landslides in the pilot study were considered when building the model of physical characteristics, the LR technique was selected. Moreover, the LR is a useful multivariate method of statistical analysis for assessing the LSM (Goyes-Penafiel et al. 2020; Lombardo and Mai, 2018), and SDM is a good tool for LR-based modeling (Neuhaser et al. 2012; Mohamed &Elmahdy 2016). The LR method and Arc-SDM tools effectively create an LSM for Hofu City. Also, this technique is well applied worldwide, such as in Italy (Lombardo and Mai, 2018), India (Feby et al. 2020), Colombia (Goyes-Penafiel et al. 2020), and China (Zhao et al. 2019). On the other hand, SDM had some limitations; for instance, training data on landslides were provided as points (up to 1,000), but the scarp centroid may not be sufficient to describe scarp circumstances for massive landslides. Also, a distinct condition table with up to six thousand circumstances serves as the foundation for the LR, reducing the amount of environmental parameters that can be input.

In this chapter, rainfall-triggered shallow landslides provided the basis for the predicted landslide susceptibility because the landslide spots in the inventory map were triggered by an equitable distribution of rainfall. Moreover, the destabilizing effects of the rainfall intensity may be affecting the soil cohesion; this is mainly in the case of increasing rainfall from its lowland minimum to its highland maximum. Hofu city was hit by intense rainfall in late July 2009, and on 21st July 2009, the rainfall reached the highest intensity, up to 274 mm/day, then triggered many shallow landslides, resulting in 928 landslides scattered in the hilly area. Additionally, previous studies revealed that incorporating rainfall into the analyses improves the landslide susceptibility models' accuracy (Zhao et al. 2020), and the combination of event inventory and spatial rainfall data could help researchers better understand the factors that determine landslide vulnerability (Smith et al. 2023).

Topography and lithology are static terrain factors that influence susceptibility. However, the development of landslides altered the set of circumstances that existed during the triggering events. Moreover, the LR coefficient result indicates that lithology is the physical element that has the most influence on this shallow landslide occurrence. Since lithological and structural variations frequently result in differences in the durability and permeability of rocks and soils, it is well known that geology is one of the physical elements that strongly influences the likelihood of landslides (Pradhan and Lee 2010). Moreover, type-specific lithology, physical and chemical properties, and mineralogy are essential key factors, together with slope aspect, angle, and land cover. Some literature (Chen et al. 2011; Moragues et al. 2023; Machay et al. 2023; Zhao et al. 2023) most commonly consider lithological conditions.

2.5 Conclusion

This chapter used LR models with 10 m grid sizes to assess landslide susceptibility. The elements selected as controlling factors were aspect, elevation, slope, rainfall, lithology, and land cover. Based on the weight calculation results of LR models, the physical parameters of elevation ranging between 50 and 350 m with a slope range of 5° to 50° were favorable for the likelihood of landslides. Also, north and northeast slope directions were conducive to the incidence of landslides. Moreover, the land coverings conducive to the likelihood of landslides included forests, bare soil, agricultural fields, and urban areas. Meanwhile, granodiorite were mostly lithologically suitable for the landslides. Further, according to the LSM, of the 928 landslide spots, 916 were located in very high susceptibility zones, which revealed that 98.7% of the LSM estimations were close to the inventory/reference dataset. The results of the proposed method show great promise for the spatial prediction of landslides, and their land susceptibility maps could serve as a first evaluation for any planning project involving municipalities to carry out more in-depth research in areas previously determined to be highly susceptible. Ideally, the susceptibility models should consider many physical variables, including curvature, geomorphology, distance to the geological fault, distance to the road, and flow density. Still, our research was limited to six variables due to local site data availability. In the future, it is necessary to add more variables that can be generated from DEM, such as curvature, flow density, and terrain roughness index, and combine the LR models and Principal Component Analysis to obtain the most influential factors among many proxies of the model.

2.6 Author Contribution Statement

Berlita Nathania did data collection and analysis and wrote the initial draft manuscript. Martiwi Diah Setiawati did data curation, supervised, and revised the manuscript. Both authors have equal contributions.

References

Agterberg FP, Bonham-Carter GF, Cheng Q, Wright DF (1993) Weights of evidence modeling and weighted logistic regression for mineral potential mapping. In: Davis JC, Herzfeld UC (eds) Computers in geology, 25 years of progress. Oxf Univ Press, Oxford, pp 13–32

Bonham-Carter GF (1994) Geographic Information Systems for Geoscientists: modelling with GIS, Vol. 13 of Computers Methods in Geosciences, Pergamon Press

Bouvet M, Goryl P, Chander G, Santer R, Saunier S (2007) Preliminary radiometric calibration assessment of ALOS AVNIR-2. In: Geoscience and Remote Sensing Symposium, 2007. IGARSS 2007. IEEE International, 2673–2676

Burnett C, Blaschke T (2003) A multi-scale segmentation/object relationship modeling methodology for landscape analysis. Ecol Model 168:233–249

Burrough PA (1986) Principles of geographical information systems for land resources assesment. Clarendon Press, Oxford

Chang M, Dou X, Su F, Yu B (2023) Remote sensing and optimized neural networks for landslide risk assessment: Paving the way for mitigating Afghanistan landslide damage. Ecol Ind 156:111179. https://doi.org/10.1016/j.ecolind.2023.111179

Chen J, Zhu X, Imura H, Chen X (2010) Consistency of accuracy assessment indices for soft classification: Simulation analysis. ISPRS J Photogramm Remote Sens 65:156–164

Chen H, Lin G, Lu M, Shih TY, Horng MJ, Wu SJ, Chuang B (2011) Effects of topography, lithology, rainfall and earthquake on landslide and sediment discharge in mountain catchments of southeastern Taiwan. Geomorphol 133:132–142

Chen CW, Saito H, Oguchi T (2015) Rainfall intensity–duration conditions for mass movements in Taiwan. Prog Earth Planet Sci 2:1–13. https://doi.org/10.1186/s40645-015-0049-2

Chen L, Guo H, Gong P, Yang Y, Zuo Z, Gu M (2021) Landslide susceptibility assessment using weights-of-evidence model and cluster analysis along the highways in the Hubei section of the Three Gorges Reservoir Area. Comput Geosci 56:104899

Chen W, Pourghasemi HR, Kornejady A, Xie X (2018) GIS-based landslide susceptibility evaluation using certainty factor 455 and index of entropy ensembled with alternating decision tree models, In book: Natural Hazards GIS-Based Spatial Modeling Using Data Mining Techniques. Adv Nat Technol Hazards Res, 48

Cheng C, Yang Y, Zhong F, Song C, Zhen Y (2022) An optimization of statistical index method based on gaussian process regression and GeoDetector, for higher accurate landslide susceptibility modeling. Appl 12:10196. https://doi.org/10.3390/app122010196

Chigira M, Mohamad Z (2011) Landslides in weathered granitic rocks in Japan and Malaysia.Bulletin of the Geological Society of Malaysia 57: 1–6 . Access at https://gsm.org.my/wp-content/uploads/gsm_file_2/702001-100367-PDF.pdf

Chow VT, Maidment DR, Mays LW (1988) Applied hydrology (p. 564). McGraw-Hill, Inc. https://wecivilengineers.files.wordpress.com/2017/10/applied-hydrology-ven-te-chow.pdf

Congalton RG (1991) A review of assessing the accuracy of classifications of remotely sensed data. Remote Sens Environ 37:35–46

Feby B, Achu A, Jimnisha K, Ayisha VA, Reghunath R (2020) Landslide susceptibility modelling using integrated evidential belief function based logistic regression method: A study from Southern Western Ghats, India. Remote Sens Appl: Soc Environ 20:100411

Froude MJ, Petley DN (2018) Global fatal landslide occurrence from 2004 to 2016. Nat Hazards Earth Syst Sci 18(8):2161–2181. https://doi.org/10.5194/nhess-18-2161-2018

GSI (2014) Geological map. Access on 3 June 2014 at www.gsi.go.jp

Hong H (2023) Assessing landslide susceptibility using combination models. For Ecol Manage 545:121288. https://doi.org/10.1016/j.foreco.2023.121288

ICHARM (2020) Collection of critical situations during flood emergency response. Access on October 18, 2023 at https://www.pwri.go.jp/icharm/special_topic/20200625_flood_response_collection_e/flood_response_collection_A3_e.pdf

JAXA (2008) alos data users handbook revision C. [Cited 2013, April 17]. Available from: URL: http://www.eorc.jaxa.jp/ALOS/en/doc/fdata/ALOS_HB_RevC_EN.pdf

Jensen JR (2005) Introductory digital image processing: a remote sensing perspective. John R Jensen 3rd ed, Pearson Education Inc., Upper Saddle River

JMA (Japan Meteorological Agency). (2023) Monthly rainfall dataset from 1990–2022. Access on December 1, 2023 at https://www.data.jma.go.jp/risk/obsdl/index.php

Kohno M, Higuchi Y (2023) Landslide susceptibility assessment in the japanese archipelago based on a landslide distribution map. ISPRS Int. J. Geo-Inf. 12:37. https://doi.org/10.3390/ijgi12020037

Laura Zangmene F, Nsangou Ngapna M, Christian Balla Ateba M, Marie Monespérance Mboudou G, Pascal Wabo Defo L, Tetang Kouo R, Kagou Dongmo A, Owona S (2023) Landslide susceptibility zonation using the analytical hierarchy process (AHP) in the Bafoussam-Dschang region (West Cameroon). Adv Space Res, 71:5282–5301. https://doi.org/10.1016/j.asr.2023.02.014

Lombardo L, Mai PM (2018) Presenting logistic regression-based landslide susceptibility results. Eng Geol 244:14–24. https://doi.org/10.1016/j.enggeo.2018.07.019

Machay F, Moussaoui SE, Talibi HE (2023) Insights into large landslide mechanisms in tectonically active agadir, morocco: the significance of lithological, geomorphological, and soil characteristics. Sci Afr 22:e01901. https://doi.org/10.1016/j.sciaf.2023.e01901

Mirus BB, Jones ES, Baum RL, Godt JW, Slaughter S, Crawford MM, Lancaster J, Stanley T, Kirschbaum DB, Burns WJ et al (2020) Landslides across the USA: Occurrence, susceptibility, and data limitations. Landslides 17:2271–2285

Mohamed MM, Elmahdy SI (2016) Remote sensing and information value (IV) model for regional mapping of fluvial channels and topographic wetness in the Saudi Arabia. Giscience & Remote Sens 53:520–541

Mohammadi A, Torkashvand A, Irani J, Sorur, (2014) The preparation of landslide map by landslide numerical risk factor (LNRF) model and geographic information system (GIS). Egypt J Remote Sens Space Sci 17(2):159–170. https://doi.org/10.1016/j.ejrs.2014.08.001.ISSN1110-9823

Moragues S, Lenzano MG, Jeanneret P, Gil V, Lannutti E (2023) Landslide susceptibility mapping in the Northern part of Los Glaciares National Park, Southern Patagonia, Argentina using remote sensing, GIS and frequency ratio model. Quat Sci Adv. 100146. https://doi.org/10.1016/j.qsa.2023.100146

Panchal S, Shrivastava AK (2021) Landslide hazard assessment using analytic hierarchy process (AHP): A case study of National Highway 5 in India. Ain Shams Eng J 13:101626. https://doi.org/10.1016/j.asej.2021.10.021

Panchal S, Shrivastava AKr (2022) Landslide hazard assessment using analytic hierarchy process (AHP): a case study of National Highway 5 in India. Ain Shams Eng J 13. https://doi.org/10.1016/j.asej.2021.10.021

Pei Y, Qiu H, Yang D, Liu Z, Ma S, Li J, Cao M, Wufuer W (2023) Increasing landslide activity in the Taxkorgan River Basin (eastern Pamirs Plateau, China) driven by climate change. CATENA 223:106911. https://doi.org/10.1016/j.catena.2023.106911

Persichillo MG, Bordoni M, Meisina C, Bartelletti C, Barsanti M, Giannecchini R, D'Amato Avanzi G, Galanti Y, Cevasco A, Brandolini P, Galve JP (2016) Shallow landslides susceptibility assessment in different environments. Geomatics, Nat. Hazards Risk 8(2), 748–771. https://doi.org/10.1080/19475705.2016.1265011

Pradhan B, Lee S (2010) Landslide Susceptibility assessment and factor effect analysis: back propagation artificial neural networks and their comparison with frequency ratio and bivariate logistic regression modeling. Environ Model Softw 25:747–759. https://doi.org/10.1016/j.envsoft.2009.10.016

Qiu HJ, Cui YF, Pei YQ, Yang DD, Hu S, Wang XG, Ma SY (2019) Temporal patterns of nonseismically triggered landslides in Shaanxi Province. China Catena 187:1043–1054. https://doi.org/10.1016/j.catena.2019.104356

Roccati A, Paliaga G, Luino F, Faccini F, Turconi L (2021) GIS-based landslide susceptibility mapping for land use planning and risk assessment. Land 10(2), 162. https://doi.org/10.3390/land10020162

Sah A, K, Sah BP, Honji K, Kubo N, Senthil S (2012) Semi-automated cloud/shadow removal and land cover change detection using satellite imagery. Int Arch Photogramm, Remote Sens Spat Inf Sci, XXXIX-B7, XXII ISPRS Congress, Melbourne, Australia, 25 August–01 September 2012, pp 335−340

Smith HG, Neverman AJ, Betts HD, Spiekermann RI (2023) The influence of spatial patterns in rainfall on shallow landslides. Geomorphology 437:108795. https://doi.org/10.1016/j.geomorph.2023.108795

World Health Organization (WHO) (2022) Landslides. https://www.who.int/news-room/fact-sheets/detail/landslides

Yamashita K,Hattanji T, Tanaka Y,Doshida S, Matsushima T (2017) Topographic characteristics of rainfall-induced shallow landslides on granitic hillslopes: A case study in Hofu City, Yamaguchi Prefecture, Japan. Tsukuba Geoenvironmental Sciences, 13, pp 23–29. https://tsukuba.repo.nii.ac.jp/record/44942/file_preview/TGS_13_23.pdf

Yu L, Zhou C, Wang Y, Cao Y, Peres DJ (2022) Coupling Data- and Knowledge-Driven methods for landslide susceptibility mapping in Human-Modified environments: a case study from Wanzhou County, three gorges reservoir area. China Remote Sens 14:774. https://doi.org/10.3390/rs14030774

Yu X, Zhang K, Song Y et al (2021) Study on landslide susceptibility mapping based on rock–soil characteristic factors. Sci Rep 11:15476. https://doi.org/10.1038/s41598-021-94936-5

Zhao Y, Wang RT, Jiang Y, Liu H, Wei Z (2019) GIS-based logistic regression for rainfall-induced landslide susceptibility mapping under different grid sizes in Yueqing. Southeast China Eng Geol 259:105147. https://doi.org/10.1016/j.enggeo.2019.105147

Zhao S, He S, Li X, Deng Y, Liu Y, Yan S, Bai X, Xie Y (2021) The Xinmo rockslide-debris avalanche: An analysis based on the three-dimensional material point method. Eng Geol 287:106109. https://doi.org/10.1016/j.enggeo.2021.106109

Zhao S, Dai F, Deng J, Wen H, Li H, Chen F (2023) Insights into landslide development and susceptibility in extremely complex alpine geoenvironments along the western Sichuan-Tibet Engineering Corridor. China CATENA 227:107105. https://doi.org/10.1016/j.catena.2023.107105

Chapter 3
Landslide Hazard Risk and Vulnerability Monitoring—GIS Based Approach

Vipin Upadhyay

Abstract Recent technological strides have ushered in a transformative era for disaster assessment, monitoring, mitigation, and management. Amid this progress, Geographic Information System (GIS) has emerged as a pivotal tool, assuming a pivotal role in monitoring landslide hazards, risks, and vulnerabilities. Through the fusion of diverse spatial data sources—encompassing topography, land cover, rainfall, geology, and infrastructure details—GIS enables comprehensive data integration and interpretation. This integration bears multifaceted applications, spanning Landslide Hazard and Risk Assessment, Vulnerability Monitoring and mapping, Real-time Information Collection, Early Warning Systems, Decision Support, Management, and Community Awareness. With GIS as a cornerstone, a wealth of benefits unfurl—from bolstering landslide hazard, risk, and vulnerability monitoring to seamlessly incorporating this intelligence into land-use planning and infrastructure development. Particularly significant in Himalayan regions, GIS empowers stakeholders with actionable insights, guiding judicious choices that foster resilience and safety. This chapter provides an introduction to landslide risk and vulnerability and also navigates through the need and methodologies for the preparation of an inventory of historical landslide records, provides an understanding of various factors, their importance, and methodologies of evaluation of these factors as driving factors for the occurrence of landslide. This chapter also provides insight into statistical and machine learning technologies that have been used for landslide susceptibility mapping, provides knowledge about implications and prospective avenues of GIS technologies primed for landslide monitoring, and elucidates the prospective applicability of landslide hazard risk and vulnerability monitoring.

Keywords Landslide · Geo-hazard · Geographic information system · Disaster · Geography

V. Upadhyay (✉)
School of Life Sciences, Central University of Himachal Pradesh, Dharamshala, HP, India
e-mail: vipinupadhaya@hpcu.ac.in

G. K. Panda et al. (eds.), *Landslide: Susceptibility, Risk Assessment and Sustainability*, Advances in Natural and Technological Hazards Research 52,
https://doi.org/10.1007/978-3-031-56591-5_3

3.1 Introduction

3.1.1 Understanding Landslide as a Potential Hazard

Landslides, which involve the downward movement of rock, debris, or soil due to the force of gravity, are complex phenomena (Cruden and Varnes 1996). They occur as a result of interactions between factors such as slope geometry, soil, and rock properties, as well as surface and groundwater dynamics (Bogaard and Greco 2016). Landslides represent significant natural disasters that predominantly occur on natural hilly slopes in mountainous regions, influenced by a variety of conditioning factors, including elevation, slope angle, aspect, curvature of the terrain, soil characteristics, lithology, streams, drainage patterns, land and vegetation cover. Additionally, landslides can be triggered by various factors such as extreme rainfall events and earthquakes (Yang et al. 2003). These preexisting characteristics have a significant influence on weathering processes and may lead to a reduction in shear strength (Skilodimou et al. 2018). When this decrease in shear strength is coupled with triggering factors like precipitation, seismic activity, snowmelt, or human interventions, it can culminate in potentially devastating landslides (Petley 2012; Haque et al. 2019). The rapid expansion of the human population, combined with human-made alterations for developmental purposes (Mukane 2014) and the impacts of climate change on mountainous terrains (Mandal and Mandal 2019a), exert immense stress on our natural environment. This situation renders these areas highly susceptible to severe geological and hydrological natural disasters, including cloud bursts, debris flows, and landslides in various regions across the globe (Hulme 2016). Construction activities in hilly regions, driven by anthropogenic factors, further expose these terrains to heightened slope instability (Panchal and Shrivastava 2020). Consequently, this heightened instability contributes to an increased susceptibility to landslides. Conditioning factors, which are pivotal to understanding the landslide process, can be broadly categorized as geological (Bhandari and Dhakal 2020a, b; Tofani et al. 2017; Tsou et al. 2018; Yalcin et al. 2011; Yalcin 2008; Gerrard 1994), topographical (Tsou et al. 2018; Zhang et al. 2016), and hydrological (Dahal et al. 2008) conditions. Weak and fragile geological formations stand out as one of the primary factors contributing to the initiation of landslides (Devkota et al. 2013; Regmi et al. 2014; Gadtaula and Dhakal 2019; Bhandari and Dhakal 2020a). Similarly, the specific geological properties of the soil on slopes play a crucial role in triggering landslides (Bhandari and Dhakal 2020b; Yalcin et al. 2011; Tofani et al. 2017; Bicocchi et al. 2019).

3.1.2 Potential Impact of Landslide

These phenomena not only result in loss of life and property but also pose substantial risks to physical infrastructure, including buildings, roads, bridges, and communication networks. They can disrupt socioeconomic development and have adverse impacts on the natural environment, affecting land, vegetation cover, and water resources in hilly terrains worldwide (Nerella et al. 2019; Lee and Pradhan 2007). Landslide events occurring across the globe have substantial economic ramifications, both in direct and indirect terms. Direct economic losses, which can be quantified in monetary values with relative ease (Panchal and Shrivastava 2020), encompass the tangible financial costs incurred. On the other hand, indirect economic losses encompass societal disruptions, including traffic congestions, travel delays, minor injuries, and loss of life, which cannot be easily measured in monetary terms (Rawat and Pant 2023; Kamal et al. 2022). Given the economic and societal consequences of landslides, landslide susceptibility analysis has emerged as a vital tool for disaster risk reduction, with the primary goal of mitigating damages and casualties (Bragagnolo et al. 2020). The attention towards landslide hazards and their investigation has surged due to an increasing awareness of the socioeconomic impacts associated with landslides (Rai et al. 2014).

3.1.3 Most Effected Geographical Regions

Landslides and slope failures are widely recognized as some of the most destructive and threatening natural hazards in mountainous regions. Frequent landslide occurrences are particularly prevalent in the mountainous regions of the Himalaya and the Western Ghats of India. The Himalayan region of India, a zone undergoing ongoing orogeny processes, is especially susceptible to natural hazards, accounting for approximately 80% of all landslides in India (Das et al. 2023; Saha et al. 2021a, b, c, d). The Frontal Himalayan terrain, which includes the Lesser Himalaya and Siwalik Hills, faces a high risk of landslide disasters due to factors such as heavy rainfall, active tectonics, complex geology, rugged terrain with steep slopes, and evolving geomorphology (Bhandari and Dhakal 2021; Ching et al. 2018; Kayastha et al. 2013). These combined geo-environmental factors lead to numerous landslide occurrences and, consequently, significant loss of life and property (Nerella et al. 2019). For effective and sustainable development planning in hilly terrain, it is imperative to assess and monitor the susceptibility to geostructural hazards, including land subsidence, landslides, and slope failures (Rawat and Pant 2023).

3.1.4 Understanding Landslide Hazard Risk and Vulnerability Monitoring

Monitoring landslide hazard risk and vulnerability involves a comprehensive approach that integrates spatial data collection, analytical tools, and decision-making processes. This approach identifies historical events, evaluates their extent, identifies the contributing factors, predicts landslide risk, assesses exposure, and estimates landslide vulnerability within the studied landscape (Das et al. 2023; Saha et al. 2021a, b, c, d). Numerous researchers have conducted studies in the mountainous and hilly regions of the Himalaya to identify causative factors and consequences of landslide occurrences. They have also recommended mitigation practices to reduce the hazardous impacts of landslides (Bhandari and Dhakal 2021; Mandal and Maiti 2014; Mandal and Mondal 2019b; Das et al. 2023; Saha et al. 2021a, b, c, d). To accurately assess and monitor the real risk associated with any hazard in a specific location, it's crucial to comprehend three key elements: susceptibility to the hazard, exposure, and vulnerability (see Fig. 3.1). A hazard, such as landslides, floods, cyclones, tsunamis, or similar events, represents a situation that can disrupt the normal functioning of society and the environment. Risk, on the other hand, refers to the likelihood of a hazard occurring at a given location. Exposure pertains to the area and resources that are at risk under the influence of a hazard (Upadhyay and Jat 2014). Consequently, the cumulative impact of a hazard's risk on exposed resources is known as vulnerability. In the context of monitoring landslide vulnerability, the landslide event itself is the hazard, while the landscape's conditions or the spatial distribution of factors contributing to landslides are considered determinants of landslide hazard risk. The spatial distribution of resources in regions at risk of landslides provides an estimate of the resources exposed to landslide hazards. The cumulative assessment of the adverse effects that may occur on the resources within a particular area during a landslide event is what we monitor as landslide vulnerability (Dehnavi et al. 2015). Historical event data can offer valuable input for assessing landslide hazards, and evaluating landslide risk for a specific area or landscape is typically achieved through landslide susceptibility or risk mapping (LSM). This can be combined with resource inventory maps that contain information about the resources exposed and vulnerable to landslide hazards within a given region, facilitating the monitoring of landslide hazard risk and vulnerability for that specific area (Das et al. 2023).

3.1.5 Importance of Landslide Vulnerability Monitoring

Landslide vulnerability monitoring represents a conventional approach to effective land use planning and decision-making for conserving natural resources, promoting development, and mitigating the impact of landslide events in hilly regions (Dehnavi et al. 2015). In recent decades, landslide vulnerability maps have also found application in planning safe and cost-effective transportation infrastructure routes in smart cities (Panchal and Shrivastava 2020). Landslide studies can be categorized as either pre-disaster or post-disaster studies, with landslide prediction being part of the pre-disaster studies (Pal and Chowdhuri 2019; Feby et al. 2020). In landslide prediction, a specific region can be subdivided into homogeneous zones based on the likelihood of landslide occurrences, considering various conditional factors (Das et al. 2023).

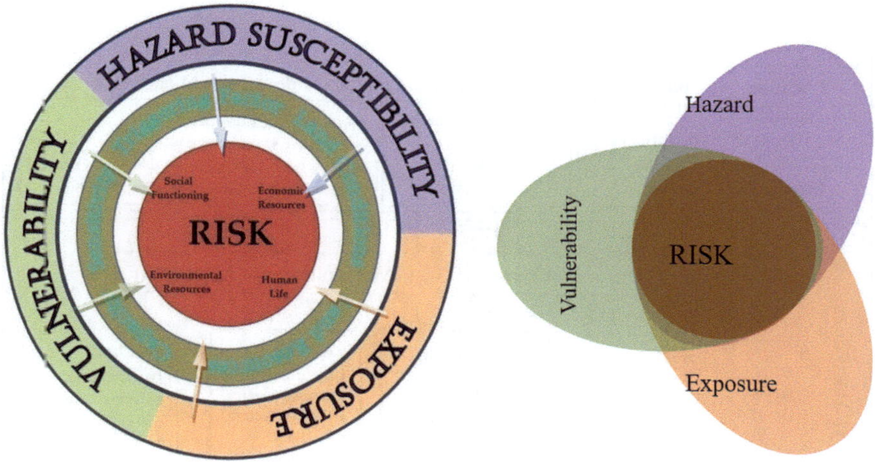

Fig. 3.1 Graphical representation of three key aspects involves in the risk of any hazard, which need to understand to monitor the landslide hazard risk

3.2 What Landslide Risk Is and How It Monitored?

In the context of disaster management, "risk" refers to the probability of a hazardous event occurring and the potential negative consequences or impacts associated with that event. It encompasses the likelihood of a disaster happening, as well as the vulnerabilities of people, property, and the environment to that specific hazard. In same manner, Landslide risk refers to the potential for landslides to occur in a particular area and the associated threat to people, property, and the environment. The risk is often expressed as a combination of the hazard, exposure, and vulnerability (Fig. 3.1). Therefore, to avoid or mitigate the risk or associated impact monitoring of this risk hazard and associated factors become crucial. Landslide hazard risk monitoring is a comprehensive process that involves mapping and categorizing a landscape or region using a continuous or categorical scale. This process estimates the potential loss of resources in terms of their importance under the influence of landslide hazard risk (Sujatha and Rajamanickam 2015). It integrates the assessment of the likelihood of landslide hazards in a given area with the distribution of resources and the vulnerability of both society and resources to cope with such hazards within the same region. This approach takes into account the value of resources, such as ranking or assigning value to assets like buildings, roads, railways, hospitals, schools, community centers, forests, wildlife, and more, which are situated within different hazard zones. This combined consideration of resource value, susceptibility to hazard occurrence (see Fig. 3.2), and hazard exposure provides a more realistic estimation of potential losses resulting from hazard risks.

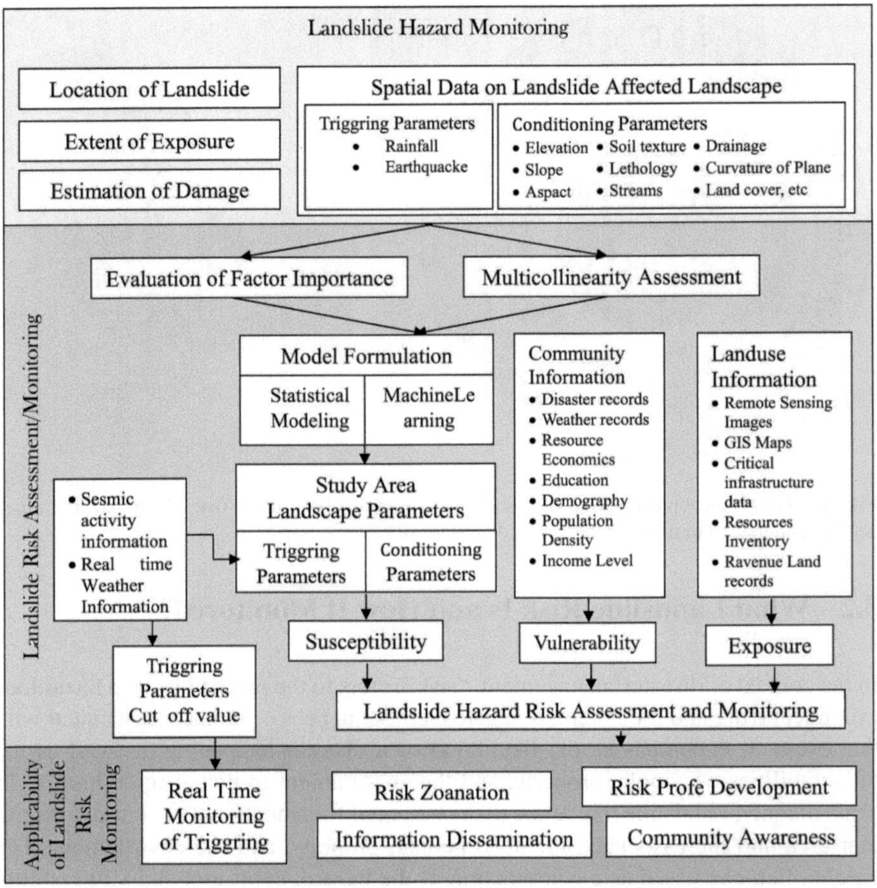

Fig. 3.2 Generic outline of data resources and their utilization for assessing and monitoring of landslide hazard risk and vulnerability was observed in related literatures

3.3 Importance of Considering Landslide Vulnerability and Exposure in Risk Assessment

While assessing hazard susceptibility alone may provide information about the probability of an event occurring in a particular area, incorporating estimates of vulnerability and exposure allows for a more comprehensive understanding of potential resource losses within different risk zones (Ram and Gupta 2022). There are situations where regions with high susceptibility may have low vulnerability and exposure, while areas with moderate or low susceptibility could have high vulnerability or risk. Therefore, the inclusion of vulnerability and exposure estimates alongside hazard susceptibility provides a more accurate assessment of risk, making it more suitable for decision-making and land-use planning compared to assessing risks in isolation (Van Westen et al. 2008).

3.3.1 *Defining Vulnerability*

Vulnerability assessment is a critical component in the evaluation of risk, encompassing the potential extent of harm to elements at risk (Ram and Gupta 2022). According to the UNISDR (2009) definition, vulnerability refers to "the characteristics and circumstances of a community, system, or asset that make it susceptible to the damaging effects of a hazard." Landslide vulnerability assessments are specifically concerned with evaluating the susceptibility of communities, infrastructure, and ecosystems to potential landslide impacts. This assessment aids in identifying realistic estimates for the resources that are most vulnerable and guides the development of targeted mitigation strategies (Van Westen et al. 2008).

3.3.2 *Types of Vulnerability*

Various types of vulnerabilities have been discussed in the literature, including Infrastructure Vulnerability, Socio-Economic Vulnerability, Community Vulnerability and Ecosystem Vulnerability, among others. Each type of vulnerability is derived by considering the exposure of various resources in each category and the estimated risk of hazard occurrence (Upadhyay and Jat 2014). In **Infrastructure Vulnerability**, the assessment focuses on the vulnerability of buildings and infrastructure. Factors such as building materials, construction quality, structural design, and proximity to hazard-prone areas are considered. **Socio-economic vulnerability** involves the analysis of socio-economic factors like income levels, education, access to healthcare, and employment opportunities among vulnerable populations. The assumption is that less equipped populations may be less capable of coping with landslide impacts. **Ecosystem Vulnerability** assesses the availability of natural resources, including forests, wetlands, and water bodies, and their importance for the surrounding environment. This assessment acknowledges the role of ecosystem resilience in reducing vulnerabilities (Ram and Gupta 2022). Each of these vulnerabilities is assessed separately in the form of spatial layers, which are then combined as weighted layers to create a composite vulnerability index for different locations. This index provides a comprehensive representation of the overall vulnerability level in each area (Upadhyay and Jat 2014).

3.3.3 *Defining Exposure*

Exposure in landslide risk assessment is a multifaceted concept encompassing human, economic, and ecological dimensions. In the context of landslide risk assessment, exposure refers to the presence and value of elements at risk within the potentially affected area i.e. Elements at risk can include people, buildings, infrastructure, environment and natural resources. Comprehensive exposure assessment provides a foundation for informed decision-making, risk prioritization, and the development of effective mitigation and preparedness measures. Therefore, evaluation of exposure becomes a crucial aspect to understand before assessment of landslide risk and its monitoring. Evaluation of exposure encompasses the identification and valuation of elements at risk within vulnerable areas. Assessing exposure involves identifying and quantifying the vulnerability of these elements to landslides. Recognizing the interconnectedness of elements at risk is essential for fostering resilient communities and safeguarding both human and environmental well-being in the face of landslide hazards. By understanding exposure, authorities and planners can prioritize areas for risk reduction measures, develop emergency response plans, and implement land-use planning strategies to minimize the potential consequences of landslides.

3.3.4 Factors to Be Considered Under Estimation of Exposure

3.3.4.1 Population Density

Human population density serves as a quantitative measure of the number of people residing in a given area. The number of people living or working in per unit area susceptible to landslides considered as critical component of exposure assessment in landslide risk assessment. The importance of this estimation lies in its direct correlation to the potential impact of landslides on communities. This information is essential for quantifying the potential risk associated with landslides. Higher population density increases the likelihood of casualties and the demand for emergency response efforts in the event of a landslide. In addition to this, areas with dense populations are more likely to have structures such as homes, schools, and businesses in close proximity, heightening the vulnerability of these elements to landslide impacts. Understanding the distribution of communities is vital for devising evacuation plans and allocating resources for emergency response. Estimates about number of people at risk allows authorities to plan and execute evacuation procedures efficiently. This includes identifying evacuation routes, establishing shelters, and allocating resources for the affected population.

3.3.4.2 Land Use

Land use information plays a crucial role in exposure assessment for landslide risk analysis. It provides a comprehensive understanding of the distribution of elements at risk, helping authorities prioritize areas for risk reduction, plan for emergencies, and implement land-use planning strategies that enhance resilience to landslide events. Understanding how land is utilized in a given area provides valuable insights into the vulnerability and potential consequences of landslides. Residential, commercial, industrial, and agricultural activities contribute differently to the vulnerability of elements at risk. For instance, residential areas, may have higher vulnerability due to the concentration of homes and infrastructure and Industrial zones may pose additional risks if hazardous materials are present. Therefore, information about type of land use exposed is become crucial for prioritization and weightage to particular area under exposure. Land use data aids in identifying critical infrastructure and valuation of area under exposure such as roads, bridges, schools, hospitals, and utilities. Understanding the location and importance of these structures is vital for assessing the potential impact of landslides on essential services and economy at risk. Commercial and industrial zones may have higher economic valuation due to businesses and manufacturing activities. Assessing the economic value of land use provides insights into the potential economic impact of landslides.

3.3.4.3 Critical Infrastructure

Critical infrastructure information is of paramount importance in exposure assessment for landslide risk. Critical infrastructure, including roads, bridges, utilities, hospitals, and other essential facilities, forms the backbone of communities. Understanding the location and vulnerability of these structures is crucial for assessing the potential impact of landslides and developing effective risk management strategies. Critical infrastructure supports the delivery of essential services and often serves as lifelines for communities. Roads and bridges facilitate transportation, while hospitals and emergency services provide critical care. Assessing the exposure of critical infrastructure to landslides helps in understanding the potential impact on public safety and the human population. The potential impact on critical infrastructure is a key consideration for minimizing disruption and ensuring effective post-landslide recovery. Assessing the exposure of such infrastructure to landslides provides insights into potential economic losses. Assessing the economic and social value of these structures aids in prioritizing risk reduction measures.

3.3.4.4 Ecological Assets

Exposure extends beyond human-centric considerations to include ecological assets such as natural resources, biodiversity, and ecosystems. Ecological assets, including diverse plant and animal species, contribute to biodiversity and ecosystem services. Landslides can have profound impacts on these elements and can disrupt these ecosystems, leading to the loss of biodiversity, environmental sustainability and provided ecosystem services, such as soil stability, water purification, and habitat provision. Assessing exposure helps understand the potential consequences for these critical ecosystem functions and contributing to a comprehensive understanding of the potential impacts on the environment. Protecting ecological assets is integral to holistic landslide risk management. Recognizing the importance of ecological assets in this context is crucial for informed decision-making and effective risk management. Additionally, vegetation and natural ecosystems play a key role in controlling erosion and stabilizing slopes. Assessing the exposure of ecological assets helps identify areas where the loss of vegetation due to landslides may compromise slope stability. This information is vital for implementing measures to protect and restore these natural stabilizers.

3.4 Identification of Landslides as Hazard

The age-old adage, "the past is the key to the future," holds true in many instances, including current practices in landslide risk monitoring. Historical data can be a valuable resource for uncovering the factors and their significance in identifying landslide-prone areas (Rawat and Pant 2023). The initial and fundamental step in assessing landslide hazard risk and vulnerability in any region involves creating an inventory based on past landslide data. This inventory offers insights into various critical attributes of past landslides, essential for understanding the occurrences of landslides and the factors responsible for them (Kamal et al. 2022). The preparation of this inventory entails the compilation of information on previous landslide events, including their locations, the extent of the affected area, landslide types, classifications, characteristics, triggers, and the date of occurrence if available.

3.4.1 Preparation of Inventory of Historical Events

Various sources can be explored to obtain information from past records, such as administrative, forest, and revenue departments (Guzzetti et al. 2012), in addition to conducting field surveys using Global Positioning System (GPS) and utilizing remote sensing technologies, be it satellite or aerial imagery (Bhandari and Dhakal 2020b). The precise location of landslides can be determined by their longitude and latitude, which can be ascertained with the assistance of GPS technology. Field surveys employing GPS devices are essential for collecting accurate coordinates and elevation data. These ground surveys help validate and complement existing datasets. Information about the size of landslides, that is, their areal extent, can be obtained from high-resolution remote sensing images captured both before and after the event using satellite or airborne systems. Satellite or aerial imagery can reveal landscape changes, like alterations in land use, erosion, and recent landslides (Bhandari and Dhakal 2020b). Nonetheless, acquiring clear imagery in hilly regions can be challenging due to issues like shadows, cloud cover, and vegetation. This can make identifying small or less destructive landslides difficult. Nevertheless, remote sensing techniques employing multi-temporal imagery are considered the most effective tool for initial landslide inventory mapping. Creating a landslide inventory through remote sensing involves a multi-step process where landslide-prone areas are identified using

Fig. 3.3 Representation of applicability of remote sensing images for collection of location and information regarding extent of landslide using pre (left) and post (right) landslide image of same location (Kangravally, Western Himalayan region; 32.185405° N, 76.217910° E) collected from google earth

high-resolution imagery from sources (Yaohui et al. 2021) like WorldView, GeoEye, QuickBird, IKONOS, SPOT-7, Sentinel, and Google Earth imagery (see Fig. 3.3). Infrared imagery, along with a comparison of SPOT, Landsat 8, and IRS-LISS satellite data, can also be used, especially for larger landslides (Fig. 3.3b). The identified landslide areas are then digitized into vector data, represented as polygons or points. Landslides identified in the infrared imagery are cross-verified in the Sentinel and Google Earth imagery. Any missing landslides in the imagery are identified through topographical means. Google Earth imagery, despite its lower spatial resolution, has been employed in various studies to identify and verify landslides (Liu et al. 2021).

3.4.2 Factors Responsible for the Occurance of Landslide

The primary driving force behind landslides is the Earth's gravity (Das and Lepcha 2019; Mangeney 2011). Intense rainfall can exacerbate this by creating prolonged wet conditions in less permeable soil slopes, ultimately initiating landslides (Tohari et al. 2007). Continuous rainfall leads to an increase in soil moisture content, potentially causing rapid soil saturation and subsequent failure. Thus, characterizing landslides while considering triggering factors such as rainfall is crucial to understanding the process (Bhandari and Dhakal 2021). However, additional factors such as adverse topography, deep and continuous erosion, and modifications in soil properties also play significant roles as conditioning factors on hilly slopes. Pore water pressure is generated in the unsaturated risk zone after heavy and continuous rainfall, contributing to landslides (Bhandari and Dhakal 2020b; Das and Lepcha 2019; Hakro and Harahap 2015; Yalcin et al. 2011). Numerous studies have explored the influence of geological factors (Henriques et al. 2015; Kim and Son 2015), rainfall (Guzzetti et al. 2007; Zêzere et al. 2015), and geomorphometric characteristics

(Vorpahl et al. 2012; Nugraha et al. 2015) in this context. The evolution of landforms is intricately dependent on the geological structure, lithology, relief configuration, climate, and biophysical processes. High relief diversity, tectonic activity, human interventions, drainage development, and intense rainfall are considered crucial factors contributing to landslide hazards. Consequently, comprehending the role of various geomorphic, geo-tectonic, and hydrologic attributes, such as elevation, slope, aspect, curvature, geomorphology, geology, lineament density, distance to lineaments, land use/land cover (LULC), soil properties, normalized differential vegetation index (NDVI), drainage density, distance to drainage, stream power index (SPI), topographic wetted index (TWI), and rainfall, becomes vital in assessing the factors responsible for landslide occurrences (Mandal and Mondal 2019b). In complex geological and topographical conditions, a single factor alone is insufficient for effective landslide characterization. A multidisciplinary approach, as noted by Ausilio and Zimmaro (2017), is a powerful tool for comprehensive landslide characterization. This approach combines various techniques to provide a thorough understanding of landslides. Therefore, it is essential to characterize landslides by broadly uncovering the mechanisms and interplay of different key determinants, including both triggering and conditioning factors (Bhandari and Dhakal 2021). While rainfall events are often considered a major triggering factor, in some cases, earthquakes can also play a role. Conditioning factors are further categorized into hydrological, topographical, and geological factors (Bhandari and Dhakal 2021).

3.4.2.1 Hydrological Factors

Given that rainfall serves as a triggering factor, the management of water retention, drainage systems, and runoff patterns has assumed increased importance. Effective water management, encompassing retention, drainage, and runoff, within a given region enhances the ability to mitigate landslide hazards by preserving soil texture and compactness on hilly slopes, even in adverse conditions brought about by heavy rainfall (Nerella et al. 2019). Therefore, collecting data related to rivers, streams, and their drainage networks (see Fig. 3.4), including parameters such as distance to drainage, drainage density, drainage intensity, drainage texture, stream frequency, stream junction frequency for calculating flow accumulation, stream power index, and topographic wetness index (refer to Table 3.1), offers valuable insights into water flow dynamics and water accumulation at specific locations. Beyond water flow and accumulation, the susceptibility of an area to landslides is also contingent upon its drainage characteristics. Drainage characteristics, in this context, are encapsulated by drainage density (Saha et al. 2021a, b, c, d). Drainage density can be defined as the length of streams per unit area of the basin. Areas where slopes are in proximity to streams or where drainage density is high are prone to soil saturation as a consequence of these drainage features. Such saturated soil becomes significantly more susceptible to failure (Panchal and Shrivastava 2020). Assessing these hydrological parameters can yield valuable insights into the soil's erosion potential (Moore et al. 1991) within

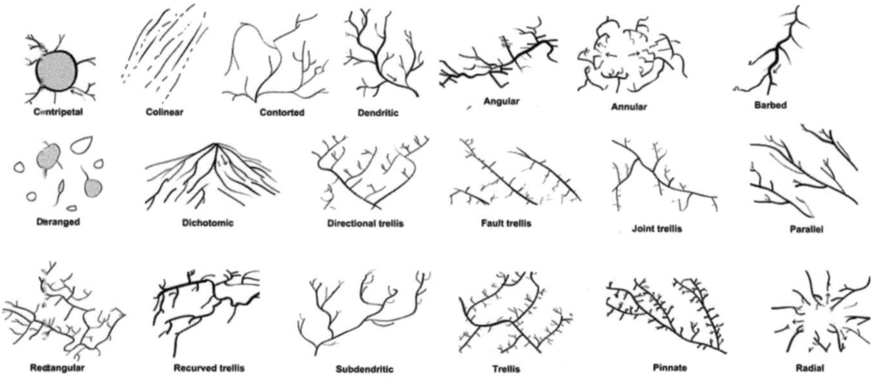

Fig. 3.4 Graphical representation of different drainage patterns available in hilly regions adopted after Drummond and Erkeling (2015)

a specific location, consequently providing a basis for estimating the likelihood of landslide occurrences (Mandal and Mondal 2019b).

3.4.2.2 Topographical Factors

As we know that the primary driving force behind landslides is the Earth's gravity, which exhibits a direct correlation with the topography of the terrain. On flat terrain, loose rock and soil have space to remain relatively stable, whereas, on uneven or sloping terrain, the tendency for loose soil and rock to move downhill is significantly increased. Data pertaining to terrain elevation, derived from various sources such as SRTM, ASTER, JAXA, LiDAR, and Cartosat, serves as foundational information. This elevation data is further used to generate various other topographical parameters, including slope, aspect, and curvature, which are essential for monitoring areas prone to landslides, especially those with steep slopes (see Table 3.1) (Meten et al. 2015).

Typically, terrain elevation is classified into different categories, such as Upper Hill, Middle Hill, Lower Hill, and Foothill, based on elevation ranges. It has been observed that lower hills within the range of 2000 m above sea level (msl) exhibit a higher potential for landslide occurrences (Mandal and Mondal 2019b). In addition to elevation distribution, the slope of the terrain also plays a crucial role in influencing the growth and development of drainage networks, surface runoff, soil erosion, and the concentration of drainage across the landscape. River basins, for instance, tend to exhibit the maximum slope angles, ranging from 0° to more than 89°. The stream power index (SPI) and topographic wetness index (TWI) are not only affected by slope steepness but also show a positive relationship with landslide frequency ratios (Mandal and Mondal 2019b). Similarly, the aspect of the terrain is another vital parameter in landslide susceptibility, as it indicates the directionality of exposure to sunlight and drying winds. This, in turn, controls soil moisture concentration, and higher soil moisture levels are positively related to landslide occurrences (Magliulo

Table 3.1 Details of various geospatial datasets, that have been used in various landslide hazard risk and vulnerability monitoring studies earlier

S. N.	GIS data	Application for landslide monitoring	Source	
1	Geological maps	Geological maps on lethology and structure contain information about geomorphic landforms, geologic structures, structure and material of rock formations, groundwater conditions, physical and mechanical properties of soils (i.e. soil type, soil texture, and soil thickness, etc.) and composition of minerals in rock and soil that contribute for erosion potential of soil, permeability, strength and weathering of rocks of any specific region. Geology also determine the water infiltration potential majorly depends upon physical and mechanical properties of constituent material of any rock and soil type such as permeability, and the soil strength (Lee et al. 2022)	Geological Departments i.e. Geological Survey of India, for Indian regions	Geological factors
2	Lineament map	Lineament maps generally include information about the location of joints, faults, foliation and bedding planes under the any geological structure. These lines are considered as lines of weakness and considered as areas of moisture accumulation which influence the terrain permeability and slope stability. Lineament map also use to calculate the distance form fault using Euclidean distance in any GIS environment (Al Brouki et al. 2023; Gomez and Kavzoglu 2005)	Information regarding lineaments can be extracted from remote sensing images obtained from satellite platforms such as LISS IV and Landsat images and aerial photography	

(continued)

Table 3.1 (continued)

S. N.	GIS data	Application for landslide monitoring	Source	
3	LULC	Land cover refers to any resource that are available on the earth and land use refers to the how these resources are used. It includes natural resources, such as river, lake, ponds, mountains, forests, grassland, rock, soil, and manmade structures like critical infrastructure (hospitals, community centres, safe houses and schools), transport facilities (road, airport and rail networks) and agriculture. Different features available on land impacted the soil mechanical behaviour and moisture in many ways. For example good vegetation cover provide mechanical anchoring to soil and provide stability to the slope by preventing soil from erosion (Masi et al. 2021; Parra et al. 2021) on other hand deforestation, steep and improper cutting of slopes on hilltops, irregularized drainage, uncontrolled settlement and excess agriculture often reduce slope stability (Chen et al. 2019)	Drived from satellite images such as LISS IV, SPOT and Landsat images	
4	Vegetation cover	Various studies have reported role of vegetation cover in landslide hazard mapping. Vegetation cover influences the soil–atmosphere interaction (i.e. rainfall interaction and, evapotranspiration) and the soil hydro-mechanical properties. Literature suggested both negative and positive influences of vegetation cover, in positive sense it provides strength to stabilize the soil by soil–root interaction and reduced pore water pressure on soil and destabilize slopes in connection with strong winds as negative impact (Kalsnes and Capobianco 2022)	Drived from satellite images such as LISS IV, SPOT and Landsat images	
5	Elevation map	Many landslide casual factors are often generated from DEMs	Obtained from various remote sensing images such as ASTER, GDEM, SRTM, ALOS and Cartosat, etc.)	Topographic factors

(continued)

Table 3.1 (continued)

S. N.	GIS data	Application for landslide monitoring	Source	
6	Slope map	Provides the information about steepness of the terrain at particular location as conditioning factor for landslide (Meten et al. 2015)	Drived from digital elevation maps using topography analysis modules of various GIS softwares i.e. spatial analysis tool from ArcGIS	
7	Aspect map	Provides the information about direction of face of the slope in respect to the face of sun. Direct involvement of aspect in landslide less prominent in the literature but aspect has indirect relation with landslide by contributing for soil moisture retention, sunlight, evaporation potential, vegetation, permeability and porosity of soil, etc. (Meten et al. 2015)		
8	Curvature	Curvature refers to the configuration of the topographical features or the shape of a line created by the change in slope angle along the curve of the hillside by the intersection of an abstract horizontal plane with the Earth's surface. Plane of curvature can provide information about flow and accumulation of water on plane and provide information about driving and resisting stresses and controls over convergence or divergence of landslide material and water in the direction of landslide motion (Meten et al. 2015)		
9	Stream network	Rivers typically have a substantial impact on shaping the terrain as they carve through various rock formations. Their contribution to creating conditions conducive to landslides is of paramount importance. Understanding the stream network in a specific region is vital for extracting valuable insights into various hydrological factors that influence landslides, including drainage density, proximity to watercourses, stream occurrence frequency, and the frequency of stream junctions. These factors are widely recognized as important considerations for landslide susceptibility (Bicocchi et al. 2019)	Drived from digital elevation maps using hydrology modules of various GIS softwares i.e. ArcHydro from ArcGIS	Hydrological factors

(continued)

Table 3.1 (continued)

S. N.	GIS data	Application for landslide monitoring	Source	
10	Stream flow	Stream flow also known as stream discharge, is defined as teh measure of the amount of the water that moves through any location in a watershed or catchment within a unit time. The flow of a stream is directly related to the amount of water moving off the watershed or catchment into the stream channel. It depends on the intensity and frequency of rainfall in the region, it is used to calculate the drainage intensity and flow accumulations (McKerchar 2003)	Calculated from incorporating the actual rainfall data with topographic information of watershed or catchment using hydrology modules of various GIS softwares i.e. ArcHydro from ArcGIS	
11	Stream power index	The stream power index (SPI) is a topographical cum hydrological component use to describe the erosion power of water that flowing through a stream in a particular catchment. SPI takes area of specific catchment and slop values in account to calculate the erosion power of water in that catchment. Stream power index is calculated as a multiplication function of specific catchment's area (A_S) and local slope gradient (tanβ) and represented as. Stream power index (SPI) = A_S * tanβ (Vijith and Dodge-Wan 2019; Singh et al. 2023; Moore et al. 1991)	Calculated from incorporating the actual rainfall data with topographic information of catchment i.e. slope and area of catchment using hydrology modules of various GIS softwares i.e. ArcHydro from ArcGIS	
	Topographic wetness index	Topographic wetness index is again a topographical cum hydrological component used to calculate soil moisture distribution under the influence of groundwater flow, which reflects the tendency of accumulated water to move through the slope of catchment under the influence of gravitational force. Various historical landslide event studies suggested that, Topographic wetness index was usually found as low in regions of steep, diverging land and high in flat, converging terrain. Topographic wetness index is calculated as a logarithm of divisional product of cumulative upslope area draining through a point (α) to the slope angle at that point (tanβ) and represented as TWI = In (α/tanβ) (Vijith and Dodge-Wan 2019; Singh et al. 2023; Moore et al. 1991)		

(continued)

Table 3.1 (continued)

S. N.	GIS data	Application for landslide monitoring	Source	
	Flow accumulation	Flow accumulation pertains to the quantity of water that can gather at a specific point along a stream or drainage system when a known amount of rainfall occurs in a particular area from which that stream or drain flows. By employing GIS software and incorporating data on the stream network and known rainfall values, it becomes possible to compute the water accumulation at a specific location based on the drainage capacity of that area. Given the significant role of water in landslides, flow accumulation and pore water pressure is utilized as a highly influential layer in landslide susceptibility assessments (Bicocchi et al. 2019)	Calculated from incorporating the actual rainfall data with topographic information of watershed i.e. slope, aspect, stream network, and Stream Flow, etc. using hydrology modules of various GIS softwares i.e. ArcHydro from ArcGIS	
12	Rainfall	Acquire historical and real-time rainfall data for the area. Rainfall is a major trigger for landslides, and analyzing precipitation patterns helps in assessing the temporal distribution of landslide events	Meteorological Department i.e. Indian Meteorological Department for Indian regions	Triggering factors
13	Stream pressure	Stream pressure can be defined as pressure caused by flow of water per unit area in any region at particular time. This information is become crucial for real time monitoring of water in streams under the effect of rainfall in the region and can provide information for identifying the potential landslide prone area during extensive rainfall situations	Calculated from real time rainfall data incorporation with stream network maps	
14	Seismic data	In seismic-prone regions, seismic data can be crucial for understanding the potential impact of earthquakes on landslide susceptibility	Realtime monitoring	

Fig. 3.5 Representation of slope (orange vertical line), positive (blue curves), negative (red curves) and flat (green horizontal line) curvature of terrain in a google earth image of kangravally, western Himalayan region; 32.22222° N, 76.37366° E

et al. 2008; Mandal and Mondal 2019b). Regarding terrain curvature, it can take one of three forms: convex, concave, or flat. Negative values represent concave terrain (Fig. 3.5), a zero value signifies flat terrain, and positive values indicate convex curvature. Terrain areas with highly positive or negative curvature values are more susceptible to slope instability (Mandal and Mondal 2019b). Additionally, other topographical information derived from digital elevation models, such as terrain roughness, terrain relief, and dissection index, has also been studied in some landslide hazard assessments.

3.4.2.3 Geological Factors

Collecting data about the geotechnical attributes of an area is imperative, and this includes aspects such as rock types, lithology, lineaments, distance to lineaments, lineament density (Mandal and Mondal 2019a, b), as well as other geological features like soil properties and vegetation cover (refer to Table 3.1). This information is vital for assessing the susceptibility of various geological formations to landslides. The region's high rainfall leads to significant weathering and erosion of rocks, contributing to frequent landslides (Magliulo et al. 2008). The susceptibility to erosion is contingent upon the type of rock and its formation at different locations. Similarly, information related to lineaments, including their density and proximity, is crucial. Lineaments describe zones of weakness and typically correspond to underlying linear geological structures, such as fault lines or fold-aligned hills in the region (Thapa et al. 2017). The distance of slopes from faults also influences slope stability, as faults indicate weaknesses in the geological structure. Therefore, they are significant considerations in susceptibility mapping (Panchal and Shrivastava 2020). It has been observed that the likelihood of a landslide event is directly proportional to the lineament density of a region. Areas with high lineament density have a greater probability of landslides, whereas regions with low lineament density exhibit lower landslide frequency ratios (Mandal and Mondal 2019b). Soil properties, including cohesion, shear strength, permeability, and the presence of vegetation cover, play crucial roles in slope stability. There is evidence in the literature that dense vegetation can stabilize slopes, underscoring the importance of information related to vegetation types, density, and health, which can be derived from satellite images and calculated using vegetation indexes like NDVI. Additionally, land use and land cover are considered critical factors in governing landslide susceptibility. Land cover pertains to the biological and physical materials on the Earth's surface, encompassing natural elements like exposed rock, soil, forests, and water bodies, as well as anthropogenic elements like agriculture, buildings, and roads (Herold et al. 2006). On the other hand, land use refers to the socio-economic utilization of the land, such as the purpose for which the land or terrain is employed by humans (Herold et al. 2006). Many studies argue that changes in land use and land cover (LULC) can significantly increase landslide susceptibility (Chen and Huang 2013).

3.5 Different Approaches Used for Evaluation of Conditioning Factors as an Important Factor in Causing Landslide Hazard

The various factors discussed in the previous section have differing impacts on the likelihood of landslide hazards. Therefore, before incorporating these factors into any further analysis for landslide hazard assessment, it is essential to determine their respective weights. Various methods and approaches have been employed to assign weightage and rank the different hydrological, topographical, and geological factors in the process of risk monitoring for landslides. Different techniques have been utilized for ranking and selecting independent factors in natural hazard assessment. One approach is the Analytic Hierarchy Process (AHP), which involves pairwise comparisons to evaluate various parameters and rank different criteria according to their influence on landslide occurrences. Another method, the Landslide Susceptibility Index (LSI), calculates weights using a weighted linear combination (WLC) technique (Panchal and Shrivastava 2020). To ensure that the predictive precision of any model is not reduced due to linear relationships between various factors, collinearity assessment is also considered. This involves conducting multiple regression analyses to explore potential interrelationships among the conditioning factors

(Arabameri et al. 2019). The process of Model Averaging (MA) is essential for planning landslide simulations. In the literature, MA has been employed to identify the correct factors contributing to landslides (Cama et al. 2017; Roy et al. 2019). To test for multicollinearity, tools such as Variance Inflation Factor (VIF), Regularized Regression (RR), and Tolerance (TOL) have been used in the past (Garosi et al. 2019). Regularized regression (RR) predictions are known for their robustness and have been introduced to address issues of overfitting instability and collinearity associated with the least squares estimator. It handles small fluctuations in the dataset used for regression and minimizes their impact on the results (Saha 2021b). Information Gain Ratio (IGR) and Chi-square Attributes Evaluation (CSAE) are some of the machine learning feature selection techniques used for selecting and classifying major Landslide Conditioning Factors. In the case of the Information Gain Ratio, historical landslide, and non-landslide data are used as training data to assess the importance of each factor. The chi-square attributes evaluator employs artificial intelligence approaches to select smaller feature training datasets, reducing the time and cost of the modeling process while achieving appropriate results (Daviran et al. 2023; Karegowda et al. 2010).

3.6 Review on Statistical and Machine Learning Techniques Used for Landslide Hazard Susceptibility Mapping

Landslide susceptibility mapping is a process that aims to identify areas in terrain that are more likely to experience slope failure by analyzing the characteristics of documented landslides (Guzzetti et al. 2005). Understanding landslide hazard susceptibility is a crucial aspect of landslide risk monitoring. The initial step in this process involves hazard identification and the estimation of the likelihood and severity of potential hazards (Huang et al. 2020). Therefore, susceptibility mapping is the most appropriate term to describe the process of estimating the spatial distribution of the probability of landslides occurring. Having explored the various factors responsible for landslide occurrences and determined their respective weights in governing landslides in specific regions in the preceding sections, we will now discuss several methods, both statistical and machine learning, that have been successfully utilized for landslide susceptibility mapping in various studies (Saha and Saha 2020; Pal and Chowdhuri 2019). These approaches are primarily classified into following categories:

1. Heuristic methods, also known as Multi-Criteria Decision Making and expert-based models, which employ expert opinions to assign weights to various triggering factors (Basu and Pal 2019; Bera et al. 2019; Mandal and Mandal 2018a; Stanley and Kirschbaum 2017). Techniques like Analytic Hierarchy Process fall under this category.

2. Deterministic methods, also known as physically-based models (Medina et al. 2021), where parameters can be determined either in the field or in a laboratory setting, such as through safety factor calculations (Aleotti and Chowdhury 1999).
3. Probabilistic methods, also known as bivariate and statistical-based models, rely on established associations between landslide distribution and causative factors in specific locations (Das et al. 2023; Rawat and Pant 2023).
4. Machine learning approaches, such as those used in data-driven techniques, which analyze landslide-controlling factors using spatiotemporal datasets of landslide events (Wang et al. 2021; Mandal and Mandal 2018b).

These various methods offer valuable tools for estimating landslide susceptibility, and their selection depends on the specific context and data available for the region of interest.

Numerous models have been employed for landslide prediction studies, offering distinct approaches to this task. Probabilistic models like Frequency Ratio (FR), Statistical Index (SI), Index of Entropy (IOE), Weight of Evidence (WoE), Evidential Belief Function (EBF), Modified Information Value Model (MIV), Logistic Regression (LR), Multiple Regression Analysis (MRA), Geographically Weighted Regression, and more have been widely used (Pal and Chowdhuri 2019; Lee and Talib 2005; Pham et al. 2015; Sahana and Sajjad 2017; Bui et al. 2011; Wang et al. 2021; Batar and Watanabe 2021; Ahmad and Dewan 2017; Chowdhuri et al. 2020; Akinci and Yavuz 2021; Aghdam et al. 2016; Zhu et al. 2014; Sujatha and Sridhar 2021).

Additionally, machine learning models have proven effective for landslide susceptibility mapping, including Artificial Neural Network (ANN), Boosted Regression Tree, MultiBoost, Least Squares Support Vector Machines, Multivariate Adaptive Regression Spline, Adaptive Neuro-Fuzzy Inference System, AdaBoost, Bagging, Dagging Random Forest (RF), Rotation Forest (RTF), Random SubSpace (RS), Decision Tree, K-Nearest Neighbor, Bayesian Logistic Regression, Logistic Model Tree, and Alternate Decision Tree (Saha et al. 2021b; Mandal and Mandal 2019c; Pham et al. 2016; Nwazelibe et al. 2023; Colkesen et al. 2016; Shirzadi et al. 2018). Furthermore, multi-criteria hybrid ensemble machine learning techniques, such as Bagging-RF (B-RF), RTFRF, and RS-RF, have been suggested to integrate multiple base classification schemes and enhance their performance in Landslide Susceptibility Mapping (LSM) (Nhu et al. 2020). Each provided methods having some merits and demerits in it, the success of these methods depends on several factors, including the classification ratio of training and validation datasets, the amount of data available for training and validation, the sampling scheme, and the size of the training data. Additionally, the accuracy of the produced landslide susceptibility maps is influenced by the accessibility of past and present landslide records in the area and the amount of training data (Roy et al. 2019; Abedini and Tulabi 2018).

3.7 Application of Remote Sensing (RS) and Geographic Information System (GIS) in Landslide Hazard Risk and Vulnerability Monitoring

Remote Sensing and Geographical Information Systems (GIS) have emerged as powerful tools for effectively managing natural disasters like landslides. In regions with challenging terrains such as the Himalayas, accessibility to landslide-prone areas can be a major issue. Geospatial techniques offer the capability to reach and analyze even the most inaccessible regions, identify various types of landslides, and establish landslide-prone zones based on significant contributing factors (Nerella et al. 2019).

Landslide hazard zonation is intrinsically linked to spatial information. GIS tools and information can be harnessed for the storage, collection, analysis, querying, management, and manipulation of data related to every facet of landslide hazard risk monitoring. This encompasses tasks from inventory collection to the creation of spatial data layers involving conditioning factors, susceptibility assessments, exposure mapping, and vulnerability evaluations (Lillesand et al. 2014; Myronidis et al. 2016). In recent times, landslide investigations have increasingly embraced the use of Remote Sensing and GIS, offering the potential for more precise results (Chawla et al. 2019; Ghosh et al. 2013; Bharadwaj and Singh 2023).

Geographic Information Systems (GIS) have proven highly effective in integrating data and information from diverse sources and methodologies concerning a study area. This integration spans various stages, providing a comprehensive understanding of the contributing and triggering factors behind landslide occurrences, risk assessment, potential impacts, vulnerability analysis of communities and infrastructure, and the finalization of risk maps (Lillesand et al. 2014; Nerella et al. 2019; Panchal and Shrivastava 2020).

One of the primary advantages of GIS is its adaptability—altering the input can lead to different outputs, making it an invaluable tool for planners. GIS is particularly effective in preparing landslide hazard zonation maps, as it considers the cumulative effects of various landscape characteristics responsible for landslide events (Shahabi and Hashim 2015; Panchal and Shrivastava 2020). The process of GIS-based landslide hazard zonation mapping involves the creation of maps that categorize regions within an area based on their susceptibility to landslides, exposure to resources, and vulnerability of communities. These maps play a crucial role in identifying potential landslide-prone zones, aiding in land-use planning, infrastructure development, and disaster risk reduction. GIS plays a vital role at various stages in landslide hazard risk and vulnerability monitoring, contributing significantly to the process (Paulin et al. 2014).

3.7.1 Data Collection and Preparation

The collection and integration of spatial data play a pivotal role in the development of precise and dependable hazard maps and risk monitoring systems (Paulin et al. 2014). Geographic Information Systems (GIS) have established themselves as invaluable platforms for executing these functions by gathering pertinent spatial data layers. These layers encompass a wide range of factors, including topography, geological information, rainfall patterns, land use, vegetation cover, soil properties, historical landslide records, population density, building characteristics, critical infrastructure such as roads and bridges, socio-economic indicators, community attributes, local environmental factors, and other assets (Zhao and Lu 2018).

3.7.2 Spatial Analysis for Data Processing and Standardization

Following the successful collection of spatial data, the next crucial step involves the analysis of this data to extract valuable insights. When dealing with spatial data, specialized tools are essential for handling and processing the information. Fortunately, most GIS platforms come equipped with a wide array of spatial analysis tools and algorithms, making the analysis of landscape data more convenient and suitable (Solari et al. 2020). Sometimes, to ensure uniformity and precision in spatial processing, it is necessary to normalize and standardize the spatial data (Mersha and Meten 2020). These processes of data normalization and standardization can also be accomplished using various GIS software tools.

3.7.3 Weightage Overlay Analysis and Integration of Data

The raw and processed spatial data collected and analyzed are organized into distinct data layers. To generate a comprehensive hazard map, these layers must be assigned values and superimposed upon one another (Wang and Li 2017). Through weighted overlays, each data layer is allocated a specific weight, reflecting its importance in the context, which leads to the creation of a unified landslide hazard index or rating for each geographic point (Vakhshoori and Zare 2016). This hazard index can then be integrated with data layers pertaining to vulnerability and exposure using GIS tools, enabling the calculation of a comprehensive risk assessment for a particular region. The significance of each layer in contributing to vulnerability should guide the assignment of weights (Ram and Gupta 2022).

3.7.4 Validation, Calibration and Accuracy Assessment

Result validation is a critical step in any assessment process, and GIS platforms offer a range of tools for validation and accuracy assessment. Among these tools are snapping tools, which enable the alignment of two or more GIS layers. Through the use of snapping and overlay tools, information gathered from real-world observations and historical landslide maps can be overlaid onto the predicted potential hazard maps (Shahabi and Hashim 2015). Overlaying and snapping multiple layers allows for the validation of hazard zonation maps by comparing them with records of historical landslide occurrences. If needed, adjustments to weights and parameters in spatial analysis tools can be made in real time to enhance the model's accuracy, along with updates to spatial information (Paulin et al. 2014).

3.7.5 Real Time Updation, Visualization and Map Designing

To ensure effective monitoring of landslide hazard and vulnerability, it is crucial to continuously update hazard zonation, exposure, and vulnerability information by incorporating changes in the landscape, new data sources, and potential modifications in hazard factors (Intrieri et al. 2017). GIS-based tools and platforms, such as web-GIS, offer users the ability to seamlessly integrate real-time updates into spatial information and their assessments related to vulnerability, exposure, and hazard zonation, accounting for shifts in the landscape, demographics, infrastructure, and other relevant factors (Sujatha and Rajamanickam 2015). GIS software also provides the capability to design hazard zonation maps, exposure maps, and vulnerability maps with clear legends, labels, and colors for effective communication of different hazard levels. Users can employ various GIS applications to depict contour lines or shading, representing various susceptibility zones (Intrieri et al. 2017).

3.8 Applicability of Landslide Hazard Risk and Vulnerability Monitoring

The monitoring of landslide hazard risk and vulnerability plays a vital role in comprehending and addressing the potential consequences of landslides on human lives, infrastructure, and the environment. It entails the continuous collection of data and the analysis of a multitude of factors that contribute to landslide risk. This monitoring effort holds diverse applications: firstly, it involves the identification of critical infrastructure, communities, assets, and vulnerable areas located within or in proximity to hazard-prone zones. Secondly, it aids in the preparation of various scenarios that assess potential hazards and their associated impacts. Additionally, it provides valuable information for establishing well-defined objectives and goals to guide decision-makers in crafting effective mitigation plans. This, in turn, supports the development of implementation plans outlining the necessary steps, timelines, responsible entities, and required resources for each mitigation strategy. Moreover, it offers critical insights for the preparation of emergency response plans, ensuring preparedness before, during, and after a landslide event. The integration of real-time data strengthens early warning systems, enhancing community safety. Furthermore, it entails an assessment of the costs and societal benefits associated with various mitigation strategies. It equips decision-makers with the information required for implementing engineering measures, land-use regulations, the relocation of vulnerable infrastructure, and the identification of suitable locations for evacuation shelters. The monitoring process also facilitates the integration and dissemination of local residents' experiences, traditional practices, and indigenous knowledge concerning historical landslide events. In addition, it harnesses hazard maps as effective tools for conveying information to stakeholders, local media, social networks, and local action groups. This contributes to increased public awareness and promotes community engagement in disaster preparedness and mitigation efforts. Finally, the monitoring process establishes a framework for conducting training and capacity-building activities for relevant stakeholders. In essence, it serves as a comprehensive foundation for informed decision-making, enabling proactive management of disaster risks and enhancing the safety and resilience of communities and their surrounding environments.

3.9 Summary

In this chapter a reviewed teh literature and studies were performed on landslide hazard, and we found landslide hazard as one of the natural disasters that impact the functioning of society throughout the globe, especially in hilly terrain like the Himalayas. The introduction section compiled the outcomes of previous studies performed on assessment of extent of landslide hazard, identification and evaluation of conditioning and triggering factors, and assessment associated risk. Where we observed that, the impact of landslide is not only in the form of direct losses such as the life of humans and livestock, and loss of physical infrastructure, property damage, but also as indirect in the form of loss to the socioeconomic development, ecology and environment. We also observe, that various factors have been identified as triggering and conditioning factors for the occurrence of an event landslide, in addition to this rapid population expansion, human-made alterations, and the impacts of climate change increase the stress on natural environments, particularly in mountainous regions, and may result in an increased probability of occurrence of the landslide as well as exposure to the resources under the threat of landslide. In one course human activities, such as deforestation and construction in hilly regions, contribute to slope instability, further increasing the risk of landslides, in another course, the construction of critical infrastructures in landslide-prone areas also increases the exposure and vulnerability to the hazard of landslide. But, as the technology grows rapidly, we have various statistical as well as machine learning algorithms that can be used to collect, evaluate, and analyze the various topographic, hydrological, and geographical data. And also can integrate real-time information about rainfall and hydrological measures along with data on various important conditioning factors according to their weightage for building effective models in the framework of GIS to monitor the hazard and vulnerability of landslide risk. In subsequent sections, we gain knowledge on various components of landslide risk monitoring such as hazard, exposure, and vulnerability, and methods of their assessment. In the process of identification of landslide hazards, we get to know about the importance of inventory on historical events and also gain the knowledge about, which resources can be explored for preparing these inventories. We also enhance our knowledge of various topographic, hydrological, and geological factors that have been considered as conditioning factors and gain knowledge about available methods for evaluating and weighing these conditioning factors in the process of assessment of landslide risk. In the end, we compile our knowledge of various statistical and machine learning algorithms and GIS tools used for landslide susceptibility mapping and hazard, risk, and vulnerability assessment and monitoring. By reading this chapter we could enhance our knowledge about, what is a landslide and how it affects the society and environment?, what are the factors responsible for its occurrence?, how humans are contributing to its occurrence and management?, which approaches are used for assessment and monitoring landslide?, How GIS and Remote Sensing contributed in assessment and monitoring? and what we can achieve by monitoring landslide hazard risk and vulnerability.

References

Abedini M, Tulabi S (2018) Assessing LNRF, FR, and AHP models in landslide susceptibility mapping index: a comparative study of Nojian watershed in Lorestan province, Iran. Environ Earth Sci 77:405. https://doi.org/10.1007/s12665-018-7524-1

Aghdam IN, Varzandeh MHM, Pradhan B (2016) Landslide susceptibility mapping using an ensemble statistical index (Wi) and adaptive neuro-fuzzy inference system (ANFIS) model at Alborz Mountains (Iran). Environ Earth Sci 75(7):1–20. https://doi.org/10.1007/s12665-015-5233-6

Ahmad B, Dewan A (2017) Application of bivariate and multivariate statistical techniques in landslide susceptibility modeling in Chittagong city corporation, Bangladesh. Remote Sens 9(4):304. https://doi.org/10.3390/rs9040304

Akinci H, Yavuz OA (2021) Landslide susceptibility mapping and hazard assessment in Artvin (Turkey) using frequency ratio and modified information value model. Acta Geophys 69:725–745. https://doi.org/10.1007/s11600-021-00577-7

Al Brouki MT, Elshaafi HZ, Mostafa MF, El Hassi MF, Khalifa AK (2023) Lineaments analysis and mapping using automatic extraction in central part of Al Jabal Al Akhdar, NE Libya. Carbonates Evaporites 38(1):4

Aleotti P, Chowdhury R (1999) Landslide hazard assessment: summary review and new perspectives. Bull Eng Geol Environ 58:21–44. https://doi.org/10.1007/s100640050066

Arabameri A, Pradhan B, Rezaei K, Sohrabi M, Kalantari Z (2019) GIS-based landslide susceptibility mapping using numerical risk factor bivariate model and its ensemble with linear multivariate regression and boosted regression tree algorithms. J Mt Sci 16:595–618

Ausilio E, Zimmaro P (2017) Landslide characterization using a multidisciplinary approach. Measurement 104:294–301

Basu T, Pal S (2019) RS-GIS based morphometrical and geological multi-criteria approach to the landslide susceptibility mapping in Gish River Basin, West Bengal, India. Adv Space Res 63:1253–1269. https://doi.org/10.1016/j.asr.2018.10.033

Batar AK, Watanabe T (2021) Landslide susceptibility mapping and assessment using geospatial platforms and weights of evidence (WoE) method in the Indian Himalayan region: recent developments, gaps, and future directions. ISPRS Int J Geo-Inf 10(3):114. https://doi.org/10.3390/ijgi10030114

Bera S, Guru B, Ramesh V (2019) Evaluation of landslide susceptibility models: a comparative study on the part of Western Ghat Region, India. Remote Sens Appl Soc Environ 13:39–52. https://doi.org/10.1016/j.rsase.2018.10.010

Bhandari BP, Dhakal S (2020a) Spatio temporal dynamics of landslide in the sedimentary terrain: a case of Siwalik zone of Babai watershed, Nepal. SN Appl Sci 2:854. https://doi.org/10.1007/s42452-020-2628-0

Bhandari BP, Dhakal S (2020b) Compositional analysis and phase relationship of soil mass from the active landslides of Babai River watershed, Siwalik zone of Nepal. Eng Geol 278:105851. https://doi.org/10.1016/j.enggeo.2020.105851

Bhandari BP, Dhakal S (2021) A multidisciplinary approach of landslide characterization: a case of the Siwalik zone of Nepal Himalaya. J Asian Earth Sci X 5:1–11. https://doi.org/10.1016/j.jaesx.2021.100061

Bhardwaj V, Singh K (2023) Landslide susceptibility using remote sensing and GIS—a review. J Min Environ 14(1):133–154

Bicocchi G, Tofani V, D'Ambrosio M et al (2019) Geotechnical and hydrological characterization of hillslope deposits for regional landslide prediction modeling. Bull Eng Geol Environ 78:4875–4891. https://doi.org/10.1007/s10064-018-01449-z

Bogaard TA, Greco R (2016) Landslide hydrology: from hydrology to pore pressure. Wiley Interdisc Rev: Water 3(3):439–459. https://doi.org/10.1002/wat2.1126

Bragagnolo L, da Silva RV, Grzybowski JMV (2020) Artificial neural network ensembles applied to the mapping of landslide susceptibility. CATENA 184:104240. https://doi.org/10.1016/j.catena.2019.104240

Bui DT, Lofman O, Revhaug I, Dick O (2011) Landslide susceptibility analysis in the Hoa Binh province of Vietnam using statistical index and logistic regression. Nat Hazards 59:1413–1444. https://doi.org/10.1007/s11069-011-9844-2

Cama M, Lombardo L, Conoscenti C, Rotigliano E (2017) Improving transferability strategies for debris flow susceptibility assessment: application to the Saponara and Itala catchments (Messina, Italy). Geomorphology 288:52–65. https://doi.org/10.1016/j.geomorph.2017.03.025

Chawla A, Pasupuleti S, Chawla S, Rao ACS, Sarkar K, Dwivedi R (2019) Landslide susceptibility zonation mapping: a case study from Darjeeling district, Eastern Himalayas, India. J Indian Soc Remote Sen 47:497–511. https://doi.org/10.1007/s12524-018-0916-6

Chen L, Guo Z, Yin K et al (2019) The influence of land use and land cover change on landslide susceptibility: a case study in Zhushan Town, Xuan'en County (Hubei, China). Nat Hazards Earth Syst Sci 19:2207–2228. https://doi.org/10.5194/nhess-19-2207-2019

Chen CY, Huang WL (2013) Land use change and landslide characteristics analysis for community-based disaster mitigation. Environ Monit Assess 185:4125–4139. https://doi.org/10.1007/s10661-012-2855-y

Ching H, Chih-Hsuan L, Chia-Ming C (2018) Numerical investigation of rainfall induced landslide in mudstone using coupled finite and discrete element analysis. Geofluids 1–15. https://doi.org/10.1155/2018/9192019

Chowdhuri I, Pal SC, Arabameri A, Ngo PTT, Chakrabortty R, Malik S, Das B, Roy P (2020) Ensemble approach to develop landslide susceptibility map in landslide dominated Sikkim Himalayan region, India. Environ Earth Sci 79:1–28. https://doi.org/10.1007/s12665-020-09227-5

Colkesen I, Sahin EK, Kavzoglu T (2016) Susceptibility mapping of shallow landslides using kernel-based Gaussian process, support vector machines and logistic regression. J Afr Earth Sci 118:53–64. https://doi.org/10.1016/j.jafrearsci.2016.02.019

Cruden DM, Varnes DJ (1996) Landslide types and processes. In: Turner AK, Shuster RL (eds) Landslides: investigation and mitigation. Transp Res Board, Washington, D.C, pp 36–75

Dahal RK, Hasegawa S, Nonomura A, Yamanaka M, Masuda T, Nishino K (2008) GIS-based weights-of evidence modelling of rainfall-induced landslides in small catchments for landslide susceptibility mapping. Environ Geol 54(2):314–324

Das G, Lepcha K (2019) Application of logistic regression (LR) and frequency ratio (FR) models for landslide susceptibility mapping in Relli Khola river basin of Darjeeling Himalaya, India. SN Appl Sci 1:1453. https://doi.org/10.1007/s42452-019-1499-8

Das J, Saha P, Mitra R, Alam A, Kamruzzaman M (2023) GIS-based data-driven bivariate statistical models for landslide susceptibility prediction in Upper Tista Basin, India. Heliyon 9(5):e16186. https://doi.org/10.1016/j.heliyon.2023.e16186. PMID: 37234665

Daviran M, Shamekhi M, Ghezelbash R, Maghsoudi A (2023) Landslide susceptibility prediction using artificial neural networks, SVMs and random forest: hyperparameters tuning by genetic optimization algorithm. Int J Environ Sci Technol 20(1):259–276

Dehnavi A, Aghdam IN, Pradhan B, Varzandeh MHM (2015) A new hybrid model using step-wise weight assessment ratio analysis (SWARA) technique and adaptive neuro-fuzzy inference system (ANFIS) for regional landslide hazard assessment in Iran. Catena 135:122–148. https://doi.org/10.1016/j.catena.2015.07.020

Devkota KC, Regmi AD, Pourghasemi HR, Yoshida K, Pradhan B, Ryu IC, Dhital MR, Althuwaynee OF (2013) Landslide susceptibility mapping using certainty factor, index of entropy and logistic regression models in GIS and their comparison at Mugling-Narayanghat road section in Nepal Himalaya. Nat Hazards 65:135–165

Drummond SA, Erkeling G (2015) Drainage pattern. In: Encyclopedia of planetary landforms. Springer, New York, NY, pp. 616–627. https://doi.org/10.1007/978-1-4614-9213-9_119-1

Feby B, Achu AL, Jimnisha K, Ayisha VA, Reghunath R (2020) Landslide susceptibility modelling using integrated evidential belief function based logistic regression method: a study from Southern Western Ghats, India. Remote Sens Appl Soc Environ 20:100411. https://doi.org/10.1016/j.rsase.2020.100411

Gadtaula A, Dhakal S (2019) Landslide susceptibility mapping using weight of evidence method in Haku, Rasuwa district, Nepal. J Nepal Geol Soc 58:163–172. https://doi.org/10.3126/jngs.v58i0.24601

Garosi Y, Sheklabadi M, Conoscenti C, Pourghasemi HR, Van Oost K (2019) Assessing the performance of GIS-based machine learning models with different accuracy measures for determining susceptibility to gully erosion. Sci Total Environ 664:1117–1132. https://doi.org/10.1016/j.scitotenv.2019.02.093

Gerrard J (1994) The landslide hazard in the Himalayas: geological control and human action. Geomorphol Nat Hazards 10(1–4):221–230. https://doi.org/10.1016/B978-0-444-82012-9.50019-0

Ghosh S, Das R, Goswami B (2013) Developing GIS-based techniques for application of knowledge and data-driven methods of landslide susceptibility mapping. Indian J Geosci 67:249–272. https://doi.org/10.1016/j.geomorph.2011.04.019

Gomez H, Kavzoglu T (2005) Assessment of shallow landslide susceptibility using artificial neural networks in Jabonosa river basin, Venezuela. Eng Geol 78:11–27

Guzzetti F, Mondini AC, Cardinali M, Fiorucci F, Santangelo M, Chang KT (2012) Landslide inventory maps: new tools for an old problem. Earth Sci Rev 112:42–66. https://doi.org/10.1016/j.earscirev.2012.02.001

Guzzetti F, Peruccacci S, Rossi M (2007) Rainfall thresholds for the initiation of landslides in central and southern Europe. Meteorol Atmos Phys 98:239–267. https://doi.org/10.1007/s00703-007-0262-7

Guzzetti F, Reichenbach P, Cardinali M, Galli M, Ardizzone F (2005) Probabilistic landslide hazard assessment at the basin scale. Geomorphology 72(1–4):272–299. https://doi.org/10.1016/j.geomorph.2005.06.002

Hakro MR, Harahap ISH (2015) Laboratory experiments on rainfall-induced flowslide from pore pressure and moisture content measurements. Nat Hazards Earth Syst Sci Discuss 3:1575–1613. https://doi.org/10.5194/nhessd-3-1575-

Haque U, da Silva PF, Devoli G, Pilz J, Zhao B, Khaloua A, Wilopo W, Andersen P, Lu P, Lee J, Yamamoto T, Keellings D, Wu JH, Glass GE (2019) The human cost of global warming: deadly landslides and their triggers (1995–2014). Sci Total Environ 682:673–684. https://doi.org/10.1016/j.scitotenv.2019.03.415

Henriques C, Zêzere JL, Marques F (2015) The role of the lithological setting on the landslide pattern and distribution. Eng Geol 189:17–31. https://doi.org/10.1016/j.enggeo.2015.01.025

Herold M, Latham JS, Di Gregorio A, Schmullius CC (2006) Evolving standards in land cover characterization. J Land Use Sci 1(2–4):157–168. https://doi.org/10.1080/17474230601079316

Huang F, Cao Z, Guo J, Jiang SH, Li S, Guo Z (2020) Comparisons of heuristic, general statistical and machine learning models for landslide susceptibility prediction and mapping. Catena 191:104580. https://doi.org/10.1016/j.catena.2020.104580

Hulme M (2016) 1.5 c and climate research after the Paris agreement. Nat Clim Change 6(3):222

Intrieri E, Bardi F, Fanti R, Gigli G, Fidolini F, Casagli N, Costanzo S, Raffo A, Massa GD, Capparelli G, Versace P (2017) Big data managing in a landslide early warning system: experience from a ground-based interferometric radar application. Nat Hazards Earth Syst Sci 17:1713–1723

Kalsnes B, Capobianco V (2022) Use of vegetation for landslide risk mitigation. In: Kondrup C et al (eds) Climate adaptation modelling. Springer Climate. Springer, Cham. https://doi.org/10.1007/978-3-030-86211-4_10

Kamal ASM, Ahmed B, Tasnim S, Sammonds P (2022) Assessing rainfall-induced landslide risk in a humanitarian context: the Kutupalong Rohingya Camp in Cox's Bazar, Bangladesh. Nat Hazards Res 2:230–248

Karegowda AG, Manjunath AS, Jayaram MA (2010) Comparative study of attribute selection using gain ratio and correlation based feature selection. Int J Inf Technol Knowl Manage 2(2):271–277

Kayastha P, Dhital MR, De Smedt F (2013) Application of the analytical hierarchy process (AHP) for landslide susceptibility mapping: a case study from the Tinau watershed, west Nepal. Comput Geosci 52:398–408. https://doi.org/10.1016/j.cageo.2012.11.003

Kim K, Son YS (2015) Geometrical and geotechnical characteristics of landslides in Korea under various geological conditions. J Mt Sci 12:1267–1280. https://doi.org/10.1007/s11629-014-3108-z

Lee S, Pradhan B (2007) Landslide hazard mapping at Selangor, Malaysia using frequency ratio and logistic regression model. Landslides 4(1):33–34

Lee S, Talib JA (2005) Probabilistic landslide susceptibility and factor effect analysis. Environ Geol 47:982–990. https://doi.org/10.1007/s00254-005-1228-z

Lee JU, Yong CC, Minseok K, Jang SJ, Lee J, Sukwoo K (2022) The effects of different geological conditions on landslide-triggering rainfall conditions in South Korea. Water 14(13):2051. https://doi.org/10.3390/w14132051

Lillesand T, Kiefer RW, Chipman J (2014) Remote sensing and image interpretation. Wiley

Liu J, Wu Z, Zhang H (2021) Analysis of changes in landslide susceptibility according to land use over 38 years in Lixian County, China. Sustainability 13:10858. https://doi.org/10.3390/su131910858b

Magliulo P, Di Lisio A, Russo F, Zelano A (2008) Geomorphology and landslide susceptibility assessment using GIS and bivariate statistics: a case study in Southern Italy. Nat Hazards 47:411–435

Mandal S, Maiti R (2014) Semi-quantitative approaches for landslide assessment and prediction. Part of the series Springer natural hazards. https://doi.org/10.1007/978-981-287-146-6

Mandal B, Mandal S (2018a) Analytical hierarchy process (AHP) based landslide susceptibility mapping of Lish river basin of eastern Darjeeling Himalaya, India. Adv Space Res 62:3114–3132. https://doi.org/10.1016/j.asr.2018.08.008

Mandal S, Mondal S (2018b) Statistical approaches for landslide susceptibility assessment and prediction. Springer International Publishing AG, Part of SpringerNature. https://doi.org/10.1007/978-3-319-93897-4

Mandal B, Mondal S (2019a) Prediction of landslide susceptibility using bivariate models. In: Geoinformatics and modelling of landslide susceptibility and risk. Environmental science and engineering, pp 121–143. https://doi.org/10.1007/978-3-030-10495-5_5

Mandal S, Mondal S (2019b) Geomorphic, geo-tectonic and hydrologic attributes and landslide susceptibility. In: Geoinformatics and modelling of landslide susceptibility and risk. Environmental science and engineering. Springer, Cham. https://doi.org/10.1007/978-3-030-10495-5_2

Mandal S, Mondal S (2019c) Machine learning models and spatial distribution of landslide susceptibility. In: Geoinformatics and modelling of landslide susceptibility and risk. Environmental science and engineering. Springer, Cham. https://doi.org/10.1007/978-3-030-10495-5_7

Mangeney A (2011) Geomorphology: landslide boost from entrainment. Nat Geosci 4(2):77

Masi EB, Segoni S, Tofani V (2021) Root reinforcement in slope stability models: a review. Geosciences 11:212. https://doi.org/10.3390/geosciences11050212

McKerchar AI (2003) Streamflow, encyclopedia of physical science and technology, vol 3, 3rd edn. Academic Press, pp 129–142. https://doi.org/10.1016/B0-12-227410-5/00741-9

Medina V, Hürlimann M, Guo Z, Lloret A, Vaunat J (2021) Fast physically-based model for rainfall-induced landslide susceptibility assessment at regional scale. Catena 201:105213. https://doi.org/10.1016/j.catena.2021.105213

Mersha T, Meten M (2020) GIS-based landslide susceptibility mapping and assessment using bivariate statistical methods in Simada area, northwestern Ethiopia. Geoenviron Disasters 7(20). https://doi.org/10.1186/s40677-020-00155-x

Meten M, PrakashBhandary N, Yatabe R (2015) Effect of landslide factor combinations on the prediction accuracy of landslide susceptibility maps in the Blue Nile Gorge of Central Ethiopia. Geoenviron Disasters 2(9). https://doi.org/10.1186/s40677-015-0016-7

Moore ID, Grayson RB, Ladson AR (1991) Digital terrain modeling—a review of hydrological geomorphological and biological application. Hydrol Process 5:3–30

Mukane P (2014) India's worst landslides, and why these might not be the last ones. http://www.dnaindiacom/india/

Myronidis D, Papageorgiou C, Theophanous S (2016) Landslide susceptibility mapping based on landslide history and analytic hierarchy process (AHP). Nat Hazards 81:245–263

Nerella SP, Alajangi S, Dhakal D (2019) Landslide susceptibility mapping using GIS-based likelihood frequency ratio model: a case study of Pakyong—Pacheykhani Area, Sikkim Himalaya. In: Rao P, Rao K, Kubo S (eds) Proceedings of international conference on remote sensing for disaster management. Springer series in geomechanics and geoengineering, Springer, Cham. https://doi.org/10.1007/978-3-319-77276-9_51

Nhu VH, Shirzadi A, Shahabi H, Chen W, Clague JJ, Geertsema M, Jaafari A, Avand M, Miraki S, Talebpour Asl D, Pham BT (2020) Shallow landslide susceptibility mapping by random forest base classifier and its ensembles in a semi-arid region of Iran. Forests 11(4):421. https://doi.org/10.3390/f11040421

Nugraha H, Wacano D, Dipayana GA, Cahyadi A, Mutaqin BW, Larasati A (2015) Geomorphometric characteristics of landslides in the Tinalah Watershed, Menoreh Mountains, Yogyakarta, Indonesia. Procedia Environ Sci 28:578–586. https://doi.org/10.1016/j.proenv.2015.07.068

Nwazelibe VE, Egbueri JC, Unigwe CO, Agbasi JC, Ayejoto DA, Abba SI (2023) GIS-based landslide susceptibility of Western Rwanda: an integrated artificial neural network, frequency ratio, and Shannon entropy approach. Environ Earth Sci 82(19):439

Pal SC, Chowdhuri I (2019) GIS-based spatial prediction of landslide susceptibility using frequency ratio model of Lachung River basin, North Sikkim, India. SN Appl Sci 1:1–25. https://doi.org/10.1007/s42452-019-0422-7

Panchal S, Shrivastava AK (2020) Landslide susceptibility mapping along highway corridors in GIS environment. In: Ahmed S, Abbas S, Zia H (eds) Smart cities—opportunities and challenges. Lecture notes in civil engineering. Springer, p 58. https://doi.org/10.1007/978-981-15-2545-2_8

Parra E, Mohr CH, Korup O (2021) Predicting Patagonian landslides: roles of forest cover and wind speed. Geophys Res Lett 48:1–10. https://doi.org/10.1029/2021GL095224

Paulin L, Bursik G, Hubp M, Mejía JL, Aceves QF (2014) A GIS method for landslide inventory and susceptibility mapping in the Río El Estado watershed, Pico de Orizaba volcano, México. Nat Hazards 71:229–241

Petley D (2012) Global patterns of loss of life from landslides. Geology 40(10):927–930. https://doi.org/10.1130/G33217.1

Pham BT, Pradhan B, Tien Bui D, Prakash I, Dholakia MB (2016) A comparative study of different machine learning methods for landslide susceptibility assessment: a case study of Uttarakhand area (India). Environ Model Softw 84:240–250. https://doi.org/10.1016/j.envsoft.2016.07.005

Pham BT, Bui DT, Indra P, Dholakia M (2015) Landslide susceptibility assessment at a part of Uttarakhand Himalaya, India using GIS-based statistical approach of frequency ratio method. Int J Eng Res Technol 4:338–344. https://doi.org/10.17577/ijertv4is110285

Rai PK, Mohan K, Kumra VK (2014) Landslide hazard and its mapping using remote sensing and GIS. J Sci Res 58:1–13

Ram P, Gupta V (2022) Landslide hazard, vulnerability, and risk assessment (HVRA), Mussoorie township, lesser Himalaya, India. Environ Dev Sustain 24:473–501. https://doi.org/10.1007/s10668-021-01449-2

Rawat PK, Pant B (2023) Geoenvironmental GIS development to investigate landslides and slope instability along frontal zone of central Himalaya. Nat Hazards Res 3(2):196–204. https://doi.org/10.1016/j.nhres.2023.03.005

Regmi AD, Devkota KC, Yoshida K, Pradhan B, Pourghasemi HR, Kumamoto T, Akgun A (2014) Application of frequency ratio, statistical index, and weights-of-evidence models and their comparison in landslide susceptibility mapping in central Nepal Himalaya. Arab J Geosci 7:725–742

Roy J, Saha S, Arabameri A, Blaschke T, Bui DT (2019) A novel ensemble approach for landslide susceptibility mapping (LSM) in Darjeeling and Kalimpong districts, West Bengal, India. Remote Sens 11(23):2866. https://doi.org/10.3390/rs11232866

Saha A, Saha S (2020) Comparing the efficiency of weight of evidence, support vector machine and their ensemble approaches in landslide susceptibility modelling: a study on Kurseong region of Darjeeling Himalaya, India. Remote Sens Appl Soc Environ 19:100323. https://doi.org/10.1016/j.rsase.2020.100323

Saha S, Arabameri A, Saha A, Blaschke T, Ngo PTT, Nhu VH, Band SS (2021a) Prediction of landslide susceptibility in Rudraprayag, India using novel ensemble of conditional probability and boosted regression tree-based on cross-validation method. Sci Total Environ 764:142928. https://doi.org/10.1016/j.scitotenv.2020.142928

Saha S, Roy J, Pradhan B, Hembram TK (2021b) Hybrid ensemble machine learning approaches for landslide susceptibility mapping using different sampling ratios at East Sikkim Himalayan, India. Adv Space Res 68:2819–2840. https://doi.org/10.1016/j.asr.2021.05.018

Saha S, Arabameri A, Saha A, Blaschke T, Ngo PTT, Nhu VH, Band SS (2021c) Prediction of landslide susceptibility in Rudraprayag, India using novel ensemble of conditional probability and boosted regression tree-based on cross-validation method. Sci Total Environ 764:142928

Saha S, Roy J, Hembram TK, Pradhan B, Dikshit A, Abdul Maulud KN, Alamri AM (2021d) Comparison between deep learning and tree-based machine learning approaches for landslide susceptibility mapping. Water 13:2664. https://doi.org/10.3390/w13192664

Sahana M, Sajjad H (2017) Evaluating effectiveness of frequency ratio, fuzzy logic and logistic regression models in assessing landslide susceptibility: a case from Rudraprayag district, India. J Mt Sci 14:2150–2167. https://doi.org/10.1007/s11629-017-4404-1

Shahabi H, Hashim M (2015) Landslide susceptibility mapping using GIS-based statistical models and remote sensing data in tropical environment. Sci Rep 5:9899. https://doi.org/10.1038/srep09899

Shirzadi A, Soliamani K, Habibnejhad M, Kavian A, Chapi K, Shahabi H, Chen W, Khosravi K, Thai Pham B, Pradhan B, Ahmad A (2018) Novel GIS based machine learning algorithms for shallow landslide susceptibility mapping. Sensors 18(11):3777. https://doi.org/10.3390/s18113777

Singh P, Sur U, Rai PK, Singh SK (2023) Landslide susceptibility prediction using frequency ratio model: a case study of Uttarakhand, Himalaya (India). Proc Indian Natl Sci Acad 89(3):600–612

Skilodimou HD, Bathrellos GD, Koskeridou E, Soukis K, Rozos D (2018) Physical and anthropogenic factors related to landslide activity in the Northern Peloponnese, Greece. Land 7(85). https://doi.org/10.3390/land7030085

Solari L, Del Soldato M, Raspini F, Barra A, Bianchini S, Confuorto P, Casagli N, Crosetto M (2020) Review of satellite interferometry for landslide detection in Italy. Remote Sens 12:1351. https://doi.org/10.3390/rs12081351

Stanley T, Kirschbaum DB (2017) A heuristic approach to global landslide susceptibility mapping. Nat Hazards 87:145–164. https://doi.org/10.1007/s11069-017-2757-y

Sujatha ER, Sridhar V (2021) Landslide susceptibility analysis: a logistic regression model case study in Coonoor, India. Hydrology 8:41. https://doi.org/10.3390/hydrology8010041

Sujatha ER, Rajamanickam GV (2015) Landslide hazard and risk mapping using the weighted linear combination model applied to the Tevankarai stream watershed, Kodaikkanal, India. Hum Ecol Risk Assess Int J 21:1445–1461. https://doi.org/10.1080/10807039.2014.920222

Thapa R, Gupta S, Gupta A, Reddy DV, Kaur H (2017) Use of geospatial technology for delineating groundwater potential zones with an emphasis on water-table analysis in Dwarka river basin, Birbhum, India. Hydrogeol J. https://doi.org/10.1007/s10040-017-1683-0

Tofani V, Bicocchi G, Rossi G, Segoni S, D'Ambrosio M, Casagli N, Catani F (2017) Soil characterization for shallow landslides modeling: a case study in the Northern Appennines (Central Italy). Landslides 14:755–770. https://doi.org/10.1007/s10346-017-0809-8

Tohari A, Nishigaki M, Komatsu M (2007) Laboratory rainfall-induced slope failure with moisture content measurement. J Geotech Geoenviron Eng 133(5):575

Tsou CY, Chigira M, Higaki D et al (2018) Topographic and geologic controls on landslides induced by the 2015 Gorkha earthquake and its aftershocks: an example from the Trishuli Valley, central Nepal. Landslides 15:953–965. https://doi.org/10.1007/s10346-017-0913-9

UNISDR (2009) UNISDR terminology on disaster risk reduction. In: United nations international strategy for disaster reduction

Upadhyay V, Jat MK (2014) Risk assessment of petroleum fire using geo-spatial techniques. Int J Remote Sens Geosci (IJRSG) 3:11–20

Vakhshoori V, Zare M (2016) Landslide susceptibility mapping by comparing the weight of evidence, fuzzy logic, and frequency ratio methods. Geomat Nat Hazards Risk 7(5):1731–1752

Vorpahl P, Elsenbeer H, Märker M, Schröder B (2012) How can statistical models help to determine driving factors of landslides? Ecol Model 239:27–39. https://doi.org/10.1016/j.ecolmodel.2011.12.007

Van Westen CJ, Castellanos E, Kuriakose SL (2008) Spatial data for landslide susceptibility, hazard, and vulnerability assessment: an overview. Eng Geol 102:112–131. https://doi.org/10.1016/j.enggeo.2008.03.010

Vijith H, Dodge-Wan D (2019) Modelling terrain erosion susceptibility of logged and regenerated forested region in northern Borneo through the analytical hierarchy process (AHP) and GIS techniques. Geoenviron Disasters 6:8. https://doi.org/10.1186/s40677-019-0124-x

Wang Q, Li W (2017) A GIS-based comparative evaluation of analytical hierarchy process and frequency ratio models for landslide susceptibility mapping. Phys Geogr 38(4):318–337

Wang Y, Fang Z, Hong H, Costache R, Tang X (2021) Flood susceptibility mapping by integrating frequency ratio and index of entropy with multilayer perceptron and classification and regression tree. J Environ Manag 289:112449. https://doi.org/10.1016/j.jenvman.2021.112449

Yalcin A (2008) GIS-based landslide susceptibility mapping using analytical hierarchy process and bivariate statistics in Ardesen (Turkey): comparisons of results and confirmations. Catena 72:1–12

Yalcin A, Reis S, Aydinoglu AC, Yomralioglu T (2011) A GIS based comparative study of frequency ratio, analytical hierarchy process, bivariate statistics and logistics regression methods for landslide susceptibility mapping in Trabzon, NE Turkey. Catena 85(3):274–287

Yang F, Kumar A, Schlesinger ME, Wang W (2003) Intensity of hydrological cycles in warmer climates. J Clim 16(14):2419–2423. https://doi.org/10.1175/2779.1

Yaohui L, Wenzhuo Z, Xiaoxian C, Mingyang Y, Yingjun S, Fei M, Xiwei F (2021) Landslide detection of high-resolution satellite images using asymmetric dual-channel network. In: IEEE international geoscience and remote sensing symposium IGARSS. Brussels, Belgium, pp 4091–4094. https://doi.org/10.1109/IGARSS47720.2021.9553683

Zêzere JL, Vaz T, Pereira S (2015) Rainfall thresholds for landslide activity in Portugal: a state of the art. Environ Earth Sci 73:2917–2936. https://doi.org/10.1007/s12665-014-3672-0

Zhang J, Liu RK, Deng W, Khanal NR, Gurung DR, Murthy MSR, Wahid S (2016) Characteristics of landslide in Koshi river basin, central Himalaya. J Mt Sci 13:1711–1722. https://doi.org/10.1007/s11629-016-4017-0

Zhao C, Lu Z (2018) Remote sensing of landslides: a review. Remote Sens 10:279

Zhu AX, Wang R, Qiao J, Qin CZ, Chen Y, Liu J, Du F, Lin Y, Zhu T (2014) An expert knowledge based approach to landslide susceptibility mapping using GIS and fuzzy logic. Geomorphology 214:128–138. https://doi.org/10.1016/j.geomorph.2014.02.003

Chapter 4
Landslide Susceptibility Mapping Methods—A Review

Evangelin Ramani Sujatha and J. S. Sudharsan

Abstract Landslides are a prevalent geological hazard in many regions globally and have a profound effect on the resources and social fabric of the affected community. Identification of landslide-prone zones is mandatory for developing land use plans in regions affected by landslides. Landslide susceptibility maps are crucial for mapping landslide hazards and evaluating the risk resulting from landslides. Zones susceptible to the geologic phenomenon of landslides are identified and mapped using various methods and models. Off-late the increase in frequency of the landslide hazards in many hill and mountain regions around the globe mandates landslide susceptibility mapping. The choice of the method depends on the factors and data available. Common methods include deterministic modelling for specific slopes and statistical & heuristic models for mapping landslide susceptibility at regional scale. Statistical and probabilistic methods like frequency ratio, conditional probability, certainty factor and logistic regression, etc. are popularly used for mapping susceptibility. Off-late machine learning and deep mining techniques are also widely adopted for assessing susceptibility. Heuristic models like analytical hierarchical and analytical network process are also common. The choice of both the model to map susceptibility and factors used to assess the susceptibility of the region are dictated by factors like the local geo-environmental set-up, quality of data used and its reliability. This chapter explores the suitability of various susceptibility models, their merits and limitations in different geo-environmental scenarios. Also, a detailed review on the various topographic, hydrologic, geological, geotechnical, environmental and anthropogenic factors is presented arguing their effectiveness in specific geologic and environmental regional set-up.

E. R. Sujatha (✉)
Centre for Advanced Research in Environment, School of Civil Engineering, SASTRA Deemed to be University, Thanjavur, India
e-mail: sujatha@civil.sastra.edu; r.evangelin@gmail.com

J. S. Sudharsan (✉)
School of Energy and Environment, NICMAR University, Pune, India
e-mail: ssudarsan@nicmar.ac.in

G. K. Panda et al. (eds.), *Landslide: Susceptibility, Risk Assessment and Sustainability*, Advances in Natural and Technological Hazards Research 52,
https://doi.org/10.1007/978-3-031-56591-5_4

Keywords Landslides · Susceptibility · Statistical models · Deterministic models · Heuristic models · Landslide causative factors

4.1 Introduction

Landslides are a frequently occurring natural hazard in many parts of the mountainous terrains globally (Froude and Petley 2018). Landslide also known as mass wasting or slope failures happen as a standalone event or in conjunction with natural hazards like seismic events, floods, cloudbursts, avalanches, forest fires, hailstorms or as a consequence of improper construction/anthropogenic activity (Bandyopadhyay and De 2018). Off-late climate change and frequent occurrence of extreme events constitute a major trigger for landslide (Dilley et al. 2005)reported that approximately 3.7 million square kilometers of terrain is prone to landslides and 300 million people are affected by it. The global damage from landslides/ mass-wasting is estimated at nearly 18 billion euros (Haque et al. 2016) and in Germany it amounts to nearly 221 million euros(Klose et al. 2016). Huge causalities are reported by landslides world-wide. The frequency of landslides is expected to increase due to various reasons like permafrost thawing (Etzeimuiier and Frauenfeider 2009), increase in frequency of heavy rainfall events (Crozier 2010), population explosion and change in land use(Promper et al. 2014). (Petley 2012) reported 80,058 fatalities due to landslides between 2004 and 2010 alone and this number is an underestimate as causalities due to landslides are poorly documented according to (Sepúlveda and Petley 2015), Countries in Asia and South and Central America are affected by landslides in particular (Petley 2012). Few examples of landslides that caused huge fatalities are the landslides triggered by Wenchuan earthquake in China and Nepal in the year 2008 and 2015 with 20,000 and 8500 fatalities respectively (Promper et al. 2014; Williams et al. n.d.). Similarly, the Uttarkashi landslide in India in the year 2013 triggered by flash flood resulted in the death of 169 confirmed deaths and 4021 missing people. This landslide affected 4200 villages and damaged 2513 houses as reported by the report of National Institute of Disaster Management (Reports|NDM India n.d.). The global landslide map shows that that south east Asia, particularly the Himalayan belt, is worst affected by landslides not only in frequency but also in magnitude and fatalities(Kirschbaum et al. 2015) and is a landslide hotspot. Also, (Kirschbaum et al. 2015) report highest fatalities in India, nearly 43% against the 15% landslide occurrences. Landslides not only lead to loss of life but also disrupt the social and economic development in the affected regions (Klose et al. 2016). Landslides also take a toll on the built environment driven by urban centric development and rapid expansion of infrastructure. For example, the Regent landslide in Sierra Leone claimed nearly 1000 lives and damaged property & infrastructure to the tune of approximately 30 million USD (Redshaw et al. 2019; World Bank Group 2017).

The increase in landslide frequency and expansion of zones affected by landslides due to unfavorable geographic and environmental settings and inadequate mitigation strategies point to the urgent need for management and mitigation of landslide hazard (Franceschini et al. 2022; Gariano and Guzzetti 2016). An efficient landslide management strategy is impeded by several challenges like limited financial resources, inaccessibility of regions and inefficiency in implementing mitigation measures, etc (Rozos 2010). The first step toward landslide management is to map the zones susceptible to landslides. Landslide susceptibility maps mark the spatial extent of landslides on various scales and enable policy makers, land use planners, disaster management cells, town and country planning authorities and authorities responsible for infrastructure planning to take into consideration the landslide hazard in developmental plans and plan effective strategies for its mitigation.

Landslide susceptibility refers to the possibility of a landslide happening in a region/area and is dictated by the local terrain (Brabb 1985) and it predicts the spatial occurrence of landslides (Guzzetti 2006). Numerous techniques are used to assess and map landslide susceptibility including mapping based on geomorphology, landslide inventory, statistical methods, knowledge based/ heuristic methods, physically based slope models and artificial intelligence based/deep learning and classification methods(Bandyopadhyay et al. 2013; Bhattacherjee et al. 2022; Kirschbaum et al. 2015). (Aleotti & Chowdhury 1999; Panchal and Shrivastava 2022) presented a detailed review on suitability of the different approaches for specific sites and scales along with their merits and limitations. They also discussed on both quantitative and qualitative approaches. (Huabin et al. 2005) reviewed the various susceptibility analysis approaches at varying scales and compared statistical, heuristic and deterministic models based on their merits and drawbacks. (Chacón et al. 2006) in their review proposed to classify susceptibility maps into three groups—(i) maps that portray the spatial occurrence of landslides without any indication of temporal occurrence (ii) maps that show spatial and temporal occurrence of landslides and (iii) maps that represent landslide risk (i.e) the expected impact of landslide on the geo-environment and society. (van Westen et al. 2008) in their review discussed on need for the relevance of environmental and thematic information used for vulnerability, susceptibility and hazard assessments and the various methods to obtain data. An associated taxonomy for landslide susceptibility classification was proposed by (Das et al. 2023) along with a scheme.

(Das et al. 2023) classified landslide as qualitative, semi-quantitative and deterministic methods. Qualitative approach includes geomorphic mapping, combination of maps and distribution analysis. Bivariate, multivariate, machine learning models and deterministic methods fall under the quantitative approach. Methods that use a combination of techniques that fall under quantitative and qualitative approaches are called semi-quantitative. Of the many methods used to map landslide susceptibility physically based slope models, data-driven methods and knowledge based/ heuristic methods are widely adopted. Heuristic method is qualitative in nature and the other two methods are quantitative. Since 2010 machine learning (ML) models are used to analyse the complex relation between landslides and factors causing landslides (Kulsoom et al. 2023; Liang et al. 2023). Statistical models can be implemented in ML environments and yield accurate susceptibility and risk maps.

Landslide models and maps require a suitable terrain/mapping unit. This terrain/ mapping unit, is a demarcated land area (portion) that differs from adjacent terrain and is characterized by ground conditions distinctly different from the surrounding area (Guzzetti 2006). Mapping unit can therefore be defined as a geographical expanse that has maximum homogeneity within the unit and heterogeneity between the surrounding units. The mapping units used in literature for landslide mapping can be broadly classified as—(i) terrain units (ii) pixels (iii) unique condition unit (iv) geo-hydrological units (v) slope units (vi) topographical units and (vii) administrative or political units. IS 14496 (Part 2), the BIS code for landslide hazard zonation used in India uses slope facets as mapping units. With the advent and wide usage of GIS pixels of different resolutions are more common in landslide susceptibility mapping.

The availability of higher resolution of satellite and remote sensing data (Abd El-Kawy et al. 2011; Bhattacharjee et al. 2021), large and accurate landslide inventories and powerful geographic information system (GIS) (Sarif et al. 2021) with higher computing power that can handle large datasets improve the predictive capacity of landslide susceptibility models (Ahmad and Goparaju 2016). The performance of the model depends on the quality and quantity of available data and sampling strategies adopted (Chang et al. 2023; Ji et al. 2023). Validation of the prepared landslide susceptibility maps is an integral part of the landslide susceptibility modelling (Brock and Purkis 2009; Chaussard and Kerosky 2016). Predictive performance of a susceptibility map can be assessed by its predictive power in retrospect (Reichenbach et al. 2018) indicating that the predictive power of the susceptibility map is validated with landslide events that occur after the map is prepared. This validation process has a higher objectivity, comprehensibility and accuracy (Fleuchaus et al. 2021) compared to the validation process outlined by (Chung and Fabbri 2003). Physically based landslide models and statistical models are widely used to map landslide susceptibility (Reichenbach et al. 2018). The choice a model to assess landslide susceptibility needs an understanding of its process flow and this study elaborates on the availability and working of these models. This chapter reviews the various factors triggering & causing landslides and methods used to map the susceptibility of landslides inorder to provide a deeper understanding of the methods that can successfully be employ a suitable susceptibility model that perform efficiently for a specific region/area. The popular models—physically based models, statistical and heuristic models are discussed in detail.

4.2 Failure Mechanism of Slopes

The stability of slopes is a critical consideration in geotechnical engineering, as slope failures can have severe consequences, ranging from property damage to loss of life. Understanding the mechanisms that lead to slope failures is essential for effective risk assessment and mitigation (Sujatha et al. 2023).

Particularly on freshly cut slopes, the slope face becomes parched and shrinks on hot, dry days. The primary determinants of extent and depth are the soil's flexibility and the kind of slope protection used. As a result of water seeping through cracks and other exposed surfaces during a downpour, the slope mass swells and becomes saturated, progressively reducing the soil's shear strength through seepage, soil particle migration, and a rise in the void ratio. At first, water seeps into the slope mostly through the fissures caused by desiccation and in reaction to the top layer of dried soil's suction pressure. Seepage occurs parallel to the slope face when the outer face of the slope swells and becomes saturated. As a result of ongoing rainfall, the slope's permeability increases. Saturation and swelling-related reduction in shear strength combined with seepage-related conditions lead to failure ultimately if the shearing resistance is equal to or less than the shearing force (Zhao et al. 2019). Understanding the forces acting on slopes—the driving force and the resisting force—is crucial to understanding soil mechanics because they keep slopes from failing. Resisting forces to driving forces are divided by the factor of safety (FoS). Slopes with lower FoS have a higher chance of failure than those with higher FOS. As a result, slopes with more FoS are more secure (Niroumand et al. 2012).

4.3 Landslide Inventory

Landslide inventories that are records of the past and present landslides is an important and vital component of landslide susceptibility analysis and mapping. (Guzzetti et al. 1999) in their review insisted the need quality landslide inventories which formed the basis for statistical analysis and information on factors used to analyse susceptibility. The landslide inventories can point to either geographic location of past/present landslides presenting landslide as a point or it can be polygon marking the complete spatial extent of the landslide. The second type of inventory where the spatial extent of landslide is presented is of primary interest though such maps are scarce. Landslide inventories can be broadly classified as follows.

(i) Geomorphological landslide inventory—the terrain/geomorphic features are used to identify the landslide. For example, the landslide can be identified from the scar of past event
(ii) Multi-temporal landslide inventory—these inventories contain the landslide event data over a period of time, say for a certain number of years for a specific area/region
(iii) Event inventories—the inventory of landslide occurrence associated with a triggering factor like a particular rainfall event, snowmelt, earthquake or for a specific season like a monsoon period in a particular year
(iv) Historical inventories—inventory of past landslide scars that dates a long period backward and may also be considered geomorphological (Reichenbach et al. 2018)

This landslide data to create the inventories can be collected from several sources like through detailed field investigation and mapping, from aerial photographs, satellite images or digital elevation models (DEM) of high resolution and through archive search (i.e.) from records that are created by various agencies involved in planning and mitigation of landslides. In general, the landslides are represented based on (i) their movement (i.e.) translation, rotation, lateral spread, flow, fall, topple, etc. and is observed to be the most common type of landslide representation (ii) landslide material like rock, soil or debris and (iii) depth, shallow or deep seated (Chirico et al. 2021).

4.4 Landslide Susceptibility, Hazard and Risk—An overview

The likelihood that specific places within the topography may cause a particular sort of landslide is known as landslide susceptibility. It can be quantitatively defined as an estimate of the chance that slope failures will occur in a specific time period, or it can be expressed quantitatively using terms like high, medium, and low, or indices like the Factor of Safety (Oldrich 2016). A description of the current or projected landslide's velocity and severity may also be included in the concept of susceptibility. Timeframe is specifically ignored in the susceptibility study, despite the fact that it is anticipated that landslides will happen more frequently in the most vulnerable places. Susceptibility to landslides includes landslides that originate locally or that may originate elsewhere and then move onto or regress into the region (Malta n.d.).

In recent decades, there has been significant advancement in the areas of landslide risk, hazard, and susceptibility to landslides. The majority of these zoning studies are qualitative in nature, but more recently, there have been instances of quantifying the dangers associated with current development and landslides by giving them a yearly likelihood (frequency).

There is seldom standard terminology used between nations, and even within a single nation. The outcomes of landslide zoning are sometimes imprecisely stated, and landslide zoning maps frequently conflate risk, danger, and susceptibility. Accuracy and dependability of these maps vary, as demonstrated by a study of recent experience with hazard zoning and landslide susceptibility (Corominas et al. 2014). These days, many different types of input data—both qualitative and quantitative—are used to generate maps. Certain nations lack the necessary statistics to properly quantify the risk of landslides. The network of literature related to keywords in landslide analysis is shown in Fig. 4.1.

4.5 Physically Based Landslide Models (PBLM)

Physically based slope/landslide models are based on the physical laws that control slope stability. Most of the PBLM follow deterministic approach while other probabilistic approach (Li et al. 2023; Sun et al. 2022). The PBLM exhibit higher prediction capacity and is a suitable quantitative assessment that takes into account the effect of

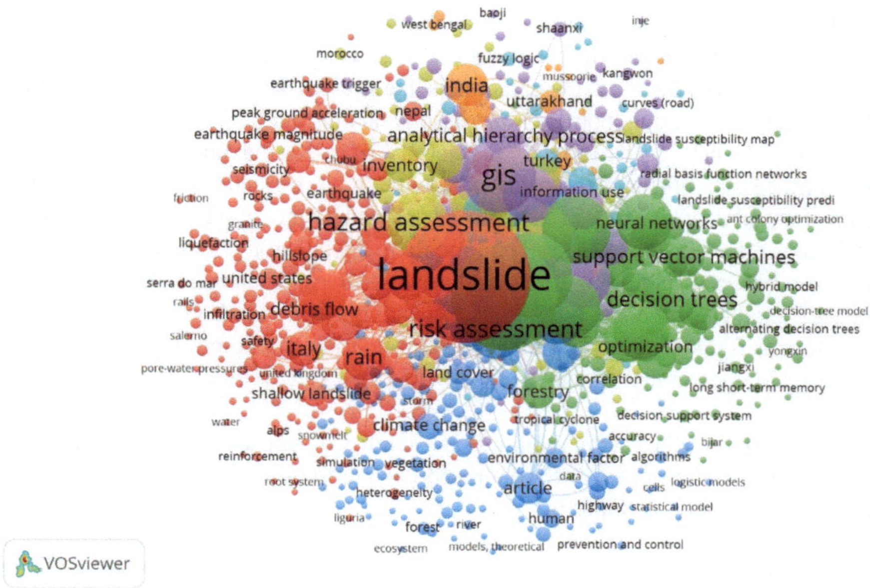

Fig. 4.1 Keywords associated with landslide susceptibility analysis

factors that contribute to initiation of landslides (Corominas et al. 2014) These models can assess the stability of a slope based on its geotechnical properties irrespective of landslide occurrences and can also be extended for analysis for wider areas owing to the availability of GIS and its general grid-based structure. These models also allow for combining with hydrological models as slope hydrology is an important factor that dictates the stability of the slope. PBLM coupled with hydrological models can effectively evaluate the stability of slopes under conditions of increasing pore pressure due to rainwater infiltration which in turn leads to a decrease in shear strength. The physical models that have been used to map slope instability in the recent past between 2010 and 2020 are SHETRAN, SLIP, SUSHI, SHALSTAB, SINMAP, SNETRAN, PISA-m, TIGRIS and GEOtop-FS and of these models SINMAP, SHALSTAB and TIGRIS have been used in nearly 93% for the assessment of the slope stability (Dietrich et al. 1998; Moreira Melo et al. 2021; Cabral et al. 2022). Of the above models, GEOtop-FS alone is a spatially distributed 3D model that calculates energy and water budgets and simulates both saturated and unsaturated subsurface flow conditions, surface runoff, channel flows and fluctuations at atmosphere—soil interface due to storms (Bain et al. 2019; König et al. 2022; Michel et al. 2014) Popular models—stability index mapping (SINMAP), shallow land sliding stability model (SHALSTAB), transient rainfall infiltration and grid-based slope stability (TIGRIS) probabilistic infinite slope stability analysis (PISA-m) and systeme hydrologique European transport (SHETRAN) use infinite slope stability analysis coupled with hydrological models.

The one-dimensional infinite slope stability analysis is an example of one of the popular PBLM (Capparelli and Versace, 2011; Thanh and De Smedt 2014) where slope stability is assessed using parameters like stresses (both normal and tangential), friction angle, cohesion, pore pressure in soil and unit weight of soil, etc. The infinite slope model can be used to analyse the stability of slopes that fail by translation (i.e.) slip surface is parallel to slope surface (ground) and is at a shallow depth. Slopes with a layered soil deposit can also be analysed using these models (Park et al. 2019). The stability of the slope is expressed commonly using a factor of safety (FoS) which is the ratio between stabilizing forces and driving forces (Thanh and De Smedt 2014). FoS of 1 indicates the slope failure is impending, while the FoS above 1 indicates the slope is stable and below 1 unstable. Considering the simplified assumptions in PBLM and uncertainty of data, a slope with a FoS greater than 1.5 is considered to be stable.

Several approaches using infinite slope stability models can be used to determine the FoS of specific slopes. A common approach used by several authors to compute the factor of safety of slopes is

$$\text{FoS} = (\text{Cs} + \text{Cr})/(\gamma\_\text{e}\ \text{D}\sin\theta) + (1 - \text{m}\gamma\_\text{w}/\gamma\_\text{e})\ \tan\varphi/\tan\theta \tag{4.1}$$

where Cs—soil cohesion (kPa); Cr—root cohesion (kPa); ϒw—unit weight of water (kN/m^3); D—thickness of soil layer above bedrock (m) and is measured in the direction perpendicular to surface topography; θ—slope angle (°); φ—friction angle of soil (°); ϒe—effective unit weight of soil (kN/m3) and is calculated as

$$\Upsilon\text{e} = \text{q}\cos\theta/\text{D} + \text{m}\gamma\_\text{s} + (1 - \text{m})\,\gamma\_\text{d} \tag{4.2}$$

where q is the surcharge on the slope/soil; γ_s—saturated unit weight of soil (kN/m^3) and γ_d —dry unit weight of soil (kN/m^3). The parameter 'm' in Eqs 4.1 and 4.2 is wetness index and reflects the relative position of ground water table in the slope and is the ratio between saturated soil thickness (h) in metres and total thickness of soil above the slope (D). Nature of soil, hydrologic & hydro-geologic conditions on the slope and terrain features influence the wetness index. Wetness index values range between 0 and 1, where 1 indicates a completely saturated soil. During a hydrologic event wetness index can reach 1 in valleys but owing to gentle gradient these slopes may not fail. On the other hand, though wetness index values on steeper slopes like cliffs, ridges and road incisions can be much lower than 1 but they are more prone to failure by virtue of their higher slope gradient. These observations point out that the capacity of wetness index to cause slope instability can be ascertained only after taking into consideration the soil properties relating to its strength and shear stress mobilized.

The choice of PBLM is limited by factors like quantity & quality of geotechnical data required, simplifying assumptions made in the process, represent only part of completely mobilized source areas & mobilized volumes and ambiguity in the interpretation of results (Corominas et al. 2014; Thanh & De Smedt 2014)Data is often limited in spatial extent and is of variable quality (van Westen et al. 2008) The strength parameters like friction angle and cohesion are inherently heterogeneous spatially as soil on the slopes is formed by natural process (Le 2014) and these uncertainties affect the PBLM. Similarly other uncertainties limiting the PBLM are uncertainties owing to geologic formation, measurement errors, spatial variability and errors associated with the model itself (Park et al. 2019).

4.6 Landslide Susceptibility using Statistical Methods

Statistical models to map landslide susceptibility depend on landslide inventories and the functional relation between known known landslide causative factors (also known as predisposing factors) with the past/present landslide distribution in a region (Guzzetti 2006; van Westen et al. 2008). Factors that cause landslide and the magnitude of their influence are intrinsic to the region studied. Statistical models are data-driven models and their performance depend on the choice and quality of data used. These statistical models are one of the early models used for analysing and mapping landslides and have been in use since mid-1970's at different scales and mapping units. (Neuland 1976) used a statistical model for landslide analysis and modelling, which is probably the first known statistical model, a bivariate discriminant analysis for landslide analysis. His study explained the relationship of the various geomechanical, lithological, morphometric and structural factors with landslides. His dataset consisted of 250 unstable and stable slopes located in Germany and he used a bivariate discriminant analysis. (Carrara 1983), then in the landmark study used a multiple regression and discriminant analysis to predict the susceptibility of landslides in two catchments Ferro, north Calabria and Buonamico, south Calabria in Italy. He used a large landslide dataset along with geological and geomorphological data for the analysis. (Carrara 1983) developed a specific software to handle thematic topography automatically which is an earlier version of the currently popular GIS (Goodchild 2010). Following the pioneering attempts made by (Neuland 1976) and (Carrara 1983) several researches were presented on landslide analysis and mapping using diverse approaches based on statistical models. The various statistical methods used to analyse landslide susceptibility include bivariate approaches (like frequency ratio, information value, weight of evidence and index of entropy, etc.), multivariate models like logistic regression and machine learning models like apply various algorithms such as random forest, decision tree, support vector machine, artificial neural network, etc. Bivariate models are more popular than other types of susceptibility mapping approaches. A brief overview of the common methods used for susceptibility mapping is discussed here.

Frequency ratio (FR) compares the landslides that are caused in a specific class of a thematic layer expressed as a ratio between the landslides in a specific class of the thematic layer to total number of landslides in the region and the spatial extent of the same class of the said thematic layer to the total area of the region under study. Usually, the area units are pixels of convenient resolution. A ratio greater than 1 indicates higher susceptibility while less than 1 means lower susceptibility to landslides. This technique can be applied at all geographic scales—global, regional and local owing to its simplicity and ease of computation but it establishes a linear relation between the causative factors and landslides which is not representative of field scenario. Information value (IV) method is an adaptation of the FR method where the natural logarithm of the FR is taken as the weight for the specific class of a particular thematic layer. If the IV of specific class is higher than 0.1, this class has higher probability to fail. The positive and negative weights provide ease of interpretation and also this non-parametric method does not require scaling of the factors. Like FR it can be applied at all geographic scales. But when IV value does not exist for a certain class, there is bias in the output and also IV method does not take into account the relative importance of the factors selected for the study.

Weight of evidence (WoE) model uses Bayesian probability in its log-linear form to estimate the weights based on the landslide events across the selected causative factors. The log-likelihood ratio is estimated based on prior and post probability and their contrast is measures the strength of the weights in a specific class of a thematic layer based on whether it is positive or negative. This is a non-parametric model with requires no tuning. The quality of the landslide controls the performance of the model. Index of entropy (IE) uses an entropy index that quantities the extent to which each causative factor affects the stability of the slopes. Entropy is a quantitative that estimates the different degrees of imbalance, disorder, uncertainty and instability within an indexing system. The IE model standardizes the entropy associated with the landslide causative factor and assigns it as relative weight for each class of a specific causative factor. This a nonparametric method with no tuning required and can be applied at all geographic scales but cannot be used with limited data as data scarcity will lead to inconsistency in prediction.

Logistic regression, a non-parametric method that is also robust considers the occurrence / non-occurrence of landslide to estimate the probability of a landslide for a set of causative factors considered and is a multivariate regression model. The logit function is assumed linear and regression coefficients are factors are estimated to define the odds. The odds are positive if the factors are significant and negative is the factor is not important. The data-fit in the form of a 'S' shaped curve is obtained using a logistic link function and is used to estimate the probability of landslide occurrence. This model is sensitive to multi-collinearity in data and does not allow for cross-checking the characteristics based on discretization. Despite the popularity and wide usage statistical models have the following limitations—(i) the quality of the models is the function of the data quality (ii) large datasets are necessary for modelling and validation (iii) interpretation of results must include the inferences from both geomorphologists and statisticians (iv) these models are region specific and do not allow for comparison with other locations and (v) the interpretation of the susceptibility map by non-specialists like policy makers or planners is difficult.

Machine learning models are in general supervised learning-based models for landslide prediction. The accuracy and performance of these models depend on the input data accuracy. Different modelling algorithms can be used to map landslide susceptibility. These models take into account the non-linear relationship between a causative factor and landslide occurrence. The output from these algorithms is flexible based on the requirements and is calibrated to a good extent. An important merit of these models is that the data type or its statistical distribution does not affect the model performance or accuracy and prior assumptions need not be considered in this approach. These models also provide the benefit of considering parallel combination function of factors causing landslide (Das et al. 2023). But these methods are data-intensive and suffers from deficiencies like under-fitting or over-fitting due to the presence of unintentional flawed data or insufficient model fitting. These models are also complex and require parameter tuning for optimal performance.

4.7 Heuristic Methods

Heuristic method relies on the opinion of experts (i.e) their subjective judgement, in this case experts in landslide analysis. Experts rank the factors based on the landslide inventory and the expected importance of the factors selected for mapping landslide susceptibility. The performance and quality of the model depends on the expert opinion and their ability in understanding the cause of instability in a specific region (Ruff and Czurda 2008). These hierarchical heuristic models also form a part of the decision support system. Analytical hierarchical process (AHP) model is one of the popularly used knowledge-driven landslide model. Weights are assigned for each class of a thematic layer based on its importance as adjudged by the expert. Pair-wise comparison is carried out for each factor with respect to other factors by subjective rating and their reliability is assessed using consistency ratio (CR) which must be less than 0.1. It is a relatively simple model and assists in clarifying the comprises in a multi-objective analysis. AHP models have been widely used for mapping landslide susceptibility in Indian Himalayas (Das et al. 2023). A lesser-known variant of the AHP models are the analytical network models (ANP) which follow the basic frame-work of the AHP models but allow for inter-dependence between factors and sub-classes defined within a factor (i.e) at all levels and between all levels inter-dependence can be allowed. Heuristic models can also be semi-quantitative by allowing for combination of statistical models to be combined for assigning the weights or rating the importance of the factors and sub-classes (Das et al. 2023).

4.8 Conclusion

Landslide susceptibility maps are vital inputs for planning any developmental activity in regions prone to landslide and also for devising mitigation strategies. The recent spate of the frequent occurrence of extreme events like heavy precipitation in short duration, flash floods, cloud burst, glacial retreat, etc. have increased the occurrence of landslides all around the world. Several methods are available for mapping landslide susceptibility. The selection of these methods for a specific region primarily depends on the available data, particularly landslide inventory, the spatial extent of the region to be mapped and the level of accuracy needed. Statistical approach is observed to be the most popular approach for modelling susceptibility. Recent literature shows an increase in the application of machine learning models with various algorithms for mapping landslide susceptibility.

Future landslide susceptibility mapping research should concentrate on building dynamic models that take temporal fluctuations in environmental conditions into account, incorporating sophisticated remote sensing data to improve precision, and refining machine learning techniques for improved performance. Predictive models would be more robust if ensemble modelling approaches were investigated and susceptibility mapping uncertainties were evaluated. The relevance and efficacy of susceptibility maps may also be improved by doing localised research that is customised to the distinct geological features of certain areas, taking climate change impacts into account, and involving the community to get pertinent insights. It is necessary to investigate the viability of real-time monitoring systems and thorough validation studies that contrast various modelling techniques in order to create standards and determine the best practises for particular situations. By offering more precise and dependable instruments for planning and reducing the danger of landslides in susceptible areas, these research avenues hope to further the topic.

References

Abd El-Kawy OR, Rød JK, Ismail HA, Suliman AS (2011) Land use and land cover change detection in the western Nile delta of Egypt using remote sensing data. Appl Geogr 31(2):483–494. https://doi.org/10.1016/J.APGEOG.2010.10.012

Ahmad F, Goparaju L (2016) Analysis of urban sprawl dynamics using geospatial technology in Ranchi City, Jharkhand, India. J Environ Geogr 9(1–2):7–13. https://doi.org/10.1515/JENGEO-2016-0002

Aleotti P, Chowdhury R (1999) Landslide hazard assessment: Summary review and new perspectives. Bull Eng Geol Environ 58(1):21–44. https://doi.org/10.1007/S100640050066

Bain RL, Hale RP, Goodbred SL (2019) Flow Reorganization in an anthropogenically modified tidal channel network: an example from the Southwestern Ganges-Brahmaputra-Meghna Delta. J Geophys Res: Earth Surf 124(8):2141–2159. https://doi.org/10.1029/2018JF004996

Bandyopadhyay S, Saha S, Ghosh K, De Kumar S (2013) Channel planform change and detachment of tributary: A study on the Haora and Katakhal Rivers, Tripura, India. Geomorphology 193:25–35. https://doi.org/10.1016/J.GEOMORPH.2013.03.024

Bandyopadhyay S, De SK (2018) Anthropogenic impacts on the morphology of the Haora River, Tripura, India. Geomorphol: Relief, Process, Environ 24(2):151–166. https://doi.org/10.4000/GEOMORPHOLOGIE.12019

Bhattacharjee S, Islam MT, Kabir ME, Kabir MM (2021) Land-use and land-cover change detection in a north-eastern wetland ecosystem of Bangladesh using remote sensing and GIS techniques. Earth Syst Environ 5(2):319–340. https://doi.org/10.1007/S41748-021-00228-3/FIGURES/15

Bhattacherjee S, Bandyopadhyay S, De SK (2022) Significance of Channel Planform Change and Morphometric Indices in the Buri River Basin, India and Bangladesh. 151–174. https://doi.org/10.1007/978-3-030-95435-2_6

Brabb EE (1985) Innovative approaches to landslide hazard and risk mapping, vol 1, pp 17–22

Brock JC, Purkis SJ (2009) The emerging role of lidar remote sensing in coastal research and resource management. J Coast Res 10053:1–5. https://www.academia.edu/64787184/The_Emerging_Role_of_Lidar_Remote_Sensing_in_Coastal_Research_and_Resource_Management
Cabral VC, Reis FAGV, Mendoza CM, de Oliveira A (2022) Model-based assessment of shallow landslides susceptibility at a petrochemical site in Brazil. Rev Bras Geomorfol 23(2):1394–1419. https://doi.org/10.20502/rbg.v23i2.2084
Capparelli G, Versace P (2011) FLaIR and SUSHI: two mathematical models for early warning of landslides induced by rainfall. 8:67–79. https://doi.org/10.1007/s10346-010-0228-6
Carrara A (1983) Multivariate models for landslide hazard evaluation. J Int Assoc Math Geol 15(3):403–426. https://doi.org/10.1007/BF01031290
Chacón J, Irigaray AC, Ferná Ndez AT, El Hamdouni AR, Chacó J, Irigaray C, El Hamdouni ÁR, Hamdouni R. El, Ferná Ndez T (2006) Engineering geology maps: landslides and geographical information systems. Bull Eng Geol Environ 65:341–411. https://doi.org/10.1007/s10064-006-0064-z
Chang Z, Catani F, Huang F, Liu G, Meena SR, Huang J, Zhou C (2023) Landslide susceptibility prediction using slope unit-based machine learning models considering the heterogeneity of conditioning factors. J Rock Mech Geotech Eng 15(5):1127–1143. https://doi.org/10.1016/J.JRMGE.2022.07.009
Chaussard E, Kerosky S (2016) Characterization of black sand mining activities and their environmental impacts in the philippines using remote sensing. Remote Sens 8(2):100. https://doi.org/10.3390/RS8020100
Chirico PG, Bergstresser SE, DeWitt JD, Alessi MA (2021) Geomorphological mapping and anthropogenic landform change in an urbanizing watershed using structure-from-motion photogrammetry and geospatial modeling techniques. Journal of Maps 17(4):241–252. https://doi.org/10.1080/17445647.2020.1746419
Chung CJF, Fabbri AG (2003) Validation of spatial prediction models for landslide hazard mapping. Natural Hazards 30(3):451–472. https://doi.org/10.1023/B:NHAZ.0000007172.62651.2B
Corominas J, van Westen C, Frattini P, Cascini L, Malet JP, Fotopoulou S, Catani F, Van Den Eeckhaut M, Mavrouli O, Agliardi F, Pitilakis K, Winter MG, Pastor M, Ferlisi S, Tofani V, Hervás J, Smith JT (2014) Recommendations for the quantitative analysis of landslide risk. Bull Eng Geol Environ 73(2):209–263. https://doi.org/10.1007/S10064-013-0538-8
Crozier MJ (2010) Deciphering the effect of climate change on landslide activity: A review. Geomorphology 124(3–4):260–267. https://doi.org/10.1016/J.GEOMORPH.2010.04.009
Das S, Sarkar S, Kanungo DP (2023) A critical review on landslide susceptibility zonation: recent trends, techniques, and practices in Indian Himalaya. Natural Hazards 115(1):23–72. https://doi.org/10.1007/S11069-022-05554-X
Dietrich WE, Real De Asua R, Coyle J, Orr B, Trso M (1998) A validation study of the shallow slope stability model, SHALSTAB, in forested lands of Northern California
Dilley M, Chen RS, Deichmann U (2005) Natural disaster hotspots: a global risk analysis, Part 611. Earth Science, 5:132. https://books.google.com/books/about/Natural_Disaster_Hotspots.html?id=X3osIdnSBdgC
Etzeimuiier B, Frauenfeider R (2009) Factors controlling the distribution of mountain permafrost in the northern hemisphere and their influence on sediment transfer. Arct, Antarct, Alp Res 41(1):48–58. https://doi.org/10.1657/1523-0430-41.1.48
Fleuchaus P, Blum P, Wilde M, Terhorst B, Butscher C (2021) Retrospective evaluation of landslide susceptibility maps and review of validation practice. Environ Earth Sci 80(15):1–15. https://doi.org/10.1007/S12665-021-09770-9/FIGURES/1
Franceschini R, Rosi A, del Soldato M, Catani F, Casagli N (2022) Integrating multiple information sources for landslide hazard assessment: the case of Italy. Sci Rep 12(1):20724. https://doi.org/10.1038/S41598-022-23577-Z
Froude MJ, Petley DN (2018) Global fatal landslide occurrence from 2004 to 2016. Hazards Earth Syst Sci 18:2161–2181. https://doi.org/10.5194/nhess-18-2161-2018

Gariano SL, Guzzetti F (2016) Landslides in a changing climate. Earth-Sci Rev 162:227–252. https://doi.org/10.1016/J.EARSCIREV.2016.08.011

Goodchild MF (2010) Twenty years of progress: GIScience in 2010. J Spat Inf Sci 1(2010):3–20. https://doi.org/10.5311/JOSIS.2010.1.2

Guzzetti F, Carrara A, Cardinali M, Reichenbach P (1999) Landslide hazard evaluation: a review of current techniques and their application in a multi-scale study, Central Italy. Geomorphology 31(1–4):181–216. https://www.academia.edu/8118340/Landslide_hazard_evaluation_a_review_of_current_techniques_and_their_application_in_a_multi_scale_study_Central_Italy

Guzzetti F (2006) Landslide hazard and risk assessment. https://bonndoc.ulb.uni-bonn.de/xmlui/handle/20.500.11811/2644

Haque U, Blum P, da Silva PF, Andersen P, Pilz J, Chalov SR, Malet JP, Auflič MJ, Andres N, Poyiadji E, Lamas PC, Zhang W, Peshevski I, Pétursson HG, Kurt T, Dobrev N, García-Davalillo JC, Halkia M, Ferri S, Keellings D (2016) Fatal landslides in Europe. Landslides 13(6):1545–1554. https://doi.org/10.1007/S10346-016-0689-3

Huabin W, Gangjun L, Weiya X, Gonghui W (2005) GIS-based landslide hazard assessment: An overview. Prog Phys Geogr 29(4):548–567. https://doi.org/10.1191/0309133305PP462RA

Ji J, Zhou Y, Cheng Q, Jiang S, Liu S (2023) Landslide susceptibility mapping based on deep learning algorithms using information value analysis optimization. Land 12(6):1125. https://doi.org/10.3390/LAND12061125

Kirschbaum D, Stanley T, Zhou Y (2015) Spatial and temporal analysis of a global landslide catalog. Geomorphology 249:4–15. https://doi.org/10.1016/J.GEOMORPH.2015.03.016

Klose M, Maurischat P, Damm B (2016) Landslide impacts in Germany: A historical and socioeconomic perspective. Landslides 13(1):183–199. https://doi.org/10.1007/S10346-015-0643-9

König T, Kux HJH, Corsi AC, König T, Kux HJH, Corsi AC (2022) Advanced models applied for the elaboration of landslide-prone maps, a review. Int J Geosci 13(3):174–198. https://doi.org/10.4236/IJG.2022.133010

Kulsoom I, Hua W, Hussain S, Chen Q, Khan G, Shihao D (2023) SBAS-InSAR based validated landslide susceptibility mapping along the Karakoram Highway: a case study of Gilgit-Baltistan, Pakistan. Sci Rep 13(1). https://doi.org/10.1038/S41598-023-30009-Z

Le TMH (2014) Reliability of heterogeneous slopes with cross-correlated shear strength parameters. Georisk 8(4):250–257. https://doi.org/10.1080/17499518.2014.966117

Li R, Huang S, Dou H (2023) Dynamic risk assessment of landslide hazard for large-scale photovoltaic power plants under extreme rainfall conditions. Water 15(15):2832. https://doi.org/10.3390/W15152832

Liang Z, Peng W, Liu W, Huang H, Huang J, Lou K, Liu G, Jiang K (2023) Exploration and comparison of the effect of conventional and advanced modeling algorithms on landslide susceptibility prediction: a case study from Yadong country, Tibet. Appl Sci 13(12):7276. https://doi.org/10.3390/APP13127276

Malta (n.d.) Environment and geoscience proceedings of the 1st wseas international conference on environmental and geological science and engineering (EG'08) energy and environmental engineering series a series of reference books and textbooks. Retrieved November 28, 2023, from www.wseas.org

Michel GP, Kobiyama M, Goerl RF (2014) Comparative analysis of SHALSTAB and SINMAP for landslide susceptibility mapping in the Cunha River basin, southern Brazil. J Soils Sediments 14(7):1266–1277. https://doi.org/10.1007/S11368-014-0886-4

Moreira Melo C, Kobiyama M, Paulo Michel G, Madruga de Brito M, Lu Z, Miguel Ferreira T (2021) The relevance of geotechnical-unit characterization for landslide-susceptibility mapping with SHALSTAB. GeoHazards 2:383–397. https://doi.org/10.3390/GEOHAZARDS2040021

Neuland H (1976) A prediction model of landslips. CATENA 3(2):215–230. https://doi.org/10.1016/0341-8162(76)90011-4

Niroumand H, Kassim KA, Nazir R (2012) Numerical modeling of geogrid reinforced sand beds by PLAXIS. Adv Sci Lett 15(1):63–65. https://doi.org/10.1166/ASL.2012.4094

Oldrich H (2016) Landslides and engineered slopes. experience, theory and practice Google Books. https://books.google.co.in/books?hl=en&lr=&id=Uh9jDAAAQBAJ&oi=fnd&pg=PA3&dq=4.+Landslide+Susceptibility,+Hazard+and+Risk+%E2%80%93+An+overview&ots=HaYQyDr9x5&sig=TZzF1p9XGIytZ5wdtBVoDKZ8xwQ#v=onepage&q=4.%20Landslide%20Susceptibility%2C%20Hazard%20and%20Risk%20%E2%80%93%20An%20overview&f=false

Panchal S, Shrivastava AK (2022) Landslide hazard assessment using analytic hierarchy process (AHP): A case study of National Highway 5 in India. Ain Shams Eng J 13(3):101626. https://doi.org/10.1016/J.ASEJ.2021.10.021

Park HJ, Jang JY, Lee JH (2019) Assessment of rainfall-induced landslide susceptibility at the regional scale using a physically based model and fuzzy-based Monte Carlo simulation. Landslides 16(4):695–713. https://doi.org/10.1007/S10346-018-01125-Z

Petley D (2012) Global patterns of loss of life from landslides. Geology 40(10):927–930. https://doi.org/10.1130/G33217.1

Promper C, Puissant A, Malet JP, Glade T (2014) Analysis of land cover changes in the past and the future as contribution to landslide risk scenarios. Applied Geography 53:11–19. https://doi.org/10.1016/J.APGEOG.2014.05.020

Redshaw P, Boon D, Campbell G, Willis M, Mattai J, Free M, Jordan C, Kemp SJ, Morley A, Thomas M (2019) The 2017 regent landslide, freetown peninsula, sierra leone. Q J Eng Geol Hydrogeol 52(4):435–444. https://doi.org/10.1144/QJEGH2018-187

Reichenbach P, Rossi M, Malamud BD, Mihir M, Guzzetti F (2018) A review of statistically-based landslide susceptibility models. Earth-Sci Rev 180:60–91. https://doi.org/10.1016/J.EARSCIREV.2018.03.001

Reports|NDM India. (n.d.). Retrieved November 28, 2023, from https://ndmindia.mha.gov.in/reports

Rozos D (2010) Comparison of the implementation of rock engineering system and analytic hierarchy process methods, upon landslide susceptibility mapping, using GIS: a case study from the Eastern Achaia County of Peloponnesus, Greece. Environ Earth Sci. https://www.academia.edu/17538897/Comparison_of_the_implementation_of_rock_engineering_system_and_analytic_hierarchy_process_methods_upon_landslide_susceptibility_mapping_using_GIS_a_case_study_from_the_Eastern_Achaia_County_of_Peloponnesus_Greece

Ruff M, Czurda K (2008) Landslide susceptibility analysis with a heuristic approach in the Eastern Alps (Vorarlberg, Austria). Geomorphology 94(3–4):314–324. https://doi.org/10.1016/J.GEOMORPH.2006.10.032

Sarif MN, Siddiqui L, Islam MS, Parveen N, Saha M (2021) Evolution of river course and morphometric features of the River Ganga: A case study of up and downstream of Farakka Barrage. Int Soil Water Conserv Res 9(4):578–590. https://doi.org/10.1016/J.ISWCR.2021.01.006

Sepúlveda SA, Petley DN (2015) Regional trends and controlling factors of fatal landslides in Latin America and the Caribbean. Nat Hazards Earth Syst Sci 15(8):1821–1833. https://doi.org/10.5194/NHESS-15-1821-2015

Sujatha ER, Sudarsan JS, Nithiyanantham S (2023) A review on sustainable reinforcing techniques to stabilize slopes against landslides. Int J Environ Sci Technol 20(12):13873–13882. https://doi org/10.1007/S13762-023-04832-W/METRICS

Sun D, Ding Y, Zhang J, Wen H, Wang Y, Xu J, Zhou X, Liu R (2022) Essential insights into decision mechanism of landslide susceptibility mapping based on different machine learning models. Geocarto Int. https://doi.org/10.1080/10106049.2022.2146763

Thanh LN, De Smedt F (2014) Slope stability analysis using a physically based model: a case study from A Luoi district in Thua Thien-Hue Province, Vietnam. Landslides 11(5):897–907. https://doi.org/10.1007/S10346-013-0437-X

van Westen CJ, Castellanos E, Kuriakose SL (2008) Spatial data for landslide susceptibility, hazard, and vulnerability assessment: An overview. Eng Geol 102(3–4):112–131. https://doi.org/10.1016/J.ENGGEO.2008.03.010

Williams JG, Rosser NJ, Kincey ME, Benjamin J, Oven KJ, Densmore AL, Milledge DG, Robinson TR (n.d.) Satellite-based emergency mapping: Landslides triggered by the 2015 Nepal earthquake. https://doi.org/10.5194/nhess-2017-273

World Bank Group (2017) Sierra Leone—Rapid damage and loss assessment of August 14th, 2017 landslides and floods in the western area. Retrieved November 29, 2023, from https://documents.worldbank.org/en/publication/documents-reports/documentdetail/523671510297364577/sierra-leone-rapid-damage-and-loss-assessment-of-august-14th-2017-landslides-and-floods-in-the-western-area

Zhao B, Dai Q, Han D, Dai H, Mao J, Zhuo L (2019) Probabilistic thresholds for landslides warning by integrating soil moisture conditions with rainfall thresholds. J Hydrol 574:276–287. https://doi.org/10.1016/J.JHYDROL.2019.04.0

Chapter 5
Mapping of Landslide Susceptibility Using State-of-the-Art Method and Geospatial Techniques in the Rangamati District in the Chattogram Hill Tracts Region of Bangladesh

K. M. Nafee, Md. Shakib Al Fahad, Md. Khayrul Islam Tuhin, Md. Sakhawat Hossen, and Md. Sofi Ullah

Abstract The chapter aims to map the susceptibility of landslides in the Rangamati district of the Chattogram Hill Tracts region in Bangladesh using state-of-the-art methods and geospatial techniques. The district has experienced a significant number of landslides that caused numerous fatalities. However, the lack of available data and proper synchronization of existing information impedes our understanding of the extent of the issue. An integrated approach combining community-based mapping, Participatory Rural Appraisal (PRA) techniques, and spatial analysis was employed to address this challenge. Generation of landslide inventory revealed 306 landslide locations spread across the entire district, with 54.58% located in Rangamati Sadar upazila. Additionally, spatial statistical analysis was conducted to find the association of landslides with 12 causative factors. Analysis revealed that areas with Boka Bill geological formation had 37.91% landslide locations, making this region highly susceptible. Similarly, areas with elevations within 50–200 m and slopes of 10°–20°

K. M. Nafee (✉) · Md. S. Al Fahad · Md. S. Hossen · Md. S. Ullah
Department of Geography and Environment, University of Dhaka, Dhaka, Bangladesh
e-mail: kmnafee15@gmail.com

Md. S. Al Fahad
e-mail: shakibalfahad742@gmail.com

Md. S. Hossen
e-mail: sakhawathossen1989@gmail.com

Md. S. Ullah
e-mail: sofiullah.geoenv@du.ac.bd

Md. K. I. Tuhin
Department of Geography and Environment, Dhaka College, Dhaka, Bangladesh
e-mail: harshenhars@gmail.com

G. K. Panda et al. (eds.), *Landslide: Susceptibility, Risk Assessment and Sustainability*, Advances in Natural and Technological Hazards Research 52,
https://doi.org/10.1007/978-3-031-56591-5_5

have recorded the highest number of landslides in the study area, with 40.85% and 50.33% of landslide locations, respectively. The study's findings also indicated that areas close to roads (0–250) m and settlements (0–25) m had the highest number of landslide locations, 61.76%, and 34.64%, respectively. Overall, findings indicated that around 43.58 km^2 area of Rangamati district are susceptible to landslides. This chapter offers a comprehensive understanding of landslide susceptibility using an integrated geospatial approach. The findings will lay the foundation for developing a landslide vulnerability model to enhance landslide risk assessment and facilitate effective mitigation measures in the future.

Keywords Landslide · Susceptibility · Rangamati · Inventory · GIS

5.1 Introduction

Landslides refer to abrupt displacement of substances such as rocks, earth or man-made materials due to the force of gravity, resulting in their movement through falling, sliding, or flowing, occurring at different velocities (Casagli et al. 2017). Landslides are frequent natural disaster in mountainous or hilly areas (Joerg et al. 2022), and they have a substantial influence on human livelihood. Landslides cause loss of life, physical degradation, income reduction, food shortages, forced displacement and destruction of infrastructure (Claudine and Ndokoye 2023). Landslides exhibit a strong negative relationship with other natural catastrophes like floods, droughts, wildfires, earthquakes, tsunamis, and volcanoes, and they can trigger a compound disaster event (Vilímek et al. 2021).

Landslides can result from factors of either natural or human origin (Sevar and Hazhir 2020). It can be triggered by various natural factors such as proximity to faults, rivers, or roads, as well as the presence of vegetation coverage (Wang et al. 2018). Additionally, slope steepness, curvature, and orientation, along with factors like rock type, yearly precipitation, soil composition, and topographic relief, also play significant roles in landslides roles in landslide occurrence (Michael and Abdul 2021). Human activities, including deforestation, land use alteration, urban sprawling (Crosta and Frattini 2008), topographic modification, and deteriorating infrastructure, are responsible for inducing landslides (Si et al. 2018). Majority of landslides are in smaller scale and slow but there are instances in which they have catastrophic consequences (Ullah 2021).

Every type of landslides is often related in tandem with earth's geological materials (Korup 2006). Hence, the stability of soil masses and rocks heavily relies on the characteristics and compaction of the materials (Mitchell et al. 2002). Landslides can be categorized into five distinct movements types: falling, toppling, sliding, spreading, and flowing (Turner and Schuster 1996). Various research (Froude and Petley 2018; Haque et al. 2016; Stanley and Kirschbaum 2017; Petley 2012; Aristizábal and

Sanchez 2020; Kirschbaum et al. 2015) have illustrated the spatial and temporal variations of landslides, highlighting the significant global risks which results in substantial human and economic losses. Due to the inconsistency of reporting and documenting of landslides in different countries, it is challenging to determine the global distribution (Sidle and Hirotaka 2006). Between 1998 and 2017, landslides had a significant impact on approximately 4.8 million individuals globally, resulting in 18,000 deaths (Sreelakshmi et al. 2022). Several nations, including New Zealand, Nicaragua, Colombia, El Salvador, Italy, Peru, Nepal, China, Brazil, and India documented a notable rise in landslide events (Gómez et al. 2023). In many parts of the world, especially in regions where there is limited capacity, infrastructure and data availability, there is a lack of attention to the timing of landslides, particularly evident in tropical areas, despite their significant and growing landslides challenges (Olivier et al. 2021). Global database of landslides offers essential information for assessing landslide prediction models, early warning systems, and for evaluating landslide susceptibility, hazards and risks (Gómez et al. 2023).

While quantifying the effects of landslides remains a challenge, the integration of satellite data with metrics like population size, traffic patterns, and structural statistics could potentially aid in estimating landslide susceptibility (Robert et al. 2020). Landslides susceptibility is the likelihood of a future landslide in a particular area, determined by assessing the correlation between responsible factors and the spatial distribution of landslides (Begum et al. 2020). Literatures covering landslide susceptibility mapping has expanded over the years leading to significant development of various techniques (Tyagi et al. 2022). Landslides susceptibility models typically operates under the assumption that it occurs regardless of time, despite the factors influencing landslide occurrence can vary with time (Jones et al. 2021). Landslide susceptibility mapping is conducted through various geospatial methods, including remote sensing, geographic information system (GIS) and the analytical hierarchy process (AHP) (Potekar et al. 2023; Nguyen et al. 2023; Mohammednur et al. 2020; Trisnawati et al. 2022). Using geospatial techniques for this purpose provides several advantages. One of these is the capacity to combine numerous elements like topography, hydrology, lithology, and land cover, all over the world (Coco et al. 2021; Ali et al. 2021). Through the amalgamation of these data sources, geospatial methods offer a holistic insight into the elements influencing landslide occurrence (Farooq 2021). Furthermore, geospatial methods facilitate in creation of precise and in-depth maps that depicts the likelihood of landslides, offering valuable insights for efficient disaster preparedness and infrastructure development (Shano et al. 2020; Cai et al. 2021).

Due to its geographical position, Bangladesh is prone to frequent natural disasters, and among these, landslides stand out as a significant concern due to their effects on the well-being and livelihoods of people residing in the in the hilly areas (Chowdhury 2023). Landslides frequently occur in the Chattogram Hill Tracts region of Bangladesh. The Chattogram Hill Tracts, comprises the districts of Chattogram, Cox's Bazar, Rangamati, Bandarban, and Khagrachari (Ara and Khan 2023). The landslides here are caused by a combination of both natural and human induced factors (Ahmed et al. 2023). Several anthropogenic factors such as cutting of hills, deforestation, malpractices in agriculture, stone and sand excavation significantly contribute to landslides (Ahmed 2021; Alam 2020). Between 2000 and 2018, there were 204 landslide incidents documented, which led to 727 deaths and 1017 injuries, averaging 19 landslides per year with a 4 percent annual growth rate (Sultana 2020). Among the 727 landslides that occurred during the time period of 2001–2018, 193 landslide occurrences were witnessed in the Rangamati (Rabby and Li 2019). During the monsoon period of 2017, due to incessant rainfall, a total of 168 landslide occurred in the entire Chattogram Hill Tracts region, with the district of Rangamati suffering the worst fate with 126 deaths and the destruction of 6000 houses, roads and telecommunication systems (Haque et al. 2018). Given the growing importance of landslide as a significant issue in Bangladesh's context, several studies have been conducted in the domain of landslide susceptibility mapping using both qualitative and quantitative techniques (Rahman et al. 2017; Ahmed and Dewan 2017). In Bangladesh, 27 different techniques have been employed for landslide susceptibility mapping, with logistic regression, random forest, weight of evidence and multiple regression being some of the commonly favored methods (Chowdhury 2023). Several techniques were used in landslide susceptibility mapping focusing in Rangamati district including Frequency Ratio (FR) and Analytical Hierarchy Process (AHP) (Wijaya et al. 2022), Cellular Automata based Artificial Neural Network and random forest (Rabby et al. 2022), Landslide Susceptibility Index (LSI) and GIS-based weighted overlay model (WOM) (Hussain et al. 2022). Past studies on landslide susceptibility relied on collecting landslide locational data from open-source databases. In Bangladesh, there is currently no established comprehensive official database for landslides (Sultana 2020) thus this issue hinders our ability to comprehend the full scope of the problem. Thus, using open-source data and generating results based on these data have reliability issues. To address this current gap of knowledge and to contribute to generation of landslide inventory for Rangamati district leading to landslide susceptibility mapping, the chapter attempts to map landslide susceptibility using an integrated approach of combining primary and secondary data sources.

The specific objectives of this chapter include,

(1) To identify landslide location by employing community-based mapping approach and PRA techniques.
(2) Landslide data acquisition from existing literatures.
(3) Generation of spatial statistics for validating the correlation between landslides to various contributing factors.

In the context of geospatial research, state-of-the-art refers to the highest level of knowledge or approaches or methodology in geospatial science. The term is relative and evolves as new advancements or milestones are reached. Upon reviewing various literature related to landslide susceptibility mapping, it was found that research was mainly focused on running simulations and models for generating results. A similar scenario is witnessed in landslide susceptibility mapping in Bangladesh. The study of Rabby and Li (2020), where landslide susceptibility mapping in Chattogram hilly areas was done using an integrated approach of combining Frequency ratio (FR) with Analytical Hierarchy Process (AHP) and Logistic Regression (LR) included a field-based component which was done through observation methods. However, only field observation is insufficient to identify landslide locations, as including local people's perceptions of landslide events can facilitate more accurate identification. The approach of investigating landslides used in this chapter stands out from most conventional approaches relying on advanced simulation for its simplicity and, most importantly, for incorporating community-based knowledge in the research framework. Community-based mapping approach and PRA techniques for landslide susceptibility are essential to ensure the incorporation of the knowledge and expertise of the local community, which can offer valuable information about the exact location and the factors that contribute to the occurrence of landslides (Basharat et al. 2023). Another advantage of using this approach is that it ensures the precision and reliability of landslide susceptibility maps by combining qualitative and quantitative analysis (Bao et al. 2022). Despite its importance and effectiveness, no previous studies tried to map landslide susceptibility using an integrated approach combining geospatial techniques, community-based mapping, and participatory rural appraisal (PRA) techniques in the Rangamati district, which this chapter will address. Studying past landslides facilitates the integration of data from multiple studies, providing a comprehensive perspective on landslide susceptibility (Martinello et al. 2022), therefore, this chapter incorporates landslide locational data from previous literatures and other databases in order to facilitate landslide susceptibility analysis. The approach used in this chapter for collecting landslide locational data from the field further enhances the existing data on landslide location in Rangamati district contributing to generating a more comprehensive landslide inventory. The generation of spatial statistics allows for the integration of various spatial data layers to identify areas at high risk of landslides (Luck and Hajnsek 2022; Woodard et al. 2023). The chapter through the generation of spatial statistics of landslides to various contributing factors focuses to validate their correlation leading to identifying landslide susceptible zones. By prioritizing the knowledge of the local community on landslides and subsequent geospatial analysis and generation of spatial statistics, this chapter provides a novel perspective on landslide susceptibility mapping. This innovative and novel approach adds to the existing development of geospatial techniques in landslide susceptibility mapping. Thus, the methods used in this chapter can be regarded as state-of-the-art.

5.2 Materials and Methods

5.2.1 Study Area

The district of Rangamati spans from 22° 27′ to 23° 44′ north latitudes and from 91° 56′ to 92° 33′ east longitudes and covers an area of 6116.13 km^2 (Hafsa et al. 2022). The district is bounded by the districts of Bandarban, Khagrachari and Chattogram in the southern and western sides respectively and by India in the northern and eastern side (Rangamati Hill District Council 2011). Previously a subdivision, it was turned into a district in 1983 and divided into 10 upazilas with Baghaichhari being the largest in terms of area and Rajasthali the smallest (BBS 2011). The total population of Rangamati is 647,587 with a density of 106 (BBS 2022). Tropical monsoon climate prevails in the district with high amount of rainfall particularly in the monsoon season (Hafsa et al. 2022). The district experiences an annual rainfall of 2540 mm, with 80% of this precipitation occurring in the monsoon season (BBS 2011). The average slope degree in the study area is 11.22° while the average elevation is 139.51 m. Tertiary and quaternary sediments being extensively fragmented by networks of waterbodies cover the entire Rangamati district. Notable geological formation in the area includes, Bhuban, Bokabil, Dupi Tila, Girujan clay and Tipam sandstone formations (Haque et al. 2018). The study area also exhibits different types of soils with most soils are sandy loam in texture that are of poor plasticity and quality (Bhuiyan et al. 2017) (Fig. 5.1).

The overall methodology in this chapter consists of three main components. (1) Landslide Location Identification (2) Preparation Landslide Causative Factors Datasets (3) Data Processing and Analysis.

5.2.2 Landslide Location Identification

The collection of landslide locational data was done using both primary and secondary sources. For the purpose of primary data collection, community-based mapping and Participatory Rural Appraisal were employed. In this process, high-resolution multi-temporal Google Earth imagery was employed to aid mapping efforts. While several attempts had been made to map landslide using Google Earth images (Sato and Harp 2009; Fisher et al. 2012; Vakhshoori and Zare 2016), this concept of mapping landslide in Bangladesh was relatively novel and had not been applied in the study area previously. High resolution 0.4-m multispectral images from Google Earth including several features such as roads, settlement and other relevant elements, were prepared for each Upazila. These images were then printed on A0 size paper and taken to the field to assist in the on-field identification of landslide location. The incorporation of local knowledge on landslide occurrence was

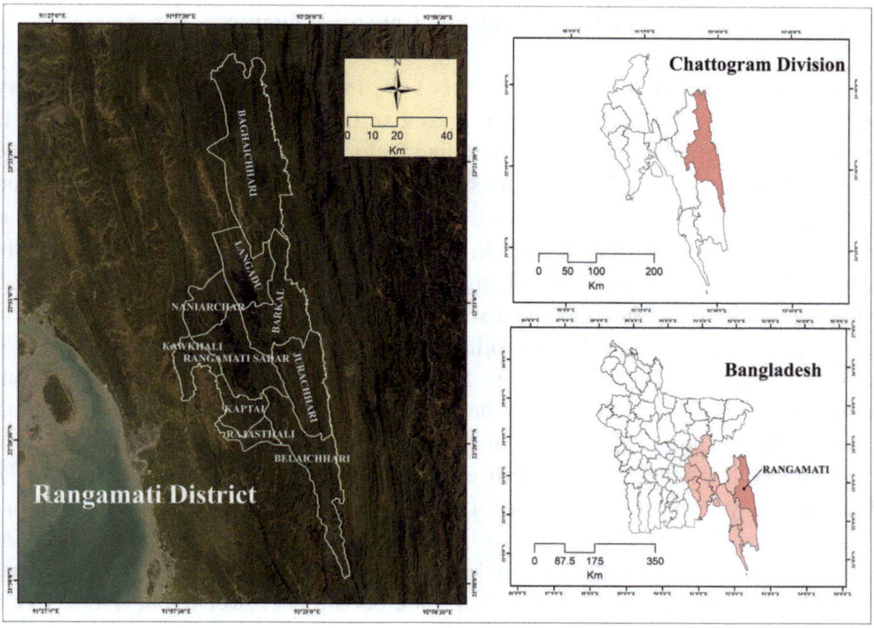

Fig. 5.1 Study area map

integrated into the mapping process through PRA technique to accurately pinpoint locations of landslide in the study area. The secondary data was collected from the District Commissioner Office of Rangamati and existing literatures pertaining to the study area. These data were plotted in the map and then taken for further analysis.

5.2.3 Preparation Landslide Causative Factors Datasets

For the preparation of different factors, Landsat 8 OLI/TIRS (Band 4 and Band 5), Geology, Fault line and Digital Elevation Model (DEM) data obtained from United States Geological Survey (USGS) website and Road network data and settlement data from Bangladesh Bureau of Statistics were collected. The collected data were converted into raster format with the same coordinate system (WGS 1984 UTM zone 46) and the spatial resolution was kept to 30-m resolution (Fig. 5.2).

5.2.3.1 Selection Criteria of Landslide Causative Factors

As mentioned earlier, 12 different factors were incorporated in this chapter. There are no fixed rules for determining the causative factors of landslides however, it is important to keep in mind that the identified factors should be practical, quantifiable and availability in the context of the specific area under investigation (Ayalew and Yamagishi 2005; Mersha and Meten 2020). The landslide conditioning factors was Typically, the most crucial factors in landslides are slope and elevation (Abedin et al. 2020; Ayalew et al. 2005; Demir et al. 2013) and therefore the computation of these factors from the DEM data was obtained. Since plan curvature and profile curvature are linked with the instability of slopes (Wang and Li 2017), these two were considered as causative factors for landslide. Another dataset obtained from calculation using the DEM data was the aspect. The aspect area influences the length of daylight, evapotranspiration, moisture retention, vegetation coverage and their distribution and hence have an indirect impact on landslides (Chen et al. 2017). Since geology represents the permeability and firmness of rocks and soil of a region, it is considered as a factor related to landslides (Ayalew and Yamagishi 2005). On the other hand, it is typically known that the areas closer to fault lines is more prone to landslides (Netra et al. 2014). In landslide occurrence, stream and rivers have a subtle impact (Rahman et al. 2017) and thus distance to drainage network was taken as a causative factor. Stream Power Index and Topographic wetness index are also contributing factors for landslides as the former one is a measurement of the capacity of a stream or water flow to cause erosion (Moore et al. 1991) and the latter one is employed to measure how topography influences hydrological process (Moazzam et al. 2020) and have a direct relationship with landslides (Lee and Min 2001). The chapter also incorporated the Normalized Difference Vegetation Index (NDVI) obtained by using the Red and Near Infrared bands to associate landslides with vegetation coverage. Apart from these geological topographical, and environmental factors, two human induced parameters namely, Roads and Settlements data were also taken as human intervention such as several anthropogenic activities including, deforestation, excavation, loads etc. play a pivotal role in slope failure (Bourenane et al. 2016) leading to landslides (Figs. 5.3 and 5.4).

5.2.4 *Data Processing and Analysis*

The primary data obtained from on-field investigation and historical landslide data sourced from secondary data sources were subsequently converted to digital format through the process of digitization. This procedure was carried out in ESRI's ArcGIS software package. The data were transformed into point data format to facilitate the subsequent analysis that followed.

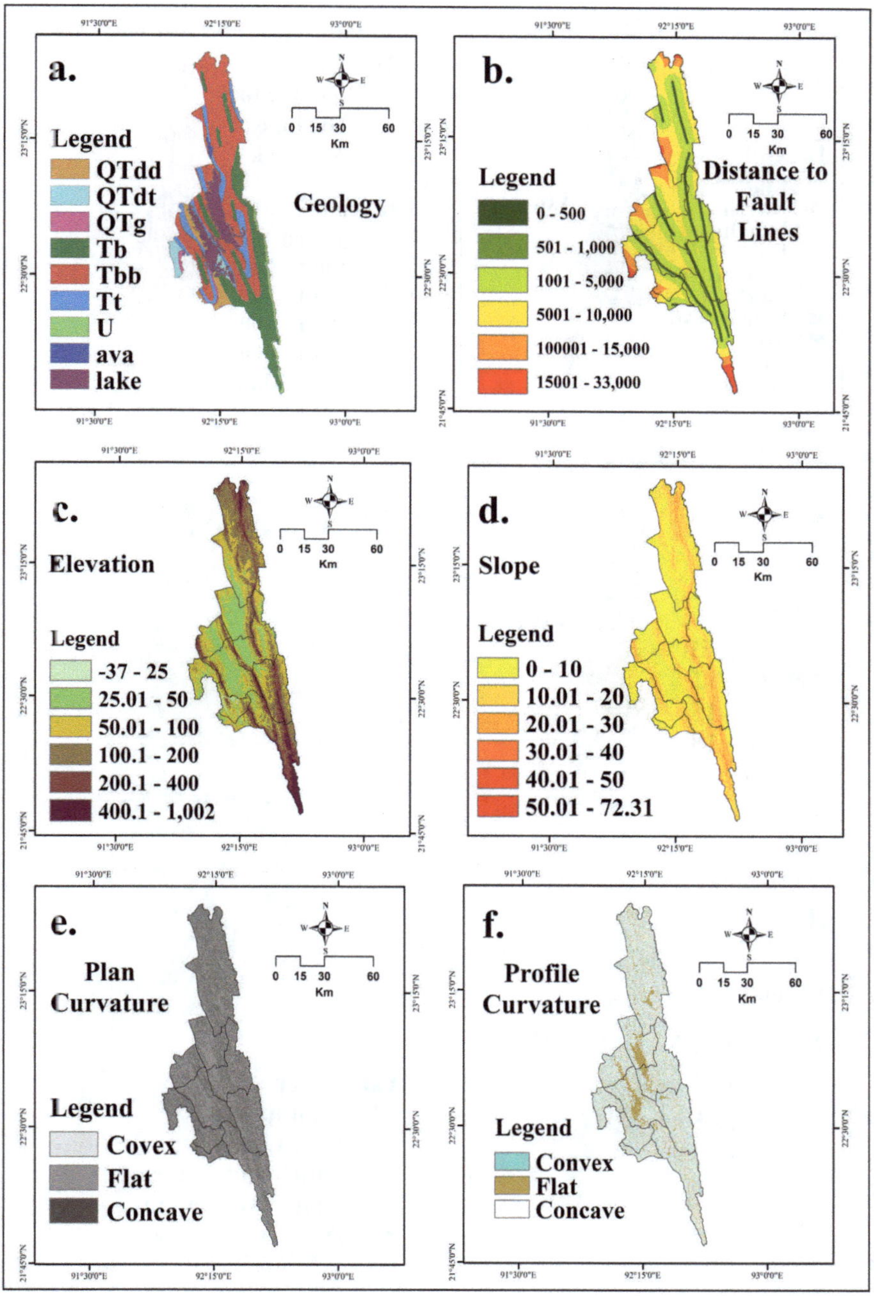

Fig. 5.2 Landslide causative factors: **a** geology Qtdd: Dihing and Dupi Tila Formation Undivided, QTdt: Dupi Tila formation (Pleistocene and Pliocene), QTg: Girujan clay (Pleistocene and Neogene), Tt: Tipam sandstone (Neogene) Tb: Bhuban formation (Miocene), Tbb: Boka Bil formation (Neogene), U: unmapped area, ava: valley alluvium and colluvium, lake: lake area; **b** distance to fault lines; **c** elevation; **d** slope; **e** plan curvature; **f** profile curvature

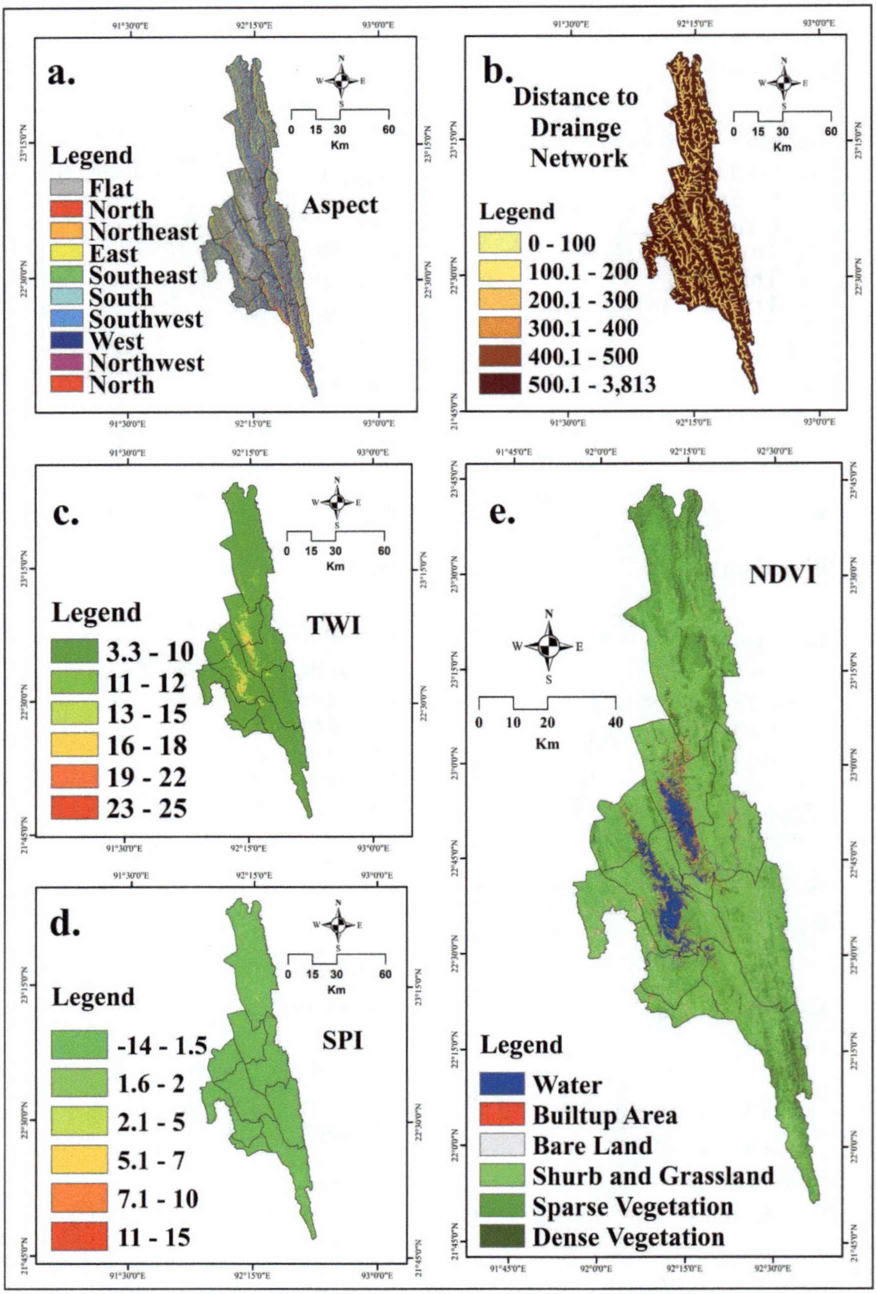

Fig. 5.3 Landslide causative factors: **a** aspect; **b** distance to drainage network; **c** TWI; **d** SPI; **e** NDVI

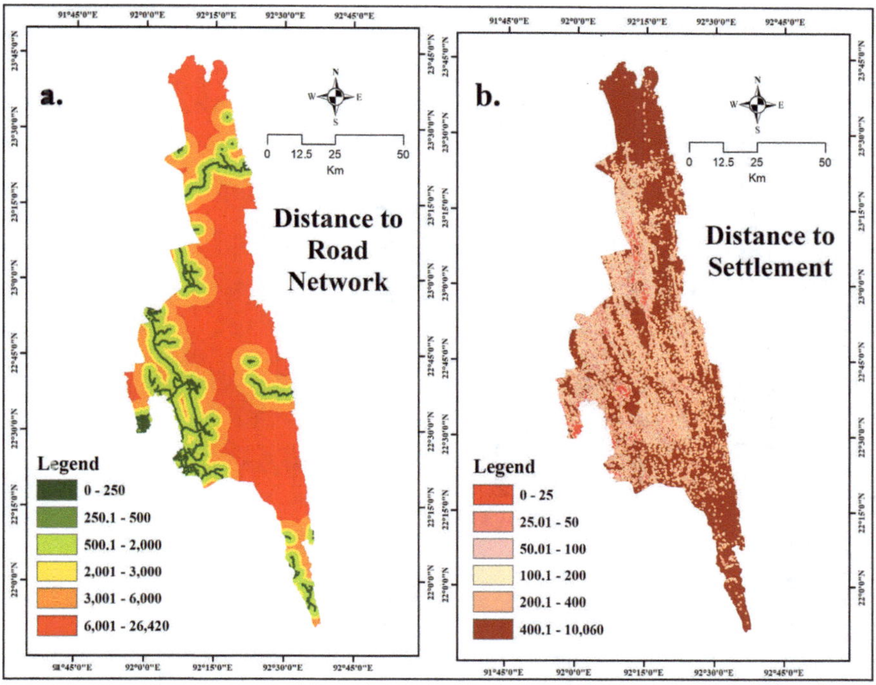

Fig. 5.4 Landslide causative factors: **a** distance to road network (meters); **b** distance to settlement (meters)

The geographical locations of landslides in the region were later on cross-referenced with twelve contributing factors associated with landslide occurrences. This association was established by generating spatial statistics to validate the relationship between landslide occurrence and several other spatial data layers eventually depicting the zones prone to landslides (Fig. 5.5).

5.3 Results and Discussion

5.3.1 Landslide Inventory

From on field investigation and secondary data sources, a total of 306 landslides locations were identified and mapped (Fig. 5.6). The upazila wise distribution of landslides in this chapter showed that, out of 306 landslides locations, 167 locations were identified in Rangamati Sadar upazila followed by Kaptai upazila with 33 landslide locations. The upazila with lowest landslide location was Langadu upazila followed by Barkal and Rajasthali with 3, 4 and 4 locations respectively (Table 5.1).

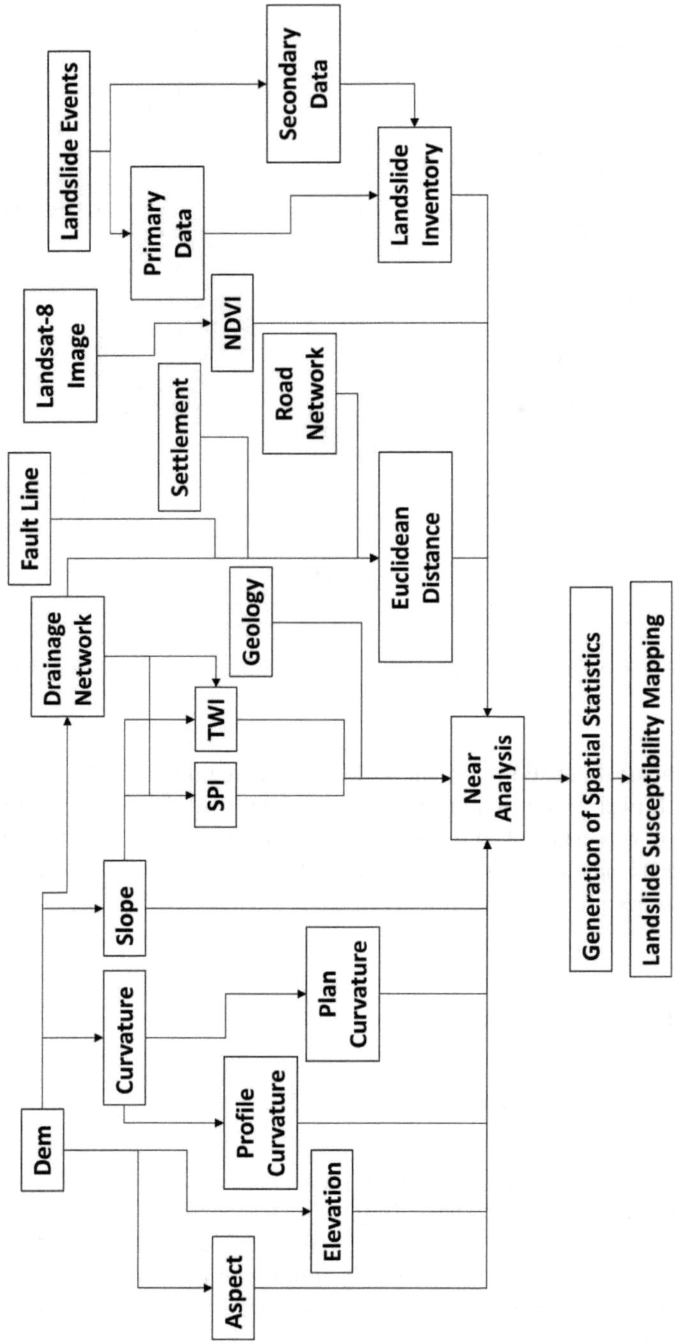

Fig. 5.5 Methodology flow chart for landslide susceptibility mapping

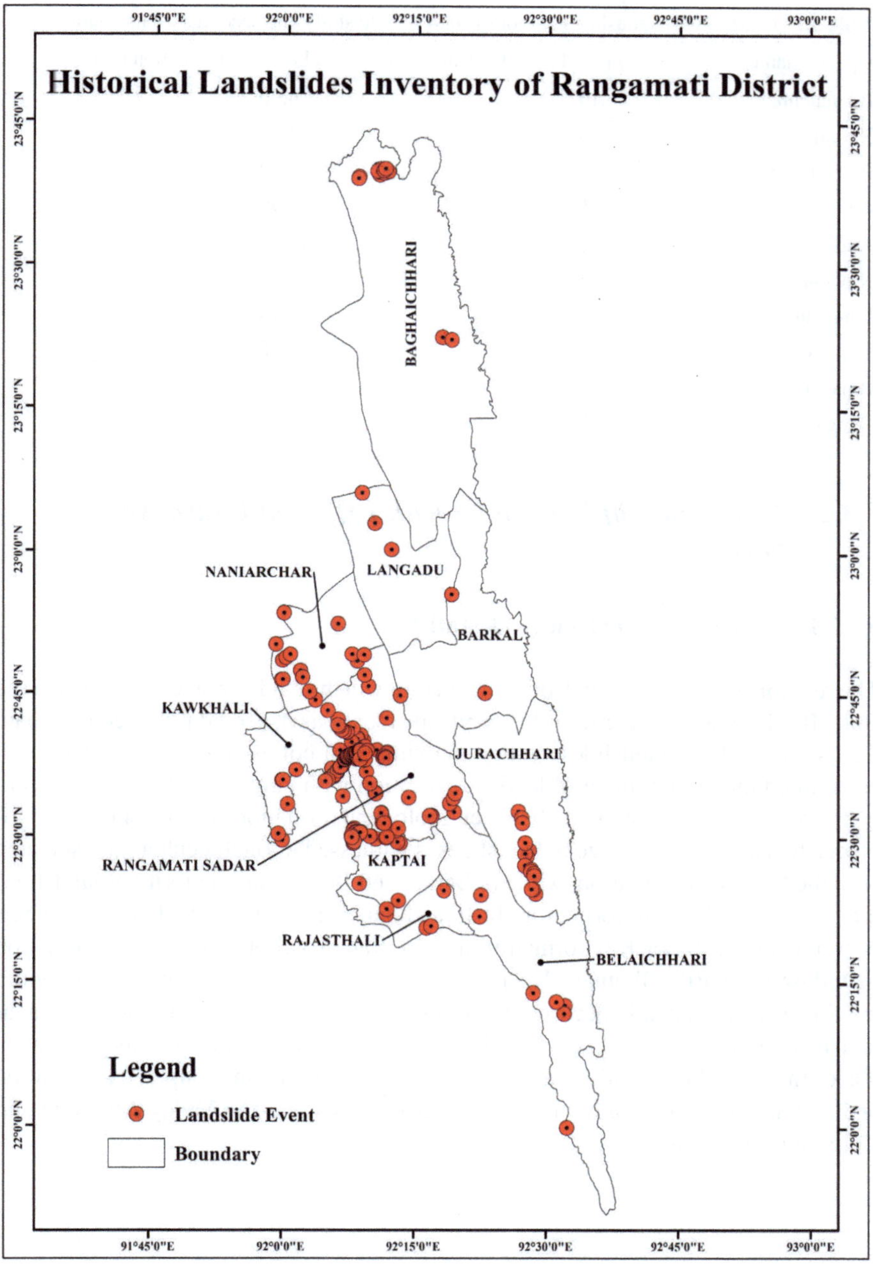

Fig. 5.6 Historical landslides inventory of Rangamati District

Table 5.1 Upazila wise landslide locations (*Source* Field study and secondary data sources)

Upazila name	Number of landslides	Percent distribution (%)
Baghaichhari	15	4.90
Barkal	4	1.31
Belaichhari	7	2.29
Jurachhari	15	4.90
Kaptai	43	14.05
Kawkhali	33	10.78
Langadu	3	0.98
Naniarchar	15	4.90
Rajasthali	4	1.31
Rangamati Sadar	167	54.58

5.3.2 Association of Landslides with Different Causative Factors

5.3.2.1 Landslide and Geological Statistics

The relationship between landslides and geology in the study area revealed that in the Boka Bill Formation there were 116 landslide locations observed followed by Tipam Sandstone with 84 landslides. Valley alluvium and colluvium on the other hand, constituted the least number of landslide location with only 1 (Table 5.2; Fig. 5.7).

From observing the statistics between geological formation and landslide events, it is seen that a high percentage of landslides is witnessed in Baghaichhari is associated with the Boka Bill formation, while in Barkal and Belaichhari majority of landslides are linked with Bhuban formation. In Juraichhari upazila all of the landslide events occurred in the Boka Bill formation and in Kaptai various geological formation including Boka Bill, Bhuman, Tipam Sandstone and Valley alluvium and colluvium are linked with landslide occurrence. Similar to that of Kaptai, in Rangamati Sadar upazila exhibits a range of geological formations where landslide events have taken place. In Langadu, landslides are associated with Dihing and Dupi Tila formation and in Naniarchar these are linked with Tipam Sandstone and Valley Alluvium and Collovium (Fig. 5.8).

Table 5.2 Relationship between geology and landslides location of the Upazilas of Rangamati District

Upazila name	Boka Bil formation (Neogene)	Bhuban formation (Miocene)	Tipam sandstone (Neogene)	Dupi Tila formation (Pleistocene and Pliocene)	Valley alluvium and colluvium	Dihing and Dupi Tila formation undivided	Lakes	Girujan clay (Pleistocene and Neogene)
Baghaichhari	14	1	–	–	–	–	3	–
Barkal	1	3	–	–	–	–	–	–
Belaichhari	1	6	–	–	–	–	–	–
Jurachhari	15	–	–	–	–	–	–	–
Kaptai	20	18	1	–	–	2	–	2
Kawkhali	21	6	–	2	1	–	–	3
Langadu	2	–	–	–	–	–	1	–
Naniarchar	8	–	7	–	–	–	–	–
Rajasthali	1	1	–	–	–	2	–	–
Rangamati Sadar	33	1	76	1	–	12	43	1

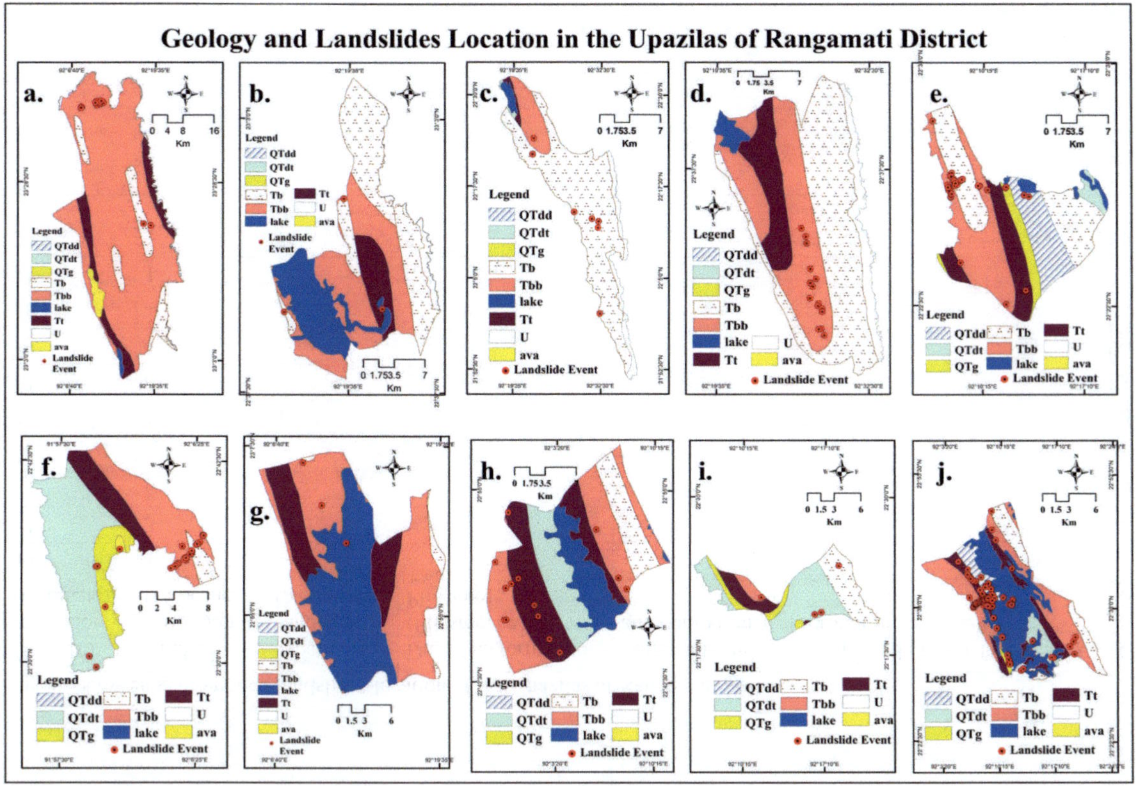

Fig. 5.7 Geology and landslides location in the Upazilas of Rangamati District **a** Baghaichhari; **b** Barkal; **c** Belaichhari; **d** Jurachhari; **e** Kaptai; **f** Kawkhali; **g** Langadu; **h** Naniarchar; **i** Rajasthali; **j** Rangamati Sadar

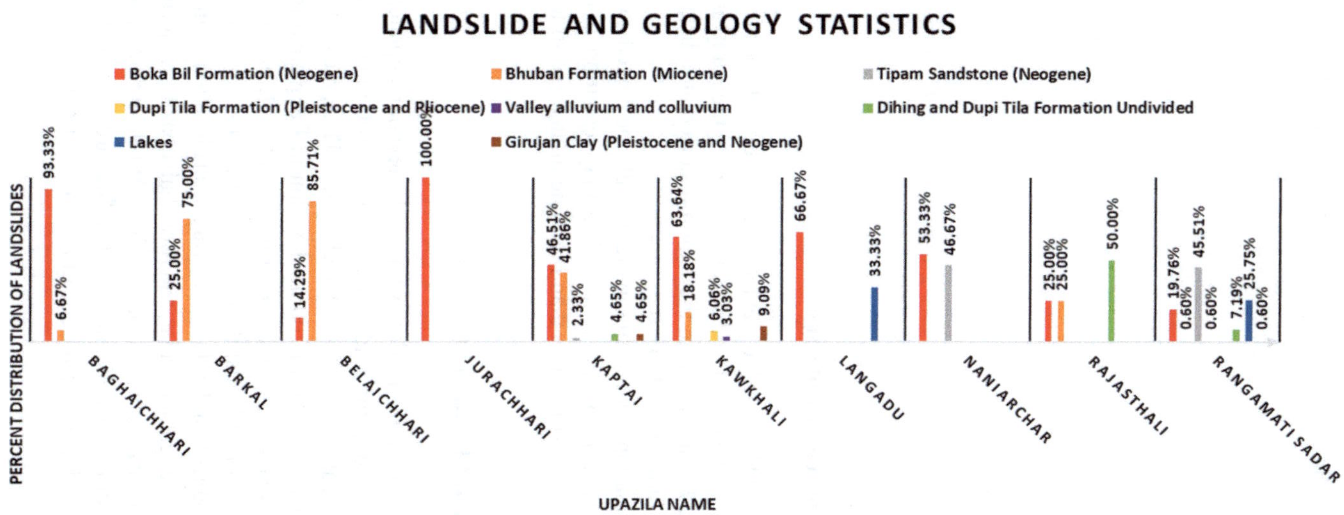

Fig. 5.8 Landslide and geology statistics

Table 5.3 Relationship between fault line and landslides location of the Upazilas of Rangamati District

Upazila name	0–100 m	100–200 m	200–300 m	300–400 m	400–500 m	More than 500 m
Baghaichhari	–	–	–	–	–	15
Barkal	–	–	–	–	–	3
Belaichhari	1	–	–	–	–	6
Jurachhari	–	–	–	–	–	15
Kaptai	–	–	–	–	–	43
Kawkhali	3	–	2	–	–	24
Langadu	–	–	–	–	–	3
Naniarchar	–	–	–	–	–	15
Rajasthali	–	–	–	–	–	4
Rangamati Sadar	–	–	–	1	1	165

5.3.2.2 Landslide and Fault Line Statistics

Contrary to the established knowledge that landslides frequently occur in areas closer to fault lines, a complete opposite trend was witnessed in the study area as most of the landslides that is 293 out of 306 in the region occurred at a distance of 500 or more meters from the fault lines (Table 5.3; Fig. 5.9).

The possible explanation behind this pattern could be linked to other factors present in areas not in close proximity to fault lines playing a more dominant role in landslide occurrences in the study area than fault line induced seismic activities.

5.3.2.3 Landslide and Elevation Statistics

Observation from associating landslides events with elevation depicted that most of the landslides that occurred in the study area was at 50–100 m elevation (125 locations) followed by 75 locations at 100–200 m elevation. Thus, most of the landslides are concentrated in areas from 50 to 200 m elevation in Rangamati indicating that these areas are more prone to landslide occurrence (Table 5.4; Fig. 5.10).

Upazila wise distribution of landslide location relating to elevation depicts that in Baghaichhari bulk amount of landslide location are at an elevation from 100 to 200 m. In Rajasthali, Langadu, Rangamati Sadar, and Kawkhali upazila, majority of the landslide location are present at an elevation of 50–100 m. While in Kaptai, at an elevation of 25–50 m, most landslides are concentrated. The Naniarchar upazila of Rangamati exhibits a diverse scenario, where the locations are distributed across 5 different

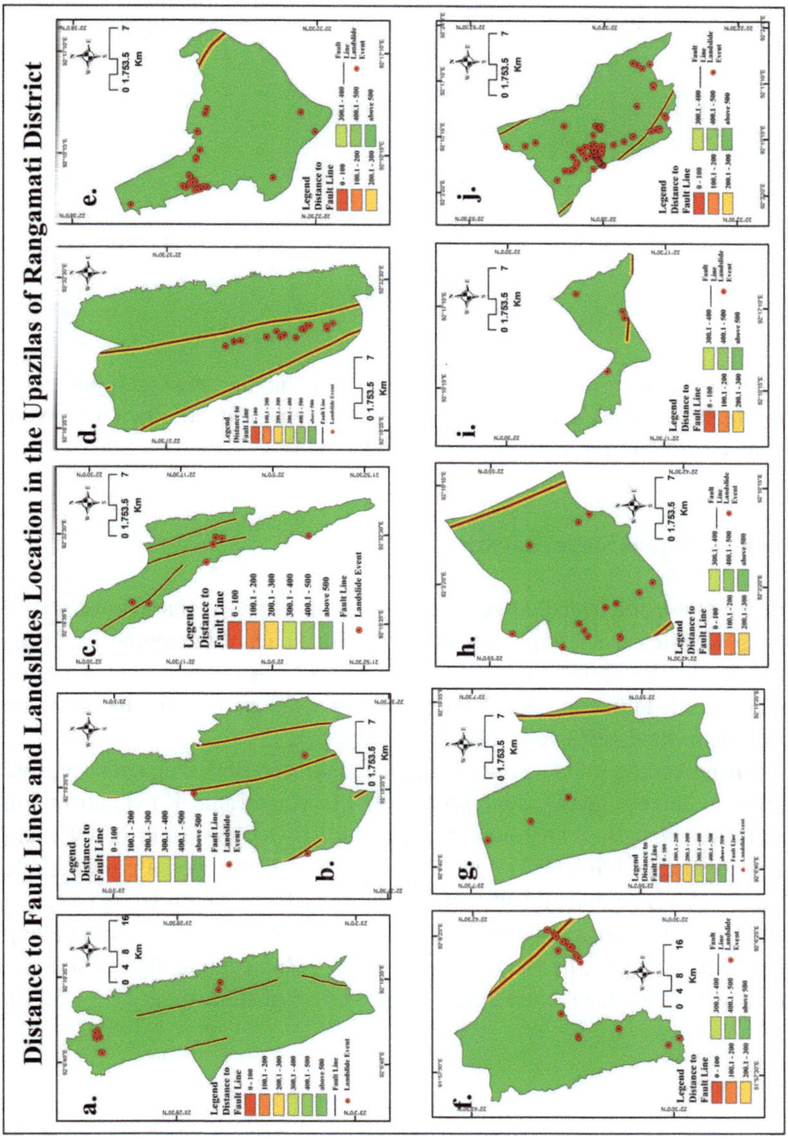

Fig. 5.9 Fault lines and landslides location in the Upazilas of Rangamati District **a** Baghaichhari; **b** Barkal; **c** Belaichhari; **d** Jurachhari; **e** Kaptai; **f** Kawkhali; **g** Langadu; **h** Naniarchar; **i** Rajasthali; **j** Rangamati Sadar

Table 5.4 Relationship between elevation and landslides location of the Upazilas of Rangamati District

Upazila name	25 or below	25–50	50–100	100–200	200–400	400 and above
Baghaichhari	–	–	–	14	1	–
Barkal	–	–	–	1	3	–
Belaichhari	–	–	–	3	2	2
Jurachhari	–	–	–	8	7	–
Kaptai	5	25	12	1	–	–
Kawkhali	–	7	22	4	–	–
Langadu	–	1	2	–	–	–
Naniarchar	–	3	6	2	3	1
Rajasthali	–	–	3	–	1	–
Rangamati Sadar	–	44	80	42	–	1

classes of elevation categories with elevation 50–100 m having most recorded landslide events. In Belaichhari and Naniarchar regions, few landslide locations were observed at an elevation of 400 m and above (Fig. 5.11).

5.3.2.4 Landslide and Slope Statistics

Being a crucial factor for landslide occurrences, an insightful scenario is observed in the study area while associating landslides with slope data of the region. Analysis shows that that most landslide locations in the Rangamati district occurred at a slope degree from 0° to 30° indicating that the areas where the slope falls in that range are more susceptible to landslide events (Table 5.5; Fig. 5.12).

A significant percentage of landslides in Baghaichhari and Barkal upazila occurred at regions with relatively gentle slope within 20°. In Kaptai and Kawkhali region majority of landslide locations are found in areas with 20°–30° slope with 67.44% and 66.67% respectively. In Juraichhari and Rangamati Sadar however, a considerable number of landslides are witnessed on areas having slope from 20° to 30°. In Langadu and Rangamati Sadar regions, the distribution is more concentrated in areas having slope from 0 to 20 degrees. The Belaichhari upazila of Ranagamati district shows a unique pattern with landslide location present in a variety of slope ranges (Fig. 5.13).

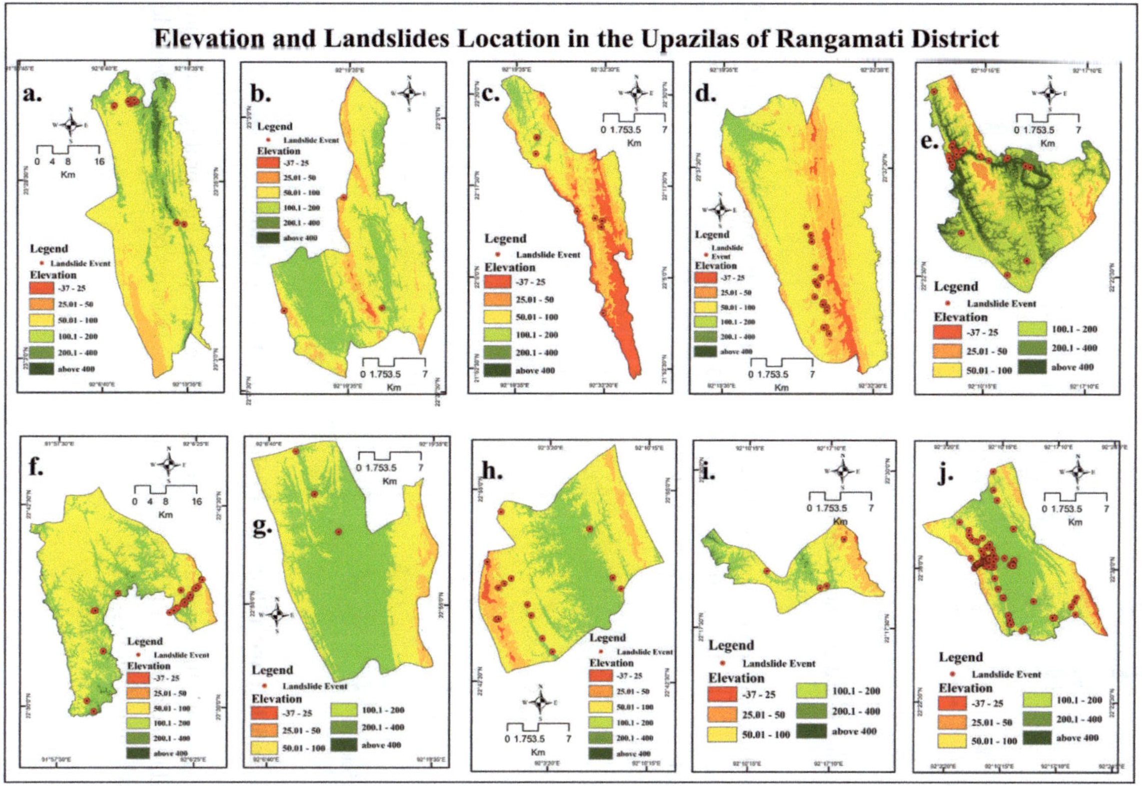

Fig. 5.10 Elevation and landslides location in the Upazilas of Rangamati District **a** Baghaichhari; **b** Barkal; **c** Belaichhari; **d** Jurachhari; **e** Kaptai; **f** Kawkhali; **g** Langadu; **h** Naniarchar; **i** Rajasthali; **j** Rangamati Sadar

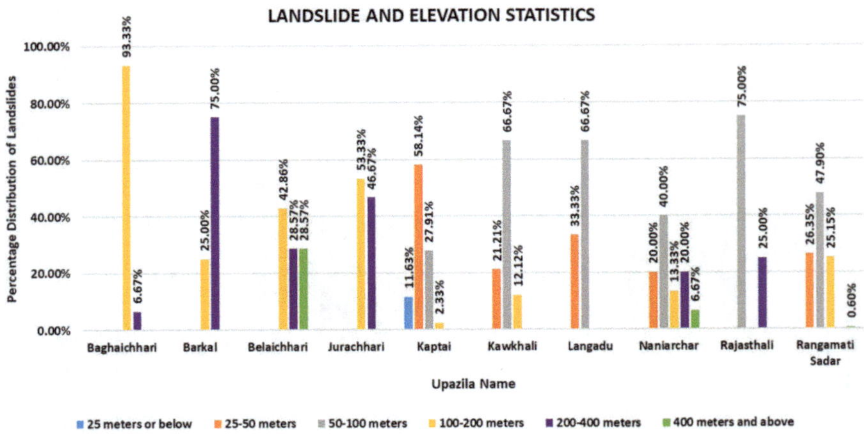

Fig. 5.11 Landslide and elevation statistics

Table 5.5 Relationship between slope and landslides location of the Upazilas of Rangamati District

Upazila name	0°–10°	10°–20°	20°–30°	30°–40°	40°–50°	50° and above
Baghaichhari	6	5	4	–	–	–
Barkal	–	3	–	1	–	–
Belaichhari	–	2	2	1	1	1
Jurachhari	2	3	9	1	–	–
Kaptai	9	29	5	–	–	–
Kawkhali	9	22	2	–	–	–
Langadu	2	1	–	–	–	–
Naniarchar	3	4	6	2	–	–
Rajasthali	–	3	1	–	–	–
Rangamati Sadar	72	82	13	–	–	–

5.3.2.5 Landslide with Plan and Profile Curvature Statistics

Observations from the analysis between landslide and plan and profile curvature reveals that in the areas with Convex and Concave curvatures occurrence of landslide is the most (Table 5.6; Fig. 5.14).

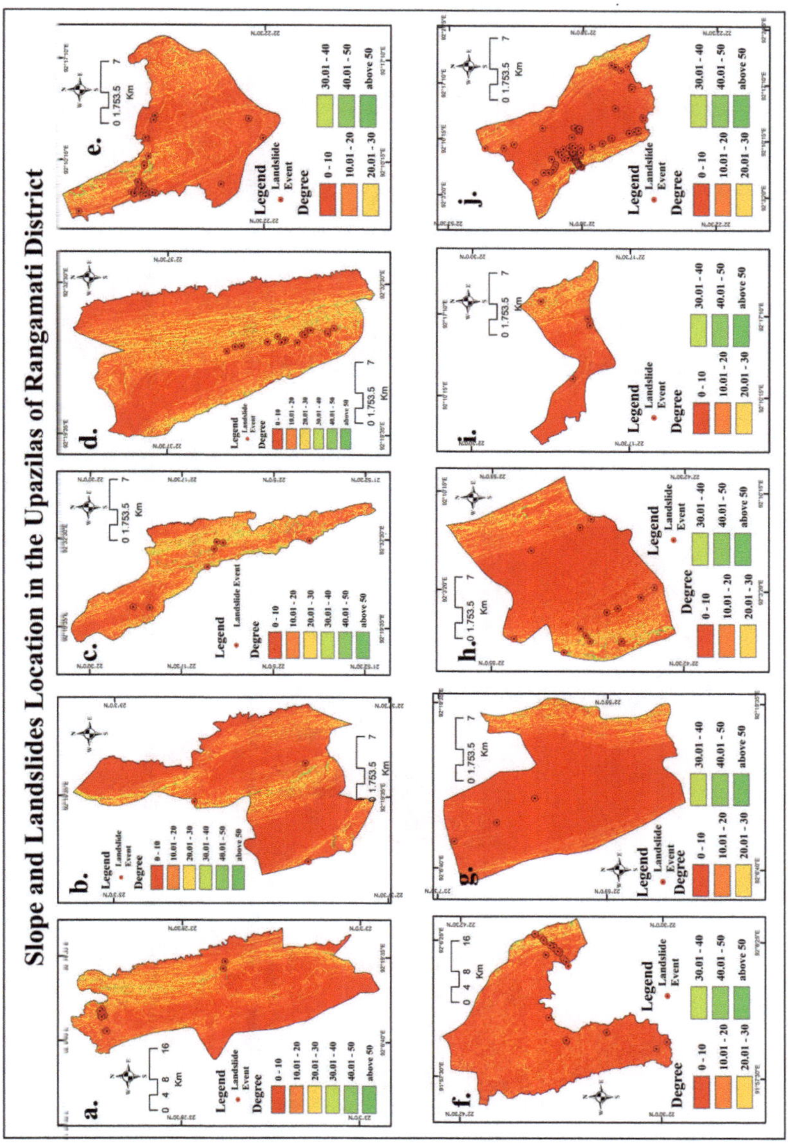

Fig. 5.12 Slope and landslides location in the Upazilas of Rangamati District **a** Baghaichhari; **b** Barkal; **c** Belaichhari; **d** Jurachhari; **e** Kaptai; **f** Kawkhali; **g** Langadu; **h** Naniarchar; **i** Rajasthali; **j** Rangamati Sadar

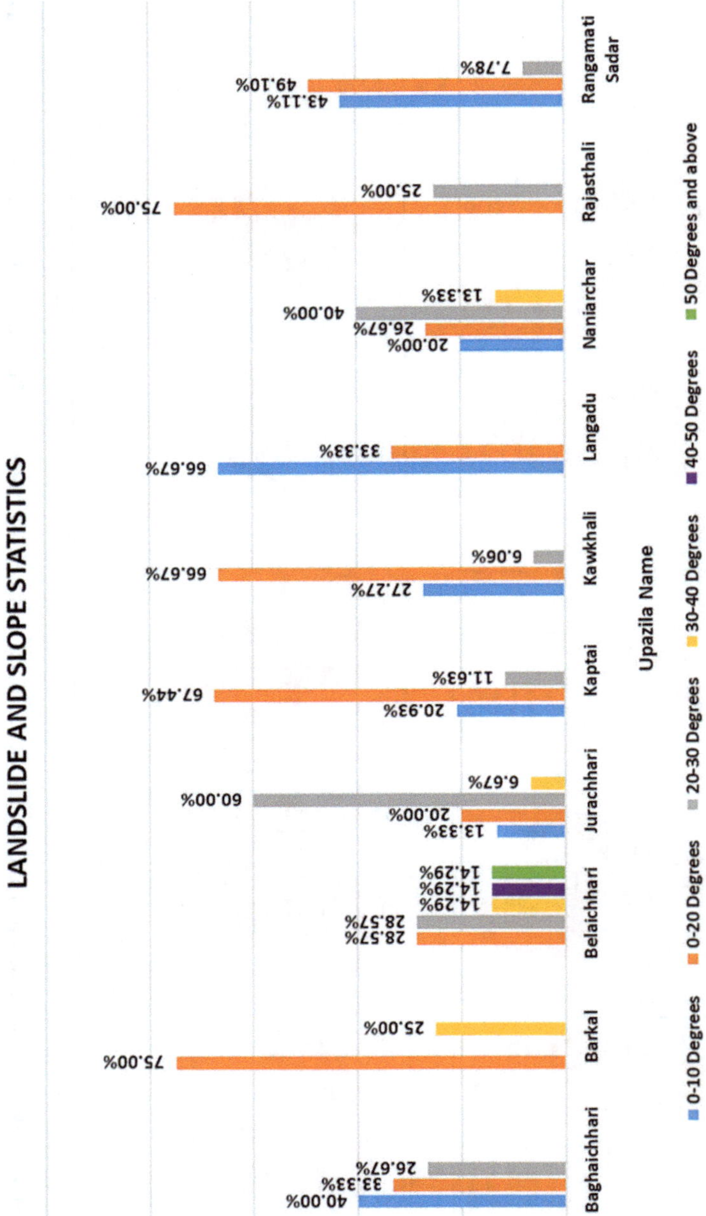

Fig. 5.13 Landuse and slope statistics

Table 5.6 Relationship between plan and profile curvature and landslides location of the Upazilas of Rangamati District

Upazila name	Convex	Flat	Concave
Baghaichhari	10	4	1
Barkal	2	2	–
Belaichhari	1	4	2
Jurachhari	2	2	11
Kaptai	12	4	27
Kawkhali	12	6	15
Langadu	1	–	2
Naniarchar	10	2	3
Rajasthali	–	2	2
Rangamati Sadar	77	16	74

The upazila wise analysis of Plan and Profile curvature with respect to landslide location in the study area provides insights into the relationship between the plan and profile curvature with landslides. In most of the upazila maximum landslides are observed in the region with convex and concave curvatures. However, few deviations from this trend were witnessed in Belaichhari upazila where most of the landslide (57.14%) occurrence and in Rajasthali upazila where 50% of landslides occurred were observed in regions with flat curvatures.

5.3.2.6 Landslide and Aspect Statistics

The analysis of data aimed at establishing association between landslides and aspect in the study area indicated that most landslides are concentrated in areas where the aspects are facing Southwest, Northeast and East. While the lowest landslide locations are North facing slopes. The flatter region has no landslide instances (Fig. 5.15).

Majority of the landslides in Baghaichhari upazila are in south aspect, while Barkal upazila these are concentrated in the east aspect. Belaichhari upazila exhibits a varied distribution of landslide location, with a notable presence in the East facing slopes. The distribution of landslide locations in Kaptai, Rangamati Sadar, Naniarchar shows a more balanced picture being present in different aspect classes. Rajasthali upazila landslide distribution are concentrated between Northeast, Southeast and West (Table 5.7).

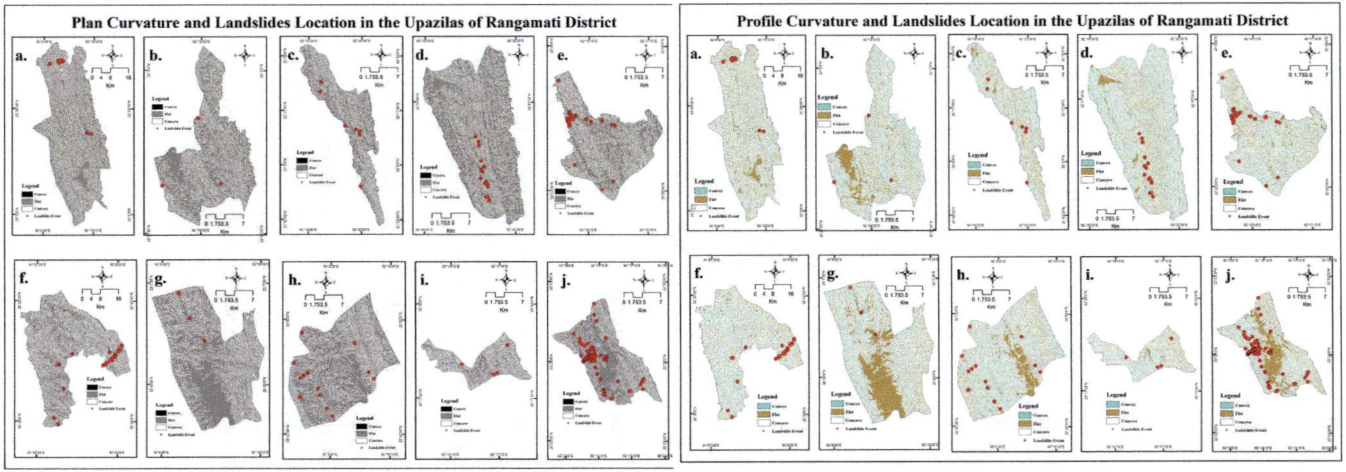

Fig. 5.14 Plan and profile curvature and landslides location in the Upazilas of Rangamati District **a** Baghaichhari; **b** Barkal; **c** Belaichhari; **d** Jurachhari; **e** Kaptai; **f** Kawkhali; **g** Langadu; **h** Naniarchar; **i** Rajasthali; **j** Rangamati Sadar

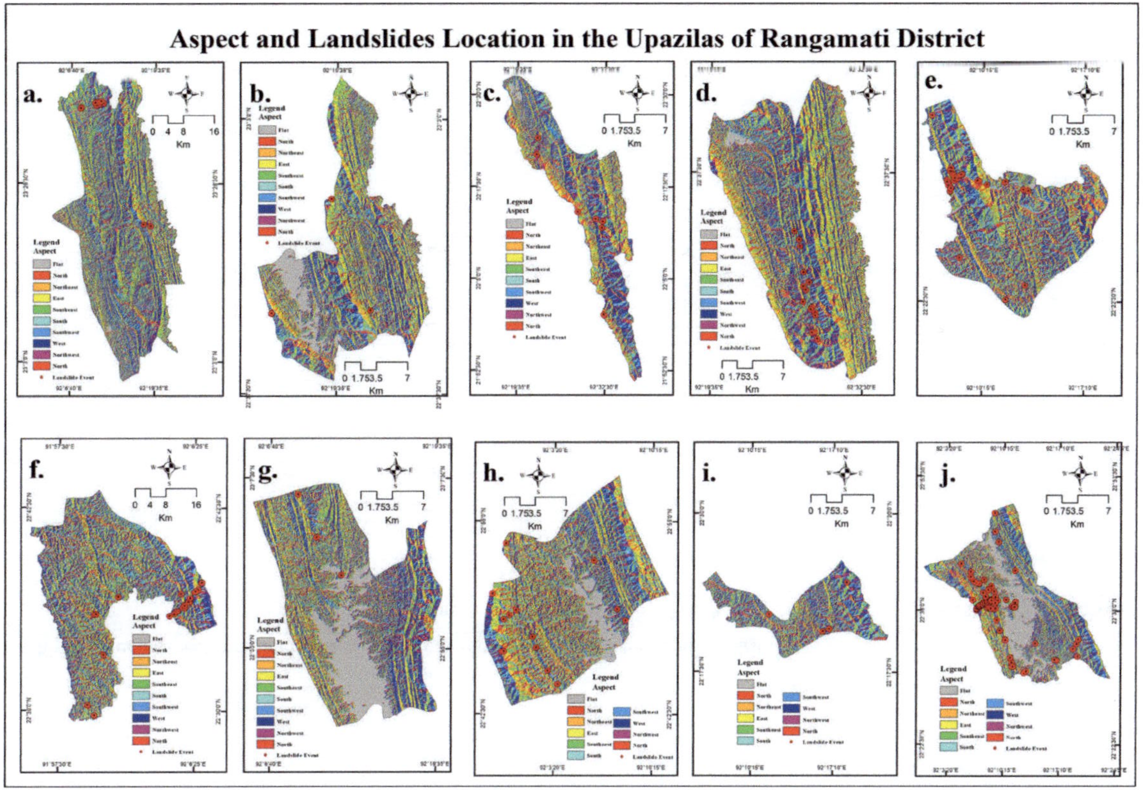

Fig. 5.15 Aspect and landslides location in the Upazilas of Rangamati District **a** Baghaichhari; **b** Barkal; **c** Belaichhari; **d** Jurachhari; **e** Kaptai; **f** Kawkhali; **g** Langadu; **h** Naniarchar; **i** Rajasthali; **j** Rangamati Sadar

Table 5.7 Relationship between aspect and landslides location of the Upazilas of Rangamati District

Upazila name	Flat	North	Northeast	East	Southeast	South	Southwest	West	Northwest	North
Baghaichhari	–	–	1	2	2	4	5	–	–	1
Barkal	–	–	–	–	1	1	2	–	–	–
Belaichhari	–	–	–	3	1	1	–	1	1	–
Jurachhari	–	–	4	3	1		1	3	3	–
Kaptai	–	1	5	5	1	1	19	9	2	–
Kawkhali	–	3	6	8	4	5	3	3	1	–
Langadu	–	–	–	1	1	–	–	–	1	–
Naniarchar	–	2	2	–	1	4	4	1	1	–
Rajasthali	–	–	1	–	2	–	–	1	–	–
Rangamati Sadar	–	6	29	24	21	29	20	18	15	5

Table 5.8 Relationship between distance to drainage network and landslides Location of the Upazilas of Rangamati District

Upazila name	0–100 m	100–200 m	200–300 m	300–400 m	400–500 m	500 m and above
Baghaichhari	5	–	3	–	1	6
Barkal	–	–	–	–	1	3
Belaichhari	–	2	–	1	1	3
Jurachhari	–	–	1	–	3	11
Kaptai	1	6	13	8	4	11
Kawkhali	9	–	5	5	3	11
Langadu	–	–	3	–	–	–
Naniarchar	–	3	1	1	1	9
Rajasthali	–	–	1	–	–	3
Rangamati Sadar	16	11	15	8	6	111

5.3.2.7 Landslide and Distance to Drainage Network Statistics

Data analysis to find the association between landslides and distance to drainage network shows an inverse relationship where most of the landslides occurred at a distance of 500 m or more from the drainage networks constituting more than 50 percent of landslide location (Table 5.8; Fig. 5.16).

The data representation shows a common trend as in majority of the upazilas in terms of landslide locations being concentrated at a distance above 500 m from drainage network with the exception of Kaptai, Kawkhali, Langadu and Baghaichhari upazila where considerable amount of landslide occurrence were witnessed in areas in close proximity to drainage networks. In Kaptai, around 30.8% of the landslide location and in Langadu all of the landslide locations are present at a distance from 200 to 300 m from the drainage network, while in Baghaicchari and Kawkhali upazila, the percent composition of landslides in closer proximity to drainage network (0–100 m) are 27.27% and 33.33% respectively (Fig. 5.17).

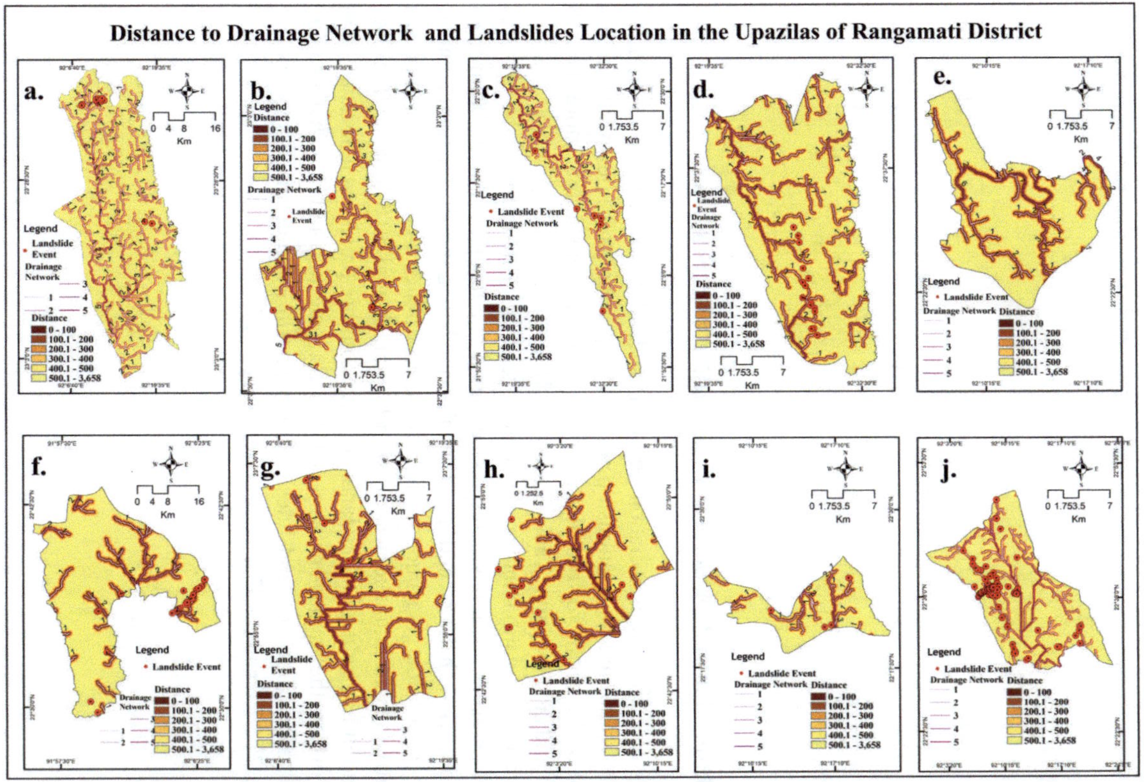

Fig. 5.16 Distance to drainage network and landslides location in the Upazilas of Rangamati District **a** Baghaichhari; **b** Barkal; **c** Belaichhari; **d** Jurachhari; **e** Kaptai; **f** Kawkhali; **g** Langadu; **h** Naniarchar; **i** Rajasthali; **j** Rangamati Sadar

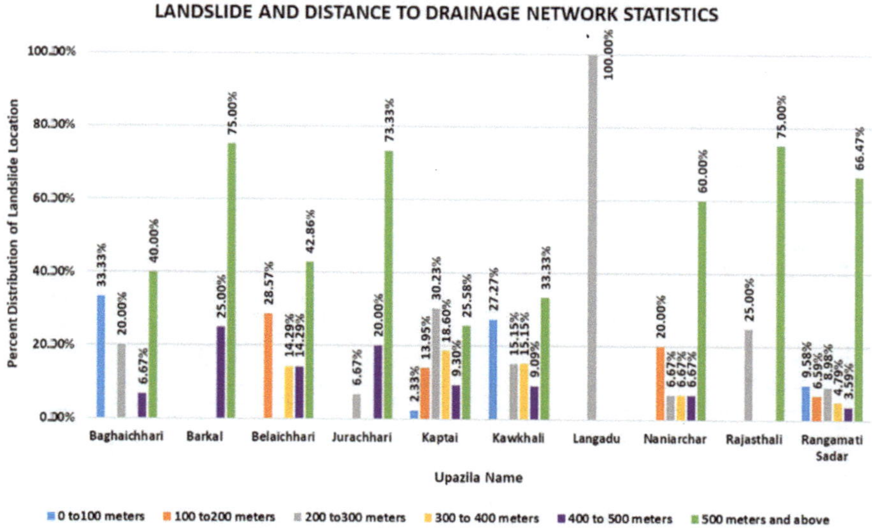

Fig. 5.17 Landslide and distance to drainage network statistics

5.3.2.8 Landslide and SPI Statistics

Stream Power Index (SPI) quantifies the erosive power of flowing water in a landscape and higher the values of SPI the chances of landslide occurrence increase. However, in the study area a complete opposite trend is witnessed as almost all of the landslides occurred in the areas where the values of SPI were lower than 1.5 (Fig. 5.18 and Table 5.9). This is indicative of the fact that though commonly, higher SPI values indicate higher probability of landslides, this relationship is much more complex than that. Other factors including vegetation cover, human intervention, geology etc. effect the relationship of SPI and Landslides.

5.3.2.9 Landslide and TWI Statistics

Topographic Wetness Index (TWI) are representation of assessing the wetness or soil moisture content of a landscape through quantification. Higher TWI values indicate the areas having more soil moisture content and potentially waterlogged condition. Thus, areas with higher TWI values are susceptible to landslides. However, in this study area, like SPI, opposite trend contrary to common knowledge is witnessed. Areas with lower TWI values constitute most of the landslide locations (Fig. 5.19 and Table 5.10). This trend is again indicative of the fact that landslides and TWI in the study area have an inverse relationship and cannot be used alone to explain

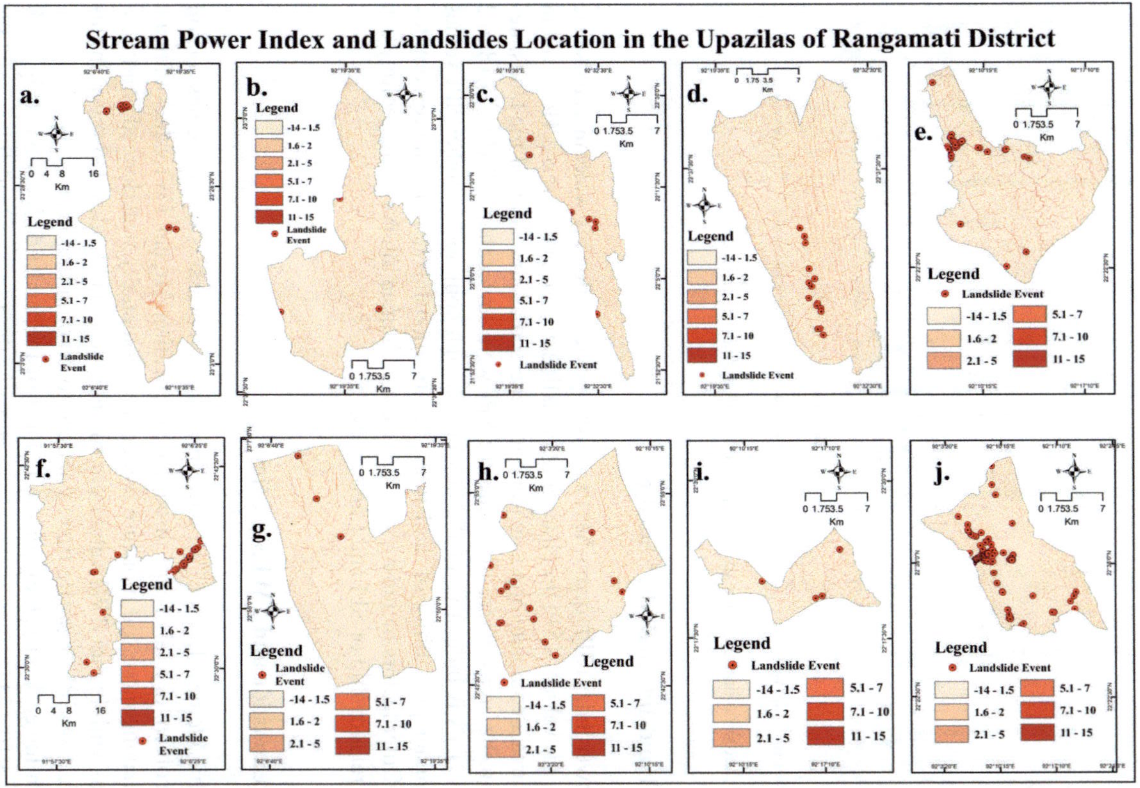

Fig. 5.18 Stream power index and landslides location in the Upazilas of Rangamati District **a** Baghaichhari; **b** Barkal; **c** Belaichhari; **d** Jurachhari; **e** Kaptai; **f** Kawkhali; **g** Langadu; **h** Naniarchar; **i** Rajasthali; **j** Rangamati Sadar

Table 5.9 Relationship between stream power index and landslides location of the Upazilas of Rangamati District

Upazila name	>1.5	1.6–2	2.1–5	5.1–7	7.1–10	10.1–15
Baghaichhari	14	–	1	–	–	–
Barkal	3	1	-	–	–	–
Belaichhari	4	2	1	–	–	–
Jurachhari	13	1	1	–	–	–
Kaptai	42	–	1	–	–	–
Kawkhali	30	–	2	–	1	–
Langadu	3	–	-	–	–	–
Naniarchar	14	–	1	–	–	–
Rajasthali	4	–	-	–	–	–
Rangamati Sadar	163	1	3	–	–	–

landslide events rather other factors should be analyzed to properly comprehend the pattern of landslide events.

5.3.2.10 Landslide and NDVI Based LULC Statistics

In this chapter, the NDVI values were categorized into 6 classes based on index values. The values from –0.16 to 0.015 corresponds to water, 0.016–0.14 corresponds to built-up areas, 0.15–0.18 represents bare land and the range 0.19–0.27, 0.28–0.36, and 0.37–0.5 are taken as shrub and grasslands, sparse vegetation and dense vegetation respectively. Overall data analysis reveals that majority of landslide locations were present in Bare Land and Shurb and Grassland class (Fig. 5.20; Table 5.11).

Notable patterns include most landslide locations were present in Shurb and Grassland Class in almost all of the upazila with the exceptiopn of Rajasthali, Baghaichhari, and Kawkhali. In Rajasthali, 50% and in Kawkhali, 45.45% of landslide locations were concentrated in Bare land area and in Baghaichhari 60% of landslide location were observed in areas having sparse vegetation. It was also observed that in Rangamati sadar and Kawkhali upazila, considerable amount of landslides were located in built-up areas (Fig. 5.21).

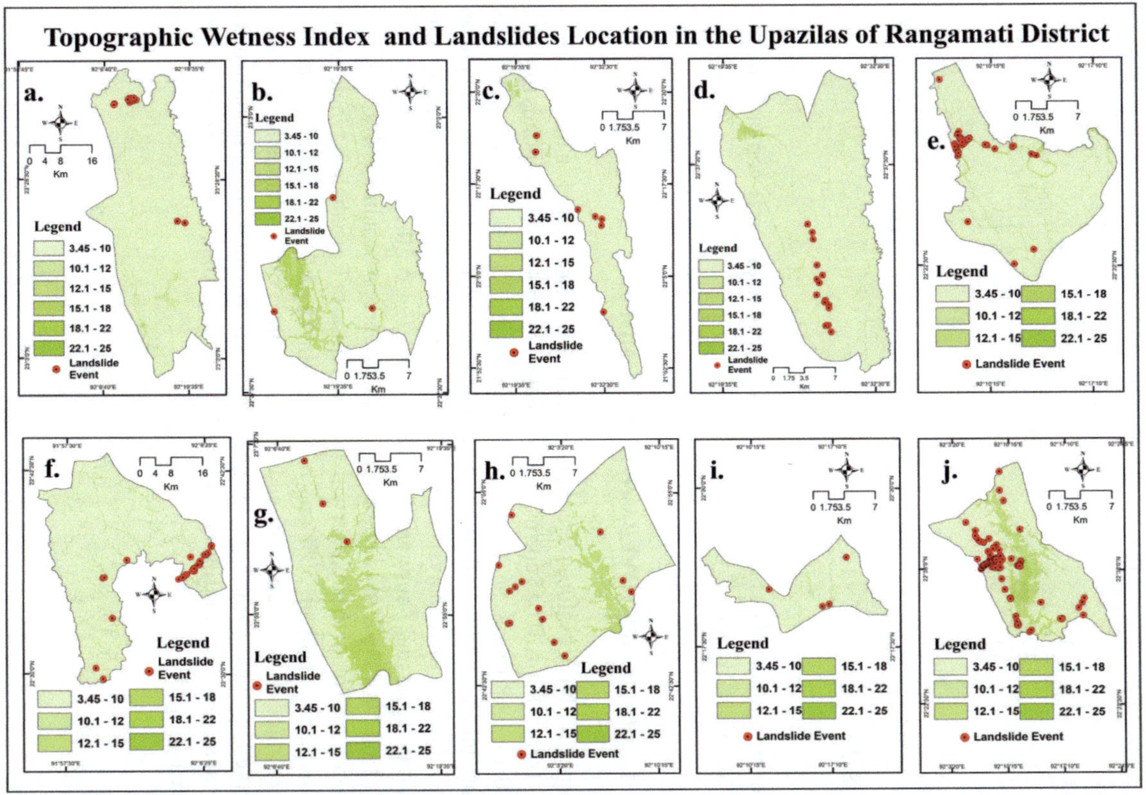

Fig. 5.19 Topographic witness index and landslides location in the Upazilas of Rangamati District **a** Baghaichhari; **b** Barkal; **c** Belaichhari; **d** Jurachhari; **e** Kaptai; **f** Kawkhali; **g** Langadu; **h** Naniarchar; **i** Rajasthali; **j** Rangamati Sadar

Table 5.10 Relationship between topographic witness index and landslides location of the Upazilas of Rangamati District

Upazila Name	3.3–10	10.1–12	12.1–15	15.1–18	18.1–22	22.1–25
Baghaichhari	15	–	–	–	–	–
Barkal	4	–	–	–	–	–
Belaichhari	7	–	–	–	–	–
Jurachhari	14	–	1	–	–	–
Kaptai	42	1	–	–	–	–
Kawkhali	32	1	–	–	–	–
Langadu	3	–	–	–	–	–
Naniarchar	15	–	–	–	–	–
Rajasthali	4	–	–	–	–	–
Rangamati Sadar	163	4	–	–	–	–

5.3.2.11 Landslide and Distance to Road Network Statistics

Road network has a close association with landslide occurrence and the areas more in close proximity to road network are more susceptible to landslides. Data analysis revealed that in the study area, most of the landslides occurred in places within 250 m from the road network (Fig. 5.22; Table 5.12).

Though most of the landslide locations were present in close proximity to road network, however in each individual upazila a different pattern is observed. In Belaichhari, Juraichhari, and Barkal, most landslide locations were concentrated in areas which were 600 m or more from the road network. In Langadu and Naniarchar these locations were concentrated in areas 2000–3000 m and 3000–6000 m from the road network (Fig. 5.23).

5.3.2.12 Landslide and Distance to Settlement Statistics

By linking the distance to settlement with landslide locational data of the study area it was found that most of the landslide locations were at places in close proximity to settlement. Areas within 25 m distance from settlement have higher concentration of landslide locations and areas from 50 to 100 m and more than 400 m from the settlement have lower concentration of landslide location (Fig. 5.24; Table 5.13).

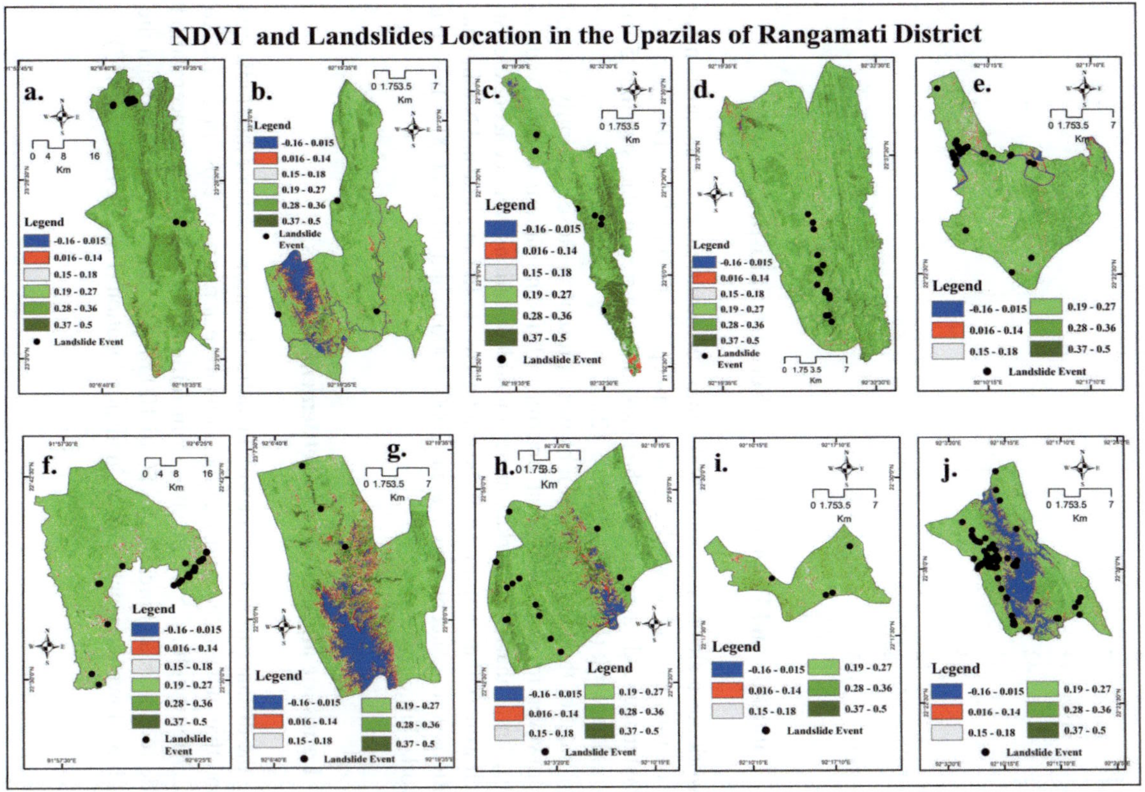

Fig. 5.20 NDVI based LULC and landslides location in the Upazilas of Rangamati District **a** Baghaichhari; **b** Barkal; **c** Belaichhari; **d** Jurachhari; **e** Kaptai; **f** Kawkhali; **g** Langadu; **h** Naniarchar; **i** Rajasthali; **j** Rangamati Sadar

Table 5.11 Relationship between NDVI based LULC and landslides location of the Upazilas of Rangamati District

Upazila name	Water (-0.16–0.015)	Builtup area (0.016–0.14)	Bare land (0.15–0.18)	Shurb and grassland (0.19–0.27)	Sparse vegetation (0.28–0.36)	Dense vegetation (0.37–0.5)
Baghaichhari	–	–	–	2	9	4
Barkal	–	–	–	3	1	–
Belaichhari	–	–	–	6	1	–
Jurachhari	–	–	–	15	–	–
Kaptai	–	2	4	36	1	–
Kawkhali	–	5	15	13	–	–
Langadu	–	–	–	3	–	–
Naniarchar	–	–	1	10	4	–
Rajasthali	–	–	2	1	1	–
Rangamati Sadar	–	25	42	90	–	10

Upazilla wise landslide locational distribution with respect to distance to settlement reveals that, in Baghaichhari, and Belaichhari upazila, most of the landslides occurred at a distance of above 400 m from the settlement. In Kaptai, Rangamati Sadar, Langadu, Rajasthali, and Kawkhali contrasting scenario was witnessed as most of the landslide locations are concentrated in the regions within 25 m distance from the settlement areas (Fig. 5.25).

The findings of the chapter associated landslide occurrence and different causative factors. One fundamental factor influencing landslides in the region was the geological formation of the landscape. The study area exhibited varying geological formation, with Boka Bill Formation and Tipam Sandstone being prominently associated with landslide events. Susceptibility to weathering and erosion of these formation are instrumental in increasing the possibility of landslide occurrences. Areas at specific elevation (50–200) m and slopes up to 30° have been identified as zones prone to landslide while, areas with convex and concave curvature in the region are at higher risk of encountering landslides events. The proximity to drainage network inversely correlates with landslide occurrence in most areas of Rangamati. The reason behind this deviation is the vegetation coverage in the areas closer to stream network plays a crucial role in aiding land stability in those areas. Additionally, low development activities in areas closer to streams could also attributed to the lower landslide occurrence in those areas. The NDVI categorization in this chapter indicates that places with bare soil and sparsely vegetated area have more chances of landslide occurrence as these regions becomes exposed to runoff and erosion compared to region with dense vegetation coverage. The association of landslide location with proximity

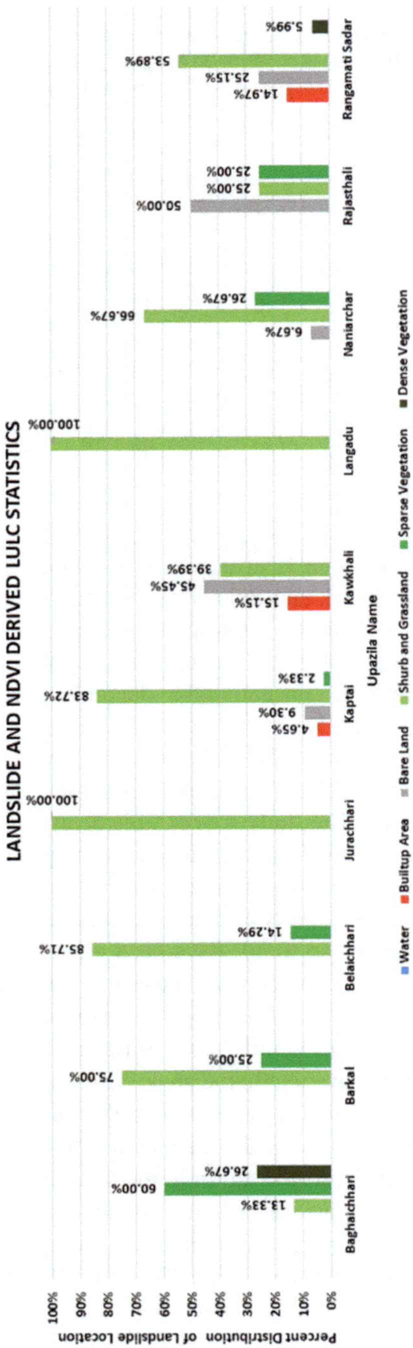

Fig. 5.21 Landslide and NDVI derived LULC statistics

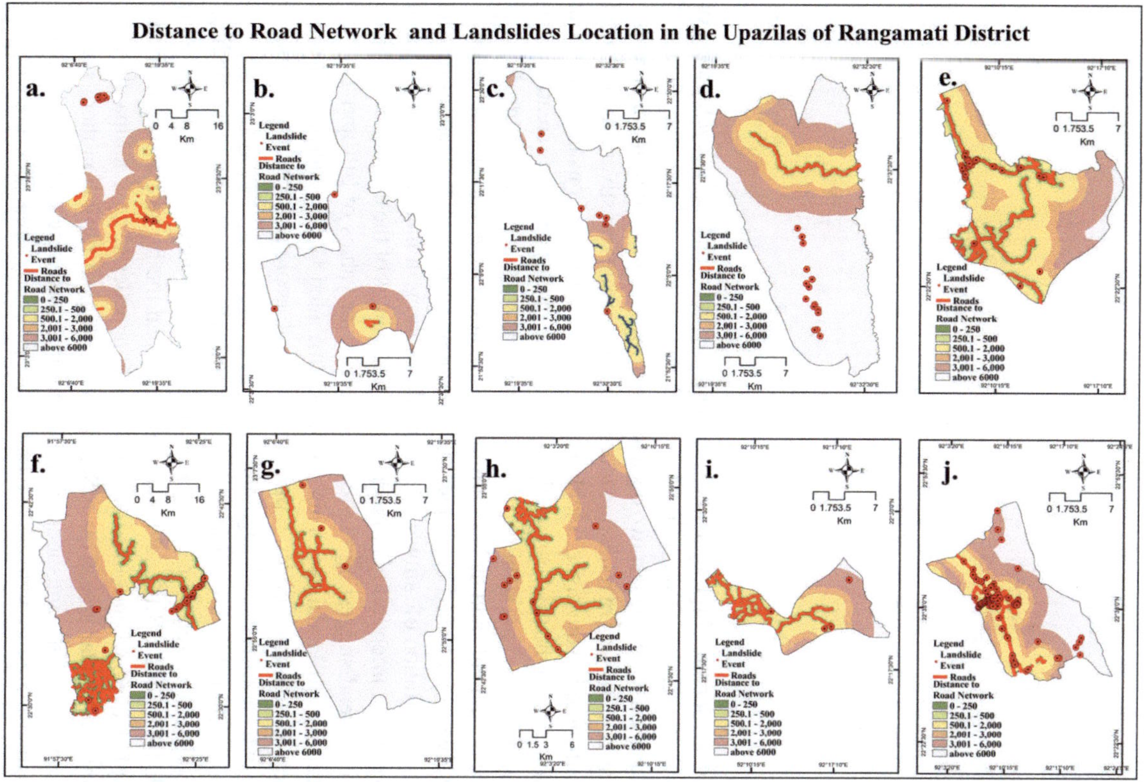

Fig. 5.22 Distance to road network and landslides location in the Upazilas of Rangamati District **a** Baghaichhari; **b** Barkal; **c** Belaichhari; **d** Jurachhari; **e** Kaptai; **f** Kawkhali; **g** Langadu; **h** Naniarchar; **i** Rajasthali; **j** Rangamati Sadar

Table 5.12 Relationship between distance to road network and landslides location of the Upazilas of Rangamati District

Upazila name	0–250 m	Above 250–500 m	Above 500–2000 m	Above 2000–3000 m	Above 3000–6000 m	Above 6000 m
Baghaichhari	2	–	–	–	–	13
Barkal	–	–	–	1	–	3
Belaichhari	–	–	–	1	1	5
Jurachhari	–	–	–	–	–	15
Kaptai	32	7	4	–	–	–
Kawkhali	29	–	1	1	2	–
Langadu	–	–	1	2	–	–
Naniarchar	5	–	1	1	8	–
Rajasthali	2	–	1	–	1	–
Rangamati Sadar	119	15	21	1	7	4

to roads and settlements highlights the role of human intervention in development of landslides as the regions closer to road network and settlement have high cluster of landslide location. Over the last two decades the growing pace of development initiatives such as construction of roads and tourism infrastructures mostly in an unplanned manner contributed to the higher incidence of landslide events. Constant movement of heavy vehicles on the road especially in areas with moderate slopes creates vibration which effects the land stability in those areas. Apart from these, cutting down of hills, faulty settlements and deforestation play a major role in landslide occurrence. Moreover, due to increased coverage of impervious surface in Rangamati district impedes the infiltration of rainwater to ground, resulting in increased rate of runoff leading to higher susceptibility of landslides. The findings of the chapter highlighted a paradoxical trend suggesting the regions with lower values of SPI and TWI are more susceptible to landslides. This could be attributed to the interplay of other factors such as vegetation cover, human intervention, geology have more influence than these two parameters in Rangamati.

Certain differences were observed while comparing the upazila-wise findings to the overall study area. For example, results in the context of the overall study area yielded that Boka Bill geological formation had the highest association with landslides; however, when looked at from a much smaller scale, i.e., upazila-wise, in the Rangamati Sadar area, Tipam Sandstone geological formation there showed higher association with landslides. Similar observations were made while associating landslide locations with aspect factors in the study area. The upazila-wise observation indicated that in the Rangamati Sadar area, the Southeast (247.5°) and

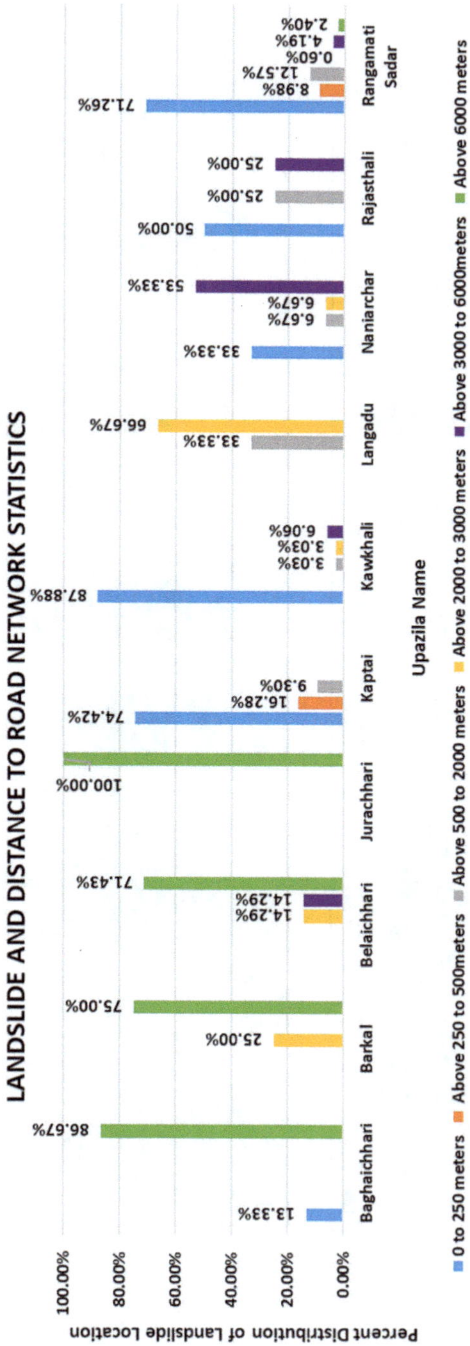

Fig. 5.23 Landslide and distance to road network statistics

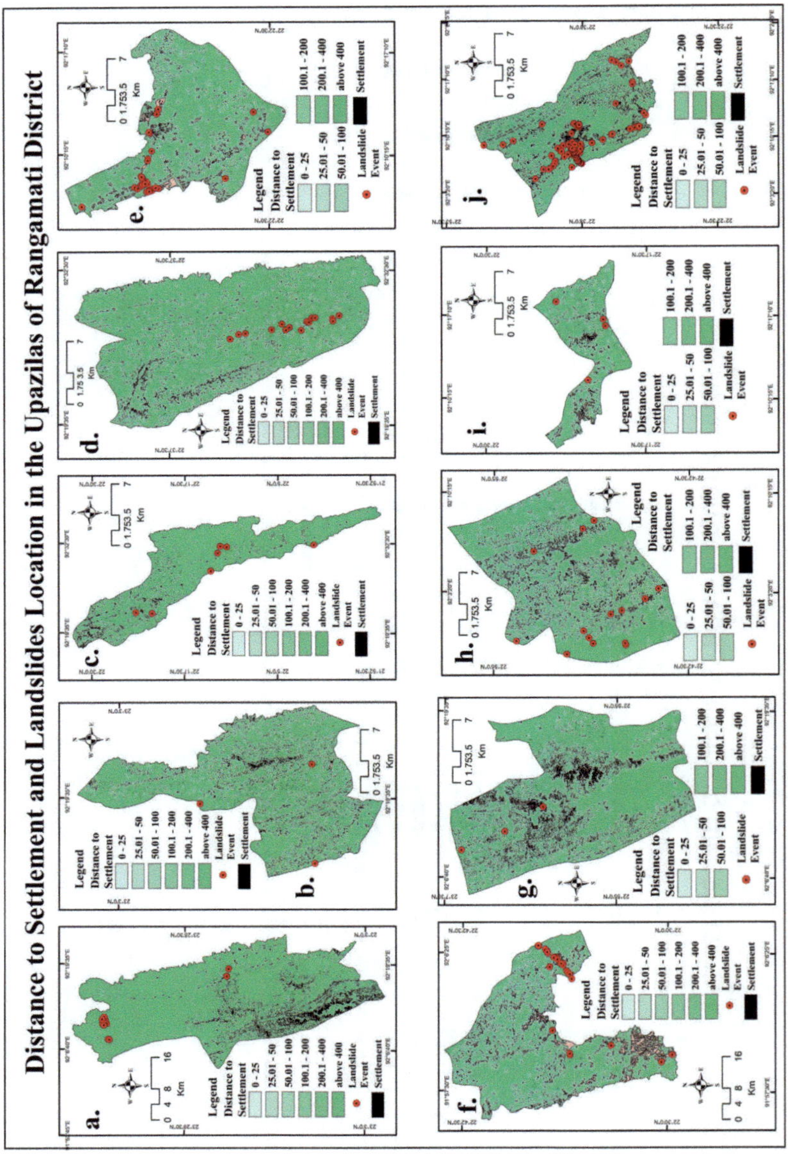

Fig. 5.24 Distance to settlement and landslides location in the Upazilas of Rangamati District **a** Baghaichhari; **b** Barkal; **c** Belaichhari; **d** Jurachhari; **e** Kaptai; **f** Kawkhali; **g** Langadu; **h** Naniarchar; **i** Rajasthali; **j** Rangamati Sadar

Table 5.13 Relationship between settlement and landslides location of the Upazilas of Rangamati District

Upazila name	25 m or less	Above 25–50 m	Above 50–100 m	Above 100–200 m	Above 200–400	Above 400 m
Baghaichhari	–	–	1	–	–	14
Barkal	–	–	–	2	1	1
Belaichhari	–	–	–	1	2	4
Jurachhari	–	–	1	2	7	5
Kaptai	27	5	6	3	1	1
Kawkhali	15	7	3	3	5	–
Langadu	1	–	1	1	–	–
Naniarchar	2	2	2	2	5	2
Rajasthali	1	–	1	1	–	1
Rangamati Sadar	60	23	19	25	30	10

Northeast (67.5°) aspects had a higher association with landslides. On the contrary, the Southwest (247.5°) aspect was more associated with landslides when the overall study area was considered. Upazila-wise susceptibility mapping accurately highlights landslide susceptible zones in the study area, reiterating the importance of the integrated approach used in the chapter. Observations of the data indicate that, the landslide locations are more concentrated in the regions where these topographical, geological and anthropogenic factors interact. Areas having certain geological (Boka Bill and Tipam Sandstone), elevation (50–200 m), slope degree (up to 20°), aspect (Northeast, Southeast and Southwest), SPI (−14 −1.5), TWI (3.45–10), distance to road (up to 250 m), settlement (up to 25 m), stream (more than 500 m) and fault lines (more than 500 m) and grasslands shrubs and sparse vegetation coverage has been identified more prone to landslides (Fig. 5.26). In the district of Rangamati, these areas cover around 43.58 square kilometers of area.

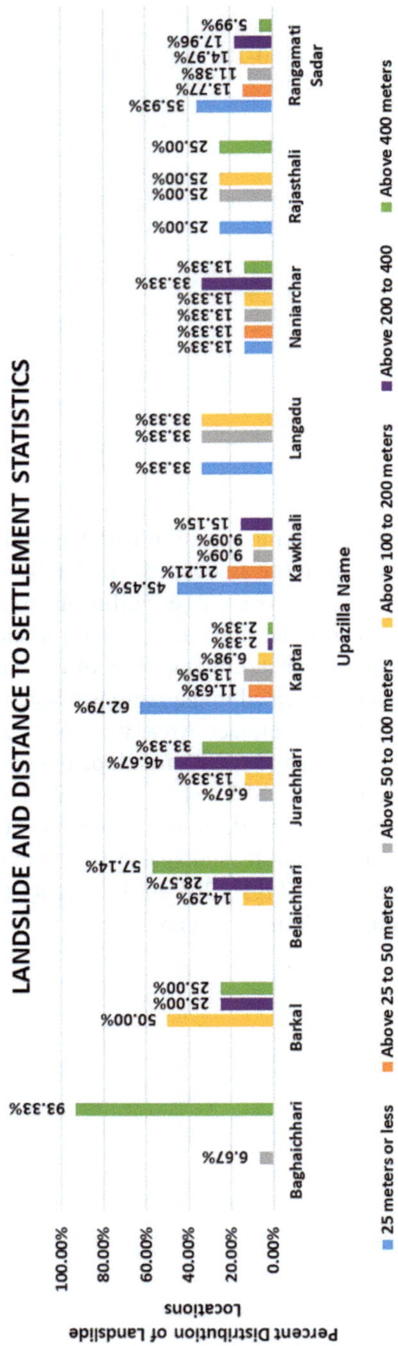

Fig. 5.25 Landslide and distance to settlement statistics

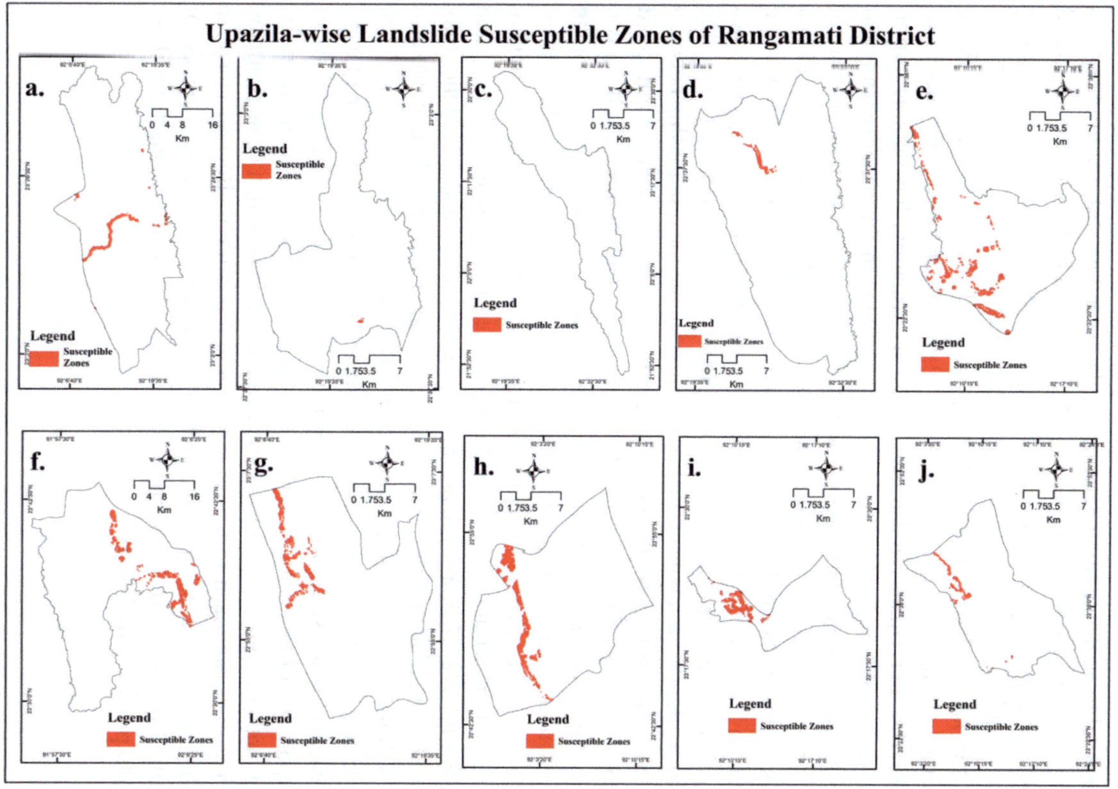

Fig. 5.26 Upazila-wise landslide susceptibility zones of Rangamati District **a** Baghaichhari; **b** Barkal; **c** Belaichhari; **d** Jurachhari; **e** Kaptai; **f** Kawkhali; **g** Langadu; **h** Naniarchar; **i** Rajasthali; **j** Rangamati Sadar

5.4 Conclusion

Combination of primary field level landslide locational data with secondary data obtained from various sources paved the way for generation of up-to-date landslide inventory covering all past landslides in Rangamati district. The findings of this chapter shed lights on the complex relationship between landslide occurrences and various 12 causative factors. Generation of spatial statistics in this chapter associated the landslide occurrences with these causative factors in each upazila and highlighted the susceptible areas for landslide occurrences in each upazila. The combined use of community knowledge, geospatial data, spatial analysis has offered a holistic understanding landslide susceptible zone and contributing factors that induces landslides. Boka Bill and Tipam Sandstone geological formation, having slope up to 20°, having elevation from 50 to 200 m and are covered with shrubs, grasslands and sparse vegetation coverage, with in close proximity to roads and settlements, certain distances from fault line and stream network and areas with certain aspect, SPI and TWI values are more susceptible to landslides in Rangamati district. Anthropogenic factors including roads and settlements also play a major role in inducing landslides as most landslides occurred in close distances from these features. Moreover, the chapter also challenges the conventional assumptions about the relationship between fault lines, SPI and TWI, highlighting the need for a more nuanced and multifaceted approach to delve deep into these issues. Further research and monitoring efforts can be built upon these findings to refine and enhance landslide susceptibility mapping and risk reduction strategies. Possible research direction in the domain of landslide susceptibility could be integration of climate change scenarios particularly changing precipitation pattern as rainfall significantly impact the soil and geological formation in an area. Furthermore, factor validation at local level for each landslide in the study area to identify which factors played a key role in inducing that landslides could also be explored. If the factors become validated then particular threshold level for each factor that initiates landslide events could be identified resulting in creation of early warning systems for landslides which can enable in taking proactive measures. Apart from incorporating climate change and factor validation, research should focus on developing methodologies to better integrate data collected through crowdsourcing and citizen science approach in landslide susceptibility studies. Beyond landslide susceptibility mapping, research should look to determine the interconnectedness of various natural hazards and their impact on landscape so that insights from these studies could provide a holistic risk assessment.

References

Abedin J, Rabby YW, Hasan I, Akter H (2020) An investigation of the characteristics, causes, and consequences of June 13, 2017, landslides in Rangamati District Bangladesh. Geoenviron Disasters 7(1). https://doi.org/10.1186/s40677-020-00161-z

Ahmed B (2021) The root causes of landslide vulnerability in Bangladesh. Landslides 18(5):1707–1720. https://doi.org/10.1007/s10346-020-01606-0

Ahmed B, Dewan A (2017) Application of bivariate and multivariate statistical techniques in landslide susceptibility modeling in Chittagong City Corporation, Bangladesh. Remote Sens 9(4):304. https://doi.org/10.3390/rs9040304

Ahmed B, Arfanul Alam SM, Ahmed I, Sammonds P (2023) The anthropogenic aggravation of landslide disasters in Bangladesh: key informants' perspectives. Prog Landslide Res Technol 1(2):385–401. https://doi.org/10.1007/978-3-031-18471-0_28

Alam E (2020) Landslide hazard knowledge, risk perception and preparedness in southeast Bangladesh. Sustainability 12(16):6305. https://doi.org/10.3390/su12166305

Ali SA, Parvin F, Vojteková J, Costache R, Linh NT, Pham QB, Vojtek M, Gigović L, Ahmad A, Ghorbani MA (2021) GIS-based landslide susceptibility modeling: a comparison between fuzzy multi-criteria and machine learning algorithms. Geosci Front 12(2):857–876. https://doi.org/10.1016/j.gsf.2020.09.004

Ara F, Khan MMR (2023) The Chittagong Hill tracts peace accord: promises and performances. In: Ranjan A, Chattoraj D (eds) Migration, regional autonomy, and conflicts in Eastern South Asia. Palgrave Macmillan, Cham. https://doi.org/10.1007/978-3-031-28764-0_12

Aristizábal E, Sanchez O (2020) Spatial and temporal patterns and the socioeconomic impacts of landslides in the tropical and mountainous Colombian Andes. Disasters 44(3):596–618. https://doi.org/10.1111/disa.12391

Ayalew L, Yamagishi H (2005) The application of GIS-based logistic regression for landslide susceptibility mapping in the Kakuda-Yahiko Mountains, Central Japan. Geomorphology 65(1–2):15–31

Ayalew L, Yamagishi H, Marui H, Kanno T (2005) Landslides in Sado Island of Japan: Part II. GIS-based susceptibility mapping with comparisons of results from two methods and verifications. Eng Geol 81(4):432–445

Bao S, Liu J, Wang L, Zhao X (2022) Application of transformer models to landslide susceptibility mapping. Sensors 22(23):9104. https://doi.org/10.3390/s22239104

Basharat M, Khan JA, Abdo HG, Almohamad H (2023) An integrated approach based landslide susceptibility mapping: case of Muzaffarabad Region, Pakistan. Geomat Nat Hazards and Risk 14(1). https://doi.org/10.1080/19475705.2023.2210255

BBS (Bangladesh Bureau of Statistics) (2011) Population and housing census 2011: national report. Bangladesh bureau of statistics, Dhaka, Bangladesh. http://www.bbs.gov.bd/WebTestApplication/userfiles/

BBS [Bangladesh Bureau of Statistics] (2022) Population and housing census 2022: preliminary report (Report No. ISBN 978-984-35-2977-0). Bangladesh Bureau of Statistics

Begum A et al (2020) Landslide susceptibility assessment: a correlation-based approach

Bourenane H, Guettouche MS, Bouhadad Y, Braham M (2016) Landslide hazard mapping in the Constantine city, Northeast Algeria using frequency ratio, weighting factor, logistic regression, weights of evidence, and analytical hierarchy process methods. Arab J Geosci 9:1–24

Bhuiyan M, Banu M, Rahman M (2017) Assessment of arbuscular mycorrhizal association in some fruit and spice plants of Rangamati Hill District. Bangladesh J Agric Res 42(2):221–232. https://doi.org/10.3329/bjar.v42i2.32810

Cai H, Chen T, Niu R, Plaza A (2021) Landslide detection using densely connected convolutional networks and environmental conditions. IEEE J Sel Top Appl Earth Obs Remote Sens 14:5235–5247. https://doi.org/10.1109/jstars.2021.3079196

Casagli N, Guzzetti F, Jaboyedoff M, Nadim F, Petley DN (2017) Hydrological risk: landslides

Chen WC, Chen H, We LW, Lin GW, Lida T, Yamada R (2017) Evaluating the susceptibility of landslide landforms in Japan using slope stability analysis: a case study of the 2016 Kumamoto earthquake. Landslides 14:1793–1801. https://doi.org/10.1007/s10346-017-0872-1.pdf. Accessed 23 June 2020

Chowdhury MS (2023) A review on landslide susceptibility mapping research in Bangladesh. Heliyon 9(7):e17972. https://doi.org/10.1016/j.heliyon.2023.e17972

Claudine IM, Ndokoye P (2023) Analysis of the impact of landslides on rural community livelihood in Rwanda: case of Ngororero district. Int J Geogr Geol 12(1):18–30. https://doi.org/10.18488/10.v12i1.3400

Coco L, Macrini D, Piacentini T, Buccolini M (2021) Landslide susceptibility mapping by comparing GIS-based bivariate methods: a focus on the geomorphological implication of the statistical results. Remote Sens 13(21):4280. https://doi.org/10.3390/rs13214280

Crosta GB, Frattini P (2008) Urban sprawling and land use alteration: impacts of human activities

Demir G, Aytekin M, Akgün A, Ikizler SB, Tatar O (2013) A comparison of landslide susceptibility mapping of the eastern part of the North Anatolian Fault Zone (Turkey) by likelihood-frequency ratio and analytic hierarchy process methods. Nat Hazards 65:1481–1506

Farooq S (2021) Comparison of data-driven landslide susceptibility assessment using weight of evidence, information value, frequency ratio and certainly factor methods. Acta Geodynamica et Geomaterialia 301–317. https://doi.org/10.13168/agg.2021.0021

Fisher GB, Amos CB, Bookhagen B, Burbank DW, Godrad V (2012) Channel widths, landslides, faults, and beyond: The new world order of high spatial resolution Google Earth imagery in the study of earth surface process. In: Whitmeyer SJ, Bailey JE, Paor DG, Ornduff T (eds) Google earth and virtual visualizations in geosciences and education research. The geological survey of America, vol 492, pp 1–22. https://doi.org/10.11301/2012.2492(01

Froude MJ, Petley DN (2018) Global fatal landslide occurrence from 2004 to 2016. Nat Hazard 18(8):2161–2181. https://doi.org/10.5194/nhess-18-2161-2018

Gómez D, García EF, Aristizábal E (2023) Spatial and temporal landslide distributions using global and open landslide databases. Nat Hazards 117:25–55. https://doi.org/10.1007/s11069-023-05848-8

Hafsa B, Chowdhury MS, Rahman MN (2022) Landslide susceptibility mapping of Rangamati District of Bangladesh using statistical and machine intelligence model. Arab J Geosci 15(15). https://doi.org/10.1007/s12517-022-10607-3

Haque DME, Sifa S, Mahmud T, Tarin MA (2018) Landslide susceptibility assessment based on satellite image processing and bi-variate statistical modeling for Rangamati district, Bangladesh. Bangladesh J Phys 23 & 24:93–106

Haque U, Blum P, da Silva PF et al (2016) Fatal landslides in Europe. Landslides 13:1545–1554. https://doi.org/10.1007/s10346-016-0689-3

Hussain N, Rizwan M, Alam MS, Suju L (2022) Remote sensing applications in landslide susceptibility index mapping of Rangamati District, Bangladesh. Re-Envis Adv Remote Sens 143–157. https://doi.org/10.1201/9781003224624-11

Joerg S, Irasema A-A, Marcus N, Stefan S (2022) Focus issue: addressing challenges of hazards, risks, and disaster management in mountain regions. Mt Res Dev 42(2):1–3

Jones JN, Boulton SJ, Bennett GL, Stokes M, Whitworth MR (2021) Temporal variations in landslide distributions following extreme events: Implications for landslide susceptibility modeling. J Geophys Res: Earth Surf 126(7). https://doi.org/10.1029/2021jf006067

Kirschbaum D, Stanley T, Zhou Y (2015) Spatial and temporal analysis of a global landslide catalog. Geomorphology 249:4–15. https://doi.org/10.1016/j.geomorph.2015.03.016

Korup O (2006) The relationship between types of landslides and geological materials

Lee S, Min K (2001) Statistical analysis of landslide susceptibility at Yongin, Korea. Environ Geol 40:1095–1113

Luck MA, Hajnsek I (2022) Integration of a graph network for landslide detection in space-borne earth observation. https://doi.org/10.1002/essoar.10512122.1

Martinello C, Mercurio C, Cappadonia C, Bellomo V, Conte A, Mineo G, Azzara G, Rotigliano E (2022) Incomplete landslide archive in landslide susceptibility assessment: a nested strategy for improving results. In: 10th International conference on geomorphology, Coimbra, Portugal, 12–16 Sep 2022, ICG2022-648. https://doi.org/10.5194/icg2022-648

Mersha T, Meten M (2020) GIS-based landslide susceptibility mapping and assessment using bivariate statistical methods in Simada area, northwestern Ethiopia. Geoenviron Disasters 7(1). https://doi.org/10.1186/s40677-020-00155-x

Michael PG, Abdul S (2021) Factors contributing to landslide susceptibility of the Kope formation, Cincinnati, Ohio. Environ Eng Geosci. https://doi.org/10.2113/EEG-D-20-00077

Mitchell NC, Masson DG, Watts AB, Gee MJR, Urgeles R (2002) The morphology of the submarine flanks of Volcanic Ocean Islands. J Volcanol Geotherm Res 115(1–2): 83–107. https://doi.org/10.1016/s0377-0273(01)00310-9

Moazzam MF, Vansarochana A, Boonyanuphap J, Choosumrong S, Rahman G, Djueyep GP (2020) Spatio-statistical comparative approaches for landslide susceptibility modeling: Case of Mae Phun, Uttaradit Province, Thailand. SN Appl Sci 2(3). https://doi.org/10.1007/s42452-020-2106-8

Mohammednur RS, Gemeda DO, Deribew KT (2020) Landslide susceptibility zonation using geospatial technologies and multi criteria evaluation techniques in Upper Didessa Sub-Basin, Southwest Ethiopia. https://doi.org/10.21203/rs.3.rs-121667/v1

Moore ID, Grayson RB, Ladson AR (1991) Digital terrain modelling: a review of hydrological, geomorphological, and biological applications. Hydrol Process 5(1):3–30

Netra P et al (2014) Proximity to fault lines and landslide susceptibility

Nguyen LD, Nguyen LM, Duong TV, Tran AN, Phung AN (2023) Landslide susceptibility zonation using geospatial techniques and analytical hierarchy process: a case study in Muong Lay Town and its vicinity. J Min Earth Sci 64(2):10–28. https://doi.org/10.46326/jmes.2023.64(2).02

Olivier D, Antoine D, Arthur D, Désiré K, Maki M, Toussaint M, Bibentyo, Judith U, Elise M (2021) Constraining landslide timing in a data-scarce context: from recent to very old processes in the tropical environment of the North Tanganyika-Kivu Rift region. Landslides. https://doi.org/10.1007/S10346-020-01452-0

Petley D (2012) Global patterns of loss of life from landslides. Geology 40(10):927–930. https://doi.org/10.1130/G33217.1

Potekar UP, Naik DD, Patil AS (2023) Geospatial approach for landslide susceptibility mapping techniques: a case study of Ratnagiri District, Maharashtra, India. Disaster Adv 16(4):34–45. https://doi.org/10.25303/1604da034045

Rabby YW, Li Y (2019) Landslide inventory (2001–2017) of Chittagong hilly areas, Bangladesh. Data 5(1):4. https://doi.org/10.3390/data5010004

Rabby YW, Li Y (2020) Landslide susceptibility mapping using integrated methods: a case study in the Chittagong hilly areas, Bangladesh. Geosciences 10(12):483. https://doi.org/10.3390/geosciences10120483

Rabby YW, Li Y, Abedin J, Sabrina S (2022) Impact of land use/land cover change on landslide susceptibility in Rangamati Municipality of Rangamati District, Bangladesh. ISPRS Int J Geo-Inf 11(2):89. https://doi.org/10.3390/ijgi11020089

Rahman MS, Ahmed B, Liping D (2017) Landslide initiation and runout susceptibility modeling in the context of hill cutting and rapid urbanization: a combined approach of weights of evidence and spatial multi-criteria. J Mt Sci 14(10):1919–1937

Rangamati Hill district Council (2011) Overview of Rangamati. http://rhdcbd.org/old/about_rangamati.php. Accessed 22 Oct 2023

Robert E, Robert E, Dalia K, Thomas S, Thomas S (2020) New global characterisation of landslide exposure. Nat Hazard. https://doi.org/10.5194/NHESS-20-3413-2020

Sato HP, Harp EL (2009) Interpretation of earthquake-induced landslides triggered by the 12 May 2008, 7.9 Wenchuan earthquake in Beichuan area, Sichuan Province, China using satellite imagery and Google Earth. Landslides 6:153–159. https://doi.org/10.1007/s10346-009-0147-6

Sevar N, Hazhir K (2020) A systematic review of GIS-based landslide hazard mapping on determinant factors from international databases. https://doi.org/10.1109/ICOASE51841.2020.9436611
Shano L, Raghuvanshi TK, Meten M (2020) Landslide susceptibility evaluation and hazard zonation techniques—a review. Geoenviron Disasters 7(1). https://doi.org/10.1186/s40677-020-00152-0
Si A, Zhang J, Tong Z, Liu X, Zhang Y, Lang Q (2018) The research of natural and human factors affect in landslide susceptibility assessment. https://doi.org/10.2991/RAC-18.2018.5
Sidle RC, Hirotaka S (2006) Challenges in determining the global distribution of landslides
Sreelakshmi S, SS VC, Shaji E (2022) Landslide identification using machine learning techniques: review, motivation, and future prospects. Earth Sci Inform 15(4):2063–2090. https://doi.org/10.1007/s12145-022-00889-2
Sultana N (2020) Analysis of landslide-induced fatalities and injuries in Bangladesh: 2000–2018. Cogent Soc Sci 6(1). https://doi.org/10.1080/23311886.2020.1737402
Stanley T, Kirschbaum DB (2017) A heuristic approach to global landslide susceptibility mapping. Nat Hazards 87:145–164
Trisnawati D, Najib Hidayatillah AS, Robbany AZ, Pellokila YA (2022) Comparative study of determination of landslide susceptibility based on weighting methods and analytical hierarchy processes in Pringapus, east Ungaran. IOP Conf Ser: Earth Environ Sci 1039(1):012024. https://doi.org/10.1088/1755-1315/1039/1/012024
Turner AK, Schuster RL (1996) Landslides can be categorized into five distinct movement types: falling, toppling, sliding, spreading, and flowing
Tyagi A, Tiwari RK, James N (2022) A review on spatial, temporal and magnitude prediction of landslide hazard. J Asian Earth Sci X 7(2022):100099. https://doi.org/10.1016/j.jaesx.2022.100099
Ullah M (2021) Geospatial modeling of landslide vulnerability and simulating spatial correlation with associated factors in Bandarban District. Dhaka Univ J Earth Environ Sci 8:51–66. https://doi.org/10.3329/dujees.v8i2.54839
Vakhshoori V, Zare M (2016) Landslide susceptibility mapping by comparing weight of evidence, fuzzy logic, and frequency ratio methods. Geomat Nat Haz Risk 7(5):1731–1752. https://doi.org/10.1080/19475705.2016.1144655.Accessed 7 Aug 2018
Vilímek V, Wang F, Strom A (2021) Catastrophic landslides and frontiers of landslide science. Landslides 18:3733–3735. https://doi.org/10.1007/s10346-021-01765-8
Wang Q, Li W (2017) A GIS-based comparative evaluation of analytical hierarchy process and frequency ratio models for landslide susceptibility mapping. Phys Geogr 38(4):318–337. https://doi.org/10.1080/02723646.2017.129452
Wang Y, Lin Q, Shi P (2018) Spatial pattern and influencing factors of landslide casualty events. J Geogr Sci 28:259–274. https://doi.org/10.1007/s11442-018-1471-3
Wijaya IP, Joshi A, Alam MN, Jayasinghe S, Laila N (2022) Climate change induced landslide susceptibility assessment—for aiding climate resilient planning for road infrastructure: a case study in Rangamati District, Chittagong Hill Tracts, Bangladesh. IOP Conf Ser: Earth Environ Sci 1091(1):012010. https://doi.org/10.1088/1755-1315/1091/1/012010
Woodard JB, Mirus BB, Crawford MM, Or D, Leshchinsky BA, Allstadt KE, Wood NJ (2023) Mapping landslide susceptibility over large regions with limited data. J Geophys Res: Earth Surf 128(5). https://doi.org/10.1029/2022jf006810

Chapter 6
Towards Artificially Intelligent Landslide Susceptibility Mapping: A Critical Review and Open Questions

Alihan Teke and Taskin Kavzoglu

Abstract Since the 1970s, the scientific community has dedicated significant efforts to the development of landslide susceptibility models through various approaches, with the current spotlight firmly on artificial intelligence techniques. Despite their unique advantages, these cutting-edge tools have introduced significant challenges, the solution of which hinges on critical user decisions. These decisions chiefly revolve around selecting landslide conditioning factors and designing the optimal configuration of internal mechanisms of susceptibility modeling approaches—both critical determinants influencing model predictive accuracy. To address origin of these issues, a systematic review of literature spanning seven years, from 2015 to 2021, was conducted. The results revealed the utilization of 151 various landslide conditioning factors, highlighting a clear dearth of consensus on the selection of geospatial covariates in the literature. Nonetheless, only about one-third of the reviewed articles considered the feature selection techniques to seek the optimal factor subset. The review also showed that 54 distinct machine learning algorithms were used, with logistic regression being the most commonly applied susceptibility modeling approach, featured in 70 articles. Notably, deep learning algorithms were marginally employed, appearing in a mere 7.08% of the reviewed articles since 2018. However, a significant proportion (64.32%) of the articles used non-optimized predictive models with default settings, while a trial-and-error approach was adopted in 10.81% of the reviewed literature. Beyond the comprehensive literature review, this chapter delves into a series of ill-explored open questions and reveals opportunities that can serve as potential research roadmaps, potentially guiding the trajectory of future studies in landslide susceptibility mapping.

A. Teke · T. Kavzoglu (✉)
Departments of Geomatics Engineering, Faculty of Engineering, Gebze Technical University, 41400 Gebze, Kocaeli, Turkey
e-mail: kavzoglu@gtu.edu.tr

A. Teke
e-mail: a.teke2020@gtu.edu.tr

G. K. Panda et al. (eds.), *Landslide: Susceptibility, Risk Assessment and Sustainability*, Advances in Natural and Technological Hazards Research 52,
https://doi.org/10.1007/978-3-031-56591-5_6

Keywords Literature review · Landslide susceptibility mapping · Machine learning · Deep learning · Feature selection · Hyperparameter optimization

6.1 Introduction

Landslide phenomena have generally been characterized as complex natural disasters due to their multi-dimensional mechanisms, unstable characteristics, and non-linear behaviors (He et al. 2023; Zhang et al. 2023), which render it difficult to discern the precursory/premonitory causes underlying their repetition. Landslides significantly engender material damage, drastic monetary assets, human losses, and disrupt the natural equilibrium by altering the natural topology of the Earth's surface system (Kalantar et al. 2018; Akinci and Yavuz Ozalp 2021; Aslam et al. 2022). Such unwarranted influences of landslides have prompted the researchers to take some proactive measures, including policy development (Alcántara-Ayala et al. 2017), early-warning system planning (Thirugnanam et al. 2020), and dam stability predictions (Shi et al. 2022). Instead, the development of landslide susceptibility maps is considered as one of the primary management and mitigation strategies to rationally avert the above-mentioned adverse influences (Merghadi et al. 2020; Khanna et al. 2021). In addition, the cost of producing a landslide susceptibility map is usually lower than its alternatives. These maps spatially characterize the landslide and non-landslide distribution in a given geographical location and can also be utilized as a base to produce emergency action plans (Pradhan and Jebur 2017; Hong et al. 2019). Therefore, the production of spatially and temporally accurate landslide susceptibility maps has consistently become a focal point of the scientific community.

Conventional statistical-based and heuristic methods, such as frequency ratio (FR), weight of evidence (WoE), and analytic hierarchy process (AHP), have long been actively applied to spatially predict the landslide susceptibility of a given zone but are hindered by subjectivity and presumptions on data distribution, limiting their effectiveness (Kavzoglu et al. 2015a; Yao et al. 2023). Also, the thematic geospatial input data obtained from imperfect in-situ observations impede the decision-making capability by eventually degrading the reliability of the prediction made by the conventional systems. They also mainly seek to construct a linear relationship between the landslide and its influencing factors, which highly decreases their generalization ability and reproducibility, since the landslides are usually driven by sophisticated internal and external conditions.

At present, the scientific literature on landslide susceptibility mapping (LSM) is populated by numerous predictive modeling schemes ranging from conventional machine learning (ML) to state-of-the-art deep learning (DL) algorithms to quantify landslide susceptibility. In addition, a set of algorithmic components within the artificial intelligence (AI) family, such as feature selection and hyperparameter optimization techniques, have gradually involved in the landslide susceptibility prediction process. These novel instruments play a critical role in not only increasing the quality of landslide susceptibility maps (Sun et al. 2021) but also making detailed inferences about landslide conditioning factors (Kutlug Sahin et al. 2017; Nirbhav et al. 2023). They make usual knowledge-based prediction procedures easier to conduct by significantly alleviating human intervention in the decision-making process and minimizing the requirement for expert knowledge. Indeed, their effectiveness and performance-enhancing contributions to landslide susceptibility maps have been proven in many recent works (Dao et al. 2020; Kavzoglu et al. 2021; Pham et al. 2021a; Pokharel et al. 2021; Sun et al. 2021).

The extensive literature on LSM has been a central theme in various survey-based studies (e.g., Dias et al. 2021; Pourghasemi et al. 2018; Tyagi et al. 2022). These studies systematically reviewed landslide susceptibility practices, examining diverse perspectives that encompass the scale of thematic data, input variables, and mapping techniques. Nonetheless, there is no thorough study that reviews algorithmic components (e.g., feature selection and hyperparameter optimization) of AI algorithms and scrutinizes the open research questions arising in their potential use in LSM practices. To the best of the authors' knowledge, there appear to be no review-based publications referring specifically to the potential effects of a series of operational elements of AI systems (e.g., feature selection and hyperparameter optimization) on LSM. To address the aforementioned problems and shed light on some obscure points of LSM practices, the recent literature was systematically reviewed by compiling a comprehensive database consisting of 210 peer-reviewed articles from 2015 to 2021. The reviewed articles in the literature database were analyzed under the five main categories: (i) geographical distribution of the case studies, (ii) landslide conditioning factors and thematic inputs, (iii) susceptibility modeling approaches, (iv) feature selection techniques, and (v) hyperparameter optimization algorithms. Lastly, a series of unresolved issues and untapped debatable points that emerged during the literature review were delved into, which shed light on various intriguing open research questions. Hence, this chapter intends to bridge these gaps by discovering the critical perspectives of the rapidly growing body of research in LSM, and is expected to be beneficial for the LSM community as well as other disaster risk-related domains.

The remainder of this chapter is organized as follows: In Sect. 6.2, recently published literature reviews were summarized to demonstrate the need for this review work and knowledge gaps. Section 6.3 is completely devoted to the theoretical background of AI tools and concepts mentioned in the review, including feature selection, ML/DL paradigms, and hyperparameter optimization algorithms. Section 6.4 represents the culmination and key findings obtained from the literature review. In Sect. 6.5, open questions and challenges identified throughout the literature survey were discussed. The last section involves a summary and conclusions drawn from this study.

6.2 Previous Literature Reviews

LSM has been of interest since the 1970s, potentially because of its nature that deeply affects human life. To rationally minimize the harsh influences of mass movement activities, a great deal of methodologies has been proposed by the global scientific communities, causing the appearance of a vast body of literature. This has further led to a practical necessity to compile these studies in a holistic context, shed light on the gaps in the current literature, and examine such rich literature on the common ground both terminologically and scientifically. Researchers did not remain unresponsive to this unavoidable need, and a series of important literature reviews were carried out that dealt with different aspects of the LSM. For instance, Reichenbach et al. (2018) systematically surveyed a total of 565 peer-reviewed articles associated with landslide susceptibility studies published between 1983 and 2016. In this wide-scope literature analysis, the authors created a comprehensive database covering 31 sub-headings, encompassing mapping techniques, input thematic variables, model type(s), validation criteria, predictive performance measure(s), scale of thematic data, and type of mapping units. They addressed that a limited number of reviewed articles considered the model uncertainty but a large proportion evaluated the predictive performance of the susceptibility models. They also applied the Susceptibility Quality Level index to rank the quality of a landslide susceptibility assessment. Huang and Zhao (2018) provided a particular review of the support vector machines (SVMs) for the spatial prediction of landslide susceptibility by explaining its underlying mechanism and fundamental concept. The authors also discussed the advantages and disadvantages of the SVMs over four different modeling approaches, including random forest (RF), artificial neural networks (ANN), logistic regression (LR), and AHP. In their study, potential applications of the SVMs together with its main hyperparameters were clarified by presenting suggestions for future works. They also explained the main steps of data preparation, sampling strategies, and some pre-processing stages (e.g., correlation analysis, transformation, feature selection) in LSM studies. Pourghasemi et al. (2018) reviewed a total of 469 studies related to landslide susceptibility published from 2005 to 2016 by benefitting from the Springer databases and Science Direct platforms. Unlike its analogous content, the compiled literature database was divided into two main parts, 2005–2012 and 2013–2016, in order to conduct a trend analysis. All articles were taken into examination by considering some article-specific information (e.g., journal's impact factor and author's name), utilized landslide-related parameters, and modeling approaches to predict landslide susceptibility. The authors reported that the LR was the most commonly utilized modeling approach in LSM and the slope gradient was selected as a dominant landslide predisposing factor that appeared in a majority of the reviewed articles.

Emphasizing the limited adoption of the ML algorithms in LSM, Merghadi et al. (2020) carried out a comparative overview of the most popular ones with the aim of providing an understandable plain language for the geoscience community. The advantages and disadvantages of these algorithms were discussed and later performed a comprehensive comparative analysis by employing 10 various ML algorithms as a case study in Algeria. They concluded that tree-based ensemble learning algorithms outperformed the others and specifically the RF algorithm yielded promising results without an intensive adjustment. The authors further broadened the scope of the literature review by including journals, authors, and their affiliations retrieved from the Web of Science platform. The study showed that Environmental Earth Sciences, Geomorphology, Landslides, Catena, and Geomatics Natural Hazard Risks were the leading journals in publishing articles focused on employing ML algorithms in landslide susceptibility research. Tyagi et al. (2022) presented a specific overview of landslide hazard prediction by concentrating on its three dimensions (i.e., susceptibility, temporal, and magnitude). To this end, the authors analyzed 236 articles published between 1972 and 2021 by using "landslide hazard mapping" and "landslide susceptibility analysis" keywords on the Google Scholar database. Within the scope of the review, a series of headings such as utilized susceptibility modeling approach, studied area, and publishing journals were examined. The authors mainly addressed that there exists general terminological confusion between "hazard" and "susceptibility" terms in a significant proportion of the reviewed articles. The review also revealed that many more articles were conducted in India, Italy, and Turkey and a geographical bias in susceptibility analysis was identified.

Lima et al. (2022) presented a bibliometric analysis of 2,585 publications from the Web of Science database and provided a review of the most recent publications from 1985 to 2020 by analyzing a series of topics, including validation techniques, classification algorithms, inventory characteristics. The authors reported that simple modeling approaches, such as FR and LR, are still in demand and a majority of the publications considered a combination of the predictors, slope angle, aspect, and geology factors. Also, the review results showed that the AUC was found to be the most popular validation criterion. Focusing on four main research subtopics (i.e., predictive models, model space data sets, prediction units, and disaster-causing factors), Yang et al. (2022) reviewed 600 documents from Scopus and Web of Science databases through bibliometric and knowledge mapping analysis in the past four decades. They highlighted that there exist numerous uncertainties (e.g., instability, fuzziness, and randomness) in landslide susceptibility prediction and concluded that future frontier research trends include neural networks, ML, and RF, and the number of published works on the uncertainty of landslide susceptibility prediction had a growing tendency. Also, review analysis demonstrated that China, Italy, and the United States were the three most influential countries in terms of the published articles.

Except for the majority of reviews investigating studies that were performed all around the world, there are also several site-specific literature surveys conducted nationwide. For instance, Dias et al. (2021) applied a systematic literature review of the publications related to LSM in Brazil using the Web of Science database. The authors selected six main topics, including the location of the test sites, publication year and journal, modeling approaches, utilized conditioning factors (i.e., thematic maps), origin of the landslide inventory, and performance assessment criterion for susceptibility map assessment. They concluded that statistical and slope stability models were frequently employed and the quality of resultant maps was evaluated based on receiver operating characteristic curves and area under the curve (AUC) metrics.

6.3 Theoretical Background

This section is entirely dedicated to the theoretical fundamentals of various concepts and techniques on specific AI components inquired in the literature review.

6.3.1 Feature Selection

With the accelerated progressions in remote sensing satellites, the internet of things, open source frameworks, and GPS-enabled applications, the last decades have witnessed a massive flux of geospatial big data from a series of sources, such as in-situ observations, high-resolution satellite images, and open-source portals, increasing the accessibility and availability of spatially and temporally important knowledge from the Earth's surface system. However, this has brought some challenging problems forcing the computational tools to manage, curate, and process these unprecedented volumes of spatial data. Another serious issue encountered in this procedure is the selection of optimal features for further computational geospatial applications. To achieve optimal performance, a high-quality input dataset proper for the nature of the problem is expected to be inputted by the learning models. This situation can be supported by an old proverb "Garbage in—garbage out", implying the poor quality of data can lead to unreliable outputs (Kilkenny and Robinson 2018). Therefore, the analyzed, collected, or acquired geospatial data should be of high accuracy, otherwise, can cast serious doubts on the reliability of the specific tasks.

From the perspective of predictive models constructed by AI, an increase in the amount of input data can sometimes accompany a decrease in their performance, which is called the Hughes phenomenon or peaking paradox. In other words, expansion of the original feature space dimension might reduce the model performance, which further increases the computational burden of the model. Similarly, another expression on the topic is "Less is more", suggesting smaller quantity can often result in higher quality.

An effective paradigm, namely feature selection, has been introduced to overcome the above-mentioned difficulties and improve the performance of the established predictive models. Considering its core philosophy, the feature selection falls into the mentioned two phrases. From these perspectives, feature selection algorithms mainly aim to identify the most relevant features by omitting the unnecessary ones. The more accurate and reliable predictive models are expected to be obtained after the feature selection process (Kavzoglu and Mather 2002; Colkesen and Kavzoglu 2018). Several significant causes for applying feature selection algorithms to input data are given as follows:

- ***Reducing Dimensionality***: Feature selection helps in reducing the dimensionality of the given dataset, thereby alleviating the computational complexity of models during the learning and prediction phases.
- ***Decreasing Processing Time***: By using a smaller set of features, feature selection contributes to decreased processing time, enhancing the efficiency of the computational process.
- ***Enhancing Model Interpretability***: Feature selection quantifies the importance of features for the specific problem under consideration, rendering models more understandable.

The feature selection techniques are generally divided into three categories: filter, wrapper, and embedded. The filter-based algorithms operate independently of a learning algorithm by adopting some statistical criteria to prioritize the variables during the feature subset evaluation (Lu 2019). Contrarily, the wrapper-based feature selection techniques evaluate the features by engaging a particular learning algorithm (El Aboudi and Benhlima 2016). Lastly, the embedded types present a combination of the filter and wrapper approaches by simultaneously evaluating features and learning models (Zhang et al. 2015).

6.3.2 Machine Learning (ML)

Machine learning (ML) is an evolving buzzword in computer science and AI fields designed to mimic the learning process of humans from their past experiences (i.e., human intelligence), encompassing a wide spectrum of computational predictive models and algorithms. One of the most overriding objectives of ML is generally to identify hidden connections and complex patterns within a given data structure. Up to now, most of the various definitions attributed to ML have generally employed the concept of "without being explicitly programming" in a common framework. What distinguishes the unique aspects of ML from the traditional programming process is this particular concept. Instead of hardly encoding a series of systematic or hierarchic rules, ML adopts the "soft coded" notion where they intrinsically automatize their inner architecture through iterations to learn a specific task.

In general, there exist two broad categories, including supervised and unsupervised learning, under the umbrella of ML on the basis of the training process. The former demands the labeled input instances under supervision to train the learning algorithm and predict unseen samples. The input dataset is generally divided into two parts, one of which is utilized to feed the predictive model. The sample subset, so-called training data, is used to fit the model with some computational process. During this period, the supervised learning algorithm tries to solve the potential hidden relationships within the data and install a function that will properly map them. The other reserved part, whose correct labels are known, is transferred to the trained model to make predictions based on the established computational structure, and the resulting model accuracy or error of the model can be quantified through various metrics. Conversely, the latter is heavily utilized in the lack of untagged instances and to label large data arrays without human supervision, making these models suitable for the clarification of the potential opaque trends.

The entrance of the ML paradigm into the field of LSM has mainly resulted from some shortcomings and weak aspects of current modeling approaches. The conventional heuristic approaches requiring expert knowledge about the area of interest struggle to keep pace with the spatial prediction of landslide susceptibility owing largely to their demanding a priori information from users that have the potential to be subjective or biased. On the one hand, landslides are too complicated and site-specific events in nature to model explicitly as a closed-form input–output relationship (Kavzoglu et al. 2015b; Hong et al. 2018; Jaafari et al. 2019). On the other hand, a series of thematic input variables from various sources are simultaneously evaluated to spatially predict the landslide susceptibility, making it difficult to establish a simple linear relationship between independent and dependent variables. Accordingly, these algorithms face the possibility of failing to model such intricate and hidden patterns, necessitating the implementation of better methods. In this context, ML algorithms have become a focal point of landslide susceptibility estimation, as in many domains, due to their improved abilities to solve real-world problems (Di Napoli et al. 2020; Pradhan 2013).

LSM and ML or more generally AI, share a long-existed history. In the early 2000s, ML ushered a new era in LSM with the application of ANN (Lee et al. 2004; Gómez and Kavzoglu 2005) and LR (Dai and Lee 2001; Dai et al. 2001) algorithms. Meanwhile, the apparent margin between ML and conventional statistical-based models still is a point of discussion. For example, the LR was originally used as a statistical-based model in the case of binary classification problems but it was adapted by ML from the domain of statistical models and is currently employed as one of the most common ML algorithms (Merghadi et al. 2020). The proliferation of these individual ML algorithms has gradually changed the trend of landslide susceptibility prediction and its procedure. In particular, they significantly decrease the labor-intensive preprocessing step and human intervention, which further alleviates the subjectivity in the decision-making process. But still, several challenging issues, such as the uncertainties in training parameters, and computational and representational problems, can emerge (Zounemat-Kermani et al. 2021), adversely affecting the efficiency of such algorithms.

To effectively handle the aforementioned issues, ensemble learning approach has been introduced and has gradually become an important tool for assorted academic fields. An ensemble learning algorithm hosts a weak (i.e., base) community of learners forming the main mechanism with the aim of boosting the prediction robustness of the final outcomes by increasing diversity within the model (Polikar 2012; Dong et al. 2020; Tonbul et al. 2022). Accordingly, ensemble ML algorithms can potentially decrease the burden on researchers in the prediction of landslide susceptibility by transforming the cumbersome prediction and parameter importance weighting process into an automated pipeline. The potential variance of an automated ML model is reduced with the core intuition behind such approaches, boosting the predictive performance of the system (Sagi and Rokach 2018). These distinct advantages have dramatically increased their practical applications for reaching the robust solution of landslide susceptibility.

6.3.3 Deep Learning (DL)

Subsumed under a broad umbrella of ML, deep learning (DL) has progressively entered into various real-world tasks from visual tracking systems to driver drowsiness detection or language translation to route optimizations. Theoretically based on conventional neural networks, the DL paradigm has long been on the agenda of many scientific communities since its first foundation in the 1950s, but has increased in popularity in the recent past. This is generally attributed to the efficiency of recent computational powers to handle the large amounts of data they demand, increasing trends of open-source programming languages and ecosystems, enhancements in hardware tools, and the emergence of cloud computing systems (Guo et al. 2016; Voulodimos et al. 2018).

DL algorithms mainly seek to extract relevant information, discover hidden relationships, and leverage a hierarchical combination of features in the given data distribution. Although there exist various DL models with different types of architectural design, their inner systems usually include similar multiple hidden layers, unlike neural networks, forming their "deep" structure and components (e.g., neurons, functions, synapses) for different tasks such as feature extraction and transformation. In DL structures, the neurons serve three different layers: (i) the input layer carrying the initial data to the subsequent layers of architecture, (ii) hidden layers performing non-linear transformation of inputs through an activation function, and (iii) output layer producing and holding the result of the problem under examination (Sildir et al. 2020).

The fundamental principle of the DL algorithms is to employ a sequence of nonlinear transformations to realize the hierarchical expression of input information in order to fit complex relationships between data (Chauhan and Singh 2018; Kavzoglu and Yilmaz 2022), which is done by using the output of the upper layer of a stacked multi-layer structure as the input of the next layer in a data-driven manner (Zhang et al. 2022). They have generally outperformed conventional algorithms in terms of predictive performance (Bui et al. 2020; Liu et al. 2022). Automatically extracting abstractions from input and showcasing robust generalization capacity, deep learning has piqued scientific interest across various domains. On the other hand, the computational complexity of DL is relatively (sometimes extremely) high, which may make its cost-efficiency in solving some problems questionable. DL algorithms usually demand a large volume of input data to efficiently produce a decision-making structure mainly because of the vast amount of internal parameters to be optimized by a learning algorithm. The trade-off between its pros and cons is usually permeated by the users by considering the nature of the problem under consideration. In brief, DL architectures build data input at multiple levels, with higher variables built as functions of lower ones (Wani et al. 2020). In contrast to the conventional approaches, DL algorithms can acquire the relevant features by themselves, with minimum human intervention (Yilmaz et al. 2023).

6.3.4 Hyperparameter Optimization

Almost all ML algorithms embody a series of internal model parameters and hyperparameters regulating their inner architecture and behavior (i.e., prediction mechanism). The former can be initialized during the learning procedure while the latter should be set by the users before training a predictive model. Having different types of presets (e.g., discrete, numerical, and categorical), hyperparameters have a significant impact on the learning algorithms' decision-making mechanism, behavior, complexity, and internal architectural design, all of which refer to their potential to affect model performance. Therefore, the adjustment procedure of predictive models requires some critical and precise tuning on the part of the user. Although the optimal hyperparameter values of learning algorithms are commonly configured by trial-and-error or some heuristic tests, such approaches require a certain level of expertise and can be impractical and time-consuming in the case of complex real-world problems (Kavzoglu and Teke 2022a). Rather, the use of automated hyperparameter optimization pipelines developed to search for optimal hyperparameters generally provides more effective and feasible results. According to Feurer and Hutter (2019), the most straightforward task is to seek optimal value from a pre-determined search space or configuration pool to get the highest achievable performance. Accordingly, optimizing the hyperparameters provides a series of further benefits (Yang and Shami 2020)

- Decreasing the manual efforts made by reducing the human intervention, especially for high-dimensional datasets.
- Boosting the predictive performance of learning algorithms.
- Increasing the reproducibility of established models and relevant scientific processes.

In summary, strategies for hyperparameter optimization provide a number of advantages, including the creation of more repeatable models, a more robust ML model, and a decrease in the need for manual effort. Therefore, there has been an increased recognition that more attention needs to be paid to the particular field of the research area.

6.4 Materials and Methods

To critically and systematically review the use of AI and its offshoots (e.g., ML, DL, hyperparameter optimization, and feature selection) in LSM literature, only peer-reviewed articles were taken into consideration by using the "Scopus" database. Initially, three criteria, including keyword, temporal constraints, and annual article volume, were considered to reach relevant literature. We limited the Boolean search criteria by using the "*landslide susceptibility*" keyword and included articles written between 2015 and 2021 to showcase contemporary research trends. After the determination of search criteria, the following search query was used: ALL ("landslide susceptibility") AND PUBYEAR > 2014 AND PUBYEAR < 2022 AND (LIMIT-TO(DOCTYPE,"ar")). Subsequently, the publications were sorted in descending order of the number of citations received, and the top 30 articles for each year were included in our literature database collection. It is noteworthy that this sorting process excluded certain irrelevant publications and survey articles, aligning with the primary objective of this chapter—reviewing real-world case experiments.

Ultimately, our literature database comprised 210 peer-reviewed articles spanning a seven-year period, setting the foundation for further analytical steps. The review design is organized into five major categories: (i) the geographical distribution of test sites vulnerable to landslide activities, (ii) thematic input data (i.e., landslide conditioning factors) used to predict landslide susceptibility, with a focus on their primary types (e.g., topographical and hydrological), (iii) landslide susceptibility modeling approaches under a broad range from conventional heuristic approaches to advanced machine and DL algorithms, (iv) feature selection (or elimination) methods used for factor prioritization, (v) hyperparameter optimization techniques considered in case of tuning the internal design of learning algorithms. The methodological flowchart that outlines the search criteria and review design for the literature survey is illustrated in Fig. 6.1.

6.5 Results and Discussion

6.5.1 Geographical Distribution

Earlier review studies (Pourghasemi et al. 2018; Merghadi et al. 2020) usually consider the authors' origins or affiliated institutions when determining the test sites or geographical locations of scientific studies on landslide susceptibility. It should be mentioned that the affiliated institutions may not necessarily be the geographical location where the study was conducted. In this chapter, the test sites of each article under investigation were independently examined in order to quantify where landslide activities are concentrated around the world. Therefore, the 'study area' in this work explicitly refers to the geographical location on the Earth where the

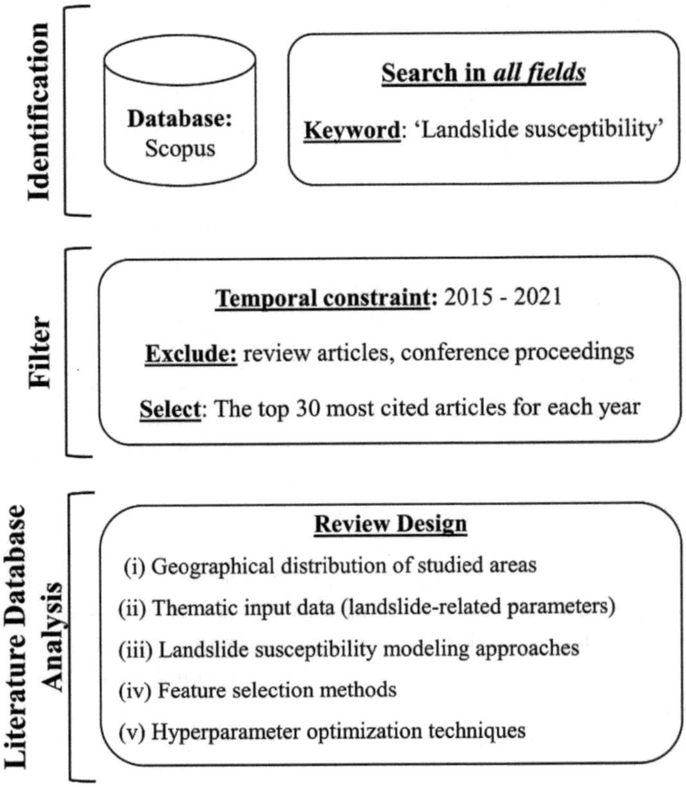

Fig. 6.1 Methodological flowchart of literature review design

susceptibility mapping work was undertaken, rather than the location of the author's affiliation.

The literature survey revealed that a significant proportion (76.67%) of the reviewed articles were conducted in Asia, implying the continent by far the most studied geographical site (with 161 articles). Within the rest of the literature database collection, the country where the most documented articles was Europe (16.67%), followed by Africa (3.81%). North America (0.95%), and Australia (0.95%) had the lowest number of published articles. On the other hand, two articles were found in the database that examined landslide susceptibility globally. All reviewed articles covered 32 various countries across the world where China had the largest number of case studies with 66 articles. The country-wise analysis also showed that approximately 84% of the articles were performed in 10 countries: China (31%), Iran (13%), India (8%), Italy (8%), Vietnam (7%), Japan (4%), South Korea (4%), Turkey (4%), Algeria (3%), and Saudi Arabia (2%) (Fig. 6.2).

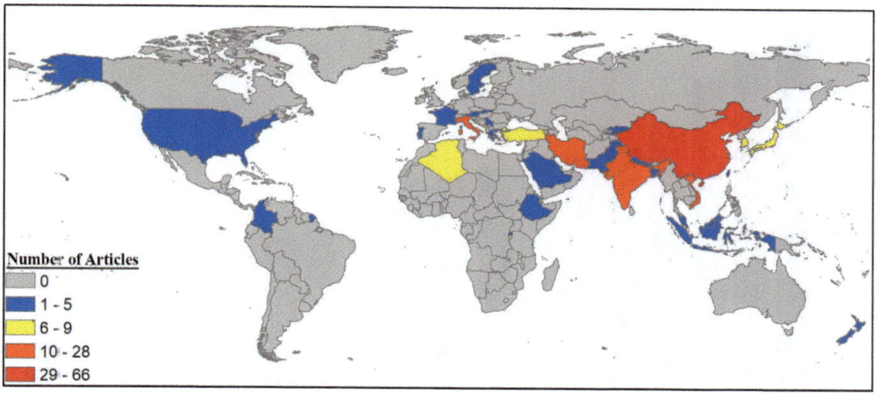

Fig. 6.2 Geographical distribution map of the test sites in reviewed 210 articles published between 2015 and 2021

6.5.2 *Landslide Conditioning Factors*

In the literature survey, landslide conditioning factors were reviewed under both the singular-context and generalized seven major categories (i.e., topographical, geological, hydrological, environmental, anthropogenic, seismic, and meteorological). It is noteworthy that the main aim was to present terminologically more generalized nomenclatures since some factors were named with different definitions but referring to the same concepts (e.g., slope aspect/aspect or slope gradient/slope), implying terminological non-uniformity and confusion in the current literature. Additionally, some parameters used in reviewed case studies, especially proximity-based factors such as distance to channels, distance to streams, and distance to rivers, can vary based on site-specific conditions, yet such factors usually correspond to almost similar settings in essence. Thus, terminologically more generalized nomenclatures were provided for these factors to provide a common taxonomy. For instance, the term "distance to water bodies" was used for the conditioning factors corresponding to the distances to the systems representing the hydrological characteristics of the test sites.

One of the most remarkable findings is that a total of 151 various landslide conditioning factors were inputted in the reviewed 210 articles. This obviously points out that there is a clear dearth of consensus on the selection of thematic input data among researchers. Figure 6.3 summarizes the mostly preferred 20 landslide-related parameters, of which the slope was considered a dominant factor in all examined studies without exception. Also, aspect (93.33%), lithology (91.90%), and elevation (81.90%) were the most commonly used parameters in the literature database. Furthermore, the results revealed that 12 factors (slope, aspect, elevation, TWI, plan curvature, profile curvature, SPI, curvature, STI, slope length, TRI, and distance to lineaments) out of the top 20 were topographical or hydrological features that are directly derived from the digital elevation model (DEM). On the other hand, findings

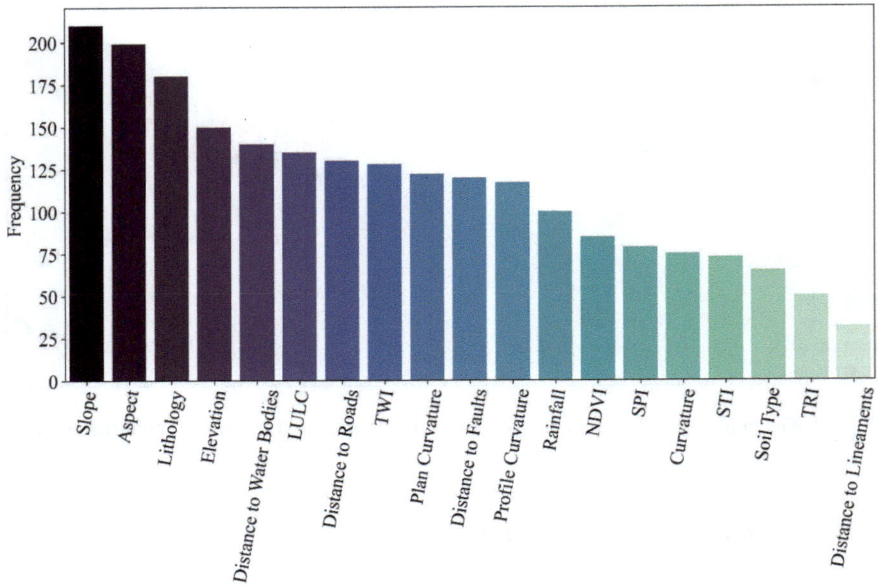

Fig. 6.3 Distribution of the top 20 landslide predisposing factors used in a total of 210 articles between 2015 and 2021

from the literature review showed that there exists a significant dominance of environmental (33.77%) and topographical-related (32.45%) factors under the general categories. Additionally, distance to water bodies and rainfall were most frequently considered hydrological conditioning factors by the authors.

6.5.3 Susceptibility Modeling Approaches

During the time period covered by the literature survey, a plethora of modeling approaches from conventional types to state-of-the-art algorithms has been continuously applied by the researchers to spatially predict landslide susceptibility. The susceptibility modeling approaches mainly refer to the computational algorithms or concepts exploited to predict landslide susceptibility in a given area. In the scope of the literature review, they were studied under three categories: heuristic (e.g., AHP), traditional statistical methods (e.g., FR), and ML/DL algorithms. Several specific approaches were expressed under the generalized topics. For instance, particular types of tree-based methods C4.5 and chi-square automatic interaction detection were represented under the decision tree.

A total of 54 different ML algorithms were identified in the database collection, of which the LR with 70 articles (e.g., Chen et al. 2017a, b; Lombardo et al. 2015; Pham et al. 2021a, b) was the most commonly applied susceptibility modeling approach (Fig. 6.4). In other words, one out of every three articles used the LR either as the

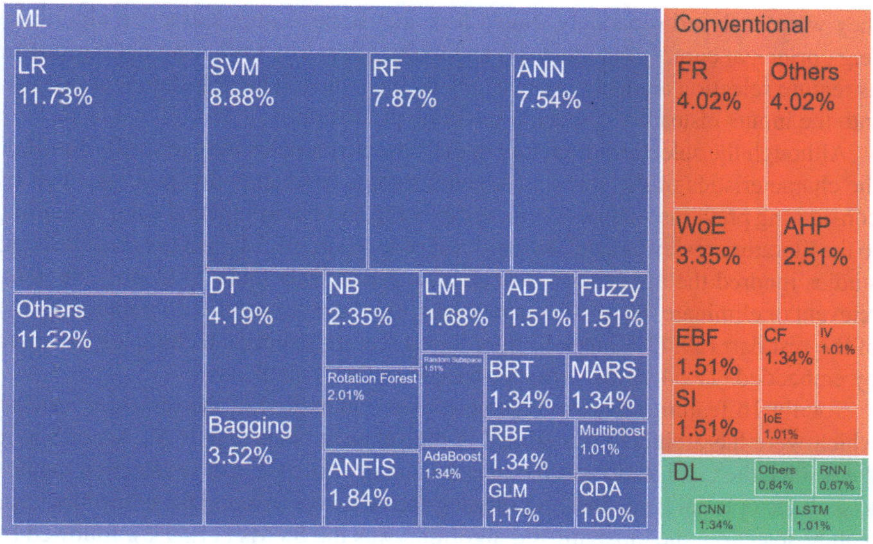

Fig. 6.4 Percentage distribution of landslide susceptibility modeling approaches (machine learning, deep learning, and conventional statistical-based) on treemap

main or benchmark technique. This wider adoption is probably due to its simplicity and structure suitable for binary problems. The SVMs with 53 articles (e.g., Chen et al. 2018; Pham et al. 2016; Phong et al. 2021), RF with 47 articles (e.g., Park and Kim 2019; Pourghasemi and Kerle 2016; Trigila et al. 2015), and ANN with 45 articles (e.g., Dou et al. 2019; Nhu et al. 2020; Youssef and Pourghasemi 2021) were other frequently utilized approaches, respectively. However, the DL algorithms were applied in only a marginal proportion (7.08%) of the reviewed articles despite their effectiveness over ML models in many studies of different domains (Abrol et al. 2021; Kratsch et al. 2021; Sujatha et al. 2021; Günen 2022). Within the compiled literature database, the emergence of the DL algorithms occurred in 2018 (Xiao et al. 2018), later followed an increasing trend. Nevertheless, it is still difficult to argue that the DL paradigm become an indispensable part of the LSM procedures, given its marginal utilization within the reviewed articles. On the other hand, traditional statistical methods were employed in relatively a modest portion (25.71%) of the reviewed articles. Among them, the FR method was the most widely utilized with 24 articles followed by the WoE, AHP, evidential belief function, and statistical index.

6.5.4 Feature Selection for Optimal Subset

Whether the feature selection techniques were employed in the reviewed articles is another subject addressed by this literature review. Since feature selection applications were surveyed under three main topics, namely, filter, wrapper, and embedded, they were analyzed separately. Noted here that some feature extraction algorithms (e.g., principal component analysis) and multicollinearity tests were not considered as feature selection algorithms since they mainly concern the translation of raw data into the inputs instead of a specific selection procedure.

Although the selection of factors in the reviewed articles was generally based on the characteristics of the study area, some studies conducted this process either by considering previous studies or utilizing existing geo-environmental data. According to the literature review results, a significant proportion (67.14%) of the analyzed case studies ignored the application of feature selection techniques to identify the most relevant or eliminate the irrelevant variables. On the other hand, a dominant majority of the rest part (59.42%) applied filter-based feature selection techniques, followed by embedded (27.54%) and wrapper-based (13.04%) methods.

Totally 16 feature selection techniques were identified in the reviewed articles, which were hierarchically visualized in the sunburst chart (Fig. 6.5). According to the results, the Gain Ratio was the most frequently utilized feature selection method to seek the optimal landslide conditioning factor subset. Among the embedded-type feature selection, the most utilized algorithm was the SVMs (11.59%), followed by RF (8.70%), and LASSO (2.90%). On the other hand, the analysis showed that the general tendency toward wrapper-based methods was less than their counterpart.

6.5.5 Hyperparameter Optimization Algorithms

Another important processing step for the ML/DL algorithms is the hyperparameter optimization that can significantly affect the quality of the established predictive models. In the ML community, a great deal of studies has been performed and a series of hyperparameter optimization algorithms has been introduced to effectively search for the optimal values of specific hyperparameters of the predictive learning methods. In this specific analysis, the articles using susceptibility modeling techniques that do not require a hyperparameter configuration were out of the review. On the other hand, if there was a method requiring an optimization process among the modeling methods applied in the articles, it was included in the scope of the review.

Considering the perspective of LSM, the literature review showed that 64.32% of the reviewed articles utilized non-optimized predictive models, and the trial-and-error concept was adopted in 10.81% of the reviewed articles (e.g., Pham et al. 2017; Wang et al. 2019; Zhou et al. 2018). These findings clearly suggested that a limited research effort was devoted to the hyperparameter tuning process. On the other hand, only 24.86% of the research works applied an automated hyperparameter optimization pipeline to get the optimal performance of learning algorithms (e.g., Al-Najjar and Pradhan 2021; Sun et al. 2020; Truong et al. 2018). According to the results, the most common algorithms were the cross-validation technique (e.g., Colkesen et al. 2016; Wang et al. 2016), followed by grid search (e.g., Hong et al. 2016, 2015) and particle swarm optimization (e.g., Mallick et al. 2021; Moayedi et al. 2019).

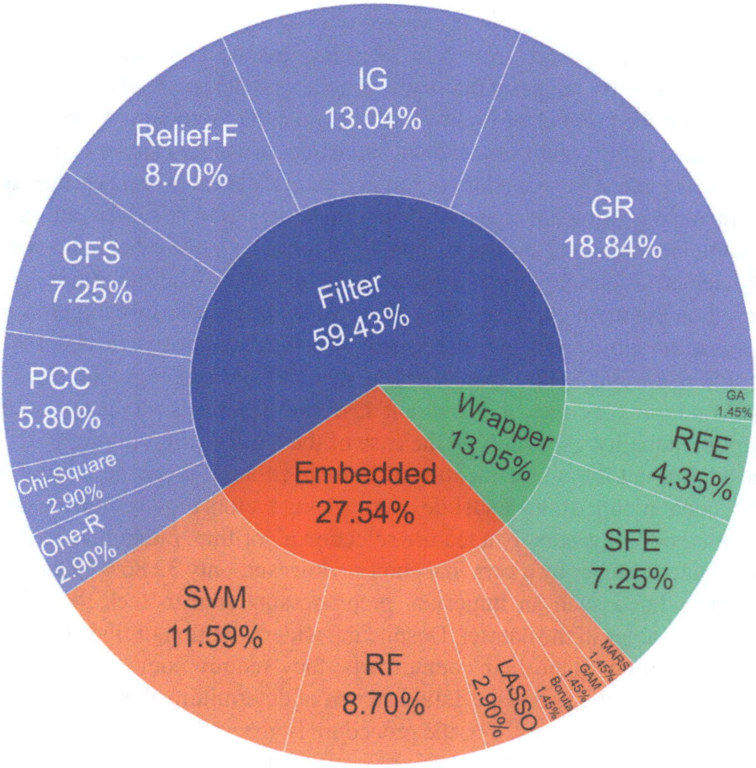

Fig. 6.5 Percentage distribution of feature selection algorithms (filter, wrapper, and embedded) on sunburst map

6.6 Open Questions and Challenges

The particular focus of this review was laid on the algorithmic elements of the AI family, including ML/DL methods, feature selection techniques, and hyperparameter optimization algorithms for landslide susceptibility mapping practices. In addition to conducting an in-depth literature review, this study aimed to enlighten several open questions that were not fully addressed in the current literature. The literature survey unveiled a series of abundant room for further progress yet rarely tapped by the scientific community, which can be also considered as potential research roadmaps and recommendations guiding the trajectory of future studies.

- **What is the optimal number of landslide conditioning factors?**

The optimal number of thematic input data for LSM has lately become a controversial issue in the community. Reviewing a total of 565 peer-review articles from 1983 to 2016, Reichenbach et al. (2018) reported that an average of nine thematic features ranging from 2 to 22 were used per single susceptibility model. Similarly, the literature survey carried out in this study showed that the number of landslide conditioning factors ranged from 5 to 29 in the reviewed studies, with an average of 12. This implicitly suggests that there is no apparent threshold concurred by the researchers on the number of input variables since the factors may vary depending on some site-specific conditions. Indeed, landslides are highly site-specific activities and may be triggered by secondary or tertiary natural disasters (e.g., earthquakes and floods) in some geographic areas, while in others they may be driven entirely by the topographic or geological structure of the local terrain. On the other hand, a majority of the reviewed articles highlighted that the selection of landslide conditioning factors was based on the main characteristics of the study area under investigation, yet this can increase the subjectivity of the procedure. Further, adopting such a kind of dataset acquisition requires a certain level of expertise in the studied geographical area. Even if these conditions are met, the resultant susceptibility maps might not necessarily be of optimal quality, especially in the case of the ML/DL algorithms.

Considering all the above-mentioned in their entirety, it can be argued that the site-dependent nature of the landslide conditioning factors requires the adoption of a specific methodology, namely a feature selection algorithm, rather than determining a critical factor number by a personal choice. However, only 32.86% of the reviewed articles considered such an important preprocessing step to seek the relevant or significant variables in the dataset. Despite the wider adoption of filter-based feature selection algorithms than their counterparts, they require some challenging decisions on the part of the user, especially in terms of determining the correct number of landslide-related parameters. Yet, the procedure is generally carried out by applying some heuristic evaluations on the basis of professional expertise or selecting the top k variables from the obtained rankings without statistically determining a critical threshold, which can restrict their effective use. However, this process needs statistically more robust approaches to determine the cut-off value instead of commonly adopted ones. In such cases, the correlation-based feature selection (CFS) strategy can be considered a reasonable alternative (Mezaal et al. 2017; Kavzoglu and Bilucan 2023). The CFS provides to identify a specific threshold value or to choose the proper number of factors in the ordered ranking list in the case of filter-based techniques. However, most studies regard all factors having merit scores over zero as significant, not eliminating any factor. This approach disregards the main philosophy of feature selection, which is the elimination of insignificant or unrelated factors.

- **Why does landslide susceptibility prediction need artificial intelligence?**

Prior to the onset of the emergence and application of AI, landslide susceptibility prediction mainly adopted conventional statistical-based models, including heuristic models based on expert knowledge (e.g., AHP), interpretative (e.g., WOE), or simple probabilistic (e.g., FR) models. These models construct an apparent relationship between landslide events and their contributing parameters by assigning weights, sometimes without requiring entire historical landslide datasets or even non-landslide samples. Even though the conventional approaches are easy-to-use and simple to develop, they mainly suffer from non-linear patterns of landslides driven by multifaceted determinants such as anthropogenic influences, climatic conditions, physical principles, and laws. Therefore, developing models through the application of such approaches might lead to suboptimal prediction results.

As a general premise, AI tools significantly minimize the requirement of prior or expert knowledge on susceptibility prediction by providing an automated decision-making process since they extract relevant knowledge by learning the underlying behavior of real-world problems with a set of historical data. Unlike the non-AI approaches, they output reproducible results by avoiding exhaustive data searching and streamlining the execution order of the program to fit the search required for a specific solution, which is especially significant for geospatial analysis involving large spatial data. They also provide computationally efficient structures and accurate models, as flexible and transferable environments.

Referring also to the previous question, AI can provide explainability of black-box models with a number of approaches (e.g., SHAP and LIME). They enable both importance scores for each feature and their positive or negative contributions to the predictive models. This explainability further provides an understanding of the underlying reason why learning algorithms result in such estimations. With the application of feature selection techniques, the triggering factors of mass movement activities can be also elaborately analyzed by prioritizing the high-impact parameters in local conditions, which significantly supports the landslide mitigation studies.

- **What is the role of ensemble learning paradigm?**

According to the literature review results, the top two most applied ML approaches were the LR and SVMs. These are single or so-called individual ML algorithms that do not adopt ensemble intelligence considering multiple models of different learning schemes. The individual learning algorithms might tend to produce biased results and fail to reach desired performance in some cases (Dong et al. 2020; Kavzoglu and Teke 2022b) since the prediction is made by only a single base model. More clearly, a susceptibility map can be produced with high accuracy by an individual learning algorithm but not by another, implying the algorithm-specific characteristics of landslide susceptibility prediction problems. Contrarily, a community of base learners participates in the prediction process and the results are combined with a set of techniques (e.g., majority voting). For instance, even if there is an ill-suited base learner in the community that produces poor performance, others can dominate it and play a decisive role in the overall performance. Within the collected database, there exists a variety of ensemble learning algorithms such as, RF (Trigila et al. 2015), AdaBoost (Pham et al. 2017), rotation forest (Chen et al. 2017b), LogitBoost (Kadavi et al. 2018), gradient boosting machines (Merghadi et al. 2018), multiboost (Pham et al. 2019), LightGBM (Merghadi et al. 2020), extreme gradient boosting machines (Can et al. 2021).

The second point of view is associated with the uncertainty about which susceptibility modeling algorithm to choose from among the wide array of available options. The ensemble learning paradigm can significantly minimize the chance of choosing the non-proper model for a particular prediction problem by strategically combining learning algorithms. This diversity both improves the predictive performance of the models and reduces the bias and influence of possible incorrect estimations. Another important role of the ensemble learning paradigm originates from computational reasons. According to Qiu et al. (2014), numerous learning algorithms, including neural networks and decision trees, include some sort of local search in their operation, which can widely lead to locally optimal solutions, yet ensemble techniques demonstrate their benefits in this case, by doing several local searches from various beginning locations.

- **Is hyperparameter optimization a prerequisite?**

The recent development of available computing systems has brought along the development of theoretically more advanced and robust artificial intelligence algorithms, often characterized by a non-parametric nature that assumes no prior knowledge. As their sophistication increased, the complexity of the configurations regulating their internal structure generally increased. In such cases, the relationship between the learning algorithm and its hyperparameters might remain uncertain. These should be set before the initialization of the learning process, otherwise, the learning algorithm can produce poor predictive performance.

The literature review showed that the optimal values for hyperparameters of specific learning algorithms were dominantly determined through a process of trial-and-error or non-optimized default settings. However, these approaches suffer from several deficiencies, generally associated with the issues faced in the training and prediction phases, which restrict their common acceptability. For instance, a trial-and-error strategy demands a certain level of prior knowledge and experience to specify the optimal hyperparameter values within a limited time (Yang and Shami 2020). Therefore, manual configuration can be impractical for the prediction of real-world scenarios under some conditions: (i) learning algorithms involving a large number of hyperparameters (e.g., ANN and ensemble models), (ii) black-box algorithms having a sophisticated inner frame, (iii) nonlinear relationship among hyperparameters (Steinholtz 2018). Application of the trial-and-error approach in such circumstances can lead to both a decrease in the time-efficiency of the models and inconsistent and poor generalization. On the other hand, using the default settings of a specific tool, such as libraries or packages embedded in programming languages or specific off-the-shelf software, may not be suitable to characterize the problem under consideration. Thus, a non-optimized model does not always guarantee the optimal prediction accuracy, casting serious doubts on the reliability of the produced landslide susceptibility maps.

Instead of the above-mentioned approaches, precise adjustment of hyperparameters needs an automated search pipeline to get optimal performance. In the literature, a variety of optimization algorithms, from metaheuristic-based to model-dependent models, has been introduced, and some of the articles compared the performances achieved by the non-optimized default and optimized models to properly measure the contribution of hyperparameter optimization algorithms to the prediction results. For example, Sun et al. (2021) applied Bayesian optimization to optimally set the hyperparameters of LR and RF algorithms in the production of landslide susceptibility maps for Fengjie County, China. The optimized RF model produced an important enhancement of up to 10% in terms of AUC score. Likewise, Rong et al.

(2021) tuned the hyperparameter of DL algorithms by adopting the tree-structured Parzen estimator. The optimization process improved the accuracy of models by up to approximately 4%. Kavzoglu and Teke (2022a) conducted a study assessing various advanced optimization algorithms, such as random search, Bayesian optimization with Gaussian Process, Bayesian optimization with Tree-structured Parzen Estimator, genetic algorithm, and the Hyperband method. Their research aimed to optimize hyperparameters within the extreme gradient boosting algorithm for better spatial prediction of landslides. The results revealed that the optimized models yielded a notable enhancement of up to 13% in overall accuracy compared to the model with default settings. There exist also conceptually similar studies (Adnan et al. 2020; Brock et al. 2020) utilizing different hyperparameter optimization techniques (e.g., grid search and random search) in the current literature. It is evident that the hyperparameter optimization techniques increase the predictive performance of the learning algorithms, and therefore, the reliability and quality of the produced susceptibility maps, albeit to different degrees. Consequently, the specific design and effective implementation of the above-mentioned algorithms are critical for the model's capabilities of generalization and learning since they have significantly affected the model's success.

- **Future roadmaps for landslide susceptibility mapping studies**

Expanding on the open questions highlighted through this extensive literature review, future studies may consider the crucial aspects noted below to increase the quality of artificially intelligent landslide susceptibility maps. First, the literature review clearly delineates the non-uniformity of the selection of landslide-related parameters in the literature, often dictated by the idiosyncratic nature of the study area(s) under scrutiny. Nonetheless, it is still important to underscore that the selection of parameters, especially within data-driven methodologies, necessitates a priori analysis. A parameter deemed pivotal in triggering or contributing to landslide initiation in practical scenarios may, at times, fail to improve the model performance as anticipated. In fact, it might even diminish it in certain instances. Conducting an initial examination via filter-based approaches proves valuable in identifying potential issues due to their lower computational complexity compared to analogous methods. For a deeper analysis, researchers might opt for wrapper or embedded feature selection techniques due to their ability to offer a nuanced and thorough understanding of feature relevance within predictive models. These methods go beyond mere feature independence evaluations seen in filter-based approaches by directly integrating the evaluation of features into the model training process. Wrappers, in particular, enable an interactive selection that considers not only individual feature importance but also their combined impact on model performance. This approach allows for a deeper exploration of feature interactions and dependencies, crucial for capturing complex relationships within data. Additionally, the model-centric nature of wrappers tailors feature selection to the intricacies of the chosen algorithm, optimizing feature subsets for improved predictive accuracy. Through an iterative process, these techniques ensure adaptability to changing datasets, providing researchers with a dynamic and precise means to enhance the quality and robustness of predictive models.

Secondly, while feature selection techniques elucidate the impact of landslide-related parameters on occurrences, they often fall short in pinpointing specific zones where these impacts manifest. In more detail, the results of feature selection can enlighten which landslide-related parameters have more impact on landslide occurrences but not in which specific zone. This kind of inquiry is also important since sometimes a landslide-related parameter has more impact on a particular landslide event, but another parameter can have more impact on another landslide phenomenon. Explainable artificial intelligence (XAI), albeit not a main concern of this review, can be highly effective in seeking an answer to this inquiry because it offers transparency in the decision-making process of black-box models. By utilizing XAI techniques such as feature importance visualization, SHAP, or LIME, researchers can delve deeper into understanding not only which parameters hold more sway in landslide occurrences but also discern the spatial distribution of their influence (Pradhan et al. 2023; Zhang et al. 2023).

Third, future research should prioritize the accurate structuring of optimal configuration design for data-driven susceptibility modeling approaches since it is evident that optimal hyperparameter values can significantly increase the resultant model performance. However, the application of exhaustive methods like grid search can be time-consuming, particularly with high-dimensional geospatial datasets. In such scenarios, researchers might consider employing random search, albeit without guaranteeing optimal predictive performance. Additionally, despite the risk of capturing local optima, evolutionary methods such as genetic algorithms, as addressed in the literature, can yield more robust results (Kavzoglu and Teke 2022a).

6.7 Review Summary and Conclusions

This review was mainly conducted in response to the need to fill a series of knowledge gaps on landslide susceptibility mapping practices and a deeper look at the literature provided a wide range of inferences on the trend of literature towards AI tools. A summary and the main conclusions of the literature survey were drawn together and summarized as follows:

- The literature survey showed that the number of landslide conditioning factors ranged from 5 to 29 in the reviewed studies, with an average of 12. The slope was found to be the most intensively used thematic input in all the case studies examined without exception, followed by aspect (93.33%), lithology (91.90%), and elevation (81.90%). On the other hand, the literature review showed how widely exploited the DEM is for the production of spatial covariates. When analyzed on a general basis, environmental (33.77%) and topographic factors (32.45%) were the most common categories in estimating landslide susceptibility.

- The literature review results explicitly highlighted that the LR was used in one of the three articles and was the most common modeling approach in the context of landslide susceptibility mapping, followed by SVMs (53 articles), RF (47 articles), and ANN (45 articles). Also, it should be noted that there has been an increasing trend recently in the adoption of DL algorithms to predict landslide susceptibility.
- The DL algorithms were considered in only 7% of the reviewed articles while the conventional methods used in 25.71% of the case studies are more inclined. This finding can be used to reach a conclusion that DL has not yet significantly penetrated the landslide susceptibility mapping practices but there is an observable upward trajectory in its utilization within LSM studies.
- A significant portion (67.14%) of the reviewed articles disregarded to apply a feature selection technique for factor optimization or prioritization. On the other hand, there was a dominant tendency for filter-based feature selection in the other parts of the database. A total of 16 various feature selection methods were identified, of which the Gain Ratio is the most common one, followed by Information Gain (13.04%), SVM (11.59%), RF (8.70%), and Relief-F (8.70%).
- Excluding non-optimized models and trial-and-error approaches, only about 25% of the reviewed studies applied hyperparameter optimization procedures, of which the cross-validation technique (19 articles) was the most frequently used one. Considering automated hyperparameter optimization pipelines, grid search was found to be the most common algorithm within the literature database. Following closely behind were particle swarm optimization, and gray wolf optimization.
- The literature review suggested that a general consensus on many significant topics on the sufficient number of conditioning factors, and the use of hyperparameter optimization and DL algorithms has not been reached, implicating available research gaps and areas in the field of landslide susceptibility mapping. Additionally, intelligence XAI tools might become highly favored, especially considering the rising popularity of algorithms featuring intricate inner architecture designs that pose challenges in directly interpreting their outcomes.

The reliance solely on the Scopus platform for literature review might inadvertently limit the perspectives presented in this study. To offer a more comprehensive outlook, future studies could adopt a broader approach by encompassing multiple database platforms, such as Web of Science, Redalyc, and REDIB. This expansion could potentially mitigate the artifact of a low representation of studies from regions like South America, Central America, and the Caribbean, which could be a consequence of the database selection process. The selection criteria focused on the top 30 articles for each year, culminating in the consideration of a total of 210 articles. However, it is crucial to note that quantitative values obtained from literature searches may not completely mirror the entirety of the landscape, and extrapolating these results to represent the entire literature may not be entirely accurate. Additionally, there exist a series of other important topics (e.g., explainability and uncertainty quantification) imposed by AI affecting the quality and interpretability of landslide susceptibility maps yet it is important to note that the focus of this review is limited to feature selection and hyperparameter optimization techniques. Future literature reviews could explore the recently emerging topic of XAI in the context of landslide susceptibility maps, specifically delving into the interpretability of the models employed for landslide susceptibility map production. Furthermore, uncertainty quantification plays a pivotal role in comprehending and communicating the reliability and robustness of susceptibility maps. While it may not inherently enhance map quality to the same degree as feature selection or hyperparameter optimization, it provides crucial insights into model predictions' limitations and variations. Hence, forthcoming literature surveys should consider incorporating discussions on uncertainty quantification as a crucial aspect of evaluating landslide susceptibility mapping methodologies.

References

Abrol A, Fu Z, Salman M et al (2021) Deep learning encodes robust discriminative neuroimaging representations to outperform standard machine learning. Nat Commun 12:353. https://doi.org/10.1038/s41467-020-20655-6

Adnan MSG, Rahman MS, Ahmed N et al (2020) Improving spatial agreement in machine learning-based landslide susceptibility mapping. Remote Sens 12:3347. https://doi.org/10.3390/rs12203347

Akinci H, Yavuz Ozalp A (2021) Landslide susceptibility mapping and hazard assessment in Artvin (Turkey) using frequency ratio and modified information value model. Acta Geophys 69:725–745. https://doi.org/10.1007/s11600-021-00577-7

Alcántara-Ayala I, Murray V, Daniels P, McBean G (2017) International Council for Science (ICSU)—on the future challenges for the integration of science into international policy development for landslide disaster risk reduction BT—advancing culture of living with landslides. In: Sassa K, Mikoš M, Yin Y (eds). Springer International Publishing, Cham, pp 143–154

Al-Najjar HH, Pradhan B (2021) Spatial landslide susceptibility assessment using machine learning techniques assisted by additional data created with generative adversarial networks. Geosci Front 12:625–637. https://doi.org/10.1016/j.gsf.2020.09.002

Aslam B, Maqsoom A, Khalil U et al (2022) Evaluation of different landslide susceptibility models for a local scale in the Chitral District, Northern Pakistan. Sensors 22:3107. https://doi.org/10.3390/s22093107

Brock J, Schratz P, Petschko H et al (2020) The performance of landslide susceptibility models critically depends on the quality of digital elevation models. Geomat Nat Hazards Risk 11:1075–1092. https://doi.org/10.1080/19475705.2020.1776403

Bui DT, Tsangaratos P, Nguyen V-T et al (2020) Comparing the prediction performance of a deep learning neural network model with conventional machine learning models in landslide susceptibility assessment. Catena 188:104426. https://doi.org/10.1016/j.catena.2019.104426

Can R, Kocaman S, Gokceoglu C (2021) A comprehensive assessment of XGBoost algorithm for landslide susceptibility mapping in the upper basin of Ataturk Dam, Turkey. Appl Sci 11:4993. https://doi.org/10.3390/app11114993

Chauhan NK, Singh K (2018) A review on conventional machine learning vs deep learning. In: 2018 International conference on computing, power and communication technologies (GUCON), Greater Noida, India, pp 347–352

Chen W, Pourghasemi HR, Zhao Z (2017a) A GIS-based comparative study of Dempster–Shafer, logistic regression and artificial neural network models for landslide susceptibility mapping. Geocarto Int 32:367–385. https://doi.org/10.1080/10106049.2016.1140824

Chen W, Shirzadi A, Shahabi H et al (2017b) A novel hybrid artificial intelligence approach based on the rotation forest ensemble and naïve Bayes tree classifiers for a landslide susceptibility assessment in Langao County, China. Geomat Nat Hazards Risk 8:1955–1977. https://doi.org/10.1080/19475705.2017.1401560

Chen W, Pourghasemi HR, Naghibi SA (2018) A comparative study of landslide susceptibility maps produced using support vector machine with different kernel functions and entropy data mining models in China. Bull Eng Geol Environ 77:647–664. https://doi.org/10.1007/s10064-017-1010-y

Colkesen I, Kavzoglu T (2018) Selection of optimal object features in object-based image analysis using filter-based algorithms. J Indian Soc Remote Sens 46:1233–1242. https://doi.org/10.1007/s12524-018-0807-x

Colkesen I, Sahin EK, Kavzoglu T (2016) Susceptibility mapping of shallow landslides using kernel-based Gaussian process, support vector machines and logistic regression. J Afr Earth Sci 118:53–64. https://doi.org/10.1016/j.jafrearsci.2016.02.019

Dai FC, Lee CF (2001) Terrain-based mapping of landslide susceptibility using a geographical information system: a case study. Can Geotech J 38:911–923. https://doi.org/10.1139/t01-021

Dai FC, Lee CF, Li J, Xu ZW (2001) Assessment of landslide susceptibility on the natural terrain of Lantau Island, Hong Kong. Environ Geol 40:381–391. https://doi.org/10.1007/s002540000163

Dao DV, Jaafari A, Bayat M et al (2020) A spatially explicit deep learning neural network model for the prediction of landslide susceptibility. Catena 188:104451. https://doi.org/10.1016/j.catena.2019.104451

Di Napoli M, Carotenuto F, Cevasco A et al (2020) Machine learning ensemble modelling as a tool to improve landslide susceptibility mapping reliability. Landslides 17:1897–1914. https://doi.org/10.1007/s10346-020-01392-9

Dias HC, Hölbling D, Grohmann CH (2021) Landslide susceptibility mapping in Brazil: a review. Geosci 11:1–15. https://doi.org/10.3390/geosciences11100425

Dong X, Yu Z, Cao W et al (2020) A survey on ensemble learning. Front Comput Sci 14:241–258. https://doi.org/10.1007/s11704-019-8208-z

Dou J, Yunus AP, Bui DT et al (2019) Evaluating GIS-based multiple statistical models and data mining for earthquake and rainfall-induced landslide susceptibility using the lidar dem. Remote Sens 11:638. https://doi.org/10.3390/rs11060638

El Aboudi N, Benhlima L (2016) Review on wrapper feature selection approaches. In: 2016 International conference on engineering and MIS (ICEMIS), Agadir, Morocco, pp 1–5

Feurer M, Hutter F (2019) Hyperparameter optimization. In: Hutter F, Kotthoff L, Vanschoren J (eds) Automated machine learning: methods, systems, challenges. Springer International Publishing, Cham, pp 3–33

Gómez H, Kavzoglu T (2005) Assessment of shallow landslide susceptibility using artificial neural networks in Jabonosa River Basin, Venezuela. Eng Geol 78:11–27. https://doi.org/10.1016/j.enggeo.2004.10.004

Günen MA (2022) Performance comparison of deep learning and machine learning methods in determining wetland water areas using EuroSAT dataset. Environ Sci Pollut Res 29:21092–21106. https://doi.org/10.1007/s11356-021-17177-z

Guo Y, Liu Y, Oerlemans A, et al (2016) Deep learning for visual understanding: a review. Neurocomputing 187:27–48. https://doi.org/10.1016/j.neucom.2015.09.116

He L, Wu X, He Z et al (2023) Susceptibility assessment of landslides in the loess plateau based on machine learning models: a case study of Xining city. Sustainability 15:14761. https://doi.org/10.3390/su152014761

Hong H, Pradhan B, Jebur MN et al (2016) Spatial prediction of landslide hazard at the Luxi area (China) using support vector machines. Environ Earth Sci 75:1–14. https://doi.org/10.1007/s12665-015-4866-9

Hong H, Liu J, Bui DT et al (2018) Landslide susceptibility mapping using J48 decision tree with AdaBoost, bagging and rotation forest ensembles in the Guangchang area (China). Catena 163:399–413. https://doi.org/10.1016/j.catena.2018.01.005

Hong H, Liu J, Zhu A-X (2019) Landslide susceptibility evaluating using artificial intelligence method in the Youfang district (China). Environ Earth Sci 78:488. https://doi.org/10.1007/s12665-019-8415-9

Hong H, Pradhan B, Xu C, Tien Bui D (2015) Spatial prediction of landslide hazard at the Yihuang area (China) using two-class kernel logistic regression, alternating decision tree and support vector machines. Catena 133:266–281. https://doi.org/10.1016/j.catena.2015.05.019

Huang Y, Zhao L (2018) Review on landslide susceptibility mapping using support vector machines. Catena 165:520–529. https://doi.org/10.1016/j.catena.2018.03.003

Jaafari A, Panahi M, Pham BT et al (2019) Meta optimization of an adaptive neuro-fuzzy inference system with grey wolf optimizer and biogeography-based optimization algorithms for spatial prediction of landslide susceptibility. Catena 175:430–445. https://doi.org/10.1016/j.catena.2018.12.033

Kadavi PR, Lee C-W, Lee S (2018) Application of ensemble-based machine learning models to landslide susceptibility mapping. Remote Sens 10:1252. https://doi.org/10.3390/rs10081252

Kalantar B, Pradhan B, Amir Naghibi S et al (2018) Assessment of the effects of training data selection on the landslide susceptibility mapping: a comparison between support vector machine (SVM), logistic regression (LR) and artificial neural networks (ANN). Geomat Nat Hazards Risk 9:49–69. https://doi.org/10.1080/19475705.2017.1407368

Kavzoglu T, Bilucan F (2023) Effects of auxiliary and ancillary data on LULC classification in a heterogeneous environment using optimized random forest algorithm. Earth Sci Inform 16:415–435. https://doi.org/10.1007/s12145-022-00874-9

Kavzoglu T, Mather PM (2002) The role of feature selection in artificial neural network applications. Int J Remote Sens 23:2919–2937. https://doi.org/10.1080/01431160110107743

Kavzoglu T, Teke A (2022a) Advanced hyperparameter optimization for improved spatial prediction of shallow landslides using extreme gradient boosting (XGBoost). Bull Eng Geol Environ 81:201. https://doi.org/10.1007/s10064-022-02708-w

Kavzoglu T, Teke A (2022b) Predictive performances of ensemble machine learning algorithms in landslide susceptibility mapping using random forest, extreme gradient boosting (XGBoost) and natural gradient boosting (NGBoost). Arab J Sci Eng 47:7367–7385. https://doi.org/10.1007/s13369-022-06560-8

Kavzoglu T, Yilmaz EO (2022) Analysis of patch and sample size effects for 2D–3D CNN models using multiplatform dataset: hyperspectral image classification of ROSIS and Jilin-1 GP01 imagery. Turk J Electr Eng Comput Sci 30:2124–2144

Kavzoglu T, Kutlug Sahin E, Colkesen I (2015a) An assessment of multivariate and bivariate approaches in landslide susceptibility mapping: a case study of Duzkoy district. Nat Hazards 76:471–496. https://doi.org/10.1007/s11069-014-1506-8

Kavzoglu T, Kutlug Sahin E, Colkesen I (2015b) Selecting optimal conditioning factors in shallow translational landslide susceptibility mapping using genetic algorithm. Eng Geol 192:101–112. https://doi.org/10.1016/j.enggeo.2015.04.004

Kavzoglu T, Teke A, Yilmaz EO (2021) Shared blocks-based ensemble deep learning for shallow landslide susceptibility mapping. Remote Sens 13:4776. https://doi.org/10.3390/rs13234776

Khanna K, Martha TR, Roy P, Kumar KV (2021) Effect of time and space partitioning strategies of samples on regional landslide susceptibility modelling. Landslides 18:2281–2294. https://doi.org/10.1007/s10346-021-01627-3

Kilkenny MF, Robinson KM (2018) Data quality: "Garbage in—garbage out." Heal Inf Manag J 47:103–105. https://doi.org/10.1177/1833358318774357

Kratsch W, Manderscheid J, Röglinger M, Seyfried J (2021) Machine learning in business process monitoring: a comparison of deep learning and classical approaches used for outcome prediction. Bus Inf Syst Eng 63:261–276. https://doi.org/10.1007/s12599-020-00645-0

Kutlug Sahin E, Ipbuker C, Kavzoglu T (2017) Investigation of automatic feature weighting methods (Fisher, Chi-square and Relief-F) for landslide susceptibility mapping. Geocarto Int 32:956–977. https://doi.org/10.1080/10106049.2016.1170892

Lee S, Ryu JH, Won JS, Park HJ (2004) Determination and application of the weights for landslide susceptibility mapping using an artificial neural network. Eng Geol 71:289–302. https://doi.org/10.1016/S0013-7952(03)00142-X

Lima P, Steger S, Glade T, Murillo-García FG (2022) Literature review and bibliometric analysis on data-driven assessment of landslide susceptibility. J Mt Sci 19:1670–1698. https://doi.org/10.1007/s11629-021-7254-9

Liu R, Yang X, Xu C et al (2022) Comparative study of convolutional neural network and conventional machine learning methods for landslide susceptibility mapping. Remote Sens 14:321. https://doi.org/10.3390/rs14020321

Lombardo L, Cama M, Conoscenti C et al (2015) Binary logistic regression versus stochastic gradient boosted decision trees in assessing landslide susceptibility for multiple-occurring landslide events: application to the 2009 storm event in Messina (Sicily, southern Italy). Nat Hazards 79:1621–1648. https://doi.org/10.1007/s11069-015-1915-3

Lu M (2019) Embedded feature selection accounting for unknown data heterogeneity. Expert Syst Appl 119:350–361. https://doi.org/10.1016/j.eswa.2018.11.006

Mallick J, Alqadhi S, Talukdar S et al (2021) Risk assessment of resources exposed to rainfall induced landslide with the development of GIS and RS based ensemble metaheuristic machine learning algorithms. Sustainability 13:457. https://doi.org/10.3390/su13020457

Merghadi A, Abderrahmane B, Tien Bui D (2018) Landslide susceptibility assessment at Mila basin (Algeria): a comparative assessment of prediction capability of advanced machine learning methods. ISPRS Int J Geo-Inf 7:268. https://doi.org/10.3390/ijgi7070268

Merghadi A, Yunus AP, Dou J et al (2020) Machine learning methods for landslide susceptibility studies: a comparative overview of algorithm performance. Earth-Sci Rev 207:103225. https://doi.org/10.1016/j.earscirev.2020.103225

Mezaal MR, Pradhan B, Sameen MI et al (2017) Optimized neural architecture for automatic landslide detection from high-resolution airborne laser scanning data. Appl Sci 7:730. https://doi.org/10.3390/app7070730

Moayedi H, Mehrabi M, Mosallanezhad M et al (2019) Modification of landslide susceptibility mapping using optimized PSO-ANN technique. Eng Comput 35:967–984. https://doi.org/10.1007/s00366-018-0644-0

Nhu V-H, Shirzadi A, Shahabi H et al (2020) Shallow landslide susceptibility mapping: a comparison between logistic model tree, logistic regression, naïve Bayes tree, artificial neural network, and support vector machine algorithms. Int J Environ Res Public Health 17:2749. https://doi.org/10.3390/ijerph17082749

Nirbhav, Malik A, Maheshwar et al (2023) Landslide susceptibility prediction based on decision tree and feature selection methods. J Indian Soc Remote Sens 51:771–786. https://doi.org/10.1007/s12524-022-01645-1

Park S, Kim J (2019) Landslide susceptibility mapping based on random forest and boosted regression tree models, and a comparison of their performance. Appl Sci 9:942. https://doi.org/10.3390/app9050942

Pham BT, Pradhan B, Tien Bui D et al (2016) A comparative study of different machine learning methods for landslide susceptibility assessment: a case study of Uttarakhand area (India). Environ Model Softw 84:240–250. https://doi.org/10.1016/j.envsoft.2016.07.005

Pham BT, Tien Bui D, Prakash I, Dholakia MB (2017) Hybrid integration of multilayer perceptron neural networks and machine learning ensembles for landslide susceptibility assessment at Himalayan area (India) using GIS. Catena 149:52–63. https://doi.org/10.1016/j.catena.2016.09.007

Pham BT, Jaafari A, Prakash I, Bui DT (2019) A novel hybrid intelligent model of support vector machines and the MultiBoost ensemble for landslide susceptibility modeling. Bull Eng Geol Environ 78:2865–2886. https://doi.org/10.1007/s10064-018-1281-y

Pham BT, Van Dao D, Acharya TD et al (2021a) Performance assessment of artificial neural network using chi-square and backward elimination feature selection methods for landslide susceptibility analysis. Environ Earth Sci 80:1–13. https://doi.org/10.1007/s12665-021-09998-5

Pham QB, Achour Y, Ali SA et al (2021b) A comparison among fuzzy multi-criteria decision making, bivariate, multivariate and machine learning models in landslide susceptibility mapping. Geomat Nat Hazards Risk 12:1741–1777. https://doi.org/10.1080/19475705.2021.1944330

Phong TV, Phan TT, Prakash I et al (2021) Landslide susceptibility modeling using different artificial intelligence methods: a case study at Muong Lay district, Vietnam. Geocarto Int 36:1685–1708. https://doi.org/10.1080/10106049.2019.1665715

Pokharel B, Althuwaynee OF, Aydda A et al (2021) Spatial clustering and modelling for landslide susceptibility mapping in the north of the Kathmandu Valley, Nepal. Landslides 18:1403–1419. https://doi.org/10.1007/s10346-020-01558-5

Polikar R (2012) Ensemble learning. In: Zhang C, Ma Y (eds) Ensemble machine learning: methods and applications. Springer US, Boston, MA, pp 1–34

Pourghasemi HR, Kerle N (2016) Random forests and evidential belief function-based landslide susceptibility assessment in Western Mazandaran Province, Iran. Environ Earth Sci 75:1–17. https://doi.org/10.1007/s12665-015-4950-1

Pourghasemi HR, Teimoori Yansari Z, Panagos P, Pradhan B (2018) Analysis and evaluation of landslide susceptibility: a review on articles published during 2005–2016 (periods of 2005–2012 and 2013–2016). Arab J Geosci 11:193. https://doi.org/10.1007/s12517-018-3531-5

Pradhan B (2013) A comparative study on the predictive ability of the decision tree, support vector machine and neuro-fuzzy models in landslide susceptibility mapping using GIS. Comput Geosci 51:350–365. https://doi.org/10.1016/j.cageo.2012.08.023

Pradhan B, Jebur MN (2017) Spatial prediction of landslide-prone areas through k-nearest neighbor algorithm and logistic regression model using high resolution airborne laser scanning data BT—laser scanning applications in landslide assessment. In: Pradhan B (ed). Springer International Publishing, Cham, pp 151–165

Pradhan B, Dikshit A, Lee S, Kim H (2023) An explainable AI (XAI) model for landslide susceptibility modeling. Appl Soft Comput 142:110324. https://doi.org/10.1016/j.asoc.2023.110324

Qiu X, Zhang L, Ren Y et al (2014) Ensemble deep learning for regression and time series forecasting. In: IEEE SSCI 2014—2014 IEEE symposium series on computational intelligence—CIEL 2014 IEEE symposium on computational intelligence in ensemble learning proceedings, pp 1–6. https://doi.org/10.1109/CIEL.2014.7015739

Reichenbach P, Rossi M, Malamud BD et al (2018) A review of statistically-based landslide susceptibility models. Earth-Sci Rev 180:60–91. https://doi.org/10.1016/j.earscirev.2018.03.001

Rong G, Li K, Su Y et al (2021) Comparison of tree-structured Parzen estimator optimization in three typical neural network models for landslide susceptibility assessment. Remote Sens 13:4694. https://doi.org/10.3390/rs13224694

Sagi O, Rokach L (2018) Ensemble learning: a survey. Wiley Interdiscip Rev Data Min Knowl Discov 8:1–18. https://doi.org/10.1002/widm.1249

Shi N, Li Y, Wen L, Zhang Y (2022) Rapid prediction of landslide dam stability considering the missing data using XGBoost algorithm. Landslides 19:2951–2963. https://doi.org/10.1007/s10346-022-01947-y

Sildir H, Aydin E, Kavzoglu T (2020) Design of feedforward neural networks in the classification of hyperspectral imagery using superstructural optimization. Remote Sens 12:956. https://doi.org/10.3390/rs12060956

Steinholtz OS (2018) A comparative study of black-box optimization algorithms for tuning of hyper-parameters in deep neural networks. MS thesis, Department of Computer Science, Electrical and Space Engineering, Luleå University of Technology

Sujatha R, Chatterjee JM, Jhanjhi NZ, Brohi SN (2021) Performance of deep learning vs machine learning in plant leaf disease detection. Microprocess Microsyst 80:103615. https://doi.org/10.1016/j.micpro.2020.103615

Sun D, Wen H, Wang D, Xu J (2020) A random forest model of landslide susceptibility mapping based on hyperparameter optimization using Bayes algorithm. Geomorphology 362:107201. https://doi.org/10.1016/j.geomorph.2020.107201

Sun D, Xu J, Wen H, Wang D (2021) Assessment of landslide susceptibility mapping based on Bayesian hyperparameter optimization: a comparison between logistic regression and random forest. Eng Geol 281:105972. https://doi.org/10.1016/j.enggeo.2020.105972

Thirugnanam H, Ramesh MV, Rangan VP (2020) Enhancing the reliability of landslide early warning systems by machine learning. Landslides 17:2231–2246. https://doi.org/10.1007/s10346-020-01453-z

Tonbul H, Colkesen I, Kavzoglu T (2022) Pixel- and object-based ensemble learning for forest burn severity using USGS FIREMON and Mediterranean condition dNBRs in Aegean ecosystem (Turkey). Adv Sp Res 69:3609–3632. https://doi.org/10.1016/j.asr.2022.02.051

Trigila A, Iadanza C, Esposito C, Scarascia-Mugnozza G (2015) Comparison of logistic regression and random forests techniques for shallow landslide susceptibility assessment in Giampilieri (NE Sicily, Italy). Geomorphology 249:119–136. https://doi.org/10.1016/j.geomorph.2015.06.001

Truong XL, Mitamura M, Kono Y et al (2018) Enhancing prediction performance of landslide susceptibility model using hybrid machine learning approach of bagging ensemble and logistic model tree. Appl Sci 8:1046. https://doi.org/10.3390/app8071046

Tyagi A, Kamal Tiwari R, James N (2022) A review on spatial, temporal and magnitude prediction of landslide hazard. J Asian Earth Sci X 7:100099. https://doi.org/10.1016/j.jaesx.2022.100099

Voulodimos A, Doulamis N, Doulamis A, Protopapadakis E (2018) Deep learning for computer vision: a brief review. Comput Intell Neurosci 2018:7068349. https://doi.org/10.1155/2018/7068349

Wang L-J, Guo M, Sawada K et al (2016) A comparative study of landslide susceptibility maps using logistic regression, frequency ratio, decision tree, weights of evidence and artificial neural network. Geosci J 20:117–136. https://doi.org/10.1007/s12303-015-0026-1

Wang Y, Fang Z, Hong H (2019) Comparison of convolutional neural networks for landslide susceptibility mapping in Yanshan County, China. Sci Total Environ 666:975–993. https://doi.org/10.1016/j.scitotenv.2019.02.263

Wani MA, Bhat FA, Afzal S, Khan AI (2020) Introduction to deep learning. In: Kacprzyk J (ed) Advances in deep learning. Springer, Singapore, pp 1–11

Xiao L, Zhang Y, Peng G (2018) Landslide susceptibility assessment using integrated deep learning algorithm along the China–Nepal highway. Sensors 18:4436. https://doi.org/10.3390/s18124436

Yang L, Shami A (2020) On hyperparameter optimization of machine learning algorithms: theory and practice. Neurocomputing 415:295–316. https://doi.org/10.1016/j.neucom.2020.07.061

Yang Z, Liu C, Nie R et al (2022) Research on uncertainty of landslide susceptibility prediction—bibliometrics and knowledge graph analysis. Remote Sens 14:3879. https://doi.org/10.3390/rs14163879

Yao J, Yao X, Zhao Z, Liu X (2023) Performance comparison of landslide susceptibility mapping under multiple machine-learning based models considering InSAR deformation: a case study of the upper Jinsha River. Geomat Nat Hazards Risk 14:2212833. https://doi.org/10.1080/19475705.2023.2212833

Yilmaz EO, Tonbul H, Kavzoglu T (2023) Marine mucilage mapping with explained deep learning model using water-related spectral indices: a case study of Dardanelles Strait, Turkey. Stoch Environ Res Risk Assess. https://doi.org/10.1007/s00477-023-02560-8

Youssef AM, Pourghasemi HR (2021) Landslide susceptibility mapping using machine learning algorithms and comparison of their performance at Abha Basin, Asir Region, Saudi Arabia. Geosci Front 12:639–655. https://doi.org/10.1016/j.gsf.2020.05.010

Zhang X, Wu G, Dong Z, Crawford C (2015) Embedded feature-selection support vector machine for driving pattern recognition. J Frankl Inst 352:669–685. https://doi.org/10.1016/j.jfranklin.2014.04.021

Zhang Y, Shi X, Zhang H et al (2022) Review on deep learning applications in frequency analysis and control of modern power system. Int J Electr Power Energy Syst 136:107744. https://doi.org/10.1016/j.ijepes.2021.107744

Zhang J, Ma X, Zhang J et al (2023) Insights into geospatial heterogeneity of landslide susceptibility based on the SHAP-XGBoost model. J Environ Manag 332:117357. https://doi.org/10.1016/j.jenvman.2023.117357

Zhou C, Yin K, Cao Y et al (2018) Landslide susceptibility modeling applying machine learning methods: a case study from Longju in the Three Gorges Reservoir area, China. Comput Geosci 112:23–37. https://doi.org/10.1016/j.cageo.2017.11.019

Zounemat-Kermani M, Batelaan O, Fadaee M, Hinkelmann R (2021) Ensemble machine learning paradigms in hydrology: a review. J Hydrol 598:126266. https://doi.org/10.1016/j.jhydrol.2021.126266

Chapter 7
Landslide Susceptibility Analysis by Frequency Ratio Model and Analytical Hierarchical Process in Mirik and Kurseong, Darjeeling Himalaya, India

Nayan Dey, Baishali Ojha, and Payel Das

Abstract Landslide is a common phenomenon in the hilly region. The Himalayan region the one of the best examples of this type of landslide-prone region. Kurseong subdivision is the study area of this chapter. Landslide is a common hazard that creates an interruption in the normal human life. With the help of Susceptibility mapping, the landslide zones can easily identified. Analytical hierarchical process (AHP) and frequency ratio model (FR) are used to prepare the susceptibility map. About 77 past landslide inventories are traced for the study. Eighteen (18) inventories from the NASA Global landslide points, six (6) inventories from the field survey (2023) and fifty-three (53) inventories from Google Earth Pro (2010–2023) in Mirik and Kurseong. Twenty (20) landslide conditioning factors are taken into account for landslide susceptibility mapping, viz., slope, elevation, aspect, curvature, relative relief, Normalized Difference Vegetation Index (NDVI), Topographic Wetness Index (TWI), Topographic Ruggedness Index (TRI), drainage density and land-use and land-cover etc. Among the two models, the Analytical Hierarchical Process (AHP) has the highest accuracy (62.27%) as per ROC-AUC. From the predicted maps, it has been traced that Kurseong is experiencing high and very high landslide-prone areas. The landscape susceptibility map helps to figure out the prime governing factors behind a landslide in the study area. In addition, the said chapter addresses the sustainable mitigation process of the landslide in Kurseong that will contribute to further sustainable development in the area.

Keywords Landslide susceptibly mapping · Topographic wetness index · Curvature · Aspect · ROC-AUC

N. Dey (✉)
Department of Geography, Pandit Ravishankar Shukla University, Raipur, India
e-mail: dey.nayanrbu@gmail.com

B. Ojha · P. Das
Department of Geography, Rabindra Bharati University, Kolkata, India

G. K. Panda et al. (eds.), *Landslide: Susceptibility, Risk Assessment and Sustainability*, Advances in Natural and Technological Hazards Research 52,
https://doi.org/10.1007/978-3-031-56591-5_7

7.1 Introduction

Landslides are mainly caused by disturbance of a natural slope. A landslide is a complex gravitational flow that initiates mass movements (Kalantar et al. 2019). A landslide is the movement of a mass of rock debris or soil down a slope. Different categories of landslides and associated triggering factors have been reported in the literature with ambiguous definitions (Lollino et al. 2015; Pradhan et al. 2017). Many of the countries in the world belong to susceptible landslide zones (De La Ville et al. 2002; Miller 2011; Prakasam et al. 2020). USA, China, Philippines, India and Indonesia belong to that category of most intensive landslide areas (Ahmed and Dewan 2017; Pisano et al. 2017; Froude and Petley 2018). The Himalayan region is the most landslide-prone area, where much devastating destruction takes place (De la Ville et al. 2002; Mandal and Mandal 2017; Chawla et al. 2019; Hirota et al. 2019). The frequent happening of landslides in the Eastern Himalayas cause huge damage to natural resources and the lives of people and animals. This disaster approximately affected a hundred million people and in another report, 66.5 billion USD were damaged over the last 5 years (Saha and Saha 2020). In June 2015, a devastating landslide and continuous heavy rainfall in Darjeeling destroyed many lives and property (Mandal and Mandal 2016). There are three types of landslides in the Darjeeling Himalayas, viz., rock falls, debris slides and earth falls. The international Disaster Database (EMDAT) suggests that landslides account for 4.9% of all natural disaster events and 1.3% of all natural hazards that happened between 1990 and 2015; and 54% of these landslide events occurred in Asia (Guha-Sapir et al. 2018; Froude and Petley 2018). The definition comes from the Latin word ‘vulnerary’ which indicates ‘to wound’. So, vulnerability means ‘the capacity to be wounded’. In the mountain region, landslides are devastating and a very common natural disaster that causes huge damage to properties and lives (Poudyal et al. 2010). Heavy rains, earthquakes or volcanic eruptions destabilize the slope and landslides occur. In the rainy season pre-monsoon or during, heavy rains dislodge the soil and rocks from the inside and they slide down the slope. In the Himalayan region, Darjeeling district is more vulnerable to its geological factors. The main task of landslide assessment is the perception of landslide susceptibility maps incorporating spatial and temporal prediction of landslides on a regular scale (Pradhan and Lee 2010). Sometimes huge landslide causes huge landslides in the form of mud. If part of the mountain is exposed on the side of the road, it remains unbalanced due to pressure, it collapses and the soil is also with it. No natural disaster can be completely prevented or the impact on society can be eliminated, but if the reason for the geological change of the soil due to the landslide is known then the damage can be reduced to a large extent. In rainy seasons landslides have a huge possibility of happening in hilly areas. According to a statistical report on the disaster in the last decade (2006–2015) produced by the International Federation of Red cross (IFRS), Landslides account for 4.7% (177 events) of the total number of reported natural hazards (Nguyan and Liu 2019).

Through the last two decades, researchers have tried to assess this hazard and try to prepare susceptibility maps of their spatial distribution (De La Ville et al. 2002; Bhattacharya 2020; Basu and Pal 2020). Van Den Eeckhaut et al. (2006), Dimri et al. (2007), Das et al. (2011), Mondal and Mandal (2017), and Mandal and Mandal (2017) are work landslide potential zones based on one model. Pal et al. (2016), Mondal and Mandal (2017), Basu and Pal (2018), Paul et al. (2019), and Saha and Saha (2020) worked on the micro geomorphological division of Darjeeling Himalaya. However, the landslide susceptibility zonation in the lower Siwalik rages of Mirik and Kurseong has not been advocated till now. According to Einstein (1988) and Mandal and Mandal (2016), analysis of landslides is mainly done by assessing their susceptibility. Many landslide susceptibility models have been developed around the world using different types of statistical methods such as bivariate analysis. Generating a landslide susceptibility map by using the AHP method has been carried out by many researchers worldwide (Nguyen and Liu 2019). The number of landslide locations in the training data set and sampling strategy are important factors that significantly affect the quality of landslide susceptibility maps (Kalantar et al. 2019). In addition, For the present study Frequency ratio (FR) model has been employed.

Mirik—Kurseong region (of Darjeeling Himalaya) is a highly populated famous tourist spot in West Bengal as well as all over India. There has been significant growth both in landslide events particularly those induced by human activities and in the number of landslide investigations in different parts of the world (Pardeshi et al. 2013). The task of cataloguing landslides to evaluate historical trends can be achieved and hazard zonation maps must be produced in this process (Chen and Chen 2021). Special prediction of landslide susceptibility zones has been the focus of research for many investigators adopting GIS-based multi-criteria analysis for the mitigation of hazards on vulnerable slopes (Pourghasemi et al. 2012; Bhattacharya 2016). The recent trend is towards the development of warning systems and land utilization regulations aimed at minimizing the loss of lives and property damage without investing in long-term, costly projects of slope stabilization (US Geological Survey 1982; Kockelman 1986; Schuster and Fleming 1986; IDNHR 1987; UNDRO 1991; Schuster 1995; Guzzetti et al. 1999). The outcome of this study will help in future sustainable development in hilly areas that are landslide and this will help them to get free from risk zones. It also has a vital economic contribution Indian economy. So it is important to make it safe and sustainable. To mitigate the problem of landslides, it is important to identify the hazardous zones first. Susceptibility mapping is the best way to mark out the landslide areas.

7.2 Study Area

Compared to the Western Ghats region, the landslides in the Himalayan region are huge. The word 'Kurseong' derived from 'Kharsang', means the land of white orchids (Mandal and Saha 2018; Bhattacharya 2020). This land was mainly formed by Darjeeling Guinness and belongs to the Dalling series of archaeology, the Gondwana

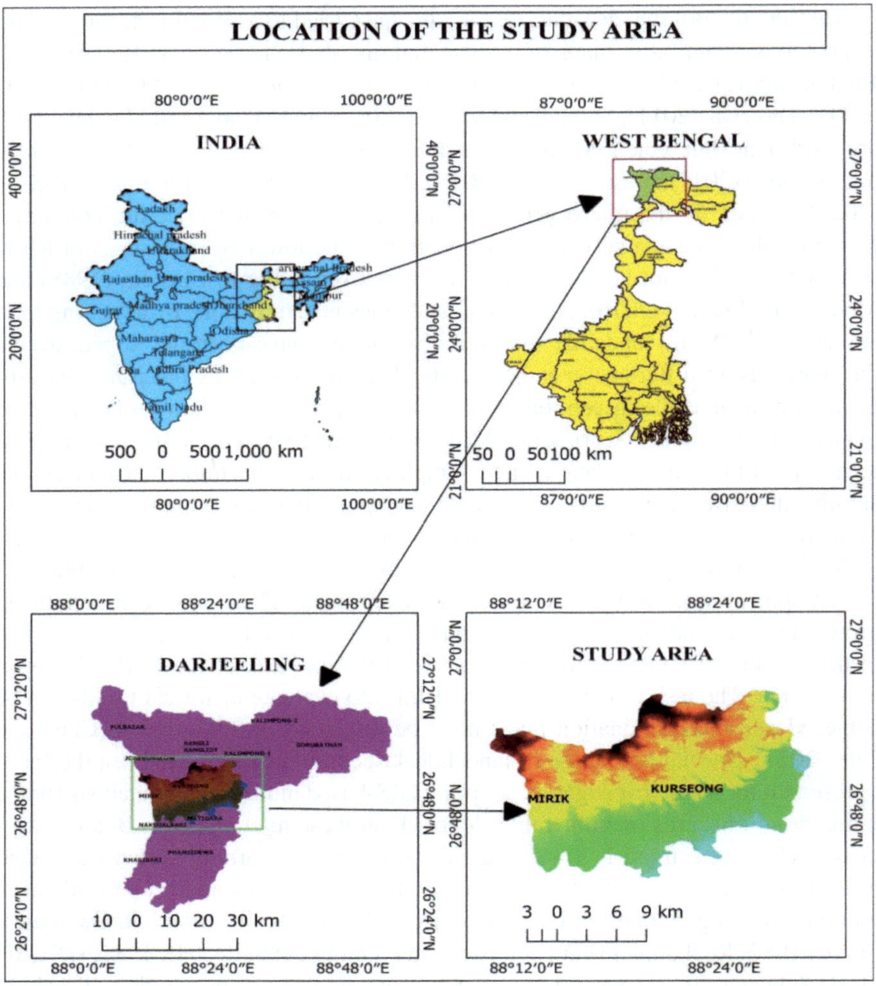

Fig. 7.1 Location map

Park rock, the Siwalik rock and the Pleistocene rise (Bhattacharya 2020). The average elevation of this land is 4858 ft (Mandal and Saha 2018). This region is sculpted by the young stage of many rivers (Saha and Saha 2020). Kurseong is a landslide-prone municipality area under Darjeeling district in the Northern Himalayas of West Bengal. It is located at an altitude of 1482.55 m. Kurseong is 32 km from Darjeeling (Fig. 7.1). According to the Municipality water record (from 1980 to 2016), Kurseong experiences about 23.69 °C average temperature and 323.5 cm average precipitation per year. According to Koppen's climatic classification, CWA type climate (subtropical humid climate) is found here. There are many tourist spots in Kurseong i.e. Dowhill deer park, Netaji Subhas Chandra Bose museum etc. The major activities of this region are horticulture, tea plantation and tourism.

7.3 Materials and Method

GIS is used in this chapter for the collection, storage, manipulation, layout and analysis of spatial data efficiently (Vijith et al. 2018; Paul et al. 2019; Mind'je et al. 2019; Debanshi and Pal 2020; Basu and Pal 2020). Remote Sensing is now a modern and scientific technique to assemble many factors such as geomorphology, land use and land cover (LULC), changes in landscape etc. (Debanshi and Pal 2020; Kanti Hembram et al. 2019; Paul et al. 2019). To mitigate the vulnerability of the landslide, firstly identification of the landslide zone is very important and mapping is the best method to mark out these zones (Table 7.1). These identifications may help to offer opportunities for their management (Pham et al. 2016; Roy and Saha 2019). With the help of landslide vulnerability modelling, there were many disasters already handled in the last few decades (Van Westen et al. 2003; Chen et al. 2017; Pisano et al. 2017; Basu and Pal 2020).

Table 7.1 Source of the data layer of the study area

Data layer	Data type	Sources of data	Data layer
Geological map	Grid	Geological Survey of India, Kolkata http://earthquake/data.gov.in (Ministry of Earth Science)	Geological map
DEM	Pixel	Alos Polsar	Altitude map, slope map, aspect map, stream power index (SPI), topographic wetness index (TWI)
Landsat 8 (Oli) (Path 139; Row41)	Pixel	USGS Earth Explorer	NDVI, POI, LULC
Soil map	Grid	NBSS and LUP Regional Centre, Kolkata	Soil map
Rainfall	Grid	CRU	Rainfall map

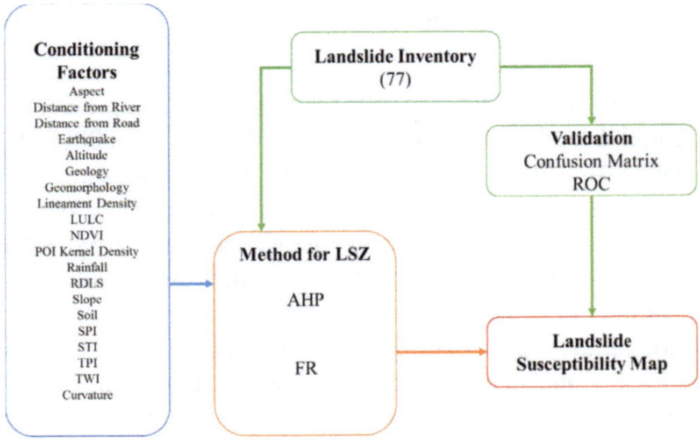

Flow Chart of Methodology

7.3.1 Landslide Experimentary Factor

Now many scientists are attracted to attention in landslide mapping (Miller 2011; Tiranti et al. 2019; Roy and Saha 2019; Basu and Pal 2020). The major influencing factors of landslides are slope, aspect, curvature, relative relief, drainage density, geology, land use land cover, NDVI etc. Thus landslide conditioning variables were obtained from the digital elevation model with the help of GIS. According to Gudiyangada Nachappa et al. (2020), He et al. (2019), Park and Kim (2019), the factors of GIS-based model depend on the availability of Dara and the expert's knowledge (Trinh 2022). However, 20 conditioning factors are selected for the landslide susceptibility mapping (LSM), viz., Aspect, altitude, slope, distance from the river, distance from road, earthquake, geology, geomorphology, lineament density etc.

7.3.1.1 Aspect

According to Yang et al. (2021) and Roy et al. (2022) topography is the key to landslide occurrence. The slope is also a dominating factor of the landslide (Trinh 2022). The slope govern the flow of azimuth and the microclimatic nature also influences the landslide (Roy et al. 2022). these factors also influence the evapotranspiration and moisture in this region in hilly region (Meten et al. 2015). Additionally, forest cover, surface temperature and humidity are pronounced by the aspect (Bilir et al. 2021; Jaksic et al. 2021). The '3D analyst' tool is used to plot the aspect map in ArcGIS (Trinh 2022) and ten categories of aspects have been recognized (Fig. 7.2a). Viz., Flat, North, Northeast, East, Southeast, South, Southwest, West, Northwest and North.

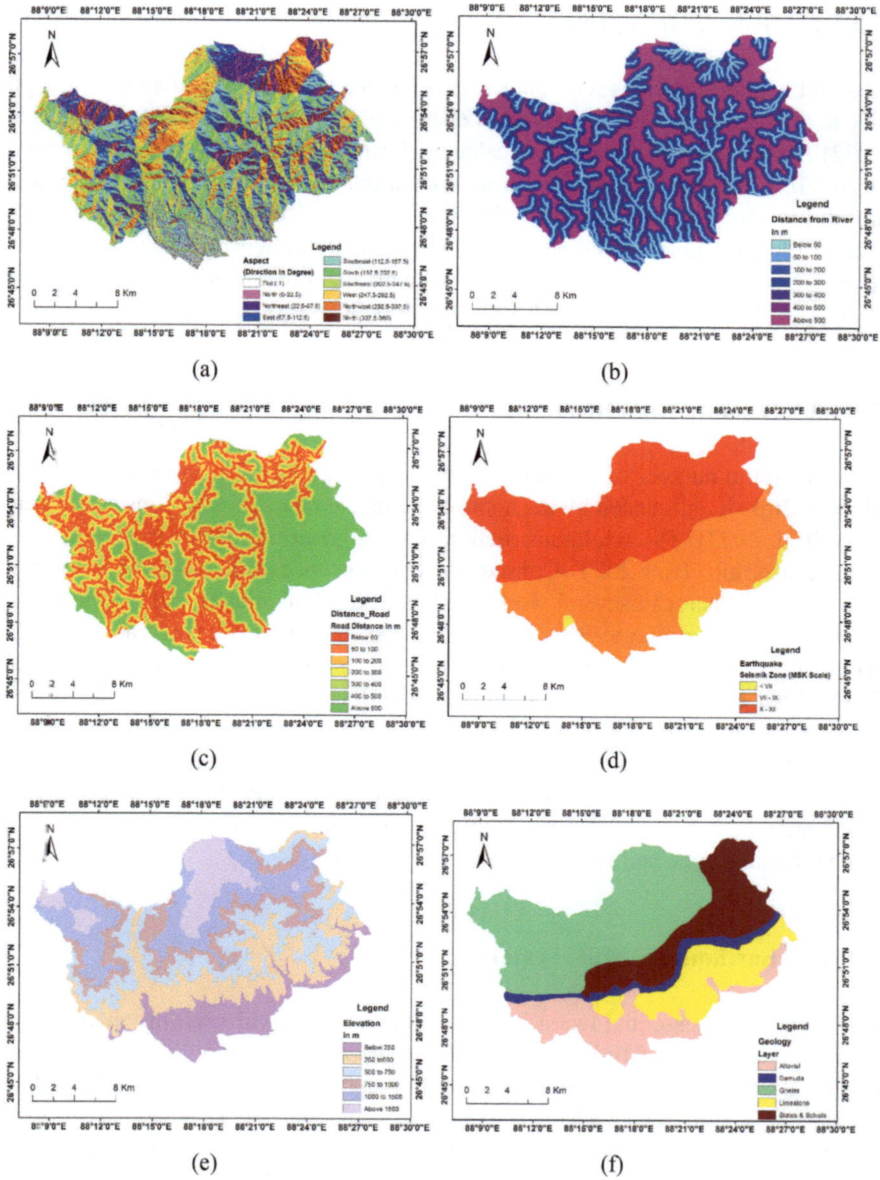

Fig. 7.2 **a** Aspect, **b** distance from the river (m), **c** distance from road, **d** earthquake, **e** altitude, **f** Geology

7.3.1.2 Distance from River (m)

One of the important causes of landslides is the deterioration of soil's shear strength due to the high pore pressure in water (Zhan et al. 2019; Liu et al. 2020; Amah et al. 2022). The distance to streams was calculated using the Euclidean distance function in ArcGIS (Park and Kim 2019). Throughout drainage lines, there are probable slope instability zones (Chueasamat et al. 2018; Xiong et al. 2019; Rodriguez et al. 2020). The Mirik and Kurseong region is covered by the high intensive drainage network. The map is manifested into six classes by the manual break method in ArcGIS, viz. <50, 50–100, 100–200, 200–300, 300–400, 400–500, >500 m (Fig. 7.2b).

7.3.1.3 Distance from Road (m)

The maximum number of landslides occurred along roads. The stabilization of the slope is directly influenced by the anthropogenic elements of road construction etc. (Das et al. 2023). The road construction plays a governing factor role in the said mountain terrain (Trinh 2022). Consequently, the distance to road factor is recognised as a causative factor in several works (Yalcin 2008; Yalcin et al. 2011; Wang et al. 2020a, b, c). According to Bourenane et al., and Road network construction by the name of development, vegetation cover clearance is common here (Bourenane et al. 2015; Abdi et al. 2021). They significantly influence the decrease in the slope's resistance (Pawluszek and Borkowski 2017). The layer distance from the road was categorized into six classes, i.e. Below 50, 50–100, 100–200, 200–300, 300–400, 400–500 and 500 m. The classes are demonstrated as very high vulnerable zone, moderately high vulnerable zone, high vulnerable zone, moderate vulnerable zone, low vulnerable zone and very low vulnerable zone (Fig. 7.2c).

7.3.1.4 Earthquake (MSK Scale)

Earthquake is an essential factor in plotting the landslide vulnerability mapping for the mountains region. The distribution of landslides has a vital relationship between earthquakes and the dip direction of the fault (Saha and Saha 2020). The Medvedev–Sponheuer–Karnik scale (MSK), is a macro-seismic intensity scale used to evaluate the intensity of the earthquake and their impact capacity on a particular region. Three classes are recoded, according to the MSK scale, viz., <VII, VII to IX and X to XII (Fig. 7.2d).

7.3.1.5 Altitude

Elevation plays a triggering factor for landslide occurrence, as a consequence of gravitational potential energy (Udin et al. 2021; Feng et al. 2022). In several previous studies, elevation is implied as a triggering factor in plotting the landslide susceptibility mapping (Kavzoglu et al. 2014; Vojteková and Vojtek 2020). Topography, aspect and slope gradient are the important elements that influenced the landslide very much (Pawluszek and Borkowski 2017; Abdi et al. 2021). The elevation of the study area directly influences the geomorphic features, types of vegetation and level of degradation (Chen et al. 2017). The 'manual breaks method' in ArcGIS to categorize into six classes, i.e. <250, 250–500, 500–750, 750–1000, 1000–1500 and >1500 m (Fig. 7.2e).

7.3.1.6 Geology

Geology plays a vital role for landslide susceptibility mapping (Yalcin 2008; Yalcin et al. 2011). The susceptibility of landslides varies with the differentiation composition of lithology. Lithology is an important factor affecting landslides, especially in mountainous areas (Shirzadi et al. 2018). The lithological characteristics of the study area are important to measure hazard susceptibility (Xiao et al. 2020; Neamat and Karimi 2020). Different lithology means that the physical and chemical properties of rocks are very different, which has an important impact on the stability of slopes (Zhou et al. 2021). The lithological structures affect the permeability and rock hardness of the earth's surface (Watakabe and Matsushi 2019; Roy et al. 2022). Generally, areas with low permeability rock masses have more potential for landslides (Zhou et al. 2021). Five geology classes are distributed in the said region of Darjeeling Himalaya, i.e. Alluvial, Damuda, Gneiss, and Limestone (Fig. 7.2f).

7.3.1.7 Geomorphology

Different types of geomorphological features often have a different levels of slope stability. For example, compared with plain landforms, mid-slope drainages and shallow valleys are more likely to have landslide disasters (Sun et al. 2020a). Micro-landform is a relatively small-scale geomorphology, and it is also the smallest unit of geomorphology (Zhou et al. 2021). In the said region twelve types of geomorphological features have been identified. The geomorphic divisions are depositional recent alluvial flood plain by the fluvial activity), alluvial fan, colluvial debris fan (occurance by the depositional activity by the gravitational pool at the foothill), fault (formed due to tectonic activity) relocated fact slope, denudational high dissected screeladen steep slope, low dissected depositional intermountain valley, intermountain structure plateau (formed due to tectonic activity), deep denudational entrenched valley, moderately denudational dissected intermountain valley, denudational niche, old depositional alluvial terrace (occurence by fluvial activity), fluvio denudational intermountain valley (Fig. 7.3a).

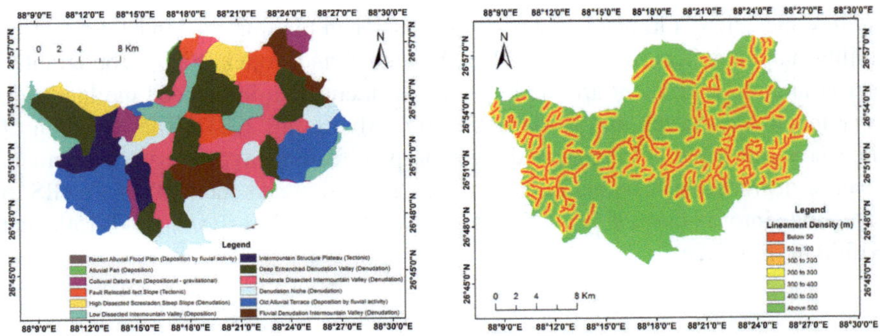

(a) (b)

Fig. 7.3 **a** Geomorphic features, **b** lineament density (m)

7.3.1.8 Lineament Density

According to Nampak et al. (2014) and Hung et al. (2016), the fault is an important factor in landslide occurrence. Lineament density is drawn by the assessment of the distance from the lineament. Six classes are plotted as follows: below 50, 50–100, 100–200, 200–300, 300–400, 400–500, and above 500 m (Fig. 7.3b).

7.3.1.9 LULC

According to Zhang et al. (2022) and Liao et al. (2022), various land use and landcover distribution pronounces to landslide differently. This factor is selected in existing susceptibility mapping (Nguyen and Liu 2019; Wang et al. 2020a, b, c; Trinh 2022). Forested areas are less landslide-prone because the roots of the plants protect the soil very much (Das et al. 2023). The absence of vegetation cover negatively pronounces the stability of the slope (Barlow et al. 2003). The plant reduces water content through evapotranspiration which helps to slope stability (Das et al. 2023). Anthropogenic activity significantly affects the frequency of landslides (Abdi et al. 2021). The land use and land cover are classified into seven classes, i.e., built-up area, waterbody or lake, river, agriculture, dense forest, sparse forest, and plantation (Fig. 7.4a).

7.3.1.10 Normalized Difference Vegetation Index (NDVI)

Normalized difference vegetation Index (NDVI) is typically applied to distinguish between areas covered in vegetation and those that are not (Mitra and Das 2022). The slope with less vegetation cover is more vulnerable to landslides. Vegetation prevents erosion on a slope by buffering the impact of rain falling on the slope, and vegetation roots increase the shear strength of the slope by increasing the shear strength of the soil (Park and Kim 2019). The vegetation cover helps the slope more stable because

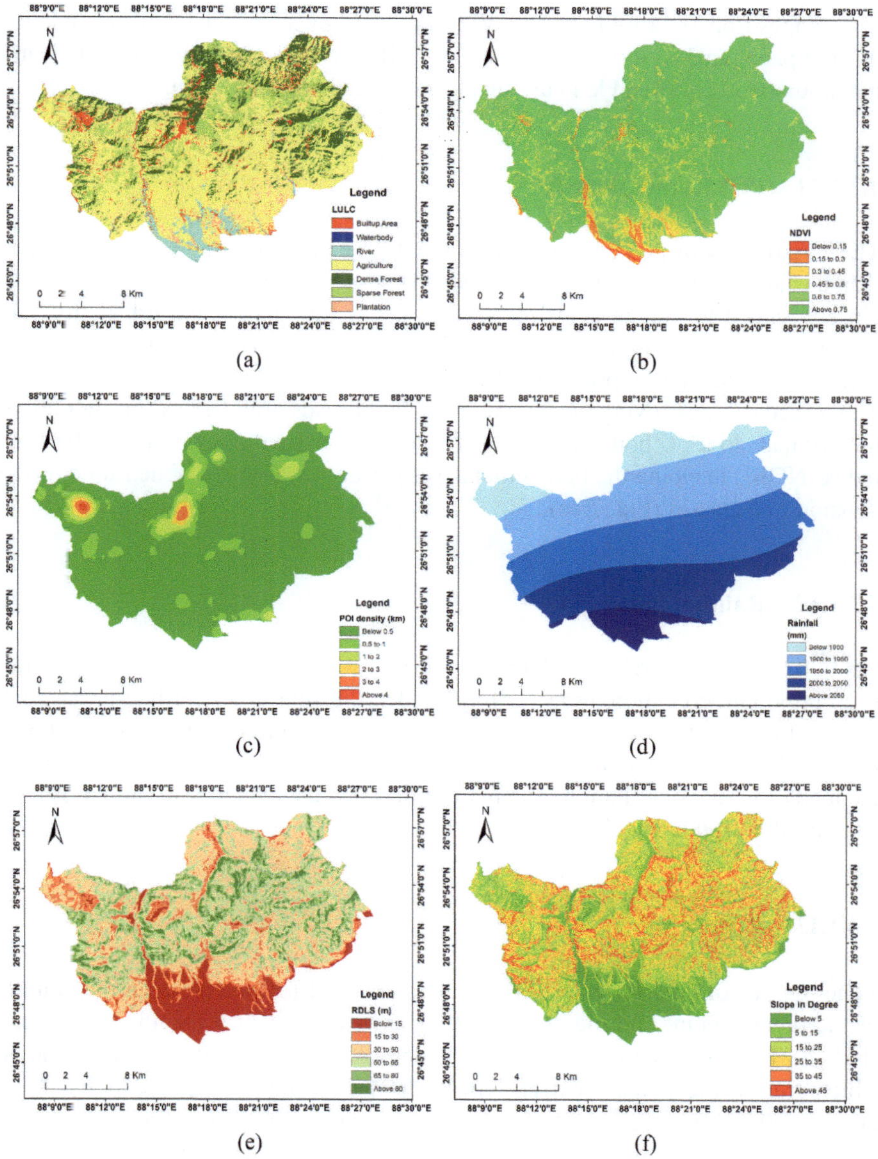

Fig. 7.4 **a** LULC, **b** NDVI, **c** POI Kernel density (m), **d** rainfall, **e** RDLS (m), **f** slope (degree)

of their soil holding capacity (Das et al. 2023). Numerous studies (Manchar et al. 2018; Rai et al. 2022; Al-Najjar et al. 2022) demonstrate the connection between landslide occurrence and NDVI (Das et al. 2023).

$$NDVI = \frac{NIR - Red}{NIR + Red} \tag{7.1}$$

It ranges from +1 to −1. The NDVI is measured by the raster calculator function of the Spatial analyst tool in ArcGIS (Fig. 7.4b). A higher positive value indicates maximum land cover and land use by the vegetation and vice versa.

7.3.1.11 POI Kernel Density

The core density of POI can well reflect the degree of human engineering activities (Zhou et al. 2021). Through the POI data, it is possible to analyze the characteristics closely related to human activities such as economic vitality and urban structure in the region (Wang et al. 2020a, b, c). Therefore, the distance from the road is a commonly used factor representing human engineering activities (Chen et al. 2021). This is because destroying the mountains to build roads will destroy the stability of the original slopes (Zhou et al. 2021). Through the normalized difference, built-up index (NDBI) mapping the POI kernel density is drawn. Density is measured by the division of 6 classes (Fig. 7.4c).

7.3.1.12 Rainfall in mm

Rainfall is intensively related to landslides (Xia et al. 2013; Avila et al. 2016). Rainfall is an important conditioning factor for the identification of a landslide-prone zone (Apostolov et al. 2021; Eitvandi et al. 2022). The annual rainfall maps are significant for landslide occurrence. Long-term average rainfall (2000–2022) is used the draw the map of spatial distribution. The map depicts that the tendency of rainfall is high from north to south (Fig. 7.4d).

7.3.1.13 RDLS (m)

Relief degree of land surface (RDLS) refers to the difference between the altitude of the highest point and the altitude of the lowest point in a specific area (Zhou et al. 2021). RDLS can be used to see the degree of surface cutting in a local area, which greatly affects the occurrence of landslides (Wang et al. 2020a, b, c). It is a macroscopic index that describes the topographic characteristics of a region (Zhou et al. 2021). The manual breaks method in ArcGIS to categorize into six classes, i.e. <15, 15–30, 30–50, 50–65, 65–80 and >80 m (Fig. 7.4e).

7.3.1.14 Slope (Degree)

Many scholars accept that the slope is a very important factor for landslide susceptibility mapping (Gupta et al. 2018; Jaksic et al. 2021; Hassangavyar et al. 2022). To create the landslide susceptibility prediction mapping, the slope of the land is a prime factor (Trinh 2022). The steepness of the land effect the landslide (Chen et al. 2019).

For analysis of the slope stability, the angle of the slope is the most crucial triggering factor (Trinh 2022). It is obvious that the slope of the land plays an important role in landslide occurrence (Yalcin 2008). The slope map is plotted from DEM by the 'spatial analyst' tool in ArcGIS (Fig. 7.4f).

7.3.1.15 Soil

According to the analysis in the previous section, soil landslides and rock landslides accounted for 75.25 and 23.04% of the landslides involved, and they were mainly induced by rainfall, human engineering activities, and river erosion (Zhou et al. 2021). In the said region four types of geomorphological features have been identified, i.e. Highly shallow Loamy-skeletal (W-004), Moderately shallow Coarse-loamy (W-002), Deep Fine Loamy (W-009), Very deep Coarse-loamy (W-006) (Fig. 7.5a).

7.3.1.16 Stream Power Index (SPI)

The stream power index (SPI) is a significant hydraulic factor that indicates the erosional power of surface water runoff and is an important factor in landslide susceptibility plotting (Abdi et al. 2021; Mitra and Das 2022; Mitra et al. 2022). The water flow's power to erode is measured by the stream power index (SPI), based on the assumption of proportionality of discharge to a catchment's specific area (Moore et al. 1991). In addition, the stream power index (SPI) employs the assessment of slope stability due to surface runoff-induced erosion (Das et al. 2023).

$$SPI = A * \tan\beta \quad \text{(Moore et al. 1991)} \tag{7.2}$$

where

A = Area
tan β = Gradient

The "Raster Calculator tool" of "Spatial Analyst Tools" was employed to plot the stream power index (SPI) map of the Kurseong. Stream power index (SPI) is divided into six classes, i.e., <50, 50–100, 100–150, 150–200 and >250 (Fig. 7.5b).

7.3.1.17 Sediment Transport Index (STI)

The sediment transport index (STI) is also often used to reflect the overland flow's power to erode (Moore and Burch 1986). The STI was calculated with their base in specific catchment areas (As) and slope maps by the following equation (Trinh 2022).

$$STI = \left(\frac{A_s}{22.13}\right)^{0.6}\left(\frac{\sin\beta}{0.0896}\right)^{1.3} \tag{7.3}$$

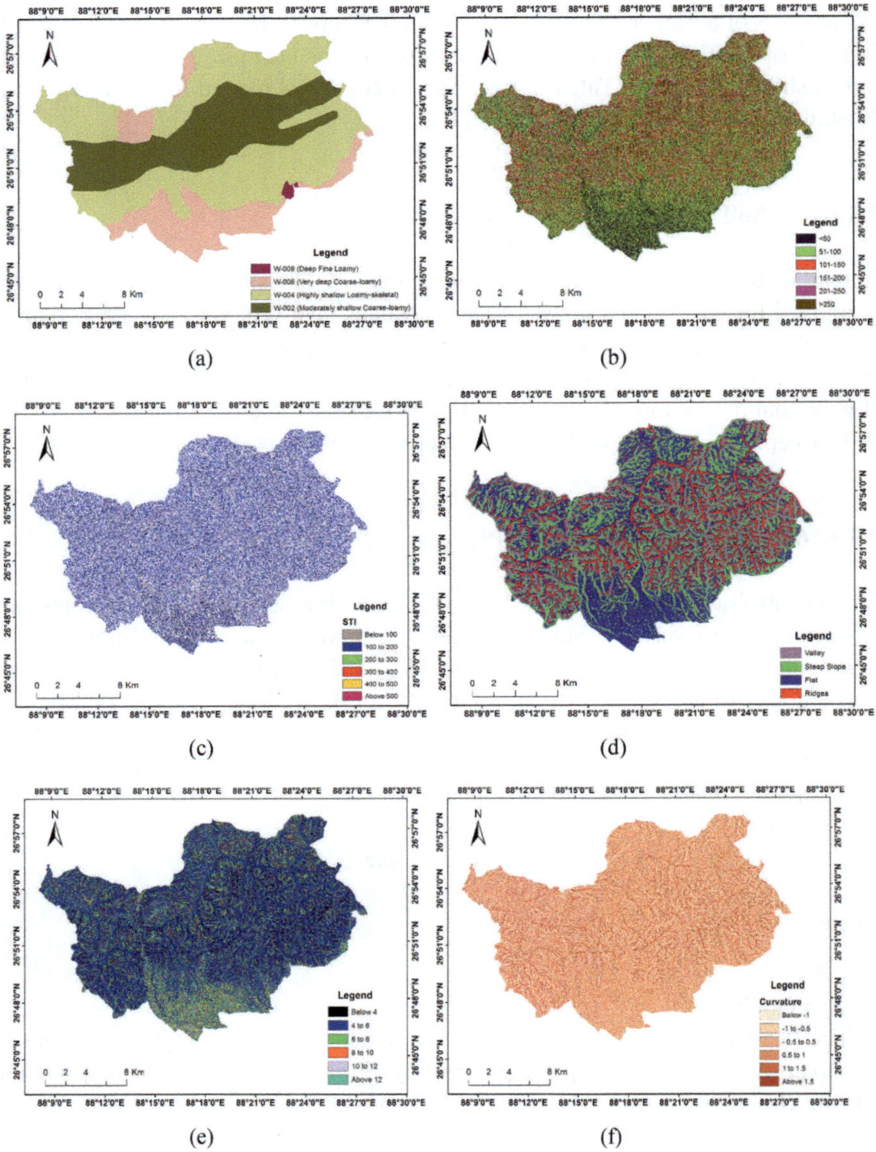

Fig. 7.5 **a** Soil, **b** SPI, **c** STI, **d** TPI, **e** TWI, **f** curvature

where
A_s = Slope in degree
β = Gradient

The larger the two values, the stronger the erosion and transportation capacity of the water flow, which will cause the slope adjacent to the river to lose stability (Zhou et al. 2021). The sediment transport index (STI) is divided into six classes, i.e., <100, 100–200, 200–300, 300–400, 400–500 and >500 (Fig. 7.5c).

7.3.1.18 Topographic Position Index (TPI)

Topographic position index (TPI) is a ratio between specific cell altitudes with their neighbour's mean cell altitude. It ranges from +1 to −1. Positive topographic position index (TPI) values indicate that the particular cell elevation is higher than the surrounding cells, that means ridges. The negative topographic position index (TPI) values indicate that the particular cell elevation is lower or depressed than the surrounding cells, that means valley. If the topographic position index (TPI) value is zero, indicates the flat region.

$$TPI = \frac{\text{Specific cell altitude}}{\text{Mean altitude of surrounding cells}} \tag{7.4}$$

In this chapter, the Mirik and Kurseong region is classified under four divisions, viz., valley, steep slope, flat and ridges (Fig. 7.5d).

7.3.1.19 Topographic Wetness Index (TWI)

Topographical Wetness index (TWI) is used as a hydrological aspect in landslides which also helps to identify the supply saturation zone (Abdi et al. 2021). Beven and Kirkby (1979) developed a topographic wetness index (TWI) that reflects water's tendency to accumulate anywhere within the catchment area, accumulations that will then tend to move downslope as a result of gravity (Poudyal et al. 2010). The topographic wetness index (TWI) is a physical indicator of the influence of regional topography on runoff flow direction and accumulation, which quantifies the control of topography on basic hydrological processes (Zhou et al. 2021). Equation (7.5) was used to calculate the TWI (Beven and Kirkby 1979; Mitra and Das 2022; Das et al. 2023).

$$TWI = \ln\left(\frac{A_s}{\tan\beta}\right) \tag{7.5}$$

A larger topographic wetness index (TWI) value means a higher soil moisture content, and the soil is more likely to reach saturation and cause slippage (Sun et al. 2020b). According to Pawluszek and Borkowski (2017), the topographic wetness index (TWI) indicates how the presence of water in the soil is pronounced the soil loss and structural degradation (Fig. 7.5e).

7.3.1.20 Curvature

Curvature, plane curvature, and profile curvature are topographical factors that reflect the geometric characteristics of the slope, which mainly affect the occurrence of landslides by controlling the erosion process and surface runoff (Moosavi and Niazi 2016; Chen et al. 2021). The curvature factor is used for the analysis of landslide susceptibility zonation by Hassangavyar et al. (2022), Bien et al. (2022), Trinh (2022) and Das et al. (2023). Positive values are related to convexity, whereas negative values are associated with concavity, and zero is represented by the flat region (Mitra and Das 2022). Both positive and negative values of curvature are highly susceptible to landslides. Carson and Kirkby (1972) and Kaur et al. (2019) researched that erosion by rainfall (by sheet erosion) and stream are pronounced by curvature and it also governs the displacement of landslide materials. A concave slope retains water after a downpour for a substantial time and helps to get soil saturated by losing its mechanical characteristics that govern the occurrence of landslides. In contrast, a landslide on a convex slope occurs by the repetition of expansion and contraction. In addition, the decomposition and disintegration of rocks pronounce the landslide occurrence on a convex slope. Curvature is plotted from the DEM by using the spatial analyst tool in ArcGIS (Fig. 7.5f).

7.3.2 Methods for Landslide Susceptibility Zonation (LSZ)

Several landslide susceptibility mapping methods have been proposed (Mahalingam et al. 2016; Jaafari 2018; He et al. 2019; Nguyen and Liu 2019). Methods could be classified into two, viz., quantitative and qualitative methods (Van Westen et al. 1997).

7.3.2.1 Analytical Hierarchical Process (AHP)

The analytic hierarchy process (AHP) model was coined and modified by Saaty (1980), Saaty and Vargas (1984). Kayastha et al. (2013), He et al. (2019), Nguyen and Liu (2019), Gudiyangada Nachappa et al. (2020), Vojteková and Vojtek (2020) used the analytic hierarchy process (AHP) by Saaty (1980) in their research as a qualitative method. The AHP method is considered as a decision-making process to employ perplexing issues (Saaty 1980). The physical process and field experiences of relevant conditioning factors to landslide occurrences are very important for the analytic hierarchy process (AHP) model by Saaty (1980). Based on the degree of impact, experts give a suitable value (ranging from 0 to 9) to the relevant factors. The analytic hierarchy process (AHP) follows the steps to draw a landslide susceptibility zone (LSZ).

Step—I

In the AHP technics, n criteria are compared and a square matrix of A = (a_{ij}). The following equation is satisfied in every matrix competent of a_{ij}.

$$a_{ij} = \frac{1}{a_{ij}} \tag{7.6}$$

In the case of the reciprocal matrix, a_{ij} follows the equality, i.e., $a_{ij} = \frac{Pi}{Pj}$; where, P_i represents the preference of the alternative i, as described below (Mitra et al. 2022).

Step—II

To construct the flood susceptibility twelve matrices and nine matrices for flood vulnerability layers are used in this paper. To calculate the AHP ratings all matrices are computed by thematic mapping. With the help of the relative scale ratio from the Pairwise reciprocal matrix, the value of weights or ratings are calculated in the following equations.

$$A = \begin{bmatrix} \frac{P_1}{P_1} \cdots \frac{P_1}{P_j} \cdots \frac{P_1}{P_n} \cdots \\ \frac{P_i}{P_1} \ldots 1 \ldots 1 \ldots \frac{P_i}{P_n} \\ \frac{P_n}{P_1} \cdots\cdots \frac{P_n}{P_j} \cdots \frac{P_n}{P_n} \end{bmatrix} \tag{7.7}$$

i. All components of j columns are summarized in matrix A:

$$\frac{P_1}{P_j} + \cdots + \frac{P_1}{P_j} + \cdots + \frac{P_n}{P_n} = \frac{\sum_{i=1}^{n} P_i}{P_j} \tag{7.8}$$

ii. Computation of normalized value by dividing the comparison, i.e., $a_{ij} = P_i/P_j$ by equation:

$$\frac{\frac{P_i}{P_j}}{\frac{\sum_{i=1}^{n} P_i}{P_j}} = \frac{P_i}{P_j} \times \frac{P_j}{\sum_{i=1}^{n} P_i} = \frac{P_i}{\sum_{i=1}^{n} P_i} \tag{7.9}$$

iii. Assigned weight or rating of *i* row (W_i), which is determined by computing the average of the components

$$W_i = \left(\frac{P_i}{\sum_{i=1}^{n} P_i} + \cdots + \frac{P_i}{\sum_{i=1}^{n} P_i} \right) \times \frac{1}{n} \tag{7.10}$$

Step—III

The consistency ratio (CR) is a mathematical expression of the pairwise matrix that is used to validate the AHP model. To compute the consistency ratio (CR) following equation is adopted step by step.

Table 7.2 Random Index for checking the consistency ratio for different matrix (Saaty and Vargas 1991)

Size of matrix	1	2	3	4	5	6	7	8	9	10
RI	0.00	0.00	0.58	0.90	1.12	1.24	1.32	1.41	1.45	1.49
Size of matrix	11	12	13	14	15	16	17	18	19	20
RI	1.51	1.48	1.56	1.57	1.59	1.6	1.61	1.61	1.62	1.63

$$CI = \frac{\lambda_{\max} - d}{d - 1} \tag{7.11}$$

where
$\lambda_{\max}$ = Large eigenvalue
d = Order of the comparison matrix

$$CR = \frac{CI}{RI} \tag{7.12}$$

where
RI = Random Index of the specific order of matrix (Table 7.2)
CI = Consistency Index

If the value of CR is $\leq$0.1, then the AHP result is accepted. If the value of CR is more than 0.1 then the process will maintain Step 2 to reconstruct the model and rejuvenate the pairwise rank of the relevant conditioning factors (Trinh 2022).

7.3.2.2 Frequency Ratio Model (FR)

The Frequency Ratio (FR) is a very popular quantitative method to use the landslide susceptibility (Lee and Pradhan 2006; Yalcin et al. 2011; Rasyid et al. 2016; Mahalingam et al. 2016). The quantitative method is used to assess the relationship between landslide and their factors (Jaafari 2018). The ratio between the landslide-occupied area and the whole study area is used for computation in the frequency ratio model (Trinh 2022). This method is used to calculate the probability of landslide occurrence and assess the importance of variables (Wang and Li 2017).

Step—I

To calculate $FR^i{}_j$, each conditioning factor *j* for different weight class *i* is calculated.

$$Fr_i = \frac{N_{pix(Si)}/N_{pix(Ni)}}{\sum N_{pix(Si)}/\sum N_{pix(Ni)}} \tag{7.13}$$

where
$N_{pix(si)}$ = Number of pixels containing side in each class *i*
$N_{pix(Ni)}$ = Total number of pixels having class (i) in the whole

$\sum N_{pi:(si)}$ = Total number of landslides pixels
$\sum N_{pi:(Ni)}$ = total number of pixels of the area of interest.

Step—II

To estimate the landslide susceptibility zone (LSZ), the frequency ratio (FR) of each class of all the landslide triggering factors was being summarized.

$$\text{LSZ} = Fr_1 + Fr_2 + \cdots + Fr_n \tag{7.14}$$

where
LSZ = Landslide Susceptibility Zonation.
Fr = Frequency Ratio of each triggering factor.

7.3.3 Landslide Inventories

To identify the landslide vulnerability zone, it is important to correlate the topography with conditioning factors and previous records of landslide occurrences (Lin and Chen 2012; Akgun and Erkan 2016; Chen et al. 2017; Roy and Saha 2019). A total of 77 landslide points have been identified. About 18 landslide incidents with locations are collected from the NASA landslide inventories map and about 53 landslide points are collected from google earth pro (between 2010 and 2023) of the said region. In addition, 6 landslide-prone areas were captured from a field survey, in February 2023.

7.3.4 Validation of Model's Performance

The initial analyses consist of the ability to create successful landslide susceptibility models with emphasis on addressing the highly susceptible area, this so-called success rate (Kalantar et al. 2019). The popular approach is to decide the accurate performance of analytical investigation (Razavi et al. 2018). The prediction rates are the ability of the test dataset to assess accurately the performances of the models' prediction strength (Kalantar et al. 2019).

7.3.4.1 Confusion Matrix

The confusion matrix includes information on model classification accuracy (Provost and Kohavi 1998), and provides useful information on model performance (Kalantara et al. 2019). This matrix helps experts determine whether the system is misidentifying the two classes (i.e. mislabelling one as the other) or correctly classifying them (Stehman 1997). The matrix columns display the instances in an estimated class while the rows show the instances in an actual class (Sammut et al. 2011). This matrix was also employed to evaluate the algorithm performance in other investigations (Naghibi and Moradi 2016). Accuracy level is measured by the confusion matrix by the following equation (Eqs. 7.15 and 7.16).

$$\text{Accuracy} = \frac{(TP + TN)}{(TP + FP + FN + TN)} \tag{7.15}$$

where
TP = True Positive
TN = True Negative
FP = False Positive
FN = False Negative

It also determined the misclassification rate (ER).

$$\text{Error rate} = \frac{(FP + FN)}{(TP + FP + FN + TN)} \tag{7.16}$$

7.3.4.2 Receiver Operating Characteristic (ROC)

The ROC curve is a graphical expression that assesses the performance of the binary classifier method (Kalantar et al. 2019). In this study model performance was conducted using receiver operating characteristic (ROC) curves (AUC) considered (Pourghasemi et al. 2013); using success rate and prediction rate for the assessments of the robustness of individual machine learning algorithms in landslide susceptibility mapping (Pham et al. 2018; Trinh 2022). The area under this curve determines the employed model accuracy in LSM (Kalantar et al. 2019). The ROC (AUC) value is categorized into scales in relation to qualitative classes 50–60% is poor, 60–70% is average, 70–80% is good, 80–90% is very good and 90–100% is exceptional (excellent) (Razavi et al. 2018; Yesilnacar and Topal 2005).

7.4 Result and Discussion

Landslides susceptibility zone assessment is very important to maintain the sustainability of the area (Saha and Saha 2020). Landslide susceptibility mapping is a very valuable method for reducing the effect of landslides (Ercanoglu and Temiz 2011; Pecoraro et al. 2019; Arabameri et al. 2019). Many methods, techniques, and approaches have been used to map the landslide vulnerability zone in previous literature (Keefer 2000; Van Westen et al. 2003; Miller 2011; Regmi et al. 2014; Mind'je et al. 2019).

The landslide susceptibility map of the Mirik and Kurseong region is sketched based on 20 prominent factors (Table 7.3). The study has revealed that among the all factors, Elevation, slope, Land use land cover, RDLS, soil, geomorphological characteristics and geological characteristics play important roles as governing factors and also help to indicate or trace the susceptible zone (Table 7.3).

The landslide susceptibility map of the Mirik and Kurseong Siwalik region is classified into 5 groups, viz. very low susceptible zone, low susceptible zone, moderate susceptible zone, high susceptible zone, and very high susceptible zone. From this research work, it has been traced that the said region is experiencing about 33% area as a high landslide susceptible zone by the AHP model and FR model (Fig. 7.6a, b). Maximum landslide occurs near the built-up region (settlement area and along the highway) due to slope destabilization by human interference.

The said chapter work quantitatively validates the landslide susceptible zonation map (LSZ) by the confusion matrix and ROC-AUC. Landslide susceptible zonation map (LSZ) by the Frequency ratio model is validated with more accuracy indicating high satisfaction. However, in ROC-AUC computation, it has been examined that the landslide susceptible zonation map (LSZ) gained a 62.27% accuracy level that indicates considerable acceptance (Table 7.4 and Fig. 7.6c).

7.5 Conclusion

Darjeeling—Kurseong region is a popular tourist destination. Every year this region faces a lot of destruction. Landslides are the most devastating natural destructions. According to Yalcin et al. (2011), Tien Bui et al. (2012), Achour et al. (2017), Zhou et al. (2020), Pham (2021), the said geological hazards is most devastating for the mountainous regions all over the world. In the eastern Himalayas, landslides are very common. Mirik municipality, Sourini Basti, Okaity Tea Estate, the south-eastern part of Bukim Tea Estate, the northern part of the Patong Tea Estate and the western part of the New Fallodi Tea Estate of Mirik region is experiencing the problem landslide. Maximum landslide occurs in the Kurseong region. Landslide occurs in the Kurseong municipality, near Sitong Forest, the southern section of Ghoomti Tea Estate, the eastern section of Rohin Tea Estate etc. of the Kurseong Himalayan region. It has an important long-term impact on human society and the natural environment. The probability of landslide due to slope destabilization is determined by the landslide susceptibility mapping. Thus, landslide susceptibility mapping can detect landslide-prone zones to establish mitigation planning. Froese and Moreno (2014), Pisano et al. (2017), Mind'je et al. (2019), urged that experts used susceptibility mapping to mitigate the nature devastating hazards. In addition, according to Chalkias et al. (2014), Hong et al. (2015), the said plotting assesses the importance of the combination of various variables in hazard propagation. Through this chapter, the landslide-prone zone of the Siwalik region of Mirik and Kurseong has been sketched by the AHP and FR models. Frequency ratio (FR) and Analytical hierarchy process (AHP) are considered as weightage methods. The west and south-facing, steep and convex slope segments are assigned with high landslide susceptibility zones. They consist of a steep

Table 7.3 Selected parameters of landslide susceptibility zonation

Classes	Number of pixel	Number of pixels (%)	Number of landslide occurrence pixel in each class	Number of landslide occurrence pixel in each class (%)	FR		AHP
					Class weight	Factor weight	Factor weight
Aspect							
Flat (−1)	27,510	1.056070389	50	0.444010301	4	1.45	4.0
North (0–22.5)	124,376	4.774620529	337	2.992629429	7		
Northeast (22.5–67.5)	300,325	11.52905633	553	4.910753929	4		
East (67.5–112.5)	364,255	13.98323954	1181	10.48752331	8		
Southeast (112.5–157.5)	393,685	15.11301604	1457	12.93846017	9		
South (157.5–202.5)	364,297	13.98485186	2213	19.65189592	16		
Southwest (202.5–247.5)	384,253	14.75093476	2754	24.45608738	19		
West (247.5–292.5)	300,302	11.52817339	1792	15.91332919	15		
Northwest (292.5–337.5)	246,836	9.4756885	751	6.669034722	8		
North (337.5–360)	99,101	3.804348661	173	1.536275642	4		
Distance from river (m)							
Below 50	235,697	9.048213349	792	7.033123168	11	1.21	5.3
50–100	223,069	8.563434848	808	7.175206465	12		
100–200	420,817	16.1548174	1770	15.71796466	14		
200–300	380,064	14.59034336	1008	8.951247669	9		
300–400	338,629	12.9996879	1092	9.697184975	11		
400–500	285,454	10.95834352	1410	12.52109049	17		
Above 500	721,171	27.68515963	4381	38.90418258	21		
Distance from road (m)							
Below 50	487,279	18.70623874	2403	21.33913507	17	1.23	3.5
50–100	296,782	11.3932161	1355	12.03267916	16		
100–200	389,141	14.9388019	1668	14.81218364	15		
200–300	277,568	10.65560649	1247	11.07361691	16		
300–400	201,173	7.722865476	674	5.985258858	12		
400–500	147,185	5.650310703	223	1.980285943	5		
Above 500	805,773	30.9329606	3691	32.77684042	16		

(continued)

Table 7.3 (continued)

Classes	Number of pixel	Number of pixels (%)	Number of landslide occurrence pixel in each class	Number of landslide occurrence pixel in each class (%)	FR		AHP
					Class weight	Factor weight	Factor weight
Earthquake (MSK scale)							
<VII	68,329	2.623058686	82	0.728176894	12	4.57	2.6
VII–IX	1,195,904	45.90915093	3515	31.21392416	29		
X–XII	1,340,703	51.46779038	7664	68.05789894	58		
Elevation (m)							
Below 250	403,931	15.50633373	621	5.514607939	5	2.09	4.4
250–500	571,614	21.94344442	1393	12.37012699	9		
500–750	486,204	18.6646766	1826	16.21525619	14		
750–1000	368,965	14.16403897	2322	20.61983838	23		
1000–1500	508,767	19.53083792	3633	32.26178847	26		
Above 1500	265,461	10.19066835	1466	13.01838203	20		
Geology							
Gneiss	1,120,637	43.02014809	6617	58.76032324	28	3.18	8.6
Slates and schists	499,029	19.15723065	2276	20.2113489	21		
Damuda	135,958	5.219293396	976	8.667081076	34		
Limestone	412,408	15.83193597	1180	10.4786431	13		
Alluvial	436,880	16.77139189	212	1.882603676	2		
Geomorphology							
Recent Alluvial flood plain	799	0.030677567	20	0.17760412	27	3.87	18.2
Alluvial fan	5234	0.200959183	0	0	0		
Colluvial debris fan	84,610	3.248596952	0	0	0		
Fault relocated fact slope	134,436	5.161663868	456	42.30055659	38		
High dissected screeladen steep slope	140,488	5.394030122	0	0	0		
Low dissected intermountain valley	187,861	7.212914219	1454	12.91181955	8		
Intermountain structure plateau	175,953	6.755707122	65	0.577213391	0		

(continued)

Table 7.3 (continued)

Classes	Number of pixel	Number of pixels (%)	Number of landslide occurrence pixel in each class	Number of landslide occurrence pixel in each class (%)	FR		AHP
					Class weight	Factor weight	Factor weight
Deep entrenched denudation valley	573,444	22.01735529	1906	16.92567268	3		
Moderate dissected intermountain valley	407,897	15.66118604	4287	38.06944321	11		
Denudation niche	346,657	13.30987914	2851	25.31746737	9		
Old alluvial terrace	380,375	14.60448015	211	1.87372347	0		
Fluvial denudation intermountain valley	166,755	6.402550346	11	0.097682266	0		
Lineament							
Below 50	141,517	5.43272086	1078	9.57286209	21	1.55	6.1
50–100	156,450	6.005986408	1282	11.38442412	23		
100–200	330,789	12.69871677	1830	16.25077702	15		
200–300	316,475	12.14921412	1027	9.119971583	9		
300–400	282,236	10.83480716	950	8.43619572	9		
400–500	238,028	9.137698515	668	5.931977622	8		
Above 500	1,139,406	43.74085618	4426	39.30379185	11		
LULC							
Builtup area	132,505	5.0920786	2108	18.71947429	46	4.68	11.0
Waterbody	170	0.006532986	0	0	0		
River	92,156	3.541493494	783	6.953201314	25		
Agriculture	1,699,612	65.31495335	6666	59.19545333	11		
Dense forest	453,577	17.430661	1118	9.928070331	7		
Sparse forest	187,083	7.189474667	586	5.203800728	9		
Plantation	37,076	1.424805903	0	0	0		
NDVI							
Below 0.15	4092	0.15725071	12	0.106562472	5	3.55	3.2
0.15–0.3	48,588	1.867179256	476	4.226978066	19		
0.3–0.45	93,933	3.609733865	1923	17.07663618	40		

(continued)

Table 7.3 (continued)

Classes	Number of pixel	Number of pixels (%)	Number of landslide occurrence pixel in each class	Number of landslide occurrence pixel in each class (%)	FR		AHP
					Class weight	Factor weight	Factor weight
0.45–0 6	217,693	8.365683991	2408	21.3835361	21		
0.6–0.75	590,487	22.69171559	2702	23.99431667	8		
Above 0.75	1,647,421	63.30843659	3740	33.21197052	4		
POI kernel density							
Below 0.5	2,233,453	85.73913411	7911	70.25130983	5	4.30	2.6
0.5–1	253,978	9.749859882	2452	21.77426516	15		
1–2	79,398	3.047978072	256	2.273332741	5		
2–3	21,160	0.812302779	213	1.891483882	16		
3–4	14,095	0.541087319	415	3.685285499	48		
Above 4	2856	0.109637842	14	0.124322884	8		
Rainfall (mm)							
Below 1900	283,533	10.88459792	600	5.328123612	13	4.89	4.9
1900–1950	757,477	29.07891701	2805	24.90897789	23		
1950–2000	827,259	31.75779041	6475	57.49933398	48		
2000–2050	588,824	22.6044675	1381	12.26356451	14		
Above 2050	147,808	5.674227159	0	0	0		
RDLS (m)							
Below 15	349,040	13.39916727	618	5.487967321	4	5.33	4.7
15–30	260,951	10.01755128	351	3.116952313	3		
30–50	868,294	33.33261675	2093	18.5862712	6		
50–65	665,583	25.55081925	2651	23.54142616	9		
65–80	331,924	12.74210749	2607	23.1506971	19		
Above 80	129,146	4.957737958	2941	26.11668591	56		
Slope (Degree)							
Below 5	306,145	11.7524825	638	5.665571441	3	5.65	8.4
5–15	439,170	16.85912799	664	5.896456798	2		
15–25	810,284	31.1056804	2443	21.69434331	5		
25–35	746,166	28.64427919	3490	30.99191901	8		
35–45	264,216	10.14288626	2782	24.70473315	19		
Above 45	38,958	1.495543658	1244	11.04697629	59		
Soil							
Highly shallow loamy-skeletal	1,350,149	51.83102647	4005	35.56522511	22	6.52	3.6
Moderately shallow coarse-loamy	762,077	29.25546229	6498	57.70357872	65		

(continued)

Table 7.3 (continued)

Classes	Number of pixel	Number of pixels (%)	Number of landslide occurrence pixel in each class	Number of landslide occurrence pixel in each class (%)	FR		AHP
					Class weight	Factor weight	Factor weight
Deep fine loamy	10,696	0.410609984	0	0	0		
Very deep coarse-loamy	481,983	18.50290126	758	6.731196164	12		
Stream Power Index (SPI)							
<50	439,774	16.88230823	1438	12.76973626	10	1.24	1.1
51–100	1,267,811	48.66948951	4491	39.88100524	11		
101–150	364,336	13.98634901	2133	18.94147944	18		
151–200	153,290	5.884588513	1044	9.270935086	21		
201–250	81,044	3.111165708	589	5.230441346	22		
>250	298,685	11.46609903	1566	13.90640263	16		
Sediment Transport Index (STI)							
Below 100	303,981	22.70594044	1078	19.18149466	23	2.72	0.9
100–200	1,004,844	75.05708585	4479	79.6975089	30		
200–300	13,498	1.008236647	26	0.462633452	12		
300–400	4801	0.358611953	6	0.106761566	8		
400–500	2480	0.18524425	1	0.017793594	2		
Above 500	9169	0.684880857	30	0.533807829	22		
Topographic Position Index (TPI)							
Valley	303,094	11.63535883	1683	14.94538673	29	1.02	1.4
Steep slope	808,107	31.02210839	3052	27.10238878	19		
Flat	1,084,705	41.64032248	4187	37.18142261	20		
Ridges	409,033	15.7022103	2339	20.77080188	30		
Topographic Wetness Index (TWI)							
Below 4	346,066	13.28499439	2220	19.71405737	30	2.27	1.3
4–6	1,401,691	53.80897595	6229	55.3148033	21		
6–8	604,000	23.18672337	2261	20.07814581	17		
8–10	158,899	6.099912512	362	3.21463458	10		
10–12	56,869	2.183122138	92	0.816978954	7		
Above 12	37,414	1.436271636	97	0.861379984	12		
Curvature							
Below −1	573,266	22.00688769	2775	24.64257171	18	1	4.3
−1 to −0.5	449,255	17.24627717	1858	16.49942279	15		
−0.5 to 0.5	555,481	21.32414617	1957	17.37856318	13		

(continued)

Table 7.3 (continued)

Classes	Number of pixel	Number of pixels (%)	Number of landslide occurrence pixel in each class	Number of landslide occurrence pixel in each class (%)	FR		AHP
					Class weight	Factor weight	Factor weight
0.5–1	451,276	17.32386056	1810	16.0731729	14		
1–1.5	326,444	12.53173299	1316	11.68635112	15		
Above 1.5	249,217	9.567095429	1545	13.7199183	23		

Source Researcher's computation

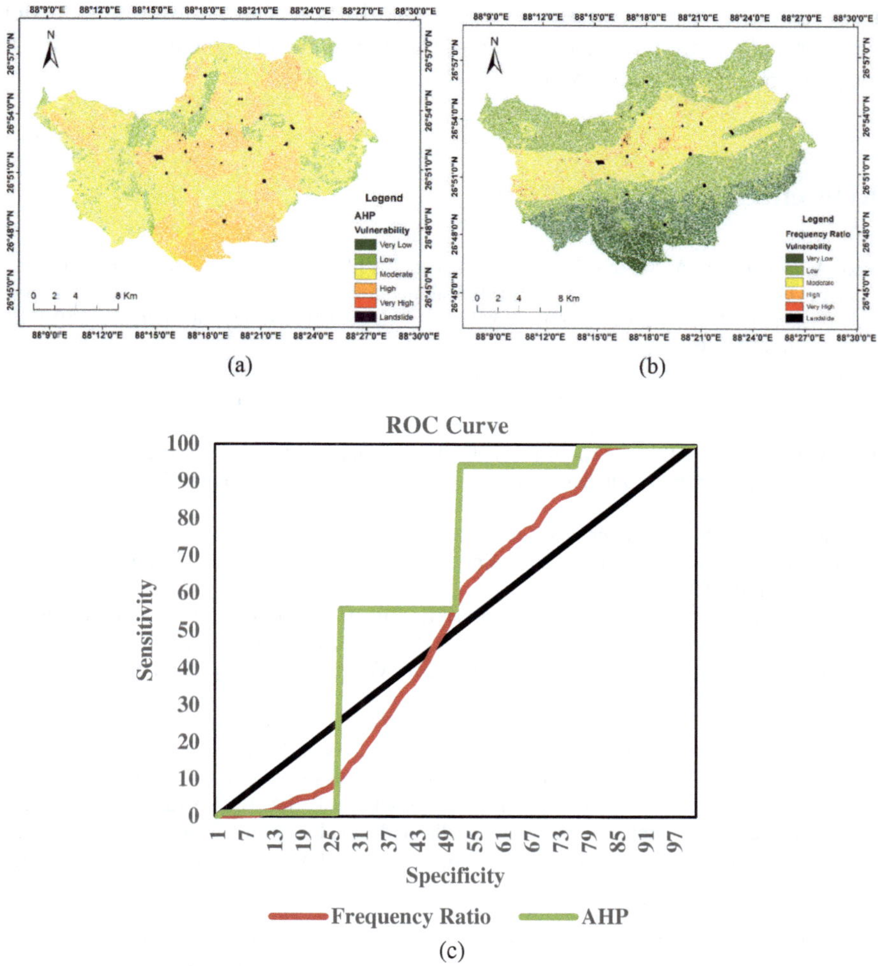

Fig. 7.6 **a** LSZ by AHP model, **b** LSZ by FR model, **c** ROC_AUC

Table 7.4 Validation of the model

Validation method		Landslide susceptibility zonation	
		AHP	FR
Confusion matrix	Accuracy	0.69	0.82
	Error rate	0.31	0.18
ROC	AUC	62.27	51.55

Source Researcher's computation

gradient slope that helps to release greater kinetic force for the increment of feasibility to transfer of mass. To prevent further landslides, it is best to avoid construction work or human interference by road construction on the aforementioned portion to keep the mountain slope stability.

References

Abdi A, Bouamrane A, Karech T, Dahri N, Kaouachi A (2021) Landslide susceptibility mapping using GIS-based fuzzy logic and the analytical hierarchical processes approach: a case study in Constantine (North-East Algeria). Geotech Geol Eng 39:5675–5691. https://doi.org/10.1007/s10706-021-01855-3

Achour Y, Boumezbeur A, Hadji R, Chouabbi A, Cavaleiro V, Bendaoud EA (2017) Landslide susceptibility mapping using analytic hierarchy process and information value methods along a highway road section in Constantine, Algeria. Arab J Geosci 10(8):194

Ahmed B, Dewan A (2017) Application of bivariate and multivariate statistical techniques in landslide susceptibility modeling in Chittagong city corporation, Bangladesh. Remote Sens 9(4):304

Akgun A, Erkan O (2016) Landslide susceptibility mapping by geographical information system based multivariate statistical and deterministic models: in an artificial reservoir area at Northern Turkey. Arab J Geosci 9(2):165

Al-Najjar HA, Pradhan B, Beydoun G, Sarkar R, Park HJ, Alamri A (2022) A novel method using explainable artificial intelligence (XAI)-based Shapley Additive Explanations for spatial landslide prediction using time-series SAR dataset. Gondwana Res.https://doi.org/10.1016/j.gr.2022.08.004

Amah EM, Katte VY, Ghogomu RT, Kamgang VK (2022) An assessment of landslides along mountain forest roads: bamenda ring road segment through bafut and befang forests on the Cameroon volcanic line. Am J Environ Stud 5:60–86. https://doi.org/10.47672/ajes.1160

Apostolov AA, Yelistratova LA, Romanciuc IF, Zakharchuk IUL (2021) Identifying potential landslide areas by employing the erosion relief index and meteorological criteria in Ukraine. Revue Roumaine de Géographie/Rom J Geogr 65:125–141. https://refhub.elsevier.com/S2405-8440(23)03393-5/sref10

Arabameri A, Pradhan B, Rezaei K, Sohrabi M, Kalantari Z (2019) GIS-based landslide susceptibility mapping using numerical risk factor bivariate model and its ensemble with linear multivariate regression and boosted regression tree algorithms. J Mt Sci 16(3):595–618

Avila A, Justino F, Wilson A, Bromwich D, Amorim M (2016) Recent precipitation trends, flash floods and landslides in southern Brazil. Environ Res Lett 11:114029. http://refhub.elsevier.com/S2405-8440(23)03393-5/sref13

Barlow J, Martin Y, Franklin SE (2003) Detecting translational landslide scars using segmentation of Landsat ETM+ and DEM data in the northern Cascade Mountains, British Columbia. Can J Remote Sens 29(4):510–517

Basu T, Pal S (2018) Identification of landslide susceptibility zones in Gish River basin, West Bengal, India, Georisk 12:14–28. https://doi.org/10.1080/17499518.2017.1343482

Basu T, Pal S (2020) A GIS-based factor clustering and landslide susceptibility analysis using AHP for Gish River Basin, India. Environ Dev Sustain 22:4787–4819. https://doi.org/10.1007/s10668-019-00406-4

Beven KJ, Kirkby MJ (1979) A physically based, variable contributing area model of basin hydrology/Un modèle à base physique de zone d'appel variable de l'hydrologie du bassin versant. Hydrol Sci J 24:43–69. https://doi.org/10.1080/02626667909491834

Bhattacharaya SK (2016) A comparison of the predicted vulnerability zones with the data based on hazard zones of landslide in the Kurseong hill subdivision, Darjeeling district, West Bengal, India. Springer, Berlin, Heidelberg. https://doi.org/10.1007/s12665-016-5729-8

Bhattacharya SK (2020) Geomorphometric analysis and terrain evaluation for environmental management in the Kurseong hill subdivision of the Darjeeling district, West Bengal, India. Environ Dev Sustain 22(2):985–1016

Bien TX, Truyen PT, Van Phong T, Nguyen DD, Amiri M, Costache R, Duc DM, Van Le H, Nguyen HBT, Prakash I, Pham BT (2022) Landslide susceptibility mapping at Sin Ho, Lai Chau Province, Vietnam using ensemble models based on fuzzy unordered rules induction algorithm. Geocarto Int 1–21. https://doi.org/10.1080/10106049.2022.2136253

Bilir TE, Fung I, Dawson TE (2021) Slope-aspect induced climate differences influence how water is exchanged between the land and atmosphere. J Geophys Res Biogeosci 126:2020JG006027. https://doi.org/10.1029/2020JG006027

Bourenane H, Bouhadad Y, Guettouche MS, Braham M (2015) GIS-based landslide susceptibility zonation using bivariate statistical and expert approaches in the city of Constantine (Northeast Algeria). Bull Eng Geol Environ 74:337–355. https://doi.org/10.1007/s10064-014-0616-6

Carson M, Kirkby M (1972) Hillslope form and process. Cambridge University Press, London

Chalkias C, Ferentinou M, Polykretis C (2014) GIS-based landslide susceptibility mapping on the Peloponnese Peninsula, Greece. Geosciences 4(3):176–190

Chawla A, Pasupuleti S, Chawla S, Rao ACS, Sarkar K, Dwivedi R (2019) Landslide susceptibility zonation mapping: a case study from Darjeeling district, Eastern Himalayas, India. J Indian Soc Remote Sens 47(3):497–511

Chen X, Chen W (2021) GIS-based landslide susceptibility assessment using optimized hybrid machine learning methods. Catena 196:104833. https://doi.org/10.1016/j.catena.2020.104833

Chen W, Panahi M, Tsangaratos P, Shahabi H, Ilia I, Panahi S, Li S, Jaafari A, Ahmad BB (2019) Applying population-based evolutionary algorithms and a neuro-fuzzy system for modeling landslide susceptibility. Catena 172:212–231

Chen CW, Oguchi T, Hayakawa YS, Saito H, Chen H (2017) Relationship between landslide size and rainfall conditions in Taiwan. Landslides 14(3):1235–1240

Cheng YS, Yu TT, Son NT (2021) Random forests for landslide prediction in Tsengwen river watershed, central Taiwan. Remote Sens 13:199. https://doi.org/10.3390/rs13020199

Chueasamat A, Hori T, Saito H, Sato T, Kohgo Y (2018) Experimental tests of slope failure due to rainfalls using 1g physical slope models. Soils Found 58:290–305. https://doi.org/10.1016/j.sandf.2018.02.003

Das I, Stein A, Kerle N, Dadhwal VK (2011) Probabilistic landslide hazard assessment using homogeneous susceptible units (HSU) along a national highway corridor in the northern Himalayas, India. Landslides 8(3):293–308

Das J, Saha P, Mitra R, Alam A, Kamruzzaman MD (2023) GIS-based data-driven bivariate statistical models for landslide susceptibility prediction in Upper Tista Basin, India. Heliyon 9:1–27. https://doi.org/10.1016/j.heliyon.2023.e16186

De La Ville N, Diaz AC, Ramirez D (2002) Remote sensing and GIS technologies as tools to support sustainable management of areas devastated by landslides. Environ Dev Sustain 4(2):221–229

Debanshi S, Pal S (2020) Assessing gully erosion susceptibility in Mayurakshi river basin of eastern India. Environ Dev Sustain 22(2):883–914

Dimri S, Lakhera RC, Sati S (2007) Fuzzy-based method for landslide hazard assessment in active seismic zone of Himalaya. Landslides 4(2):101

Einstein HH (1988) Special lecture: landslide risk assessment procedure. Proceedings 5th International Symposium on Landslides. Lausanne 2:1075–1090

Eitvandi N, Sarikhani R, Derikvand S (2022) Landslide susceptibility mapping by integrating analytical hierarchy process, frequency ratio, and fuzzy gamma operator models, case study: north of Lorestan Province, Iran. Environ Monit Assess 194:1–26. https://doi.org/10.1007/s10661-022-10206-5

Ercanoglu M, Temiz FA (2011) Application of logistic regression and fuzzy operators to landslide susceptibility assessment in Azdavay (Kastamonu, Turkey). Environ Earth Sci 64(4):949–964

Feng W, Bai H, Lan B, Wu Y, Wu Z, Yan L, Ma X (2022) Spatial–temporal distribution and failure mechanism of group-occurring landslides in Mibei village, Longchuan County, Guangdong, China. Landslides 1–14. https://doi.org/10.1007/s10346-022-01904-9

Froese CR, Moreno F (2014) Structure and components for the emergency response and warning system on Turtle Mountain, Alberta, Canada. Nat Hazards 70(3):1689–1712

Froude MJ, Petley DN (2018) Global fatal landslide occurrence from 2004 to 2016. Nat Hazards Earth Syst Sci 18:2161–2181. https://doi.org/10.5194/nhess-18-2161-2018

Ghosh S, van Westen CJ, Carranza EJM, Jetten VG, Cardinali M, Rossi M et al (2012) Generating event-based landslide maps in a data-scarce Himalayan environment for estimating temporal and magnitude probabilities. Eng Geol 128:49–62

Gudiyangada Nachappa T, Kienberger S, Meena SR, Hölbling D, Blaschke T (2020) Comparison and validation of per-pixel and object-based approaches for landslide susceptibility mapping. Geomat Nat Hazards Risk 11(1):572–600

Guha-Sapir D, Below R, Hoyois PH (2018) EM-DAT: international disaster database. Université Catholique de Louvain, Brussels, Belgium. http://www.emdat.be

Gupta SK, Shukla DP, Thakur M (2018) Selection of weightages for causative factors used in preparation of landslide susceptibility zonation (LSZ). Geomat Nat Hazards Risk 9:471–487. https://doi.org/10.1080/19475705.2018.1447027

Guzzetti F, Carrara A, Cardinali M, Reichenbach P (1999) Landslide hazard evaluation: a review of current techniques and their application in a multi-scale study, Central Italy. J Geomorphol 31:181–216

Hassangavyar MB, Damaneh HE, Pham QB, Linh NTT, Tiefenbacher J, Bach QV (2022) Evaluation of re-sampling methods on performance of machine learning models to predict landslide susceptibility. Geocarto Int 37:2772–2794. https://doi.org/10.1080/10106049.2020.1837257

He H, Hu D, Sun Q, Zhu L, Liu Y (2019) A landslide susceptibility assessment method based on GIS technology and an AHP-weighted information content method: a case study of southern Anhui, China. ISPRS Int J Geo-Inf 8(6):266

Hirota K, Konagai K, Sassa K, Dang K, Yoshinaga Y, Wakita EK (2019) Landslides triggered by the west Japan heavy rain of July 2018, and geological and geomorphological features of soaked mountain slopes. Landslides 16(1):189–194

Hong H, Pradhan B, Xu C, Tien Bui D (2015) Spatial prediction of landslide hazard at the Yihuang area (China) using two-class kernel logistic regression, alternating decision tree and support vector machines. Catena 133:266–281. https://doi.org/10.1016/j.catena.2015.05.019

Hung LQ, Van NTH, Duc DM, Ha LTC, Van Son P, Khanh NH, Binh LT (2016) Landslide susceptibility mapping by combining the analytical hierarchy process and weighted linear combination methods: a case study in the upper Lo River catchment (Vietnam). Landslides 13(5):1285–1301

IDNHR, Advisory Committee (1987) Confronting natural disasters. An international decade for natural hazard reduction. National Academy Press, WA, pp 60

Jaafari A (2018) LiDAR-supported prediction of slope failures using an integrated ensemble weights-of-evidence and analytical hierarchy process. Environ Earth Sci 77(2):42

Jaksic S, Ninkov J, Milic S, Vasin J, Zivanov M, Jaksic D, Komlen V (2021) Influence of slope gradient and aspect on soil organic carbon content in the region of Niš, Serbia. Sustainability 13:8332. https://doi.org/10.3390/su13158332

Kalantar B, Pradhan B, Naghibi S, Motevalli A, Mansor S (2018) Assessment of the effects of training data selection on the landslide susceptibility mapping: a comparison between support vector machine (SVM), logistic regression (LR) and artificial neural networks (ANN). Geomat Nat Hazards Risk. https://doi.org/10.1080/19475705.2017.1407368

Kalantar B, Ueda N, Najjar HAHAl, Gibril MBA, Lay US, Motevalli A (2019) An evaluation of landslide susceptibility mapping using remote sensing data and machine learning algorithms in Iran. ISPRS Ann Photogramm Remote Sens Spat Inf Sci 4(2):503–511. https://doi.org/10.5194/isprs-annals-IV-2-W5-503-2019

Kanti Hembram T, Paul GC, Saha S (2019) Spatial prediction of susceptibility to gully erosion in Jainti River basin, Eastern India: a comparison of information value and logistic regression models. Model Earth Syst Environ 5(2):689–708

Kaur H, Gupta S, Parkash S, Thapa R, Gupta A, Khanal GC (2019) Evaluation of landslide susceptibility in a Hill City of Sikkim Himalaya with the perspective of hybrid modelling techniques. Ann GIS 25(2):113–132

Kavzoglu T, Sahin EK, Colkesen I (2014) Landslide susceptibility mapping using GIS-based multi-criteria decision analysis, support vector machines, and logistic regression. Landslides 11(3):425–439

Kayastha P, Dhital M, De Smedt F (2013) Application of the analytical hierarchy process (AHP) for landslide susceptibility mapping: a case study from the Tinau watershed, west Nepal. Comput Geosci 52:398–408

Keefer DK (2000) Statistical analysis of an earthquake-induced landslide distribution—the 1989 Loma Prieta, California event. Eng Geol 58(3–4):231–249

Lee S, Pradhan B (2006) Probabilistic landslide hazards and risk mapping on Penang Island, Malaysia. J Earth Syst Sci 115(6):661–672

Liao M, Wen H, Yang L (2022) Identifying the essential conditioning factors of landslide susceptibility models under different grid resolutions using hybrid machine learning: a case of Wushan and Wuxi counties, China. Catena 217:106428. https://doi.org/10.1016/j.catena.2022.106428

Lin GW, Chen H (2012) The relationship of rainfall energy with landslides and sediment delivery. Eng Geol 125:108–118

Liu J, Xu Q, Wang S, Siva Subramanian S, Wang L, Qi X (2020) Formation and chemo-mechanical characteristics of weak clay interlayers between alternative mudstone and sandstone sequence of gently inclined landslides in Nanjiang, SW China. Bull Eng Geol Environ 79:4701–4715. https://doi.org/10.1007/s10064-020-01859-y

Lollino G et al (2015) Engineering geology for society and territory—volume 2: landslide processes. Lollino G, Giordan D, Crosta GB, Corominas J, Azzam R, Wasowski J, Sciarra N (eds) Engineering geology for society and territory, vol 2: landslide processes. Springer International Publishing, Cham. https://doi.org/10.1007/978-3-319-09057-3

Mahalingam R, Olsen MJ, O'Banion MS (2016) Evaluation of landslide susceptibility mapping techniques using lidar-derived conditioning factors (Oregon case study). Geomat Nat Hazards Risk 7(6):1884–1907

Manchar N, Benabbas C, Hadji R, Bouaicha F, Grecu F (2018) Landslide susceptibility assessment in Constantine region (NE Algeria) by means of statistical models. Studia Geotechnica et Mechanica 40:208–219. https://doi.org/10.2478/sgem-2018-0024

Mandal B, Mandal S (2016) Assessment of mountain slope instability in the Lish River basin of Eastern Darjeeling Himalaya using frequency ratio model (FRM). Springer International Publishing Switzerland. https://doi.org/10.1007/s40808-0169-8

Mandal S, Saha A (2018) Support vector machines for monitoring land use dynamicity and temporal variation of land surface temperature in Kurseong and surrounding of Darjeeling Himalaya. Model Earth Syst Environ 4(2):659–672

Meten M, PrakashBhandary N, Yatabe R (2015) Effect of landslide factor combinations on the prediction accuracy of landslide susceptibility maps in the Blue Nile gorge of central Ethiopia. Geoenviron Disasters 2(1):9

Miller AJ (2011) Identifying landslide activity as a function of economic development: a case study of increased landslide frequency surrounding Dominical, Costa Rica. Environ Dev Sustain 13(5):901

Mind'je R, Li L, Nsengiyumva JB, Mupenzi C, Nyesheja EM, Kayumba PM et al (2019) Landslide susceptibility and influencing factors analysis in Rwanda. Environ Dev Sustain. https://doi.org/10.1007/s10668-019-00557-4

Mitra R, Das J (2022) A comparative assessment of flood susceptibility modelling of GIS-based TOPSIS, VIKOR, and EDAS techniques in the Sub-Himalayan foothills region of Eastern India. Environ Sci Pollut Res. https://doi.org/10.1007/s11356-022-23168-5

Mitra R, Saha P, Das J (2022) Assessment of the performance of GIS-based analytical hierarchical process (AHP) approach for flood modelling in Uttar Dinajpur district of West Bengal, India. Geomat Nat Hazards Risk 13:2183–2226. https://doi.org/10.1080/19475705.2022.2112094

Mondal S, Mandal S (2017) Application of frequency ratio (FR) model in spatial prediction of landslides in the Balason river basin, Darjeeling Himalaya. Spat Inf Res 25(3):337–350

Moore ID, Burch GJ (1986) Sediment transport capacity of sheet and rill flow: application of unit stream power theory. Water Resour Res 22:1350–1360. https://doi.org/10.1029/WR022i008p01350

Moore ID, Grayson RB, Ladson AR (1991) Digital terrain modelling: a review of hydrological, geomorphological, and biological applications. Hydrol Process 5:3–30. https://doi.org/10.1002/hyp.3360050103

Moosavi V, Niazi Y (2016) Development of hybrid wavelet packet-statistical models (WP-SM) for landslide susceptibility mapping. Landslides 13:97–114. https://refhub.elsevier.com/S1674-9871(21)00075-X/h0190

Naghibi SA, Moradi DM (2016) Evaluation of four supervised learning methods for groundwater spring potential mapping in Khalkhal region (Iran) using GIS-based features. Hydrogeol J 25(1):169–189

Nampak H, Pradhan B, Manap MA (2014) Application of GIS based data driven evidential belief function model to predict groundwater potential zonation. J Hydrol 513:283–300

Neamat S, Karimi H (2020) A systematic review of GIS-based landslide hazard mapping on determinant factors from international databases. In: 2020 International conference on advanced science and engineering (ICOASE), pp 180–183. https://doi.org/10.1109/ICOASE51841.2020.9436611

Nguyen T, Liu CC (2019) A new approach using ahp to generate landslide susceptibility maps in the Chen-Yu-Lan watershed, Taiwan. Sensors 19(3):505

Pal R, Biswas SS, Mondal B, Pramanik MK (2016) Landslides and floods in the Tista Basin (Darjeeling and Jalpaiguri Districts): historical evidence, causes and consequences. J Indian Geophys Union 20(2):209–215

Pardeshi SD, Autade SE, Pardeshi SS (2013) Landslide hazard assessment: recent trends and techniques. Springer Plus 2:523. https://doi.org/10.1186/2193-1801-2-523

Park S, Kim J (2019) Landslide susceptibility mapping based on random forest and boosted regression tree models, and a comparison of their performance. Appl Sci 9(5):942

Pourghasemi HR, Pradhan B, Gokceoglu C (2012) Application of fuzzy logic and analytical hierarchy process (AHP) to landslide susceptibility mapping at Haraz watershed, Iran, Nat Hazards 63:965–996, https://doi.org/10.1007/s11069-012-0217-2

Paul GC, Saha S, Hembram TK (2019) Application of the GIS-based probabilistic models for mapping the flood susceptibility in Bansloi sub-basin of Ganga-Bhagirathi river and their comparison. Remote Sens Earth Syst Sci 2(2–3):120–146

Pawluszek K, Borkowski A (2017) Impact of DEM-derived factors and analytical hierarchy process on landslide susceptibility mapping in the region of Roznow Lake, Poland. Nat Hazards 86:919–952. https://doi.org/10.1007/s11069-016-2725-y

Pecoraro G, Calvello M, Piciullo L (2019) Monitoring strategies for local landslide early warning systems. Landslides 16(2):213–231

Pham BT, Pradhan B, Bui DT, Prakash I, Dholakia MB (2016) A comparative study of different machine learning methods for landslide susceptibility assessment: a case study of Uttarakhand area (India). Environ Model Softw 84:240–250

Pham BT, Shirzadi A, Bui DT, Prakash I, Dholakia MB (2018) A hybrid machine learning ensemble approach based on a radial basis function neural network and rotation forest for landslide susceptibility modeling: a case study in the Himalayan area, India. Int J Sediment Res 33(2):157–170. https://doi.org/10.1016/j.ijsrc.2017.09.008

Pham QB et al (2021) A comparison among fuzzy multi-criteria decision making, bivariate, multivariate and machine learning models in landslide susceptibility mapping. Geomat Nat Hazards Risk 12(1):1741–1777. https://doi.org/10.1080/19475705.2021.1944330

Pisano L, Zumpano V, Malek Ž, Rosskopf CM, Parise M (2017) Variations in the susceptibility to landslides, as a consequence of land cover changes: a look to the past, and another towards the future. Sci Total Environ 601:1147–1159

Poudyal CP, Chang C, Oh HJ, Lee S (2010) Landslide susceptibility maps comparing frequency ratio and artificial neural networks: a case study from the Nepal Himalaya. Environ Earth Sci 61:1049–1064. https://doi.org/10.1007/s12665-009-0426-5

Pourghasemi HR, Pradhan B, Gokceoglu C, Mohammadi M, Moradi HR (2013) Application of weights-of-evidence and certainty factor models and their comparison in landslide susceptibility mapping at Haraz watershed, Iran. Arab J Geosci 6(7):2351–2365. https://doi.org/10.1007/s12517-012-0532-7

Pradhan B, Lee S (2010) Landslide susceptibility assessment and factor effect analysis: back propagation artificial neural networks and their comparison with frequency ratio and bivariate logistic regression modeling. Environ Model Softw 25:747–759. https://doi.org/10.1016/j.envsoft.2009.10.016

Pradhan AM, Singh HS, Kang SL, Kim YT (2017) Spatial model integration for shallow landslide susceptibility and its runout using a GIS-based approach in Yongin, Korea. Geocarto Int 32(4):420–441. https://doi.org/10.1080/10106049.2016.1155658

Prakasam C, Aravinth R, Kanwar VS, Nagarajan B (2020) Comparative study between weighted overlay and fuzzy logic models for landslide vulnerability mapping—a case study of Rampur Tehsil, Himachal Pradesh. In: Sustainable civil engineering practices. Springer, Singapore, pp 155–171

Provost F, Kohavi R (1998) Guest editors' introduction: on applied research in machine learning. Mach Learn 30(2):127–132

Rai DK, Xiong D, Zhao W, Zhao D, Zhang B, Dahal NM, Wu Y, Baig MA (2022) An investigation of landslide susceptibility using logistic regression and statistical index methods in Dailekh district. Nepal Chin Geogr Sci 32:834–851. https://doi.org/10.1007/s11769-022-1304-2

Rasyid AR, Bhandary NP, Yatabe R (2016) Performance of frequency ratio and logistic regression model in creating GIS based landslides susceptibility map at Lompobattang Mountain, Indonesia. Geoenviron Disasters 3(1). https://doi.org/10.1186/s40677-016-0053-x

Razavi T, Vahid S, Kornejady A, Pourghasemi HR, Keesstra S (2018) Flood susceptibility mapping using novel ensembles of adaptive neuro fuzzy inference system and metaheuristic algorithms. Sci Total Environ 615:438–451. https://doi.org/10.1016/j.scitotenv.2017.09.262

Regmi AD, Devkota KC, Yoshida K, Pradhan B, Pourghasemi HR, Kumamoto T et al (2014) Application of frequency ratio, statistical index, and weights-of-evidence models and their comparison in landslide susceptibility mapping in Central Nepal Himalaya. Arab J Geosci 7(2):725–742

Rodriguez J, Macciotta R, Hendry MT, Roustaei M, Grapel C, Skirrow R (2020) UAVs for monitoring, investigation, and mitigation design of a rock slope with multiple failure mechanisms—a case study. Landslides 17(9):2027–2040. https://doi.org/10.1007/s10346-020-01416-4

Roy D, Das S, Mitra R (2022) An application of geospatial-based multi-criteria decision-making technique to identify landslide susceptibility zones in the Ragnu Khola River Basin of Darjeeling Himalayan region, India. Appl Geomat 1–19. https://doi.org/10.1007/s12518-022-00468-6

Roy J, Saha S (2019) Landslide susceptibility mapping using knowledge driven statistical models in Darjeeling District, West Bengal, India. Geoenviron Disasters 6(1):11

Saaty TL (1980) The analytic process: planning, priority setting, resources allocation. McGraw-Hill International, New York, NY, USA

Saaty T, Vargas L (1984) Inconsistency and rank preservation. J Math Psychol 28(2):205–214

Saaty T, Vargas L (1991) Prediction, projection, and forecasting: applications of the analytic hierarchy process in economics, finance, politics, games, and sports. Kluwer Academic Pub, Springer Netherlands

Saha A, Saha S (2020) Application of statistical probabilistic methods in landslide susceptibility assessment in Kurseong and its surrounding area of Darjeeling Himalayan, India: RS-GIS approach. Environ Dev Sustain. https://doi.org/10.1007/s10668-020-00783-1

Saha P, Mitra R, Chakraborty K, Roy M (2022) Application of multi-layer perceptron neural network Markov chain model for LULC change detection in the Sub-Himalayan North Bengal. Remote Sens Appl Soc Environ 26:100730. https://doi.org/10.1016/j.rsase.2022.100730

Schuster RL (1995) Keynote paper: recent advances in slope stabilization. In: Bell (ed) Landslides. Balkema, Rotterdam, pp 1715–1745

Schuster RL, Fleming RW (1986) Economic losses and fatalities due to landslides. Bull Am Assoc Eng Geol 23(0):11–28

Sammut C, Webb GI (eds) (2011) Encyclopedia of machine learning. Springer, US. https://doi.org/10.1007/978-0-387-30164-8

Shirzadi A, Soliamani K, Habibnejhad M, Kavian A, Chapi K, Shahabi H, Chen W, Khosravi K, Pham T, Pradhan B, Ahmad A, Bin Ahmad B, Bui T (2018) Novel GIS based machine learning algorithms for shallow landslide susceptibility mapping. Sensors (Basel) 18:3777. http://refhub.elsevier.com/S1674-9871(21)00075-X/h0285

Stehman SV (1997) Selecting and interpreting measures of thematic classification accuracy. Remote Sens Environ 62(1):77–89

Sun DL, Wen HJ, Wang DZ, Xu JH (2020a) A random forest model of landslide susceptibility mapping based on hyperparameter optimization using Bayes algorithm. Geomorphology 362:107201. https://refhub.elsevier.com/S1674-9871(21)00075-X/h0300

Sun DL, Xu JH, Wen HJ, Wang Y (2020b) An optimized random forest modeland its generalization ability in landslide susceptibility mapping: application in two areas of Three Gorges Reservoir, China. J Earth Sci (China) 31:1068–1086. https://refhub.elsevier.com/S1674-9871(21)00075-X/h0310

Tiranti D, Nicolò G, Gaeta AR (2019) Shallow landslides predisposing and triggering factors in developing a regional early warning system. Landslides 16(2):235–251

Tien Bui D, Pradhan B, Lofman O, Revhaug I, Dick OB (2012) Landslide susceptibility assessment in the Hoa Binh province of Vietnam: a comparison of the Levenberg-Marquardt and Bayesian regularized neural networks. Geomorphology 171–172:12–29

Trinh T (2022) A comparative analysis of weight-based machine learning methods for landslide susceptibility mapping in Ha Giang area. Big Earth Data 1–30. https://doi.org/10.1080/20964471.2022.2043520

US Geological Survey (1982) Goals and tasks of the landslide part of a ground failure hazard reduction program. US Geological Survey Circular, vol 880, pp 44

UNDRO (1991) Mitigating natural disasters. Phenomena, effects and options. United Nations, New York, pp 164

Udin WS, Yahaya NN, Shariffuddin SIM (2021) Landslide susceptibility assessment using geographic information system in Aring, Gua Musang, Kelantan. In: IOP conference series: earth and environmental science, vol 842, p 012008. https://refhub.elsevier.com/S2405-8440(23)03393-5/sref115

Van Den Eeckhaut M, Vanwalleghem T, Poesen J, Govers G, Verstraeten G, Vandekerckhove L (2006) Prediction of landslide susceptibility using rare events logistic regression: a case-study in the Flemish Ardennes (Belgium). Geomorphology 76(3–4):392–410

Wang Q, Li W (2017) A GIS-based comparative evaluation of analytical hierarchy process and frequency ratio models for landslide susceptibility mapping. Phy Geogr 38(4):318–337

Van Westen CJ, Rengers N, Terlien MTJ, Soeters R (1997) Prediction of the occurrence of slope instability phenomenal through GIS-based hazard zonation. Geol Rundsch 86(2):404–414

Van Westen CJ, Rengers N, Soeters R (2003) Use of geomorphological information in indirect landslide susceptibility assessment. Nat Hazards 30(3):399–419

Vijith H, Seling LW, Dodge-Wan D (2018) Estimation of soil loss and identification of erosion risk zones in a forested region in Sarawak, Malaysia, Northern Borneo. Environ Dev Sustain 20(3):1365–1384

Vojteková J, Vojtek M (2020) Assessment of landslide susceptibility at a local spatial scale applying the multi-criteria analysis and GIS: a case study from Slovakia. Geomat Nat Hazards Risk 11(1):131–148

Wang G, Lei X, Chen W, Shahabi H, Shirzadi A (2020a) Hybrid computational intelligence methods for landslide susceptibility mapping. Symmetry 12(3):325

Wang Y, Fang Z, Wang M, Peng L, Hong H (2020b) Comparative study of landslide susceptibility mapping with different recurrent neural networks. Comput Geosci 138:104445

Wang Y, Sun DL, Wen HJ, Zhang H, Zhang FT (2020) Comparison of random forest model and frequency ratio model for landslide susceptibility mapping (LSM) in Yunyang County (Chongqing, China). Int J Environ Res Public Health 17(12):4206. https://refhub.elsevier.com/S1674-9871(21)00075-X/h0355

Watakabe T, Matsushi Y (2019) Lithological controls on hydrological processes that trigger shallow landslides: observations from granite and hornfels hillslopes in Hiroshima, Japan. Catena 180:55–68. https://doi.org/10.1016/j.catena.2019.04.010

Xia M, Ren GM, Ma XL (2013) Deformation and mechanism of landslide influenced by the effects of reservoir water and rainfall, Three Gorges, China. Nat Hazards 68:467–482. https://doi.org/10.1007/s11069-013-0634-x

Xiao T, Segoni S, Chen L, Yin K, Casagli N (2020) A step beyond landslide susceptibility maps: a simple method to investigate and explain the different outcomes obtained by different approaches. Landslides 17:627–640. https://doi.org/10.1007/s10346-019-01299-0

Xiong X, Shi Z, Xiong Y, Peng M, Ma X, Zhang F (2019) Unsaturated slope stability around the Three Gorges Reservoir under various combinations of rainfall and water level fluctuation. Eng Geol 261:105231. https://doi.org/10.1016/j.enggeo.2019.105231

Yalcin A (2008) GIS-based landslide susceptibility mapping using analytical hierarchy process and bivariate statistics in Ardesen (Turkey): comparisons of results and confirmations. Catena 72(1):1–12

Yalcin A, Reis S, Aydinoglu A, Yomralioglu T (2011) A GIS-based comparative study of frequency ratio, analytical hierarchy process, bivariate statistics and logistics regression methods for landslide susceptibility mapping in Trabzon, NE Turkey. Catena 85(3):274–287

Yang D, Qiu H, Hu S, Pei Y, Wang X, Du C, Long Y, Cao M (2021) Influence of successive landslides on topographic changes revealed by multitemporal high-resolution UAS-based DEM. Catena 202:105229. https://doi.org/10.1016/j.catena.2021.105229

Yesilnacar E, Topal T (2005) Landslide susceptibility mapping: a comparison of logistic regression and neural networks methods in a medium scale study, Hendek Region (Turkey). Eng Geol 79(3–4):251–266. https://doi.org/10.1016/j.enggeo.2005.02.002

Zhan J, Wang Q, Zhang W, Shangguan Y, Song S, Chen J (2019) Soil-engineering properties and failure mechanisms of shallow landslides in soft-rock materials. Catena 181. https://doi.org/10.1016/j.catena.2019.104093

Zhang W, Liu S, Wang L, Samui P, Chwala M, He Y (2022) Landslide susceptibility research combining qualitative analysis and quantitative evaluation: a case study of Yunyang County in Chongqing, China. Forests 13:1055. https://doi.org/10.3390/f13071055

Zhou C, Cao Y, Yin K, Wang Y, Shi X, Catani F, Ahmed B (2020) Landslide characterization applying sentinel-1 images and in sar technique: the muyubao landslide in the three Gorges Reservoir Area, China. Remote Sens 12(20):3385

Zhou X, Wen H, Zhang Y, Xu J, Zhang W (2021) Landslide susceptibility mapping using hybrid random forest with Geo detector and RFE for factor optimization. Geosci Front 12:101211. https://doi.org/10.1016/j.gsf.2021.101211

Chapter 8
Suitability Analysis of Landslide Susceptibility Model of Uttarkashi District in Uttarakhand, India: A Comparative Approach Between Weighted Overlay and Multi-criteria Decision Analysis

Asutosh Goswami, Suhel Sen, and Priyanka Majumder

Abstract Natural disasters are can take place anytime and anywhere. When a natural disaster strikes, it lays a severe effect on the natural and social environment. Sometimes the intensity of damage goes to such an extent that it becomes almost difficult for man to cope with the situation and come out of the losses. Hence, it is very much necessary to adopt proper mitigation strategies so that the severity of the disaster can be reduced and colossal loss to life and property can be averted. The present study makes an attempt to perform a comparative study of landslide susceptibility of Uttarkashi district of Uttarakhand prepared using weighted overlay technique and multi-criteria decision analysis technique by applying GIS and Remote Sensing tools and also tries to point out the suitable model out of the two. The study results revealed five landslide susceptibility zones and also found that the both the models were 'Good'. However, AUC value of success rate curve of model prepared using weighted overlay (79.7%) is greater than that of multi-criteria decision analysis (78.9%) and thus is considered to more applicable for the future scenario. The study has also delved into the assessment of outrageous values of landslide events through the computation of stand error. The standard error values are widely scattered giving a result of non-homogenous distribution of the landslide points. The multi-spectral behavior also indicate the identical result as majority of the landslide events are detected near the higher elevated areas.

Keywords Susceptibility modelling · Landslide · Weighted overlay · Multi criteria decision · Predictive rate

A. Goswami (✉) · P. Majumder
Department of Earth Sciences and Remote Sensing, JIS University, Agarpara, West Bengal, India
e-mail: asutosh.goswami@jisuniversity.ac.in

S. Sen
Department of Geography, Bhairab Ganguly College, Belgharia, West Bengal, India

G. K. Panda et al. (eds.), *Landslide: Susceptibility, Risk Assessment and Sustainability*, Advances in Natural and Technological Hazards Research 52,
https://doi.org/10.1007/978-3-031-56591-5_8

8.1 Introduction

In spite of being technologically blessed, man is a mere puppet in the hands of nature even at present times. Nature plays a vital role in shaping the life of man and no one can go beyond the limit set by nature. It is a true fact that man in many ways have become successful in taming the nature and have established his existence dictating the rules of the environment. However, in no way man can free himself from the controls posed by environment upon him. Nature selects its own way. Man through his activities can modify this way, but cannot control it. This scenario becomes clearer when the issue of natural disaster comes into consideration (Krishnan et al. 2021; Hembram et al. 2022; Hussain et al. 2022; Gao et al. 2023; Ma et al. 2023). When a natural disaster takes place, it causes colossal loss of human lives and property. Man fights against the losses, copes with the situation and tries to get back to normalcy. There is no doubt of the fact that disasters act as a major obstacle in the path of development. But, occurrence of disasters and the losses caused by it also helps man to identify his weaknesses and thus gives him a chance to develop strategies to fight against it. For instance, earthquakes cause severe damage to human society and at the same time it also helps man to understand the reasons responsible for such losses. Thus, man will try to develop earthquake resistant structures. In other words, it is nature who sets all the obstacles and at the same time it is nature who gives the hint to bring the solution of those problems. So, the positive insight of a disaster should also be considered keeping in mind the negative consequences.

Landslide is one very well-known natural hazard. In simple words, landslide can be defined as the movement of rocks, debris and earth down the slope under the influence of gravity. When the shear stress present within a hill-slope exceeds the shear strength, landslide takes place. Landslides can be triggered by Earthquakes or due to heavy rainfall in mountainous areas. In higher altitudes, snow avalanches are also common. The Himalayan Mountains in India is one of the young fold mountain systems of the world and have been experiencing severe landslide events since time immemorial. Uttarkashi in Uttarakhand state in India is one such landslide prone districts located in the lap of the lofty Garhwal Himalayas. The Varunavat Hill landslide which took place 23rd September, 2003 is one such widely known landslide event (Sarkar et al. 2011). Slope instability and heavy rainfall were considered to be the primary reasons responsible for the landslide. People residing in the mountainous areas encounter several problems when a landslide takes place. Transport networks are totally destroyed and thus the link between the affected area and lower plains are disturbed. This in turn poses serious problems in providing relief and medical care to the victims. There is loss of crop which they cultivate on hill terraces. Communication network gets badly affected as telephone towers and internet service often comes to a halt. Landslides are also a major threat to the livelihood of many people and this in turn often stimulates livelihood transformation (Goswami et al. 2022a, b). Uttarkashi is considered to be one of the best spots for religious tourism and is enriched with a number of religious tourist spots. Besides, the district is also well known for its natural beauty. Owing to these, tourists from all over India as well as from different parts of the world come here and this becomes a source of income for the local people. But when landslide takes place, the tourism industry gets severely affected as tourists are not allowed to travel at higher altitudes for their own security. This in turn affects the employment structure of the tourist spots and many of them are compelled to adopt some other means to earn their daily breads stimulating a short-term livelihood transformation. In other words, landslide acts as an hindrance in the path of development of the district and thus needs proper management and mitigation strategies.

There is no doubt of the fact that adoption and implementation of landslide mitigation is not an easy task and demands lot of time and investment. Mitigation techniques have to be scientific and appropriate and then only the losses and damages can be reduced. The advancing technologies of GIS and Remote Sensing has made this much easier and have helped in preparation of Landslide Susceptibility Models using different techniques. These models have become indispensable weapons in the hands of the planners and researchers to adopt proper mitigation strategies. Several techniques like Shannon entropy, frequency ratio, analytical hierarchy process, multi criteria decision analysis and weighted overlay can be used for preparing landslide susceptibility maps. However, the present research aims to prepare and make a comparative analysis of landslide susceptibility models prepared using multi criteria decision analysis and weighted overlay analysis. It also aims to perform suitability analysis of the two models so that its application for future can be done.

Modelling has become a very important tool in disaster mitigation. This has been accelerated with the advance of geospatial technology. However, if the models are not validated and do not achieve necessary accuracy, they become futile (Wubalem and Meten 2020). Several research works have been done on modelling for mitigation of losses of disasters. Goswami et al. (2022a, b) is of the opinion that GIS and Remote Sensing in present times has become an important tool in preparing disaster susceptibility maps like landslides, floods, drought etc. and this maps play an important role in decision making (Goswami et al. 2022a, b). Mersha and Meten (2020) applied GIS and bi-variate statistical technique in preparing landslide susceptibility map of Simda in north-western Ethiopia. Frequency Ratio (FR) and Weight of Evidence techniques (WoE) were applied for modelling and 576 landslide events were used as data inventory (Mersha and Meten 2020). The study revealed five landslide susceptibility zones and identified predictive rate of 88.2% and 84.8% for FR and WoE respectively (Mersha and Meten 2020). Thus, FR model was considered to be more applicable. Fayez et al. (2018) considered Frequency Ratio to be an ideal technique for landslide susceptibility modelling while conducting landslide susceptibility assessment of a part of Uttarakhand Himalaya (Fayez et al. 2018). The study aimed for the preparation of landslide susceptibility map of a part of Uttarakhand Himalaya using Frequency Ratio model and took into account six landslide conditioning factors and 430 landslide locations of which 70% were used for modelling and 30% were used for validation (Pham et al. 2015). The AUC under success curve is 75% and thus implies that the model is good and can be an important tool in decision making (Pham et al. 2015). Roy and Saha (2019) aimed for making landslide susceptibility model of Darjeeling district in West Bengal using Landslide Numerical Risk Factor (LNRF) and Analytical Hierarchy Process (AHP) (Roy and Saha 2019). The study revealed that about 34% and 22% of the total study were highly susceptible to landslides in case of LNRF and AHP respectively and LNRF model was found to be promising as it enjoyed a higher AUC value of 91% than AHP with AUC value of 90% (Roy and Saha 2019). Shahabi and Hashim (2015) used GIS technique for preparation of landslide susceptibility map of Cameroon Highlands of Malaysia. For the completion of the work, spatial multi-criteria evaluation (SMCE), analytical hierarchy process (AHP) and weighted linear combination (WLC) were used. Model prepared using SMCE was found to be more appropriate than AHP and WLC as SMCE was found to be 96% accurate while for AHP and WLC, the accuracy was 91% and 89% respectively (Shahabi and Hashim 2015). Batar and Watanabe (2021) aimed for the preparation of landslide susceptibility map of Rudraprayag district using Weight of Evidence method and achieved 85.7% accuracy. Sonker et al. (2021) also prepared landslide susceptibility map of Sikkim Himalaya using AHP technique and achieved 73% accuracy and hence was considered good. Arsyad and Hamid (2020) also achieved 91.3% accuracy in making landslide susceptibility map of West Sulawesi using AHP process.

In view of the above mentioned issues, the main objective of the study is to prepare a landslide susceptibility map of Uttarkashi district of Uttarakhand state using weighted overlay and multi-criteria decision analysis techniques and also to determine the suitability of both the maps prepared using two different techniques. The novelty of this research lies in the detection of standard error prelude of the landslide events which are having vociferous impact on the growth of economic activities in the affected areas.

8.2 Methodology

The entire research work has been done using geospatial techniques and primary observations (Fig. 8.1). In order to accomplish the work, five major factors which trigger landslide in an area were taken into account and they were considered as 'conditioning factors'. The five conditioning factors taken in case of present research work were elevation, aspect, slope, rainfall and lineament density. About 488 landslide events were used as 'Landslide Data Inventory' and they were downloaded from Bhukosh website of Geological Survey of India. For the preparation of elevation, slope and aspect map, digital elevation model (DEM) was downloaded from USGS Earth Explorer website. Rainfall data for the year 2021 was obtained in gridded format from India Meteorological Department, Pune. Lineament data was also downloaded from Bhukosh website of GSI.

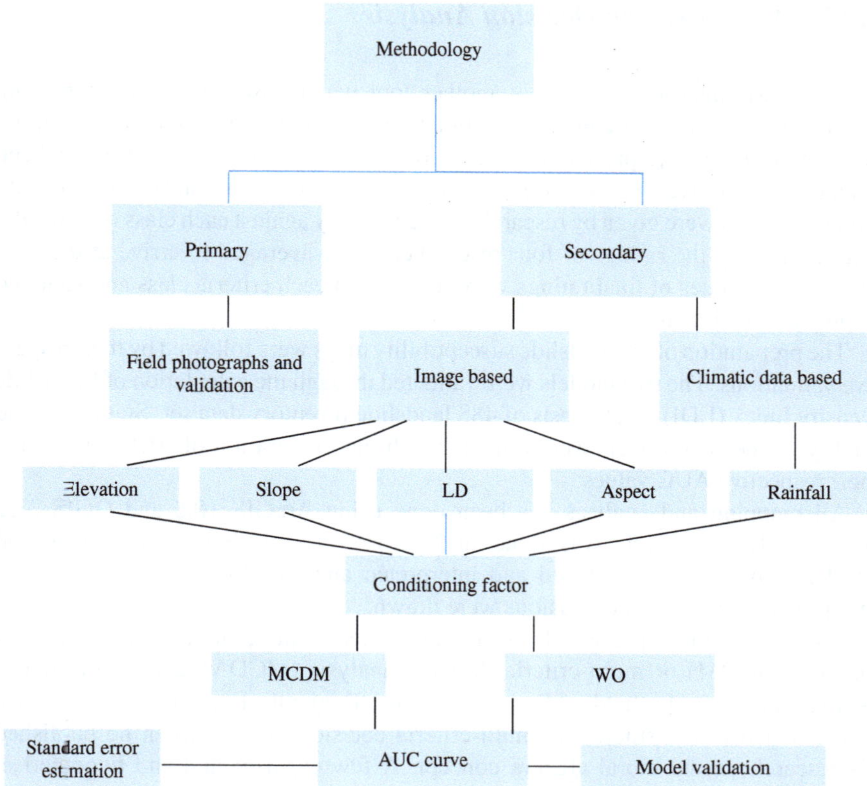

Fig. 8.1 Methodological flow-chart

8.2.1 *Weighted Overlay Technique*

Weighted overlay is a tool which is used for solving problems having more than one criteria. All the conditioning factors taken into consideration are divided into equal number of classes and are rated from 1 to 5 based on their susceptibility. 1 stands for very low susceptibility while 5 stands for very high susceptibility. On the basis of susceptibility of each parameter, weightage is given in percentage by the researchers after field visit and detailed discussion with the experts and extensive literature review. Thus, elevation was assigned the weightage value of 9%, slope 27%, aspect 9%, rainfall 27% and lineament density 27%. All these weightages and the criteria were overlayed to prepare the final landslide susceptibility map.

8.2.2 Multi Criteria Decision Analysis

Multi criteria decision analysis is another tool widely used for decision making and for arriving at rigid conclusions. Final decision is taken on the basis of more than one criteria. The present study considers five criteria. Each criteria has been categorized into five classes. Based on detailed literature review and discussion with experts, ratings were given by researchers individually against each class of a specific criteria. Finally the ratings of four researchers were averaged to arrive at the final ratings. The values of final ratings were replaced in each criteria class and summed to prepare the final map.

The preparation of the landslide susceptibility maps were followed by their respective validations. The two models were validated through the calculation of Landslide Density Index (LDI) on the basis of 488 landslide inventory data set. Suitability and quality of the two models were done through the preparation of AUC curves and their respective AUC values.

All mapping and analysis has been done using ArcGIS 10.8 and QGIS 3.22 software. Data analysis was done in Microsoft Excel. Finally, all maps and graphs prepared were analyzed and interpreted and based on those analysis and interpretation, the final conclusions were drawn.

One of the most precise decision-making techniques, multi-criteria decision making (MCDM) or multi-criteria decision analysis (MCDA) can be referred to as a revolution in the field of Remote Sensing. Benjamin Franklin developed one of the first research studies on multi-criteria decision-making when he published his research on the moral algebra concept. A few experimental and hypothetical researchers have dealt with MCDM strategies to inspect the numerical demonstrating ability of these strategies since 1950s to give a system that can assist with organizing dynamic issues and create inclinations from choices. The various methods that make up MCDM differ from one another in various ways have been discussed in the following sections. In order to find the best solution, this method takes into account various qualitative and quantitative criteria that must be fixed. For instance, cost and nature of the cycles are among the most widely recognized rules in numerous dynamic issues. Furthermore, in these issues, master bunches give various loads to the standards that depend on the significance of every standard in that particular case. MCDM can be applied to problems that people face on a daily basis. However, the evaluation of criteria is an important issue when the problem is based on more significant subjects, such as capital levels. In this way dynamics should be in view of legitimate organizing and express assessment of the relative multitude of measures utilizing fitting programming and devices. Practically, MCDM is used to deal with the steps of structuring, making decisions, and planning when the domain has many criteria for finding the best solution based on the preferences of the decision-makers. Over the past few decades, a variety of authors have come up with or improved a number of different kinds of MCDM methods. The primary distinctions between these techniques are connected with the intricacy level of calculations, the weighting strategies for rules, the way of addressing inclinations assessment measures, unsure information probability, lastly, information conglomeration type. There are various ways of interpreting tackling a MCDM issue. The process can be thought of as selecting the most desirable (best) alternative from a set of alternatives. It can also be described as grouping options into multiple preference sets and selecting a small set from those sets. Additionally, the definition of non-dominated or efficient alternatives is the goal of these problems. This point can assist decision-makers in selecting a solution-set from the set of non-dominated ones; however, there is no way to move from a non-dominated solution to another solution without sacrificing at least one of the criteria. A mathematical definition of an MCDM problem is as follows:

$$D = \{Di\,l\,i = 1, 2, \ldots, m\} \tag{8.1}$$

where D will be a particular and limited set of choices, and m addresses the quantity of them.

$$A = \{Ai\,l\,I = 1, 2, 3, \ldots, n\} \tag{8.2}$$

where n is the total number of the criteria that are used to evaluate A in C.

Although this is not required for the criteria, the alternatives naturally share the same characteristics. That is to say, criteria can have distinct units that do not have any relationships with one another and can have distinct objectives that are at odds with one another (minimizing objectives in some cases and maximizing them in others).

$$W = \{W_n\,l\,I = 1, 2, 3, \ldots, n\} \tag{8.3}$$

where W is a set of weights that are normalized and assigned to each criterion according to how important they are. The numerical type of sets examined above is a basic method for characterizing a MCDM issue, and the acquired data normally is coordinated as a network structure. Additionally, the merits and disadvantages of each type of MCDM will likely be explained specifically based on the approaches used. For instance, the analytic hierarchy process (AHP) is simple to use but has difficulties as a result of the interdependence between alternatives and criteria. On the other hand, using imprecise input is possible in Fuzzy Set Theory (FST); however, developing this method is difficult. All MCDM approaches have the advantages of taking into account the disproportionate and contradictory effects of decisions. Negative aspects include the fact that, due to the nature of the problem, the solutions produced by these techniques compromise between multiple objectives.

8.3 Results and Discussions

8.3.1 A Brief Idea About the Study Area

Uttarkashi district came into existence on 24th February, 1960. The district owes its name from the district headquarter Uttarkashi. The word 'Uttarkashi' actually means 'Kashi of the north'. It is called so because of its almost same standard of veneration of the Kashi of the plain region i.e. Varanasi located in Uttar Pradesh. It is considered to be one of the main abodes of Lord Shiva and have many temples where the Lord is addressed as Vishwanath. The district has the total area of about 7534.28 km^2. It has its latitudinal extension from 30° 28′ 34″N to 31° 27′ 19″N while its longitudinal extension is from 77° 49′ 22″E to 79° 24′ 49″E. In the north of Uttarkashi, there is Shimla and Kinnaur districts of Himachal Pradesh. The southern part of the district is occupied by Tehri Garhwal district of Uttarakhand. In the east and west, the district is bounded by Chamoli and Dehradun districts respectively (Fig. 8.2). According to 2011 census, Uttarkashi had a total population of about 330,086 and the population growth rate for the period 2001–2011 was 11.75%. About 98.42% of the total population are Hindus. The district has the sex ratio of 959 females per 1000 males and has got a literacy rate of 75.98%.

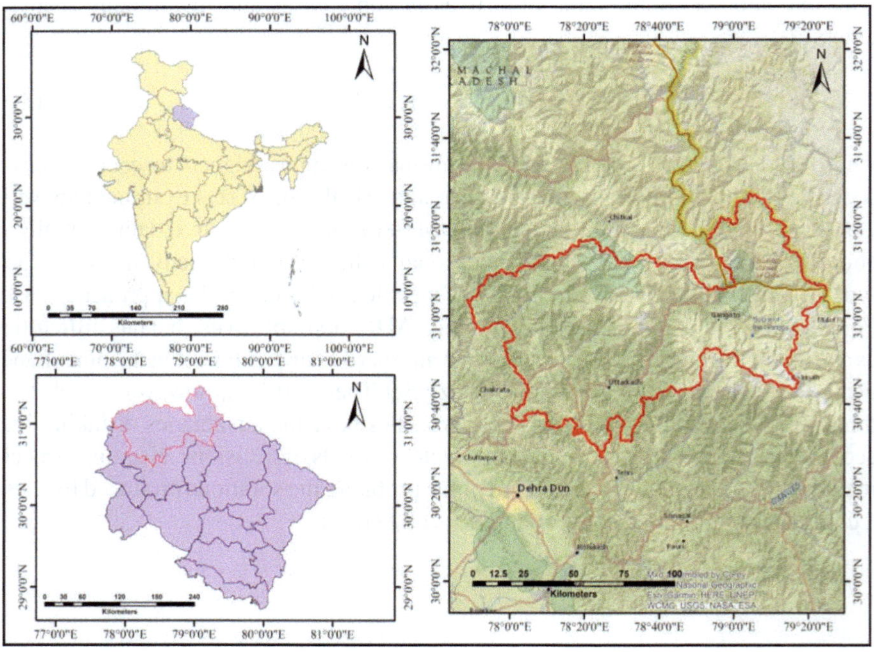

Fig. 8.2 Location map of the study area

8.3.2 Analysis of Conditioning Factors Used for Modelling

Landslide is triggered by a number of factors. Hence, it is very much necessary to have a detailed idea about the conditioning factors which are responsible for the occurrence of landslide in a particular region. The present study has taken into account five conditioning factors for performing landslide susceptibility analysis and they are discussed below.

8.3.2.1 Elevation

Elevation is one of the most important factors which affect occurrence of landslides. Landslides do not occur on plain areas and is very common in high altitude areas. The highest and the lowest elevation of the district are 7122 m and 688 m respectively. Elevation exhibits a rising trend from the western to the eastern part of the district (Fig. 8.3).

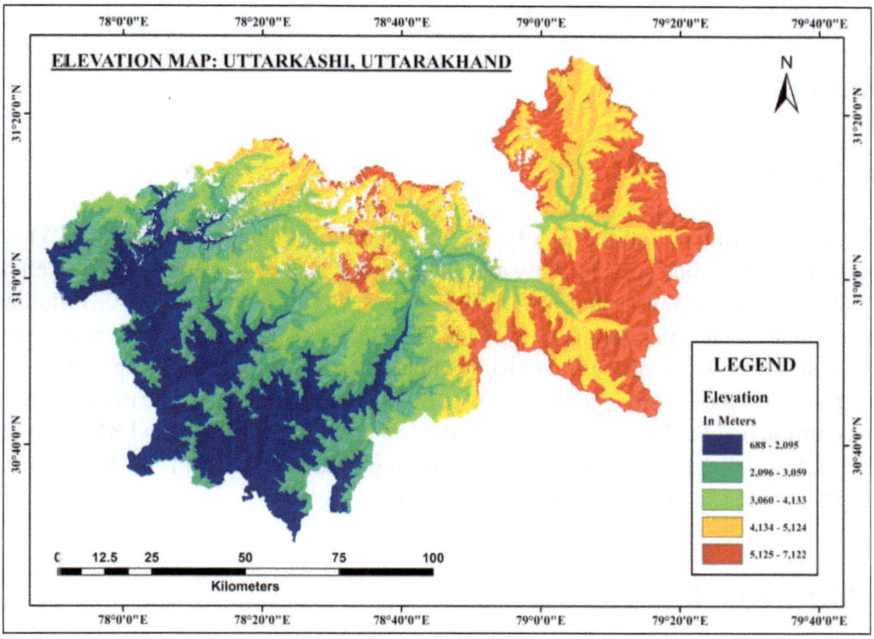

Fig. 8.3 Elevation map of the study area

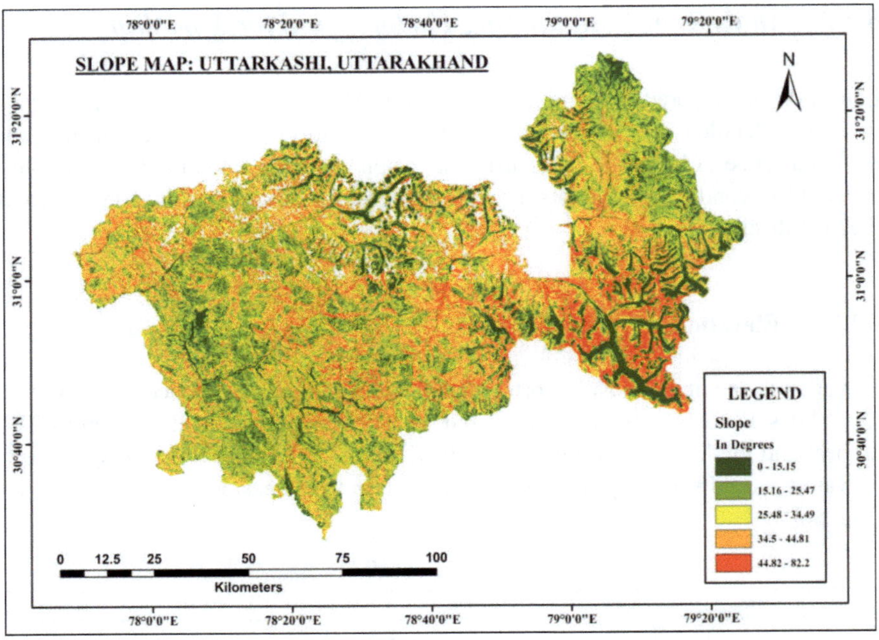

Fig. 8.4 Slope map of the study area

8.3.2.2 Slope

Slope is considered to be a key factor determining landslide activity. While performing landslide susceptibility analysis, slope is also given a high weight. Çellek (2022) is of the opinion that out of 293 studies related to landslides, 281 studies used slope as an input parameter (Çellek 2022). It is usually stated that areas with steep slope are more prone to landslides than the areas with gentle slopes. Critical slopes experience more landslides (Doke 2017). However, some researchers are of the opinion that landslides can also occur at low slope angles (Hammad Khaliq et al. 2023). In the present study, the highest slope angle value is found to be 82.2°. Gentle slopes are seen in the western part of the district (Fig. 8.4).

8.3.2.3 Rainfall

Rainfall can be considered as a triggering factor of landslide. Generally, slope instability is considered to be the primary reason for the occurrence of landslides. But chances of landslide increase to a number of times when there is heavy rainfall. Landslides occur when the amount of rainfall exceeds a threshold value (Abraham et al. 2020). In simple words, when heavy rainfall takes place, the debris, earth and rock on the slopes of the mountains become viscous. Since the slope surface also gets slippery, these materials easily come downslope under the influence of gravity

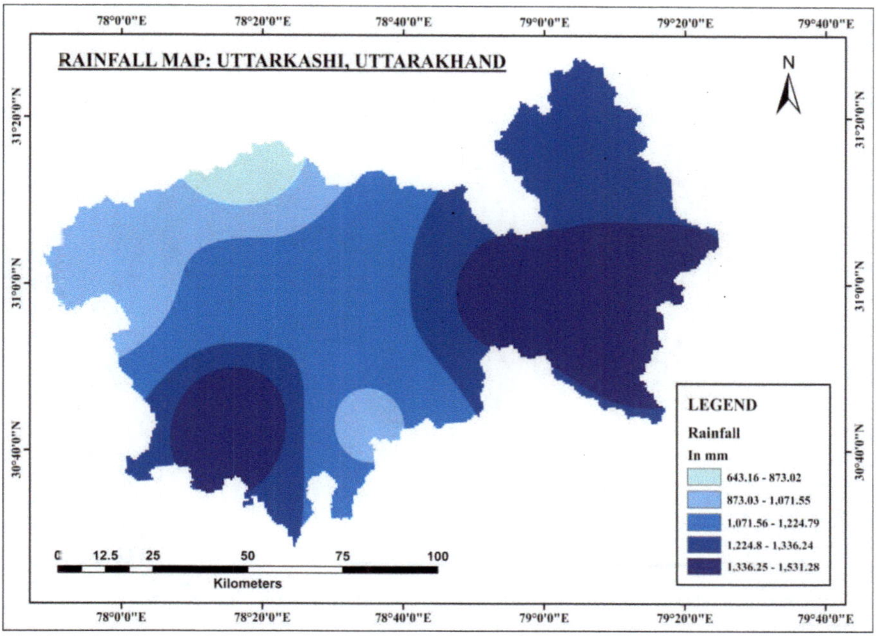

Fig. 8.5 Rainfall map of the study area

and thus cause landslides. In the present study, the highest amount of annual rainfall obtained by the district was 1531.28 mm while the lowest amount of rainfall was 643.16 mm. Two zones of very high rainfall can be seen in the south western and eastern part of Uttarkashi district (Fig. 8.5).

8.3.2.4 Lineament Density

Lineament is defined as the zone of weakness. In order to prepare landslide hazard maps, mapping of lineaments is an essential task (Ramli et al. 2010). The strength of slope material along the lineaments is reduced and any slight disturbance along these lineament zones may compel the slope materials to come down the slope and cause catastrophic landslides (Champati Ray et al. 2016). In the present study, the highest lineament density value is 0.84 km per square kilometer. Concentration of maximum lineaments can be seen in the western part of the district while number of lineaments in the eastern part is very low (Fig. 8.6).

8.3.2.5 Aspect

Aspect is defined as the direction of slope. Frequency of landslides have no direct relation with aspect. Landslides tend to occur more in areas where slopes are faced

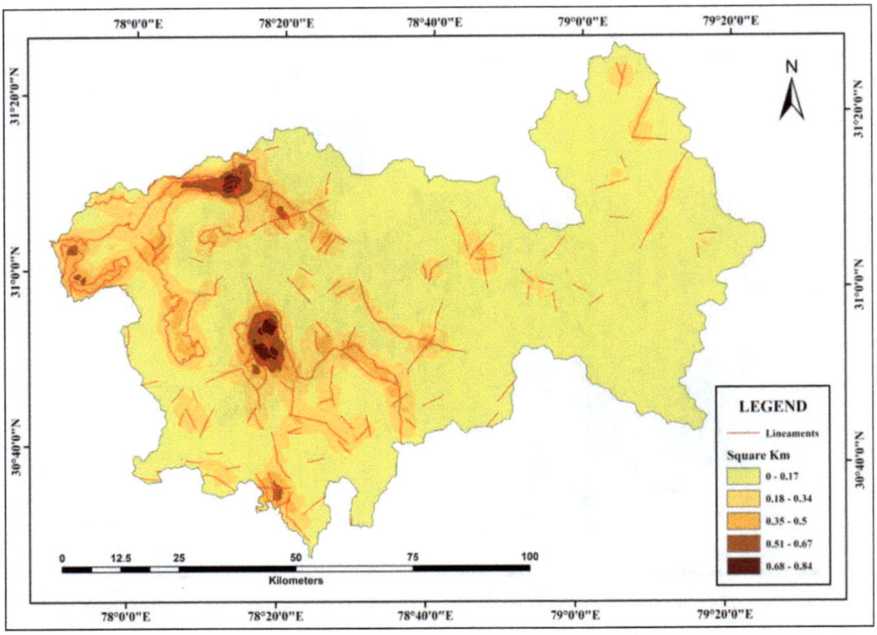

Fig. 8.6 Lineament density map of the study area

in more than one direction because the areas having slope in more than one direction are under the influence of other parameters (Çellek 2022). While considering aspect as a conditioning factor, the number of landslide occurring in a specific class of aspect has been taken into account. For instance, out of 488 landslide events taken for study, about 135 landslide events were found to occur in the aspect class of 69.78–143.37 (Fig. 8.7). For this reason, it has been assigned the highest weightage and susceptibility rating value in Weighted Overlay and MCDA respectively.

8.3.3 Landslide Susceptibility Analysis of Uttarkashi

Landslide susceptibility has been performed using weighted overlay and multi-criteria decision analysis (Tables 8.1 and 8.2). In both cases, five classes were made namely very low, low, moderate, high and very high. The map shows that the high and very high susceptibility areas have remained concentrated in the south-western part of the district (Fig. 8.8).

Landslide susceptibility map prepared using Multi-criteria decision analysis also reveals almost the same scenario. From the figure (Fig. 8.9), it is clearly seen that the areas of high and very high landslide susceptibility remained concentrated in the south-western part of the study area.

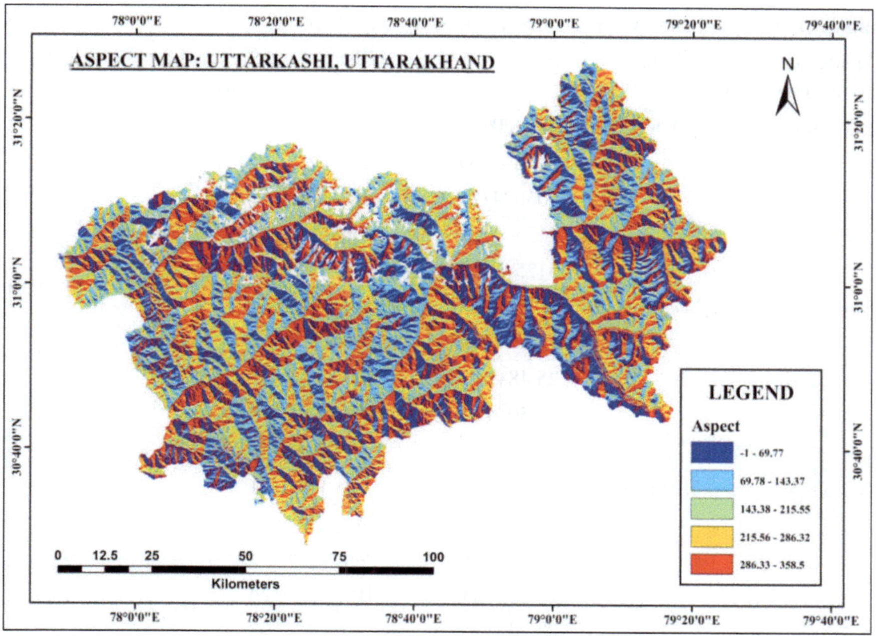

Fig. 8.7 Aspect map of the study area

8.3.4 Comparative Analysis of the Areal Coverage of Susceptibility Zones

For a better understanding of the scenario, a comparative analysis of the area coverage of the susceptibility zones is done. It is seen that in case of the model prepared using MCDA, area under very low susceptibility zone is 5.80% of the total study area while in case of the model prepared using weighted overlay, it is only 0.02% of the total study area. Low susceptibility zone covers about 19.79% of total study area in MCDA while in case of weighted overlay, it is much lower and covers about 10.75% of the total study area. About 45.9% of the total study area is seen under moderate susceptibility zone in the map prepared using weighted overlay technique. This is relatively higher than the map prepared using MCDA where it is seen that about 35.47% of the total area is in moderate susceptibility zone. Using MCDA, high susceptibility zone covers about 38.51% of the total area while in case of weighted overlay, it is comparatively lower and covers about 20.83% of the total study area. Very high susceptibility zone occupies about 18.11% of the total study area in case of MCDA and only 4.81% in case of weighted overlay model. The comparative analysis shows that in both the models, most part of Uttarkashi district is located in moderate landslide susceptibility zone (Fig. 8.10).

Table 8.1 Calculation of weights using weighted overlay (WO)

Landslide controlling factor	Units	Range	Class	Susceptibility rating	Weightage
Elevation	Meters	688–2095	Very low	1	9
		2096–3059	Low	2	
		3060–4133	Moderate	3	
		4133–5124	High	4	
		5125–7122	Very high	5	
Slope	Degrees	0–15.15	Very low	1	27
		15.16–25.47	Low	2	
		25.48–34.49	Moderate	3	
		34.50–44.81	High	4	
		44.82–82.2	Very high	5	
Aspect	Direction	0–69.77	Very low	1	9
		69.78–143.37	Very high	5	
		143.38–215.55	Moderate	3	
		215.56–286.32	High	4	
		286.33–358.5	Low	2	
Rainfall	Millimeters	643.16–873.02	Very low	1	27
		873.03–1071.55	Low	2	
		1071.56–1224.79	Moderate	3	
		1224.80–1336.24	High	4	
		1336.25–1351.28	Very high	5	
Lineament density	Square km	0–0.17	Very low	1	27
		0.18–0.34	Low	2	
		0.35–0.50	Moderate	3	
		0.51–0.67	High	4	
		0.68–0.84	Very high	5	

8.3.5 Model Validation

Models play a vital role in scientific decision making. Landslide susceptibility models are also not an exception in this regard. The present study validated its two models through the calculation of Landslide Density Index (LDI). Landslide Density Index is defined as the ratio between the percentage of number of landslide events and the percentage of class pixels in every class of landslide susceptibility model (Quan and Lee 2012; Aditian et al. 2018; Fayez et al. 2018; Akinci 2022; Goswami et al. 2022a, b).

In both the models (Tables 8.3 and 8.4), it can be seen that the lowest LDI values can be seen in the very low and low susceptibility zones. The LDI values rise as we

Table 8.2 Calculation of weights using multi-criteria decision analysis

	Class	Value	No. of landslides	Weight	Weight	Weight	Average
Elevation in meters	1	688–2095	427	5	5	1	3.7
	2	2096–3059	60	4	3	2	3.0
	3	3060–4133	1	3	4	3	3.3
	4	4133–5124	0	2	1	4	2.3
	5	5125–7122	0	1	2	5	2.7
	Total		488				
Aspect	1	0–69.77	51	1	1	1	1.0
	2	69.78–143.37	135	5	5	2	4.0
	3	143.38–215.55	93	3	4	3	3.3
	4	215.56–286.32	132	4	3	5	4.0
	5	286.33–358.5	77	2	2	4	3.0
	Total		488				
Slope in degrees	1	0–15.15	46	2	1	1	1.3
	2	15.16–25.47	136	4	2	2	2.7
	3	25.48–34.49	153	5	4	3	4.0
	4	34.50–44.81	115	3	3	4	3.3
	5	44.82–82.2	38	1	5	5	3.7
	Total		488				
Rainfall in millimeters	1	643.16–873.02	11	1	1	1	1.0
	2	873.03–1071.55	37	2	2	2	2.0
	3	1071.56–1224.79	206	5	3	4	4.0
	4	1224.80–1336.24	132	4	4	5	4.3
	5	1336.25–1351.28	102	3	5	3	3.7
	Total		488				
Lineament density square km	1	0–0.17	179	1	1	4	2.00
	2	0.18–0.34	255	2	2	5	3.00
	3	0.35–0.50	45	3	3	1	2.33
	4	0.51–0.67	9	4	4	2	3.33
	5	0.68–0.84	0	5	5	3	4.33
	Total		488				

gradually move towards the very high susceptibility zone. In both the models, the highest value of LDI is seen in case of the very high susceptibility zone and it is 5.2 in weighted overlay model and 3.3 in MCDA.

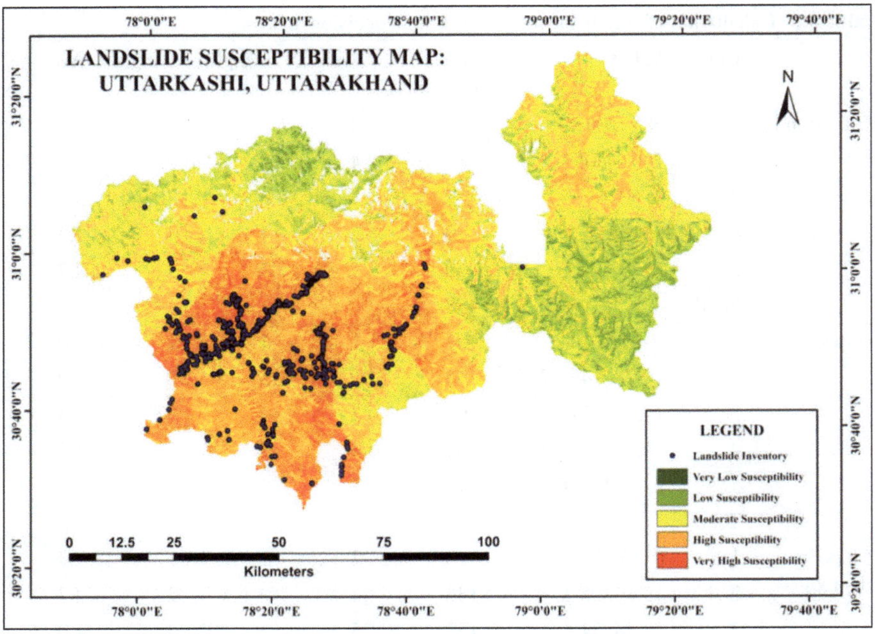

Fig. 8.8 Landslide susceptibility map of Uttarkashi using weighted overlay technique

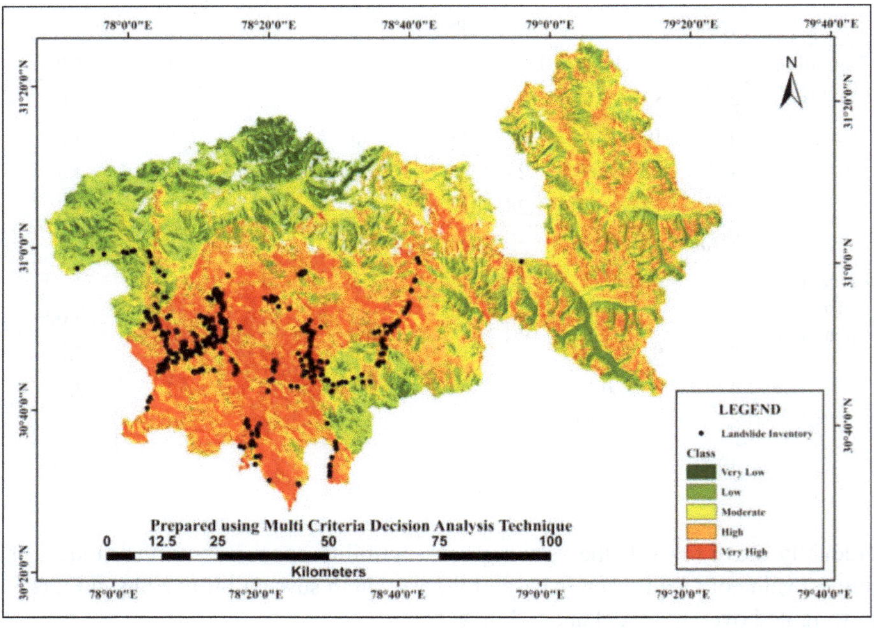

Fig. 8.9 Landslide susceptibility map of Uttarkashi using multi criteria decision analysis

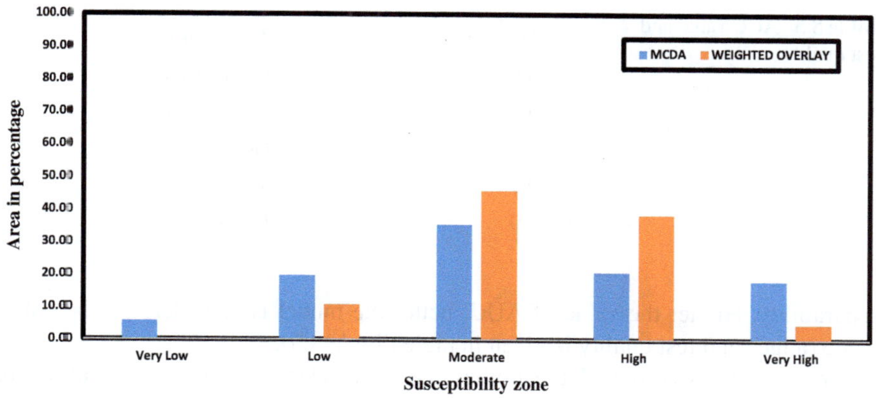

Fig. 8.10 Comparative analysis of areal coverage of landslide susceptibility zones

Table 8.3 Validation of model prepared using weighted overlay technique

Class	Susceptibility	No. of landslides	Percentage	Total pixels	Percentage	LDI
1	Very low	0	0	1974	0.0	0.0
2	Low	0	0	899,869	10.7	0.0
3	Moderate	69	14.14	3,842,513	45.9	0.3
4	High	296	60.66	3,224,226	38.5	1.6
5	Very high	123	25.20	402,844	4.8	5.2
Total		488	100	8,371,426	100	

Table 8.4 Validation of model prepared using MCDA

Class	Susceptibility	No. of landslides	Percentage	Total pixels	Percentage	LDI
1	Very low	0	0	483,427	5.8	0.0
2	Low	21	4.3	1,656,549	19.8	0.2
3	Moderate	84	17.2	2,969,272	35.5	0.5
4	High	93	19.1	1,743,988	20.8	0.9
5	Very high	290	59.4	1,516,434	18.1	3.3
Total		488	100	8,369,670		

8.3.5.1 Success Rates and Suitability Analysis of the Models

It can be easily concluded from the above discussions that both the models used for modelling of landslide are scientific. Now, the question arises which of these two modes will be more appropriate and suitable for application. The decision is taken through the preparation of an AUC or area under curve and receiver operating characteristics or ROC (Al-Sababhah 2022; Alsabhan et al. 2022; Ray et al. 2015). ROC is considered to be the probability while AUC is referred to as the degree of

Table 8.5 AUC value and test quality

AUC value range	Test quality
0.5–0.6	Unsatisfactory
0.6–0.7	Satisfactory
0.7–0.8	Good
0.8–0.9	Very good
0.9–1.0	Excellent

separability. Higher the value of AUC, better the model is considered to be. AUC value ranges and test quality is given in table (Table 8.5).

On preparing AUC of both the models, it is seen that the AUC of success rate curve for MCDA is 78.9% (Fig. 8.11a) and for weighted overlay, it is 79.7% (Fig. 8.11b). On the basis of test quality in Table 8.5, both the values fall between 0.7 and 0.8 and hence can be considered 'Good'. Since the AUC value of weighted overlay is slightly higher than that of MCDA, the landslide susceptibility model prepared using Weighted Overlay technique enjoys more suitability and applicability than that of MCDA.

The following figures represent different aspects of landslides that has taken place in different spots within the study area in different times (Fig. 8.12). At different spots, rock-slide, mud-slide and debris slide events were noticed. Steep slope and unplanned anthropogenic activities are solely responsible for such occurrences. Construction of protected walls along the steep slope by the side of the roads on steep hill side slope is considered to be one of the ideal ways of mitigating the disastrous impact of landslide in the study area.

The study also highlights the landslide event variations in the different spectrum of pixel behavior. The pixels have been segregated as per the landuse and land cover variations. Landslides events are more concentrated near the uplands with barren topography while near the river beds or the valley fill portion due to higher adhesive capacity by the soil the events of landslide phenomena are found to be very less (Fig. 8.13). While we have estimated the spatial expansion of landslide events on the panchromatic imageries, the spectral reflectance of the events are much greater near the river tributaries due to the water excavation in the form of rills and gullies. More is the distance from the river cross-section, lesser is the occurrences of landslide events (Fig. 8.14). On depicting the complex relationship between landslide events and altitude parameters on multispectral imageries a linear positive relation has been noted in this regard. While majority of the landslide events are found to concentrate near the water-bodies, the shallow seated vacant lands are also the major exceptions. It is hypothesized that the vacant lands are having the gentle slope to promote the settle-down of the debris materials (Fig. 8.15). The stand error functions of the landslide events have indicated that the smaller areas have the higher potentiality to drag the greater amount of debris flow. To validate our study, the events have been segregated into some pockets, which clearly shows that the events are more biased towards lower spatially extended areas (Fig. 8.16a). Even the fishnet wise segregation of the major events bring out the fact that the landslide events have disproportionate

Fig. 8.11 AUC of landslide susceptibility models

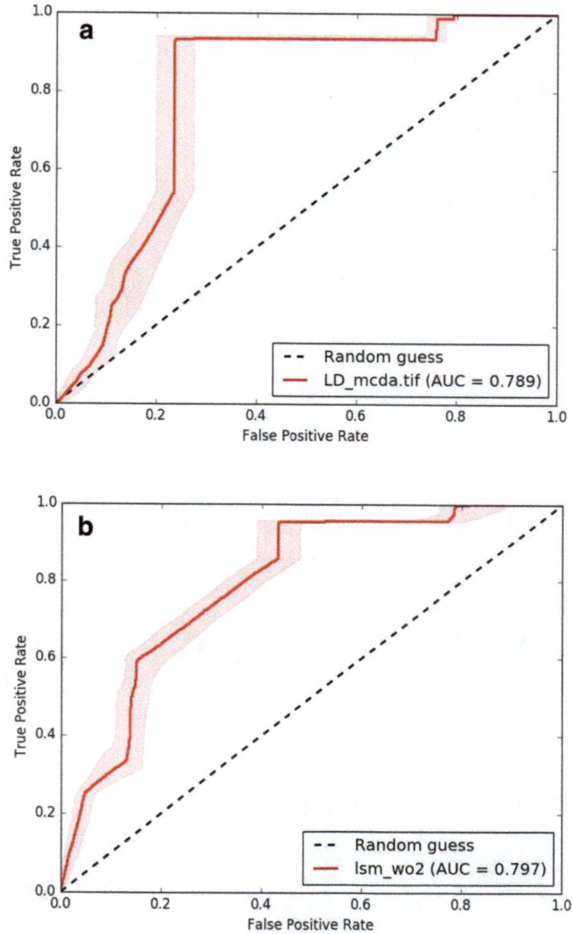

relationship with the areal extension (Fig. 8.16b); same is also noted for the circular distribution of the events (Fig. 8.16c).

Fig. 8.12 Various aspects of landslide events at Uttarkashi (*Source* Field photograph, 2022)

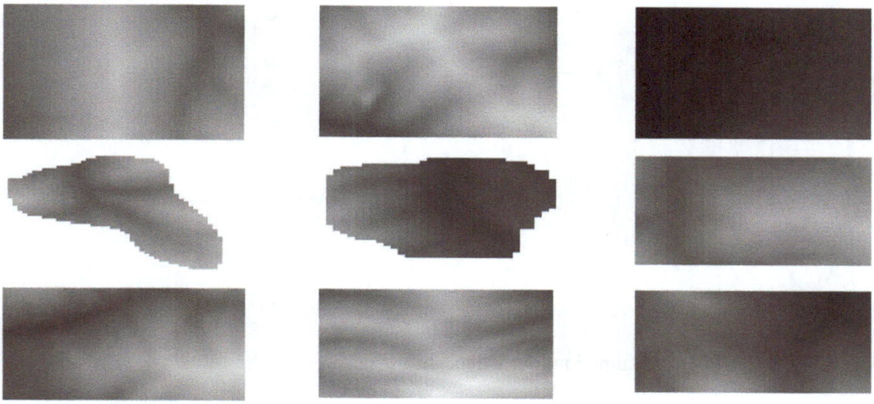

Fig. 8.13 Pixel behavior for landslide

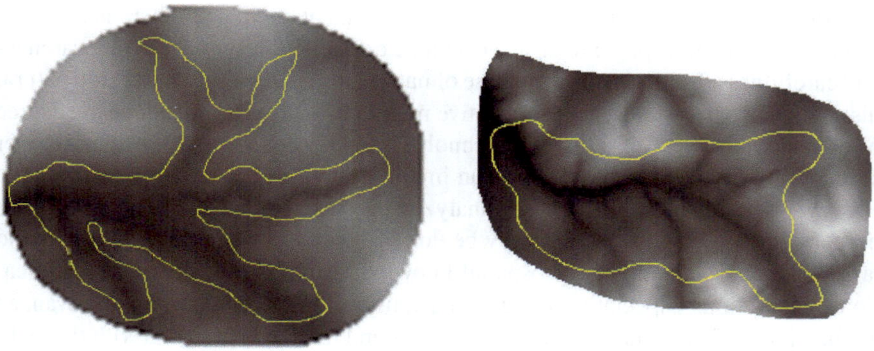

Fig. 8.14 Panchromatic signatures for landslide events

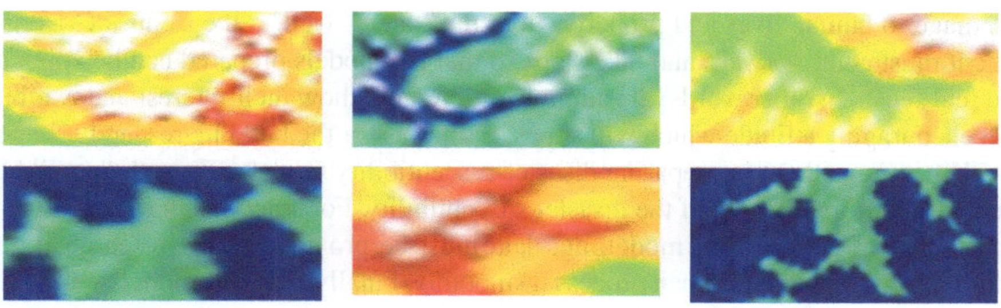

Fig. 8.15 Multispectral behavior for landslide events

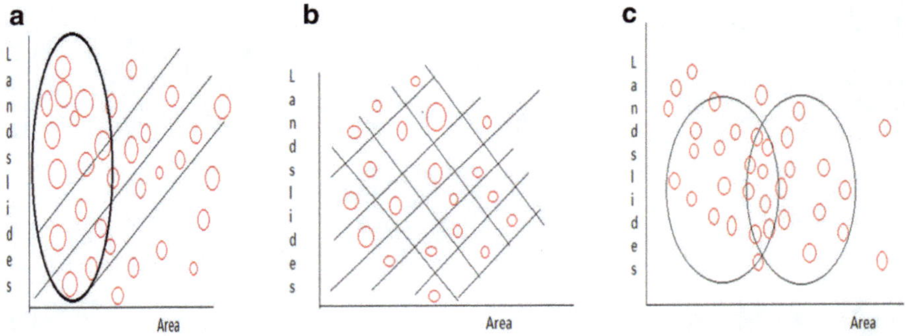

Fig. 8.16 Standard error estimation for the landslide events

8.4 Conclusion

Disasters are inevitable and can happen anytime and at any place. There is no doubt of the fact that anthropogenic disasters can be controlled through human awareness and carefulness. But it is not true in case of natural disasters. Man cannot stop natural disasters, but can certainly take preventive measures in advance to reduce the degree of damage. In the era of advanced technology of GIS and remote sensing, disaster susceptibility modelling has become an important tool in the hands of researchers. However, models prepared must be analyzed properly so that its suitability can be properly assessed. All these can only be done when there will be an expansion and large scale dissemination of geospatial knowledge among the researchers. Educational institutions especially colleges and universities must incorporate advanced application of GIS technology in their syllabus and train their students with the technology so that they can utilize it not only for upliftment of their career but also for the welfare of the nation. Being a developing nation, India must undertake this strategy because landslides are very common in the entire Himalayan belt and cause severe damage to human lives and property. Both the state as well as the Central Government must come forward and take steps so that the models prepared can be applied in real sense. In other words, the administration must show their interest and exhibit an encouraging attitude which will act as a stimulus for the budding researchers and will encourage them in preparing disaster susceptibility models. Thus, there will be a continuous monitoring of the disaster susceptibility. For smooth and rapid accomplishment of the work of modelling, data must be provided to the researchers in digital format so that they can process it smoothly. Finally, all these models will only be effective once they are used for taking important decisions in disaster mitigation and thus is introduced in the real world for looking into the human welfare. Once these models are successful as a decision-making tool, then only one of the major goals of disaster mitigation will be achieved and the administrative authority must play their respective roles for achieving the goal.

Declarations

Funding No funding source is available.

Conflict of Interest There is no conflict of interest regarding the publication of this article.

References

Abraham MT, Satyam N, Pradhan B, Alamri AM (2020) Forecasting of landslides using rainfall severity and soil wetness: a probabilistic approach for Darjeeling Himalayas. Water (Switzerland) 12(3):1–19. https://doi.org/10.3390/w12030804

Aditian A, Kubota T, Shinohara Y (2018) Comparison of GIS-based landslide susceptibility models using frequency ratio, logistic regression, and artificial neural network in a tertiary region of Ambon, Indonesia. Geomorphology 318:101–111. https://doi.org/10.1016/j.geomorph.2018.06.006

Akinci H (2022) Assessment of rainfall-induced landslide susceptibility in Artvin, Turkey using machine learning techniques. J Afr Earth Sci 191:104535. https://doi.org/10.1016/J.JAFREARSCI.2022.104535

Al-Sababhah NMA (2022) Development of landslide susceptibility mapping using GIS modeling in Jordan's Northern Highlands. Environ Ecol Res 10(6):701–727. https://doi.org/10.13189/eer.2022.100607

Alsabhan AH, Singh K, Sharma A, Alam S, Pandey DD, Rahman SAS, Khursheed A, Munshi FM (2022) Landslide susceptibility assessment in the Himalayan range based along Kasauli–Parwanoo road corridor using weight of evidence, information value, and frequency ratio. J King Saud Univ Sci 34(2):101759. https://doi.org/10.1016/j.jksus.2021.101759

Arsyad A, Hamid W (2020) Landslide susceptibility mapping along road corridors in west Sulawesi using GIS-AHP models. IOP Conf Ser: Earth Environ Sci 419(1). https://doi.org/10.1088/1755-1315/419/1/012080

Batar AK, Watanabe T (2021) Landslide susceptibility mapping and assessment using geospatial platforms and weights of evidence (WoE) method in the Indian Himalayan region: recent developments, gaps, and future directions. ISPRS Int J Geo-Inf 10(3). https://doi.org/10.3390/ijgi10030114

Çellek S (2022) Effect of the slope angle and its classification on landslides. Himalayan Geol 43(1):85–95

Champati Ray PK, Chattoraj SL, Bisht MPS, Kannaujiya S, Pandey K, Goswami A (2016) Kedarnath disaster 2013: causes and consequences using remote sensing inputs. Nat Hazards 81(1):227–243. https://doi.org/10.1007/s11069-015-2076-0

Doke A (2017) Malin village landslide: a geological hazard. Indian Streams Res J 7(5):4–11

Fayez L, Pazhman D, Pham BT, Dholakia MB, Solanki HA, Khalid M, Prakash I (2018) Application of frequency ratio model for the development of landslide susceptibility mapping at part of Uttarakhand State, India. Int J Appl Eng Res 13(9):6846–6854. http://www.ripublication.com

Gao Y, Fan LC, Deng J, Zhao S, Dai W, Zhu J, He Z, Yang Z, Lane SN (2023) Primary and potential secondary risks of landslide outburst floods. Nat Hazards. https://doi.org/10.1007/s11069-022-05775-z

Goswami A, Sen S, Sanyal R (2022a) Delineation of landslide hazard zones of a part of Sutlej Basin in Himachal Pradesh through frequency ratio model. In: Convergence of deep learning and artificial intelligence in internet of things. CRC Press, Taylor and Francis, pp 211–230

Goswami A, Sen S, Sanyal R (2022b) Convergence of deep learning and artificial intelligence in internet of things. In: Rana AA, Rana AK, Dhawan S, Sharma S, Elngar (eds) Convergence of deep learning and artificial intelligence in internet of things, 1st edn. CRC Press, pp 211–230. https://doi.org/10.1201/9781003355960

Hammad Khaliq A, Basharat M, Talha Riaz M, Tayyib Riaz M, Wani S, Al-Ansari N, Ba Le L, Thi Thuy Linh N (2023) Spatiotemporal landslide susceptibility mapping using machine learning models: a case study from district Hattian Bala, NW Himalaya, Pakistan. Ain Shams Eng J 14(3):101907. https://doi.org/10.1016/J.ASEJ.2022.101907

Hembram M, Sen S, Chakraborty SD (2022) Analysis of correlation between built up scenario and urban heat island effect of Bangalore urban area of Karnataka. Int J Res Eng Sci 10(10):364–372

Hussain MA, Chen Z, Wang R, Shah SU, Shoaib M, Ali N, Xu D, Ma C (2022) Landslide susceptibility mapping using machine learning algorithm. Civ Eng J (Iran) 8(2):209–224. https://doi.org/10.28991/CEJ-2022-08-02-02

Krishnan AR, Kasim MM, Hamid R, Ghazali MF (2021) A modified critic method to estimate the objective weights of decision criteria. Symmetry 13(6). https://doi.org/10.3390/sym13060973

Ma W, Dong J, Wei Z, Peng L, Wu Q, Wang X, Dong Y, Wu Y (2023) Landslide susceptibility assessment using the certainty factor and deep neural network. Front Earth Sci 10(January):1–14. https://doi.org/10.3389/feart.2022.1091560

Mersha T, Meten M (2020) GIS-based landslide susceptibility mapping and assessment using bivariate statistical methods in Simada area, northwestern Ethiopia. Geoenviron Disasters 7(1). https://doi.org/10.1186/s40677-020-00155-x

Quan HC, Lee BG (2012) GIS-based landslide susceptibility mapping using analytic hierarchy process and artificial neural network in Jeju (Korea). KSCE J Civ Eng 16(7):1258–1266. https://doi.org/10.1007/s12205-012-1242-0

Ramli MF, Yusof N, Yusoff MK, Juahir H, Shafri HZM (2010) Lineament mapping and its application in landslide hazard assessment: a review. Bull Eng Geol Env 69(2):215–233. https://doi.org/10.1007/s10064-009-0255-5

Ray PKC, Chattoraj SL, Bisht MPS, Kannaujiya S, Pandey, Goswami A (2015) Kedarnath disaster 2013: causes and consequences using remote sensing inputs. Nat Hazards 81:227–243. https://doi.org/10.1007/s11069-015-2076-0

Roy J, Saha S (2019) Landslide susceptibility mapping using knowledge driven statistical models in Darjeeling District, West Bengal, India. Geoenviron Disasters 6(1). https://doi.org/10.1186/s40677-019-0126-8

Sarkar S, Kanungo DP, Chauhan PKS (2011) Varunavat landslide disaster in Uttarkashi, Garhwal Himalaya, India. Q J Eng Geol Hydrogeol 44(1):17–22. https://doi.org/10.1144/1470-9236/09-029

Shahabi H, Hashim M (2015) Landslide susceptibility mapping using GIS-based statistical models and remote sensing data in tropical environment. Sci Rep 5:1–15. https://doi.org/10.1038/srep09899

Sonker I, Tripathi JN, Singh AK (2021) Landslide susceptibility zonation using geospatial technique and analytical hierarchy process in Sikkim Himalaya. Quat Sci Adv 4:100039. https://doi.org/10.1016/j.qsa.2021.100039

Thai Pham B, Khuc T, Xuan T, Noi H, Nam Dieu Tien Bui V, Indra P (n.d.) Landslide susceptibility assessment at a part of Uttarakhand Himalaya, India using GIS—based statistical approach of frequency ratio method. http://www.ahec.org.in/wfw/maps.htm

Wubalem A, Meten M (2020) Landslide susceptibility mapping using information value and logistic regression models in Goncha Siso Eneses area, northwestern Ethiopia. SN Appl Sci 2(5):1–19. https://doi.org/10.1007/s42452-020-2563-0

Chapter 9
Determining Land Induced Factors for Landslide Susceptibility in Indian Cities

Apurv Bhogibhai Patel and Vibhore Bakshi

Abstract India has witnessed humongous growth in the frequency of landslides in the past 40–50 years. The process of urbanization, concretization, and climate change has imposed various challenges for Indian cities; out of which landslide is one. Landslides inflict substantial damage to human lives, infrastructure, and the environment. The book chapter tries to identify the relevance of land use and land cover as major determinants for triggering natural disasters; however, this book chapter identifies the role of geotechnical characteristics in landslide susceptibility. The next section of the book chapter addresses various cases from Indian cities: Shimla, Himachal Pradesh deciphers the phenomenon of landslides majorly caused by soil erosion; the classical case of Western Ghats in India that is prone to repeated landslides. Moreover, the case of Jharkhand has been triggered in the last few years due to mining activities. Lastly, the chapter embarks on mitigation by the adoption of resilient strategies for landslide susceptibility and envisions the approach for reducing landslides in the near future.

Keywords Landslides · Land use · Land cover · Geotechnical characteristics · Mitigation strategies · Resilience

A. B. Patel (✉) · V. Bakshi
Department of Urban and Regional Planning, School of Planning and Architecture, Bhopal, India
e-mail: apurvab1112@gmail.com

V. Bakshi
e-mail: vibhore.bakshi@spabhopal.ac.in

G. K. Panda et al. (eds.), *Landslide: Susceptibility, Risk Assessment and Sustainability*, Advances in Natural and Technological Hazards Research 52,
https://doi.org/10.1007/978-3-031-56591-5_9

9.1 Introduction

9.1.1 Landslide as Emerging Issue

Landslides, geological phenomena characterized by the downward movement of soil, rocks, and debris on slopes, pose significant risks to human lives and infrastructure worldwide (Hong 2023; Huang et al. 2023a, b, c). Understanding the factors that contribute to landslide occurrence and magnitude is crucial for effective risk assessment and mitigation strategies (Badola et al. 2023; Gariano and Guzzetti 2016). Among the various factors influencing landslides, the role of land use and land cover has gained increasing attention in geotechnical research (Gómez and Kavzoglu 2005; Tanyas et al. 2019; Thi Ngo et al. 2021). This study aims to provide a clear and analytical exploration of the relationship between land use, land cover, and landslide occurrences, shedding light on their interdependencies and implications. In India, landslides pose a significant threat to both rural and urban areas, impacting communities and infrastructure. Recent studies underscore the gravity of the landslide issue in the country. Western Ghats region highlights the susceptibility of hilly terrains to landslides, emphasizing the need for comprehensive risk assessment in such geologically sensitive areas. The Himalayan region underscores the complex interaction between geological factors and human activities, revealing how deforestation and improper land use practices contribute to heightened landslide risks.

9.2 Paradox of Land Use; Land Cover and Occurrence of Landslide

Land use and land cover play crucial roles in shaping the conditions that lead to landslides (Dehnavi et al. 2015; Fang et al. 2021). Human activities and their associated land use practices, such as urbanization, agriculture, and mining, significantly alter the natural landscape and modify slope characteristics. Additionally, land cover elements, including vegetation, impervious surfaces, and water bodies, contribute to slope stability or vulnerability (Dehnavi et al. 2015; Fang et al. 2021; Zhu et al. 2019). Rapid urban development altered the landscape, leading to elevated landslide risks; it was identified as a key factor contributing to increased landslide susceptibility in the Three Gorges Reservoir area in China (Tang et al. 2019). Human activities, including urban expansion and infrastructure development, were linked to the occurrence of landslides in mountainous regions of the European Alps (Gariano and Guzzetti 2016; Plat et al. 1983). A comprehensive global analysis of landslide distribution revealed a clear correlation between areas undergoing rapid development and an increased susceptibility to landslides (Wang et al. 2018). This underscores the urgent need for implementing sustainable land use practices and effective risk mitigation strategies in regions experiencing intense development (Gariano and Guzzetti 2016; Plat et al. 1983; Tang et al. 2019; Wang et al. 2018). Understanding the influence of land use and land cover on landslide susceptibility is essential for effective land management, hazard assessment, and mitigation planning (Roy et al. 2023; Tyagi et al. 2022).

9.3 Definition Understanding the Land Use and Land Cover

In the context of this study, land use and land cover are fundamental concepts that influence landslide susceptibility and provide insights into the human-environment interactions shaping slope stability (Addis 2023; Fang et al. 2020; Rabby et al. 2022). It is essential to establish a clear understanding of these terms:

9.3.1 Land Use

Land use refers to the activities and purposes for which land is utilized by human societies (Rabby et al. 2022). It encompasses various forms of human interventions and practices on the land, including urban areas, agricultural fields, industrial zones, transportation infrastructure, and mining sites. Land use choices and management practices have direct implications for the natural and physical characteristics of slopes and their vulnerability to landslides (Chang et al. 2022; Dhungana et al. 2023).

9.3.2 Land Cover

Land cover, on the other hand, describes the physical coverage of the Earth's surface by natural or artificial elements (Bonnesoeur et al. 2019; Chang et al. 2022). It encompasses the observable features and materials present on the land, such as vegetation, water bodies, bare soil, rocks, and artificial surfaces like buildings, roads, and pavement. The composition and arrangement of these land cover elements play a crucial role in determining the stability or susceptibility of slopes to landslides (Ali et al. 2021; Bonnesoeur et al. 2019; Rabby et al. 2022). The understanding of land use and land cover dynamics provides a foundation for assessing the influence of human activities, environmental changes, and natural processes on slope stability, enabling informed decision-making for land management, hazard assessment, and mitigation strategies (Ojala et al. 2019; Saha et al. 2023; Saikh and Mondal 2023).

9.4 Identifying the Relevant Parameters that Impact Landslide Occurrence

9.4.1 Factors Affecting the Occurrence of Landslides

This study adopts an analytical research approach to investigate the relationship between land use, land cover, and landslide occurrences. A systematic analysis of relevant literature, including research papers, case studies, and empirical studies, was conducted. The selected sources were critically evaluated to extract valuable data, identify trends, and analyze the interplay between land use, land cover, and landslide events. This research aims to provide a comprehensive understanding of the complex dynamics and implications of land use and land cover changes on landslide occurrences.

The interactions between land use, land cover, and landslides are multifaceted and exhibit varying degrees of influence depending on the specific context (Saravanan et al. 2009; Sutoris 2018). Urbanization, characterized by the conversion of natural landscapes into built environments, often leads to increased landslide risks (Nightingale and Richmond 2022). The construction of infrastructure, alteration of slopes, and removal of vegetation can weaken slope stability and enhance landslide susceptibility (Dao et al. 2020; Dou et al. 2019). On the other hand, agricultural activities can also contribute to landslide occurrences. Improper land management practices, excessive soil disturbance, and inadequate water management in agricultural areas can exacerbate soil erosion and slope instability, thereby increasing landslide hazards (Infante et al. 2019). Furthermore, mining activities have been identified as significant factors influencing landslides. The excavation of earth materials, such as open-pit mining, can alter slope geometry and weaken natural slopes, making them prone to landslides (Wei et al. 2022). Improper waste disposal practices and the accumulation of mine tailings can further exacerbate slope instability, increasing the likelihood of landslides in mining regions (Chen and Li 2020; Lima et al. 2023).

9.5 The Impact of Land Use on Landslide Susceptibility

9.5.1 Dichotomous Urbanization and Increasing Landslide

Urbanization is a significant land use change that involves the transformation of natural or rural areas into urban settlements. The rapid expansion of cities and towns often leads to the alteration of slopes, removal of vegetation, excavation activities, and construction of infrastructure (Saikh and Mondal 2023). These activities can modify the natural slope stability and increase the susceptibility to landslides. Urban areas are particularly vulnerable to shallow landslides triggered by intense rainfall events due to the reduced infiltration capacity of impervious surfaces and disrupted drainage systems (Bonnesoeur et al. 2019). The process of urbanization, characterized by the expansion of urban areas, construction of buildings, and infrastructure development, often leads to increased landslide risks (Addis 2023). The removal of vegetation cover during construction activities, modification of slopes for building purposes, and the introduction of impervious surfaces disrupt the natural drainage patterns, resulting in increased surface runoff and reduced infiltration (Achu et al. 2023; Paryani et al. 2021). These changes weaken slope stability and can trigger landslides, especially during intense rainfall events.

9.5.2 Impact of Agricultural Land Holding on Landslide Occurrence

Agricultural activities can also contribute to landslide occurrences. Improper land management practices, such as excessive soil disturbance, inadequate water management, and improper irrigation techniques, can lead to soil erosion and loss of soil cohesion (Dhungana et al. 2023). Steep slopes and the removal of natural vegetation for agricultural purposes increase the likelihood of landslides, especially during periods of heavy rainfall (Pánek 2019). This weakened soil structure makes slopes more susceptible to landslides, especially in hilly or mountainous agricultural regions (Chen et al. 2018).

9.5.3 Impact of Deforestation on Landslide Occurrence

Deforestation involves the removal or clearing of forests for various purposes, such as agriculture, logging, or urbanization. Vegetation plays a crucial role in binding soil particles, enhancing soil cohesion, and reducing surface erosion (Hammad Khaliq et al. 2023; Puente-Sotomayor et al. 2021). The removal of trees and vegetation cover reduces slope stability, increases surface runoff, and exposes the soil to erosion. Deforested areas are highly susceptible to landslides, particularly in regions with steep slopes and heavy precipitation (Lombardo et al. 2020; Sonker et al. 2022).

9.5.4 Illegal Mining Activities Trigger the Landslide Occurrence

Mining operations often involve the excavation of large quantities of earth materials, altering the natural topography and geotechnical conditions (Guo et al. 2023; Huang et al. 2023a, b, c). Open-pit mining and underground mining can weaken slope integrity, disrupt natural drainage patterns, and create unstable slopes (Canlas

2023; Horn 2015). Moreover, the accumulation of mine waste and tailings can further increase the susceptibility to landslides. Mining activities in hilly or mountainous areas require careful planning, monitoring, and reclamation measures to minimize the potential for landslides (Chang et al. 2023a, b; Huang et al. 2022). Effective land use planning and management strategies can help mitigate landslide risks. Implementing appropriate zoning regulations, considering slope stability assessments in urban planning, and integrating measures to preserve vegetation cover in agricultural and forestry practices are essential steps in reducing landslide susceptibility (Badola et al. 2023; Chen et al. 2023; Gariano and Guzzetti 2016; Thi Ngo et al. 2021). Additionally, maintaining natural drainage patterns, implementing erosion control measures, and adopting sustainable land management practices contribute to enhancing slope stability and minimizing landslide hazards.

Understanding the influence of land use on landslide susceptibility is vital for informed decision-making and the development of effective mitigation strategies (Hong 2023; Huang et al. 2023a, b, c; Pennington et al. 2015). By considering the interactions between human activities, slope dynamics, and environmental factors, researchers and policymakers can work towards sustainable land management practices that minimize the impact of landslides on both human lives and the environment (Hong 2023; Huang et al. 2023a, b, c; Pennington et al. 2015).

9.6 Factors Impacting Landslide Susceptibility

Land cover, encompassing the physical elements present on the Earth's surface, exerts a significant influence on landslide susceptibility (Abdulkareem et al. 2019). The composition and arrangement of land cover elements play a crucial role in shaping slope stability and the occurrence of landslides (Rabby et al. 2022).

9.6.1 Key Concerns for Landslide Susceptibility

Vegetation Cover: Vegetation cover plays a crucial role in slope stability and landslide prevention. The roots of plants and trees bind soil particles, increase soil cohesion, and enhance slope reinforcement (Chen et al. 2019; Guo et al. 2023). Dense vegetation cover acts as a barrier to surface runoff, reducing erosion and maintaining the stability of slopes (Abdulkareem et al. 2019). Forests, grasslands, and other types of natural vegetation provide effective protection against landslides by intercepting rainfall, promoting infiltration, and reducing soil moisture content. Deforestation or removal of vegetation cover can weaken slopes, increase soil erosion, and escalate landslide risks (Tanyas et al. 2019).

Soil Erosion: Soil erosion is a natural process influenced by rainfall, topography, and land management practices (Dehnavi et al. 2015; Gómez and Kavzoglu 2005).

It involves the detachment and transport of soil particles by water or wind, leading to the loss of fertile topsoil and increased slope instability (Fang et al. 2021; Zhu et al. 2019). Eroded slopes are more susceptible to landslides, as the removal of soil weakens the overall slope strength (Roy et al. 2023). Controlling soil erosion through proper land management techniques, such as contour plowing, terracing, or reforestation, can mitigate landslide risks (Li and Mo 2019; Pradhan et al. 2023; Tyagi et al. 2022).

Impervious Surfaces: The presence of impervious surfaces, such as concrete pavements, roads, or buildings, significantly alters the natural hydrological processes and increases the likelihood of landslides. Impervious surfaces reduce the infiltration capacity of the soil, resulting in increased surface runoff. This excess runoff accumulates on slopes, saturates the soil, and reduces slope stability (Dao et al. 2020; Dou et al. 2019; Infante et al. 2019). Urban areas with extensive impervious surfaces are prone to landslide events, particularly during heavy rainfall or when combined with other triggering factors. The excess water adds pressure to the slopes, reducing their stability and raising the potential for landslides (Chang et al. 2023a, b; Es-smairi et al. 2023).

Water Bodies: The presence of water bodies, such as rivers, lakes, and ponds, can significantly influence slope stability. Water bodies increase pore water pressure within the slopes, reducing the effective stress and weakening soil strength. Saturation of soil due to prolonged rainfall or the rise in water levels can trigger landslides, especially in areas with high groundwater levels or in slopes adjacent to water bodies (Chen and Li 2020; Lima et al. 2023; Wei et al. 2022).

Bare Soil and Exposed Rocks: Unvegetated or bare slopes, devoid of any land cover elements, are highly susceptible to erosion and landslides (Ali et al. 2021; Rabby et al. 2022). Without the protective covering of vegetation or other land cover, bare soil is exposed to erosive forces of wind and water, leading to soil detachment and slope instability. Similarly, exposed rocks, particularly in areas with weathering or fracturing, are prone to weathering processes, which can weaken the rocks and trigger landslides (Ojala et al. 2019; Saha et al. 2023; Saikh and Mondal 2023).

Land Cover Heterogeneity: The spatial arrangement and diversity of land cover elements within a landscape influence landslide susceptibility (Rabby et al. 2022). Heterogeneous landscapes, with a mix of vegetation, water bodies, and different land cover types, often exhibit a more complex pattern of slope stability. Variations in land cover can create different moisture conditions, soil properties, and erosion patterns, resulting in varied susceptibility to landslides (Abdulkareem et al. 2019; L. Chen et al. 2019; Guo et al. 2023; Rabby et al. 2022).

Understanding the impact of land cover on landslide susceptibility is crucial for assessing and mitigating landslide hazards (Guo et al. 2023). By recognizing the role of vegetation, water bodies, impervious surfaces, and the overall composition of land cover, researchers and practitioners can develop strategies for sustainable land management, erosion control, and effective land use planning. Incorporating measures to preserve or restore vegetation cover, implementing appropriate drainage systems, and minimizing the introduction of impervious surfaces are essential steps in reducing landslide susceptibility and promoting slope stability (Guo et al. 2023; Rabby et al. 2022).

9.7 The Role of Geotechnical Characteristics in Landslide Susceptibility

Geotechnical characteristics, which encompass the physical and mechanical properties of soil and rock materials, play a critical role in landslide susceptibility (Huang et al. 2023a, b, c; Pennington et al. 2015). These characteristics determine the stability and behavior of slopes, influencing their vulnerability to landslides, and hydrological conditions (Gariano and Guzzetti 2016). This section provides a clear and analytical exploration of the role of geotechnical characteristics in landslide susceptibility:

Soil Composition: The composition and properties of soil significantly influence its stability and susceptibility to landslides (Badola et al. 2023). Soils with high clay content are more prone to landslides due to their low shear strength and high water-holding capacity. Sandy soils, on the other hand, are generally more stable but can become susceptible to landslides when saturated (Thi Ngo et al. 2021). Cohesive soils, such as clays and silts, have a higher susceptibility to landslides due to their low permeability and potential for high pore water pressure buildup (Chen et al. 2023; Huang et al. 2022). Coarse-grained soils, such as sands and gravels, generally exhibit better drainage and shear strength properties, reducing their landslide susceptibility (Patil and Panhalkar 2023; Tempa et al. 2021). The presence of geological discontinuities, such as faults or bedding planes, can also affect slope stability by creating potential slip surfaces. Understanding the composition and texture of soils within a slope is crucial for assessing their stability and susceptibility to landslides (Abdulkareem et al. 2019; Chang et al. 2023a, b).

Slope Gradient and Landslide Susceptibility: Slope gradient plays a crucial role in landslide susceptibility. Steeper slopes are inherently more susceptible to landslides as the gravitational forces acting on the slope increase (Canlas 2023; Horn 2015). The stability of a slope is influenced by the balance between driving forces (e.g., gravity) and resisting forces (e.g., soil strength). High slope gradients can exceed the resisting forces of the soil, leading to slope failures and landslides (Canlas 2023; Guo et al. 2023). As the slope angle approaches the angle of repose, the potential for slope failure and landslides increases. The critical angle of slope stability depends on various factors, including soil properties, vegetation cover, and water content (Huang et al. 2023a, b, c).

Hydrological Conditions and Landslide Hazard: Hydrological conditions, such as rainfall intensity and groundwater levels, have a significant impact on landslide susceptibility (Sonker et al. 2022). Prolonged or intense rainfall events can saturate the soil, reducing its shear strength and increasing pore water pressure (Lombardo et al. 2020). Elevated pore water pressure reduces the effective stress in the soil, making it more susceptible to landslides (Chen et al. 2019). Groundwater levels that rise close to the ground surface can also contribute to landslide occurrence by increasing pore water pressure and reducing slope stability (Hammad Khaliq et al. 2023). High porosity facilitates water storage within the soil, potentially leading to increased pore water pressure and reduced shear strength. Low permeability restricts water movement, resulting in increased saturation and reduced drainage, both of which can contribute to slope instability and landslides (Puente-Sotomayor et al. 2021). Evaluating the porosity and permeability characteristics of soils is essential for understanding their response to rainfall and groundwater conditions (Chen et al. 2018; Pánek 2019).

Shear Strength: Shear strength is a fundamental geotechnical property that determines the resistance of soil or rock to sliding along potential failure planes. It depends on factors such as soil cohesion, frictional resistance, and effective stress (Chang et al. 2022; Dhungana et al. 2023). Soils with low shear strength, either due to their inherent properties or changes induced by external factors like water content, are more susceptible to landslides (Achu et al. 2023; Paryani et al. 2021). Assessing the shear strength parameters of soils is crucial for evaluating slope stability and landslide susceptibility (Bonnesoeur et al. 2019; Ojala et al. 2019; Saikh and Mondal 2023).

Rock Mass Strength: In the case of rock slopes, the strength and integrity of the rock mass are critical factors in landslide susceptibility (Rabby et al. 2022; Saha et al. 2023). Rock mass properties, such as intact rock strength, joint spacing, and orientation, influence the stability of rock slopes (Chen and Li 2020; Lima et al. 2023). Weakened or weathered rock masses are more prone to slope failures and landslides. Evaluating the rock mass characteristics through geological and geotechnical investigations is crucial for assessing rock slope stability and landslide susceptibility (Es-smairi et al. 2023; Tanyas et al. 2019; Wei et al. 2022).

9.8 Case Studies and Research Findings

9.8.1 Case Study 1: Land Use Change and Landslide Occurrence in India

In a compelling case study conducted in a region in India that experienced significant land use changes, researchers sought to explore the relationship between urbanization and landslide occurrence (John and Rajendran 2009; Panchal and Shrivastava 2022; Sharma et al. 2018). The study revealed a clear correlation, demonstrating the adverse consequences of converting natural landscapes into urban areas. The findings shed light on the increased frequency and magnitude of landslides observed as a result of extensive urban development (John and Rajendran 2009; Panchal and Shrivastava

2022; Sharma et al. 2018). The case study focuses on the city of Shimla, located in the northern state of Himachal Pradesh, India (Fig. 9.1). Shimla, once renowned for its pristine natural beauty and forested hills, has undergone rapid urbanization and population growth over the years (John and Rajendran 2009; Panchal and Shrivastava 2022; Sharma et al. 2018). The transformation of the region from a relatively undisturbed landscape to a bustling urban center provided an ideal context to investigate the impact of land use change on landslide susceptibility (Dai et al. 2023; Pangapanga-Phiri et al. 2023). One of the primary factors identified in this case study was the removal of vegetation cover during the process of urbanization (Panchal and Shrivastava 2022). In Shimla, as urban areas expanded, the dense forest cover that once adorned the hillsides was significantly reduced. Trees were cleared to make way for residential and commercial constructions (Sharma et al. 2018). This loss of vegetation cover weakened the natural stability of the slopes, leaving them exposed to erosion and increasing the likelihood of landslides during heavy rainfall events (John and Rajendran 2009).

The alteration of slopes during urban development played a crucial role in exacerbating landslide susceptibility (Alsabhan et al. 2022). In Shimla, the rapid growth of the city led to extensive hillside excavations for constructing buildings and infrastructure.

Slopes were reshaped and graded to accommodate the expanding urban footprint (John and Rajendran 2009; Sharma et al. 2018). However, these alterations disrupted the natural equilibrium of the hills, compromising their stability. As a result, even moderate rainfall events could trigger landslides due to the weakened slopes being unable to withstand the additional pressure (Panchal and Shrivastava 2022; Thacker et al. 2023). The case study also identified inadequate stormwater

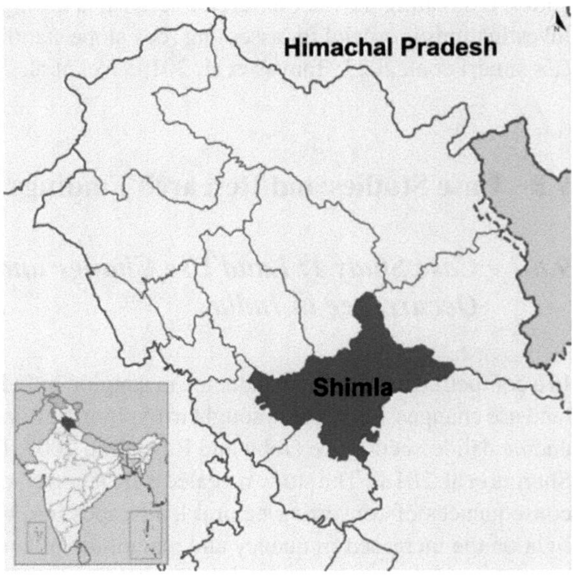

Fig. 9.1 Location of the Shimla in India

management systems as a contributing factor to increased landslide occurrences. With the expansion of urban areas in Shimla, impermeable surfaces such as roads, pavements, and buildings became more prevalent (Panchal and Shrivastava 2022). During heavy rainfall, the water runoff from these impervious surfaces accumulates and flows over the land surface instead of being absorbed into the soil. The excessive surface runoff added pressure to the slopes, further reducing their stability and augmenting the susceptibility to landslides. To exemplify the impact of land use change on landslide occurrence, a specific area in Shimla serves as a relevant example. The hillsides near a newly developed residential area experienced a series of landslides during the monsoon season (Davis and Friedmann 2005; Santosh' et al. 2003). The removal of vegetation cover and alteration of slopes for construction purposes significantly weakened the stability of the hillsides, making them prone to landslides triggered by intense rainfall (Manna and Maiti 2018; Mehrotra et al. 1999). These landslides resulted in property damage, and disruptions to daily life, and posed risks to the safety of the residents.

The findings of this case study emphasize the critical importance of considering land use changes and their associated risks in urban planning and development. It highlights the need for sustainable land management practices, preservation of vegetation cover, and the implementation of effective stormwater management systems (John and Rajendran 2009; Panchal and Shrivastava 2022). By incorporating these measures into land use policies, cities like Shimla can mitigate the adverse impacts of urbanization on landslide susceptibility and strive for sustainable development (Panchal and Shrivastava 2022). This case study provides valuable insights for researchers, policymakers, and practitioners involved in urban planning, geotechnical engineering, and disaster risk management in India and other regions experiencing similar land use changes. It underscores the significance of adopting comprehensive approaches that integrate environmental considerations, slope stability assessments, and community engagement to minimize the risks associated with landslides (John and Rajendran 2009; Panchal and Shrivastava 2022; Sharma et al. 2018). By taking proactive measures and implementing appropriate strategies, cities can strike a balance between development and environmental preservation, ensuring the safety and well-being of their residents in landslide-prone areas (Dhungana et al. 2023; Li and Mo 2019; Lima et al. 2023; Puente-Sotomayor et al. 2021; Tempa et al. 2021).

9.8.2 Case Study 2: Vegetation Cover and Slope Stability

A noteworthy case study conducted in India focused on investigating the critical role of vegetation cover in maintaining slope stability (Li and Mo 2019). The research aimed to understand the impact of deforestation on landslide occurrences and highlight the importance of intact forests in mitigating slope instability. The study took place in the Western Ghats, (Fig. 9.2) a mountain range that stretches along the western coast of India (Patil and Panhalkar 2023; Thacker et al. 2023). The region is characterized by diverse and dense forest cover, making it an ideal setting to explore the relationship between vegetation and slope stability (Patil and Panhalkar 2023). The research revealed a clear correlation between intact forests and a lower frequency of landslides. Areas that had experienced deforestation showed a higher susceptibility to slope failures (Revadekar et al. 2018). This disparity demonstrated the protective role of vegetation cover in preventing landslides and maintaining the stability of slopes.

One of the key mechanisms identified in this case study was the contribution of tree roots to slope stability (Bonnesoeur et al. 2019). In areas with intact forests, the intricate network of tree roots acted as natural reinforcements, anchoring the soil and rock layers together (Chen et al. 2018; Segura-Millán and Perez-Verdin 2023). The roots provided mechanical strength to the slopes, reducing the risk of

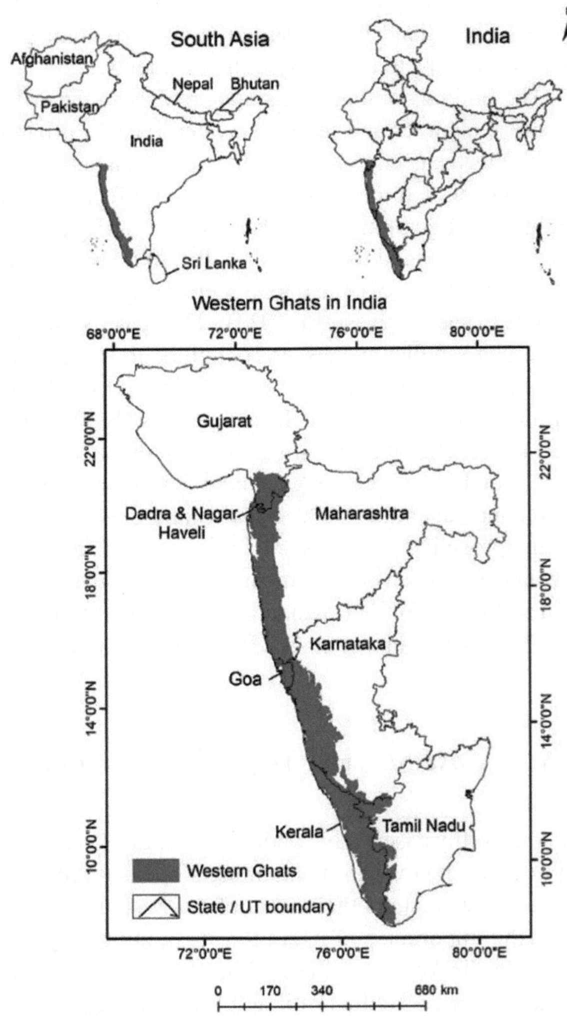

Fig. 9.2 Location of the Western Ghats in India

soil erosion and slope failure. Moreover, the fibrous nature of the roots increased soil cohesion, enhancing the overall stability of the slopes (Bonnesoeur et al. 2019; Chen et al. 2018; Dou et al. 2019; Segura-Millán and Perez-Verdin 2023). Additionally, the dense canopy formed by the forest vegetation played a significant role in intercepting rainfall (Segura-Millán and Perez-Verdin 2023). When rainwater fell on the forest canopy, it was absorbed and evaporated, reducing the amount of water reaching the ground surface. This interception process effectively regulated the amount of water infiltrating into the soil, preventing excessive saturation and reducing the likelihood of landslides triggered by saturated slopes (Bonnesoeur et al. 2019; W. Chen et al. 2018; Dou et al. 2019; Segura-Millán and Perez-Verdin 2023; Tanyas et al. 2019; Thacker et al. 2023). The case study also highlighted the role of vegetation in reducing surface runoff. Forests with a dense cover of vegetation acted as natural sponges, absorbing rainfall and slowing down the movement of water across the slopes. By slowing down the flow of water, the vegetation helped to reduce the erosive force of runoff, preventing soil erosion and maintaining the stability of the slopes.

To exemplify the impact of vegetation cover on slope stability, a specific area in the Western Ghats can be considered. A forested region with intact vegetation exhibited minimal landslide occurrences during heavy rainfall events (Manna and Maiti 2018; Pradhan et al. 2023). The dense forest canopy intercepted a significant portion of the rainfall, and the extensive network of tree roots held the soil in place, preventing erosion and slope failures (Infante et al. 2019). In contrast, adjacent areas that had undergone deforestation witnessed a higher frequency of landslides due to the loss of vegetation cover and the subsequent destabilization of the slopes (Chang et al. 2022; Li and Mo 2019; Tao et al. 2020). The findings of this case study emphasize the crucial importance of preserving and restoring forest ecosystems to maintain slope stability and mitigate landslide risks. It underscores the need for sustainable forest management practices and highlights the benefits of afforestation and reforestation initiatives (Tanyas et al. 2019). By promoting the growth of vegetation and ensuring the conservation of forests, communities can effectively reduce landslide susceptibility and protect the surrounding slopes. This case study provides valuable insights for researchers, policymakers, and practitioners involved in land management, environmental conservation, and disaster risk reduction in India and other regions with similar forested landscapes (Chang et al. 2022). It reinforces the significance of incorporating vegetation-based solutions and promoting sustainable practices to safeguard slope stability and mitigate the impacts of landslides. By recognizing the essential role of vegetation cover in maintaining slope stability, proactive measures can be taken to ensure the long-term sustainability and resilience of these ecosystems (Bonnesoeur et al. 2019; Chen et al. 2018; Dou et al. 2019; Segura-Millán and Perez-Verdin 2023; Tanyas et al. 2019; Thacker et al. 2023).

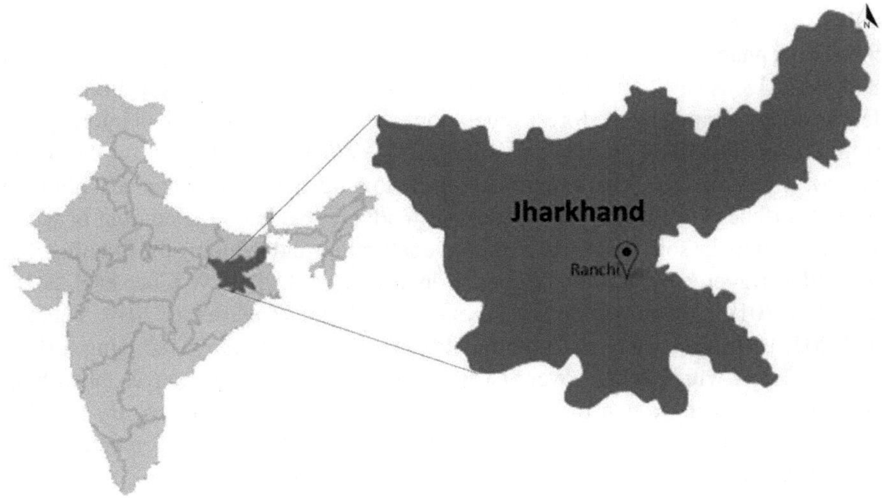

Fig. 9.3 Location of the Jharkhand

9.8.3 *Case Study 3: Mining Activities and Landslide Risk*

A significant case study conducted in mining regions of India shed light on the elevated landslide risk associated with mining activities. The investigation aimed to explore the impact of mining operations on slope stability and highlight the importance of implementing proper management practices to mitigate landslide hazards (Manna and Maiti 2018; Tao et al. 2020). The case study focused on a mining region in Jharkhand, (Fig. 9.3) a state in eastern India that is rich in mineral resources (Rao and Das 2021). The study area encompassed open-pit mining sites where minerals such as coal, iron ore, and limestone were extracted (Tao et al. 2020). The examination of this region provided valuable insights into the relationship between mining activities and landslide occurrences (Sharma et al. 2018).

One of the key factors identified in this case study was the excavation of large volumes of soil and rock during open-pit mining operations (Mishra et al. 2013). The removal of these materials weakened the natural stability of the slopes surrounding the mining sites. The exposed slopes became vulnerable to erosion and mass movements, increasing the risk of landslides (Manna and Maiti 2018). The excavation process altered the geomorphological characteristics of the terrain, disrupting the equilibrium and leading to slope failures. Improper waste disposal practices associated with mining activities further exacerbated the landslide risk (Tao et al. 2020). Mining operations generate substantial amounts of waste materials, such as overburden and mine tailings. In some instances, these waste materials were not adequately managed or disposed of, leading to their accumulation on the slopes (Fang et al. 2020; Paryani et al. 2021). The presence of large volumes of unstable mine waste intensified the susceptibility to landslides. Rainfall events and natural erosional processes could trigger mass movements, causing significant slope failures and posing threats to nearby communities and infrastructure (Chen and Li 2020; Wei et al. 2022).

The case study also highlighted the importance of implementing proper slope management measures and reclamation practices in mining areas. By adopting proactive measures, such as slope stabilization techniques and erosion control measures, the risks associated with mining-induced landslides can be effectively reduced (Chang et al. 2023a, b; Es-smairi et al. 2023). For instance, the installation of retaining structures, such as retaining walls and geotechnical reinforcements, can enhance slope stability and prevent soil erosion (Infante et al. 2019). Moreover, implementing appropriate reclamation practices, including the re-vegetation of disturbed areas and the restoration of natural landforms, can help restore the stability and ecological balance of the mining sites (Manna and Maiti 2018; Mishra et al. 2013; Tao et al. 2020). To illustrate the impact of mining activities on landslide risk, a specific mining site in Jharkhand can be considered. The open-pit mining operations in this area resulted in extensive excavation, leaving behind steep and unstable slopes (Manna and Maiti 2018; Mishra et al. 2013; Tao et al. 2020). Improper waste disposal practices led to the accumulation of mine waste on the slopes, further compromising their stability. As a consequence, landslides frequently occur, endangering the lives of nearby communities and disrupting infrastructure (Rabby et al. 2022; Saha et al. 2023).

The findings of this case study underscore the importance of responsible mining practices and effective environmental management in mining regions. It emphasizes the need for robust regulations and monitoring systems to ensure proper waste disposal and slope stabilization in mining operations (Manna and Maiti 2018; Mishra et al. 2013; Tao et al. 2020). By implementing sustainable mining practices and adhering to reclamation guidelines, the risk of landslides can be significantly reduced, ensuring the safety of workers and local communities (Alsabhan et al. 2022; Revadekar et al. 2018). It provides valuable insights for researchers, mining authorities, and policymakers involved in the mining sector and environmental management in India and other regions with similar mining activities (Sharma et al. 2018). It emphasizes the significance of incorporating slope management measures and reclamation practices into mining operations to minimize landslide hazards. By integrating these measures, mining regions can mitigate the adverse impacts on slope stability and work towards sustainable mining practices that prioritize environmental protection and the well-being of local communities.

9.9 Mitigation Strategies and Best Practices

Landslides pose a significant risk to communities and infrastructure in many parts of the world (Gómez and Kavzoglu 2005; Tanyas et al. 2019). As such, implementing effective mitigation strategies and adopting best practices is crucial to minimize the impact of landslides and enhance overall resilience (Chen et al. 2023; Dehnavi et al.

2015). This section will go through some of the most effective mitigation techniques and safeguards for landslides and susceptible locations.

Geotechnical Investigations and Mapping: To comprehend the underlying geological conditions and recognize possible risks, thorough geotechnical investigations and mapping of landslide-prone regions are required (Chen and Li 2020; Sonker et al. 2022). This involves assessing the soil composition, slope characteristics, groundwater conditions, and other relevant factors (Ali et al. 2021; Saikh and Mondal 2023). The collected data can inform the development of accurate landslide susceptibility maps, which are valuable tools for land-use planning and infrastructure development (Ali et al. 2021; Chen et al. 2018; Chen and Li 2020; Patil and Panhalkar 2023; Saikh and Mondal 2023; Sonker et al. 2022).

Land Use Planning and Zoning: Proper land use planning and zoning regulations play a vital role in preventing landslide (Abdulkareem et al. 2019)s. The probability of landslides can be considerably decreased by recognizing and avoiding high-risk regions, such as steep slopes, unstable terrains, or areas with documented geological weaknesses (Rabby et al. 2022). Land use policies should prioritize the preservation of natural landscapes, discourage development in hazard-prone areas, and promote sustainable practices that minimize disturbance to the environment (Chen et al. 2019). Limiting urban expansion on steep slopes, promoting green infrastructure, and preserving natural areas can help mitigate landslide risks. Comprehensive land use plans should incorporate risk assessments and prioritize sustainable development practices (Abdulkareem et al. 2019; Chen et al. 2019; Rabby et al. 2022).

Slope Stabilization Measures: Implementing slope stabilization measures is crucial to reinforce slopes and prevent slope failures (Li and Mo 2019). Various methods can be used, based on the unique circumstances of the site. These may include installing retaining walls, soil nailing, rock bolting, terracing, or bioengineering methods that utilize vegetation and erosion control measures (Chang et al. 2022). To ensure the efficacy and long-term stability of these measures, proper engineering design and construction techniques should be used (Chang et al. 2022; Li and Mo 2019; Tanyas et al. 2019).

Vegetation Management: Maintaining and enhancing vegetation cover is an effective strategy for landslide prevention. Vegetation, particularly dense forests, plays a vital role in stabilizing slopes by providing root reinforcement, intercepting rainfall, reducing soil erosion, and increasing soil cohesion (Bonnesoeur et al. 2019; Chen et al. 2018). Afforestation, reforestation, and implementing erosion control measures, such as the establishment of vegetative barriers and bioengineering techniques, can significantly reduce landslide susceptibility and contribute to overall slope stability (Abdulkareem et al. 2019; Bonnesoeur et al. 2019; Chen et al. 2018; Dou et al. 2019).

Stormwater Management: Proper stormwater management is critical for landslide prevention, as excessive water infiltration can significantly increase slope instability (Nepal et al. 2019). Implementing effective drainage systems, including the construction of channels, culverts, and retention ponds, helps control the flow of water and prevents the saturation of slopes (Abdulkareem et al. 2019). Additionally, measures like sedimentation basins and check dams can reduce sediment transport and prevent erosion, further enhancing slope stability (Hong 2023).

Early Warning Systems: Developing and implementing early warning systems can provide valuable time for evacuation and emergency preparedness in landslide-prone areas (Pennington et al. 2015). These systems utilize real-time monitoring of slope movements, rainfall data, and other relevant parameters to detect potential landslide triggers (Gariano and Guzzetti 2016; Huang et al. 2023a, b, c). When threshold levels are exceeded, timely alerts can be issued, allowing authorities and communities to take necessary actions to minimize the risks and ensure the safety of the affected population (Badola et al. 2023; Thi Ngo et al. 2021).

Education, Awareness, and Capacity Building: Promoting education, awareness, and capacity-building initiatives is crucial for building resilience against landslides (Van Assche et al. 2020). This involves educating communities, stakeholders, and decision-makers about the risks associated with landslides, the importance of adherence to regulations, and the adoption of best practices (Segura-Millán and Perez-Verdin 2023). Training programs, workshops, and awareness campaigns can empower individuals and communities to take proactive measures and make informed decisions regarding land use, construction practices, and emergency response (Segura-Millán and Perez-Verdin 2023; Van Assche et al. 2020).

Sustainable Land Management Practices: Implementing sustainable land management practices is essential to maintain slope stability and reduce landslide hazards (Pangapanga-Phiri et al. 2023). Measures such as afforestation, reforestation, contour plowing, and terracing can enhance soil conservation, improve infiltration capacity, and minimize erosion (Gholkar et al. 2022). Promoting responsible agricultural practices, such as crop rotation and agroforestry, can also contribute to soil stability and reduce landslide susceptibility (Kolapo et al. 2022; Mcharo and Maghenda 2021).

By implementing these mitigation strategies and best practices, the vulnerability to landslides can be significantly reduced (Issahaku and Abdul-Rahaman 2019; Martínez-Mena et al. 2020; Seid et al. 2022). Governments, policymakers, engineers, and communities need to collaborate and prioritize the adoption of these measures in both urban and rural settings. By investing in preventive measures, monitoring systems, and public awareness, we can build more resilient societies that are better equipped to mitigate the risks posed by landslides and safeguard lives and infrastructure (Dehnavi et al. 2015; Gómez and Kavzoglu 2005; Roy et al. 2023; Tanyas et al. 2019; Thacker et al. 2023; Zhu et al. 2019).

9.10 Conclusion

The environment, infrastructure, and human lives are all significantly at hazards from landslides. Understanding the factors that contribute to landslide susceptibility and implementing effective mitigation strategies are crucial for enhancing resilience and reducing the impact of these natural hazards. This study examined how many elements, including human activity, geotechnical characteristics, land use, and cover, affect landslide vulnerability.

It has been established via extensive research and case studies that changes in land use, especially urbanization, are directly related to an increase in the frequency of landslides. The conversion of natural landscapes into urban areas disrupts the delicate balance of slopes and alters the hydrological regime, leading to higher landslide risks. Similarly, the removal of vegetation cover, alteration of slopes, and inadequate stormwater management systems further contribute to heightened susceptibility to landslides. The presence of intact vegetation cover has been shown to play a critical role in maintaining slope stability. Forested areas with dense canopies and well-established root systems exhibit lower landslide occurrence compared to deforested regions. The roots of trees reinforce slopes, intercept rainfall, and reduce soil erosion, thereby enhancing the overall stability of the slopes. Activities related to mining pose significant landslide risks. Open-pit mining operations and improper waste disposal practices weaken natural slopes and increase the susceptibility to landslides. However, implementing proper slope management measures and reclamation practices can mitigate these risks and reduce landslide hazards in mining areas. To enhance landslide resilience, it is crucial to adopt comprehensive strategies that encompass geotechnical investigations, land use planning, slope stabilization measures, vegetation management, stormwater management, early warning systems, and education initiatives. By integrating these strategies, governments, communities, and stakeholders can effectively mitigate landslide risks and protect vulnerable areas.

Geotechnical investigations and mapping enable a better understanding of landslide-prone areas, aiding land-use planning and infrastructure development. Proper land use planning and zoning regulations help identify and avoid high-risk areas, minimizing landslide potential. Implementing slope stabilization measures, such as retaining walls or bioengineering techniques, reinforces slopes and prevents slope failures. Maintaining and enhancing vegetation cover through afforestation and reforestation plays a crucial role in slope stability. Effective stormwater management helps control water infiltration and prevents the saturation of slopes. Early warning systems provide timely alerts, allowing for evacuation and emergency preparedness. Education, awareness, and capacity-building initiatives empower individuals and communities to make informed decisions and take proactive measures against landslides. Preventing and mitigating landslides requires a multidimensional approach that addresses the underlying factors contributing to landslide susceptibility. By implementing comprehensive strategies, we can reduce the impact of landslides, protect lives and infrastructure, and create more resilient communities. Governments, researchers, engineers, and communities need to collaborate, prioritize preventive measures, and raise awareness to effectively tackle the challenges posed by landslides. With coordinated efforts, we can build a more secure and sustainable future where landslide hazards are reduced and both human and environmental well-being are protected.

Reference

Abdulkareem JH, Pradhan B, Sulaiman WNA, Jamil NR (2019) Prediction of spatial soil loss impacted by long-term land-use/land-cover change in a tropical watershed. Geosci Front 10(2):389–403. https://doi.org/10.1016/j.gsf.2017.10.010

Achu AL, Aju CD, Di Napoli M, Prakash P, Gopinath G, Shaji E, Chandra V (2023) Machine-learning based landslide susceptibility modeling with emphasis on uncertainty analysis. Geosci Front 101657. https://doi.org/10.1016/j.gsf.2023.101657

Addis A (2023) olGIS—based flood susceptibility mapping using frequency ratio and information value models in upper Abay river basin, Ethiopia. Nat Haz Res. https://doi.org/10.1016/j.nhres.2023.02.003

Ali SA, Parvin F, Vojteková J, Costache R, Linh NTT, Pham QB, Vojtek M, Gigović L, Ahmad A, Ghorbani MA (2021) GIS-based landslide susceptibility modeling: a comparison between fuzzy multi-criteria and machine learning algorithms. Geosci Front 12(2):857–876. https://doi.org/10.1016/j.gsf.2020.09.004

Alsabhan AH, Singh K, Sharma A, Alam S, Pandey DD, Rahman SAS, Khursheed A, Munshi FM (2022) Landslide susceptibility assessment in the Himalayan range based along Kasauli—Parwanoo road corridor using weight of evidence, information value, and frequency ratio. J King Saud Univ Sci 34(2). https://doi.org/10.1016/j.jksus.2021.101759

Badola S, Mishra VN, Parkash S, Pandey M (2023) Rule-based fuzzy inference system for landslide susceptibility mapping along national highway 7 in Garhwal Himalayas, India. Quat Sci Adv 11:100093. https://doi.org/10.1016/j.qsa.2023.100093

Bonnesoeur V, Locatelli B, Guariguata MR, Ochoa-Tocachi BF, Vanacker V, Mao Z, Stokes A, Mathez-Stiefel SL (2019) Impacts of forests and forestation on hydrological services in the Andes: a systematic review. In: Forest ecology and management, vol 433. Elsevier BV, pp 569–584. https://doi.org/10.1016/j.foreco.2018.11.033

Canlas IP (2023) Three decades of disaster risk reduction education: a bibliometric study. Nat Haz Res. https://doi.org/10.1016/j.nhres.2023.02.007

Chang L, Ma J, Xing G, Zhang R, Zhao N, Yin H, Fan L, Huang F (2023a) Landslide susceptibility evaluation and interpretability analysis of typical loess areas based on deep learning. Nat Haz Res. https://doi.org/10.1016/j.nhres.2023.02.005

Chang Z, Huang F, Huang J, Jiang SH, Liu Y, Meena SR, Catani F (2023b) An updating of landslide susceptibility prediction from the perspective of space and time. Geosci Front 14(5). https://doi.org/10.1016/j.gsf.2023.101619

Chang Z, Catani F, Huang F, Liu G, Meena SR, Huang J, Zhou C (2022) Landslide susceptibility prediction using slope unit-based machine learning models considering the heterogeneity of conditioning factors. J Rock Mech Geotech Eng. https://doi.org/10.1016/j.jrmge.2022.07.009

Chen J, Li L, Xu C, Huang Y, Luo Z, Xu X, Lyu Y (2023) Freely accessible inventory and spatial distribution of large-scale landslides in Xianyang City, Shaanxi Province, China. Earthq Res Adv 100217. https://doi.org/10.1016/j.eqrea.2023.100217

Chen L, Guo Z, Yin K, Pikha Shrestha D, Jin S (2019) The influence of land use and land cover change on landslide susceptibility: a case study in Zhushan Town, Xuan'en County (Hubei, China). Nat Hazard 19(10):2207–2228. https://doi.org/10.5194/nhess-19-2207-2019

Chen W, Li Y (2020) GIS-based evaluation of landslide susceptibility using hybrid computational intelligence models. CATENA 195. https://doi.org/10.1016/j.catena.2020.104777

Chen W, Zhang S, Li R, Shahabi H (2018) Performance evaluation of the GIS-based data mining techniques of best-first decision tree, random forest, and naïve Bayes tree for landslide susceptibility modeling. Sci Total Environ 644:1006–1018. https://doi.org/10.1016/j.scitotenv.2018.06.389

Dai F, Zhu W, Ren M, Niu L, Hou C (2023) Experimental and numerical simulation on the sliding process of roof rock blocks triggered by dynamic disturbance. Underground Space. https://doi.org/10.1016/j.undsp.2022.10.007

Dao DV, Jaafari A, Bayat M, Mafi-Gholami D, Qi C, Moayedi H, Phong TV, Ly HB, Le TT, Trinh PT, Luu C, Quoc NK, Thanh BN, Pham BT (2020) A spatially explicit deep learning neural network model for the prediction of landslide susceptibility. CATENA 188. https://doi.org/10.1016/j.catena.2019.104451

Davis GA, Friedmann SJ (2005) Large-scale gravity sliding in the Miocene Shadow Valley supradetachment basin, Eastern Mojave Desert, California. Earth Sci Rev 73(1–4):149–176. https://doi.org/10.1016/j.earscirev.2005.04.008

Dehnavi A, Aghdam IN, Pradhan B, Morshed Varzandeh MH (2015) A new hybrid model using step-wise weight assessment ratio analysis (SWARA) technique and adaptive neuro-fuzzy inference system (ANFIS) for regional landslide hazard assessment in Iran. CATENA 135:122–148. https://doi.org/10.1016/j.catena.2015.07.020

Dhungana G, Ghimire R, Poudel R, Kumal S (2023) Landslide susceptibility and risk analysis in Benighat Rural Municipality, Dhading, Nepal. Nat Haz Res 3(2):170–185. https://doi.org/10.1016/j.nhres.2023.03.006

Dou J, Yunus AP, Tien Bui D, Merghadi A, Sahana M, Zhu Z, Chen CW, Khosravi K, Yang Y, Pham BT (2019) Assessment of advanced random forest and decision tree algorithms for modeling rainfall-induced landslide susceptibility in the Izu-Oshima Volcanic Island, Japan. Sci Total Environ 662:332–346. https://doi.org/10.1016/j.scitotenv.2019.01.221

Es-smairi A, Elmoutchou B, Mir RA, Ouazani Touhami A El, Namous M (2023) Delineation of landslide susceptible zones using Frequency Ratio (FR) and Shannon Entropy (SE) models in northern Rif, Morocco. Geosyst Geoenviron 2(4). https://doi.org/10.1016/j.geogeo.2023.100195

Fang Z, Wang Y, Peng L, Hong H (2020) Integration of convolutional neural network and conventional machine learning classifiers for landslide susceptibility mapping. Comput Geosci 139. https://doi.org/10.1016/j.cageo.2020.104470

Fang Z, Wang Y, Peng L, Hong H (2021) A comparative study of heterogeneous ensemble-learning techniques for landslide susceptibility mapping. Int J Geogr Inf Sci 35(2):321–347. https://doi.org/10.1080/13658816.2020.1808897

Gariano SL, Guzzetti F (2016) Landslides in a changing climate. In: Earth-science reviews, vol 162. Elsevier BV, pp 227–252. https://doi.org/10.1016/j.earscirev.2016.08.011

Gholkar M, Thombare P, Koli U, Kumbhar N (2022) Techno-economic assessment of agricultural land remediation measures through nutrient management practices to achieve sustainable agricultural production. Environ Challenges 7. https://doi.org/10.1016/j.envc.2022.100492

Gómez H, Kavzoglu T (2005) Assessment of shallow landslide susceptibility using artificial neural networks in Jabonosa River Basin, Venezuela. Eng Geol 78(1–2):11–27. https://doi.org/10.1016/j.enggeo.2004.10.004

Guo Z, Ferrer JV, Hürlimann M, Medina V, Puig-Polo C, Yin K, Huang D (2023) Shallow landslide susceptibility assessment under future climate and land cover changes: A case study from southwest China. Geosci Front 14(4). https://doi.org/10.1016/j.gsf.2023.101542

Hammad Khaliq A, Basharat M, Talha Riaz M, Tayyib Riaz M, Wani S, Al-Ansari N, Ba Le L, Thi Thuy Linh N (2023) Spatiotemporal landslide susceptibility mapping using machine learning models: A case study from district Hattian Bala, NW Himalaya, Pakistan. Ain Shams Eng J 14(3). https://doi.org/10.1016/j.asej.2022.101907

Hong H (2023) Assessing landslide susceptibility based on hybrid Best-first decision tree with ensemble learning model. Ecol Ind 147. https://doi.org/10.1016/j.ecolind.2023.109968

Horn DP (2015) Storm surge warning, mitigation, and adaptation. In: Coastal and marine hazards, risks, and disasters. Elsevier Inc, pp 153–180. https://doi.org/10.1016/B978-0-12-396483-0.00006-6

Huang F, Teng Z, Guo Z, Catani F, Huang J (2023a) Uncertainties of landslide susceptibility prediction: influences of different spatial resolutions, machine learning models and proportions of training and testing dataset. Rock Mech Bull 2(1):100028. https://doi.org/10.1016/j.rockmb.2023.100028

Huang F, Xiong H, Yao C, Catani F, Zhou C, Huang J (2023b) Uncertainties of landslide susceptibility prediction considering different landslide types. J Rock Mech Geotech Eng. https://doi.org/10.1016/j.jrmge.2023.03.001

Huang W, Ding M, Li Z, Yu J, Ge D, Liu Q, Yang J (2023c) Landslide susceptibility mapping and dynamic response along the Sichuan-Tibet transportation corridor using deep learning algorithms. CATENA 222. https://doi.org/10.1016/j.catena.2022.106866

Huang F, Yan J, Fan X, Yao C, Huang J, Chen W, Hong H (2022) Uncertainty pattern in landslide susceptibility prediction modelling: Effects of different landslide boundaries and spatial shape expressions. Geosci Front 13(2). https://doi.org/10.1016/j.gsf.2021.101317

Infante D, Di Martire D, Confuorto P, Tessitore S, Tòmas R, Calcaterra D, Ramondini M (2019) Assessment of building behavior in slow-moving landslide-affected areas through DInSAR data and structural analysis. Eng Struct 199. https://doi.org/10.1016/j.engstruct.2019.109638

Issahaku G, Abdul-Rahaman A (2019) Sustainable land management practices, off-farm work participation and vulnerability among farmers in Ghana: is there a nexus? Int Soil Water Conserv Res 7(1):18–26. https://doi.org/10.1016/j.iswcr.2018.10.002

John B, Rajendran CP (2009) Evidence of episodic brittle faulting in the cratonic part of the Peninsular India and its implications for seismic hazard in slow deforming regions. Tectonophysics 471(3–4):240–252. https://doi.org/10.1016/j.tecto.2009.02.022

Kolapo A, Didunyemi AJ, Aniyi OJ, Obembe OE (2022) Adoption of multiple sustainable land management practices and its effects on productivity of smallholder maize farmers in Nigeria. Resour Environ Sustain 10. https://doi.org/10.1016/j.resenv.2022.100084

Li Y, Mo P (2019) A unified landslide classification system for loess slopes: a critical review. In: Geomorphology, vol 340. Elsevier BV, pp 67–83. https://doi.org/10.1016/j.geomorph.2019.04.020

Lima P, Steger S, Glade T, Mergili M (2023) Conventional data-driven landslide susceptibility models may only tell us half of the story: potential underestimation of landslide impact areas depending on the modeling design. Geomorphology 108638. https://doi.org/10.1016/j.geomorph.2023.108638

Lombardo L, Opitz T, Ardizzone F, Guzzetti F, Huser R (2020) Space-time landslide predictive modelling. In: Earth-science reviews, vol 209. Elsevier BV. https://doi.org/10.1016/j.earscirev.2020.103318

Manna A, Maiti R (2018) Geochemical contamination in the mine affected soil of Raniganj Coalfield—a river basin scale assessment. Geosci Front 9(5):1577–1590. https://doi.org/10.1016/j.gsf.2017.10.011

Martínez-Mena M, Carrillo-López E, Boix-Fayos C, Almagro M, García Franco N, Díaz-Pereira E, Montoya I, de Vente J (2020) Long-term effectiveness of sustainable land management practices to control runoff, soil erosion, and nutrient loss and the role of rainfall intensity in Mediterranean rainfed agroecosystems. CATENA 187. https://doi.org/10.1016/j.catena.2019.104352

Mcharo M, Maghenda M (2021) Cost-benefit analysis of sustainable land and water management practices in selected highland water catchments of Kenya. Sci Afr 12. https://doi.org/10.1016/j.sciaf.2021.e00779

Mehrotra RC, Awasthi N, Dutta SK (1999) Study of fossil wood from the upper Tertiary sediments (Siwalik) of Arunachal Pradesh, India and its implication in palaeoecological and phytogeographical interpretations. In: Review of palaeobotany and palynology, vol 107. www.elsevier.com/locate/revpalbo

Mishra DP, Sugla M, Singha P (2013) Productivity improvement in underground coal mines—a case study. J Sustain Min 12(3):48–53. https://doi.org/10.7424/jsm130306

Nepal N, Chen J, Chen H, Wang X, Pangali Sharma TP (2019) Assessment of landslide susceptibility along the Araniko Highway in Poiqu/Bhote Koshi/Sun Koshi Watershed, Nepal Himalaya. Progr Disast Sci 3. https://doi.org/10.1016/j.pdisas.2019.100037

Nightingale E, Richmond CAM (2022) Building structures of environmental repossession to reclaim land, self-determination and Indigenous wellness. Health Place 73. https://doi.org/10.1016/j.healthplace.2021.102725

Ojala AEK, Mattila J, Markovaara-Koivisto M, Ruskeeniemi T, Palmu JP, Sutinen R (2019) Distribution and morphology of landslides in northern Finland: an analysis of postglacial seismic activity. Geomorphology 326:190–201. https://doi.org/10.1016/j.geomorph.2017.08.045

Panchal S, Shrivastava AK (2022) Landslide hazard assessment using analytic hierarchy process (AHP): A case study of National Highway 5 in India. Ain Shams Eng J 13(3). https://doi.org/10.1016/j.asej.2021.10.021

Pánek T (2019) Landslides and quaternary climate changes—the state of the art. In: Earth-science reviews, vol 196. Elsevier BV. https://doi.org/10.1016/j.earscirev.2019.05.015

Pangapanga-Phiri I, Mungatana ED, Pangapanga L, Nkoka FS (2023) Understanding the impact of sustainable landscape management on farm productivity under intensifying tropical cyclones in Southern Malawi. Trop Cycl Res Rev. https://doi.org/10.1016/j.tcrr.2023.02.002

Paryani S, Neshat A, Pradhan B (2021) Improvement of landslide spatial modeling using machine learning methods and two Harris hawks and bat algorithms. Egypt J Rem Sens Space Sci 24(3):845–855. https://doi.org/10.1016/j.ejrs.2021.08.006

Patil AS, Panhalkar SS (2023) Remote sensing and GIS-based landslide susceptibility mapping using LNRF method in part of Western Ghats of India. Quat Sci Adv 11:100095. https://doi.org/10.1016/j.qsa.2023.100095

Pennington C, Freeborough K, Dashwood C, Dijkstra T, Lawrie K (2015) The National Landslide Database of Great Britain: acquisition, communication and the role of social media. Geomorphology 249:44–51. https://doi.org/10.1016/j.geomorph.2015.03.013

Plat JP, Van Den Eeckhout B, Janzen E, Konert G, Simon OJ, Weijermars R (1983) The structure and tectonic evolution of the Aguil6n fold-nappe, Sierra Alhamilla, Betic Cordilleras, SE Spain. J Struct Geol 5(5)

Pradhan B, Dikshit A, Lee S, Kim H (2023) An explainable AI (XAI) model for landslide susceptibility modeling. Appl Soft Comput 142. https://doi.org/10.1016/j.asoc.2023.110324

Puente-Sotomayor F, Egas A, Teller J (2021) Land policies for landslide risk reduction in Andean cities. In: Habitat international, vol 107. Elsevier Ltd. https://doi.org/10.1016/j.habitatint.2020.102298

Rabby YW, Li Y, Abedin J, Sabrina S (2022) Impact of land use/land cover change on landslide susceptibility in Rangamati Municipality of Rangamati District, Bangladesh. ISPRS Int J Geo-Inf 11(2). https://doi.org/10.3390/ijgi11020089

Rao MRK, Das MK (2021) Malaria among the Sauria Paharia, a primitive and vulnerable tribe of Jharkhand state, India. Clin Epidemiol Glob Health 11. https://doi.org/10.1016/j.cegh.2021.100778

Revadekar JV, Varikoden H, Murumkar PK, Ahmed SA (2018) Latitudinal variation in summer monsoon rainfall over Western Ghat of India and its association with global sea surface temperatures. Sci Total Environ 613–614:88–97. https://doi.org/10.1016/j.scitotenv.2017.08.285

Roy D, Sarkar A, Kundu P, Paul S, Chandra Sarkar B (2023) An ensemble of evidence belief function (EBF) with frequency ratio (FR) using geospatial data for landslide prediction in Darjeeling Himalayan region of India. Quat Sci Adv 11:100092. https://doi.org/10.1016/j.qsa.2023.100092

Saha S, Majumdar P, Bera B (2023) Deep learning and benchmark machine learning based landslide susceptibility investigation, Garhwal Himalaya (India). Quat Sci Adv 10. https://doi.org/10.1016/j.qsa.2023.100075

Saikh NI, Mondal P (2023) Gis-based machine learning algorithm for flood susceptibility analysis in the Pagla river basin, Eastern India. Nat Haz Res. https://doi.org/10.1016/j.nhres.2023.05.004

Santosh M, Yokoyama K, Biju-Sekhal S, Rogers JJW (2003) Multiple tectonothermal events in the granulite blocks of Southern India revealed from EPMA dating: implications on the history of supercontinents. Int Assoc Gondwana Res 6(I)

Saravanan S, Chandrasekar N, Hentry C, Rajamanickam M, Immanuel JL, Subramanian PS (2009) Post-tsunami assessment in the coastal region between Kanyakumari and Ovari, Tamil Nadu-a case study. Dixue Qianyuan/earth Sci Front 16(6):129–137. https://doi.org/10.1016/S1872-5791(08)60111-6

Segura-Millán K, Perez-Verdin G (2023) The effect of payments for ecosystem services on forest cover, land use, and capacity building in northern Mexico. Trees Forests People 12. https://doi.org/10.1016/j.tfp.2023.100387

Seid M, Simane B, Teferi E, Azmeraw A (2022) Determinants of farmers' multiple-choice and sustainable use of indigenous land management practices in the Southern Rift Valley of Ethiopia. Curr Res Environ Sustain 4. https://doi.org/10.1016/j.crsust.2022.100158

Sharma A, Gupta AK, Ganguly R (2018) Impact of open dumping of municipal solid waste on soil properties in mountainous region. J Rock Mech Geotech Eng 10(4):725–739. https://doi.org/10.1016/j.jrmge.2017.12.009

Sonker I, Tripathi JN, Swarnim. (2022) Remote sensing and GIS-based landslide susceptibility mapping using frequency ratio method in Sikkim Himalaya. Quat Sci Adv 8. https://doi.org/10.1016/j.qsa.2022.100067

Sutoris P (2018) Elitism and its challengers: Educational development ideology in postcolonial India through the prism of film, 1950–1970. Int J Educ Dev 60:1–9. https://doi.org/10.1016/j.ijedudev.2017.10.017

Tang H, Wasowski J, Juang CH (2019) Geohazards in the three Gorges Reservoir Area, China – Lessons learned from decades of research. Eng Geol 261. Elsevier BV. https://doi.org/10.1016/j.enggeo.2019.105267

Tanyas H, Rossi M, Alvioli M, van Westen CJ, Marchesini I (2019) A global slope unit-based method for the near real-time prediction of earthquake-induced landslides. Geomorphology 327:126–146. https://doi.org/10.1016/j.geomorph.2018.10.022

Tao Z, Shu Y, Yang X, Peng Y, Chen Q, Zhang H (2020) Physical model test study on shear strength characteristics of slope sliding surface in Nanfen open-pit mine. Int J Min Sci Technol 30(3):421–429. https://doi.org/10.1016/j.ijmst.2020.05.006

Tempa K, Peljor K, Wangdi S, Ghalley R, Jamtsho K, Ghalley S, Pradhan P (2021) UAV technique to localize landslide susceptibility and mitigation proposal: a case of Rinchending Goenpa landslide in Bhutan. Nat Haz Res 1(4):171–186. https://doi.org/10.1016/j.nhres.2021.09.001

Thacker M, Limaye RB, Padmalal D, Rajaguru SN, Kumaran KPN, Punekar SA, Karthick B (2023) Holocene climate dynamics and ecological responses in Kaas Plateau, Western Ghats, India: Evidence from lacustrine deposits. Quat Sci Adv 11. https://doi.org/10.1016/j.qsa.2023.100087

Thi Ngo PT, Panahi M, Khosravi K, Ghorbanzadeh O, Kariminejad N, Cerda A, Lee S (2021) Evaluation of deep learning algorithms for national scale landslide susceptibility mapping of Iran. Geosci Front 12(2):505–519. https://doi.org/10.1016/j.gsf.2020.06.013

Tyagi A, Kamal Tiwari R, James N (2022) A review on spatial, temporal and magnitude prediction of landslide hazard. J Asian Earth Sci X 7. Elsevier Ltd. https://doi.org/10.1016/j.jaesx.2022.100099

Van Assche K, Gruezmacher M, Deacon L (2020) Land use tools for tempering boom and bust: strategy and capacity building in governance. Land Use Policy 93. https://doi.org/10.1016/j.landusepol.2019.05.013

Wang J, Zhang Q, Gou T, Mo J, Wang Z, Gao M (2018) Spatial-temporal changes of urban areas and terrestrial carbon storage in the Three Gorges Reservoir in China. Ecol Ind 95:343–352. https //doi.org/10.1016/j.ecolind.2018.06.036

Wei R, Ye C, Sui T, Ge Y, Li Y, Li J (2022) Combining spatial response features and machine learning classifiers for landslide susceptibility mapping. Int J Appl Earth Obs Geoinf 107. https://doi.org/10.1016/j.jag.2022.102681

Zhu AX, Miao Y, Liu J, Bai S, Zeng C, Ma T, Hong H (2019) A similarity-based approach to sampling absence data for landslide susceptibility mapping using data-driven methods. CATENA 183. https://doi.org/10.1016/j.catena.2019.104188

Chapter 10
Moisture-Driven Landslides and Cascade Hazards in the Himalayan Region: A Synthesis on Predictive Assessment

Danish Monga and Poulomi Ganguli

Abstract Moisture-driven landslides (MDL), mainly caused by rain, seriously threaten lives and property, resulting in catastrophic damages and considerable economic losses. In areas with steep topography, the temporal and spatial clustering of long-duration, moderate-to-short-duration high-intensity rainfall contributes to landslides. Shifting precipitation patterns due to climate change, alterations to sub-surface conditions, such as pore-water pressures, retreating glaciers, and permafrost, further increase the risk of landslides. This synthesis apprises the physical drivers associated with MDLs and their complex interplay in triggering MDLs. The added value of this synthesis is multifaceted: (1) to uncover moisture-driven landslide trends and associated cascading hazards across the high-mountain areas of the globe (2) To explore the probability of rain-driven landslide-related mortality rate across different continents using the archived landslide information. (3) To present a systematic review of available physical and statistical tools to identify MDL triggers and highlight the need for updating rain thresholds using observed and projected climate information to address nonstationarities related to climate change and climate variability. Our analysis of archived rainfall-triggered landslides of the global landslide catalogue between 2007 and 2022 shows that global precipitation contributes to more than 60% of landslides. Further, our observational assessments of the fatality versus landslide frequency curves show that high-mountain Asia experiences the most frequent landslides, and its populations are especially vulnerable to such catastrophic events. The contribution of this synthesis paper is to inform scientists and practitioners of the latest developments in MDLs, aiding the translation of scientific understanding into developing resilience policies and adaptation efforts.

D. Monga · P. Ganguli (✉)
Agricultural and Food Engineering Department, Indian Institute of Technology Kharagpur, Kharagpur, West Bengal 721302, India
e-mail: pganguli@agfe.iitkgp.ac.in

D. Monga
e-mail: davinder.fdk@kgpian.iitkgp.ac.in

G. K. Panda et al. (eds.), *Landslide: Susceptibility, Risk Assessment and Sustainability*, Advances in Natural and Technological Hazards Research 52,
https://doi.org/10.1007/978-3-031-56591-5_10

Keywords Moisture-driven landslides · Climate change · Multi-hazard risk · Resilience

10.1 Introduction

A significant risk to human life and infrastructure in mountainous areas is landslides, including debris flows, rock falls, avalanches, mudslides, and other downward movements of soil, gravel, or debris under gravity (Cruden and Varnes 1996; Hungr et al. 2014). The Emergency Events Database (EM-DAT 2022), estimated that landslides affect 48 million people globally, making up about 5% of all natural hazards (WMO 2018). Precipitation is a primary causal driver for landslides (Petley et al. 2005; Johnston et al. 2021). Moisture-driven Landslides (MDL) are triggered by increased water infiltration, which results from increased rainfall, snowmelts, and their combinations, leading to elevated sub-surface moisture content or an increased volume of water (Whiteley et al. 2019). Several assessments across the globe, such as in California, USA (Handwerger et al. 2022), Kashmir (northwestern) Himalayas (Shah et al 2023), and Southern Hiroshima, Japan (Neto and Bhandary 2023), highlighted the pivotal role of precipitation-driven landslides. While Shah et al. (2023) illuminated the direct impact of precipitation magnitude, intensity, and frequency on landslides, Neto and Bhandary (2023) showed the causal link between landslide activation to peak-intensity rainfall and accumulated precipitation. Handwerger et al. (2022) showed that despite significant differences in hydroclimate across California, landslides show similar characteristics and hydrologic sensitivity in response to precipitation: wet areas of northwestern California and dry regimes of southwestern California showed similar sensitivity to seasonal and multi-year changes in precipitation.

The slope instability due to MDL involves several interconnected drivers (Reichenbach et al. 2018; Whiteley et al. 2019; Greco et al. 2023; Arnone et al. 2016): (1) hillslope processes, such as evapotranspiration, infiltration and percolation, runoff and groundwater recharge; (2) geohydrological characteristics, such as soil properties, root strength, antecedent moisture content, precipitation characteristics, evaporative demand; and (3) morphological features, such as topography, slope, lithology, and land use that shape the evapotranspiration, controlling rainfall or snowmelts, contributing sub-surface moisture and/or changes in groundwater level, *i.e.*, "effective infiltration" throughout the climate cycle. The key precipitation attributes, such as intensity and duration, as well as their frequency, mediate the likelihood of landslides. Shallow to deep-seated landslides are caused by either low-intensity, prolonged rain episodes or high-intensity, short-duration extreme precipitation events (Tsunetaka 2021). Therefore, understanding the various factors contributing to MDLs is crucial for effective hazard management and risk mitigation. Figure 10.1 provides a comprehensive overview of global landslide dynamics. The pie chart (Fig. 10.1a) portrays environmental controls triggering landslides, with rainfall being the most prominent driver, contributing around 60.7% of landslides. Figure 10.1b shows global distribution of MDLs based on available Global landslide catalogue, Cooperative Open

Online Landslide Repository (COOLR) archive maintained by the NASA (Juang et al. 2019) during the 16-year time window 2007–2022. The Himalayan region is one of the landslide 'hotspots' due to the complex interplay of hydroclimate, topography, geology, and land use and cover (Fig. 10.1b). The distributions of MDLs also shows distinct variations in landslide frequency and its magnitude across different continents, highlighting the interdependencies of environmental and socio-economic factors. Finally, the Fatality (*F*) versus Frequency (*N*) curves (Fig. 10.1c), identify high-mountain areas in Asia and South America as highly risk-prone in terms of population exposure to landslides. This can be explained by the intensity of the hazard, exposed population and the adaptation practices developed in such countries (Hock et al. 2019; Merzdorf 2020). Interestingly, we find between 2007 and 2022, the high mountain Asia, North America and South America contribute to around 50%, 19% and 11.50% of large (1000–1,000,000 m^3) to catastrophic (>1 million m^3) slide volumes of MDLs globally (Fig. 10.1b). These areas have a high frequency of severe landslides linked to factors such as precipitation, seismicity, deforestation, urbanization, economic disparities and low adaptation strategies (Li et al. 2022). In summary, the visual representations emphasize landslides as a global issue that requires an in-depth synthesis on its drivers, enriching robust adaptation and mitigation strategies. Furthermore, Fig. 10.1 highlights the importance of the Himalayan region in advancing landslide research and developing countermeasures.

Landslide research has made significant progress, yet it continues to face numerous challenges (Dash et al. 2023; Ward et al. 2020). Current generation predictive models often struggle to credibly capture the complex interplay of physical and environmental drivers that influence landslide initiation and behavior (Wang et al. 2022; Kirschbaum and Stanley 2018). Conventional approaches often oversimplifies or fails to thoroughly evaluate the entire hazard chain associated with landslides, including interdependent causal drivers and the resulting cascading impacts (Chen et al. 2021). Recent studies have developed "physics-based statistical learning predictive frameworks" for real-time monitoring (Tehrani et al. 2022; Lei et al. 2023) and high-resolution susceptibility mapping (Sharma et al. 2024) of catastrophic landslides. These advancements offer a more nuanced understanding of landslide dynamics and suggest potential solutions for improved forecasts of such events with high confidence and significantly low false alarms (Tehrani et al. 2022). However, the effectiveness of landslide projections in future climate is often hindered by various sources of epistemic and aleatory uncertainties, ranging from measurement inaccuracies, scale-related challenges, model assumptions, and lack of adequate knowledge of the intricate physical processes. This is particularly true for present-generation scenario-based Earth system models, which often fail to capture fine-scale topographical features and hence underlying physical processes, such as orographic rainfall (Tehrani et al. 2022; Tengtrairat et al. 2021). The objective of this synthesis is to strengthen our knowledge in hydrometeorological aspect of MDL processes by adding value to the following aspects: (a) to present a systematic review on physical processes for MDLs; (b) To explore the probability of rain-driven landslide-related mortality rate across different continents using archived landslide information; (c) to present a synthesis on recent advances in physically-based and statistical

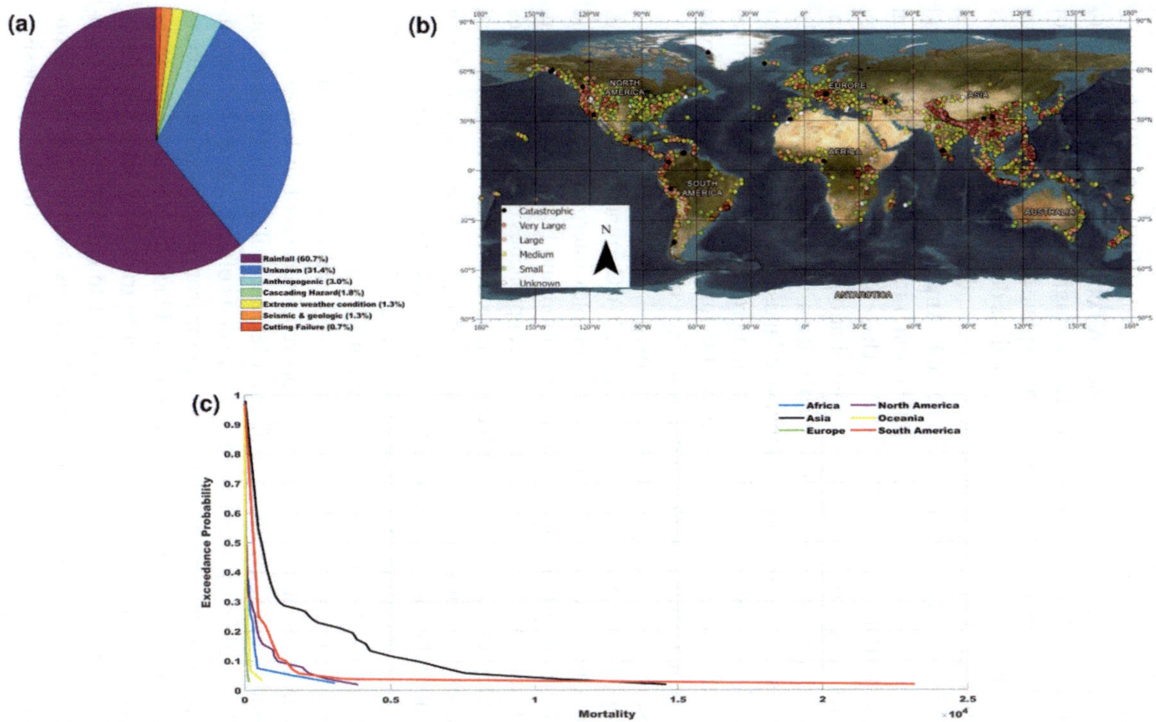

Fig. 10.1 Global landslide dynamics: environmental triggers, geographical distribution of MDLs and frequency-versus-mortality (*FN*) relations of MDLs. **a** Contribution of different environmental controls on landslide genesis **b** spatial distribution of rain-induced landslides and their severity levels as depicted by landslide volume. The landslide events are categorized to small (<10 m^3), medium (10 to 1000 m^3), large (1000 to <100,000 m^3), very large (100,000 to 1 million m^3) and catastrophic based on their volume, severity of impact and the spatial coverage. While small landslide affects one hillslope or small area, the catastrophic landslide comprises multiple interrelated events forming a hazard cascade with small recovery period impacting larger spatial area. **c** Frequency versus exceedance probability curves relating mortality rate of MDLs to its frequency of exceedance (annual rate of exceedance) across different continents. The exceedance probability $(\overline{P})$ is determined using by rank-order transformation (*i*) of *N* observations and then following the equation: $\overline{P}_i = 1 - P_i$, where $P_i = P(X < x) = \frac{i}{N+1}$. The mortality records were obtained from NASA COOLR archive available at: https://gpm.nasa.gov/landslides/index.html

approaches on determining rain trigger on landslides in mountainous terrains that consequently contribute to a deeper understanding of the complex processes in landslides mechanics (d) to provide an overview of cascading hazard chains associated with MDLs, and finally, (e) to discuss new avenues and future research perspectives for landslides research.

10.2 Rainfall-Induced Landslides: Mechanisms, Triggers, and Complex Interplay

Extreme rainfall characteristics, such as its magnitude and duration can significantly impact slope stability. Figure 10.2 shows possible physical mechanisms and forcings for MDLs in high-elevated terrains. Rainfall alters the geohydrological properties of soil and rock formations, influencing pore-water distribution and pressure, shear strength, and overall slope weight. Pore water pressure is the force that water exerts within soil and rock fractures, whereas shear strength is the resistance to sliding or displacement. The weight of the slope, influenced by gravity, is also a crucial factor (DMR 2010). Landslides typically occur in areas with a threshold slope >10°, and the likelihood rises noticeably as the slope angle increases (Ozturk et al. 2022; Modugno et al. 2022). When rainfall or snowmelts exceed a certain threshold, it leads to increased water infiltration, raising the groundwater table—the boundary between saturated and unsaturated zones. This increase reduces effective stress, (*i.e.*, the difference of total stress versus pore water pressure), thereby reducing shear strength. A landslide, characterized by the rapid downward movement of large soil and rock masses occurs when shear strength falls below shear stress (BGS 2023; Kim et al. 2017). The mechanism of MDL-driven slope failure is complex and mediated by soil types, slope angle, and rainfall intensity. The key factors leading to slope failure during rain events are the infiltration of water into soil slopes, leading to elevated pore-water pressure (Rahardjo et al. 2019). The other driving factors that trigger landslide response include, impoundment of lakes, rivers and reservoirs due to large-scale landslides and debris flow in mountainous regions, leading to formation of natural 'landslide dams'(Chen et al. 2019; Fomenko et al. 2020; Tacconi Stefanelli et al. 2015), river erosion resulting in changes in bank morphology and hence slope stability (Lévy et al. 2012), deforestation on steep hillslopes (Lehmann et al. 2019), and glacial retreat (Liu et al. 2021).

Dry spells and heatwaves can significantly increase the risk of MDLs. In soils rich in clay, preceding dry spells can make slopes more susceptible to sliding as soil desiccation reduces its stability. When heavy rainfall follows a dry spell, the soil may not quickly absorb enough water due to the formation of dense surface crust, leading to elevated pore-water pressure and depletion in shear strength, triggering landslides (Tichavský et al. 2019; Sidle and Bogaard 2016). Precipitation 'whiplash', characterized by an abrupt shift between dry and wet extremes, can lead to concurrent landslides and soil erosion and pose little time for recovery, amplifying the individual

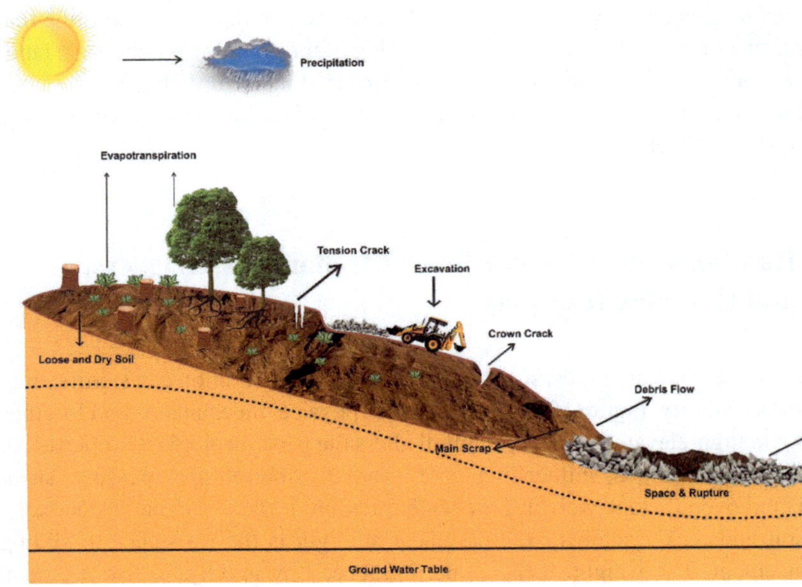

Fig. 10.2 Unveiling the causal drivers of landslides. Charting hydrometeorological and geological factors for landslides. The dashed line in black indicates the elevated transient groundwater table due to prolonged or short-duration intense rainfall or snowmelts

impacts of the isolated occurrence of either extreme (Tan et al. 2023). The climate model projection shows a concurrent increase in precipitation intensity and lengthening dry spells in response to anthropogenic climate change, confirming elevated landslide susceptibility in a warming climate (Giorgi et al. 2011; Tan et al. 2023).

The role of seismic factors in MDLs is complex and multifaceted. An earlier assessment (Lacroix et al. 2020) suggests that seismic shaking can increase shear stress on hillslopes, affecting the factor of safety (FoS) and the likelihood of landslide movement. Seismic activity can also cause soil liquefaction, reducing shear strength of the soil, and increasing pore-water pressure, thereby making slopes more vulnerable. Notably, slow-moving landslides, which differ from rapid landslides characterized by sudden catastrophic collapse, can be influenced by preceding seismic events (Iverson 2005; Iverson and George 2016). Slow-moving landslides can persist over extended periods, often hydrometeoroloically driven, and are influenced by a complex interplay of several drivers, such as soil properties, seismicity, river incision, and human activities (Rutter and Green 2011; Riva et al. 2018; Handwerger et al. 2022). However, rapid or high-speed landslides result from sudden increases in shear stress, often driven by heavy rainfall or seismic events over steep slopes, causing substantial damage (Lacroix et al. 2020).

In conclusion, understanding the mechanisms and triggers of rainfall-induced landslides is crucial for predicting these events and mitigating their impacts. This involves assessing various environmental drivers, such as rainfall patterns, lithology and topographical features, soil properties, and seismic activities. This understanding becomes even more critical as shifting climate and human activities continue to alter our landscapes, impacting steep-slope agriculture worldwide (Wang et al. 2022). The interplay of these drivers in Anthropocene underscores the complexity of landslide dynamics. Further research and development of predictive models are required to enhance our understanding of physical processes, enabling credible assessment of rain thresholds in triggering landslides.

10.3 Determining Moisture Thresholds for Landslides

This section provides a concise exploration of the methodologies used to determine moisture thresholds, a crucial element for developing a credible 'Landslide Early Warning Systems (LEWS)'. A wide range of approaches are in use to determine moisture (mainly rain) thresholds (See Fig. 10.3). Each method, with its unique strengths and applications, significantly enhances our understanding and prediction of MDLs. This review covers a range of approaches, including Bayesian and frequentist methods, empirical techniques, physically-based strategies, and hybrid models.

While most of the MDLs impact the uppermost three meters of the slope, causing shallow slides, deep-seated landslides are frequently associated with the attenuated hydrological response following extreme rain events, for instance, time-lagged groundwater level variations (Greco et al. 2023). Therefore, several studies stressed

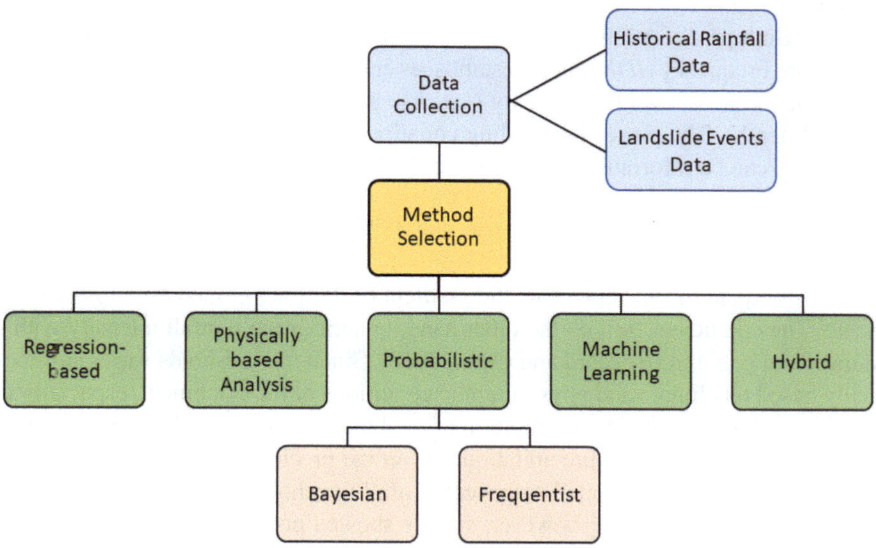

Fig. 10.3 Methodologies for determining rainfall thresholds in landslide prediction using different approaches

the importance of analyzing extreme rainfall attributes, such as intensity and duration, as viable approach for assessing landslide hazards (Tsunetaka 2021; Johnston et al. 2021). Landslide triggering rainfall initiates sliding movement of sediment layer when the driving force is greater than the resistive force. In this case, rain threshold is typically obtained as the lower envelop of rain events that potentially lead to landslide initiation (Tsunetaka 2021; Bogaard and Greco 2018). Process-based deterministic modelling approaches relies on mechanical rules for slope failures and determine infiltrations and eventually variations in groundwater table due to rainfall forcing (Crosta and Frattini 2003; Meena et al. 2022). A coupled distributed hydrological stability model can evaluate the risk of shallow MDLs at a catchment scale by simulating slope failures as a result of interdependent spatiotemporal dynamics of hydrological and land-surface processes. These methods leverage topography, geology, and geotechnical parameters alongside rainfall data to provide insight into landslide occurrence (Aristizábal et al. 2016). They utilize limit equilibrium theory to assess slope safety by considering time-invariant and transient information (Ortiz-Giraldo et al. 2023). However, most of the process-based spatially distributed models involve a large number of parameters; a robust calibration of such models at a natural watershed scale is often not feasible, primarily due to spatially sparse and non-existent hydrogeologic information. Second, the uncertainty in model parameters and associated forcings, such as hydraulic conductivity, and soil layer depth used in process-based models, contribute significantly to the discrepancy between observed and simulated catchment-scale landslide distribution (Burton and Bathurst 1998). Furthermore, uncertainty predominates in estimated soil hydrological properties, particularly in unsaturated conditions (Arnone et al. 2016).

Considering such operational limitations in process-based distributed models, empirical rainfall threshold curves are widely been used for assessing landslide hazard. Both precipitation Intensity (I)–Duration (D) thresholds and the Intensity–Duration–Frequency (IDF) curves establishes empirical relationships linking rainfall duration (D) to its average intensity (*i.e.,* $I = M/D$, where M is the event-specific rainfall depth). The power-law scaling equation, which exhibits a negative scaling exponent, can be reformulated as follows:

$$I = m * d^{-n} \tag{10.1}$$

In this equation, 'n' represents the gradient of the line, plotted on a logarithmic scale. The coefficient 'm', on the other hand, quantifies the rainfall intensity with a duration of one unit (Bogaard and Greco 2018). Since ID thresholds varies substantially based on climate and physiographic conditions of the catchment, especially in absolute values, a scaled representation of ID thresholds are often been proposed by normalizing mean annual precipitation to average precipitation intensity (Guzzetti et al. 2007). Most documented empirical rainfall thresholds lie between 10 and 100 mm of cumulative rainfall; however, studies showed precipitation volumes of less than 10 mm and more than 1000 mm can also contribute to landslides (Guzzetti et al. 2007). Under soil field capacity, in normal antecedent moisture conditions, accumulated precipitation <10 mm, in general, fails to trigger landslides. However, it is challenging to determine whether precipitation volumes greater than 1000 mm or periods longer than 100 h can cause landslides or debris flows across different hydrogeological settings, often misleading the use of rainfall ID threshold (Bogaard and Greco 2018).

"Antecedent soil moisture" has a critical role in landslide initiation, primarily mediated by the water content in the soil before the landslide triggering extreme rain events (Lazzari et al. 2018; Marino et al. 2020). The antecedent moisture conditions determine soil response to precipitation, potentially leading to landslides. In areas prone to MDL-driven shallow landslides, antecedent saturation is a crucial preconditioned driver (Marino et al. 2020). In contrast, the actual landslide trigger is associated with the hydrological response to recent and preceding rain events (Marino et al. 2020). A dry spell of at least 24 h typically precedes the landslide triggering rain events (Kim et al. 2021). The n-day ($n = 1, 2, \ldots, d$) cumulative rainfall magnitude leading up to a triggering precipitation event is termed antecedent rainfall, and its accurate measurement is crucial for assessing landslide risk (Kim et al. 2021). The role of cumulative antecedent rain events on *E-D* thresholds is investigated using the Crozier (1989) equation, which is given by (Kim et al. 2021)

$$ARx_n = \delta R_1 + \delta^2 R_2 + \cdots + \delta^n R_n \tag{10.2}$$

Here, ARx_n in Eq. 10.2 shows the n-day antecedent rainfall responsible for landslide triggering rain event x. δ denotes the "decay constant" that ranges from 0.8 to 0.9 and is an empirical constant representing the drainage capacity of sub-surface soil and rocks and the hydrological properties of the location (Capecchi and Focardi 1988). However, Eq. (10.2) requires optimization of additional parameters, such as lag and decay constants representing a "temporal distance" based on historical rainfall conditions. Mean soil moisture conditions averaged over large spatial domains are often used to derive empirical thresholds for triggering landslides. Some regional-scale assessments (Segoni et al. 2018; Abraham et al. 2021) have used antecedent soil moisture as a proxy for antecedent rainfall accumulated over long periods to derive landslide early warning systems. However, this approach has potential pitfalls as it relies on modelled soil moisture and credible soil moisture measurements are often not available for vast spatial domains. At a catchment scale, an index based on daily catchment-averaged precipitation is often used to quantify antecedent soil moisture condition and has been used for regional-scale flood risk assessments (Blöschl et al. 2019; Betrola et al. 2021)

Rainfall Event-Duration (ED) thresholds representing the relation of rainfall event duration (D) versus the landslide initiating accumulated moisture (E) are pivotal tools in determining the minimum precipitation required for landslides. These thresholds, based on rainfall records, serve as a quantitative tool for predicting landslide occurrence, valuable for developing LEWS and formulating risk management strategies, especially in landslide-prone regions (Guzzetti et al. 2007). The equation is given as follows (Guzzetti et al. 2007):

$$E = m.D^n \tag{10.3}$$

where, E denotes the cumulative rainfall, D indicates the event duration, the intercept m and exponent n are parameters determined through empirical analysis (Marra 2019). The E versus D (Event vs Duration) approach is an improvement over conventional Intensity-Duration (*ID*) techniques for landslide prediction. It emphasizes the non-linear relationship of accumulated rainfall (E) and the corresponding event duration (D), in shaping landslide frequency and often used as a proxy for soil wetness. If antecedent conditions are wetter, a small magnitude of rainfall can be sufficient to trigger a landslide. This is in contrast to the *ID* threshold approach, which relies on power law equations and often poorly calibrated. Therefore, the intricate relationships between rainfall and slope stability may not be considered in the ID threshold approach. Sengupta et al. (2010) found that a predictive threshold of 250 mm of normalized cumulative rainfall over 15 days was associated with landslide triggers in the Sikkim Himalayan region. This suggests that the debris zone becomes saturated and fails after a certain period of rainfall, regardless of the storm event intensity. However, *E vs D* enhances predictability, eliminating the requirement for conversion factors used in intensity determination as observed in *ID* threshold approach. Furthermore, it establishes a more direct link between rainfall parameters and landslides (He et al. 2021). The integration of *ED* and *ID* methods enhances landslide prediction models, strengthening the capability of LEWS for disaster management as noted in earlier assessments (Marra 2019; Huang et al. 2022). Statistical tools, such as the least-square method, quantile regression, logarithmic regression, and log–log linear regression, are frequently employed to estimate rain thresholds for landslides (He et al. 2021; Tyralis and Papacharalampous 2021; Oosterbaan et al. 1994; Chen et al. 2020). These empirical tools provide valuable insights into MDLs. However, to effectively address the complexity of hydrogeological hazard cascades, they should be supplemented with other analytical and modelling techniques to effectively address the complexity of landslide hazards.

The Kinetic Energy versus Duration (*KE* vs D) approach is a significant advancement in the study of MDLs. It calculates the kinetic energy of rainfall, considering its erosive power, and establishes a robust correlation between rainfall kinetic energy and landslide occurrence. This method stands out for its simplicity and effectiveness, enabling the estimation of landslide initiation thresholds based on soil properties and rainfall characteristics. A notable refinement by Ferro et al. (2020) linked rainfall event duration with the rainfall kinetic energy for MDLs, facilitating a proactive landslide hazard prediction. Unlike empirical relationship, which often lacks precision in defining threshold coefficients and relies on empirically defined thresholds (Guzzetti et al. 2008), the *KE* versus D approach offers quantitative rigor. It enhances the precision and reliability in assessing and predicting rain-induced landslides. Considering kinetic energy as a rainfall parameter complements traditional rainfall intensity-duration parameters and enriches landslide forecasting efforts (Lin and Chen 2012). High rainfall kinetic energy generated from large-sized raindrops with strong intensity accelerates soil erosion, resulting in a landslide when the response time is short (Bradford et al. 1987). In extreme conditions, landslide events may even be triggered by kinetic energy greater than 2000 J/m^2 (Lin and Chen 2012), underscoring the critical role of kinetic energy in landslide prediction.

Under Probabilistic methodologies, both frequentist and Bayesian approaches, offer robust statistical tools for assessing the susceptibility of landslides induced by rainfall. The frequentist approach is a statistical method used to establish critical probability levels for landslide occurrence across large regions. The method is based on a rainfall frequency analysis, responsible for landslides. The process can be summarized as follows (Brunetti et al. 2010): first, a best-fit line based on linear regression estimates of mean rainfall intensity responsible for landslides is obtained. Second, the probability density function (PDF) of the residual error is obtained following a standard normal distribution, assuming residual errors are Gaussian distributed with a zero mean and constant variance. Finally, the shift of different threshold levels from the best-fit line is derived, representing $x\%$ exceedance, where x varies for thresholds ranging from 1%, 5%, 10%, 20%, and 50%. Globally, frequentist rainfall thresholds are instrumental in developing probabilistic frameworks within the LEWS (Gariano et al. 2023; Larsen and Simon 1993) and in South Asia (Harilal et al. 2019; Kanungo and Sharma 2014). In India, the LANDSLIP project (https://www.landslip.org/) has focused on creating a LEWS for predicting rainfall-induced landslides in Darjeeling and Nilgiris. Gariano et al. (2023) developed rainfall thresholds at different exceedance probabilities using the frequentist method and an automated tool for these pilot locations. The Bayesian approach integrates prior knowledge and uncertainty, manages intricate models, and provides probabilistic estimates of landslide susceptibility (Brunetti et al. 2010). Nie et al. (2019) used a dynamic Bayesian network to evaluate slope stability over time, linking key drivers, rainfall intensity, cumulative infiltration, pore-water pressure, and the FoS. This method improves predictive performance by weighing observational value against uncertainty.

The hybrid methodology integrates physical and hydrological techniques to establish MDL thresholds (D'Ippolito et al. 2023). Based on the Bayesian probabilistic approach, the newly developed Combined Rainfall Thresholds and Landslide Events Tool (CTRL-T) is a significant tool in this domain. Leveraging continuous rainfall time series and landslide information, it computes probabilistic rainfall thresholds, thereby improving landslide prediction (Santangelo et al. 2023). Further, integrating rain thresholds into real-time hydrologic simulation systems, such as Delft - Flood Early Warning System (Delft-FEWS) modelling platforms, offers a dynamic analysis of hydrologic responses. This dynamic analysis can be further facilitated by forcing precipitation nowcasts and forecasts at specific lead times to improve the predictability of the rainfall-triggered landslide model (Hsu et al. 2018). Due to the complexity of landslides genesis, combining different approaches, as demonstrated by the hybrid methods, offers more robust and reliable estimates of rainfall thresholds. Real-time monitoring and warning systems are pivotal in mitigating MDLs. Leveraging continuous hydrologic simulations, they offer timely warnings, reducing casualties and property damage (Zhang et al. 2023). Combining Kalman filter, fast Fourier transform, support vector machine and more recently, deep learning-based approaches provides real-time landslide alerts. Machine learning, increasingly applied in hydrology, excels in improving predictability of landslides, often outperforming conventional least-square based approaches (Tehrani et al. 2022). Methods, such as logistic regression, adaboost tree, support vector machine, random forest, long-short term memory (LSTM), and multilayer perceptron capture characteristics of rainstorm impact on landslide susceptibility. While empirical rainfall thresholds relies on preceding and antecedent moisture conditions for triggering landslides, the process-based approaches mostly emphasize on local conditions by simulating hydrological and mechanical processes acting on slope stability. Recently, hybrid process-based machine learning approaches considering hourly rainfall information and geomorphic basin characteristics, such as topographic distance, local morphological information, and the difference in elevation between landslides and rain gauges, were proposed in the literature (Mondini et al. 2023) to overcome the drawbacks of both approaches.

The Table 10.1 presents a comprehensive synthesis of local and national scale studies on MDLs. These studies leverage rainfall records with varying temporal and spatial resolutions and employ a diverse array of statistical methods. The derivation of rain thresholds is significantly impacted by the temporal resolutions of rainfall time series. The table succinctly emphasizes key findings drawn from each study. Given the recent evolution in precipitation patterns due to changing climatic conditions, it emphasizes the need for updating rain thresholds. While empirical threshold-based relations are commonly used to predict MDLs in various parts of the world, the quality and quantity of available rainfall observations and landslide data limit the effectiveness of such approaches. Recently, Gariano et al. (2020) showed that considering different time resolutions of rain records has resulted in considerable uncertainty in the shape and estimated rainfall *ED* thresholds derived from the frequentist-based approach. They further highlighted the necessity for higher temporal resolution rainfall time series as rainfall with coarser temporal resolution results in systematic bias (underestimation) in derived thresholds especially at short durations, leading to large false alarm rates in operational landslide prediction systems. However, the lack of reliable high-resolution rainfall time series, developing countries continue to rely on daily time series for threshold determination. To this end, Gariano et al. (2020) recommended that the rainfall thresholds derived using the daily temporal resolution should be shown using equations portraying duration with the units of days instead of hours.

10.4 Multi-Hazard Risk Framework for Moisture-Driven Landslides

Extremes often show synchronicity over space and time. Hazard occurs in a sequence of specific order, in which a preceding event or hazard typically triggers one or multiple following hazards, creating feedback loops of cascading hazard chains (Culler et al. 2023; Rusk et al. 2022). A multi-hazard event may overlap one another temporally or spatially, causing various hazard processes to interact (Gill and Malamud 2014). Two different hazard interactions may occur (Mani et al. 2022; Zscheischler et al. 2020; Schauwecker et al. 2019): (1) A single driver triggers a

Table 10.1 Synthesis of MDLs studies based on empirical rainfall or moisture thresholds—key insights and implications

Author name	Region	Periods	Resolution	Rain-thresholds equation	Type of statistics	Key findings
Akcali et al. (2010)	Trabzon Province, Turkey	1998–2008	Hourly	$I = 0.69\ D^{-0.99}$ *for 0.01 h* < D < *1000 h*	Least-square method	Landslide risk increases with modest rainfall, and often influenced by drivers, such as slope angle, lithology, cover type, agriculture, and landslide nature. Notably, gentler slopes and vegetated areas have higher rainfall thresholds
Sengupta et al. (2010)	Lanta Khola, Sikkim, India	1998–2006	Daily	$E = 0.012D^{1.54}$ *20 h* < *D* < *600 h*	Logistic regression analysis	The study proposes a new rainfall threshold for landslide prediction at Lanta Khola, outperforming the existing approach by considering unique geological conditions and a 15-day cumulative rainfall of ≥ 250 mm. It emphasizes the role of fine-grained debris in landslide genesis
Saito et al. (2010)	Japan	2006–2008	hourly	$I = 2.18\ D^{-0.26}$ *for 3 h* < D < *537 h* $I_{MAP} = 0.0007\ D^{-0.21}$ *for 3 h* < D < *537 h*	Quantile regression method	The rainfall threshold for initiating shallow landslides in Japan is lower than elsewhere due to the country's high-relief topography, geologic conditions, and rainfall characteristics

(continued)

Table 10.1 (continued)

Author name	Region	Periods	Resolution	Rain-thresholds equation	Type of statistics	Key findings
Zhou and Tang (2014)	Wenchuan southwestern China	2008–2012	hourly	$I = 66.36D^{-0.79}$ *for* $2\,h \leq D \leq 15\,h$ $I_{MAP} = 0.036D^{-0.61}$ *for* $2\,h \leq D \leq 15\,h$ $E_{MAP} = 0.0376D^{0.37}$ *for* $2\,h \leq D \leq 15\,h$	Least-square regression	The research discerns elevated precipitation thresholds for triggering debris flows in regions affected by the Wenchuan earthquake. Such thresholds, being generalized, hold potential for application in early warning systems across geologically analogous areas
Mathew et al. (2014)	National Highway-58 from Rishikesh to Mana in the Garhwal Himalaya, India	1998–2006 and 1987–2006	3-hourly and daily	$I = 58.7D^{-1.12}$ for $D \leq 12\,h$ $N_I = 0.0612D^{-1.17}$ for $D \leq 12\,h$	Binary logistic regression	(1) In the studied area, landslides can be caused by long-lasting rainfall of about 4 mm/*h* for duration ≤12 h (2) Daily, 3-daily cumulative, and monthly antecedent rainfall significantly affect slope stability (3) The insights highlight the need for an updated landslide warning system, incorporating the information of antecedent rainfall in an optimal lag-time in an intensity-duration rain threshold model

(continued)

Table 10.1 (continued)

Author name	Region	Periods	Resolution	Rain-thresholds equation	Type of statistics	Key findings
Kanungo and Sharma (2014)	Chamoli-Joshimath region, Garhwal Himalayas, India	2009–2012	Daily	$I = 1.82\,D^{-0.23}$ for $24\,h < D < 336\,h$	Frequentist approach	To trigger landslides, minimum of 10-day antecedent rainfall of 55 mm and 20-day antecedent rainfall of 185 mm are suffice to trigger landslides. The obtained threshold can be used in a specific point locations of Garhwal Himalaya region during monsoon season
Chang et al. (2017)	Taiwan	2006–2012	minute-scale	$E_{kn} = 32.19*(1–0.725e^{-0.029I})$ $E_{ks} = 32.23*(1–0.643e^{-0.022I})$ Empirical rain threshold relations: $I_N = 13.81\,D^{-0.31}$ $I_S = 66.44\,D^{-0.57}$	Rainfall Kinetic energy and least-square regression	In northern Taiwan, landslides are triggered during both monsoon and typhoon periods. In contrast, in southern Taiwan, they predominantly occur during typhoons. Notably, in the north, landslides typically occur when the duration *D* exceeds 10 h. In contrast, in the south, landslides are observed when *D* ranges between 16 and 20 h. This suggests that prolonged rainfall events leading to landslides are primarily cause by rainfall infiltration. Conversely, for shorter storm events, the driving factor is the kinetic energy of intense rainfall

(continued)

Table 10.1 (continued)

Author name	Region	Periods	Resolution	Rain-thresholds equation	Type of statistics	Key findings
Do and Yin (2018)	Ha Giang city and the surroundings, Vietnam	1957–2013	Daily	$E = 40.8 - 0.201D_3$	Bayesian Probability approach	Identified that conditional landslide probability is significantly influenced by rainfall intensity of >40 mm and duration of >0.3 day or approximately 7-h
Nolasco-Javier and Kumar (2018)	Baguio City and neighbouring municipalities in northern Luzon, Philippines	1970–2013	Daily	$I = 6.46D^{-0.28}$ *for* $24\ h \le D \le 264\ h$ $N_I = 0.002D^{-0.28}$ *for* $24\ h \le D \le 264\ h$	Least-square regression	Highlighted the substantial role of antecedent rainfall in triggering landslides, with rainfall within a 24-*h* window exerting a more pronounced effect than earlier rainfall
Dikshit and Satyam (2018)	Kalimpong, West Bengal	2010–2016	Daily	$I = 3.52\ D^{-0.41}$ for $24\ h < D < 720\ h$	Frequentist approach	Rainfall characteristics significantly influence landslide hazards in Kalimpong. Specifically, a 24-h rainfall event with an intensity of 0.95 mm/h can trigger a high risk of landslides. Furthermore, 10 to 20 days preceding antecedent moisture content of around 88.37–133.5 mm is adequate to trigger landslides
Abraham et al. (2019)	Idukki district, Kerala, India	2010–2018	Daily	$I = 0.9D^{-0.16}$ for $24\ h < D$	Frequentist approach	Landslide events are more influenced by antecedent rainfall conditions rather than the rainfall magnitude on the day of the event

(continued)

Table 10.1 (continued)

Author name	Region	Periods	Resolution	Rain-thresholds equation	Type of statistics	Key findings
Teja et al. (2019)	Kalimpong, Darjeeling Himalayas	2010–2016	Daily	$E = (4.2 \pm 1.3)$ $D^{(0.56 \pm 0.05)}$ $for\ 24\ h \leq D \leq 288\ h$	Bayesian method implemented in CTRL-T	The rainfall threshold for triggering landslides in the Kalimpong Region is around 37 mm over a duration of 48 h
Kim et al. (2020)	South Korea	1963–2018	hourly	$I = 10.40D^{-0.31}$ $I_{MAP} = 0.006D^{-.26}$ $for\ 4\ h \leq D \leq 84\ h$	Quantile regression analysis	(1) In South Korea, rainfall thresholds for shallow landslides decrease with increasing antecedent rainfall (2) Thresholds vary regionally, typically lower in the southern region, where rainfall is relatively more intense
Roy et al. (2022)	Cox's Bazar district, Bangladesh	1997–2021	3-hourly	For 5% exceedance probability $I = 3.63D^{-0.1313}$ $E = 3.63D^{0.8687}$ For 1% exceedance probability $I = 2.78D^{-0.1313}$ $E = 2.78D^{0.8687}$	Frequentist approach	(1) Landslides can be triggered by both short and long rainfall events and require a mean intensity of 2.39 mmh^{-1} or 57.4 mm of rainfall in 24 h (2) Antecedent rainfall plays a crucial role, with a minimum of 130 mm over 72 h and 210 mm over 120 h, potentially can cause catastrophic slope failure. These thresholds, lower than the global averages, reflect Cox's Bazar area's distinct geoclimatic conditions

* The symbols denote the following- I_{MAP}: Normalized rainfall, determined by normalizing rainfall intensity to mean annual precipitation; E_{MAP}: Normalized event rainfall(in percent); N_I: Normalized intensity, per hour; E_{kn} & E_{ks}: Rainfall kinetic energy, where subscripts, *n* denotes northern Taiwan and *s* denotes southern Taiwan; I_N & I_S: Rainfall intensity, where subscripts *N* denotes northern Taiwan and *S* denotes southern Taiwan; D_3: 3-day antecedent rainfall

temporally clustered sequence of similar processes or events that are concurrent in space or sequential with a short time lag. An illustration includes cloudbursts, temporal clustering of early precipitation, and movement of the southwest monsoon, which resulted in the excessive rise in the water level of the Chorbari glacial lake in Uttarakhand, resulting in over 6000 fatalities in June 2013 (Gupta and Agrawal 2016). (2) A hazard process sets off a series of cascading effects, which increase in progression over time, resulting in unprecedented secondary event chains with catastrophic consequences. An example of a rapid sequence of events include the Rain on Snow (hereafter ROS) episodes, when raindrops fall on a pre-existing snowpack, triggering floods, the loss of woodlands and landslides, and contributing to soil erosion in high-altitude regions (Ombadi et al. 2023). Floods and/or fallen trees over roads disrupt transportation networks (e.g., railways), leading to blackouts and damaging agriculture, clean water supply, communication, and other sectors (Schauwecker et al. 2019). In contrast, an example of the delayed response include wildfires in mountainous terrain, elevates likelihood of MDLs and debris flow. The sequence of process chain involves weakening of roots, reduced evapotranspiration, changes in vegetative cover, elevated runoff and flood risk due to alteration in water-budget, leading to changes in soil mechanical and hydraulic properties, triggering landslides (Belongia et al. 2023; Abdollahi et al. 2023). Often, MDL's occur when hillslopes are in unsaturated conditions. The unsaturated soil develops negative pore pressure, which can diminish due to rainfall or snowmelt infiltrations. However, the infiltration capacity of soils in burned landscapes recovers after a few years of post-wildfire events, elevating reduction in pore pressure, soil strengths and the stability of hillslopes due to moisture (Abdollahi et al. 2023).

In multi-hazard interactions, while each event in isolation may not be extreme, the response time between two or multiple consecutive extremes dictates the aggregated hazard potential, overtaxing disaster mitigation (Leonard et al. 2014; Zscheischler et al. 2020). The two types of hazard interactions were identified in the literature (Kappes et al. 2012): (1) cascade, domino or triggering effects occur as a hazard chain, triggering one hazard by another instantaneously with a short response or delay time between processes, eventually leading to catastrophic consequences. (2) Hazard disposition or delay: a preceding hazard alters the subsequent hazard with a relatively extended period between two processes, providing some time for preparing an intervention. However, in the former, the time between two events is so concise that the consecutive extremes offer little time for human preparedness for adaptation; hence, their joint impact exacerbates the individual effects. Figure 10.4 presents schematics of different types of hazard interactions that may exacerbate the societal impact.

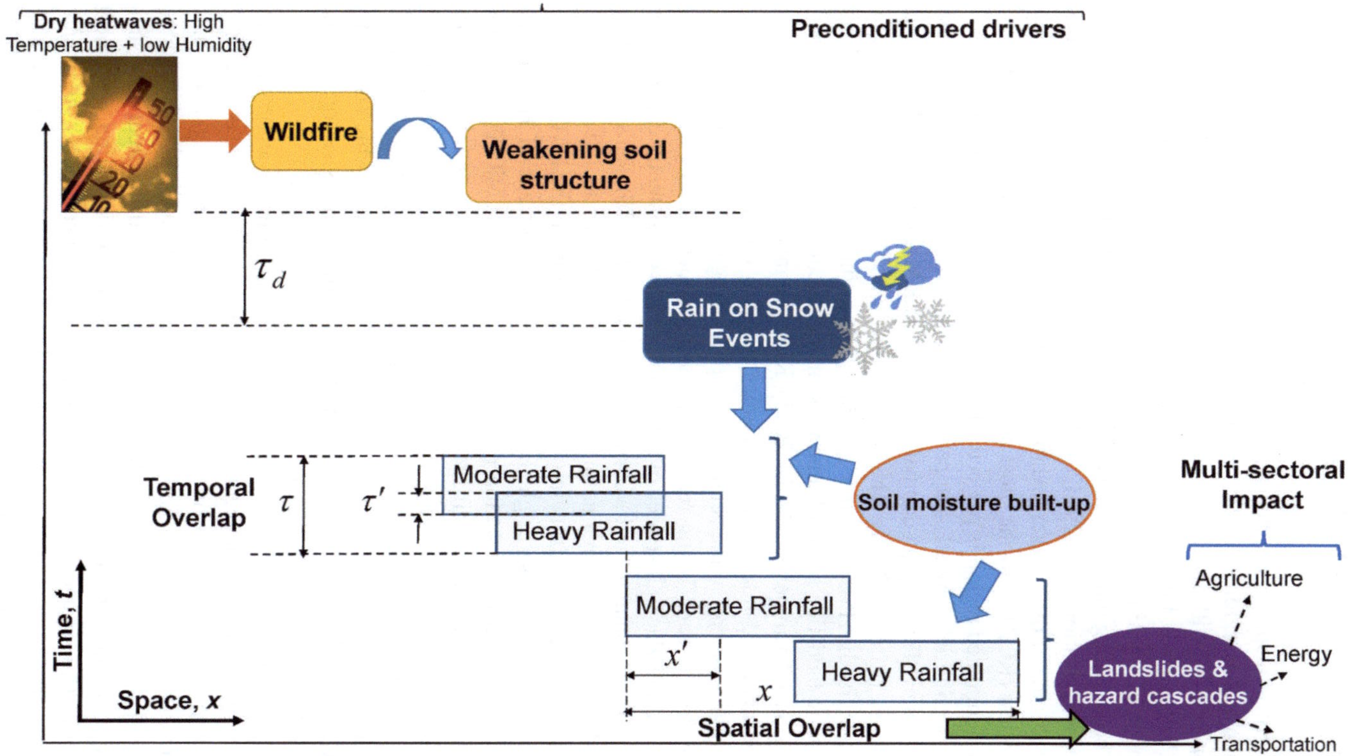

Fig. 10.4 Multi-hazard interactions over space and time. τ_d indicates delay or hazard disposition. The symbols τ and τ' show temporal overlaps between the events for which the joint probability fields can be modelled considering (i) events are mutually exclusive, i.e., 'OR' case when one/more events occurs and (ii) following the 'AND' concept, i.e., when both/all events occurs together. Likewise, *x* and *x*' show the spatial overlaps of events. The arrows indicate causal chain of events

10.5 Moisture-Driven Landslides in Anthropocene across High Mountain Asia

Over the northern hemisphere, the landslide peaks in the summer, with a notable inter-annual variation across regions. 77–93% of MDLs worldwide occur during the south-west monsoon season, with India and Nepal contributing 16% and 10% of MDLs (Froude and Petley 2018). Leveraging observed and re-analyses-based precipitation products, Jiang et al. (2023) showed that while the southeast of High Mountain Asia (HMAS) has been drying out over the past 20 years, the northern HMAS showed a steady rise in precipitation from 1951 to 2020. The climate models, participating in the Fifth Assessment of the Intergovernmental Panel on Climate Change (IPCC AR5) with modest (RCP4.5) to the most aggressive (RCP8.5) emission scenarios, show the rain-dominated region is projected to expand to around half of the HMAS with a robust decline in snowfall is apparent over eastern HMAS, including East Tienshan and eastern Tibet region compared to western HMAS covering Pamir and Karakoram ranges (Li et al. 2020). This shift in precipitation regime have altered the landslide activity in HMAS region. Forcing global climate model simulations in a near-real-time landslide hazard framework, namely, Landslide Hazard Assessment for Situational Awareness (LHASA 2023), Kirschbaum et al. (2020) showed a projected increase in landslide activity over areas currently covered by glacial lakes and glaciers around the end of the century, potentially triggering cascading hazards for population downstream. Using remotely-sensed products, Zhang et al. (2022) demonstrated that the glacial lake area over the HMAS has consistently expanded at a rate of $0.58 \pm 0.21\%$/annum in the current climate (1990–2020), with the fastest expansion evident in the lake area at an elevation of $\geq$5000 m a.s.l. Temperature increase, glacial thinning, and shifts in precipitation patterns in the HMAS are the possible causes for the increase in glacial lake area. The shrinkage of Himalayan glaciers have contributed to formation of more glacial lakes (CWC 2015a). The obstructions created across the drainage path due to landslides or infrastructure development, cause the formation of glacial lakes from 'moraines' (*i.e.*, glacial ice) impounding water. The melt water may accumulate in the basin, whereas the large impoundment of water turns the lake unstable, leading to flash floods, which is widely known as Glacial Lake Outburst Flood (*i.e.*, GLOF), resulting in massive soil erosion and changes in landscape (CWC 2015a).

The HMAS region has experienced significant changes due to anthropogenic warming and climate change. The recent IPCC AR6 (Masson-Delmotte et al. 2021) assessment report has projected a high confidence in decline in the glaciers and permafrost area and a medium confidence in an increase in annual and summer monsoon rainfall over South Asia. Climate change has altered the frequency and peak intensity of rainfall, shifting towards the ROS events and reducing snowpack in the late summer (Maina and Kumar 2023). Rainfall and snowmelt over the mountains are the two primary factors that increase the antecedent moisture content and play a crucial role in triggering landslides in the Himalayas (Fayaz et al. 2022).

Land use changes and anthropogenic activities, such as deforestation and construction activities also contribute to landslide severity and its frequency. Anthropogenic activities can disturb the natural slope stability, making them more susceptible to landslides (Tripathi 2021; Singh et al. 2022). Moreover, human-induced alterations, such as mining, hill cutting, and constructions, installation of hydropower facilities trigger landslides (Singh et al. 2022). Climate change, together with urbanization and infrastructure development on these world's water towers situated over steep topography, are likely to exacerbate such hazard cascades due to warming-induced glacial mass loss (Hugonnet et al. 2021; Immerzeel et al. 2020). A recent study (Lee et al. 2021a, b) has claimed that the Himalayas are experiencing a ten-fold increase in ice loss, which is greater than the rate of ice loss recorded elsewhere. For instance, in February 2021, catastrophic landslides and debris flow descended in Rishiganga and Dhauliganga valleys in Chamoli, Uttarakhand, have caused widespread damage and devastated hydropower projects. As identified by Shugar et al. (2021) the key drivers for the catastrophic cascading hazards were: (1) The installation of several hydroelectric plants across the course of a river; (2) A notable ratio of rock to ice mass that have caused glacier ice to melt and increased debris flow mobility, and finally; (3) the unprecedented fall height of mass supported substantial gravitational potential energy. Likewise, Central Water Commission (CWC 2015b; CWC 2011) identified 16 glacial lakes out of 294 lakes located across the Brahmaputra catchment as critical based on their water spread area of greater than 50 hectares across Sikkim, the northeast state of India, which included the recent flash flood due to GLOF in South Lhonak lake in October 2023.

10.6 Discussion and Conclusions

Globally the highest counts of fatalities from landslides occur in high mountain Asia followed by North America and steep topographies of Southeast Asia (Perkins et al. 2023). In 2010, globally, around 1543 rain-triggered landslide events were detected, which have caused over 5000 fatalities; among these, 48% (736/1543) of events were identified in Asia, followed by 30% (462/1543) in North America (Perkins et al. 2023). The shifting climate and rapid urbanization, particularly in low to lower-middle-income economy countries in tropical regions, will expose more of the population to landslide hazards. Despite this, a long-term historical database, for example, a long-term reliable archive of infrastructure and/or agricultural damage data are often missing or poorly maintained in low and lower-middle-income economy countries. Further, climate change consequences and other environmental changes constrain the usefulness of historical moisture thresholds triggering landslides. This emphasizes the need for updating rain thresholds to address nonstationarity due to climate change and develop climate-resilient infrastructure. Moreover, the complex topography of high mountains and simulation of atmospheric circulation patterns require improved parameterizations and the development and deployment of 'hyper-resolution' land surface and climate models (Xue et al. 2021) for impact assessment, which is often lacking in the present generation earth system models, putting additional challenges for future projection of precipitation extremes that mediates landslides.

This review underscores the significance of MDLs in managing hazards and mitigating risks, considering multi-hazard interactions and the evolving climate and human activities. It emphasizes the global relevance of understanding and predicting moisture-driven landslides, facilitated by advanced "physics-guided machine learning"-based landslide prediction systems considering causal attributes. The key added value of our synthesis is two-fold: (1) Uncovering moisture-driven landslide trends, synthesizing compounding and cascading hazards, and the multi-hazard interactions causing catastrophic landslides across the Himalayas, which is globally relevant due to the high landslide frequency, exposure to low-economy populations, and associated impacts. (2) Presenting a systematic review of available physical and statistical tools to identify MDL triggers and highlight the need for updating rain thresholds using observed and projected climate information to address nonstationarities related to climate change and climate variability. The knowledge gained from this review can enhance the effort to set up efficient LEWS and risk reduction strategies worldwide. Landslide-prone landscapes are, in general, steep and are not readily accessible post-catastrophic hazard chains with either field equipment or real-time sensors. While some of the direct drivers of cascade hazard chains, such as rainfall, landslides size, and volume, can be accessed and monitored in near real-time through space, based on compilation of available information from openly-sourced archives, developing data-driven and probabilistic hazard framework, and vulnerability maps can serve as a "giant leap" towards climate adaptation and mitigation efforts in low- and lower middle economies, where deployment of expensive sensors might not be feasible to map soil instability at a large spatial domain. The study of MDLs is not merely an academic pursuit but a vital endeavour to protect lives, property, and the environment. The development of medium to long-lead predictive models and interdisciplinary collaborations between stakeholders, engineers, and climate modeller communities can develop resilient strategies to confront the devastating consequences of such unprecedented cascading hazard chain.

References

Abdollahi M, Vahedifard F, Tracy FT (2023) Post-wildfire stability of unsaturated hillslopes against rainfall-triggered landslides. Earth's Future 11(3):e2022EF003213

Abraham MT, Pothuraju D, Satyam ND (2019) Rainfall thresholds for prediction of landslides in Idukki, India: an empirical approach. Water 11(10):2113. https://doi.org/10.3390/w11102113

Abraham MT, Satyam N, Rosi A, Pradhan B, Segoni S (2021) Usage of antecedent soil moisture for improving the performance of rainfall thresholds for landslide early warning. CATENA 200:105147

Akcali E, Arman H, Firat S, Saltabas L, Gunduz Z (2010) Rainfall thresholds for the initiation of landslides in Trabzon province of Turkey. Int J Eng Appl Sci 1(1):1–9

Aristizábal E, Vélez JI, Martínez HE, Jaboyedoff M (2016) SHIA_Landslide: a distributed conceptual and physically based model to forecast the temporal and spatial occurrence of shallow landslides triggered by rainfall in tropical and mountainous basins. Landslides 13:497–517

Arnone E, Dialynas YG, Noto LV, Bras RL (2016) Accounting for soil parameter uncertainty in a physically based and distributed approach for rainfall-triggered landslides. Hydrol Process 30(6):927–944

Belongia MF, Hammond Wagner C, Seipp KQ, Ajami NK (2023) Building water resilience in the face of cascading wildfire risks. Sci Adv 9(37):eadf9534

Bertola M, Viglione A, Vorogushyn S, Lun D, Merz B, Blöschl G (2021) Do small and large floods have the same drivers of change? A regional attribution analysis in Europe. Hydrol Earth Syst Sci 25(3):1347–1364

Blöschl G, Hall J, Viglione A, Perdigão RA, Parajka J, Merz B, ... , Živković N (2019) Changing climate both increases and decreases European river floods. Nature 573(7772):108–111

Bogaard T, Greco R (2018) Invited perspectives: Hydrological perspectives on precipitation intensity-duration thresholds for landslide initiation: proposing hydro-meteorological thresholds. Nat Hazard 18(1):31–39

Bradford JM, Ferris JE, Remley PA (1987) Interrill soil erosion processes: I. Effect of surface sealing on infiltration, runoff, and soil splash detachment. Soil Sci Soc Am J 51(6):1566–1571

British Geological Survey (2023) Understanding Landslides. Available at https://www.bgs.ac.uk/discovering-geology/earth-hazards/landslides/, Accessed on September 2023.

Brunetti MT, Peruccacci S, Rossi M, Luciani S, Valigi D, Guzzetti F (2010) Rainfall thresholds for the possible occurrence of landslides in Italy. Nat Hazard 10(3):447–458

Burton A, Bathurst JC (1998) Physically based modelling of shallow landslide sediment yield at a catchment scale. Environ Geol 35:89–99

Capecchi F, Focardi P (1988) Rainfall and landslides: research into a critical precipitation coefficient in an area of Italy. In: Bonnard C (ed) Landslides, Proceedings of the 5th International Symposium on Landslides. Balkema, Roterdam, pp 1131–1136.

Chang JM, Chen H, Jou BJD, Tsou NC, Lin GW (2017) Characteristics of rainfall intensity, duration, and kinetic energy for landslide triggering in Taiwan. Eng Geol 231:81–87

Chen B, Ma C, Krajewski WF, Wang P, Ren F (2020) Logarithmic transformation and peak-discharge power-law analysis. Hydrol Res 51(1):65–76

Chen KT, Chen XQ, Hu GS, Kuo YS, Chen HY (2019) Effects of river flow velocity on the formation of landslide dams. J Mt Sci 16(11):2502–2518

Chen X, Li D, Tang X, Liu Y (2021) A three-dimensional large-deformation random finite-element study of landslide runout considering spatially varying soil. Landslides 18(9):3149–3162

Crosta GB, Frattini PJNH (2003) Distributed modelling of shallow landslides triggered by intense rainfall. Nat Hazard 3:81–93

Cruden DM, Varnes DJ (1996) Landslides: investigation and mitigation. Chapter 3-Landslide types and processes. Transportation research board special report, Technical Report no. 247, pp 36–75

Culler ES, Livneh B, Rajagopalan B, Tiampo KF (2023) A data-driven evaluation of post-fire landslide susceptibility. Nat Hazard 23(4):1631–1652

Central Water Commission (2011) Inventory and Monitoring of Glacial Lakes/Water Bodies in the Himalayan Region of Indian River Basins Report. Retrieved from cwc.gov.in

Central Water Commission (2015a). advisory sheet glacial lake outburst flood South Lhonak System in Teesta River Basin. Retrieved from cwc.gov.in

Central Water Commission (2015b) Lake outburst flood (GLOF) study of South Lhonak Lake in Sikkim. Retrieved from cwc.gov.in

Dash RK, Samanta M, Kanungo DP (2023) debris flow hazard in India: current status, research trends, and emerging challenges. In: Landslides: detection, prediction and monitoring: technological developments. Springer, pp 211–231

D'Ippolito A, Lupiano V, Rago V, Terranova OG, Iovine G (2023) An integral assessment of landslide dams generated by the occurrence of rainfall-induced landslide and debris flow hazard chain. Water 15(2):2771

Department of Mineral Resources (2010) Knowledge of Landslides (in Thai). Available at: www.dmr.go.th, Retrieved on: October 31, 2021

Dikshit A, Satyam DN (2018) Estimation of rainfall thresholds for landslide occurrences in Kalimpong, India. Innov Infrastruct Solut 3:1–10

Do HM, Yin KL (2018) Rainfall threshold analysis and Bayesian probability method for landslide initiation based on landslides and rainfall events in the past. Open J Geol 8(7):674–696

EM-DAT (2022) International emergency disaster database. Available at: emdat.be. Accessed on: September, 2022

Fayaz M, Khader SA, Rafiq M (2022) Landslides in the Himalayas: Causes, Evolution, and Mitigation—A Case Study of National Highway 44, India. In: Disaster management in the complex Himalayan terrains: natural hazard management, methodologies and policy implications. Springer International Publishing, pp 43–58

Ferro V, Carollo FG, Serio MA (2020) Establishing a threshold for rainfall-induced landslides by a kinetic energy–duration relationship. Hydrol Process 34(16):3571–3581

Fomenko I, Strom A, Zerkal O (2020, July) Possibility of landslide damming in the Vakhsh River catchment and its effect on the hydraulic schemes and population. In: IOP conference series: materials science and engineering, vol 883, no 1. p 012069

Froude MJ, Petley DN (2018) Global fatal landslide occurrence from 2004 to 2016. Nat Hazard 18(8):2161–2181

Gariano SL, Melillo M, Peruccacci S, Brunetti MT (2020) How much does the rainfall temporal resolution affect rainfall thresholds for landslide triggering? Nat Hazards 100:655–670

Gariano SL, Melillo M, Brunetti MT, Kumar S, Mathiyalagan R, Peruccacci S (2023) Challenges in defining frequentist rainfall thresholds to be implemented in a landslide early warning system in India. In: Progress in landslide research and technology, vol 1, no 1, 2022. Springer, pp 409–416

Gill JC, Malamud BD (2014) Reviewing and visualizing the interactions of natural hazards. Rev Geophys 52(4):680–722

Giorgi F, Im ES, Coppola E, Diffenbaugh NS, Gao XJ, Mariotti L, Shi Y (2011) Higher hydroclimatic intensity with global warming. J Clim 24(20):5309–5324

Greco R, Marino P, Bogaard TA (2023) Recent advancements of landslide hydrology. Wiley Interdiscip Rev: Water:e1675

Gupta A, Agrawal R (2016) Management issues in Kedarnath Tragedy. Available at: https://www.preventionweb.net/publication/, Accessed on October, 2023

Guzzetti F, Peruccacci S, Rossi M, Stark CP (2007) Rainfall thresholds for the initiation of landslides in central and southern Europe. Meteorol Atmos Phys 98:239–267

Guzzetti F, Peruccacci S, Rossi M, Stark CP (2008) The rainfall intensity–duration control of shallow landslides and debris flows: an update. Landslides 5:3–17

He Z, Zuo R, Zhang D, Ni P, Han K, Xue Z, Wang J, Xu D (2021) A least squares method for identification of unknown groundwater pollution source. Hydrol Res 52(2):450–464

Harilal GT, Madhu D, Ramesh MV, Pullarkatt D (2019) Towards establishing rainfall thresholds for a real-time landslide early warning system in Sikkim India. Landslides 16(12):2395–2408

Handwerger AL, Fielding EJ, Sangha SS, Bekaert DP (2022) Landslide sensitivity and response to precipitation changes in wet and dry climates. Geophys Res Lett 49(13):e2022GL099499

Hock R, Rasul G, Adler C, Cáceres B, Gruber S, Hirabayashi Y, ... , Steltzer H (2019) high mountain areas. IPCC special report on the ocean and cryosphere in a changing climate. In: The intergovernmental panel on climate change. IPCC, Geneva

Hsu Y, Chang Y, Chang C, Yang J, Tung Y (2018) Physical-based rainfall-triggered shallow landslide forecasting. Smart Water 3(3):1–16. https://doi.org/10.1186/s40713-018-0011-8

Huang F, Chen J, Liu W, Huang J, Hong H, Chen W (2022) Regional rainfall-induced landslide hazard warning based on landslide susceptibility mapping and a critical rainfall threshold. Geomorphology 408:108236

Hugonnet R, McNabb R, Berthier E, Menounos B, Nuth C, Girod L, ... , Kääb A (2021) Accelerated global glacier mass loss in the early twenty-first century. Nature 592(7856):726–731

Hungr O, Leroueil S, Picarelli L (2014) The Varnes classification of landslide types, an update. Landslides 11:167–194

Immerzeel WW, Lutz AF, Andrade M, Bahl A, Biemans H, Bolch T, ..., Baillie JEM (2020) Importance and vulnerability of the world's water towers. Nature 577(7790):364–369
Iverson RM (2005) Regulation of landslide motion by dilatancy and pore pressure feedback. J Geophys Res: Earth Surf 110(F2). https://doi.org/10.1029/2004JF000268
Iverson RM, George DL (2016) Modelling landslide liquefaction, mobility bifurcation and the dynamics of the 2014 Oso disaster. Géotechnique 66(3):175–187
Jiang J, Zhou T, Qian Y, Li C, Song F, Li H, Chen X, Zhang W, Chen Z (2023) Precipitation regime changes in High Mountain Asia driven by cleaner air. Nature, 1–6. https://doi.org/10.1038/s41586-023-06619-y
Juang CS, Stanley TA, Kirschbaum DB (2019) Using citizen science to expand the global map of landslides: Introducing the Cooperative Open Online Landslide Repository (COOLR). PLoS ONE 14:e0218657
Johnston EC, Davenport FV, Wang L, Caers JK, Muthukrishnan S, Burke M, Diffenbaugh NS (2021) Quantifying the effect of precipitation on landslide hazard in urbanized and non-urbanized areas. Geophys Res Lett 48(16):e2021GL094038
Kanungo DP, Sharma S (2014) Rainfall thresholds for prediction of shallow landslides around Chamoli-Joshimath region, Garhwal Himalayas India. Landslides 11(6):629–638
Kappes MS, Keiler M, von Elverfeldt K, Glade T (2012) Challenges of analyzing multi-hazard risk: a review. Nat Hazards 64:1925–1958
Kim J, Kim Y, Jeong S, Hong M (2017) Rainfall-induced landslides by deficit field matric suction in unsaturated soil slopes. Environ Earth Sci 76:1–17. https://doi.org/10.1007/s12665-017-7127-2
Kim J-H, Kim D-J, Lee J-H, Lee J-W (2020) Effect of antecedent rainfall conditions and their variations on shallow landslide-triggering rainfall thresholds in South Korea. Nat Hazard 20(6):1505–1516
Kim SW, Chun KW, Kim M, Catani F, Choi B, Seo JI (2021) Effect of antecedent rainfall conditions and their variations on shallow landslide-triggering rainfall thresholds in South Korea. Landslides 18:569–582
Kirschbaum D, Stanley T (2018) Satellite-based assessment of rainfall-triggered landslide hazard for situational awareness. Earth's Future 6(3):505–523
Kirschbaum D, Kapnick SB, Stanley T, Pascale S (2020). Changes in extreme precipitation and landslides over High Mountain Asia. Geophys Res Lett 47(4):e2019GL085347
Lacroix P, Handwerger AL, Bièvre G (2020) Life and death of slow-moving landslides. Nat Rev Earth & Environ 1(8):404–419
Larsen MC, Simon A (1993) A rainfall intensity-duration threshold for landslides in a humid-tropical environment, Puerto Rico. Geogr Ann Ser B 75(1–2):13–23
Lazzari M, Piccarreta M, Manfreda S (2018) The role of antecedent soil moisture conditions on rainfall-triggered shallow landslides. Nat Hazards Earth Syst Sci Discuss 2018:1–11. https://doi.org/10.5194/nhess-2018-371
Lee E, Carrivick JL, Quincey DJ, Cook SJ, James WH, Brown LE (2021) Accelerated mass loss of Himalayan glaciers since the Little Ice Age. Sci Rep 11(1):24284. Provide https://doi.org/10.1038/s41598-021-03805-8
Lei Q, Sornette D, Yang H, Loew S (2023) Real-time forecast of catastrophic landslides via dragon-king detection. Geophys Res Lett 50(6):e2022GL100832
Lehmann P, von Ruette J, Or D (2019) Deforestation effects on rainfall-induced shallow landslides: Remote sensing and physically-based modelling. Water Resour Res 55(11):9962–9976
Leonard M, Westra S, Phatak A, Lambert M, van den Hurk B, McInnes K, ... , Stafford-Smith M (2014) A compound event framework for understanding extreme impacts. Wiley Interdiscip Rev: Clim Chang 5(1):113–128
Lee WL, Martinelli M, Shieh CL (2021b) An investigation of rainfall-induced landslides from the pre-failure stage to the post-failure stage using the material point method. Front Earth Sci 9:764393

Lévy S, Jaboyedoff M, Locat J, Demers D (2012) Erosion and channel change as factors of landslides and valley formation in Champlain Sea Clays: The Chacoura River, Quebec, Canada. Geomorphology 145:12–18

Li BV, Jenkins CN, Xu W (2022) Strategic protection of landslide vulnerable mountains for biodiversity conservation under land-cover and climate change impacts. Proc Natl Acad Sci 119(2):e2113416118

Li Y, Chen Y, Wang F, He Y, Li Z (2020) Evaluation and projection of snowfall changes in High Mountain Asia based on NASA's NEX-GDDP high-resolution daily downscaled dataset. Environ Res Lett 15(10):104040

Lin GW, Chen H (2012) The relationship of rainfall energy with landslides and sediment delivery. Eng Geol 125:108–118

Liu J, Wu Y, Gao X (2021) Increase in occurrence of large glacier-related landslides in the high mountains of Asia. Sci Rep 11(1):1635. https://doi.org/10.1038/s41598-021-81212-9

Maina FZ, Kumar SV (2023) diverging trends in rain-on-snow over high Mountain Asia. Earth's Future 11(3):e2022EF003009

Mani PA, Allen SK, Evans SG, Kargel J, Mergili M, Petrakov DA, Stoffel M (2022) Geomorphic process chains in high-mountain regions – A review and classification approach for natural hazards assessment. Rev Geophys. https://doi.org/10.1029/2022RG000791

Marino P, Peres DJ, Cancelliere A, Greco R, Bogaard TA (2020) Soil moisture information can improve shallow landslide forecasting using the hydrometeorological threshold approach. Landslides 17:2041–2054

Marra F (2019) Rainfall thresholds for landslide occurrence: systematic underestimation using coarse temporal resolution data. Nat Hazards 95(3):883–890

Masson-Delmotte V, Zhai P, Pirani A, et al (2021) Climate change 2021: the physical science basis. Contribution of working group I to the sixth assessment report of the intergovernmental panel on climate change. Cambridge University Press, UK

Mathew J, Babu DG, Kundu S, Kumar KV, Pant CC (2014) Integrating intensity–duration-based rainfall threshold and antecedent rainfall-based probability estimate towards generating early warning for rainfall-induced landslides in parts of the Garhwal Himalaya, India. Landslides 11:575–588

Meena V, Kumari S, Shankar V (2022, June) Physically based modelling techniques for landslide susceptibility analysis: a comparison. In: IOP conference series: earth and environmental science, vol 1032, no 1. IOP Publishing, p 012033

Merzdorf J (2020, February 11) Climate change could trigger more landslides in high Mountain Asia. In: NASA's Goddard Space Flight Center, Available at climate.nasa.gov, Accessed on September 2023

Modugno S, Johnson SCM, Borrelli P, Alam E, Bezak N, Balzter H (2022) Analysis of human exposure to landslides with a GIS multiscale approach. Nat Hazards 112(1):387–412

Mondini AC, Guzzetti F, Melillo M (2023) Deep learning forecast of rainfall-induced shallow landslides. Nat Commun 14(1):2466

LHASA (2023) LHASA. GitHub Repository. Available at: https://github.com/nasa/LHASA, Accessed from: https://gpm.nasa.gov/landslides/projects.html#LHASA

Neto JMDSR, Bhandary NP (2023) Influence of localized rainfall patterns on landslide occurrence—a case study of Southern Hiroshima with eXtended Radar Information Network Data during the July 2018 Heavy Rain Disasters. Geosciences 13(8):245

Nie W, Feng D, Lohpaisankrit W, Li C, Yuan J, Chen Y (2019) A dynamic Bayesian network-based model for evaluating rainfall-induced landslides. Bull Eng Geol Env 78(5):2069–2080

Nolasco-Javier D, Kumar L (2018) Deriving the rainfall threshold for shallow landslide early warning during tropical cyclones: a case study in northern Philippines. Nat Hazards 90:921–941

Ombadi M, Risser MD, Rhoades AM, Varadharajan C (2023) A warming-induced reduction in snow fraction amplifies rainfall extremes. Nature 619(7969):305–310

Oosterbaan RJ, Nijland HJ, Ritzema HP (1994) Drainage principles and applications. International Institute for Land Reclamation and Improvement (ILRI, Wageningen, The Netherlands), pp 1–1124

Ortiz-Giraldo L, Botero BA, Vega J (2023) An integral assessment of landslide dams generated by the occurrence of rainfall-induced landslide and debris flow hazard chain. Front Earth Sci, 1–11. https://doi.org/10.3389/feart.2023.1157881

Ozturk U, Bozzolan E, Holcombe EA, Shukla R, Pianosi F, Wagener T (2022) How climate change and unplanned urban sprawl bring more landslides. Nature 608(7922):262–265

Perkins L, Kirschbaum DB, Ng J (2023) Global rainfall-triggered landslides from 2007 through 2015. Available at: svs.gsfc.nasa.gov. Retrieved on October 27, 2023

Petley DN, Higuchi T, Petley DJ, Bulmer MH, Carey J (2005) Development of progressive landslide failure in cohesive materials. Geology 33(3):201–204

Rahardjo H, Kim Y, Satyanaga A (2019) Role of unsaturated soil mechanics in geotechnical engineering. Int J Geo-Eng 10:1–23. https://doi.org/10.1186/s40703-019-0104-8

Reichenbach P, Rossi M, Malamud BD, Mihir M, Guzzetti F (2018) A review of statistically-based landslide susceptibility models. Earth Sci Rev 180:60–91

Riva F, Agliardi F, Amitrano D, Crosta GB (2018) Damage-based time-dependent modeling of paraglacial to postglacial progressive failure of large rock slopes. J Geophys Res Earth Surf 123(1):124–141

Rusk J, Maharjan A, Tiwari P, Chen THK, Shneiderman S, Turin M, Seto KC (2022) Multi-hazard susceptibility and exposure assessment of the Hindu Kush Himalaya. Sci Total Environ 804:150039

Rutter EH, Green S (2011) Quantifying creep behaviour of clay-bearing rocks below the critical stress state for rapid failure: Mam Tor landslide, Derbyshire England. J Geol Soc 168(2):359–372

Roy R, Haque DME, Sifa SF, Tasnim S, Mahmud T, Hayat T (2022) Empirical approach based rainfall threshold estimation for landslide occurrence in Cox's Bazar district, Bangladesh. Dhaka Univ J Earth Environ Sci 11(1):81–94

Saito H, Nakayama D, Matsuyama H (2010) Relationship between the initiation of a shallow landslide and rainfall intensity—duration thresholds in Japan. Geomorphology 118(1–2):167–175

Santangelo M, Althuwaynee O, Alvioli M, Ardizzone F, Bianchi C, Bornaetxea T., ... , Fiorucci F (2023) Inventory of landslides triggered by an extreme rainfall event in Marche-Umbria, Italy, on 15 September 2022. Sci Data 10(1):427. https://doi.org/10.1038/s41597-023-02336-3

Schauwecker S, Gascón E, Park S, Ruiz-Villanueva V, Schwarb M, Sempere-Torres D, ... , Rohrer M (2019) Anticipating cascading effects of extreme precipitation with pathway schemes-Three case studies from Europe. Environ Int 127:291–304

Segoni S, Rosi A, Lagomarsino D, Fanti R, Casagli N (2018) Brief communication: Using averaged soil moisture estimates to improve the performances of a regional-scale landslide early warning system. Nat Hazard 18(3):807–812

Sengupta A, Gupta S, Anbarasu K (2010) Rainfall thresholds for the initiation of landslide at Lanta Khola in north Sikkim, India. Nat Hazards 52:31–42

Shah B, Alam A, Bhat MS, Ahsan S, Ali N, Sheikh HA (2023) Extreme precipitation events and landslide activity in the Kashmir Himalaya. Bull Eng Geol Env 82(8):328

Shugar DH, Jacquemart M, Shean D, Bhushan S, Upadhyay K, Sattar A, ... , Westoby MJ (2021) A massive rock and ice avalanche caused the 2021 disaster at Chamoli, Indian Himalaya. Science 373(6552):300–306

Sharma N, Saharia M, Ramana GV (2024) High resolution landslide susceptibility mapping using ensemble machine learning and geospatial big data. CATENA 235:107653

Sidle RC, Bogaard TA (2016) Dynamic earth system and ecological controls of rainfall-initiated landslides. Earth-Sci Rev 159:275–291

Singh K, Kumar V (2021) Rainfall thresholds triggering landslides: a review. In: Sustainable environment and infrastructure: proceedings of EGRWSE 2019. pp 455–464

Singh S, Joshi A, Sahu A, Arun Prasath R, Sharma S, Dwivedi CS (2022) Himalayan landslides–causes and evolution. In: Disaster management in the complex Himalayan terrains: natural hazard management, methodologies and policy implications. Springer, pp 33–42

Tacconi Stefanelli C, Catani F, Casagli N (2015) Geomorphological investigations on landslide dams. Geoenvironmental Disasters 2(1):1–15. https://doi.org/10.1186/s40677-015-0030-9

Tan X, Wu X, Huang Z, Fu J, Tan X, Deng S, ... , Liu B (2023) Increasing global precipitation whiplash due to anthropogenic greenhouse gas emissions. Nat Commun 14(1):2796. https://doi.org/10.1038/s41467-023-38510-9

Tehrani FS, Calvello M, Liu Z, Zhang L, Lacasse S (2022) Machine learning and landslide studies: recent advances and applications. Nat Hazards 114(2):1197–1245

Teja TS, Dikshit A, Satyam N (2019) Determination of rainfall thresholds for landslide prediction using an algorithm-based approach: Case study in the Darjeeling Himalayas India. Geosciences 9(7):302. https://doi.org/10.3390/geosciences9070302

Tengtrairat N, Woo WL, Parathai P, Aryupong C, Jitsangiam P, Rinchumphu D (2021) Automated landslide-risk prediction using web GIS and machine learning models. Sensors 21(13):4620. https://doi.org/10.3390/s21134620

Tichavský R, Ballesteros-Cánovas JA, Šilhán K, Tolasz R, Stoffel M (2019) Dry spells and extreme precipitation are the main trigger of landslides in Central Europe. Sci Rep 9(1):14560. https://doi.org/10.1038/s41598-019-51148-2

Tripathi S (2021) What causes landslides? Are they linked to climate change? India Today. Available at: www.indiatoday.in, Accessed on August 2023

Tsunetaka H (2021) Comparison of the return period for landslide-triggering rainfall events in Japan based on standardization of the rainfall period. Earth Surf Proc Land 46(14):2984–2998

Tyralis H, Papacharalampous G (2021) Quantile-based hydrological modelling. Water 13(23):3420

Wang W, Pijl A, Tarolli P (2022) Future climate-zone shifts are threatening steep-slope agriculture. Nature Food 3(3):193–196

Ward PJ, Blauhut V, Bloemendaal N, Daniell JE, de Ruiter MC, Duncan MJ, ... , Winsemius HC (2020) Natural hazard risk assessments at the global scale. Nat Hazards Earth Syst Sci 20(4):1069–1096

Whiteley JS, Chambers JE, Uhlemann S, Wilkinson PB, Kendall JM (2019) Geophysical monitoring of moisture-induced landslides: a review. Rev Geophys 57(1):106–145

World Meteorological Organization (2018) 2018 Annual Report: WMO for the Twenty-first Century. Retrieved from: public.wmo.int. Technical report number WMO-No. 1229, pp 1–32

Xue Y, Houser PR, Maggioni V, Mei Y, Kumar SV, Yoon Y (2021) Evaluation of High Mountain Asia-Land Data Assimilation System (Version 1) From 2003 to 2016, Part I: A Hyper- Terrestrial Resolution Modeling System. J Geophys Res: Atmos 126(8):e2020JD034131.

Zhang M, Chen F, Guo H, Yi L, Zeng J, Li B (2022) Glacial Lake area changes in high mountain asia during 1990–2020 using satellite remote sensing. Research. https://doi.org/10.34133/2022/9821275

Zhang W, Liu H, Wang L, Zhu X, Zhang Y (2023) Real-Time Monitoring and Early Warning of Landslide. In: Application of machine learning in slope stability assessment. Springer, pp 31–43

Zhou W, Tang C (2014) Rainfall thresholds for debris flow initiation in the Wenchuan earthquake-stricken area, southwestern China. Landslides 11:877–887

Zscheischler J, Martius O, Westra S, Bevacqua E, Raymond C, Horton RM, ..., Vignotto E (2020) A typology of compound weather and climate events. Nat Rev Earth Environ 1(7):333–347

Chapter 11
Landslide Susceptibility Map Showing the Spatial Relationship Between Various Landslide Factors and Landslide Using Remote Sensing and GIS-Based Frequency Ratio Method in Chamoli District, Uttarakhand, India

Subrata Kundu Paul, Ershad Ali, and Bipul Chandra Sarkar

Abstract Landslides are common in Uttarakhand state due to many geo-environmental variables. The main objective of this study is to produce a landslide susceptibility map of Chamoli district, Uttarakhand. To assess the influence of geo-environmental parameters on the occurrence and distribution of landslides, the Frequency Ratio (FR) model using remote sensing and GIS-based techniques is applied in the present study. The FR model is developed from the landslide inventory map and geo-environmental parameters including slope aspect, altitude, slope, TRI, TWI, SPI, rainfall, earthquake, soil, NDVI, geomorphology, geology, LULC, distance to rivers, distance to roads, and distance to faults. The model is validated using the AUC curve method. The landslide susceptibility map is divided into five zones, viz. Very high (7.92% area), high (15.78% area), moderate (21.13% area), low (25.02% area), and very low (30.15% area). The result shows that about 7.92% of the study area is a high landslide potential area where almost 55% of total landslide events are occurred. we can trace about 90% of total landslide occurrences found within 23.7% of the study area.

Keywords Landslide · Landslide susceptibility mapping · Frequency ratio method · GIS software · AUC method

S. K. Paul · E. Ali (✉) · B. C. Sarkar
Department of Geography, Ananda Chandra College, Jalpaiguri, West Bengal, India
e-mail: aliershad.geo@gmail.com

G. K. Panda et al. (eds.), *Landslide: Susceptibility, Risk Assessment and Sustainability*, Advances in Natural and Technological Hazards Research 52,
https://doi.org/10.1007/978-3-031-56591-5_11

11.1 Introduction

Decayed and fragmented rock materials owing to the mechanism of weathering processes (mechanical, chemical, biotic, and biochemical) are called rock debris. Generally, mass movement of rock debris is the movement of rock debris down a slope or is simply referred to as mass movement. Mass movement is the detachment and gravitational transport of soil and rock materials under the influence of gravity. The term paper is used for a variety of processes in which the material of the earth's surface is pulled down a slope by gravity. Two forces act on a mass standing on a slope; One is the driving force, and the other is the resisting force. The weight of the standing mass is the primary driving force. When the driving force exceeds the drag force, the mass becomes unstable and moves downward. When the gravitational force acting on a slope exceeds its resisting force, failure of the slope occurs in holding the mass, and mass wasting may occur at a very slow rate, particularly in regions that are very dry or those areas that receive sufficient rainfall such that vegetation has stabilized the surface, and it may also occur at extremely high speed, such as in rock slides or landslides, with disastrous impacts.

The term landslide is considered a North American word and is equivalent to the English word landslip in the Oxford English Dictionary (Onions 1993) The term landslip was first introduced by Lyell in his book Principles of Geology (1st Edition, Volume 3), in which landslide was described as a part of the land that has been destroyed as a result of, or because of, the disturbance of an earthquake, has sunk undermined by water that has washed away the lower strata that supported it. The first utilization of the word landslide was in the Jeffersonian Magazine, published in Albany on March 10, 1838, under the headline "Town of Natchez Struck by Landslide". The term landslide refers to the movement of a mass of rock, debris, or earth down a slope. A landslide is a process of variations in the stress–strain state of a slope, or ground mass that results in mass separation and downslope soil movement while maintaining continuous contact between the sliding mass and the underlying undisturbed soil (Kazeev and Postoev 2017). Landslide describes a variety of processes that result in the downward movement of slope-forming material, including rock, soil, debris, crops, or a combination thereof. The materials can move by falling, sliding, spreading, or flowing. Recently, several landslides have occurred in the mountainous regions of India, damaging infrastructure, causing fatalities, and causing various transportation difficulties ranging from a very minor inconvenience to a major economic and social disaster (Prakesh 2015).

In India, approximately 0.42 million km^2 of land area, or 12.6% of the land area (excluding snow-covered areas), is at risk of landslides. Of this, 0.18 million km^2 of the landmass is in the Northeast Himalayas, including Darjeeling and Sikkim Himalayas; 0.14 million km^2 fall in the northwestern Himalayas (including the states of Uttarakhand, Himachal Pradesh, and Jammu & Kashmir); The economic loss is estimated at 0.09 million km^2 in the Western Ghats and Konkan Hills (including the states of Tamil Nadu, Kerala, Karnataka, Goa, and Maharashtra) and 0.01 million km^2 in the Eastern Ghats of the Aruku region of Andhra Pradesh due of landslides can be between 1 and 2% of gross national product in many developing countries. Landslide hazard and risk assessment and mitigation is a major challenge for technocrats and policymakers in developing countries, as 80% of reported landslide deaths occur in developing countries (Landslide Hazard, Geological Survey of India).

The Uttarakhand region has been hit by several landslides and some of the best-researched and most devastating landslides are mentioned here. Some major landslide events in the Uttarakhand region occurred in the Okhimath area in the Mandakini valley of Rudraprayag district. On September 14, 2012, a series of landslides triggered by torrential rain caused extensive property damage and the deaths of 51 people. In August 1998, a heavy rain event resulted in a total of 466 landslides, killing 103 people and damaging 47 villages across the Mandakini Valley (NRSC 2012). The worst disasters have recently occurred in the districts of Uttarkashi, Chamoli, Rudraprayag, Bageshwar, and Pithoragarh in the Uttarakhand Himalayas. Due to the geological, structural, and climatic conditions, natural disasters have repeatedly occurred in the Chamoli district. In the recent past, there have been many instances where these disasters have caused severe loss of life and property. These include damage from an earthquake (1803 AD) at the Badrinath Temple (Smith 1843), a landslide that blocked the Birahi Ganga in 1868 (Juyal et al. 1996), and the rupture of the artificial lake Rishi Ganga in 1970 and 1996 (Raina et al. 1980). Lake Patal Ganga eruption in 1970 (Chansarkar 1975), Natural hazards around Gopeshwar in 1991 (Bhatt 1998), Cloudbursts in Bhimtala and Chamoli in 1995 (Joshi and Maikhuri 1997), Alaknanda floods in the 1995 and Chamoli earthquakes in 1999 (Pandey et al. 2002), landslides in Badrinath in 2004, debris flows through the local Nala near Govind Ghat in 2005 and landslides in 2010 (NIDM 2015). In June 2013, Uttarakhand's Chamoli and Rudraprayag districts experienced further heavy rainfall of 350 mm in five days (Martha et al. 2015). Approximately 30% of landslides were observed due to geotechnical and geomorphological activities in the Himalayan regions. Garhwal Himalaya is the most landslide-prone region in the Indian subcontinent because of physical factors of the terrain viz immature and unstable geological structures, rugged topography along with initiating factors. Seismic forces and heavy rainfall during the monsoon season cause severe landslide phenomena in the Indian Himalayan region (Kanungo and Sharma 2013).

With population growth, modern economic activity in the mountainous region is also increasing, paving the way for landslides (Pal and Chowdhuri 2019). The greatest damage caused by landslides affects the transportation system, i.e. roads and highways, which directly or indirectly affect many people in hilly terrain (Singh et al. 2017). Aside from this loss, other vital facilities such as water supply, electricity, and communications are also affected by the landslide activities. As landslides are a natural phenomenon that while uncontrollable, loss of life and property can be minimized if predicted in advance. Therefore, proper landslide susceptibility mapping is being conducted in this area to minimize infrastructure collapse and loss of life. For this reason, the use of landslide susceptibility mapping is considered the first step in landslide planning, assessment, and remediation.

Landslide hazard assessment is a complex and time-consuming task that involves the science of geology, geomorphology, geotectonics, statistics, and hydrology

(Glade et al. 2005). Landslide susceptibility is the probability that a landslide will occur in an area, depending on the terrain (Brabb 1984). Landslide susceptibility mapping serves not only to identify the factors that had the greatest impact on the landslides that occurred in the region, but also to estimate each factor's relative contribution to slope failure, and it is also important to establish a correlation between the factors and landslides predict the landslide risk in the future (Chen and Wang 2007). It shows the level of vulnerability of the area to landslide events. Remote sensing and geographic information systems are used to demarcate areas at risk of landslides. For better analysis and interpretation of surface landscape features, remote sensing is a valuable tool due to its spatial, temporal, and repeated coverage (Mersha and Meten 2020). In geographic information systems (GIS), we can collect georeferenced data from remote sensing and other resources for analysis, modeling, simulation, and visualization, helping to generate informed results (Roy and Saha 2019; Li et al. 2021). The study area is in the Himalayan region, where many locations are physically inaccessible and prone to landslides. Therefore, remote sensing and GIS are very useful tools to create a landslide inventory map as well as thematic maps related to landslide events in the study region. In addition, GIS, remote sensing, and advanced computer systems are reducing the need for manual fieldwork to identify vulnerable areas (Gupta et al. 2022; Anbalagan et al. 2015; Ilanloo 2011; Velayudham, et al. 2021).

In recent years, statistical methods (quantitative approach) have been widely used for landslide hazard assessment or hazard mapping due to the advantages of remote sensing and geographic information system (GIS) techniques (Van Westen et al. 1997). The statistical methods (quantitative approach) have gained importance because they can reduce the inherent subjectivity in the choice of input data and can be applied in small to large areas (Van Westen et al. 2006). The different statistical methods, which can be roughly divided into three types: bivariate, multivariate, and probabilistic prediction models, were applied and compared to test the validity and efficiency and to find the most suitable model to assess landslide susceptibility mapping (Sarkar 2004; Pradhan 2010). Statistical methods commonly used to map landslide susceptibility include the information value method (IV) (Bai et al. 2010) and the logistic regression method (LR) (Bai et al. 2010; Das et al. 2012). the fuzzy logic method (FL) (Wang et al. 2013), the artificial neural network (ANN) method (Poudyal et al. 2010; Bui et al. 2012), and the frequency ratio method (FR). (Choi et al. 2011; Park, et al. 2012), Bayes' theorem is based on the Weights of Evidence (WoE) method (Pradhan et al. 2010), and bivariate statistical analysis (Armas 2012; Nandi and Shakoor 2010).

Of these methods, frequency ratio is often used to assess landslide susceptibility with good performance. Several researchers have used the frequency ratio model for landslide studies (Bahrain et al. 2014; Meten et al. 2015; Hong et al. 2015; Pham et al. 2015a, b; Pirasteh and Li 2017; Fayez et al. 2018).; Khan et al. 2019).

11.2 Database and Methodology

11.2.1 Objective of the Study

The main objective of this study is to evaluate the landslide vulnerability map generated from remote sensing satellite imagery and the effectiveness of a GIS-based modeling technique in predicting the possibility of landslide vulnerability in the study area of Chamoli District, Uttarakhand State, India.

11.2.2 Study Area

The study area chosen for this research is Chamoli District, established on February 20, 1960. It was previously part of Garhwal District as Chamoli Subdivision and it occupied the central part of the Uttrakhand Division until 20 December 1968 when that division was abolished and the area occupied by the current district was incorporated into the newly developed Garhwal Division. Chamoli used to be a district of the Garhwal Division of Uttarakhand State, but after the creation of the Gairsain Division in March 2021, Chamoli is now an important district of this division. Thedistrict Chamoli is surrounded by Uttarkashi in North-West, Pithoragarh in the South-West, Almora in South East, Rudrprayag in the South-West, and Tehri Garhwal in the West. The district lies between 29 55 37″N to 31 27 3″N latitude and 78 54 3″E to 80 2 3″E longitude. The administrative headquarters of Chamoli district is in Gopeshwar. The district is split into six tehsils namely Joshimath, Chamoli, Karnaprayag, Pokhari, Gairsain, and Tharali, and the district is split into nine development blocks Joshimath, Dasoli, Pokhari, Ghat, Karnaprayag, Tharali, Narayanbagar, Dewal and Gairsain (Bhatt et al. 2014). The district has a geographical area of 8030 km^2. The district in its present form occupies over 15% of the total area of the State. The physiography of the district is hilly with a moderate to high angle of inclination. The altitude of the district varies from 600 to 7800 m. The district lies beneath the impressive Himalayan ridges amidst the breathtaking beauty of the upper Himalayas. The district also stretches across snow-free valleys to sky-high peaks with perpetual snow and glaciers. Chamoli District is located in a region with a tectonically folded and thrustable mountain range. Geologic formations are characterized by complex reverse faults, folds, thrusts, and nappes of large, variable dimensions (Krishnan 1982). The most important lithological units in this region are the Vaikrita group, the Musiari, Tejam, Berinag, and Damtha formations. The Vaikrita rock group is part of the higher Himalayas and consists of heavily metamorphic rocks of granite and gneisses. The Precambrian is exposed in the area north of Chamoli. The Ramgarh Group consists of crystalline rocks, predominantly quartzite and sedimentary, exposed south of the Chamoli district. The crystalline rocks of metasediments and acidic intrusions are exposed near Siraupani and north of Chamoli. The Almora Cliff Zone is moderately metamorphic, covering much of the area and forming the boundary south of

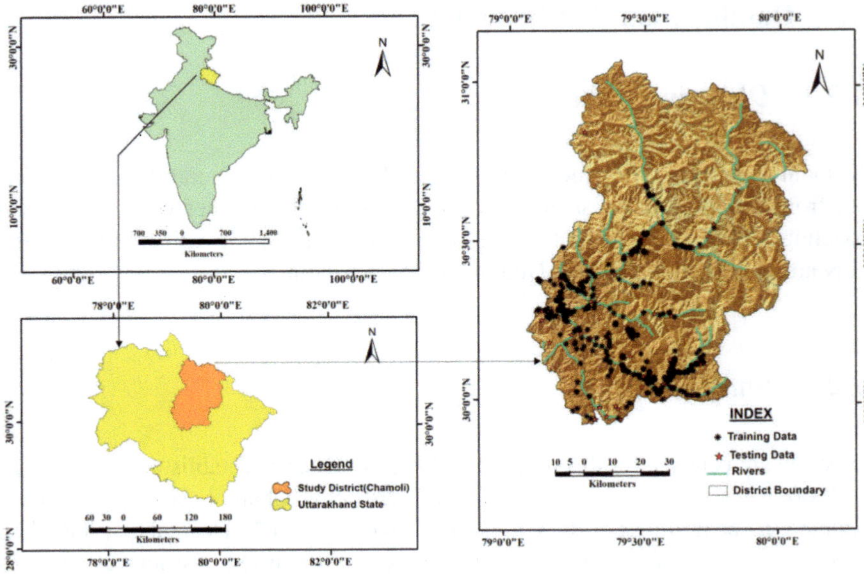

Fig. 11.1 Location map of the study area (**a** India, **b** Uttarakhand, **c** Chamoli district)

the Vaikrita Group. The boundary between the high-grade metamorphic rocks of the Vaikrita Group and the low-grade metamorphic rocks of the Almora Group is the Vaikrita Thrust, defined as the Main Central Thrust (MCT) (Fig. 11.1).

The central crystalline rocks are well exposed in the higher Himalayas of the Alaknanda Valley in the Chamoli district. The Garhwal group forms the largest rock group in the Chamoli district. The Garhwal Group is composed of quartzite, phyllite, shale and limestone. Geologically, the area belongs to the Little Himalayas and is located in a tectonically active zone. The climatic conditions in the district vary depending on the altitude and terrain. Therefore, hot and cold weather conditions as well as rainfall amounts vary from area to area. Most of the region lies on the southern slopes of the outer Himalayas, the monsoon flows through valleys, with rainfall being heaviest from June to September when the southern half of the district accounts for 70–80% of the annual precipitation and 55–65% of the annual precipitation. in the northern half. The main rivers of this area are Alaknanda, Pinder, Rishi Ganga, and Dhauli Ganga.

11.2.3 History of Landslides

In terms of deadly landslides, India ranks first in the world (Martha et al. 2021). Large vulnerable areas (0.42 million km^2), high population densities, and monsoon rains complicate India's task of managing and containing landslides (Martha et al.

2021). According to official figures from the United Nations International Strategy for Disaster Reduction (UN/ISDR) and the Center for Research on Disaster Epidemiology (CRED) in 2006, landslides ranked third among the ten most common natural disasters in terms of number of deaths. Located in the lap of the Himalayas, Chamoli has a long history of landslides. Table 11.1 lists the location and timing of the major landslides in this county and their impact in terms of loss of life and livestock, and damage to the economy and infrastructure (Table 11.1).

11.3 Methodology

The development of the susceptibility to landslides in the present study area takes place in six steps. These steps are (5.1) selection of landslide conditioning parameters, (5.2) data collection and organization, (5.3) construction of a landslide inventory map, (5.4) construction of various thematic maps of landslide conditioning parameters, (5.5) application of the frequency ratio (FR) model, and (5.6) development a landslide susceptibility map.

11.3.1 Selection of Landslide Conditioning Parameters

Any type of mass movement of weathered debris causes landslides, whether on a slope or in a valley. The relationship between shear forces and the resistance of materials to shear forces is called the safety factor (F_S) (Popescu 2002). Landslides can occur when either shear forces increase or the shear strength of materials decreases. The occurrence of landslides is influenced by the interaction of various parameters, such as B. Initiating parameters (precipitation, earthquakes, etc.); topo geomorphological parameters (elevation, dip angle, orientation, topographic roughness index, drainage frequency, distance to rivers, geomorphology factor, etc.); hydrological parameters (stream transport index, stream performance index, topographic wetness index, etc.); lithological parameters (fault line, distance to faults, soil depth, soil properties, geology, gravity anomaly, etc.) man-made parameters (roads, land use and land cover factor, etc.). The factors influencing landslides are the main criteria for mapping landslide susceptibility. Therefore, the selection of causative factors is considered a fundamental step in susceptibility modeling. In this study, based on an analysis of the history of landslides in the study area and the underlying geomorphometric conditions, a total of sixteen factors such as orientation, elevation, tilt, precipitation, soil, geomorphology, geology, earthquakes and Normalized Difference Vegetation Index (NDVI), Topographical Robustness Index (TRI), Topographic Wetness Index (TWI), Sream Power Index (SPI), Land Use and Land Cover (LULC), Distance to Rivers, Distance to Roads, Distance to Faults are taken into account in detecting landslide-prone areas. All of these factors are extracted from

Table 11.1 Major landslide events in the study area (Chamoli district, Uttarakhand)

Serial no.	Location	Time (month/ year)	Consequences	References
1	Joshimath Landslide	1842	Damaged roads and blocked traffic	Parkash (2014)
2	Alaknanda River, Chamoli	1868	The landslide blocked the Alaknanda River, swept through two villages, and killed 70 pilgrims	Rautela and Pande (2005)
3	Gohna, Birahi Ganga, Chamoli	October 1893	The Gohna landslide filled the bed of the Birahi River to a height of 350 m. The landslide dam was partially breached in 1894, and the final breach occurred in 1970, raising the water level in Srinagar by 50 m. Two days later, the water level of the river in Haridwar rose by 4 m	Prakesh (2015)
4	Alaknanda, Chamoli	893	Floods in the Birehi Ganga River near its confluence with the Alaknanda River triggered landslides, causing major blockage of the river with a 10–13 m afflux. A girder bridge was bypassed and another one was destroyed	Pandey and Mishra (2015)
5	A landslide dam broke near Gohna	1894	The rupture of Lake Gohna caused the Birehi disaster in Alaknanda Valley	Rautela and Pande (2005)
6	Around Badrinath, Alaknanda River, Chamoli	1930	The Alaknanda was blocked in 1930, causing the water level to rise to 9 m and flooding many homes	Prakesh (2014)
7	Dhauli Ganga, near Bhapkund, Chamoli	1957	An avalanche fell along the Dronagiri Creek, a tributary of the Dhauli Ganga, in 1957, blocking the river near Bhapkund and creating a 3.0 km long lake that was later filled with debris	Prakesh (2014)
8	Rishi Ganga, near Reni village, Chamoli	May 1968	The Rishi Ganga in Chamoli district was blocked to a height of 40 m due to a landslide in Reni village. The dam collapse in 1970 caused significant damage	Gulia (2007), NDMA (2009)

(continued)

Table 11.1 (continued)

Serial no.	Location	Time (month/ year)	Consequences	References
9	DhakNala, near Tapovan, Chamoli	20th July 1970	The Dhak Nala has dug its narrow bed deeply and the resulting toe cuttings have reactivated the prehistoric landslide at the Kuari Pass. A 300 m section of the Joshimath-Malari Road near the Dhak Nala plunged 40 m and the landslide debris temporarily blocked the Dhauliganga with a 15–20 m high embankment	Gulia (2007)
10	Near Helong, Chamoli	July 1970	The current in the Karmanasa Nadi reactivated another prehistoric slide, tearing away part of the road and blocking the Alaknanda River, although residents of several villages near the Helong Nala prayed for their lives throughout the night	Gulia (2007)
11	Hanuman Chatti, near Badrinath, Chamoli	20th July 1970	About 275 mm of rain (in a few hours) caused a landslide that blocked the river and created temporary lakes. Alaknanda and Dhauli Ganga rose 30–60 and 15–20 m respectively and erupted, causing significant loss of life	Gulia (2007), NDMA (2009)
12	Patal Ganga, Chamoli	July 1970	The Patal Ganga, a tributary of the Alaknanda River got choked and a reservoir more than 60 m deep was created. The bursting of this choked reservoir resulted in flash floods in the Alaknanda River, triggering many landslides. Belakuchi village was washed away and communications were disrupted due to a landslide that occurred on the left bank of the Alaknanda River	Pandey and Mishra (2015)
13	Bamni Village near Badrinath, Chamoli	April1979	Avalanches blocked the Alaknanda River and a lake formed which later erupted over the village of Bamni near Badrinath	Bisht et al. (2002)

(continued)

Table 11.1 (continued)

Serial no.	Location	Time (month/ year)	Consequences	References
14	Landslides at Jakholi in Tehri Garhwal and Devaldhar in Chamoli	1986	32 lives were lost	Prakesh (2015)
15	Gopeshwar, Chamoli	16th Aug 1991	In 6 villages 36 people were killed and 26 animals lost	Prakesh (2015)
16	Phata and Byung Gad Landslides, Chamoli	17th July 2001	21 were killed and several houses were damaged	Prakesh (2015)
17	Saigari Village, Chamoli	2002	The Gankhwi River was blocked by sediment and debris. In 2002 and 2003, a glacial lake flooded, destroying the village of Saigari in the Dhauli Ganga valley	Bisht et al. (2011)
18	Badrinath, Chamoli	July 2004	16 people were killed, 200 pilgrims were stranded, 800 shopkeepers and 2300 villagers were trapped when a downpour triggered massive landslides that almost washed away the Badrinath Strait and severed the Badrinath area. 200 m of road flooding	Pandey and Mishra (2018)
19	Govind Ghat, Chamoli	29th–30thJune 2005	A cloudburst &landslide occurred, throwing many debris and boulders along a seasonal nala. Eleven people were killed and property was lost	Pandey and Mishra (2018)
20	Chamoli	2007	23 people died	Pandey and Mishra (2018)
21	Chamoli	26th June 2008	8 killed, 2 injured	Prakesh (2015)
22	Joshimath, Chamoli	22nd Sept. 2010	2 people died	Prakesh (2015)
23	Chamoli	30th June 2011	10 people dead, and Rishikesh-Badrinath closed	Prakesh (2015)
24	Pithoragarh and Chamoli Districts	1st July 2016	A cloudburst killed 30 people and landslides buried several houses in remote villages	Pandey and Mishra (2018)
25	Tharali Ghat, Chamoli	16th July 2018	Several houses & vehicles damaged	Pandey and Mishra (2018)

(continued)

Table 11.1 (continued)

Serial no.	Location	Time (month/ year)	Consequences	References
26	Joshimath, Chamoli	19th July 2018	2 people died. 150 m the Joshimath Highway was washed away. Several farmland and roads were damaged	Pandey and Mishra (2018)
27	Ronti Gad, Rishiganga and Dhauliganga valleys in Chamoli	7th Feb. 2021	More than 72 people died in the incident when a huge mass of rock and snow erupted from a left valley of Raunthi Gad, crashing into the Rishiganga valley and causing a deadly flood in the Rishiganga River	Mao et al. (2022)

respective spatial databases and using the Frequency Ratio (FR) method using GIS software to produce the final vulnerability map.

11.3.2 Data Collection and Organization

The main source of this research is secondary data. As already mentioned, a total of 16 causal parameters are used to create the landslide susceptibility map for the present study; Therefore, different geodata are collected from different secondary sources. Data for the development of the landslide susceptibility map was collected and extracted from the SRTM Digital Elevation Model (DEM), LandSat Images, Geological Survey of India (GSI), Google Earth imagery, and CRU TS (Climatic Research Unit Gridded Time Series). Table 11.2). Arc GIS 10.5 version and QGIS 3.16.0 with GRASS 7.8.4 software version are used to develop various thematic layers and Google Earth Pro software is used to test the accuracy of the LULC map. Microsoft Office 7 is used to format, organize, and calculate data generated from open attribute tables of various thematic levels. (Table 11.2 and Fig. 11.2).

11.3.3 Construction of a Landslide Inventory Map

The occurrence of landslides is based on the principle that slope failure is more likely under the conditions that led to previous instability (Guzzetti et al. 2000). Therefore, tracking past landslides is important to predict future landslide events. The data on previous landslides that occurred at different times and locations is called the landslide inventory (Wu et al. 2016), and landslide inventories are the simplest and most direct method of mapping landslide locations (Guzzetti et al. 2012). It can be used for high predictive power in landslide vulnerability analysis and landslide

Table 11.2 Different sources of information for the various landslide factors utilized in the mapping of landslide

Serial no	Causative parameters	Description and sources
1	Aspect	Used SRTM DEM (1 Are-Second Global) with 3% cloud cover Spatial Resolution is 30 m The acquisition date is February 11, 2000, and the publication date is September 23, 2014 *Source* https://earthexplorer.usgs.gov/
2	Altitude	
3	Slope	
4	Topographic ruggedness index (TRI)	
5	Topographic wetness index (TWI)	
6	Stream power index (SPI)	
7	Rainfall	Used high-resolution gridded time-series dataset of Climatic Research Unit (CRU) The resolution of the dataset is 0.5 × 0. 5 degrees The version used is CRU TS v. 4.07 The release date is April 19, 2023 *Source* https://crudata.uea.ac.uk/cru/data/hrg/
8	Earthquake	Long-term data are used for making earthquake maps *Source* https://bhukosh.gsi.gov.in/
9	Soil	The FAO digital soil map of the world is used. The scale is 1:5,000,000 *Source* https://www.fao.org/soils-portal/data-hub/soil-classification/en/
10	Normalized difference vegetation index (NDVI)	Band 4 and Band 5 of the Landsat-8 OLI/TIRS satellite image were used. The spatial resolution is 30 m The acquisition date of the image is 16th October 2020 *Source* https://earthexplorer.usgs.gov/
11	Geomorphology	Data are collected from https://bhukosh.gsi.gov.in/
12	Geology	Data are collected from https://www.gsi.gov.in/
13	Land use and land cover (LULC)	Used Sentinel-2 satellite image for Land Use/Land Cover The spatial resolution of this image is 10 m *Source* https://livingatlas.arcgis.com/landcover/
14	Distance to rivers	The river network is collected from https://www.diva-gis.org/
15	Distance to roads	The road network is collected from https://www.diva-gis.org/
16	Distance to faults	Data are collected from https://www.gsi.gov.in/

vulnerability mapping (Sonker et al. 2021). For this work, past landslide data is collected by the Geological Survey of India (https://www.gsi.gov.in/). A total of 616 landslide points were used as a landslide inventory for analysis and validation of landslide susceptibility. Of these landslides, 36.85% came from debris, 56.17% from rock material, 6.33% from a mixture of rock and debris, and only 0.65% from soil. Multiple variations in the type of movement can be observed in landslides at

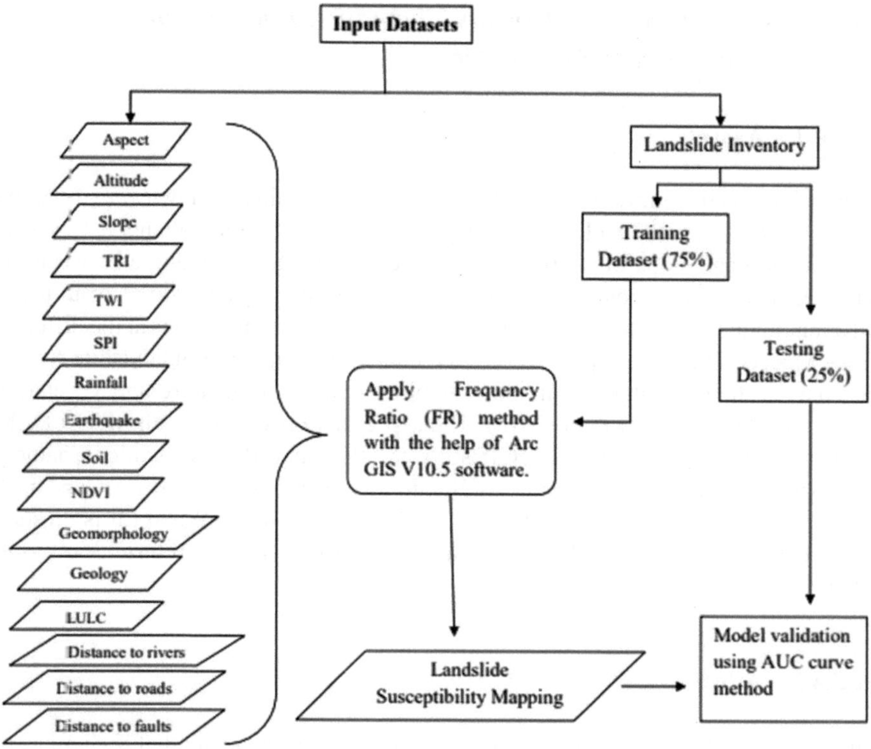

Fig. 11.2 Flowchart showing methodology for constructing LSM

different locations with different environmental conditions. According to the analysis of past landslide events, different types of movement of landslides are observed in the study area. These are landslides (83.77%), falls (4.38%), streams (3.25%), sinks (1.46%), and falls (0.49%). The geometry type of all landslides is the point. All landslide data is randomly split into two datasets using Arc GIS 10.5v software. For example, 75% of the entire dataset (429 landslides) is used for training purposes for building frequency ratio models and 25% for testing purposes. The full dataset (187 landslides) is used to verify the result. In general, training data is used to develop the LSZ model, while test data is used to validate the model (Chen et al. 2016).

11.3.4 Construction of Various Thematic Maps of Landslide Conditioning Parameters

11.3.4.1 Aspect

Aspect is defined as the direction of the slope, which explains the downward direction of the maximum rate of elevation change. An aspect map simultaneously shows the direction and slope of a terrain. The slope aspect affects the angle between the surface, the wind direction, and the solar radiation received by the ground, resulting in different light, temperature, and soil physical and chemical properties at the different slope aspects. The slope aspect is considered an important factor in landslide studies and plays a fundamental role in slope stability due to temperature and vegetation variations (Mohammady et al. 2012). The aspect map is created with SRTM DEM (30 m spatial resolution) using ArcGIS software. We found 10 classifications, namely Flat, North, Northeast, East, Southeast, South, Southwest, West, Northwest, and back north. If the cell value is −1 (flat), it means there is no slope. When tilt is present, the orientation is measured clockwise, starting at 0 to the north. It returns 360° north again.

11.3.4.2 Altitude

The elevation factor affects landslide susceptibility in such a way that higher elevations are generally more susceptible to landslides. Elevation is typically used to identify local relief, giving the minimum and maximum elevation values of the area (Chen et al. 2020). In this study, an elevation was created using data from SRTM DEM. The altitude value is divided into five classes, namely: under 2040, 2040–3072, 3072–4187, 4187–5190 m, and over 5190 m.

11.3.4.3 Slope

The slope is of significant importance for landslide formation, development, and susceptibility and is defined as an input parameter in susceptibility studies. The inclination is the angle between each surface segment and a horizontal reference point that measures the rate of elevation change and supports the flow of water and other materials in the direction of the incline in terms of the steepest incline drop for elevation (Dehnavi et al. 2015). As the incline of the slope increases, so does the instability of the slope, and with it the risk of landslides. In this study, a slope map was created to show the direct association with the occurrence of landslides. SRTM DEM (30 m spatial resolution) is used to create a slope map and is divided into five classes,i.e. below 20°, 20°–40°, 40°–60°, 60°–80°, and above 80°.

11.3.4.4 Topographic Ruggedness Index (TRI)

TRI is a geomorphic parameter that shows whether the topography of the study area is flat or undulating and represents the unevenness of a land surface. In general, it is used to measure the variation in height difference between adjacent cells of a DEM. To create the TRI layer, SRTM DEM (30 m spatial resolution) is used to make three focal statistical layers by using the neighborhood tool (spatial analyst tools) and then applying the formula (Mukherjee and Singh 2020) with the help of the Raster calculator Map algebra tools).TRI layer is divided into five classes, i.e. below 0.2667, 0.2667–0.4222, 0.4222–0.5778, 0.5778–0.7333, and above 0.7333 (Fig. 11.3).

$$\mathrm{TRI} = (\mathrm{FS}_{\mathrm{mean}} - \mathrm{FS}_{\mathrm{min}})/(\mathrm{FS}_{\mathrm{max}} - \mathrm{FS}_{\mathrm{min}}) \tag{11.1}$$

11.3.4.5 Topographic Wetness Index (TWI)

TWI is an index commonly used in hydrological analysis to describe an area's tendency for water accumulation. The TWI is an important parameter to characterize the hydrological similarity of landslide-prone areas by quantifying the influence of local topography on the hydrological process and showing the spatial distribution of soil moisture and surface saturation. The wetness of the soil material in the various topographical areas is measured by the TWI and has a direct impact on the slope. SRTM DEM (30 m spatial resolution) and the following formula are used to calculate the TWI.

$$\mathrm{TWI} = \ln\,(\alpha/\tan\beta) \tag{11.2}$$

where α is the local upslope area per unit contour length and $\tan\beta$ is the local gradient. With the help of ArcGIS software the TWI layer is created and it is divided into five classes ranging from below 4.67 to above 12.52.

11.3.4.6 Stream Power Index (SPI)

The upstream catchment area, the erosive force of the flowing water, and the slope of each grid cell are used to determine the SPI. Potential sites for soil conservation measures can be identified to mitigate the effects of concentrated surface runoff. Possible flow erosion and associated landscape processes can be described with SPI. As the specific catchment area and slope steepness increase, the water volume of the slope areas, the rate of water flow and the risk of erosion also increase. This directly affects the likelihood of landslides in the area. The SPI is calculated using the below equation (Digital Terrain Analysis Manual: Zumbro River Watershed Restoration Prioritization 2014):

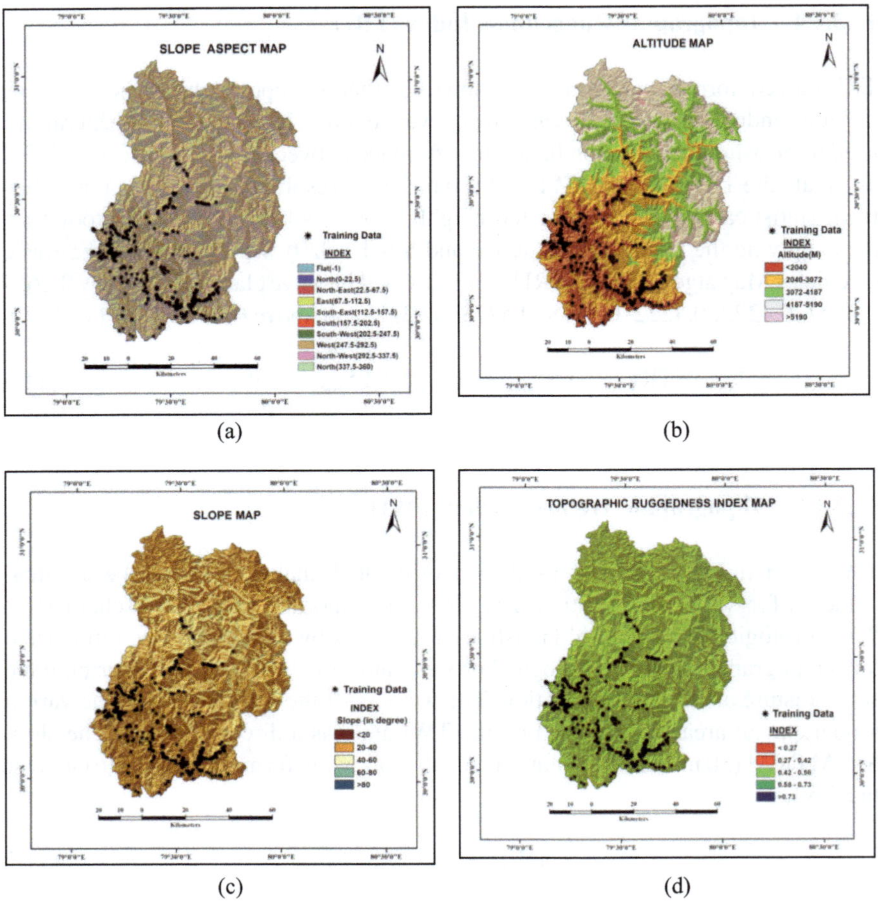

Fig. 11.3 Showing different maps of important factors related to landslides: **a** Slope Aspect map. **b** Altitude map. **c** Slope map. **d** TRI map

$$\text{SPI} = \ln(''\text{flow\_accumulation}'' + 0.001) * ((''\text{slope\_percent}''/100) + 0.001) \tag{11.3}$$

SRTM DEM (30 m spatial resolution) is used to get the necessary output layers,i.e. flow direction, flow accumulation, and slope map using the hydrology tool (spatial analyst tool) of ArcGIS software.

11.3.4.7 Rainfall

In the Himalayan region, precipitation is the main trigger for landslides, which play an important role in their occurrence throughout the rainy season. The consequences of precipitation in geological catastrophes are as follows: first, the soil becomes heavier and more hydrated, which increases gliding power, second, precipitation weakens the structural integrity of the soil by softening it, and third, water intrusion reduces friction by lubricating the sliding surface (Sui et al. 2022). Therefore, an average annual precipitation map was created to show the variation of precipitation in the study area using the interpolation method in ArcGIS software.

11.3.4.8 Earthquake

Landslides are caused by earthquakes because gravitational and seismic acceleration increase short-lived stress to levels higher than the cohesive and frictional strengths of the underlying rocks and soils (Newmark 1965). While the frequency of earthquakes in the studied area is relatively high, their magnitudes range from low to moderate. This weakens the strength and stiffness of the rock mass and causes landslides. An earthquake zone map of the study area is created using the interpolation method of the ArcGIS software. (Fig. 11.4).

11.3.4.9 Soil

The continuous deterioration of the soil through erosion is one of the main causes of the landslide. Wind activity, flowing water, the beating action of raindrops, the nature of the land surface, and the nature of soil properties are some of the reasons soil material separates and moves from one place to another. Assessing the stability of soil and the susceptibility of land to slipping depends largely on the depth of the soil, its texture, its erosion, its stoniness, and its surface finish. The increase in soil depth increases the rate of infiltration while decreasing the rate of runoff. With flat ground, the above situation is reversed. The nature of the soil is also important for the stability of the soil and its susceptibility to landslides. Clay-rich soils exhibit stable, separation-resistant aggregates. Therefore, the most common features of a landslide, including a sandy bottom, a steep gradient, and large falls, cause severe damage to land (Sonker et al. 2021). Therefore, studying the soil properties is essential to be able to assess the probability of a landslide in the study area. For this reason, a soil map is created from FAO soil data using ArcGIS software.

11.3.4.10 Normalized Difference Vegetation Index (NDVI)

The Normalized Difference Vegetation Index (NDVI) measures how green or active vegetation is. The NDVI is a simple numerical indicator that provides a measure of ground surface vegetation cover in large regions (Zhang et al. 2009). It is associated with photosynthetically active radiation (PAR). The multispectral remote sensing data approach is used by NDVI to analyze the remote sensing data and determine the vegetation index, land cover categorization, water bodies, open regions, bush areas, hilly areas, agricultural areas, heavy forests, and thin forests (Gandhi et al. 2015).

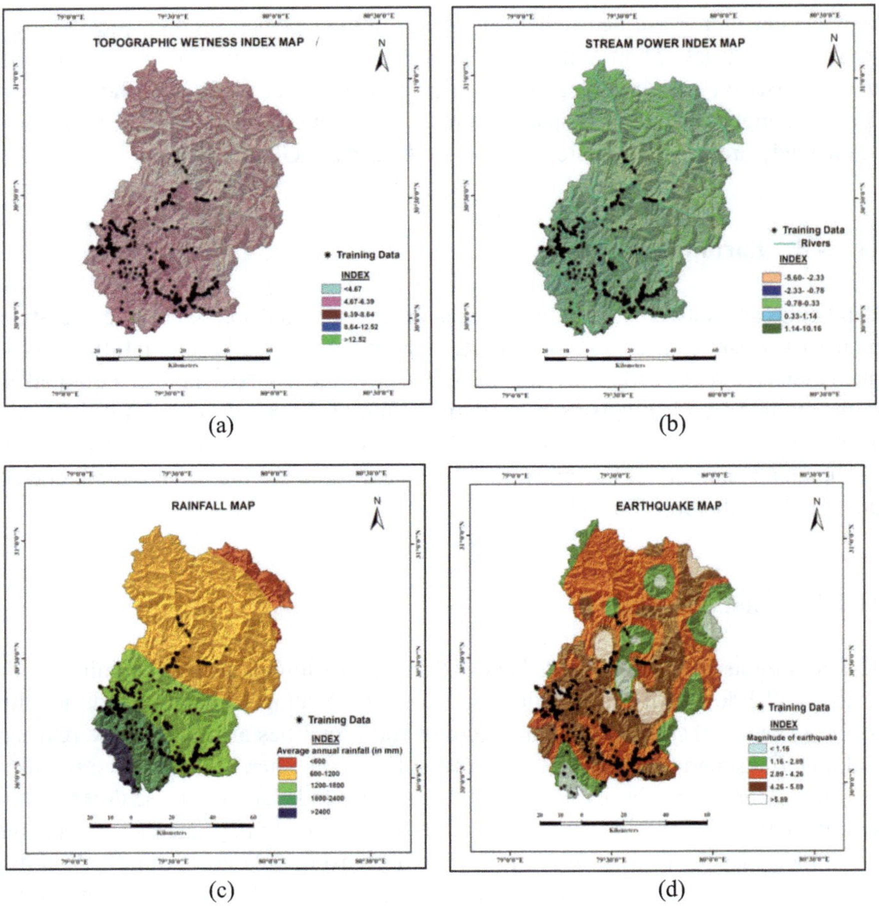

Fig. 11.4 Showing different maps of important factors related to landslides: **a** TWI map, **b** SPI map, **c** Rainfall map, **d** Earthquake map

Increased vegetative growth reduces the likelihood of landslides and vice versa in mountainous areas. In this work, Landsat 8 OLI/TIRS imagery was used to create an NDVI map to represent low to high vegetation zones using the following equation (Tucker 1980):

$$\text{NDVI} = (\text{NIR} - \text{RED})/(\text{NIR} + \text{RED}) \tag{11.4}$$

Here, NIR stands for the near-infrared band, and RED for the red band. The vegetation cover of the research region is analyzed using bands 5 and 4 of Landsat 8 images.

11.3.4.11 Geomorphology

Geomorphology is the scientific study of the properties of the Earth's surface, its environment, and the processes that gave the planet its current shape. The surface landscape elements responsible for landslides are shown on the geomorphological map. Landslides can occur for various morphological reasons. These include tectonic or volcanic uplift, glacial recoil, river, wave, or glacial erosion at the top or lateral edges of the slope, depositional loading of the slope or its crest, weathering by freeze and thaw, etc. The main geomorphological landforms in the region consist of various hills, mountains, and valleys. The geomorphological map is created using data from the Geological Survey of India using ArcGIS software.

11.3.4.12 Geology

Understanding the geological structure of a region helps determine the types and properties of rocks and slopes, the physical and chemical properties of soils, the availability of minerals, and surface and subsurface water resources. As this is a mountainous region, studying the geology of the study area is critical to understanding the nature of landslides. For this reason, the geological map is made using the data from the Geological Survey of India (Fig. 11.5).

11.3.4.13 Land Use and Land Cover (LULC)

Landslides may occur more frequently and severely due to anthropogenic interventions such as deforestation and deforestation of slopes. Land use and land cover (LULC) data, representing both naturally occurring and artificially modified earth surfaces, can be used to assess the impact of human activities on slope stability (Chen et al. 2019). Hydrological processes and the physical structure of the soil are affected by LULC, which increases the likelihood of landslides. For example, slope stability can be affected by road or building improvements, deforestation, or changes in soil moisture or surface water flow. In this study, the Sentinel-2 satellite imagery is used to create a LULC map using ArcGIS software.

11.3.4.14 Distance to Rivers

Drainage plays an important role in the likelihood of a landslide, as water runoff is one of the main causes of landslides (Berhane et al. 2021). The drainage distance map for this research area shows the proximity of drainage and is divided into five distinct buffer zones. The Euclidean distance method is used to generate these buffer zones using ArcGIS software.

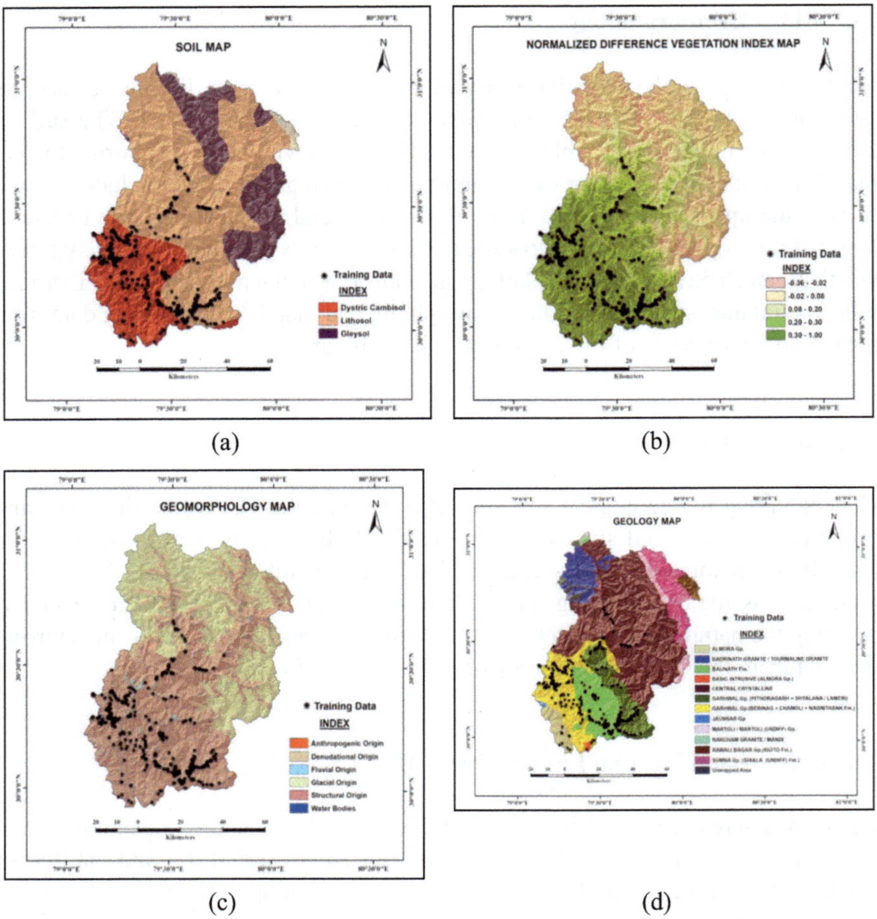

Fig. 11.5 Showing different maps of important factors related to landslides: **a** Soil map. **b** NDVI map. **c** Geomorphology map. **d** Geology map

11.3.4.15 Distance to Roads

Drainage plays an important role in the likelihood of a landslide, as water runoff is one of the main causes of landslides (Berhane et al. 2021). The drainage distance map for this research area shows the proximity of drainage and is divided into five distinct buffer zones. The Euclidean distance method is used to generate these buffer zones using ArcGIS software.

11.3.4.16 Distance to Faults

Due to tectonic fractures, the strength of the rock around geological fault zones decreases, making them vulnerable to landslides (Chen et al. 2017). This factor is taken into account when mapping landslides and the fault buffers are created to generate the distance map to faults using the Euclidean distance method in ArcGIS software (Fig. 11.6).

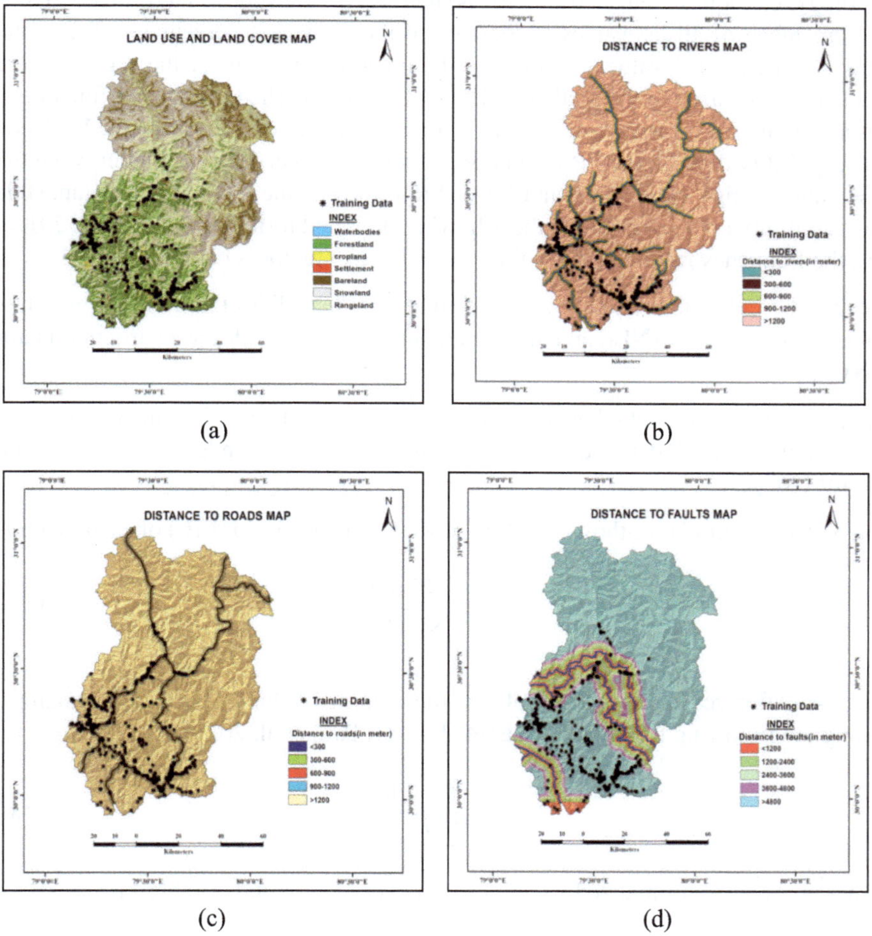

Fig. 11.6 Showing different maps of important factors related to landslides: **a** LULC map. **b** Distance to rivers map. **c** Distance to roads map. **d** Distance to faults map

11.3.4.17 Application of the Frequency Ratio (FR) Model

In this study, the frequency ratio (FR) was used to assess landslide susceptibility. A bivariate statistical technique with reliable results is the frequency ratio (FR), which is easy to use. An observation-based method for creating landslide susceptibility maps is the FR model. The FR is often used to map landslide susceptibility. For a reclassified factor or categorical factor, FR was used to calculate the ratio of cell to landslide incidence in each class (aspect, elevation, slope, precipitation, soil, geomorphology, geology, earthquake, etc.). So the FR is the ratio of landslides in a desired class as a percentage of all landslides to the area of the class as a percentage of the total map. Thus, the FR value 1 represents the mean of the ratio of the region where landslides occurred to the total area (Ehret et al. 2010). The GIS overlay function is used to combine the many parameters assigned to the ratio to construct the landslide susceptibility through the FR technique. Landslide vulnerability is higher when the FR value is high (value >1), and lower FR values (value <1) mean less landslide susceptibility in the area. Using the following steps and formulas (Pham et al. 2015a, b), the frequency ratio model for the classes of each factor is determined:

Step 1: Calculate the ratio between the number of landslides that fall in every class of a particular factor (**NL_i**) and the overall number of landslides in a particular factor (**NL_t**).

Step 2: Calculate the ratio between the number of cells of landslides in every class of a particular factor (**NC_i**) and the overall number of cells of landslides in a particular factor (**NC_t**).

Step 3: The formula for the calculation of the Frequency ratio (**FR_i**) of each factor is

$$FR_i = \frac{NL_i/NL_t}{NC_i/NC_t} \tag{11.5}$$

Step 4: After getting the values of FR, the obtained values of FR are normalized using the following formula (Sonker et al. 2021; Patil et al. 2022).

$$FR_n = \frac{FR}{\sum \text{Class FR}} \tag{11.6}$$

where FR_n is the normalized value of FR of every factor class (Table 11.3).

11.3.4.18 Development of a Landslide Susceptibility Map

Landslide Susceptibility Indices (LSI) are calculated and ranked for the entire research region to produce a map of landslide susceptibility. The level of a region's vulnerability to landslide events is indicated by the LSI. Areas with lower LSI are less likely to experience landslides and vice versa. The following equation (Pham

Table 11.3 The spatial relationship between different landslide factors and landslides using the frequency ratio

Factors	Classes	No. of landslide (NL_i)	No. of landslides in percentage	No. of cells in classes (NC_i)	Percentage of cells area in classes	FR (FR_i)	Normalization (FR_n)
Aspect	Flat	43	10.02	880,435	10.15	0.99	0.77
	North	36	8.39	874,840	10.09	0.83	0.65
	North–East	36	8.39	876,798	10.11	0.83	0.65
	East	34	7.93	886,084	10.22	0.78	0.61
	South–East	44	10.26	883,667	10.19	1.01	0.79
	South	56	13.05	883,992	10.19	1.28	1
	South–West	51	11.89	917,345	10.58	1.12	0.88
	West	47	10.96	843,604	9.73	1.13	0.88
	North–West	39	9.09	796,156	9.18	0.99	0.77
	North	43	10.02	828,599	9.56	1.05	0.93
Altitude	<2040	378	88.11	1,910,593	22.03	4.00	1
	2040–3072	51	11.89	2,255,401	26.01	0.46	0.11
	3072–4187	0	0	2,318,307	26.73	0	0
	4187–5190	0	0	2,114,076	24.38	0	0
	>5190	0	0	73,143	0.84	0	0
Slope	<20	0	0	6406	0.07	0	0
	20–40	332	77.39	6,378,591	73.56	1.05	1
	40–60	96	22.38	2,109,495	24.33	0.92	0.87
	60–80	1	0.23	177,145	2.04	0.11	0.11
	>80	0	0	43	0.00	0	0

(continued)

Table 11.3 (continued)

Factors	Classes	No. of landslide (NL_i)	No. of landslides in percentage	No. of cells in classes (NC_i)	Percentage of cells area in classes	FR (FR_i)	Normalization (FR_n)
TRI	<0.27	0	0	13,744	0.16	0	0
	0.27–0.42	54	12.59	1,013,838	11.67	1.08	1
	0.42–0.58	343	79.95	6,752,669	77.74	1.03	0.95
	0.58–0.73	32	7.46	897,314	10.33	0.72	0.67
	>0.73	0	0	9204	0.11	0	0
TWI	<4.67	145	33.80	2,970,168	34.25	0.99	0.89
	4.67–6.39	195	45.45	3,539,288	40.82	1.11	1
	6.39–8.64	61	14.22	1,531,826	17.67	0.80	0.72
	8.64–12.52	22	5.13	505,164	5.83	0.88	0.79
	>12.52	6	1.40	125,074	1.44	0.97	0.87
SPI	−5.60 to −2.33	30	6.99	742,421	8.56	0.82	0.69
	−2.33 to −0.78	42	9.79	1,079,556	12.45	0.79	0.66
	−0.78–0.33	131	30.54	2,815,857	32.47	0.94	0.79
	0.33–1.14	186	43.36	3,158,858	36.43	1.19	1
	1.14–10.16	40	9.32	874,828	10.09	0.92	0.78
Rainfall (mm/year)	< 600	0	0	433,001	4.99	0	0
	600–1200	47	10.96	4,784,910	55.18	0.20	0.06
	1200–1800	227	52.91	2,236,387	25.79	2.05	0.65
	1800–2400	149	34.73	949,557	10.95	3.17	1
	>2400	6	1.40	267,635	3.09	0.45	0.14

(continued)

Table 11.3 (continued)

Factors	Classes	No. of landslide (NL_i)	No. of landslides in percentage	No. of cells in classes (NC_i)	Percentage of cells area in classes	FR (FR_i)	Normalization (FR_n)
Earthquake's magnitude	<1.16	10	2.33	434,621	5.01	0.47	0.31
	1.16–2.89	19	4.43	1,426,078	16.44	0.27	0.18
	2.89–4.26	181	42.19	3,588,102	41.37	1.02	0.67
	4.26–5.89	206	48.02	2,751,349	31.72	1.51	1
	>5.89	13	3.03	473,006	5.45	0.56	0.37
Soil	Dystric Cambisol	233	54.31	1,863,923	21.74	2.50	1
	Lithosol	196	45.69	5,106,660	59.56	0.77	0.31
	Gleysol	0	0	1,602,723	18.69	0	0
NDVI	−0.36 to −0.02	1	0.23	687,628	7.93	0.03	0.02
	−0.02–0.08	41	9.56	2,879,828	33.21	0.29	0.15
	0.08–0.20	129	30.07	1,525,597	17.59	1.71	0.90
	0.20–0.30	179	41.72	1,909,771	22.02	1.89	1
	0.30–1.00	79	18.41	1,668,276	19.24	0.96	0.51
Geomorphology	Anthropogenic origin	1	0.23	602	0.01	0.06	0.01
	Denudational origin	12	2.80	369,314	4.26	0.66	0.06
	Fluvial origin	48	11.19	88,741	1.02	10.93	1
	Glacial origin	3	0.70	3,012,448	34.76	0.02	0.00
	Structural origin	345	80.42	5,078,462	58.60	1.37	0.13
	Water bodies	20	4.66	116,536	1.34	3.47	0.32
Geology	ALMORA Gp	9	2.10	248,131	2.86	0.73	0.23

(continued)

Table 11.3 (continued)

Factors	Classes	No. of landslide (NL_i)	No. of landslides in percentage	No. of cells in classes (NC_i)	Percentage of cells area in classes	FR (FR_i)	Normalization (FR_n)
	BADRINATH GRANITE/ TOURMALINE GRANITE	0	0	555,411	6.41	0	0
	BAIJNATH Fm	110	25.64	858,493	9.91	2.59	0.82
	BASIC INTRUSIVE (ALMORA Gp.)	2	0.47	23,775	0.27	1.70	0.54
	CENTRAL CRYSTALLINE	32	7.46	3,954,930	45.65	0.80	0.25
	GARHWAL Gp. (PITHORAGARH = SHYALANA/LAMERI)	94	21.91	808,258	9.33	2.35	0.75
	GARHWAL Gp. (BERINAG = CHAMOLI = NAGNITHANK Fm.)	182	42.42	1,167,184	13.47	3.15	1
	JAUNSAR Gp	0	0	13,924	0.16	0	0
	MARTOLI/MARTOLI (UNDIFF) Gp	0	0	182,432	2.11	0	0
	RAKCHAM GRANITE/MANDI	0	0	25,627	0.30	0	0
	RAWALI BAGAR Gp. (KIOTO Fm.)	0	0	117,107	1.35	0	0
	SUMNA Gp. (SHIALA (UNDIFF) Fm.)	0	0	628,586	7.26	0	0
	Unmapped area	0	0	79,096	0.91	0	0
LULC	Waterbodies	4	0.93	23,041	0.27	3.51	1
	Forestland	182	42.42	2,328,791	26.86	1.58	0.45
	Cropland	3	0.70	37,205	0.43	1.63	0.46
	Settlement	0	0	17,680	0.20	0	0
	Bareland	58	13.52	1,323,457	15.26	0.89	0.25

(continued)

Table 11.3 (continued)

Factors	Classes	No. of landslide (NL_i)	No. of landslides in percentage	No. of cells in classes (NC_i)	Percentage of cells area in classes	FR (FR_i)	Normalization (FR_n)
	Snowland	0	0	1,869,308	21.56	0	0
	Rangeland	182	42.42	3,071,981	35.43	1.20	0.34
Distance to rivers (M)	<300	107	24.94	400,327	4.62	5.40	1
	300–600	64	14.92	376,765	4.34	3.43	0.64
	600–900	57	13.29	374,551	4.32	3.08	0.57
	900–1200	29	6.76	372,417	4.29	1.57	0.29
	>1200	172	40.09	7,147,430	82.426	0.49	0.09
Distance to roads (M)	<300	70	16.32	293,510	3.38	4.82	1
	300–600	46	10.72	273,072	3.15	3.40	0.71
	600–900	37	8.62	268,013	3.10	2.79	0.58
	900–1200	18	4.20	263,085	3.03	1.38	0.29
	>1200	258	60.14	7,573,782	87.34	0.69	0.143
Distance to faults (M)	<1200	23	5.36	810,335	9.34	0.57	0.52
	1200–2400	26	6.06	619,505	7.14	0.85	0.77
	2400–3600	21	4.90	500,481	5.77	0.85	0.77
	3600–4800	22	5.13	401,849	4.63	1.11	1
	>4800	337	78.55	6,339,320	73.11	1.07	0.52

et al. 2015a, b) is used to construct a landslide susceptibility map of the study area using normalized FR values (Table 11.3).

$$\mathrm{LSI} = \sum_{i=1}^{16} \mathrm{FRn} \tag{11.7}$$

where FRn is the normalized frequency ratio values of seventeen parameters that have been selected for LSI. The classification of the landslide susceptibility indexes (LSI) in this study has been done using the natural break method. Landslide susceptibility indexes are divided into five classes using this approach. The classes are Very Low (4.99–8.05), Low (8.05–9.33), Moderate (9.33–10.7), High (10.7–12.1), and Very High (12.1–16.6) (Fig. 11.7).

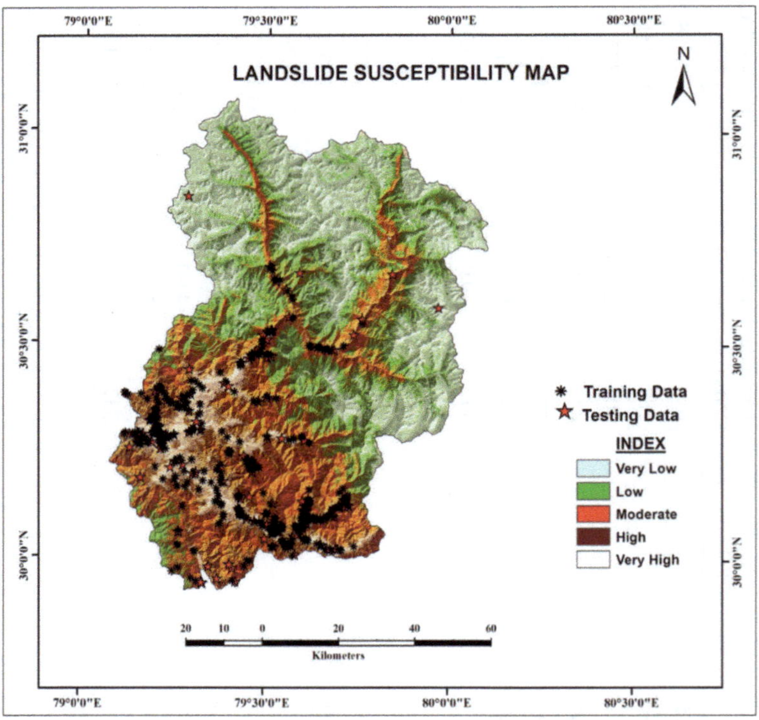

Fig. 11.7 Landslide susceptibility map of Chamoli district, Uttarakhand State

11.4 Result and Discussion

11.4.1 LSZ in the Study Region

In this study, LSI is divided into five categories using ArcGIS software version 10.5, as described in the previous part. Looking at the classes, we discovered five classes with different proportions of regions with low to high amounts of landslide events. To the southwest of this district is the very high zone, as this region accounts for 55.94% of landslide events. Although only 7.92% of the region's total landmass lies in this very high zone, this area is extremely sensitive to landslides. With regard to the probability of landslides, the high zone is also considered a danger zone. This region has the second-highest landslide inventory points after the Very High Zone, which collectively accounted for approximately 34% of all historically visible landslides in this area. This zone occupies only 15.78% of the total land area. The main rivers of the study region are severely affected by regular landslide events in the south and south-west, in the middle and in the east. If we look at the research map, the probable landslide zone almost entirely covers the southern and southwestern regions. Approximately 89.74% of the total landslide events in this region were observed in a region only 1829.42 km^2 (23.7%). The number of landslide events has decreased significantly, with only about 9.32% of all documented landslide events occurring in the moderately affected landslide zone, which has a total area of 1630.81 km^2, larger than the very high and high landslide susceptibility zones. In the central and northern regions of this district, there are zones of low and very low susceptibility to landslides. Since both zones occupied 4258.34 km^2 of this district, <1% of all landslide events occurred there, accounting for more than half (55.17%) of the total landmass of the study area. Many previously mentioned parameters, lithological, hydrological, topo-geomorphological, man-made, and initiating parameters (earthquakes, rainfall), are responsible for the uneven distribution of landslide events in this study region (Fig. 11.8).

Southwest areas in this district are very prone to landslides. Most of them can be seen along the major rivers of the district. Compared to other sections of this district, this region is not significantly above mean sea level, with an elevation of <2040 m. The 1719.53 km^2 of this area contains around 88.11% of all landslide sites. Looking at the second class, which includes altitudes between 2040 and 3072 m, only 11.89% of the total landslide area (2029.86 km^2) occurred there. Since slope gradient is related to landslides, the altitude of a location above mean sea level is not the only factor that triggers a landslide; The incline or incline of a surface is also important. Compared to other slope groups where landslides occur less frequently, the 2040 slope group accounts for over 77.40% of all landslides and covers 74% of the research area or 5740.73 km^2. Only 22.38% of the landslides occur on slopes between 40 and 60°. It can thus be seen that the majority of the areas in the study area have slopes between 20° (low angle) and 60° (medium angle). The unevenness of the surface roughness and the surface height values are closely related. The research area generally has a moderate roughness ranging from 0.42 to 0.58. Nearly 80% of

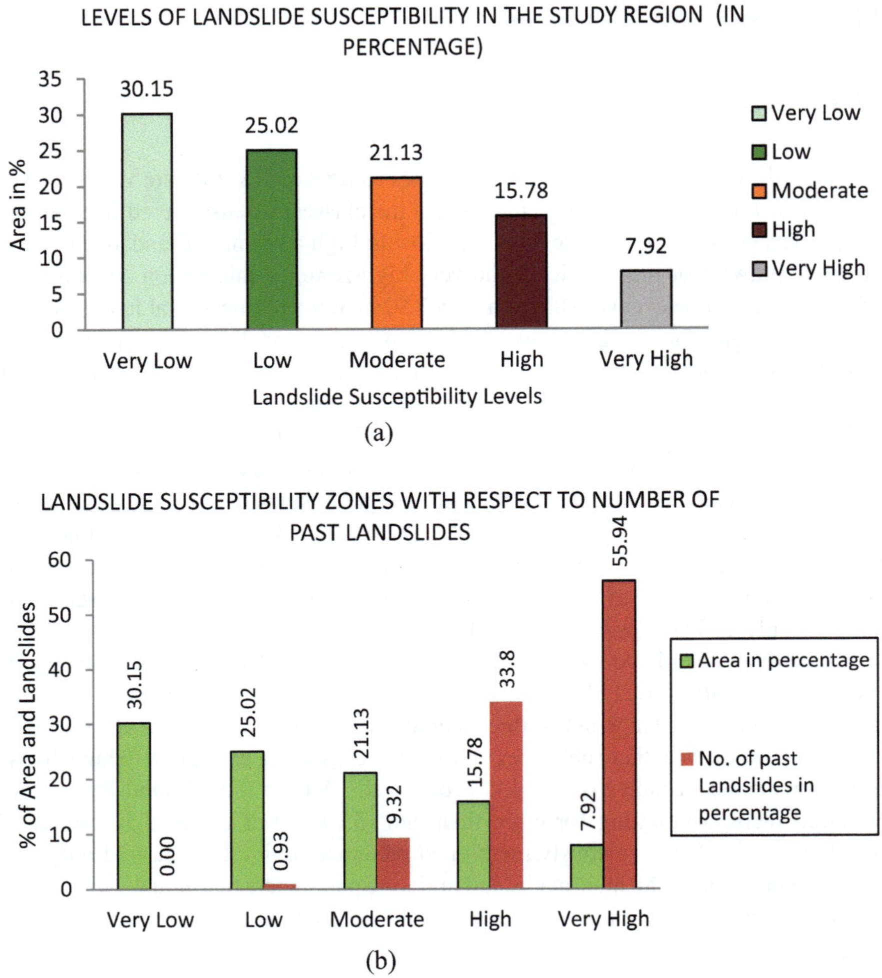

Fig. 11.8 **a** Area of different landslide susceptibility zones, and **b** Percentage of past landslide events in different LSZ

the total number of landslide sites were found in the study area shown on the map, however the bulk is concentrated in the south and southwest of the district. According to the calculated values for the different classes, the number of landslides decreased as the TWI value increased, with the exception of the second class. A significant number of landslides (33.80%) were identified in the first class, while about 46% of all landslides are found in the second class, which covered an area of 3185.36 km^2 of the district. The erosive power of flowing water is quantified by the SPI. Negative values are shown for topographically prone areas for deposits and positive values for potentially erosive areas. The lowest values reflect relatively flat sites, which affect susceptibility to sludge deposition and accumulation, while the highest values are associated with steep slope gradients. In the research area, SPI classes ranging from medium to high show positive values in the 5,377.24 km^2 zone where nearly 74% of the landslides occurred. Therefore, most of the land (about 69%) is potentially erosive in character. As a triggering factor, precipitation changes surface and groundwater dynamics, which reduces slope stability and leads to landslides. Where the average rainfall is moderate to high, several landslides are caused by rainfall each year. The calculation table shows that rainfall in this area triggers landslides. About 52.91 percent of the landslides occur in this location, where the average annual rainfall ranges from 1200 to 1800 mm. These landslides affect an area of 2021.74 km^2, but most of them are in the western part. About 35% of the total occurs in just 10.95% of the region, where average annual rainfall is high enough to accelerate the occurrence of landslides. Rainfall totals of more than 1800 mm but <2400 mm are recorded here and there is a strong suspicion of landslides in this region. Almost half of all landslides are recorded in the most significant earthquake zone, which ranges from magnitude 4.26 to magnitude 5.89. However, zone three, which lies between magnitude 2.89 and magnitude 4.26, is also significant as 42.19% of landslides are reported here. The three main soil types in the study area are gleysol (18.48%), lithosol (58.89%) and dystriancambisol (21.49%). Dystric cambisol soils are the most prone to landslides, accounting for more than 50% of all recorded landslides.It is often observed that the susceptibility of hills to landslides increases significantly when forest or scrub cover is removed and replaced with grass or crops. In this area, landslides of varying magnitude are observed on bare and wet areas with no vegetation (−0.36–0.08), areas with very little to little vegetation cover (0.08–0.3), and forests (0.3–1) to observe. According to the NDVI calculation table, we found that 72% of all landslides occurred in areas with low to extremely low vegetation cover, which is not enough to avoid landslides. This area consists of a variety of landforms, including anthropogenic, denuded, fluvial, glacial, structural, and water bodies, depending on geomorphological origin. The largest landmass by area is the landform of structural origin. Here are the records for about 81% of all landslides, but the landform of fluvial origin has the highest probability of landslides since only 1% of the area occupied 11% of all landslides. The lithological and geological formation also contributes significantly to the occurrence of landslides in this region. A large number of landslides are caused by the Baijnath and Garhwal formations; these two locations account for 68.08% of all landslides in this area. Landslides can happen more often in the Garhwal Formation. Forestland and rangeland, which make up about 4860.69 km^2 of this region, account for around 85% of all documented landslides. The proportion of LSD within a 300 m radius of rivers is around 24.94%. This percentage is greater than the percentage of the area within a 300 m radius of rivers, which is 4.62%. Landslides are a serious threat in this area. The second potential zone is located between 300 and 600 m from rivers. Distance to roads is also a significant factor because road construction and different types of developmental work related to roads can destroy the stability of slopes. Maintaining the stability of road cuttings and road embankments is crucial where roads and highways cross steep terrain. Additionally, it is necessary to redirect the flow of water from slopes so that the road does not hinder the water's path and that the groundwater table does not descend too far below the road's surface. In the study area, we have created five zones to find out which zone is more likelihood of landslides. To determine which section of the research area has a higher chance of landslides, we have divided it into five zones. Due to the fact that 16.32% of all landslide incidents have been documented within an area of 264.159 km^2 (3.38%), the first zone, which extends 300 m from roads, is the most vulnerable to landslides. In this area, landslides are quite likely to occur. Due to its high likelihood of landslides, the second zone, which extends from 300 to 600 m, is also crucial. Approximately 11% of all landslides were contained in just 3.15% of this zone. A significant component influencing the geographical distribution of landslides is faulting. A key element is how far a location is from the faultline. Locations in the research region that are between 3600 and 4800 m from the faultlines are susceptible to slides. Landslides can occur more frequently in the Garhwal Formation. Forest and rangeland, which make up about 4860.69 km^2 of this region, are responsible for about 85% of all documented landslides. The proportion of LSD within a radius of 300 m from rivers is around 24.94%. This proportion is larger than the proportion of land within 300 m of rivers, which is 4.62%. Landslides are a serious threat in this area. The second potential zone is between 300 and 600 m from rivers. Distance to roads is also an important factor, as road works and various road-related works can destroy the stability of slopes. Maintaining the stability of road cuts and road embankments is critical when roads and highways traverse steep terrain. In addition, it is necessary to divert the flow of water from embankments so that the road does not obstruct the path of water, and the water table does not drop too far below the road surface. In the study area, we created five zones to find out which zone has a higher probability of landslides. To determine which section of the research area has a higher risk of landslides, we divided it into five zones. Since 16.32% of all landslide events were documented over an area of 264.159 km^2 (3.38%), the first zone, extending 300 m from roads, is the most prone to landslides. In this area, the risk of landslides is very high. Due to the high probability of landslides, the second zone, which extends from 300 to 600 m, is also crucial. Approximately 11% of all landslides were contained in just 3.15% of this zone. A major component affecting the geographic distribution of landslides is faulting. A key element is how far a site is from the fault line. Sites in the research area that are between 3600 and 4800 m from the fault lines are prone to landslides.

11.5 Accuracy Assessment of the Landslide Susceptibility Map (LSM)

A very useful approach to assess the accuracy of the LSZ mapping produced by the FR model and the landslide inventory point is the Receiver Operating Characteristic (ROC) curve. In this analysis, the total number of landslide inventory points (616) was randomly split between two types of data sets: training data, which made up 75% (429), and test data, which made up 25% (187). In this study, the FR model is used to calculate the probability of landslide susceptibility. Area under the curve (AUC) is used to assess the performance of this model. The percentage of consecutive incidence and predictive incidence of the method is measured using this curve. The compatibility of the susceptibility models was evaluated using the training data set and their prediction probability was then also evaluated using the test data set. The AUC value ranges from 0.5 to 1. A higher value indicates the model is working great, while a lower value indicates the model is not working well. The AUC values show consecutive rates and validation rates of approximately 91.5 and 87.2%, respectively. This is shown in Fig. 11.9. The AUC curve values for both incidences are above 0.8, showing the good performance of the model (Fig. 11.9).

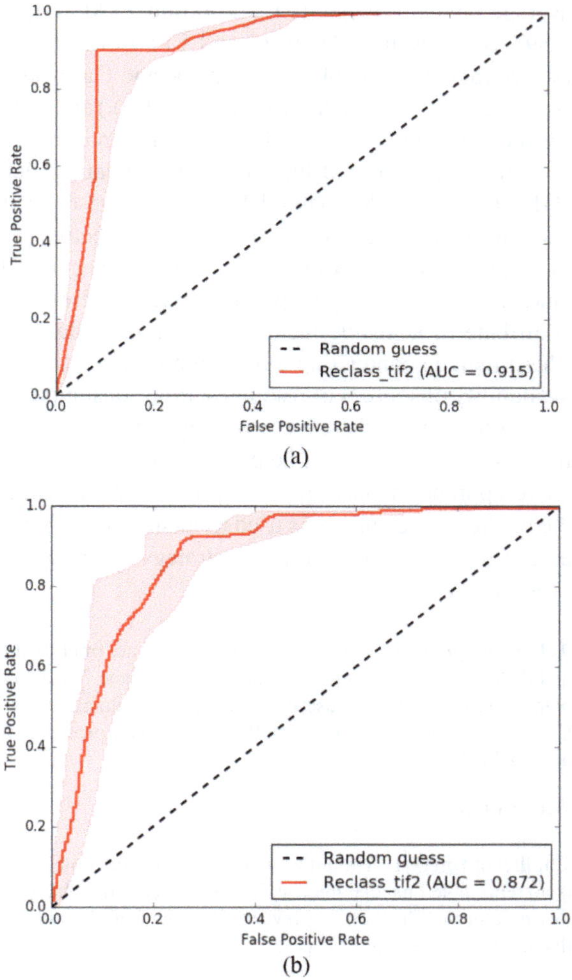

Fig. 11.9 The AUC analysis: **a** Performance of success rate curve, and **b** Performance of predicted curve

11.6 Conclusion

Using ArcGIS software version 10.5, the landslide susceptibility zoning (LSZ) map for the research region is generated by considering 16 conditioning factors and applying the frequency ratio approach. There are five classifications for the LSZ map: very low, low, medium, high, and very high. According to the LSZ map, there are extremely high landslide-prone regions in the southern, southeastern, and southwestern directions of Chamoli district, accounting for about 611.52 km^2 or about 56% of all landslides. Due to the high risk of landslides in this area, special care must be taken to minimize or reduce the risk of landslides. The combined effects of altitude (<2040 m), slope (20–40), TRI (0.42–0.58), TWI (<6.39), SPI (0.78–1.14), average annual precipitation (1200 mm-2400 mm), and seismic magnitude (2.89–5.89), soil (Dystric Cambisol and Lithosol), NDVI (0.08–0.30), geomorphology (structural origin), geology (Baijnath- and Garhwal formation), LULC, distance to rivers (<1200 m), distance to roads (<300 m).) and distance to faults (3600 m–4800 m) are the main contributors to landslide susceptibility in this study area.

A thorough study of the numerous regulating elements is necessary since landslides are a complex natural hazard that arises from a variety of endogenic and exogenic sources. Therefore, a number of effective landslide mitigation and management programs are required to lower the danger of landslides, and in this regard, landslide susceptibility mapping is essential. Because there are many variables that contribute to landslide incidents, some procedures need to be carefully followed. These include conducting a thorough investigation, gathering the appropriate data, availability of sources of resources, time consumption, and accessibility of the study area. Additionally, there are two categories into which different methods of assessing the risk of landslides can be placed: qualitative and quantitative. While the quantitative approach is used to determine hazard zonation and landslide susceptibility, the qualitative technique is used to identify possible landslide zones. Therefore, it is essential to select an appropriate strategy for the given situation, as each method has pros and cons.

Acknowledgements All data used in this study are freely available. We would like to thank USGS and ISRO for the various datasets used in this article. We also thank Mr. Chiranjib Sarkar for their help in preparing this manuscript. Finally, special thanks to the Department of Geography, Ananda Chandra College, Jalpaiguri, West Bengal, India for providing the data needed for the study, as well as valuable help and great support.

Declarations

Conflict of Interest The authors have reviewed the content of the manuscript and agree to it. There is no reportable financial interest. The author confirms that the submission is an original work and that no other publication is a review. On behalf of all authors, the corresponding author declares that no conflict of interest exists.

Availability of Data and Material On reasonable request,, the corresponding author will provide the data sets generated as part of the ongoing investigation.

References

Anbalagan R, Kumar R, Lakshmanan K, Parida S, Neethu S (2015) Landslide hazard zonation mapping using frequency ratio and fuzzy logic approach. A case study of Lachung Valley, Sikkim. https://doi.org/10.1186/s40677-014-0009-y

Armas I (2012) Weights of evidence method for landslide susceptibility mapping. Prahova Subcarpathians, Romania. Nat Hazards 60:937–950

Baharin A, Farshid S, Javad M, Barat M (2014) Using frequency ratio method for spatial landslide prediction. Res J App Sci Eng Technol. 7:3174–3180

Bai SB, Wang J, Lü GN, Zhou PG, Hou SS, Xu SN (2010) GIS-based logistic regression for landslide susceptibility mapping of the Zhongxian segment in the Three Gorges area, China. Geomorphology 115:23–31

Berhane G, Kebede M, Alfarrah N (2021) Landslide susceptibility mapping and rock slope stability assessment using frequency ratio and kinematic analysis in the mountains of Mgulat area, Northern Ethiopia. Bull Eng Geol Environ 80(1):285–301. https://doi.org/10.1007/s10064-020-01905-9

Bhatt O (1998) Living in the shadows of death: Landslides in Uttrakhand. He calamity-prone Central Himalayas, Lok Soochana Evam Sahayata Kendra, Gopeshwar, Chamoli, Uttarakhand, India, pp 1–4

Bhatt GD, Sinha K, Deka PK, Kumar A (2014) Flood hazard and risk assessment in chamoli district, uttarakhand using satellite remote sensing and GIS techniques. Int J Innov Res Sci Eng Technol 3(8), August 2014

Bisht MPS, Mehta M, Nautiyal SK (2002) A report on geomorphic hazards around Badrinath Area (Uttaranchal) and innovative control measures proposed, NDBR project. Department of Geology, HNB Garhwal University, Uttarakhand

Bisht MPS, Mehta M, Nautiyal SK (2011) Impact of depleting glaciers on the Himalayan biosphere reserve—a case study of Nanda Devi Biosphere Reserve, Uttarakhand Himalaya. In: Bisht MPS, Pal D (eds) Mountain resource management: application of remote sensing and GIS. Transmedia Publication, Srinagar, Uttarakhand, pp 17–31

Brabb EE (1984) Innovative approaches to landslide hazard mapping. In: Proceedings 4th international symposium on landslides, vol 1, Toronto, pp 307–324

Bui DT, Pradhan B, Lorfman O, Revhaug I, Dick OB (2012) Landslide susceptibility assessment in the HoaBinh province of Vietnam: a comparison of the Levenberg–Marquardt and Bayesian regularized neural networks. Geomorphology 171:12–29

Chansarkar RA (1975) Geological and geomorphic factors in landslide investigations. In: Proceedings of seminar on landslides and toe erosion problems with special reference to Himalayan region, Gangtok, pp 54–66

Chen Z, Wang J (2007) Landslide hazard mapping using a logistic regression model in Mackenzie Valley, Canada. Nat Hazard 42(1):75–89

Chen W Panahi M, Pourghasemi HR (2017) Performance evaluation of GISbased new ensemble data mining techniques of adaptive neuro-fuzzy inference system (ANFIS) with genetic algorithm (GA), differential evolution (DE), and particle swarm optimization (PSO) for landslide spatial modeling. CATENA 157:310–324

Chen L, Guo Z, Yin K, Shrestha DP, Jin S (2019) The influence of land use and land cover change on landslide susceptibility: a case study in Zhushan Town, Xuan'en County (Hubei, China). Nat Hazard 19:2207–2228

Chen W, Li W, Chai H, Hou E, Li X, Ding X (2016) GIS-based landslide susceptibility mapping using analytical hierarchy process (AHP) and certainty factor (CF) models for the Baozhong Region of Baoji City, China. Environ Earth Sci 75:63. https://doi.org/10.1007/s12665-015-4795-7

Chen Z, Ye F, Fu W, Ke Y, Hong H (2020) The influence of DEM spatial resolution on landslide susceptibility mapping in the Baxie River basin, NW China. Nat Hazards 101(3):853–877. https://doi.org/10.1007/s11069-020-03899-9

Choi J, Oh HJ, Lee HJ, Lee C, Lee S (2011) Combining landslide susceptibility maps obtained from frequency ratio, logistic regression, and artificial neural network models using ASTER images and GIS. Eng Geol 124:12–23

Das I, Stein A, Kerle N, Dadhwal VK (2012) Landslide susceptibility mapping along road corridors in the Indian Himalayas using Bayesian logistic regression models. Geomorphology 179:116–125

Dehnavi A, Aghdam IN, Pradhan B, Varzandeh MHM (2015) A new hybrid model using step-wise weight assessment ratio analysis (SWARA) technique and adaptive neuro-fuzzy ınference system (ANFIS) for regional landslide hazard assessment in Iran, Catena 135:122–148. https://doi.org/10.1016/j.catena.2015.07.020

Digital Terrain Analysis Manual: Zumbro River Watershed Restoration Prioritization (2014). [Minnesota] :: Zumbro Watershed Partnership, Print. https://mnpals-lrl.primo.exlibrisgroup.com/discovery/openurl?institution=01MNPALS_LRL&vid=01MNPALS_LRL:LRL&rft.oclcnum=957518973. Accessed 16 June 2023

Ehret D, Rohn J, Dumperth C, Eckstein S, Ernstberger S, Otte K, Rudolph R, Wiedenmann J (2010) Frequency ratio analysis of mass movements in the Xiangxi Catchment, Three Gorges Reservoir Area, China. J Earth Sci 21:824–834

Fayez L, Pham BT, Solanki HA, Pazhman D, Dholakia MB, Khalid M, Prakash I (2018) Application of frequency ratio model for the development of landslide susceptibility mapping at part of Uttarakhand state, India. Int J Appl Eng Res 13:6846–6854

Gandhi GM, Parthiban S, Thummalu N, Christy A (2015) Ndvi: vegetation change detection using remote sensing and Gis—a case study of Vellore District. Procedia Comput Sci 57:1199–1210. https://doi.org/10.1016/j.procs.2015.07.415

Glade T, Anderson M, Crozier MJ (2005) Landslide hazard and risk. Wiley. https://doi.org/10.1002/9780470012659

Gulia KS (2007) Discovering Himalaya: tourism of Himalayan region. ISHA Books, Delhi

Gupta N, Pal SK, das J (2022) GIS-based evolution and comparisons of landslide susceptibility mapping of the East Sikkim Himalaya. Spatial Sci 3:359–384. https://doi.org/10.1080/19475683.2022.2040587

Guzzetti F, Cardinali M, Reichenbach P, Carrara A (2000) Comparing landslide maps: a case study in the Upper Tiber River Basin, Central Italy. Environ Manag 25(3):247–363. https://doi.org/10.1007/s002679910020

Guzzetti F, Mondini AC, Cardinali M, Fiorucci F, Santangelo M, Chang KT (2012) Landslide inventory maps: new tools for an old problem. Earth Sci Rev 112(1–2):42–66. https://doi.org/10.1016/j.earscirev.2012.02.001

Hong H, Xu C, Bui DT (2015) Landslide susceptibility assessment at the Xiushui area (China) using frequency ratio model. Procedia Earth Planet Sci 15:513–517

Ilanloo M (2011) A comparative study of fuzzy logic approach for landslide susceptibility mapping using GIS: an experience of Karaj dam basin in Iran. Procedia Soc Behav Sci 19:668–676. https://doi.org/10.1016/j.sbspro.2011.05.184

Joshi V, Maikhuri RK (1997) Cloudburst: A natural calamity-A case study from Garhwal Himalaya. Up J Indian Build Congr 4:207–217

Juyal N, Pant RK, Bhatt OP (1996) He calamity-prone Central Himalayas. Report of Seminar on Natural Calamities, Dasholi Gram Swaraj Mandal (DGSM), Gopeshwar, Chamoli, Uttarakhand, India

Kanungo DP, Sharma S (2013) Rainfall thresholds for prediction of shallow landslides around Chamoli Joshimath region, Garhwal Himalayas, India. https://doi.org/10.1007/s10346-013-0438-9

Kazeev A, Postoev G (2017) Landslide investigations in Russia and the former USSR. Nat. Hazards 1–21

Khan H, Shafique M, Khan M, Bacha M, Shah S, Calligaris C (2019) Landslide susceptibility assessment using Frequency Ratio, a case study of northern Pakistan. Egyptian J Remote Sens Space Sci 22(1):11–24. https://doi.org/10.1016/j.ejrs.2018.03.004

Krishnan MS (1982) Geology of India and Burma, 6th edn. CBS Publishers, Delhi

Li B, wang N, Chen J (2021) GIS-based landslide susceptibility mapping using information, frequency ratio, and artificial neural network methods in qinghai province, northwestern China. Adv Civ Eng 14–4758062. https://doi.org/10.1155/2021/4758062

Mao W, Wu L, Singh RP, Qi Y, Xie B, Liu Y, Ding Y, Zhou Z, Li J (2022) Progressive destabilization and triggering mechanism analysis using multiple data for Chamoli rockslide of 7 February 2021. Geomat Nat Haz Risk 13(1):35–53. https://doi.org/10.1080/19475705.2021.2013960

Martha TR, Roy P, Govindharaj KB, Kumar KV, Diwakar PG, Dadhwal VK (2015) Landslides triggered by the June 2013 extreme rainfall event in parts of Uttarakhand state, India. Landslides 12:135–146. https://doi.org/10.1007/s10346-014-0540-7

Martha TR, Roy P, Jain N, Khanna K, Mrinalni K, Kumar KV, Rao PVN (2021) Geospatial landslide inventory of India—an insight into occurrence and exposure on a national scale. Landslides. https://doi.org/10.1007/s10346-021-01645-1

Mersha T, Meten M (2020) GIS-based landslide susceptibility mapping and assessment using bivariate statistical methods in Simada area, northwestern Ethiopia. Geoenviron Disasters 7:20. https://doi.org/10.1186/s40677-020-00155-x

Meten M, Bhandary NP, Yatabe R (2015) GIS-based frequency ratio and logistic regression modeling for landslide susceptibility mapping of Debre Sina area in Central Ethiopia. J Mt Sci 12(6). https://doi.org/10.1007/s11629-015-3464-3

Mohammady M, Pourghasemi HR, Pradhan B (2012) Landslide susceptibility mapping at Golestan Province, Iran: A comparison between frequency ratio, Dempster-Shafer, and weights-of-evidence models. J Asian Earth Sci 61:221–236. https://doi.org/10.1016/j.jseaes.2012.10.005

Mukherjee I, Singh UK (2020) Delineation of groundwater potential zones in a drought-prone semi-arid region of east India using GIS and analytical hierarchical process techniques. CATENA 194:104681. https://doi.org/10.1016/j.catena.2020.104681

Nandi A, Shakoor A (2010) A GIS-based landslide susceptibility evaluation using bivariate and multivariate statistical analyses. Eng Geol 110:11–20

NDMA (2009) National Disaster Management Guidelines—Management of landslides and snow avalanches. Publication of National Disaster Management Authority, Government of India. New Delhi

Newmark NM (1965) Effects of earthquakes on dams and embankments. Géotechnique 15(2):139–160. https://doi.org/10.1680/geot.1965.15.2.139

NIDM (2015) Uttarakhnd Disaster 2013. National Institute of Disaster Management Ministry of Home affairs, Government of India, New Delhi, India

Okhimath landslides in 2012—satellite based study—Bhuvan—Google Search (NRSC)

Onions CT (1993) The Oxford English dictionary. Oxford University Press, Oxford, p P31

Pal SC, Chowdhuri I (2019) GIS-based spatial prediction of landslide susceptibility using frequency ratio model of Lachung River basin, North Sikkim, India. SN Appl Sci 1(5):416

Pandey VK, Mishra A (2015) Causes and disaster risk reduction measures for hydrometeorological disaster in Uttarakhand, India: an overview. Int J Curr Res Sci Technol 1(3):61–80

Pandey VK, Mishra A (2018) Trends of Hydro-Meteorological Disaster in Uttarakhand, India. Int J Curr Res Sci Technol 12(4):1–7

Pandey Y, Dharmaraju R, Chauhan PKS (2002) Estimation of Source parameters of Chamoli Earthquake. In: Nainwal HC, Prasad C (eds) Geodynamics & environmental management of Himalaya, pp 99–105

Park S, Choi C, Kim B, Kim J (2012) Landslide susceptibility mapping using frequency ratio, analytic hierarchy process, logistic regression, and artificial neural network methods at the Inje area, Korea. Environ Earth Sci 68:1443–1464

Parkash S (2014) Some socio-economically significant landslides in Uttarakhand Himalaya: events, consequences and lessons learnt. Disaster Risk Reduction 211–232. https://doi.org/10.1007/978-4-431-55242-0_12

Patil AS, Panhalkar SS, Shinde D (2022) Frequency Ratio approach for landslide susceptibility mapping of Phonda Ghat of Maharashtra. In: Remote sensing and geographic information system for policy decision support, pp 17–43

Pham BT, Tien Bui D, Prakash I, Dholakia M (2015a) Landslide susceptibility assessment at a part of Uttarakhand Himalaya, India using GIS–based statistical approach frequency ratio method. Int J Eng Res Technol 4:338–344

Pham BT, Tien Bui D, Prakash I, Dholakia M (2015b) Landslide susceptibility assessment at a part of Uttarakhand Himalaya, India using GIS-based statistical approach of frequency ratio method. Int J Eng Res Technol 4:338–344

Pirasteh S, Li J (2017) Probabilistic frequency ratio (PFR) model for quality improvement of landslide susceptibility mapping from LiDAR-derived DEMs. Pirasteh Li Geoenviron Dis 4:19. https://doi.org/10.1186/s40677-017-0083-z

Popescu ME (2002) Landslide causal factors and landslide remediatial options. Accessed 16 June 2023

Poudyal CP, Chang C, Oh HJ (2010) Lee S (2010) Landslide susceptibility maps comparing frequency ratio and artificial neural networks: a case study from the Nepal Himalaya. Environ Earth Sci 61:1049–1064

Pradhan B (2010) Landslide Susceptibility mapping of a catchment area using frequency ratio, fuzzy logic and multivariate logistic regression approaches. J Indian Soc Remote Sens 38:301–320

Pradhan B, Oh HJ, Buchroithner M (2010) Weights-of-evidence model applied to landslide susceptibility mapping in a tropical hilly area. Geomat Nat Hazards Risk 1:199–223

Prakesh S (2014) Geo-environmental characteristics, natural resources and disasters in Uttarakhand State. In: Retrospect and prospects of natural resource and disaster management in Uttarakhand Himalaya, pp 51–99

Prakesh S (2015) Some socio-economically landslides in Uttarakhand Himalaya: events, consequences and lessons learnt. In: Nibanupudi HK, Shaw R (eds) Mountain hazards and disaster risk reduction. Springer, pp 1–271

Raina BN, Hukku BM, Rao RVC (1980) Geological features of the Himalayan region with special reference to their impact on environmental appreciation and environmental management. In: Proceedings of national seminar on resource development and environment in the Himalayan region, New Delhi, India

Rautela P, Pande RK (2005) Traditional inputs in disaster management: the case of Amparav, North India. Int J Environ Stud 62(5):505–515. Schuster, R.L.

Roy J, Saha S (2019) Landslide susceptibility mapping using knowledge driven statistical models in Darjeeling District, West Bengal, India. Geoenviron Disasters 6(1). https://doi.org/10.1186/s40677-019-0126-8

Sarkar S, Kanungo DP (2004) An integrated approach for landslide susceptibility mapping using remote sensing and GIS. Photogramm Eng Remote Sens 70:617–625

Singh A, Pal S, Kanungo DP, Pareek N (2017) An overview of recent developments in landslide vulnerability assessmentpresentation of a new conceptual framework. In: Workshop on World Landslide Forum. Springer, pp 795–802

Smith LRB (1843) Memoir of Indian earthquakes. J Asiatic Soc Bengal 12:1029–1059

Sonker I, Tripathi JN, Singh AK (2021) Landslide susceptibility zonation using geospatial technique and analytical hierarchy process in Sikkim Himalaya. Quat Sci Adv 4:100039. https://doi.org/10.1016/j.qsa.2021.100039

Sui H, Su T, Hu R, Wang D, Zheng Z (2022) Study on the risk assessment method of rainfall landslide. Water 14(22):3678. https://doi.org/10.3390/w14223678

Tucker CJ (1980) Remote sensing of leaf water content in the near infrared. Remote Sens Environ 10(1):23–32. https://doi.org/10.1016/0034-4257(80)90096-6

Van Westen CJ, Rengers N, Terlien MTJ (1997) Prediction of the occurrence of slope instability phenomena through GIS-based hazard zonation. Geol Rundsch 86:4004–4414

Van Westen CJ, van Asch TWJ (2006) Soeters R (2006) Landslide hazard and risk zonation—Why is it still so difficult? Bull Eng Geol Environ 65:167–184

Velayudham J, Kannaujiya S, Sarkar T, Ray CPK, Taloor AK, Singh B, MP, Chawla S, Pal SK (2021) Comprehensive study on evaluation of Kaliasaur landslide attributes in Garhwal Himalaya by the execution of geospatial, geotechnical and geophysical methods. Quat Sci Adv 3:100025, November 2020. https://doi.org/10.1016/j.qsa.2021.100025

Wang LJ, Kazuhide S, Shuji M (2013) Landslide susceptibility analysis with logistic regression model based on FCM sampling strategy. Comput Geosci 57:81–92

Wu Y, Li W, Wang Q, Liu Q, Yang D, Xing M, Pei Y, Yan S (2016) Landslide susceptibility assessment using frequency ratio, statistical index and certainty factor models for the Gangu County, China. Arab J Geosci 9(2):1–16. https://doi.org/10.1007/s12517-015-2112-0

Zhang X, Hu Y, Zhuang D, Oi Y, Ma X (2009) NDVI spatial pattern and its differentiation on the Mongolian Plateau. J Geog Sci 19:403–415

Chapter 12
Landslide Susceptibility Using Weighted Regression Model: A Geo-spatial Approach

R. M. Yuvaraj and Bhagyasree Dolui

Abstract Steep geographical terrain is very vulnerable to the trend of landslip all around the world. Landslides take place regularly and on a yearly basis throughout India's diverse hill and mountain ranges. Tamil Nadu's Nilgiris district is especially susceptible to landslides because of the region's abundant rainfall for the duration of the South and North East monsoons. The focus of this study is to define high-risk areas and pinpoint features that make landslides vulnerable. It is essential to use the landslide hazard zonation maps appropriately and conduct a thorough analysis of each slope that is vulnerable to landslides. Development, the risk of landslides today, and projected future slopes where we can place the early warning system for slope failure based on past landslide locations should all be represented on planning-level maps. Using the weighted regression model, a straightforward statistical method has been used to calculate the proximity of their relationship. Additionally, weighted regression models were useful in validating the chosen causal factors based on their capacity to prevent a landslide episode because they could explain in detail the variance in scores between the causative factors for each class of landslides as well as the distribution of landslides using geographic information systems (GIS). The outcome indicates a relationship between the likelihood of a landslip occurring and the slope, with steeper slopes having higher landslip probability and a slope prediction rate of 23.82. The research area's western and eastern halves are heavily populated with high-susceptibility areas.

Keywords Landslide · Weighted regression model · Nilgiri district · Landslide susceptibility map · GIS

R. M. Yuvaraj (✉)
Department of Geography, University of Madras, Chennai, Tamil Nadu, India
e-mail: yuvaraj1023@gmail.com

B. Dolui
Department of Geography, Queen Mary's College, Chennai, Tamil Nadu, India
e-mail: bshree40@gmail.com

G. K. Panda et al. (eds.), *Landslide: Susceptibility, Risk Assessment and Sustainability*, Advances in Natural and Technological Hazards Research 52,
https://doi.org/10.1007/978-3-031-56591-5_12

12.1 Introduction

The frequent occurrence of landslides, which are common geohazards that cause monetary losses and fatalities (Haque et al. 2016; Papoutsis et al. 2020), is one of the primary causes of terrain modification (Farooq et al. 2015). Between 2008 and 2017, landslides claimed over 3 million lives worldwide, resulting in US economic losses of over $2.7 billion (Chen et al. 2020). Regarding natural disaster classification based on natural processes, landslides rank third among the most hazardous disasters (Tsironi et al. 2022). The region with the highest frequency of landslides is Asia, however there is a wide geographical dispersion of these events (Froude and Petley 2018). The probable of landslip databases for comprehending the global effects of landslip phenomena has been demonstrated by several writers. For instance, Kirschbaum et al. (2010) presented a global analysis of rainfall-related landslides that were documented in three catalogues for the years 2003, 2007, and 2008, with around 230, 400, and 550 reports, correspondingly. These assessments, which identified the Asian continent as the area most impacted by the event of landslides, were compiled using data on landslide dangers found in scholarly papers, newspaper articles, and other sources. Petley (2012) and Froude and Petley (2018) conducted an analysis of a global database of fatal non-seismic landslides that covered the years 2004–2016. The database contained 4862 different landslide incidents. The findings highlighted a recent rise in the frequency of landslides brought on by human activity and revealed that Asia had the largest number of recorded deadly landslides globally. In hilly terrain, landslides pose a far greater geologic risk. They typically occur during or after periods of intense rainfall, which results in the death of humans and other animals as well as the destruction of man-made infrastructure and the natural ecology (Yuvaraj and Dolui 2021). Numerous predisposing factors, including topography, geology, hydrology, and land cover, influence the evolution of landslides. As an outcome, the spatial probability of landslides occurring in different places differs due to the nonlinear coupling impacts of these elements (Conforti and Letto 2020).

Furthermore, other temporal possibilities are linked with the possibility of landslides due to external sources such as earthquakes, poor engineering practices, and heavy rainfall (Huang et al. 2021). Around the world, landslides brought on by rainfall are common and severely hinder social and economic advancement. This work emphasizes the detailed investigation of the landslide hazard caused by regional rainfall, which is explained as forecasting the spatial–temporal probability of future landslides conditionally influenced by triggering and predisposing factors.

Different natural hazard categories may be paired in many areas, and their combined effects may intensify considerably (Marjanović 2013). Under such conditions, determining the impact of each danger becomes exceedingly difficult (Kanwal et al. 2017). Since landslides are unpredictable, it is critical to comprehend their contributing components and map the regions that are vulnerable to landslides in the future. Landslip susceptibility (LS) mapping is done for this purpose, utilizing data from geographic information systems (GIS) and remote sensing. The LS map is separated into various classes based on vulnerability (very low, low, moderate, high, and very high) without considering the occurrence rate (Ozdemir and Altural 2013).

Several techniques for LS mapping have been developed and deployed globally, classed as qualitative as well as quantitative approaches (Guzzetti et al. 1999; Tariq et al. 2023). The qualitative way is a knowledge-driven methodology on the basis of an specialist's understanding. It is a heuristic or somewhat subjective method. According to the researcher's area of expertise, it ranks and weighs the landslide-influencing components to evaluate landslide vulnerability (Shah et al. 2022). Certain qualitative methods are regarded as semi-quantitative since they employ analytical instruments for rating and weighing (Kouli et al. 2010). Simple additive weighting (Feizizadeh and Blaschke 2013), frequency ratio model (Yuvaraj and Dolui 2023), and ordered weighted average (Ayalew et al. 2004) are among the often used subjective techniques. The weighted linear combination (Yuvaraj and Dolui 2021), TOPSIS (Arabameri et al. 2019), analytical network process (Neaupane and Piantanakulchai 2006), and analytical hierarchy process (Alexakis et al. 2014; Dolui et al. 2019). This chapter proposes employing a weighted regression model to map landslip susceptibility. Several investigations have demonstrated that these models produce superior results than other traditional methods (Hong et al. 2017; Li et al. 2020). The most often used local regression format, geographically weighted regression (GWR), demonstrates excellent ability to handle spatially non-stationary correlations (Wheeler and Páez 2009). The model's regression coefficients are calculated for every geographic zone, allowing the relationship between the independent and dependent variables to vary over time (Düzgün and Kemeç 2008). Numerous academic disciplines, including social economics, geography, and meteorology, have used this methodology (Paez 2006; Tu 2011). Spatial proximity is one issue preventing the GWR model from being applied to assessing landslip susceptibility. It is the fundamental input condition and central challenge of the GWR model, and the task of finding a suitable formulation for the spatial propinquity at various points directly affects the model's capacity to simulate. The Nilgiris district, located in the Western Ghats, is among the areas in India with a high risk of landslips. The Nilgiris district served as the study site for numerous landslip hazard investigations (Rajarathnam and Ganapathy 2006; Jaiswal and Westen 2009; Gomathi et al. 2013; BhagyasreeDolui and Geetha 2020; Dolui et al. 2023). The Nilgiri District's land use planning urgently needed a landslip susceptibility assessment, which is why this study was carried out. This study aims to (1) enhance the predictors of landslides for mapping susceptibility and (2) assess the landslide susceptibility of geographically weighted regression (GWR) models.

12.2 Study Area

The Nilgiris district has a extensive history of landslides in Tamil Nadu's Western Ghats. In the district, landslides occur virtually exclusively during certain seasons. Landslides are common in October and November. In the Nilgiris district's history, landslides have occurred in 1902, 1978, 1979, 1993, 2001, 2006, and 2009. Heavy rains in November 1891 damaged the Kotagiri—Metuppalayam road and produced numerous landslips on the Coonoor Ghat (Ganapathy and Rajawat 2015). Across a 250 km^2 area, the district had approximately 100 landslides in 1978 due to exceptional rains. On November 5, 1978, landslides, home collapses, and drownings claimed several lives in Ooty. Around 200 landslides were reported in 1979, with several fatalities and significant property damage. Although landslides are known to happen in highland areas like the Nilgiri, there have never been any reports of such widespread occurrences. There was constant, severe rain from November 12 to November 19, 1979. In Coonoor, there was a tremendous downpour of 102.2 mm; in Selas, there was a significant landslip that entirely buried a house with three children and couple of women. Another "cloud burst" took place at the upper extreme of Coonoor Taluk's Marappalam on November 11, 1993. Over eighteen cottages under the road were damaged by this one, and it washed out Coonoor MTP Ghat Road for almost 11/2 km. The number of landslide-related deaths and property damage in the Nilgiri Hills has recently increased. Intense rainfall in 2009 caused landslides in the Nilgiris regions of Ooty, Coonoor, and Kotagiri. Within 48 h on November 10, 2009, 42 persons passed away. In the village of Acchanakal, close to Ooty, a family of seven perished.

Additionally, the slides shut off access to the Nilgiris via Mettupalayam and destroyed trees. One among the districts in Tamil Nadu, is the Nilgiri district which is generally a hilly region, rising between 1,000 and 2,600 m above sea level in the western regions of south India's Ghats. The study area is 2,552.50 km^2 an overall geographic area. The district lies between 11° 12 min N and 11° 37 min N in latitude and between 76° 30 min E and 76° 55 min E in longitude. The Karnataka districts of Wayanad, Malappuram, Palakkad, and Mysore lie to the west of the Nilgiris, while Coimbatore (Tamil Nadu), Erode (Karnataka), and Chamarajanagar (Karnataka) are to the south and north of the Nilgiris (Fig. 12.1). Of the thirty-two revenue districts in Tamil Nadu, only the Nilgiris district is completely contained inside the Nilgiri Plateau, which is the meeting point of the Western and Eastern Ghats mountain ranges. This district, which is 250 km^2, comprises the governmental regions of Ooty, Kundah, Kotagiri, Coonoor, and Pandalur. Landslides have caused significant damage in the past, and some of the triggering elements have been briefly covered. Landslides in the Nilgiris are a result of human involvement in the form of development projects that have wreaked havoc on the environment. The flood of people from the surrounding areas has resulted in deforestation and habitat damage in the Nilgiris. Environmental damage, depletion, and degradation, as well as the depletion and deterioration of natural resources, have all been exacerbated by intensive falling. Construction operations such as road construction have resulted in landslides and slope destabilization.

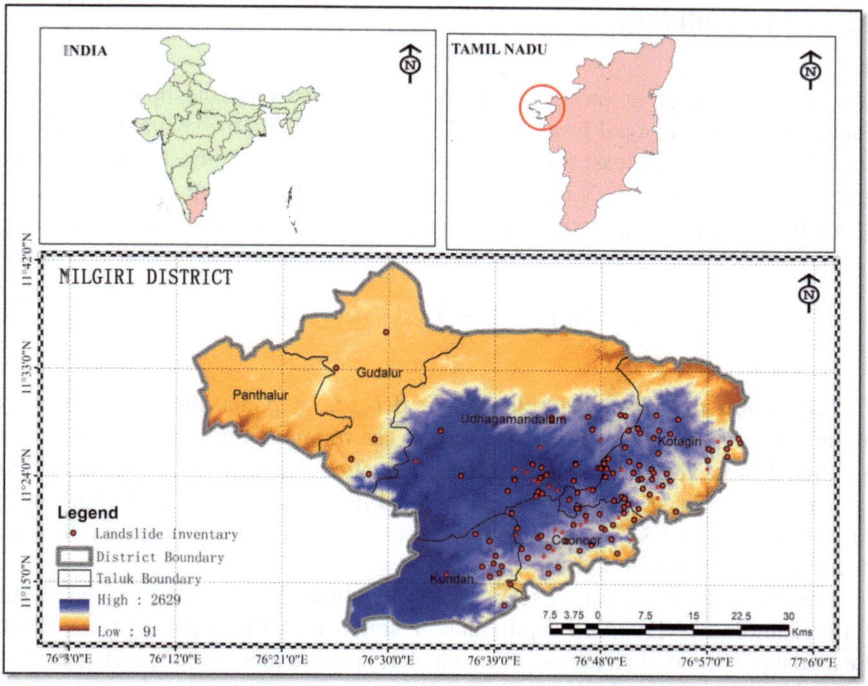

Fig. 12.1 Location of the study area

12.3 Data

The USGS provided the Landsat 8 OLI data satellite image taken in March 2019. The State Ground and Surface Water Board at Taramani is where the rainfall data for each month for the various stations in the Nilgiri district has been gathered. The Niligiri district's land use and land cover have been retrieved from EOSDIS. USGS data with an SRTM 30 m resolution has been acquired. We have collected information on Lithology, soil depth, and soil from the Indian Geological Survey. For risk management and land use planning in landslide-prone locations, assessing landslide hazards is essential. The landslide hazard zonation maps are useful not only for monitoring landslides but also for forecasting future slope collapses in a specific area. Numerous studies assessing the risk of landslides have been conducted for the Nilgiris district since 1979. The Geological Survey of India (GSI), the State Geology branch, and the Tamil Nadu government began conducting in-depth geological investigations of the landslides in August 1979.

12.4 Methodology

The Nilgiris district in India was the first to receive the landslip susceptibility map created on a regional level. The GWR model weights the spatial dependency, extending the ordinary least squares regression (OLS). The established spatial proximity estimates the model's coefficients for each unit; the coefficients' values and symbols vary depending on the unit. Equation (12.1) is the format of this model.

$$y_t = \beta_0(\mu_t, v_t) + \sum_{j=1}^{p} \beta_{tj}(\mu_t, v_t)x_{tj} + \epsilon_t \tag{12.1}$$

The number of independent elements is denoted by p, and the coordinates of a tth unit in space are represented by (μ_t, v_t). The tth unit's jth independent variable is denoted as x_{tj}. The intercept parameter at location t is $\beta_0(\mu_t, v_t)$, and the random error is ϵ_t. The local regression coefficient, $\beta_{tj}(\mu_t, v_t)x_{tj}$, for the jth explanatory variable at position t is a crucial parameter for the spatial non-stationary model since it changes as the spatial position does.

Thus, ten influencing parameters (i.e., aspect, slope, lineament density, lineament buffer, Lithology, NDVI, soil depth, rainfall, soil, and LULC) were used to estimate the landslip of the studied region. A weighted regression model was used to calculate the weight of every assessment factor. IDW interpolation in ArcGIS software has been used to create the yearly average rainfall spatial distribution. From satellite imagery, the Normalized Difference Vegetation Index (NDVI) can be obtained by applying the following association: $NDVI = NIR - R/NIR + R$. On the other hand, in a satellite image, R stands for the red band and NIR for the near infra-red band. PCI Geomatics and ArcGIS software have been used to restore lineament from a Landsat image. The lineament density has been prepared using the density tools in ArcGIS programme. Using ArcGIS, slope and aspect have been produced from the SRTM DEM data (Fig. 12.2).

12.5 Discussion

To build a model, it is challenging to select the appropriate conditioning variables that have a large impact on the frequency of landslides (Ayalew and Yamagishi 2005). There aren't many fundamental rules of thumb for choosing factors in landslip susceptibility mapping. Ayalew and Yamagishi (2005) state that the characteristics selected for the assessment of landslip susceptibility in a GIS-based study have to be operational, comprehensive, non-uniform, fundamental, and measurable. Eleven potential predisposing factors were considered in this study: lithology, rainfall, slope, aspect, soil depth, NDVI, and characteristics related to land use and land cover.

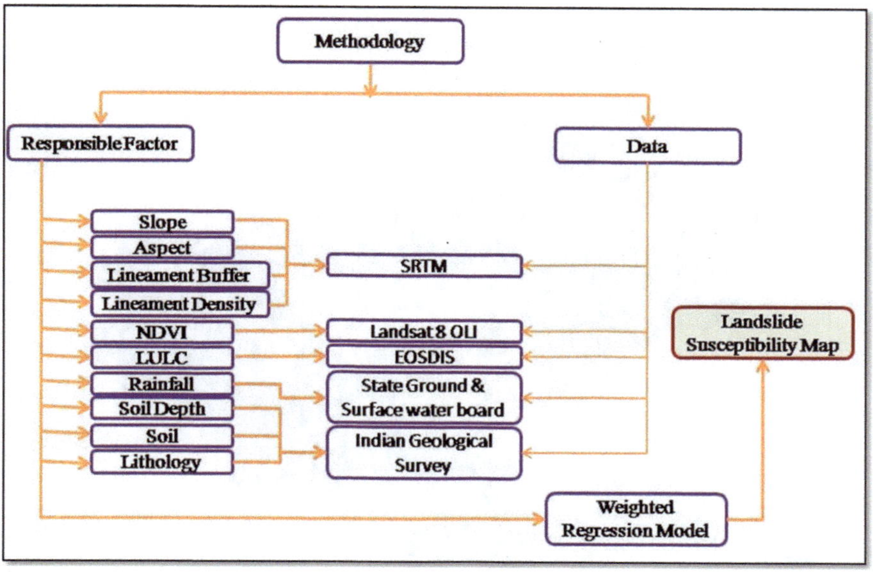

Fig. 12.2 Methodological flowchart

12.6 Result

12.6.1 Lineament

Any linear feature in a given area is referred to as a lineament. It can be a fault, shearing, fracture, discontinuity in the structure, or weaker zones in a geological formation. Because of the substantial changes that their occurrence or non-occurrence causes in landslide vulnerability, lineaments are considered one of the important and crucial geological elements that determine the incidence of landslides (Ayalew and Yamagishi 2005). Lineament density is divided into three categories: low, moderate, and high. Except for the central region and a few areas in the north and southwest, the majority of the Nilgiri district has a high lineament density (Fig. 12.3). The lineament buffer is displayed in Fig. 12.4.

12.6.2 NDVI

The NDVI plays a significant influence on the likelihood of landslip events. In mountainous areas, a decrease in vegetation serves as a contributing factor to the likelihood of landslides. The soil's erodibility on a slope is reduced by the dense vegetation. A district with the highest vegetation coverage will have lower erosion, lessening the risk of landslides. The NDVI map shown in Fig. 12.5 shows that the maximum NDVI

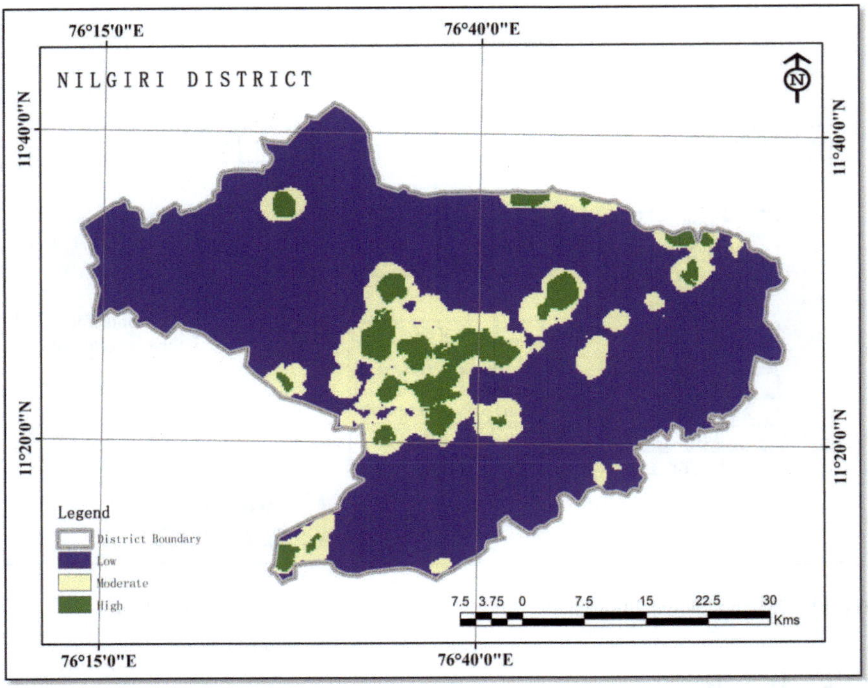

Fig. 12.3 Lineament density

is located in the north and south-west of Nilgiri, whereas the extreme northern region has moderate NDVI. In the research region's east, west, and south, the minimum NDVI is subtracted.

12.6.3 Slope

Slope is a significant and noticeable portion of the terrain, especially in hilly and high-altitude areas. An angular inclination between two elevations is known as a slope. The viewpoint and angle of slopes determine the earth's stability or instability. The degree and intensity of landslides are influenced by significant slope steepness, which rises with height, gradient, and slope concavity. In response to gravity, the material (in the form of boulders or other debris) might slowly slide down a slope over many years or devastatingly swiftly in seconds. Because of the various forms of progression that go into their development and the kind, quantity, and state of the slope materials, slope characteristics vary from location to location. The slope has been divided into five subfactors for this study: 0–500, 50–150, 15–300, 30–500, and >500. Figure 12.6 shows that the preponderance of the northern portion of the study district has a 0–50 slope. The slope ranges from 5 to 150 in the central and western

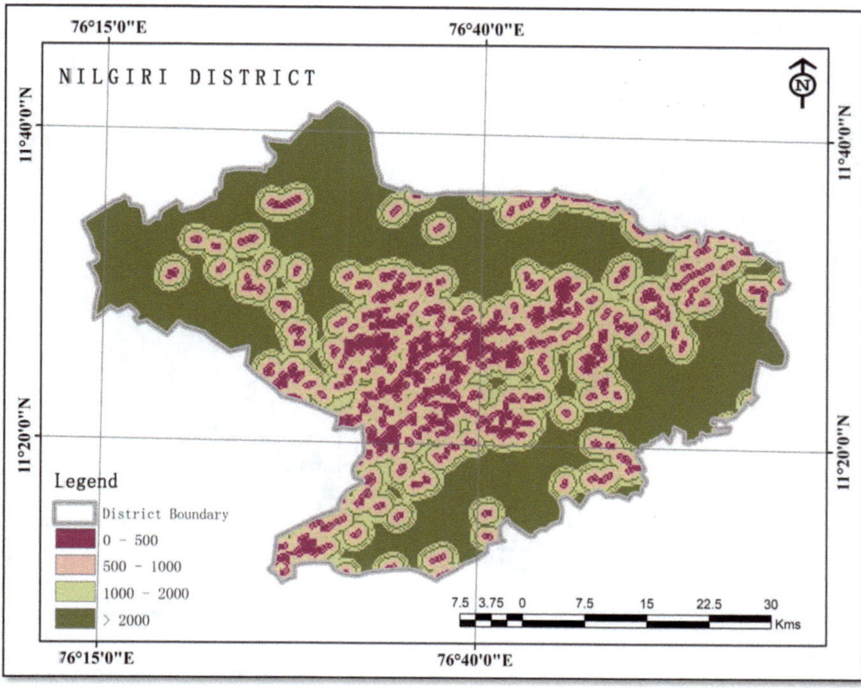

Fig. 12.4 Lineament buffer

sections. There is a 15–300 slope in the eastern, southern, and western sections and a small amount in the central areas. Most of the art in the northeastern region and the southeast and southwest corners depicted in Fig. 12.6 is mostly covered by a small number of artists on the west since most of the slope in rocky places runs from 30 to 500. There is very little Land in Nilgiri, with a slope of more than 500. These slopes are identifiable in stony areas in the south-west and south-east.

12.6.4 Aspect

Aspect is a significant contributing component to landslip causes. Rainfall, lineament, wind, and sun exposure are examples of aspects that affect the slope's contour and trajectory. The essential element that promotes slope stability is the aspect. Nine groupings, including flat, north, northeast, east, south-east, south, south-west, west, and northwest, are used to classify aspects. The aspect's path through the Nilgiri district is depicted on the map in Fig. 12.7. The Nilgiri district is situated in the Western Ghats, hence there isn't much level land there. The aspect is flat in the middle. Most of the district's north and northeast elements are located in the centre,

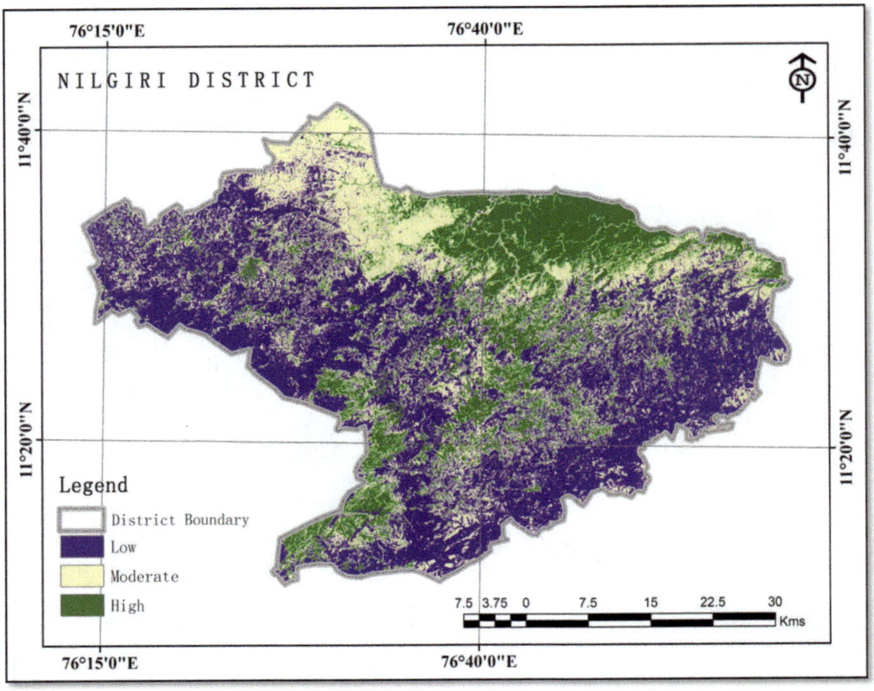

Fig. 12.5 Vegetation indices

east, and far west of Nilgiri. The northwest and south-eastern regions of the study area have an east-south aspect.

12.6.5 Rainfall

Rainy seasons occur in the Nilgiris District during the South West and North East monsoons. Gudalur, Pandalaur, and Kundah taluks receive rain during the South-West monsoon. Rainfall falls in Kotagiri, Coonoor, and Udhagamandalam taluk during the Northeast monsoon. The Nilgiris region has an standard yearly precipitation of 1500 to 3000 mm. Rainfall has a significant influence and is a major factor in "triggering" landslides (Anis et al. 2019; Karim et al. 2019). When it rains, rainfall penetrates into the soil on the majority of naturally existing hill slopes in tropical regions, diminishing the soil's shear strength and causing landslides. Possible landslip events can be predicted using an early warning or signalling system on the basis of rainfall data and thresholds. If the amount of rainfall in that specific location can be reliably predicted, the same threshold value can be used to send early warnings for landslides. If predicted rainfall exceeds threshold limits, an immediate warning alert signal should be sent to the local community, the government, and the search and rescue

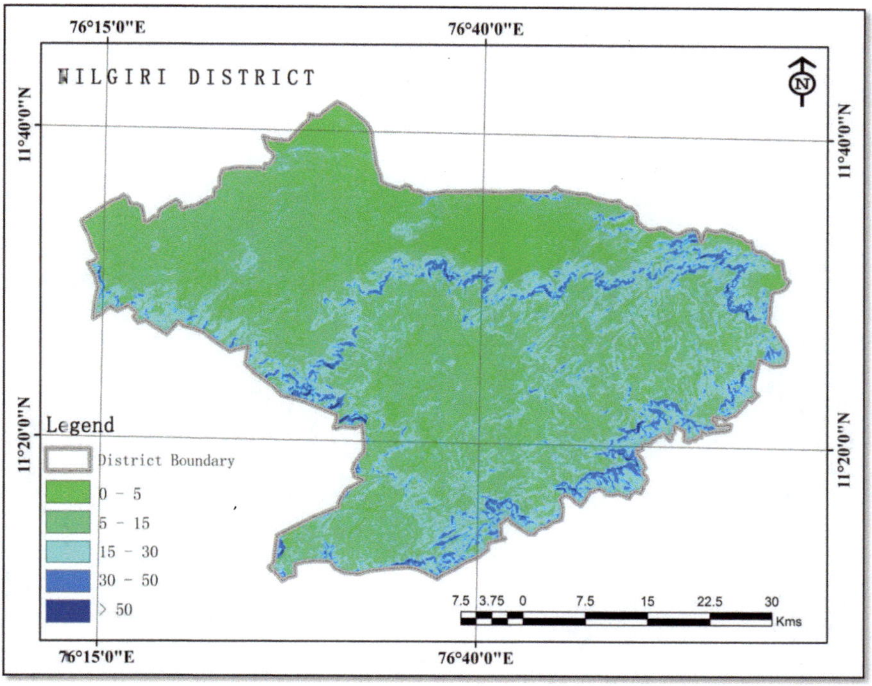

Fig. 12.5 Slope

team. Therefore, it is essential for accurate and timely rainfall forecasts. The map in Fig. 12.8 shows that most of the northeast, south-east, and east regions receive little rainfall. The regions with the highest rainfall are in the west and south.

12.6.6 Lithology

The study of lithology is crucial to understanding landslides because various lithological units express distinct geomorphological action, including landslides, that are now taking place (Manchar et al. 2018). The creation has been examined in this study. Lithology can be divided into archcan, mesozoic, and genozoic. Genozoic rock is found in the north, northwest, and west, as seen in Fig. 12.9. In Nilgiri, Mesozoic rocks are extremely rare. There are a few places in the northeast and western regions where Mesozoic rock can be found. Archean rocks cover the whole Nilgiri district's east, south, central, and south-west sections.

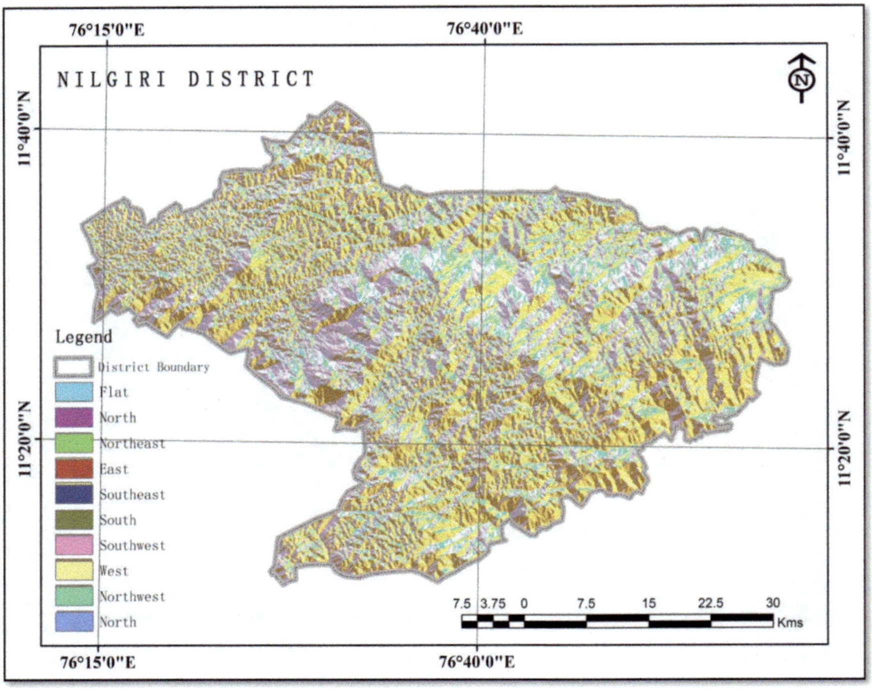

Fig. 12.7 Aspect

12.6.7 Soil

Another important causal factor for landslides is the soil. More sand in the soil, an unstable slope, and heavy rains—all of which are major contributors to landslides—can seriously harm the Land. The relative amounts of clay loam, rock land, sandy clay, and sandy clay loam/sandy loam content are revealed by the texture of the soil. A discernible chain of rock land extends from the study region's western portion to its middle and northeastern regions. It is also discernible in certain areas of the south-east. Only small percentages of rock land produce a stable, complete structure resistant to separation in Nilgiri soils.

On the other hand, sandy clay loam soil covers the whole south-west region and most of the south and eastern regions. Because they contain less organic stuff, they are easier to break off and are less able to form particularly stable aggregates. There is sandy, clayey soil in the far north. Figure 12.10 shows that sandy clay loamy soil covers much of the Nilgiri districts.

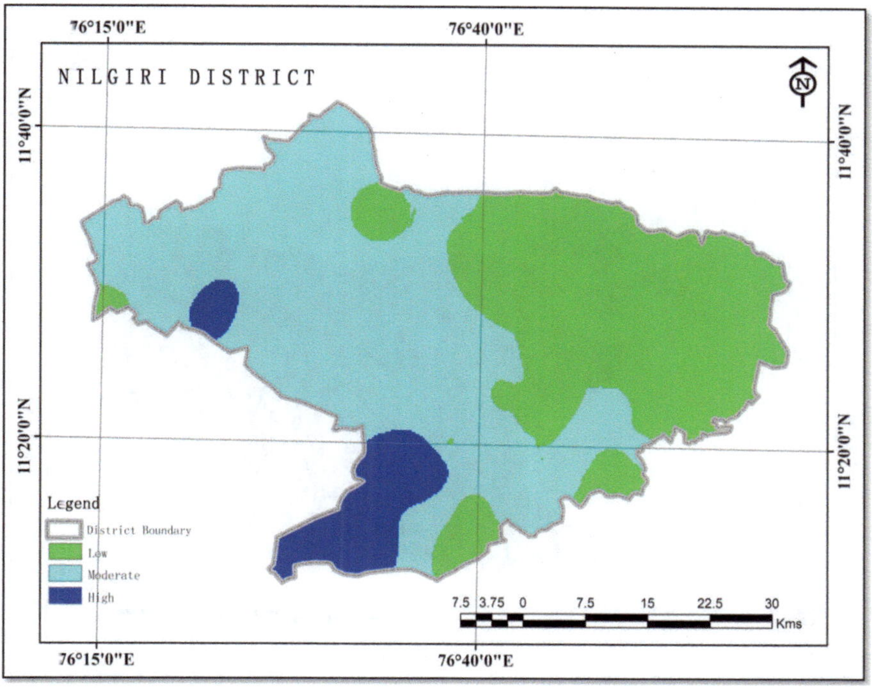

Fig. 12.8 Rainfall

12.6.8 Soil Depth

One important metric for determining soil stability and landslide-prone locations is soil depth. An increase in soil depth also increases the soil's capacity to absorb surface water, reducing the runoff quantity. For this reason, it is thought that shallow soil is more prone to landslides and instability than deep soil. The majority of the region is thought to have exceptionally deep soil, measuring more than 150 cm, based on Fig. 12.11. The centre region, significant portions of the western and southern regions, the northwest, and a few north places are visible at soil depths of more than 150 cm. At the same time, there are a few identifiable regions in the west and north of Rock Land. It is also observed in a few southern regions. The depth of the soil in the far north is between 100 and 150 cm.

12.6.9 Land Use Land Cover

Landslip incidence is thought to be significantly influenced by land use/land cover. Land use and land cover elements are the main determinants of a slope material's ability to manage the water content of the slope and prevent sliding. Plant roots are

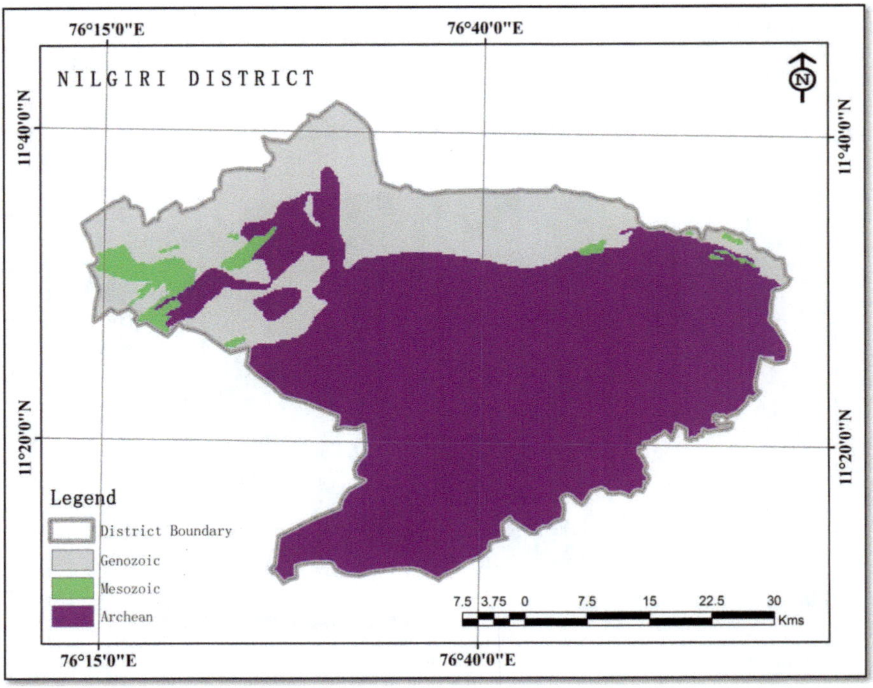

Fig. 12.9 Lithology

often considered reinforcements since they strengthen the sand at slopes. Land cover reduces the likelihood of landslides by allowing the soil to absorb water. The research region's land use and land cover have been broken down into six categories for this study: built-up Land, water bodies, barren Land, Forests and grassland. Water bodies can be found in the territories' south-west and centre sections, as seen in Fig. 12.12. Most of the western region has agricultural fields, while nearly all of the middle, eastern, and southern sections are also home to agricultural areas; parts of the west and south-west are covered in grasslands. There is forest in the northern and northeast parts of the district and certain areas in the west and southwestern regions. There is barren Land in the southern part, north and centre. The regions in the centre and east-south-east are covered with built-up areas. The Nilgiri district's western sections also contain a tiny build-up area.

12.7 Landslide Susceptibility

Table 12.1 explains the GWR-based regression analysis. The adj R^2 score of 0.82 implies that independent variables of 10 conditioning factors account for 82% of the variation in landside. If the probability of the F-statistic is less than the crucial

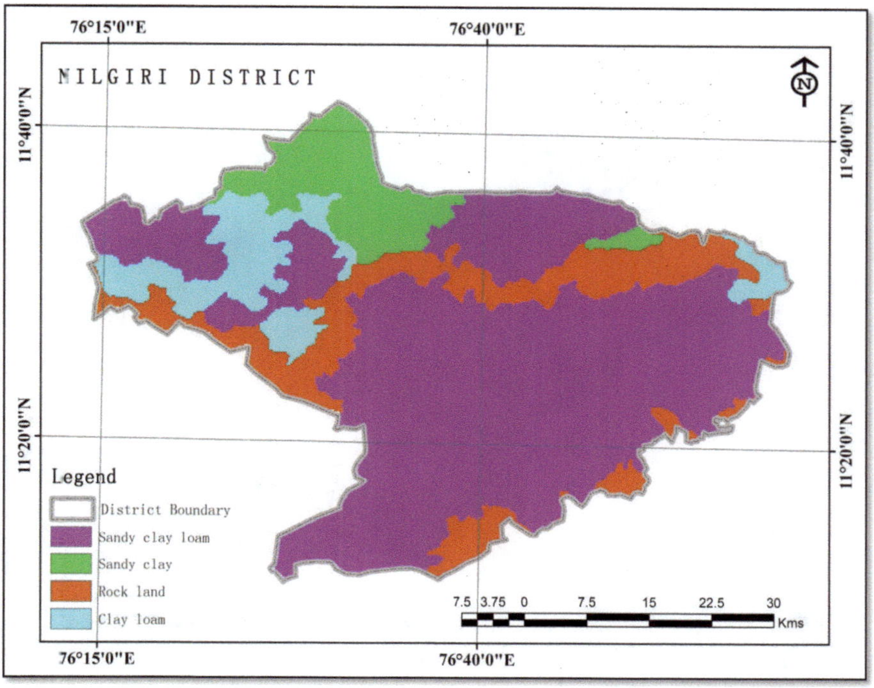

Fig. 12.10 Soil

value of 0.01, it suggests that the suggested model is meaningful and well-fit. The forecast is not biased because the Jarque–Bera Statistics Significant value is more than 0.01. The GWR summary includes a range of scales for assessing the model. The coefficient denotes the potency of the association among each explanatory variable and the dependent variable. According to the Variance Inflation Factor (VIF), a big VIF value—more than 7.5—indicates redundancy among the explanatory variables. The statistical significance of each independent variable explains the susceptibility related to the Nilgiri district.

Table 12.2 explains that out of the ten conditioning factors, the slope has the highest coefficient at 23.82, followed by land use and land cover (21.22) and, respectively, Lithology (6.19) and soil depth (6.32). Maps showing the vulnerability of landslides to future events provide information on the most likely areas for landslides to occur. Where landslip events could occur in a given area is the primary goal of susceptibility mapping. Numerous influencing elements, such as climate, hydrology, Lithology, structure, and geomorphic history, are linked to the landslip phenomena. However, it is only feasible to incorporate some of these characteristics for some susceptibility mapping studies. In this study, many parameters (like slope; soil; rainfall; Lithology; etc.) are believed as prompting layers for landslip vulnerability zoning, depending on the characteristics of the location. Using the landslide occurrence variables by GWR, landslide risk zones are analyzed and mapped for this study.

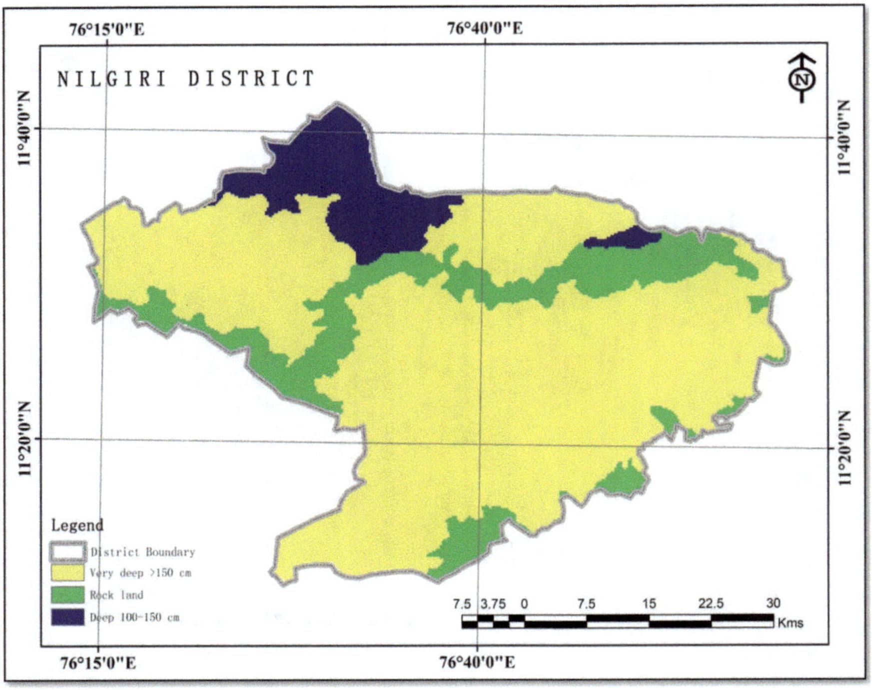

Fig. 12.11 Soil depth

Every factor contributes to the incidence of landslides, some large and others small. The landslip susceptibility map for the entire Nilgiri district is displayed in Fig. 12.13. Three susceptibility classes were identified: high, moderate, and low landslip-prone zones. The Nilgiri district's western and northern backbones, as depicted in Fig. 12.13, have low susceptibility. The research region is largely covered by moderate susceptibility. Most of the Nligiris's high susceptibility regions are in its west, south-west, and southern regions. Table 12.3 provided us with the area or percentage statistics of the susceptible classes to landslides. According to the results, 510.11 km^2 (19.99%) of the region was classified as a great danger zone, 1863.53 km^2 (73.02%) was classified as a moderate zone, and 178.36 km^2 (6.99%) as a zone susceptible to landslides.

12.8 Limitation

Although information theory-based techniques, like the one this study suggests, have demonstrated significant promise in various predictive spatial modelling scenarios, they are flexible. Although the suggested methodology's use as an objective weighting system does not rely on the experience or judgment of the decision

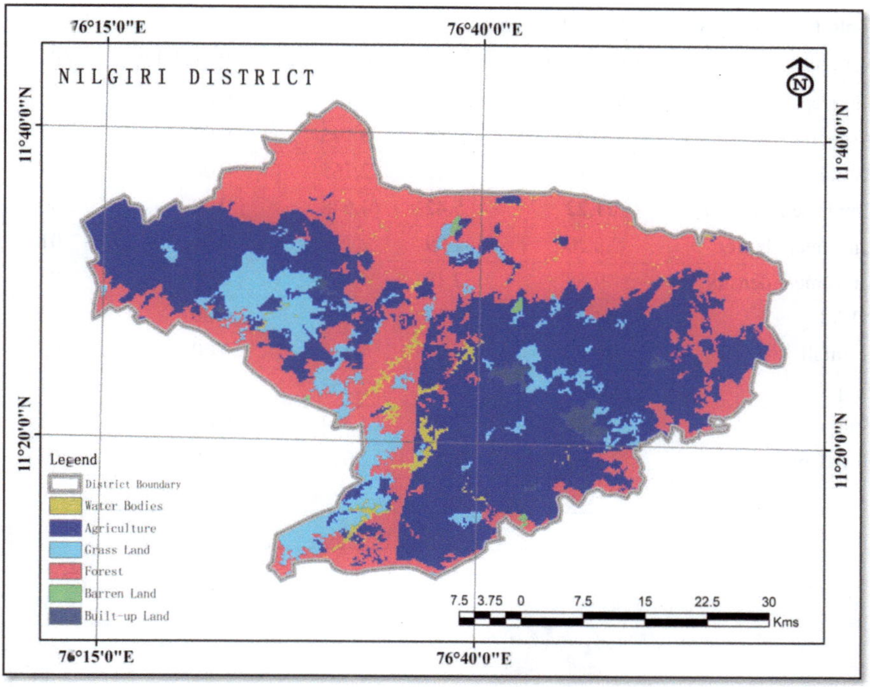

Fig. 12.12 Land use land cover

Table 12.1 Ordinary least square diagnostics

Input features	Landslide hotspot	Dependent variable	Susceptibility
No. of observations	136	Akaike's information criterion (AICc)	673.218
Multiple R-squared	0.848	Adjusted R-squared	0.840
Joint F-statistic	107.88	Prob (>F), (6,193) degrees of freedom	0.001
Jarque–Bera statistic	0.352	Prob (>chi-squared), (2) degrees of freedom	0.742

maker, it does depend on the measurement of specific features of landslip data points through a series of mathematical calculations. This is dependent upon a succinct and representative database being present. For this study, having a thorough and easily accessible database of landslip inventories was very helpful in reaching the intended result. A further drawback of the methodology is evident, especially in the eastern portion of the research area, where there are false alarms due to low rainfall regions classified as high susceptibility classes (extremely few pixels in a class that is highly sensitive). This is mostly because rainfall alone is not considered a major factor influencing the spatial pattern of landslide occurrence.

Table 12.2 Summary of ordinary least square

Variable	Coefficient	SE	t-Statistic	Probability	VIF
Intercept	3394.48	819.67	4.14	0.00	–
Aspect	10.06	2.89	3.143	0.00	1.04
Lithology	6.19	1.12	2.765	0.00	1.04
Land use land cover	21.22	5.42	4.648	0.00	1.03
Lineament buffer	8.20	1.99	2.153	0.01	1.01
Lineament density	9.58	2.18	4.109	0.00	1.01
NDVI	14.23	3.63	2.235	0.00	1.03
Rainfall	20.56	5.24	4.421	0.00	1.03
Slope	23.82	5.78	2.253	0.01	1.01
Soil	11.36	2.22	4.324	0.00	1.01
Soil depth	6.32	1.13	2.012	0.00	1.03

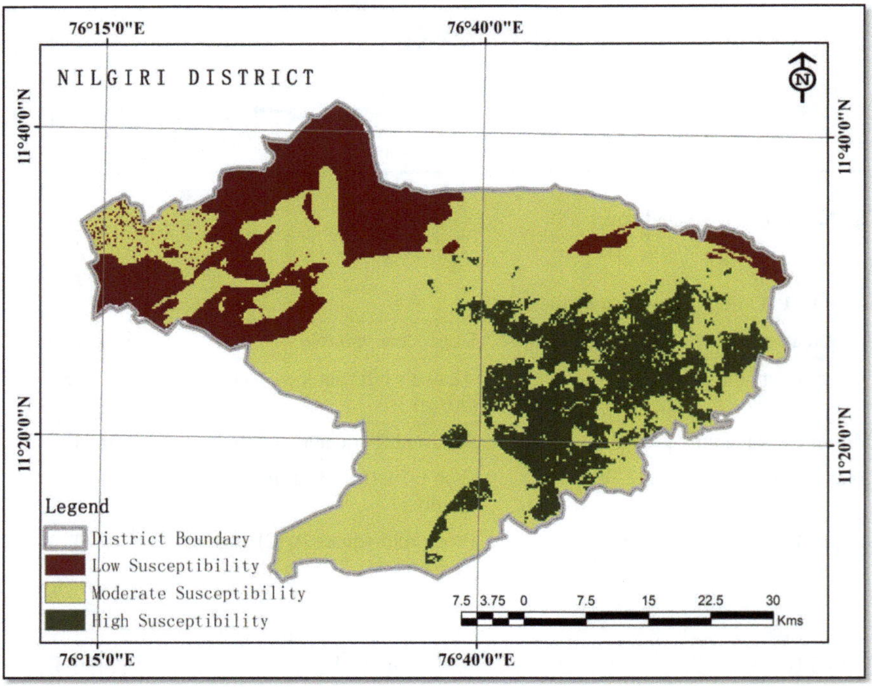

Fig. 12.13 Landslide susceptibility

Table 12.3 Area of landslide susceptibility zone of Nilgiri district

Category	Area (km^2)	Percentage
Low susceptibility	0510.11	19.99
Moderate susceptibility	1863.53	73.02
High susceptibility	0178.36	06.99

12.9 Conclusion

Landslides can be caused by humans or by nature. It is one of the biggest natural disasters, with a high death toll and extensive damage to communication networks, farming and forested areas, and human settlements. Creating and investigating the area's landslip zonation is crucial for efficient disaster management planning and can help with landslip hazard reduction strategy. This study aimed to determine the main cause of the landslip process in the Nilgiris district and to make a map of the susceptibility to landslips, an essential first step towards managing hazards comprehensively. This investigation was carried out using the GWR susceptibility mapping approach. Ten factors are shown on this landslide map as the causes of landslides. As a result, we have identified the primary causes of landslides in the Nilgiri district.

Due to excessive rainfall, most of the high mountain areas in the Nilgiri district are at risk of experiencing slope failures. We have identified three sensitive zones in this susceptibility assessment: low, moderate, and high. Most of the Nilgiris are high susceptibility regions in the west, south-west, and southern regions. According to the findings, 510.11 km^2 (19.99%) of the region was classified as a great danger zone; 1863.53 km^2 (73.02%) was classified as a moderate zone; and 178.36 km^2 (6.99%) was classified as a zone susceptible to landslides. Because it shows areas that are more vulnerable to landslides, the Nilgiri district's landslip sensitivity map will be advantageous to engineering projects, regional development, the tourism sector, and travelers. These treatments may lessen the causes and consequences of landslides. Moving a small stream away from a landslide-prone area can enhance surface drainage, and silt needs to be dredged out of the drainage to prevent landslides. In areas susceptible to landslides, barriers made of concrete should be built alongside rivers and roadways. Tree planting is required in the places where landslides occur.

12.9.1 Future Scope

When comparing all landslip areas to highly or very highly vulnerable areas, most LSM techniques result in astronomically large false positive rates. Since this issue extends beyond the scope of current study, it would have to catch the interest of the community of physical geographer, especially methodological development specialists. This will promote to investigate the new strategies for minimizing the issue of overestimated susceptibility in subsequent research by incorporating cutting-edge developments like artificial intelligence.

References

Alexakis DD, Agapiou A, Tzouvaras M, Themistocleous K, Neocleous K, Michaelides S, Hadjimitsis DG (2014) Integrated use of GIS and remote sensing for monitoring landslides in transportation pavements: the case study of Paphos area in Cyprus. Nat Hazards 72:119–141

Anis Z, Wissem G, Riheb H, Biswajeet P, Essghaier GM (2019) Effects of clay properties in the landslides genesis in flysch massif: case study of Aïn Draham, North Western Tunisia. J Afr Earth Sci 151:146–152

Arabameri A, Pradhan B, Rezaei K, Conoscenti C (2019) Gully erosion susceptibility mapping using GIS-based multi-criteria decision analysis techniques. CATENA 180:282–297

Ayalew L, Yamagishi H (2005) The application of GIS-based logistic regression for landslide susceptibility mapping in the Kakuda-Yahiko Mountains, Central Japan. Geomorphology 65(1–2):15–31

Ayalew L, Yamagishi H, Ugawa N (2004) Landslide susceptibility mapping using GIS-based weighted linear combination, the case in Tsugawa area of Agano River, Niigata Prefecture, Japan. Landslides 1:73–81

Bhagyasree Dolui YR, Geetha G (2020) Statistical analysis on the interrelationship between socio-demographic and anxiety of landslide among the tribal people

Chen W, Chen Y, Tsangaratos P, Ilia I, Wang X (2020) Combining evolutionary algorithms and machine learning models in landslide susceptibility assessments. Remote Sens 12(23):3854

Conforti M, Ietto F (2020) Influence of tectonics and morphometric features on the landslide distribution: a case study from the Mesima Basin (Calabria, South Italy). J Earth Sci 31(2):393–409

Dolui B, Yuvaraj RM, Geetha G (2019) Landslide susceptibility mapping using AHP model in Nilgiri District. Themat J Geogr 8(12):189–208

Dolui B, Yuvaraj RM, Geetha G (2023) Risk perception of landslide among the tribal population: a case study of Kotagiri Taluk of Nilgiri District, Tamil Nadu, India. Indigenous knowledge and disaster risk reduction: insight towards perception, response, adaptation and sustainability. Springer International Publishing, Cham, pp 209–225

Düzgün HS, Kemeç S (2008) Spatial and geographically weighted regression

Farooq K, Rogers JD, Ahmed MF (2015) Effect of densification on the shear strength of landslide material: a case study from Salt Range, Pakistan. Earth Sci Res 4(1):113

Feizizadeh B, Blaschke T (2013) GIS-multicriteria decision analysis for landslide susceptibility mapping: comparing three methods for the Urmia lake basin, Iran. Nat Hazards 65:2105–2128

Froude MJ, Petley DN (2018) Global fatal landslide occurrence from 2004 to 2016. Nat Hazards 18(8):2161–2181

Ganapathy GP, Rajawat AS (2015) Use of hazard and vulnerability maps for landslide planning scenarios: a case study of the Nilgiris, India. Nat Hazards 77:305–316

Gomathi R, Mohan RV, Backiaraj S (2013) Landslide susceptibility zonation in Kukalthurai Halla watershed, Moyar sub-basin in Nilgiris mountains, South India using Remote Sensing and GIS. Int J Geomat Geosci 4(2):366

Guzzetti F, Carrara A, Cardinali M, Reichenbach P (1999) Landslide hazard evaluation: a review of current techniques and their application in a multi-scale study, Central Italy. Geomorphology 31(1–4):181–216

Haque U, Blum P, Da Silva PF, Andersen P, Pilz J, Chalov SR et al (2016) Fatal landslides in Europe. Landslides 13:1545–1554

Hong H, Pradhan B, Sameen MI, Chen W, Xu C (2017) Spatial prediction of rotational landslide using geographically weighted regression, logistic regression, and support vector machine models in Xing Guo area (China). Geomat Nat Haz Risk 8(2):1997–2022

Huang F, Ye Z, Jiang SH, Huang J, Chang Z, Chen J (2021) Uncertainty study of landslide susceptibility prediction considering the different attribute interval numbers of environmental factors and different data-based models. CATENA 202:105250

Jaiswal P, Van Westen CJ (2009) Probabilistic landslide initiation hazard assessment along a transportation corridor in the Nilgiri area, India. In: EGU general assembly conference abstracts, p 2854
Kanwal S, Atif S, Shafiq M (2017) GIS-based landslide susceptibility mapping of northern areas of Pakistan, a case study of Shigar and Shyok Basins. Geomat Nat Haz Risk 8(2):348–366
Karim Z, Hadji R, Hamed Y (2019) GIS-based approaches for the landslide susceptibility prediction in Setif Region (NE Algeria). Geotech Geol Eng 37(1):359–374
Kirschbaum DB, Adler R, Hong Y, Hill S, Lerner-Lam A (2010) A global landslide catalog for hazard applications: method, results, and limitations. Nat Hazards 52:561–575
Kouli M, Loupasakis C, Soupios P, Vallianatos F (2010) Landslide hazard zonation in high risk areas of Rethymno Prefecture, Crete Island, Greece. Nat Hazards 52:599–621
Li Y, Liu X, Han Z, Dou J (2020) Spatial proximity-based geographically weighted regression model for landslide susceptibility assessment: a case study of Qingchuan area, China. Appl Sci 10(3):1107
Manchar N, Benabbas C, Hadji R, Bouaicha F, Grecu F (2018) Landslide susceptibility assessment in Constantine region (NE Algeria) by means of statistical models. Studia Geotechnica et Mechanica 40(3):208–219
Marjanović M (2013) Advanced methods for landslide assessment using GIS. Doctoral dissertation, Palacký University Olomouc
Neaupane KM, Piantanakulchai M (2006) Analytic network process model for landslide hazard zonation. Eng Geol 85(3–4):281–294
Ozdemir A, Altural T (2013) A comparative study of frequency ratio, weights of evidence and logistic regression methods for landslide susceptibility mapping: Sultan Mountains, SW Turkey. J Asian Earth Sci 64:180–197
Paez A (2006) Exploring contextual variations in land use and transport analysis using a probit model with geographical weights. J Transp Geogr 14(3):167–176
Papoutsis I, Kontoes C, Alatza S, Apostolakis A, Loupasakis C (2020) InSAR Greece with parallelized persistent scatterer interferometry: a national ground motion service for big Copernicus sentinel-1 data. Remote Sens 12(19):3207
Petley D (2012) Global patterns of loss of life from landslides. Geology 40(10):927–930
Rajarathnam S, Ganapathy GP (2006) Landslide hazard zonation of India, a GIS approach. In: Proceedings of the first India disaster management congress, New Delhi, vol 29, p 30
Shah SHIA, Jianguo Y, Jahangir Z, Tariq A, Aslam B (2022) Integrated geophysical technique for groundwater salinity delineation, an approach to agriculture sustainability for Nankana Sahib Area, Pakistan. Geomat Nat Haz Risk 13(1):1043–1064
Tariq A, Yan J, Gagnon AS, Riaz Khan M, Mumtaz F (2023) Mapping of cropland, cropping patterns and crop types by combining optical remote sensing images with decision tree classifier and random forest. Geo-Spat Inf Sci 26(3):302–320
Tsironi V, Ganas A, Karamitros I, Efstathiou E, Koukouvelas I, Sokos E (2022) Kinematics of active landslides in Achaia (Peloponnese, Greece) through InSAR time series analysis and relation to rainfall patterns. Remote Sens 14(4):844
Tu J (2011) Spatially varying relationships between land use and water quality across an urbanization gradient explored by geographically weighted regression. Appl Geogr 31(1):376–392
Wheeler DC, Páez A (2009) Geographically weighted regression. In: Handbook of applied spatial analysis: software tools, methods and applications. Springer, Berlin, Heidelberg, pp 461–486
Yuvaraj RM, Dolui B (2021) Statistical and machine intelligence-based model for landslide susceptibility mapping of Nilgiri district in India. Environ Chall 5:100211
Yuvaraj RM, Dolui B (2023) Geographical assessment of landslide susceptibility using statistical approach. Quat Sci Adv 11:100097

Chapter 13
Assessment of Climate Change Impact on Landslides in Darjeeling District of West Bengal: A Geospatial, Geostatistical and Ecosystem Service Based Approach

Anusha Mondal, Ishika Chowdhury, Sayani Mukherjee, and Asutosh Goswami

Abstract The climate system of the earth is definitely warming. That environment changes influence the steadiness of regular and designed slants and have results on avalanches, is likewise undisputable. Less clear is the sort, degree, greatness and heading of the progressions in the security conditions, and on the area, overflow, movement and recurrence of avalanches because of the projected climate change. The present study delves into the assessment of climo-geomorphic role on the occurrences of landslides through the investigation of ecosystem service evaluation by employing remote sensing technology. Environment and avalanches act at just somewhat covering spatial and transient scales, confusing the assessment of the environment influences on avalanches. We found a skewed distribution in the geographical spreading of the landslide-climate studies that have been published, with significant portions of the globe not covered. The present study has identified the change in temperature and precipitation through the climatic based analysis and indices computation. The present study utilizes a probabilistic avalanche danger model to evaluate provincial avalanche alterations. The study advocates for developing outfits of projections in light of a scope of discharges situations, and to utilize cautiously results from most pessimistic scenario situations that may over/under-gauge avalanche dangers and hazard. Additionally, we have computed frequency ratio model (FRM) for landslide susceptibility zonation. The study concludes with the finding of regulating services in the steep terrain to nullify the impact of landslides.

Keywords Climate change · Landslide · Darjeeling Himalaya · Hazard · Ecosystem services

---

A. Mondal · I. Chowdhury · S. Mukherjee · A. Goswami (✉)
Department of Earth Sciences and Remote Sensing, JIS University, Agarpara, West Bengal, India
e-mail: goswamiasutosh@gmail.com

G. K. Panda et al. (eds.), *Landslide: Susceptibility, Risk Assessment and Sustainability*, Advances in Natural and Technological Hazards Research 52,
https://doi.org/10.1007/978-3-031-56591-5_13

13.1 Introduction

Landslides are a significant natural hazard that can have devastating impacts on human communities, infrastructure, and the environment (Chawla et al. 2018; Abraham et al. 2020; Dikshit et al. 2020; Ali et al. 2021; Alsabhan et al. 2022). Understanding landslide vulnerability is crucial for effective risk assessment, mitigation, and disaster management. This chapter aims to provide a comprehensive discussion on landslide vulnerability, covering various aspects, including the vulnerability, factors contributing to landslide vulnerability, methods for assessing vulnerability, and strategies for mitigating landslide risk. Landslides are the downward movement of rock, soil, and debris on slopes due to various factors such as rainfall, earthquakes, volcanic eruptions, or human activities (Anis et al. 2019; Berhane et al. 2020; Araújo et al. 2022; Bansal et al. 2022; Asmare 2023). These natural events can result in loss of life, property damage, and disruption of vital services. The vulnerability of an area to landslides is determined by a complex interplay of geological, environmental, and human factors. Understanding landslide vulnerability is crucial to develop effective strategies for risk reduction and disaster management. Landslide vulnerability refers to the degree of susceptibility of an area or community to landslides and the potential consequences of such events (Bhatta 2011; Biswas and Pal 2016; De 2017; Demir 2019; Chen and Zhang 2021). Vulnerability is a multifaceted concept that encompasses not only the physical and environmental aspects of a location but also social, economic, and institutional factors (Meghanadh et al. 2022; Mekonnen et al. 2022; Villuri et al. 2022; Hammad Khaliq et al. 2023). Landslides are complex geological phenomena characterized by the downward movement of earth material, often triggered by various factors such as heavy rainfall, earthquakes, and human activities. These events can have devastating consequences, causing loss of life, property damage, disruption of infrastructure, and long-term environmental impacts (Korup et al. 2010; Kahlon et al. 2014; Khan et al. 2019; Huangfu et al. 2021; Hussain et al. 2022). As climate change and human activities continue to alter the landscape, understanding the relationship between landslides and ecological services is becoming increasingly important. This chapter aims to explore the interplay between landslides and ecological services, demonstrating the crucial role of ecosystems in mitigating landslide risks and supporting recovery efforts. Climate change can significantly influence the occurrence and intensity of landslides through various mechanisms. The relationship between climate change and landslides is complex and involves several interconnected factors. Climate change is associated with an increase in the frequency and intensity of extreme weather events such as heavy rainfall, storms, and prolonged periods of precipitation which can saturate soil, leading to increased water content and reduced stability, making slopes more susceptible to landslides. Climate change can alter precipitation patterns, leading to changes in the timing, duration, and intensity of rainfall. Changes in precipitation can affect soil moisture levels and contribute to the triggering of landslides particularly in the areas formed by tertiary sedimentary. While not a direct cause, the changing climate may contribute to the perception of an increase in landslide incidents due to the amplification of predisposing factors.

It is important to note that the relationship between climate change and landslides is context-specific and can vary depending on geographical, geological, and environmental conditions. Mitigation and adaptation strategies to reduce landslide risk often involve a combination of land-use planning, early warning systems, and sustainable environmental management practices. The Darjeeling Himalaya, nestled in the northeastern part of India, is a region renowned for its breathtaking landscapes and rich biodiversity. However, it is also notorious for its susceptibility to landslides. Landslides in the Darjeeling Himalaya have been a recurring problem, causing considerable damage to infrastructure, loss of life, and disruption to livelihoods. This comprehensive essay explores the causes, impacts, and mitigation strategies associated with landslides in the Darjeeling Himalaya, delving into the geological, environmental, and human factors that contribute to this phenomenon. This chapter aims to provide a comprehensive understanding of landslides in the Darjeeling Himalaya. It will explore the geological, environmental, and human factors contributing to landslides, their impacts on society and the environment, and the various mitigation strategies employed to manage this recurring problem. The research for this essay is based on an extensive review of scientific literatures (Doke 2017; Fayez et al. 2018; Dolui and Chakraborty 2022; Goswami et al. 2022a, b; Gao et al. 2023), government reports, and case studies related to landslides in the Darjeeling Himalaya. Field surveys and expert interviews have also been conducted to gain insights into the ground realities and the effectiveness of mitigation strategies. The information gathered has been synthesized to provide a comprehensive overview of the subject.

The region is characterized by its steep slopes, deep valleys, and the presence of numerous rivers and streams. The geological foundation of the Darjeeling Himalaya plays a crucial role in the occurrence of landslides. The region is primarily composed of a complex mix of rocks, including sedimentary rocks, schists, gneisses, and granites. The presence of fault zones and fractures in the rocks further increases its susceptibility to landslides. The Darjeeling Himalaya is located at the convergence of the Indian plate and the Eurasian plate. The ongoing collision and interaction of these tectonic plates have given rise to the formation of the Himalayan mountain range. This tectonic activity, along with the complex geology of the region, contributes to the instability of slopes and the occurrence of landslides. The climate in the Darjeeling Himalaya is characterized by distinct wet and dry seasons. The monsoon, which occurs between June and September, brings heavy rainfall to the region. This seasonal variation in weather patterns is a significant factor contributing to landslides in the area. Landslides can be classified into several types based on their movement and characteristics. Common landslide types in the Darjeeling Himalaya include rockfalls, debris flows, and translational slides. Each type has its specific triggers and causes. Rockfalls are sudden, rapid movements of individual rocks or boulders down steep slopes. They are often triggered by heavy rainfall, seismic activity, or freeze–thaw cycles. Debris flows, also known as mudslides, involve a mixture of water, soil, and loose rocks moving downhill at high speeds. They are prevalent in the monsoon season. Translational slides are characterized by the movement of a mass of rock or soil along a well-defined plane. They can be triggered by heavy rainfall, saturation of the ground, or seismic activity. The type of rock in a particular location has a significant influence on landslide susceptibility. In the Darjeeling Himalaya, rocks with high weathering potential are more likely to disintegrate and lead to landslides. For example, shale, schist, and phyllite are highly weathered rock types prevalent in the region. The constant process of weathering weakens the rock, making it more susceptible to erosion and landslides. The presence of geological faults and fractures in the rocks of the Darjeeling Himalaya further exacerbates the landslide risk. These fractures provide pathways for water infiltration, which can weaken the surrounding rocks. Changes in precipitation patterns, increased frequency and intensity of extreme weather events, and alterations in temperature can influence landslide dynamics. It is crucial for researchers and policymakers to consider these climate change factors in vulnerability assessments and mitigation strategies. Constructing retaining walls and check dams can help stabilize slopes and control soil erosion. These structures are strategically placed to divert water and prevent saturation of the soil, reducing landslide risks. Slope stabilization techniques, including the use of geogrids, soil nails, and rock bolts, are employed to reinforce unstable slopes.

Landslide vulnerability is a complex issue influenced by geological, environmental, social, and economic factors. Understanding and assessing vulnerability is crucial for effective risk reduction and disaster management. By employing methods such as GIS, remote sensing, and vulnerability indices, we can better identify areas at risk and implement mitigation strategies to protect communities and infrastructure. Landslide risk reduction is a shared responsibility that requires collaboration between government agencies, communities, and experts in various fields. By working together, we can reduce the impacts of landslides and create safer, more resilient communities. Ecological services, also known as ecosystem services, are the various benefits that ecosystems provide to humans and the environment. These services can be broadly categorized into four types. Provisioning services encompass the tangible goods that ecosystems offer, such as food, water, timber, and medicine. Forests, wetlands, and agricultural landscapes are primary sources of provisioning services. The role of these services in landslide mitigation lies in their ability to stabilize the landscape, reduce soil erosion, and maintain soil quality. Regulating services involve the control and moderation of natural processes. Examples include climate regulation, flood control, and disease regulation. Ecosystems can play a significant role in landslide risk reduction through regulating services by controlling erosion, stabilizing slopes, and reducing the likelihood of mass movements. Supporting services are essential for the production of other ecosystem services. They include nutrient cycling, soil formation, and pollination. Healthy soils and vegetation, promoted by supporting services, can enhance the stability of landscapes and reduce the risk of landslides. Cultural services are the non-material benefits people obtain from ecosystems, such as spiritual and recreational experiences. These services contribute to human well-being, promote a sense of community, and can encourage sustainable land management practices that prioritize ecosystem conservation.

Ecosystems play a crucial role in landslide mitigation through their ability to stabilize slopes, control erosion, and regulate water flow. By harnessing the potential of ecosystems and implementing ecosystem-based approaches, we can reduce landslide risk and promote sustainable land management practices that prioritize ecosystem conservation. However, addressing challenges related to land tenure, knowledge dissemination, climate change, and economic considerations is essential to ensure the success of such approaches. In a changing world, the preservation and restoration of ecosystems are keys to enhancing landslide resilience and securing the well-being of communities and the environment. The novelty of the present study lies in its integration of the ecosystem services valuation and landslide distributions.

13.2 Materials and Methods

13.2.1 Study Area

Darjeeling district is situated in the northern part of West Bengal, bordering the countries of Nepal and Bhutan (Fig. 13.1). It is part of the Eastern Himalayas and is known for its proximity to the Sikkim region. The district's terrain is characterized by a diverse topography, with elevations ranging from the low-lying foothills to the towering peaks of the Himalayas. The district is home to some of the highest mountains in India, including Kanchenjunga, the third-highest peak in the world. The district features a series of hills and valleys, with several small and large rivers crisscrossing the landscape. The Darjeeling Himalayan region is renowned for its tea plantations that cover many of the slopes. The climate in Darjeeling district varies with elevation. The lower regions have a subtropical climate, while the higher elevations experience a more temperate and alpine climate. The district is known for its cool and pleasant weather, making it a popular destination for tourists seeking respite from the heat in the plains. Darjeeling district is famous for its tea gardens, which are among the best in the world. The tea plantations are primarily located in the hilly areas, and the district's tea is renowned for its unique flavor and aroma. The district is home to lush forests with a variety of flora and fauna. These forests are part of the Eastern Himalayan biodiversity hotspot and harbor a wide range of wildlife, including red pandas, leopards, and various bird species. Several rivers flow through the district, including the Teesta, Rangeet, and Balasun. These rivers are important for both local agriculture and hydropower generation. The predominant land use in the district includes agriculture, with the cultivation of tea, maize, paddy, and ginger being significant. Additionally, tourism plays a crucial role in the local economy. The district is known for its cultural and ethnic diversity, with various communities, including Gorkhas, Lepchas, Bhutias, and Bengalis, residing in the region. Geography of the district not only makes it a beautiful and scenic place but also contributes to its cultural richness and economic significance, primarily through tea production and tourism.

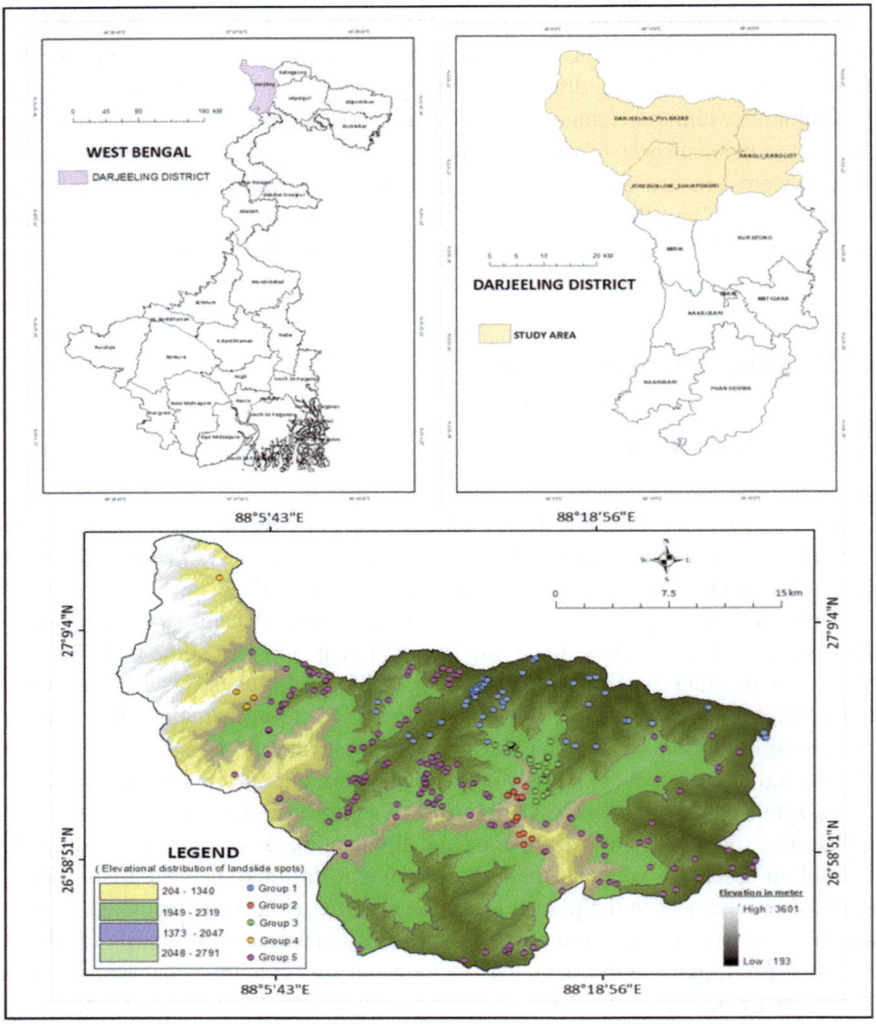

Fig. 13.1 Location map of the study area

13.2.2 *Road Contributing Area*

Road contributing area in GIS refers to the portion of the landscape that directs water runoff toward a specific road or highway. It considers factors such as topography and surface characteristics to determine the area from which water flows toward the road. Analyzing road-contributing areas in GIS involves delineating and measuring the surface area that contributes runoff to a particular road segment or network. This analysis takes into account factors like slope, land cover, and hydrological properties to determine the extent of the landscape that contributes water to the road.

The concept of road contributing area holds significance in hydrological modeling, flood management, and transportation planning. Understanding the road-contributing area helps assess the potential flood risks, identify areas prone to erosion and sedimentation, and design appropriate drainage systems. By delineating the road-contributing area, GIS enables valuable insights for various purposes. Analyzing road-contributing areas assists in identifying the catchment area that contributes runoff to the road. This information supports the design of drainage systems, culverts, and infrastructure to effectively manage stormwater runoff. Determining the road-contributing area helps evaluate the likelihood of flooding along road networks. It aids in assessing road vulnerability during heavy rainfall events and helps develop strategies for flood mitigation. Understanding the road contributing area allows the identification of erosion and sedimentation-prone areas. It helps prioritize erosion control measures like sediment basins, vegetative buffers, and contour plowing techniques. Analyzing the road contributing area helps identify areas where runoff can affect road safety, such as locations susceptible to pooling or hydroplaning. It supports planning maintenance activities and implementing measures to ensure road functionality and safety. GIS software offers tools and techniques for delineating road-contributing areas based on terrain analysis, hydrological modeling, and spatial data integration (Fig. 13.2). These tools enable the visualization, analysis, and integration of road-contributing area information with other relevant datasets. This empowers informed decision-making in transportation and water management planning.

13.2.3 Frequency Ratio Modeling

FRM is a widely used statistical method for assessing landslide susceptibility. It operates on the premise that specific factors or variables contribute to landslide occurrences. This model determines the ratio of landslide frequency to the occurrence frequency of a particular factor within a given area. To employ the model, factors influencing landslides, such as slope steepness, geological characteristics, land cover, soil properties, and rainfall intensity, are identified and mapped. Each factor is assigned a rating based on its relative importance in triggering landslides. The frequency ratio is then computed by dividing the number of landslides in areas with a specific factor by the total occurrences of that factor across the study area. This ratio serves as a measure of the likelihood or probability of landslides associated with the factor. By applying the frequency ratio model, a susceptibility map can be generated, delineating areas with different levels of landslide susceptibility. This map aids land managers, planners, and decision-makers in identifying high-risk zones and implementing appropriate mitigation measures. The accuracy and reliability of the model hinge on the quality and availability of data used for factor identification, mapping, and the specific geological and environmental conditions of the area under study.

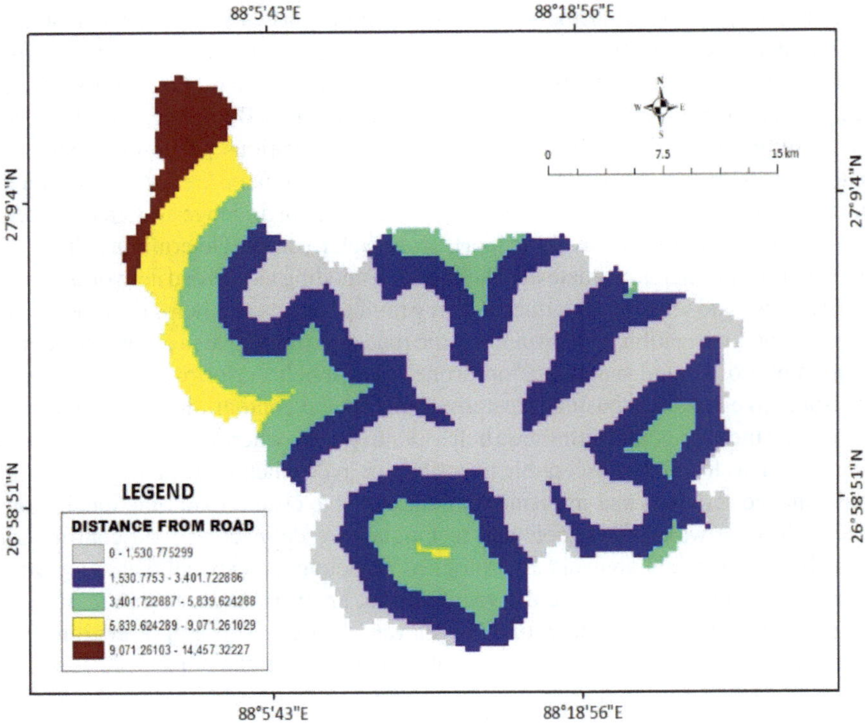

Fig. 13.2 Distance from road mapping

The frequency ratio model is a preferred method for calculating landslide susceptibility zonation maps due to its numerous advantages in assessing landslide risk.

The frequency ratio (FR) is computed for each factor category using the formula:

$$\begin{aligned} \text{FR} = {} & (\text{Number of landslides in a category/Total number of landslides}) \\ & \div (\text{Number of non - landslide points in a category/Total number of} \\ & \text{non - landslide points}) \end{aligned} \tag{13.1}$$

The following detailed equation may be utilized to calculate frequency ratio (FR).

$$\text{R} = \text{f1/f2} \tag{13.2}$$

where

f1 is the first frequency.

f2 is the second frequency.

Angular frequency is often used in the context of oscillations and waves. The angular frequency, denoted as ω, is related to the frequency f by the equation $\omega = 2\pi f$. The angular frequency ratio is then:

$$R\omega = \omega 2\omega 1 = 2\pi f 2 2\pi f 1 = f2f1 \tag{13.3}$$

The period T is the reciprocal of frequency (T = f1). The period ratio is the reciprocal of the frequency ratio:

$$RT = T1/T2 = 1/f1 = 1/f2 \tag{13.4}$$

This is essentially the inverse of the frequency ratio.

13.2.3.1 ROC Curve

ROC (receiver operating characteristic) curves are commonly used for model validation, particularly in the context of binary classification problems. These problems involve the classification of spatial features into two categories, such as land cover classes (e.g., forest or non-forest) or the presence and absence of a certain phenomenon (e.g., ESV). ROC curves are widely employed to assess the performance of predictive models in GIS applications. ROC curves illustrate the trade-off between sensitivity (true positive rate) and specificity (true negative rate) at various classification thresholds. In GIS, this could relate to the accuracy of classifying land cover types or identifying areas prone to a specific environmental condition. GIS models are often used for tasks like land cover classification, habitat suitability modeling, or risk assessment. Model validation is crucial to ensure the reliability of these models and their outputs, especially when applied to real-world scenarios. For instance, a land cover classification model may classify each pixel as either vegetation or non-vegetation. The ROC curve is generated by plotting the true positive rate (sensitivity) against the false positive rate (1-specificity) at various classification thresholds. The area under the ROC curve (AUC-ROC) is a summary measure of the model performance, where a value of 0.5 indicates random classification and a value of 1.0 represents perfect classification. We can use ROC curves to assess how well a model discriminates between different classes. An ROC curve that hugs the upper-left corner of the plot indicates high sensitivity and specificity, suggesting a good model performance. This is the proportion of actual positive instances correctly classified by the model. It is calculated as:

$$\text{Sensitivity} = \text{True Positives}/(\text{True Positives} + \text{False Negatives}) \tag{13.5}$$

False positive rate (1 − Specificity) is the proportion of actual negative instances incorrectly classified as positive by the model. It is calculated as:

$$\text{False Positive Rate} = \text{False Positives}/(\text{False Positives} + \text{True Negatives}) \tag{13.6}$$

In a GIS model, the predicted probabilities are converted into class labels using a threshold. The ROC curve is generated by varying this threshold, and at each threshold, the true positive rate and false positive rate are calculated. The AUC-ROC is a scalar value that represents the area under the ROC curve. A model with better discriminatory power will have a higher AUC-ROC value. A random classifier will have an AUC-ROC of 0.5, while a perfect classifier will have an AUC-ROC of 1.

$$\text{AUC-ROC} = 0.5 : \text{Random classifier.}$$
$$\text{AUC-ROC} > 0.5 : \text{Better than random.}$$
$$\text{AUC-ROC} = 1 : \text{Perfect classifier.}$$

13.2.4 Precipitation Concentration Index (PCI)

To calculate the PCI based on the ratio of the sum of the squared monthly precipitation values to the square of the sum of the monthly precipitation values. The monthly precipitation data set has been collected from the India Meteorological Department daily gridded directory. The data range is from 1981 (May, June, July, August) to 2021 (May, June, July, August). The year-wise monsoon season data has been taken to show the accurate precipitation concentration index (PCI) for our study area.

The formula is as follows:

$$\text{PCI} = \left(\sum \text{Sp}/\text{A}^2\right) * 33.33 \tag{13.7}$$

where

Sp represents the sum of monthly precipitation values.

A indicates the monthly annual precipitation value.

13.2.5 LST Mapping

A land surface temperature (LST) map is a representation of the temperature of the Earth's land surface at a given point in time. It provides information about the thermal characteristics of the Earth surface. By analyzing the emitted thermal radiation the temperature of the land surface can be determined. To assess the health of vegetation, analyze energy consumption patterns, and understand climate change impacts on land surfaces, a land surface temperature map has been prepared. A high value generally refers to a relatively warmer temperature, it indicates areas where the land surface is experiencing higher thermal energy. LST can provide valuable information about the thermal conditions of the Earth's surface, which can indirectly contribute to the potential occurrence of landslides. Land surface temperature influences vegetation health and can indirectly impact landslide occurrence. Vegetation plays a crucial role in stabilizing slopes by reinforcing soil and reducing surface erosion. In remote sensing, the Land Surface Temperature (LST) is an important parameter that represents the temperature of the Earth's surface. LST can be estimated using thermal infrared remote sensing data. One common method for computing LST is based on the radiative transfer equation and involves the use of thermal bands from satellite sensors.

The following is a simplified formula for estimating LST from remote sensing data:

$$\mathrm{LST} = [\mathrm{K2}/\{\ln(\mathrm{TB\ K1} + 1)\}] - 273.15 \tag{13.8}$$

where

LST is the Land Surface Temperature in degrees Celsius.

TB is the brightness temperature in Kelvin obtained from the thermal band of the satellite sensor.

K1 and K2 are constants specific to the satellite sensor and need to be provided by the sensor manufacturer. It is important to note that different satellite sensors may have different calibration constants, and the formula may be adjusted accordingly. Additionally, atmospheric correction may be applied to the thermal infrared data to account for the effects of the Earth's atmosphere on temperature readings.

13.2.6 SEVI

The calculation of the Shadow Eliminated Vegetation Index (SEVI) involves several essential steps within the SCP tool in QGIS. Initially, atmospheric correction algorithms available in SCP are employed to eliminate atmospheric interference, such as haze and scattering, and obtain accurate reflectance values. This correction is crucial for ensuring reliable vegetation analysis. Following the atmospheric correction, two intermediate indices are derived. Those are namely the Ratio Vegetation Index (RVI) and the Simple Vegetation Index (SVI). The RVI is computed by taking the ratio between the Near Infrared (NIR) band and the Red band of the Landsat image, providing valuable information on vegetation health and density. Meanwhile, the SVI is obtained by calculating the reciprocal of the Red band, focusing on the vegetation response in the red spectral region and indicating vegetation presence and vigor. SEVI is a spectral index commonly used in remote sensing and image processing to detect and analyze vegetation in satellite or aerial imagery while accounting for the presence of shadows. Shadows can create challenges when trying to accurately assess vegetation health or cover because they can reduce the reflectance values in the affected areas. SEVI is designed to mitigate the impact of shadows on vegetation analysis by combining multiple spectral bands in a way that reduces shadow-related distortions. It is typically calculated using the following formula:

$$\text{SEVI} = (\text{NIR} - \text{Red})/(\text{NIR} + \text{Red} + \text{L}) \tag{13.9}$$

where

NIR (Near-Infrared) is the reflectance in the near-infrared band.

Red is the reflectance in the red band.

L is a constant or an adjustable parameter introduced to reduce the shadow effect. The value of L can be determined based on the specific characteristics of the imagery or the research being conducted.

SEVI helps to enhance the visibility of vegetation, even in areas affected by shadows. By normalizing the reflectance values in the red and near-infrared bands, it reduces the differences caused by varying illumination conditions. This index is particularly useful in applications like land cover classification, vegetation health monitoring, and environmental studies where accurate vegetation detection is crucial, and shadows can pose challenges to analysis.

13.2.7 Assignment of Ecosystem Service Valuation (ESV)

Through their capabilities, environments straightforwardly or by implication give various administrations that are essential to life on the planet and to human endurance. Different biological systems give many structures like arrangement (food creation, unrefined components, water), guideline (environment control, disintegration control, water guideline), social administrations (culture and diversion), and backing administrations (soil development and nourishment cycling), which contrasts in amount and quality. These structures support biological cycles and assist with supporting life on the planet. Environment administrations are straightforwardly connected with the kind of biological system, e.g., landuse and land cover (LULC) type in a specific region. Because of colossal populace development, financial turn of events, and urbanization, LULCs have gone through significant changes all over the planet in late many years, especially from forests to agrarian land, building area, and brushing land. Dynamic changes in LULC are perceived to be quite possibly of the main figure environment misfortune.

The complete ESV upsides of chosen scene have been assessed utilizing the accompanying equation:

$$\text{ESV} = \sum(\text{Ak} \times \text{V Ck}) \tag{13.10}$$

where ESV is the assessed environment administration esteem; AK is the region (ha) of LULC classification k and V; CK esteem coefficient (US \$ ha^{-1} year^{-1}) for LULC classification k.

The progressions in ESV esteem are determined by the result assessed upsides of distinction for each LULC classification reference year. The level of ESV has been determined through the accompanying equation:

$$\text{Level of ESV change} = \left(\left(\text{ESV}_{\text{final year}} - \text{ESV}_{\text{initial year}}\right)/\text{ESV}_{\text{initial year}}\right) \times 100 \tag{13.11}$$

where ESV means the complete assessed environment administration esteem. Positive qualities demonstrate an increment and negative qualities mean a diminishing in amount. Besides, we have likewise determined the singular environment capability of the study area utilizing the accompanying equation:

$$\text{ESV} = \sum \text{Ak} \times \text{V C f k} \tag{13.12}$$

ESV f = Environment administration worth of biological system capability ‘f’, Ak is the region (ha) and V C f k is the worth coefficient of capability ‘f’ (US \$ ha^{-1} year^{-1}) for LULC classification ‘k’.

The entire methodological framework has been reflected in the following figure (Fig. 13.3).

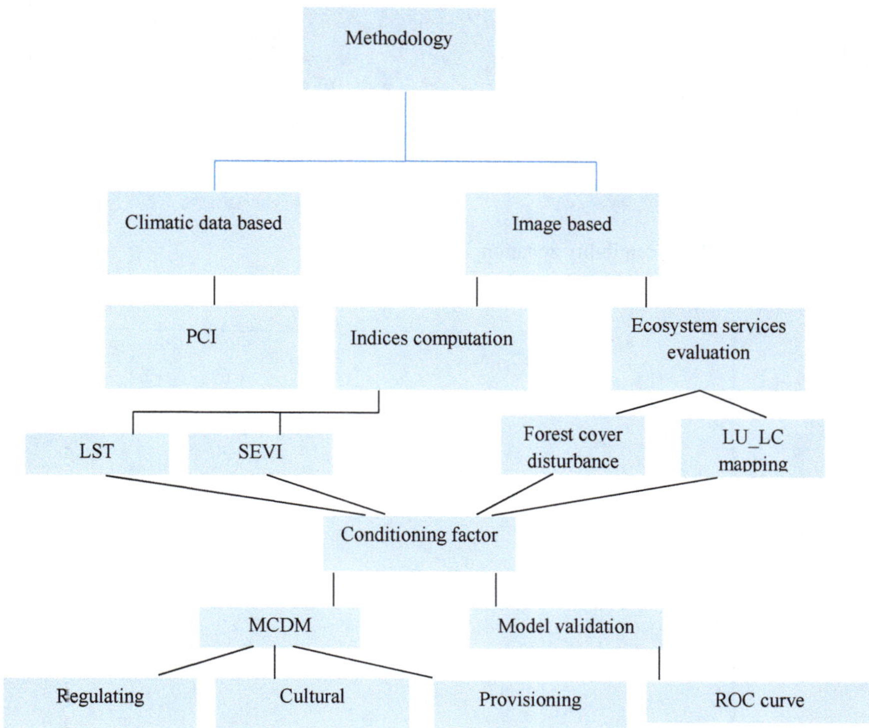

Fig. 13.3 Methodological flowchart

13.3 Results and Discussions

In the present research, the recurrence proportion model grouping has been utilized (Fig. 13.4) and ensembled with the ROC bend (Fig. 13.5). The avalanche molding factors to be specific, slant, precipitation, geomorphology, soil, and vegetation record are utilized as the contribution of the FRM characterization.

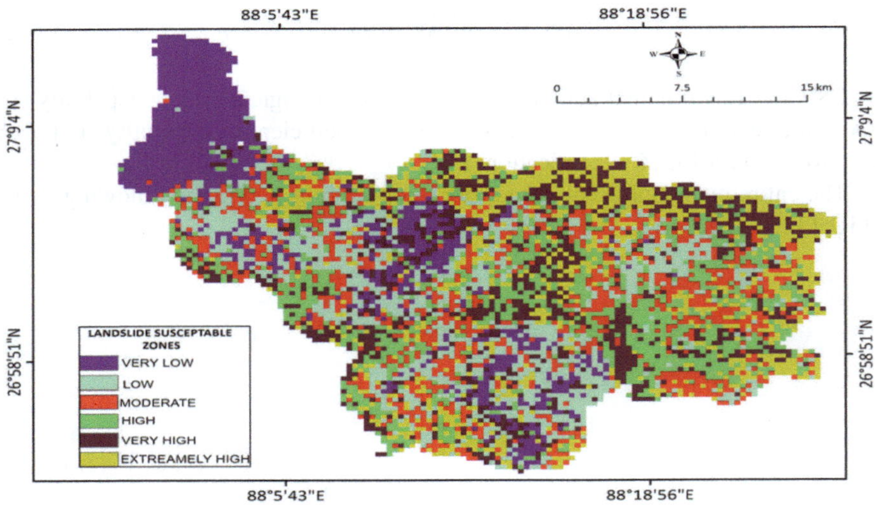

Fig. 13.4 Landslide susceptibility zonation

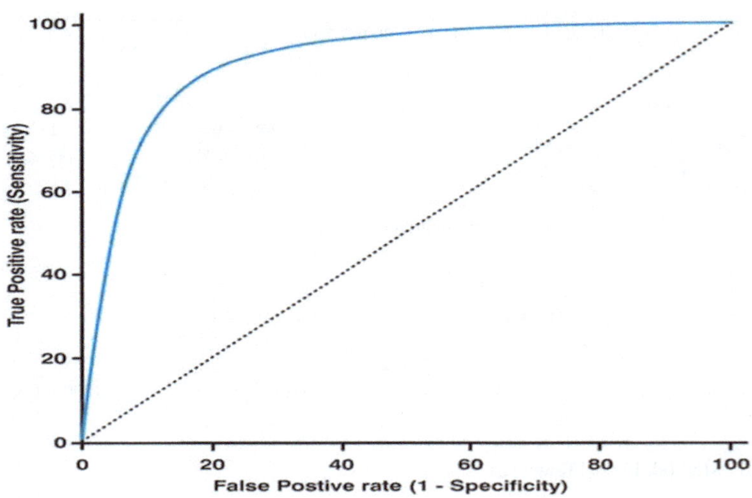

Fig. 13.5 ROC curve to validate the predicted landslide susceptible zones

These landslide vulnerability maps (LSMs) have been arranged into six classifications to be specific, exceptionally low, low, moderate, high, exceptionally high, and very high defenselessness to avalanches, involving the regular breaks order strategy in the GIS environ. The four avalanche susceptible classes of low, medium, high, and extremely high covered 1071 km^2 (34%), 813 km^2 (25.8%), 635 km^2 (20.2%), and 630 km^2 (20%) region of the areas, separately. The regions in these distributions nearer to streets, for example, Pulbazar-jorethang Street, slope truck street, shopping center street, and Rishi Street, are exceptionally impacted via sudden avalanches. The other basic avalanche regions are the 6th mile Bazar region, Pubong Kashmahal, Naya Basti region, Kolbong, Limbic, and Sukhiapokhri regions. The concentrated elements deciding avalanche risk in these sectors are vegetation degradation, weighty precipitation, slant, and soil. However, the landslide risk zones of Pulbazar, Sukhiapokhri, and Rangliot community development blocks are meticulously drawn out in the study. The government will be able to lessen the impact of the landslide with the assistance of this study, which will also raise public awareness of sustainable development.

A ROC bend is a visual portrayal that shows how well a twofold classifier framework performs. In this specific circumstance, it is being utilized to assess the precision of a defenselessness zonation model by contrasting it with a realized avalanche zone. The x-pivot of the ROC bend addresses the misleading positive rate, which demonstrates the extent of non-avalanche regions that are inaccurately named powerless. The proportion of actual landslide areas correctly identified as susceptible is represented by the true positive rate on the y-axis.

The ROC bend investigates the connection between the genuine positive rate and the bogus positive rate, giving experiences into the model presentation. The region under the ROC bend (AUC) is a measurement used to evaluate the general classifier execution. The AUC esteem goes from 0 to 1, with higher qualities showing better execution. For this situation, the bend esteem shows that the AUC esteem is moving toward 1, proposing that the weakness zonation model exhibits high particularity. At the point when the AUC esteem is near 1, it infers that the model actually distinguishes regions that are truly inclined to avalanches while limiting misleading up-sides. The ROC bend and AUC values have been created to survey and approve the exactness and explicitness of the defenselessness zonation model in distinguishing avalanche inclined regions when contrasted with a current avalanche zone. Over the 41-year period, the average annual rainfall varied between approximately 2126.5–4203.2 cm. The highest annual rainfall occurred in 1989 with a value of 4203.2 cm, while the lowest was recorded in 2004 with 2126.5 cm. Concentrating at individual months, the highest rainfall was observed in the month of June, reaching a maximum of 6371.1 cm in 1995. This was followed by the month of July, which experienced a maximum rainfall of 2959.1 cm in 1995. August recorded a maximum rainfall of 1394.5 cm in 1995. Analyzing the specific years, 1995 stands out as a year with extremely high rainfall, experiencing a total precipitation of 108,474.0 cm. Other notable years in terms of total rainfall include 1983 with 30,474.4 cm, 1987 with 17,746.8 cm, and 1988 with 28,849.9 cm. Analyzing the precipitation concentration index (PCI), which measures the evenness of rainfall distribution, the values ranged

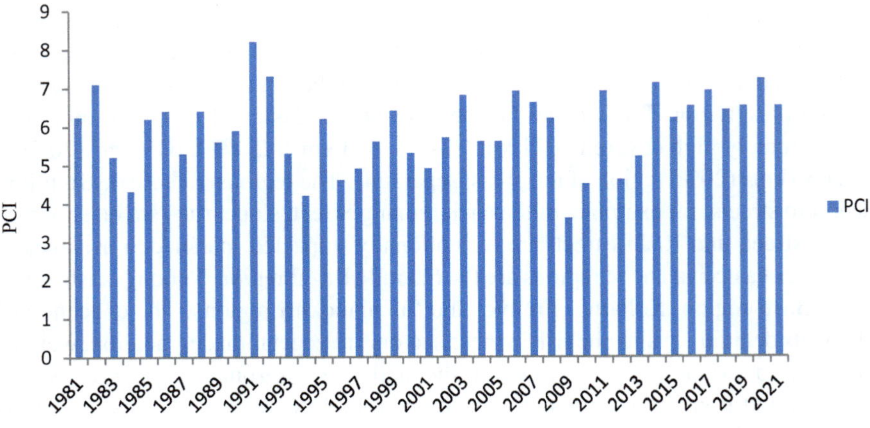

Fig. 13.6 PCI distribution for the study area

from 3.901 to 8.1626. A lower PCI suggests a more evenly distributed rainfall, while a higher PCI indicates a greater concentration of rainfall in certain periods. In this dataset, the highest PCI values have been recorded in 1991 (8.1626), 2004 (7.4659), and 2012 (7.6472), indicating a more concentrated pattern of rainfall during those years. Conversely, the lowest PCI value was observed in 2010 (3.901), suggesting a more evenly distributed rainfall pattern during that year (Fig. 13.6). Overall, the 41-year rainfall data demonstrate considerable variation in annual rainfall amounts and the concentration of precipitation. The fluctuations in rainfall amounts and the varying PCI values reflect the dynamic nature of rainfall patterns over the years, highlighting the importance of studying long-term trends to understand the climatic characteristics of a region.

The land surface temperature map of the study area has been demonstrated in figure (Fig. 13.7). The maps are revealing that the high temperature is 26.2507 °C and the low temperature is −2.39035 °C. North-western part of the study area contains very low-temperature frequency, on the other hand south and eastern part of the study area contains very high-temperature frequency while northern and central part contain high to moderate temperature frequency.

The SEVI map has revealed that the higher values indicate areas with denser and healthier vegetation cover (Fig. 13.8). High SEVI frequency values correspond to areas with high vegetation density, such as dense forests, and agricultural fields. The highest SEVI frequency value is 55.5455 units and the lowest SEVI frequency value is found to be 3.04947 units. The north-western part of the study area contains high vegetation index values along with some areas of the eastern part also revealing the same nature of vegetation canopy coverage.

The generated map (Fig. 13.9) represents the ecological impact of forest degeneration caused by landslides. The researchers have identified the need to understand the ecological consequences of such hazards and focuses on studying the vegetation

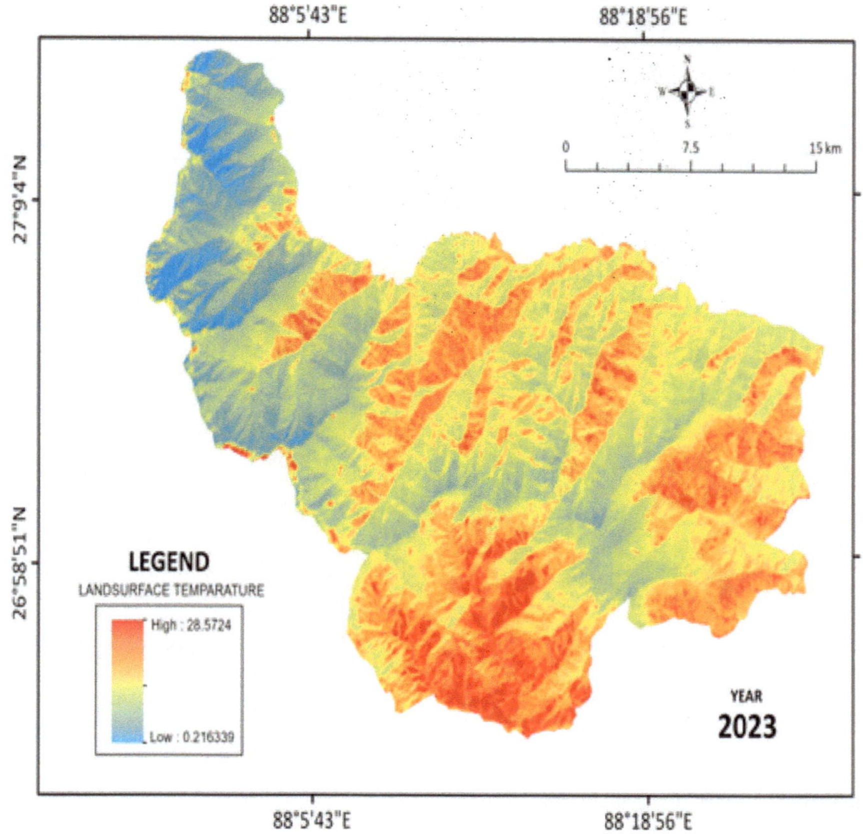

Fig. 13.7 LST mapping of the study area

dynamics as an indicator. The primary objective is to map the distribution of forests in the study area over time.

By comparing the temporal forest patches, we have calculated the areas that have undergone degeneration or deforestation. It is observed that the existing landslide spots coincide with areas where the forests are degrading. This observation suggests that landslides have a significant impact on the forest ecosystem.

The LULC map (Fig. 13.10) provides insights into the changes in land cover over time. It indicates the persistence of forest cover, a slight expansion of water bodies, and the growth of both cultivated and built-up land areas. This information is valuable for monitoring land use trends, assessing environmental impacts, and informing land management and planning decisions.

Ecosystem services valuation allow us to understand the contributions of different land cover categories in terms of provisioning services. Forests play a significant role in food production, raw materials, genetic resources, and water supply. Cultivated areas primarily contribute to food production, while water bodies provide food

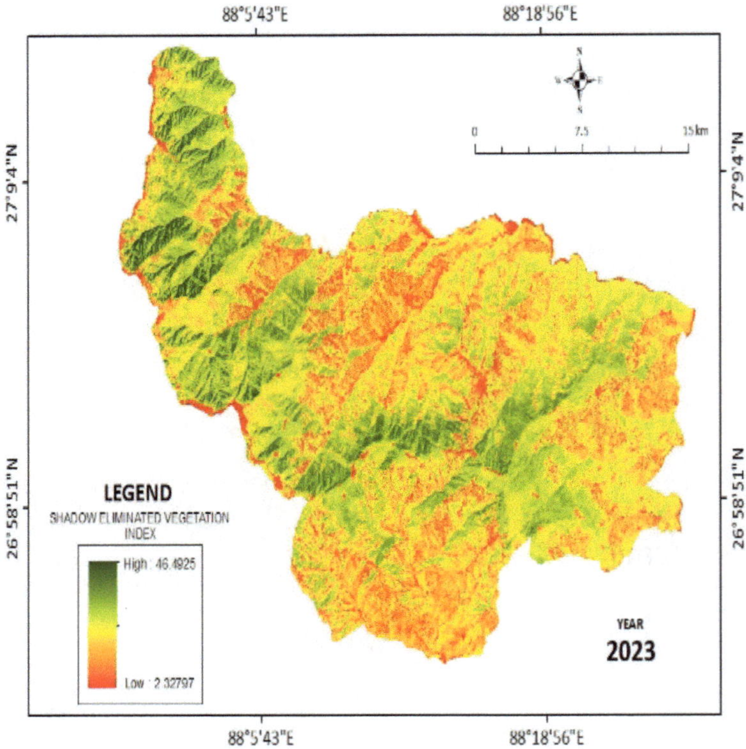

Fig. 13.8 SEVI map of the study area

production and water supply services. Built-up land, in this dataset, does not report any provisioning services. Recognizing and quantifying the provisioning services provided by different land cover categories is crucial for sustainable land management, resource planning, and decision-making processes related to food security, material production, and water resource management (Figs. 13.11, 13.12 and 13.13).

These data allow us to understand the extent of different land cover categories and their contributions to regulating services (Fig. 13.14). Forests play a crucial role in providing climatic regulation, waste treatment, erosion control, and water regulation services, which are vital for maintaining ecosystem balance and resilience. Cultivated areas contribute to pollination and biological control services, supporting agricultural productivity and pest management. Water bodies provide waste treatment services, while built-up land does not report any regulating services in this dataset. The heights of existing landslides vary across different categories. Ranging from the very low category heights between 204 and 1340 m represents minor disturbance in the landscape. Moving up the scale, the low category includes heights from 1341 to 1949 m. The moderate category spans from 1950 to 2047 m and the high category covers heights between 2048 to 2319 m. And finally, heights ranging from 2320 to 2791 m are the highest in elevation among the categories provided, indicating a

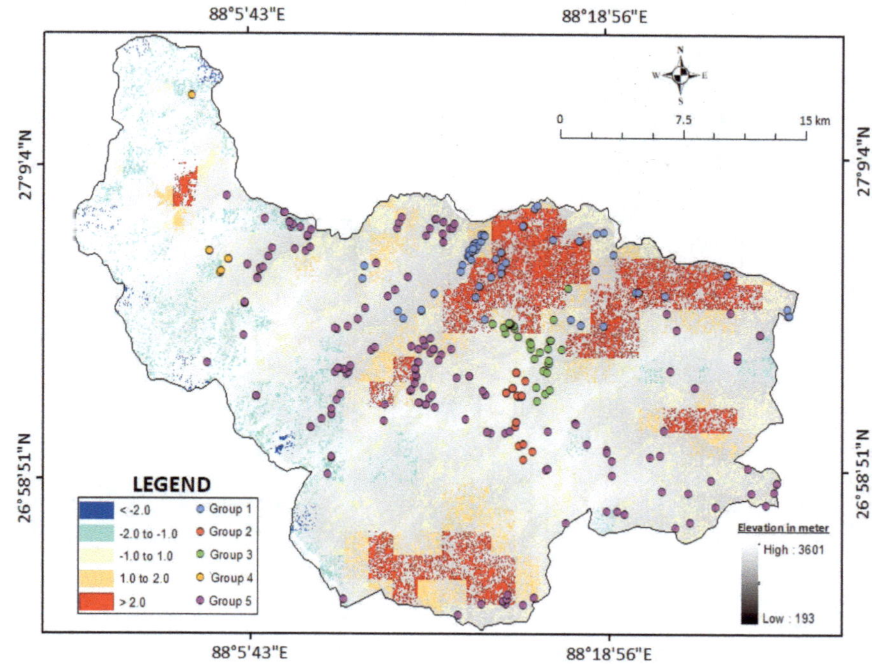

Fig. 13.9 Forest cover disturbance mapping

significant vertical displacement. The amount intensity duration and concentration of rainfall can significantly affect the stability of slope and trigger landslides in our study area. The position or topography of our study area influences landslide occurrence about rainfall concentration. Slopes that face the direction of prevailing rain storms or received higher amounts of rainfall due to their geographical location are more susceptible to landslides in our study area. Here we find an increasing trend of rainfall in our study area which made this region more susceptible to landslides. The area with high forest distribution, forest loss, and unauthorized construction is very much prone to landslide susceptible zones. The landslides susceptible zones are varying from 1340 to 2791 m. These areas continuously become more and more landslide susceptible zones. There is a significant impact on the forest ecosystem's service value regarding provisioning service.

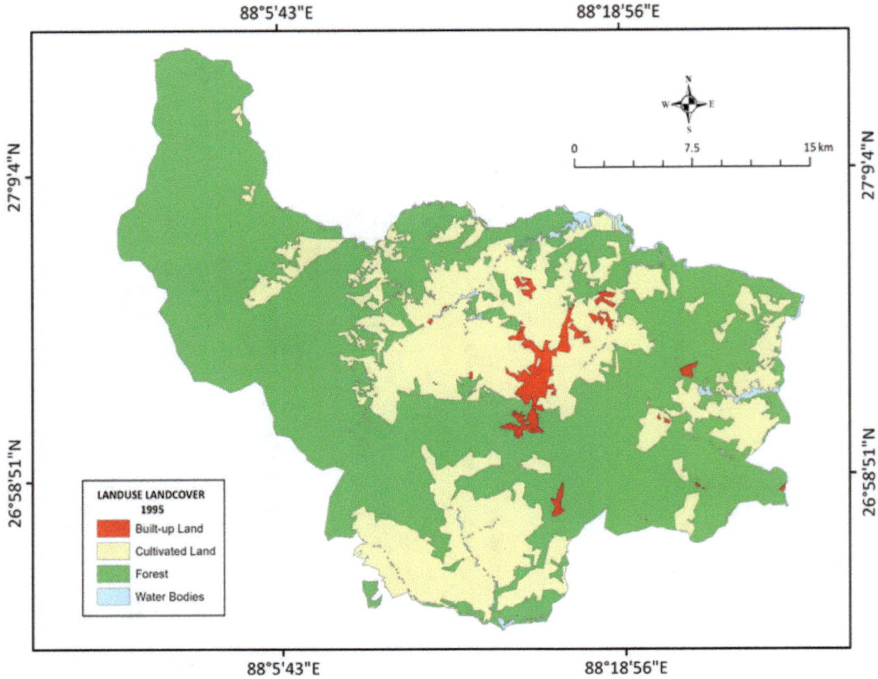

Fig. 13.10 LULC map of the study area

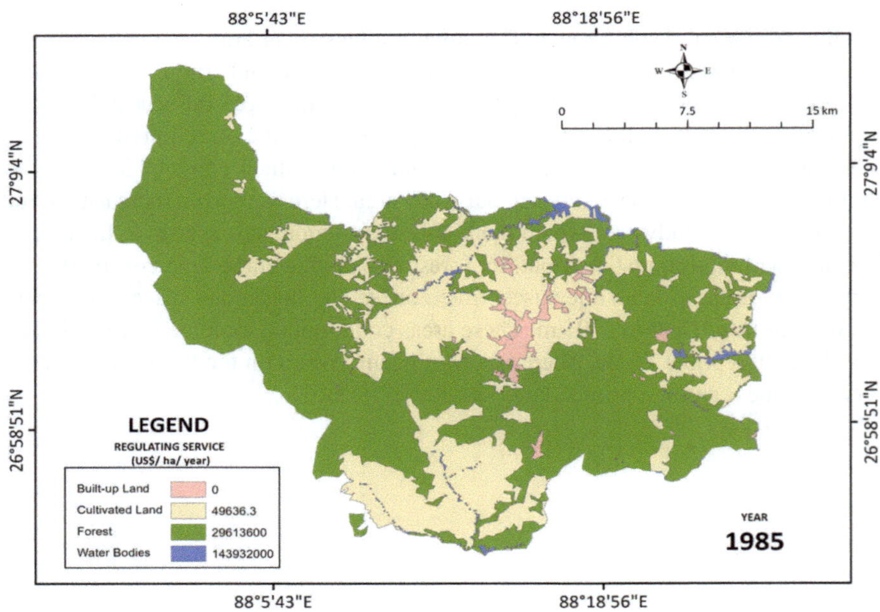

Fig. 13.11 Provisioning service value in the year 1985

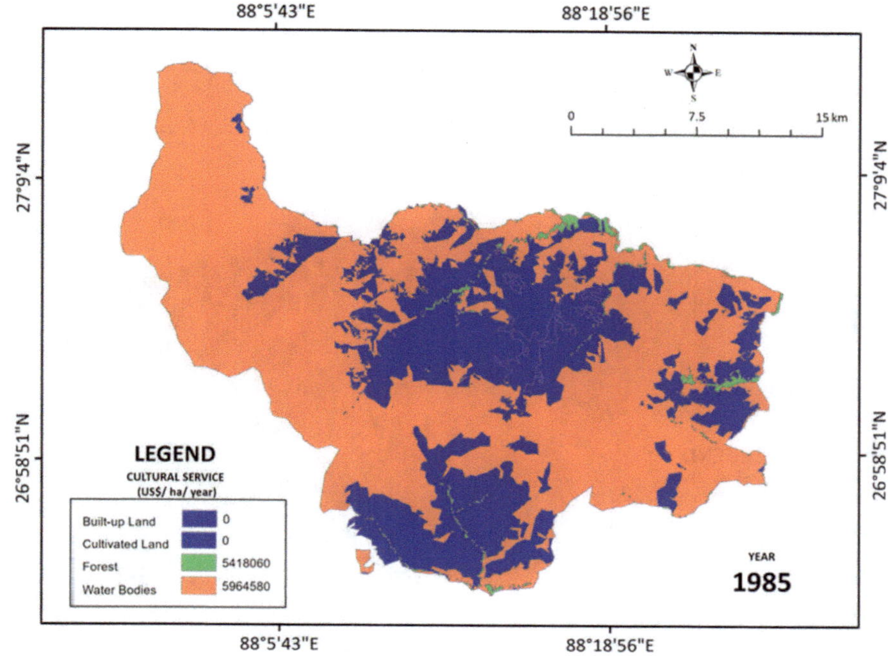

Fig. 13.12 Cultural service value in the year 1985

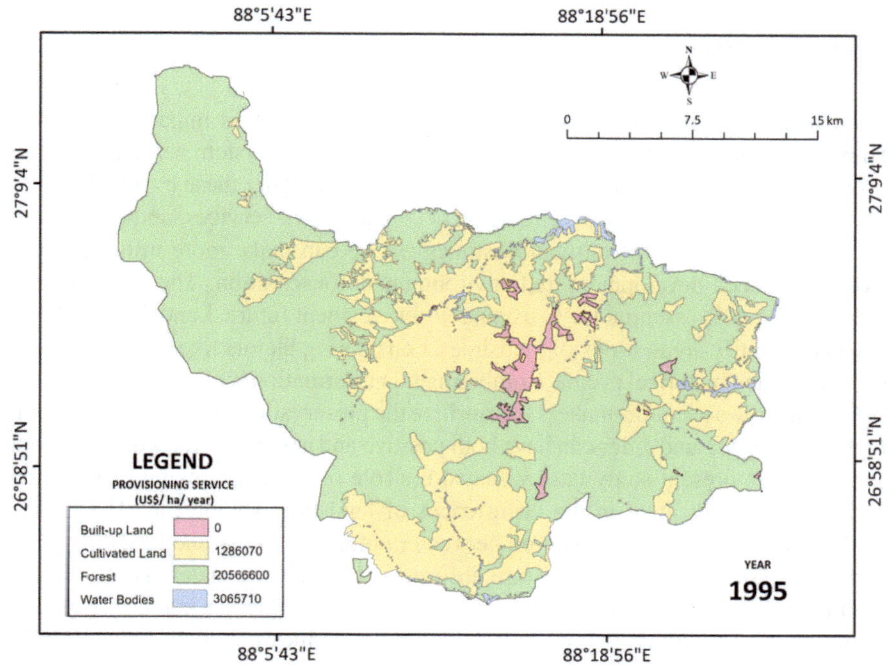

Fig. 13.13 Provisioning service value in the year 1995

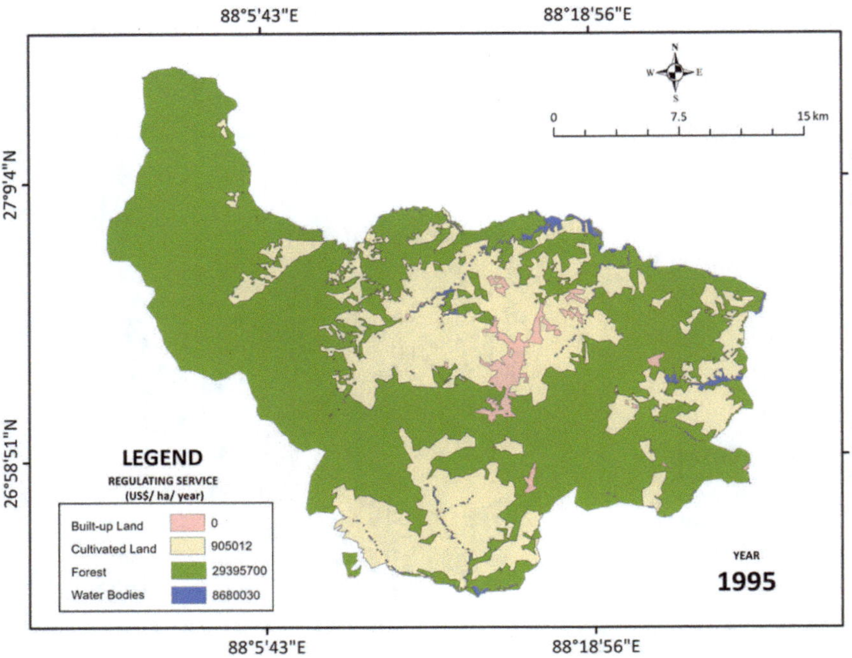

Fig. 13.14 Regulating service value in the year 1995

13.4 Conclusion

Landslides are complex geological events that have profound impacts on ecosystems and society. Combining landslide modeling and ecosystem service valuation provides a holistic approach to understanding and managing these events. By identifying vulnerable areas, assessing impacts on ecosystem services, and quantifying the economic benefits of mitigation strategies, we can make more informed decisions to balance development and environmental conservation. The integration of these two fields is essential for a sustainable and resilient future. Landslide modeling can help identify areas at risk of landslides, considering factors like topography, soil properties, and historical events. Combining this information with ecosystem service valuation allows decision-makers to prioritize the preservation of valuable services in high-risk zones. Landslides can have both positive and negative effects on ecosystem services. The present study has evaluated the role of LULC, climatic extremeness and human involvement in the occurrences of landslide phenomena. The present study identifies a reciprocal relation between rainfall variability and landslide as the phenomena depends on the moisture absorption by the rock materials. The places with better ecosystem protection measures are marked by lesser intensity landslide occurrences. The computation of ROC curve determines the spatial auto correlation in the time dependent landslide occurrences. Decision-makers can evaluate the consequences of land development or infrastructure projects on the landscape and its services. Ecosystem service valuation is the process of assigning economic, social, or ecological values to the benefits provided by ecosystems. Though the procedure is time consuming but it will have long endurance and reprimand the unscientific developmental projects. This valuation is essential for decision-makers, as it helps to recognize the trade-offs between land use choices and the preservation of ecosystem services.

References

Abraham MT, Satyam N, Pradhan B, Alamri AM (2020) Forecasting of landslides using rainfall severity and soil wetness: a probabilistic approach for Darjeeling Himalayas. Water (Switzerland) 12(3):1–19. https://doi.org/10.3390/w12030804

Ali SA, Parvin F, Vojteková J, Costache R, Linh NTT, Pham QB, Vojtek M, Gigović L, Ahmad A, Ghorbani MA (2021) GIS-based landslide susceptibility modeling: a comparison between fuzzy multi-criteria and machine learning algorithms. Geosci Front 12(2):857–876. https://doi.org/10.1016/j.gsf.2020.09.004

Alsabhan AH, Singh K, Sharma A, Alam S, Pandey DD, Rahman SAS, Khursheed A, Munshi FM (2022) Landslide susceptibility assessment in the Himalayan range based along Kasauli–Parwanoo road corridor using weight of evidence, information value, and frequency ratio. J King Saud Univ Sci 34(2):01–12. https://doi.org/10.1016/j.jksus.2021.101759

Anis Z, Wissem G, Vali V, Smida H, Mohamed Essghaier G (2019) GIS-based landslide susceptibility mapping using bivariate statistical methods in North-western Tunisia. Open Geosci 11(1):708–726. https://doi.org/10.1515/geo-2019-0056

Araújo JR, Ramos AM, Soares PMM, Melo R, Oliveira SC, Trigo RM (2022) Impact of extreme rainfall events on landslide activity in Portugal under climate change scenarios. Landslides 19(10):2279–2293. https://doi.org/10.1007/s10346-022-01895-7

Asmare D (2023) Application and validation of AHP and FR methods for landslide susceptibility mapping around choke mountain, Northwestern Ethiopia. Sci Afr 19:e01470. https://doi.org/10.1016/j.sciaf.2022.e01470

Bansal BK, Verma M, Gupta AK, Prasath RA (2022) On mitigation of earthquake and landslide hazards in the eastern Himalayan region. In: Natural hazards, vol 114, Issue 2. Springer Science and Business Media B.V., pp 1079–1102. https://doi.org/10.1007/s11069-022-05448-y

Berhane G, Kebede M, Alfarah N, Hagos E, Grum B, Giday A, Abera T (2020) Landslide susceptibility zonation mapping using GIS-based frequency ratio model with multi-class spatial data-sets in the Adwa-Adigrat mountain chains, northern Ethiopia. J Afr Earth Sci 164:103795. https://doi.org/10.1016/J.JAFREARSCI.2020.103795

Bhatta B (2011) Remote sensing and GIS, 2nd edn. Oxford University Press

Biswas SS, Pal R (2016) Causes of landslides in Darjeeling Himalayas during June-July, 2015. J Geogr Nat Disasters 6(2). https://doi.org/10.4172/2167-0587.1000173

Chawla A, Chawla S, Pasupuleti S, Rao ACS, Sarkar K, Dwivedi R (2018) Landslide susceptibility mapping in Darjeeling Himalayas, India. Adv Civ Eng 2018. https://doi.org/10.1155/2018/6416492

Chen W, Zhang S (2021) GIS-based comparative study of Bayes network, Hoeffding tree and logistic model tree for landslide susceptibility modeling. CATENA 203(March):105344. https://doi.org/10.1016/j.catena.2021.105344

De SK (2017) Landslides and human interference in Darjiling Himalayas, India. Revista de Geomorfologie 19(1):44–57. https://doi.org/10.21094/rg.2017.014

Demir G (2019) GIS-based landslide susceptibility mapping for a part of the North Anatolian fault zone between Reşadiye and Koyulhisar (Turkey). Catena 183(October 2018):104211. https://doi.org/10.1016/j.catena.2019.104211

Dikshit A, Sarkar R, Pradhan B, Segoni S, Alamri AM (2020) Rainfall induced landslide studies in Indian Himalayan region: a critical review. Appl Sci (Switz) 10(7). https://doi.org/10.3390/app10072466

Doke A (2017) Malin village landslide: a geological hazard. Indian Streams Res J 7(5):4–11

Dolui S, Chakraborty S (2022) Identification of preferable ecotourism destinations in Purulia District, West Bengal (India): AHP and GIS approach. J Geogr Stud 6(2):73–93. https://doi.org/10.21523/gcj5.22060203

Fayez L, Pazhman D, Pham BT, Dholakia MB, Solanki HA, Khalid M, Prakash I (2018) Application of frequency ratio model for the development of landslide susceptibility mapping at part of Uttarakhand State, India. Int J Appl Eng Res 13(Issue 9). http://www.ripublication.com

Gao Y, Fan LC, Deng J, Zhao S, Dai W, Zhu J, He Z, Yang Z, Lane SN (2023) Primary and potential secondary risks of landslide outburst floods. Nat Hazards. https://doi.org/10.1007/s11069-022-05776-z

Goswami A, Sen S, Sanyal R (2022a) Delineation of landslide hazard zones of a part of Sutlej Basin in Himachal Pradesh through frequency ratio model. In: Convergence of deep learning and artificial intelligence in internet of things. CRC Press, Taylor and Francis, pp 211–230

Goswami A, Sen S, Chakraborty PK (2022b) Examination of vegetation health and its relation with normalized difference built-up index a study on Rajarhat block of North 24 Parganas District of West Bengal, India. In: Gupta R, Rana AK, Dhawan S, Cengiz K (ed) Advanced sensing in image processing and IoT, 1st edn. CRC Press, Taylor and Francis, pp 255–276. https://doi.org/10.1201/9781003221333

Hammad Khaliq A, Basharat M, Talha Riaz M, Tayyib Riaz M, Wani S, Al-Ansari N, Ba Le L, Thi Thuy Linh N (2023) Spatiotemporal landslide susceptibility mapping using machine learning models: a case study from district Hattian Bala, NW Himalaya, Pakistan. Ain Shams Eng J 14(3):101907. https://doi.org/10.1016/J.ASEJ.2022.101907

Huangfu W, Wu W, Zhou X, Lin Z, Zhang G, Chen R, Song Y, Lang T, Qin Y, Ou P, Zhang Y, Xie L, Huang X, Fu X, Li J, Jiang J, Zhang M, Liu Y, Peng S et al (2021) Landslide geo-hazard risk mapping using logistic regression modeling in Guixi, Jiangxi, China. Sustainability (Switzerland) 13(9):01–20. https://doi.org/10.3390/su13094830

Hussain MA, Chen Z, Wang R, Shah SU, Shoaib M, Ali N, Xu D, Ma C (2022) Landslide susceptibility mapping using machine learning algorithm. Civ Eng J (Iran) 8(2):209–224. https://doi.org/10.28991/CEJ-2022-08-02-02

Kahlon S, Chandel VBS, Brar KK (2014) Landslides in Himalayan Mountains: a study of Himachal Pradesh, India. Int J IT Eng Appl Sci Res Explor Int Res J Consort 3(9):2319–4413

Khan H, Shafique M, Khan MA, Bacha MA, Shah SU (2019) Landslide susceptibility assessment using Frequency Ratio, a case study of northern Pakistan. Egypt J Remote Sens Space Sci 22(1):11–24. https://doi.org/10.1016/j.ejrs.2018.03.004

Korup O, Densmore AL, Schlunegger F (2010) The role of landslides in mountain range evolution. Geomorphology 120(1–2):77–90. https://doi.org/10.1016/j.geomorph.2009.09.017

Meghanadh D, Kumar Maurya V, Tiwari A, Dwivedi R (2022) A multi-criteria landslide susceptibility mapping using deep multi-layer perceptron network: a case study of Srinagar-Rudraprayag region (India). Adv Space Res 69(4):1883–1893. https://doi.org/10.1016/J.ASR.2021.10.021

Mekonnen AA, Raghuvanshi TK, Suryabhagavan KV, Kassawmar T (2022) GIS-based landslide susceptibility zonation and risk assessment in complex landscape: a case of Beshilo watershed, northern Ethiopia. Environ Chall 8:100586. https://doi.org/10.1016/J.ENVC.2022.100586

Villuri VGK, Saha A, Bhardwaj A (2022) Development and assessment of GIS-based landslide susceptibility mapping models using ANN, Fuzzy-AHP, and MCDA in Darjeeling Himalayas, West Bengal, India. Land 11(10):01–28. https://doi.org/10.3390/land11101711

Chapter 14
Landslide Vulnerability Analysis of Tourist Spots Through Shannon Entropy Model: A Case Study on Rudraprayag, Uttarakhand

Saswati Roy and Suhel Sen

Abstract Landslides are very common phenomena along the full stretch of Himalayan Mountain belt of northern India. Every year the foothill states like Himachal Pradesh, Uttarakhand, West Bengal, Sikkim, Meghalaya and Arunachal Pradesh experience severe landslides especially during the monsoon season. Higher frequency and intensity of landslide imparts a negative impact on the tourism industry which is one of the major economic sectors of this region. Rudraprayag district of Uttarakhand is endowed with a number of religious tourist spots like Gauri Kund, Madmaheswar temple, Kalimath, Omkareswar temple, Trigunayan temple of which Kedarnath derives special mention. Every year huge influx of tourists visit these places of Rudraprayag which generates good revenue. The study area lies within the landslide vulnerability zone and experiences the aftermath of landslide every year which affects the tourism industry in turn. The present research work attempts to perform landslide susceptibility using Shannon Entropy Model, a widely used decision making tool coupled with Geospatial techniques and applied on some selected tourist spots of Rudraprayag district in Uttarakhand. The outcome of the research work would reveal the validation of the Shannon entropy model in landslide susceptibility assessment of a few tourist spots' of Rudraprayag. This paper would help in prognosticating the future of the region's vulnerability towards landslide and henceforth in judicious management planning for attaining sustainable tourism in the study area.

Keywords Rudraprayag · Tourism industry · Landslide susceptibility · Shannon entropy · Geospatial techniques · Sustainable tourism

---

S. Roy (✉)
Assistant Professor, Department of Geography, Sarsuna College, Kolkata, West Bengal 700061, India
e-mail: saswatiroy2008@yahoo.co.in

S. Sen
Department of Geography, Lovely Professional University, Phagwara, Punjab 144411, India

G. K. Panda et al. (eds.), *Landslide: Susceptibility, Risk Assessment and Sustainability*, Advances in Natural and Technological Hazards Research 52,
https://doi.org/10.1007/978-3-031-56591-5_14

14.1 Introduction

Natural disasters act as an obstacle in the path of development of a society on all perspectives (Goswami et al. 2022). Landslide is also not an exception in this regard. Landslide is defined as a situation under which the rocks and debris resting on hill slope moves downward under the influence of gravity (Singh 2011). Landslide is considered to be a manifestation of slope instability and is often triggered by external factors like heavy rainfall, large scale deforestation, injudiciously unplanned human activity and large scale urban development on the hill slope region (Pal et al. 2016). The Himalayan Mountains also known as 'Abode of Snow' is located in the northern periphery of Indian sub-continent and runs for a total length of 2400 km from the west to the east (Khullar 2014). The Himalayan range acts as a barrier between India and Tibetan Plateau. Being one of the young fold mountains of the world, it is considered to be rising even at present and thus has not yet achieved stability. As a result, the mountain region is considered to be prone to seismic activity and experience earthquakes. Owing to such earthquakes, loose rocks and debris on the slope often gets disturbed and comes down-slope causing landslides. In higher altitudes, earthquakes often cause avalanches. The Nepal earthquake of magnitude 7.8 which occurred on 25th April, 2015 caused avalanche in Everest Base camp and killed about 24 people (Goswami et al. 2022).

Landslides in Himalayas are often stimulated by heavy rainfall. Since Himalayas block the monsoon winds and cause heavy rainfall on its south facing slope, the rock and debris on the steep slope become viscous and cannot stay on the slope surface and comes down in the form of rock fall, earth flow, mud flow and debris flow (Singh 2011). Besides, sudden landslides may also occur due to outbreak of cloudbursts (Araújo et al. 2022). The Kedarnath disaster is a manifestation of such an event (Goswami et al. 2022). Besides, all these natural factors, landslides also occur in this mountain belt due to unplanned human activity. Since the mountain is endowed with some of the world's most beautiful hill-stations and tourist spots, man is undertaking developmental activities without taking into account the vulnerability of the spots (Bianchini et al. 2017; Chaudhary et al. 2022). Man is constructing and expanding roads by blasting hill slopes indiscriminately. Owing to the increasing popularity of the tourist spots, they are constructing expansive multi-storied buildings and hotels. While doing all these activities, they are totally ignoring the natural environment. As a result, anthropogenic landslides have become very common (Fayez et al. 2018).

Tourism industry in India has emerged as a major source for sustainable economic development because of its strong forward and backward linkage with other prominent sectors of economy. India ranked 17th in world in term of international tourist receipts and 10th in Asia Pacific in term of international tourist arrival and 7.7 million tourists arrived in India during 2014. It showed a growth rate of 10.6% where as previously it was 5.6% in 2013. This growth rate attracts and motivates many researchers to obtain in depth knowledge of tourism sector. According to the published reports of World Travel and Tourism Council (WTTC), tourism in India is able to generate 6.3% of India's total employment and it increases the level of economic activities in developed or underdeveloped region by increasing the tourist inflow. It helps the local communities and government for generating the revenues by various sources (taxes) like luxury transportation, and income from tourist protected areas (TPA) via accommodation facilities, road network expansion, communication facilities, medical and health facilities, banking facilities, proper vehicle parking facilities, safety and security etc. In terms of growth rate of domestic tourists during 2013 Jammu & Kashmir was an exceptional case. On the contrary the state of Uttarakhand shows −25.67% growth rate and there may be so many reasons for negative growth rate like less government support and policy initiatives, pollution, climate degradation, lack of management (Tourism and Disaster) at government and private level and 2013 Himalayan T-Sunami was the major reason of this negative growth rate (Rana and Kumar 2017).

Landslides in mountains affect tourism industry the most particularly the Himalayan region. Since landslides cause large scale disruption of the transport sector due to damage of road-network, it severely affects the tourism industry. For instance, when tourists travel to Bhutan, they enter Bhutan from Jaigaon in Alipurduar District in West Bengal. However, before entering into Bhutan from Jaigaon, the information about weather condition of the upper altitudes is collected. If there are chances of heavy rainfall, then car drivers of Bhutan do not take the risk of driving the tourists to Thimpu as the route is highly susceptible to landslide. Shopkeepers of Jaigaon market are of the opinion that their income gets highly affected during monsoon season as tourists do not prefer to come to the area due to the threat of getting trapped in a landslide (Goswami et al. 2022). Unplanned human activity is also responsible for making many tourist spots landslide prone as a result of which they have lost their popularity and laid a negative impact on the Eco-tourism industry. The home stay owner in Roghi Village of Himachal Pradesh opined that they do not get any tourists during rainy season for which they engage themselves in some other means for earning their daily bread and thus short-term livelihood transformation takes place. As tourism industry of the Himalayas contributes to the entire GDP of the country, proper management and planning is highly essential (Goswami et al. 2022).

The objectives of this work are firstly to identify the major conditioning factors affecting landslide within the study area, secondly to prepare thematic maps of all the selected conditioning factors, thirdly assign criteria weights on the basis of past landslide events using Shannon entropy model, fourthly preparing landslide susceptibility map by validating the model and finally using the same to analyse landslide susceptibility of some selected popular tourists spots of Rudraparayg.

The entropy concept, introduced by Shannon, measures the uncertainty of a random variable in the information theory. The expected value of the information contained in a message is quantified using this concept. Shannon entropy uses a mathematical function for explaining the expected value of alternative conditions of a variable. The entropy function always has a positive value. Due to many attributes, the entropy concept is accepted as an objective criterion that can be used in measuring information content of any statistical process.

This paper has been prepared through the application of Shannon Entropy in which weights are assigned by selecting five major conditioning factors of landslide. The Shannon entropy model is validated by calculating the Landslide Density Index (LDI) by taking into account 327 past landslide events. Shannon Entropy and geospatial technique has been applied to perform a vulnerability analysis of a few selected tourist spots i.e. to analyze the degree of chances of occurrence of landslides on those spots so that proper mitigation and safety measures can be implemented. It is expected that the present research would not only reveal the vulnerability of tourist spots to landslides but would also help in evaluating judicious management planning for attaining sustainable tourism in the study area.

14.2 Methodology

In the era of rapidly advancing technology, GIS and Remote Sensing has become an indispensable weapon in the hands of the researchers (Fayez et al. 2018; Sonker et al. 2021). Using remote sensing technology, data about any place of the world can be collected easily even without going to the field. GIS is used to prepare maps which act as tools for quick and proper decision making. The rapidly advancing technology has all become successful in providing in the hands of budding researchers ample data in digital format which can be fed into the GIS environment and can be processed to generate maps (Sarif et al. 2023). Geospatial techniques and statistical tools at present has become very popular in preparing landslide susceptibility maps (Alsabhan et al. 2022; Mersha and Meten 2020). A landslide susceptibility map is very important for bringing about balanced development in fragile Himalayan environment and it is a key which ensures proper urban planning, land management and safe human occupation (Aditian et al. 2018; Akinci 2022; Ghorbanzadeh et al. 2019; Huangfu et al. 2021).

Fayez et al. (2018) applied geospatial technique and statistical technique of Frequency ratio to demarcate the landslide susceptibility zones of a part of Uttarakhand state in India and the study revealed that the northern part of the study area is more susceptible to landslide than the southern part (Fayez et al. 2018). Elmoulat and Brahim (2018) used Weights of Evidence (WoE) technique coupled with GIS to prepare the landslide susceptibility zones of Tetouan Ras-Mazari area of northern Morocco of area 590 km^2 and used 126 past landslide events as landslide data inventory (Elmoulat and Brahim 2018). Results of the study showed that areas which are more prone to landslides are having conglomerate, mudstones and marlstone outcrops, have high altitudes and slope of 57° approximately (Elmoulat and Brahim 2018). Sadiki et al. (2022) in their work aimed for the identification of landslide prone areas in Ziz Upper Watershed of Morocco by the application of GIS based Fuzzy Analytical Hierarchy Process (FAHP) and used nine geo-environmental factors namely elevation, precipitation, slope, aspect, lithology, landuse/landcover, streams, fault and roads for the landslide prone area identification (Sadiki et al. 2022). The study revealed that about 16.7% of the total study area falls in high susceptibility zone and the western part of the upstream falls in high susceptible area (Sadiki et al. 2022). Sababhah used geology, slope, flood susceptibility, rainfall, run-off, aspect, elevation and landuse/landcover for preparation of landslide susceptibility map of Northern Highlands in Jordan and it shows that landslides are a serious threat to the road network of the study area as about 53.8% of the total roads were located in very high and high landslide susceptibility zone. Goswami et al. (2022) used Frequency Ratio (FR) technique for determining landslide susceptibility zones of a part of Sutlej basin in Himachal Pradesh and the study showed the northern portion of the basin falls under very high landslide susceptible zone in comparison to the southern part (Goswami et al. 2022). Although this paper mainly focuses on the delineation part but a thorough description of the factors is lacking. Villuri et al. (2022) performed a comparative analysis to determine the degree of performance of landslide susceptibility models prepared using Artificial Neural Network (ANN), Analytical Hierarchy Process (AHP) and Fuzzy Analytical Hierarchical Process (FAHP) and it was seen that Artificial Neural Network (ANN) turned out to be the best with the highest Area Under Curve (AUC) value of 88.1% against AHP and FAHP with AUC values of 86.1 and 85.4% respectively (Villuri et al. 2022). Similarly, Nohani et al. (2019) in their work made an attempt to prepare the landslide susceptibility map of Klijanrestagh Basin of Northern Iran using GIS based bivariate models. The four models used were Frequency Ratio Model (FR), Shannon Entropy (SE), Weights of Evidence (WoE) and Evidential Belief Function (EBF) and it was seen that out of four models used for the preparation of landslide susceptibility map, weights of evidence technique turned out to be relatively more accurate with the highest prediction rate of 0.84 (Nohani et al. 2019). After reviewing all the literature it was evident that implications for future risk was not elaborate and mitigation strategies pertinent to the specified study area were not explicit.

After analyzing all these models, the present study has been done aiming to prepare landslide susceptibility map of Rudraprayag District of Uttarakhand. This is done by applying Shannon Entropy and geospatial technique and a vulnerability analysis of seven selected tourist spots has been performed i.e. to analyze the degree of chances of occurrence of landslides on those spots. It is done so that landslide susceptible zones could be prepared in the long run which would provide a framework for the implementation of judicious planning and executing proper mitigation measures. It is expected that the present research would not only reveal the vulnerability of tourist spots to landslides but would also help in evaluating sensible management planning for attaining sustainable tourism in the study area.

14.2.1 Conditioning Factors

The present research article takes into account five conditioning factors namely elevation, slope, rainfall, drainage density and Normalized Difference Vegetation Index (NDVI) as landslide controlling factors. Five conditioning factors are selected for analysis in this paper as these are the major controlling aspects of landslide in the study area. Moreover database of these factors are available for the relatable pursuance of this paper. Landslides tend to occur more in the areas of higher elevation especially in the mountainous areas. Steep and unstable slopes are considered to be more prone to landslides than the gentle slopes. However, the issue of anthropogenic interference cannot be ignored in this aspect because large scale, unplanned and unscientific human activities on fragile hill slope may trigger landslides even in the areas of gentle slope. Heavier rainfall always stimulates landslides as they make the weathered debris resting on the hillslope viscous and thus compels them to come down under the gravitational influences. Areas with high drainage density also have chances of more landslide as water from the smaller streams percolate through the rock cracks and crevices and trigger undercutting. As a result, the debris resting on it becomes unstable and have more chances to come down as landslides. Vegetation health is also considered to be another important landslide controlling factor as vegetation with good health has more grip on the soil than the vegetation having weak health. Thus, chances of landslide are less in areas with good amount of vegetation with good health and thus both the quantitative and qualitative aspect of vegetation plays a vital role in determining landslide susceptibility. NDVI is one band index that helps in understanding the vegetation health as NDVI values ranging from 0.3 to 0.6 indicate stressed and weak vegetation health while those exceeding 0.6 indicate healthy vegetation (Sarif et al. 2023).

Thematic map of all the conditioning factors were prepared in GIS environment and the 'Natural Breaks' method was used to classify all the thematic layers into five classes. Criteria weights were assigned to the conditioning factors on the basis of number of past landslide events in each class of the conditioning factor.

14.2.2 Database

The landslide map of Rudraprayag has been prepared by using Shannon Entropy model developed by Claude Shannon in 1948. In order to accomplish the present task, five major factors which trigger landslide events are taken into consideration and are termed as 'Landslide Conditioning Factors'. They are elevation, slope, rainfall, drainage density and Normalized Difference Vegetation Index (NDVI). Elevation, slope and drainage density map has been prepared using SRTM Digital Elevation Model downloaded from USGS Earth Explorer website. NDVI map has been prepared using near-infrared band (NIR) and visible red band of Landsat 9 satellite image downloaded from USGS Earth Explorer. The formula used for calculating NDVI is given in Eq. (14.1).

$$\mathrm{NDVI} = \frac{\mathrm{NIR} - \mathrm{Red}}{\mathrm{NIR} + \mathrm{Red}}. \tag{14.1}$$

In Landsat 9, band number 5 is the NIR band while band number 4 is the Red band. Substituting the band number in Eq. (14.1), the final equation for computing

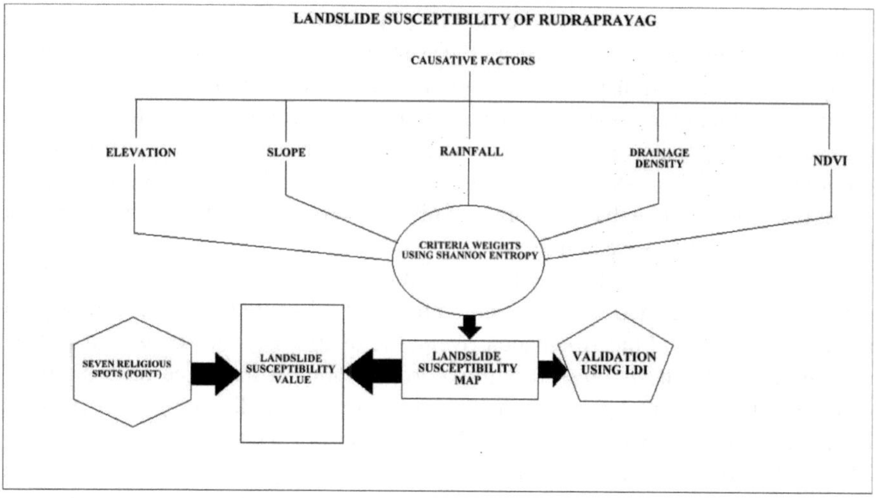

Fig. 14.1 Methodological flowchart

NDVI for Landsat 9 is given in Eq. (14.2).

$$\text{NDVI} = \frac{\text{Band 5} - \text{Band 4}}{\text{Band 5} + \text{Band 4}}. \tag{14.2}$$

The annual rainfall map of the study area for the year 2021 has been prepared using rainfall data obtained from India Meteorological Department. For preparing the model, about 327 past landslide events within the district have been used as 'Landslide Data Inventory' and they have been obtained from Bhukosh website from Geological Survey of India (Fig. 14.3). The detailed methodology has been explained in the methodological flow shown below (Fig. 14.1).

The present research selects Rudraprayag district of Uttarakhand district as the study area because the district has a long history of occurrence of landslides. Studies conducted by Indian Space Research Organization (ISRO) reveal the fact that Rudraprayag is the most landslide prone area of India. Both natural as well as human activities are considered to be responsible for this large scale occurrences of landslides. This region also experiences moderate earthquake events as the Main Central Thrust which separates Eurasian plate from Indian plate lies on this region (Singh 2011). When earthquakes take place, the loose debris resting on unstable hill slope comes down under the effect of gravity to cause debris flow and rock slide. Besides, heavy rainfall during the monsoon season is also responsible for lubrication of debris and weathered materials on the slope which slides down causing devastating landslides (Goswami et al. 2022). As Rudraprayag is a holy place for the Hindus, many people visit the place to pay homage to Lord Shiva. Hence, it has turned out to be a place of religious tourism. The natural beauty of the place also attracts many tourists not only from different parts of the India but also from different parts of the world. Hence, it can be said that the economic backbone of this region primarily rests on religious and nature tourism. However, when landslide takes place disrupting the major transport and communication system, religious tourism sector gets collapsed and face large scale losses and thus the economy the study area gets weakened. The present research work analyzes the vulnerability of seven selected religious tourist spots which will enable the decision makers in adoption of proper strategies which will enable to adopt sustainable tourism strategies keeping in mind the issue of landslide susceptibility.

Rudraprayag is one of the most important districts of Uttarakhand state in India and came into existence on 16th September 1997. Located at the meeting point of the Mandakini and the Alkananda rivers, the district is endowed with a number of religious spots and is blessed with nature's bounty. It has three blocks namely Ukhimath, Jakholi and Agastyamuni. The total area of the district is about 1986.20 square kilometers. It extends from 30.176° N to 30.811° N latitude and from 78.815° E to 79.355° E longitude (Fig. 14.2). It is surrounded by Uttarkashi district in the north, Chamoli district in the east, Tehri Garhwal district in the west and finally Pauri Garhwal district in the south. Out of 13 districts in Uttarakhand, Rudraprayag has the lowest population according to census 2011 and the total population of the district is 242,285. Population growth rate of the district for the year 2001–2011 was only 4.14%. In terms of population, Rudraprayag is in the 585th position out of total 640 districts in India.

14.2.3 Shannon Entropy Model

The concept of information entropy was first formulated by renowned American mathematician Claude Elwood Shannon in 1948 in his famous work '*A Mathematical Theory of Communication*' and hence is also known as Shannon Entropy. The concept of entropy is widely applied in the field of physical and social sciences. It is also used for estimation of weights of several parameters taken by a researcher under consideration. The procedure by which weights can be assigned to a certain attribute is done through completion of three steps.

In the first step, normalization of arrays of the decision matrix are performed to obtain the outcome of the project (P_{ij}) and it is done by using the Eq. (14.3)

$$P_{ij} = \frac{X_{ij}}{\sum_{i=1}^{m} X_{ij}}. \tag{14.3}$$

The second step involves the calculation of entropy (E_j) of the project outcomes and is done by using Eq. (14.4).

$$E_j = -k \sum_{i=1}^{m} P_{ij} \cdot \log P_{ij}. \tag{14.4}$$

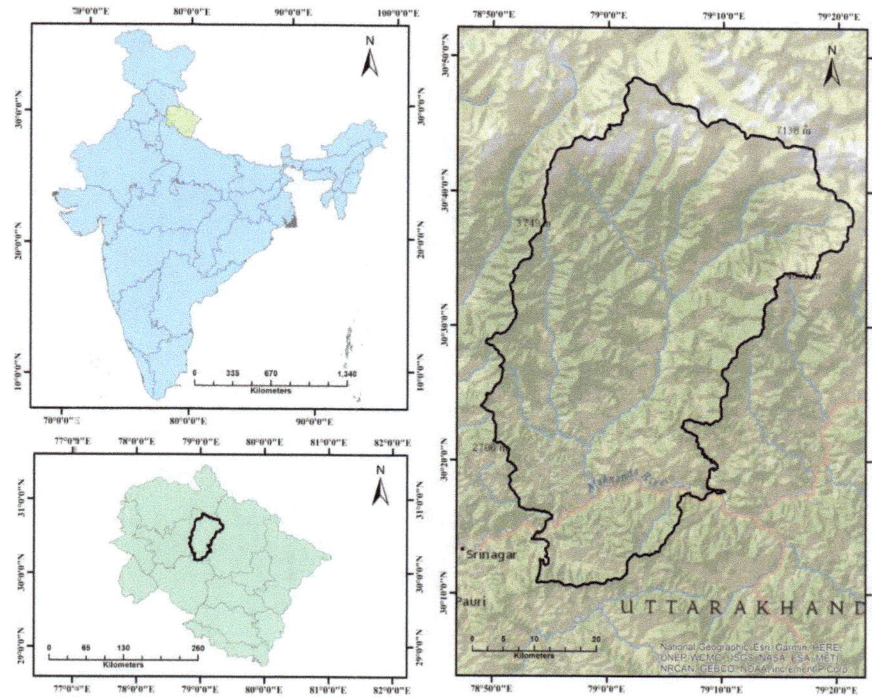

Fig. 14.2 Location map of the district

where k = 1/log. Number of values (m).

In the final step, weights (W_j) of respective attributes are determined on the basis of entropy concept and it is given in Eq. (14.5).

$$W_j = \frac{1 - E_j}{\sum_{j=1}^{n}(1 - E_j)}. \quad (14.5)$$

Through the application of Shannon Entropy, weights were assigned to the five conditioning factors. Thus, elevation was assigned the weight of 0.377, slope was given the weight of 0.065, rainfall 0.316, drainage density 0.09 and finally NDVI 0.151.

Validation of the Model

The landslide susceptibility model generated has been validated by calculating the Landslide Density Index (LDI) and have taken into account all the 327 past landslide events. The equation used for the purpose is mentioned below:

$$\text{LDI} = \frac{\text{Percentage of number of landslides present in a susceptible zone}}{\text{Percentage of pixels in each class on the susceptibility map}} \quad (14.6)$$

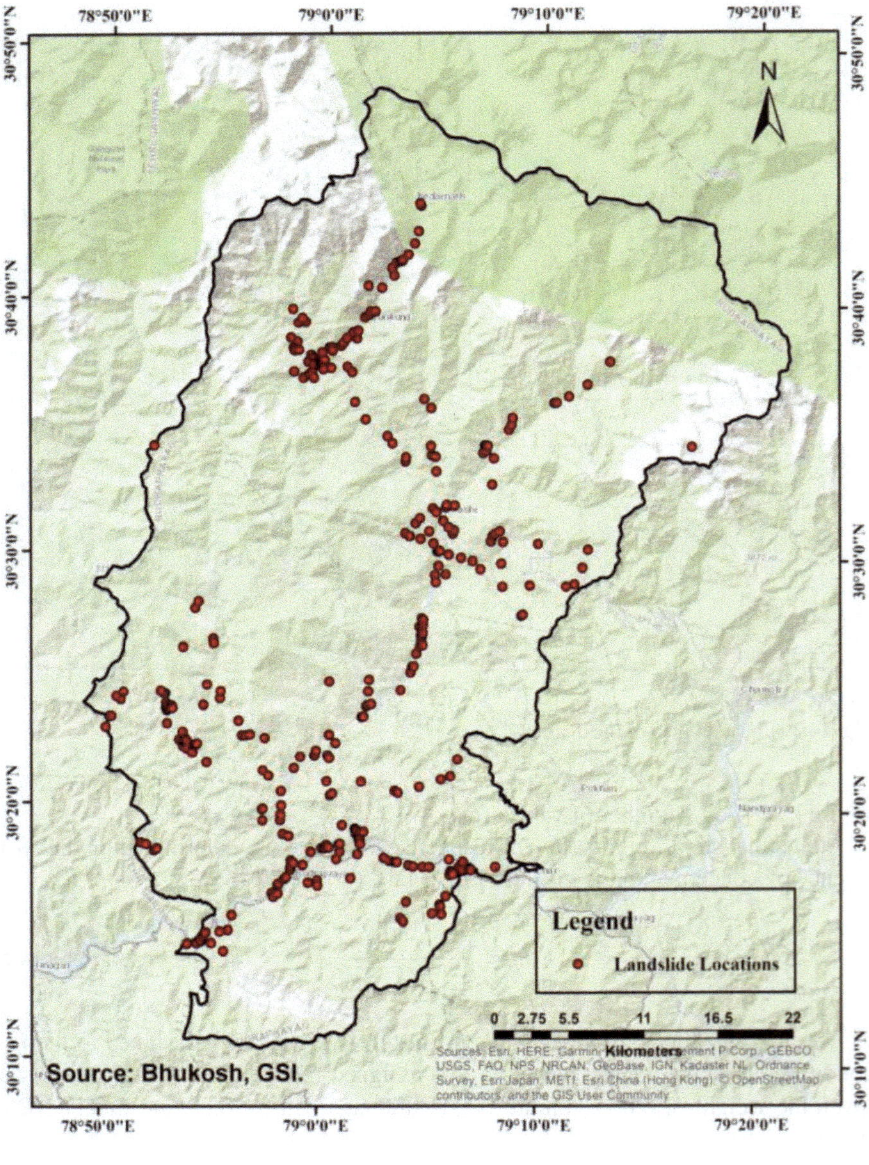

Fig. 14.3 Landslide inventory (*Source* Bhukosh, GSI)

14.2.4 Calculation of Landslide Susceptibility Value of Different Religious Tourist Spots

Locations of seven popular religious tourist spots were obtained from Google Earth Pro and then they were transferred in GIS environment as point layers. The landslide susceptibility map generated was overlaid on those points and then 'Extract values to points' tool was used to get the landslide susceptibility pixel value of the respective points as explained and interpreted in Sect. 14.3.4.

14.2.5 Processing and Analysis

All data were processed and maps were prepared using ArcGIS 10.8 and ERDAS Imagine 2015 software. Validation of the model was done through computation of Landslide Density Index (LDI) using MS Excel. Finally, maps and graphs prepared were analyzed and interpreted to arrive at necessary conclusions.

14.3 Results and Discussions

14.3.1 Analysis of the Conditioning Factors Affecting Landslides

For successful modelling and its applicability in the real world, it is very much necessary to have an insight on the factors affecting or triggering the specific event under consideration. In the context of present study, it is very much necessary to understand about the factors which affect or act as a triggering factor for the occurrence of landslide in the study area. The present study thus takes into account five conditioning factors.

14.3.1.1 Elevation

Elevation is an important factor that affect landslide. It is usually seen that several environmental factors like temperature, rainfall, vegetation cover and humidity exhibits variations with elevation and thus affects the degree of landslide susceptibility (Mekonnen et al. 2022). Rudraprayag has the lowest elevation of 569 m while the highest elevation of the study area is 6976 m. Elevation exhibits a rising trend from the south to the north (Fig. 14.4). Out of the past 327 landslide events, about 210 landslides have occurred in the elevation zone of 569–1594 m (Table 14.1). This could be attributed to the large scale human activities over the area in the form of urbanization, construction and expansion of roads.

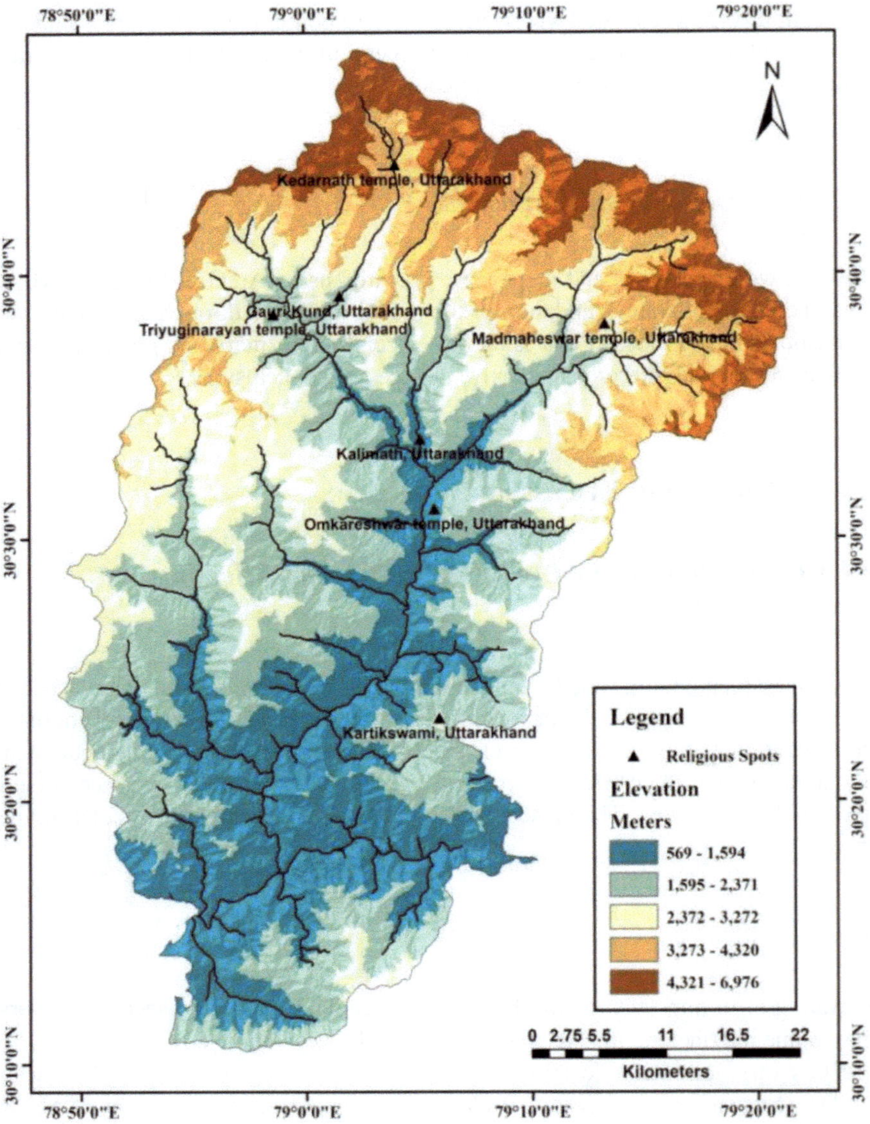

Fig. 14.4 Elevation map of Rudraprayag

14.3.1.2 Slope

Since landslide is a manifestation of slope instability, slope is considered to be one of the most important conditioning factor of landslide (Goswami et al. 2022). Usually, occurrence of landslide in an area depends upon slope angle whereby it is noticed that steep slopes are prone to landslides than gentle slopes (Crosta and Frattini 2007). It is

Table 14.1 Calculation of criteria weights using Shannon entropy

Parameter	Units	Range	Class	No. of landslide	Percentage	Total pixels	Percentage	Pii (normalization)	Pii x log (Pii)	E ii	1_E ii	Wii
Elevation	Meters	569–1594	1	210	64.22	558,229	25.27	0.64	−0.123513689	0.535059	0.464941	0.377
		1595–2371	2	95	29.05	651,286	29.49	0.29	−0.155958086			
		2372–3272	3	17	5.2	483,843	21.91	0.05	−0.066757432			
		3273–4320	4	5	1.53	305,085	13.81	0.02	−0.027761128			
		4321–6976	5	0	0	210,377	9.52	0	0			
				327	**100**	**2,208,820**	**100**		**−0.373990333**			
Slope	Degrees	0–17.25	1	24	7.34	323,396	14.64	0.07	−0.083254056	0.919452	0.080548	0.065
		17.26–26.98	2	72	22.02	593,494	26.87	0.22	−0.144707946			
		26.99–36.08	3	100	30.58	632,922	28.65	0.31	−0.157354053			
		36.09–46.12	4	98	29.97	468,448	21.21	0.3	−0.156836466			
		46.13–79.69	**5**	**33**	**10.09**	**190,558**	**8.63**	**0.1**	**−0.100517174**			
Rainfall	Millimeter	943–1119	1	11	3.36	50,751	5.34	0.03	−0.049555675	0.609623	0.390377	0.316
		1120–1294	2	23	7.03	166,508	17.51	0.07	−0.081085193			
		1295–1470	3	232	70.95	143,121	15.05	0.71	−0.105754942			
		1471–1646	4	38	11.62	444,397	46.74	0.12	−0.108627027			
		1647–1821	5	23	7.03	145,992	15.36	0.07	−0.081085193			
				327	**100**	**950,769**	**100**		**−0.42610803**			
Drainage density	km/km^2	0–0.73	1	77	23.55	151,064	17.79	0.24	−0.147891104	0.888399	0.111601	0.09
		0.74–1.5	2	112	3425	474,352	55.86	0.34	−0.15937899			
		1.6–2.2	3	91	27.83	152,511	17.96	0.28	−0.154590455			
		2.3–2.9	4	33	10.09	57,470	6.77	0.1	−0.100517174			

(continued)

Table 14.1 (continued)

Parameter	Units	Range	Class	No. of landslide	Percentage	Total pixels	Percentage	Pii (normalization)	Pii x log (Pii)	E ii	1_E ii	Wii
		3.0–3.6	5	14	4.28	13,832	1.63	0.04	−0.005858678			
				327	**100**	**849,229**	**100**		**−0.6209645**			
NDVI	Value	Below 0.07	1	2	0.61	254,702	11.53	0.011	−0.013538335	0.813638	0.186362	0.151
		0.08–0.17	2	65	19.88	258,921	11.72	0.2	−0.139468611			
		0.18–0.26	3	143	43.73	552,099	25	0.44	−0.157086469			
		0.27–0.34	4	78	23.85	563,319	25.5	0.24	−0.148475063			
		0.34–0.62	5	39	11.93	579,771	26.25	0.12	−0.110140192			
				327	**100**	**2,208,812**	**100**		**−0.568708669**		**1.233828**	

seen that Rudraprayag is having gentle and moderate slope in the southern part while the northern periphery of the district is having steep slopes. The highest value of slope of the study area was found to be 79.69° (Fig. 14.5). Out of 327 past landslide events, about 100 landslide events were seen in the slope zone of 26.99°–36.08° (Table 14.1) as this slope region was mainly vulnerable to human activities and it experienced maximum instability (Crosta and Frattini 2007).

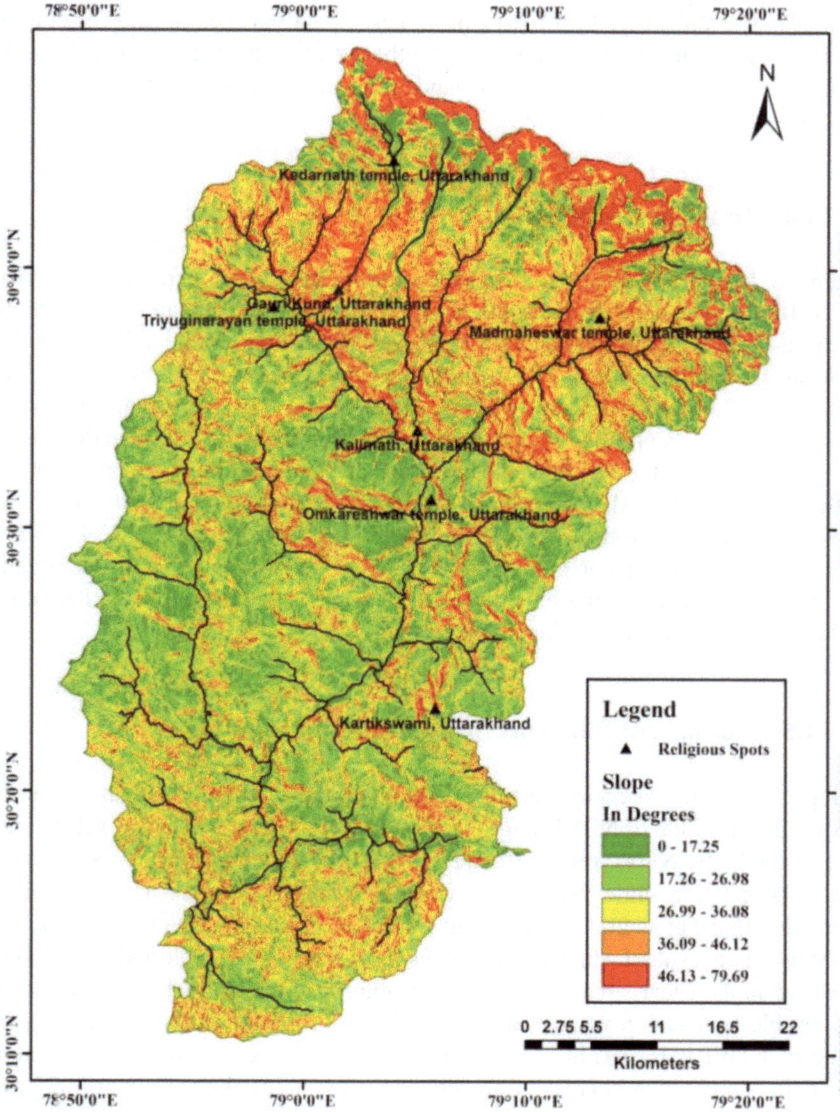

Fig. 14.5 Slope map of Rudraprayag

14.3.1.3 Rainfall

Rainfall act as triggering factor of landslide. Heavy rainfall turns the rock debris on the slope viscous and thus stimulate them to come down-slope and bring about landslides. Most of the landslides which the Himalayan belt experiences occurs mainly during the rainy season which extends from June to September because India gets its 80% of its rainfall during this time period (Khullar 2014). The highest amount of annual rainfall obtained by the district is 1821 mm while the lowest amount of rainfall is 943 mm (Fig. 14.6). Area of highest rainfall is seen in the western part while the area of lowest rainfall is in the southern part of the district (Fig. 14.6). Out of 327 past landslide events, about 232 landslide events were seen in the rainfall zone of 1295–1470 mm (Table 14.1).

14.3.1.4 Drainage Density

Drainage density is one of the most important morphometric parameter of a river basin. It is obtained by dividing the total length of all streams and rivers present within a basin with the total area of the basin (Mandal and Mandal 2016). During a rainstorm, the hydrograph of river undergo significant modifications. Rivers with high value of drainage density will have steep falling limb in its hydrograph and will have high susceptibility of landslides (Mandal and Mandal 2016). In the present study, highest drainage density value is 3.6 km/km^2 and is mainly seen on either side of the Mandakini and Alkananda rivers (Fig. 14.7). Out of 327 past landslide events, about 112 landslide events were seen in the drainage density zone of 0.74–1.5 km/km^2 (Table 14.1) as this area have more number of first and second order streams and water from these streams enter the cracks and crevices and conduct undercutting and thus make the debris resting above unstable.

14.3.1.5 Normalized Difference Vegetation Index (NDVI)

Using NDVI, the vegetation health condition of area under investigation can be analyzed. The value of NDVI ranges from -1 to $+1$ whereby negative values indicate water body. NDVI value ranging between 0.3 and 0.6 indicate existence of stressed vegetation while NDVI values exceeding 0.6 marks the existence of healthy vegetation (Sarif et al. 2023). Weak vegetation has weak roots which fail to hold the soil tightly as a result of which the soil cohesion also becomes feeble and brings about a decline in the shear strength of the soil profile. This in turn makes the area more susceptible to landslides (Goswami et al. 2022). In the present study, the highest value of NDVI is 0.61 (Fig. 14.8) and thereby indicates the presence of healthy vegetation and is seen in areas of moderate to heavy rainfall areas. Out of 327 past landslide events, about 78 landslide events were seen in the NDVI zone of 0.27–0.34 which represent the very stressed vegetation while only 39 past landslide events were seen in the NDVI zone of 0.34–0.62 which represent moderately stressed and

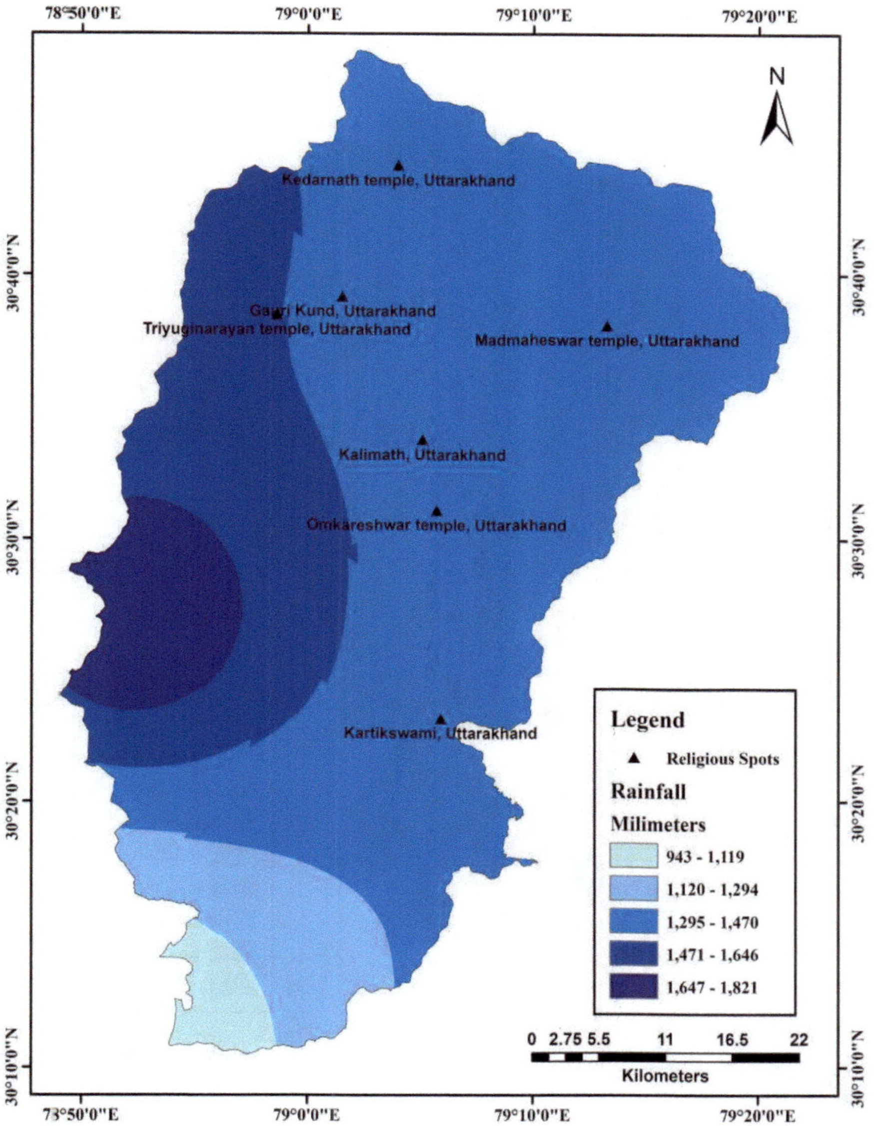

Fig. 14.6 Rainfall map of Rudraprayag (*Source* IMD)

healthy vegetation (Table 14.1). Maximum number of past landslides (143 events) were seen in the NDVI zone of 0.18–0.26 which indicate barren lands and built up areas (Table 14.1).

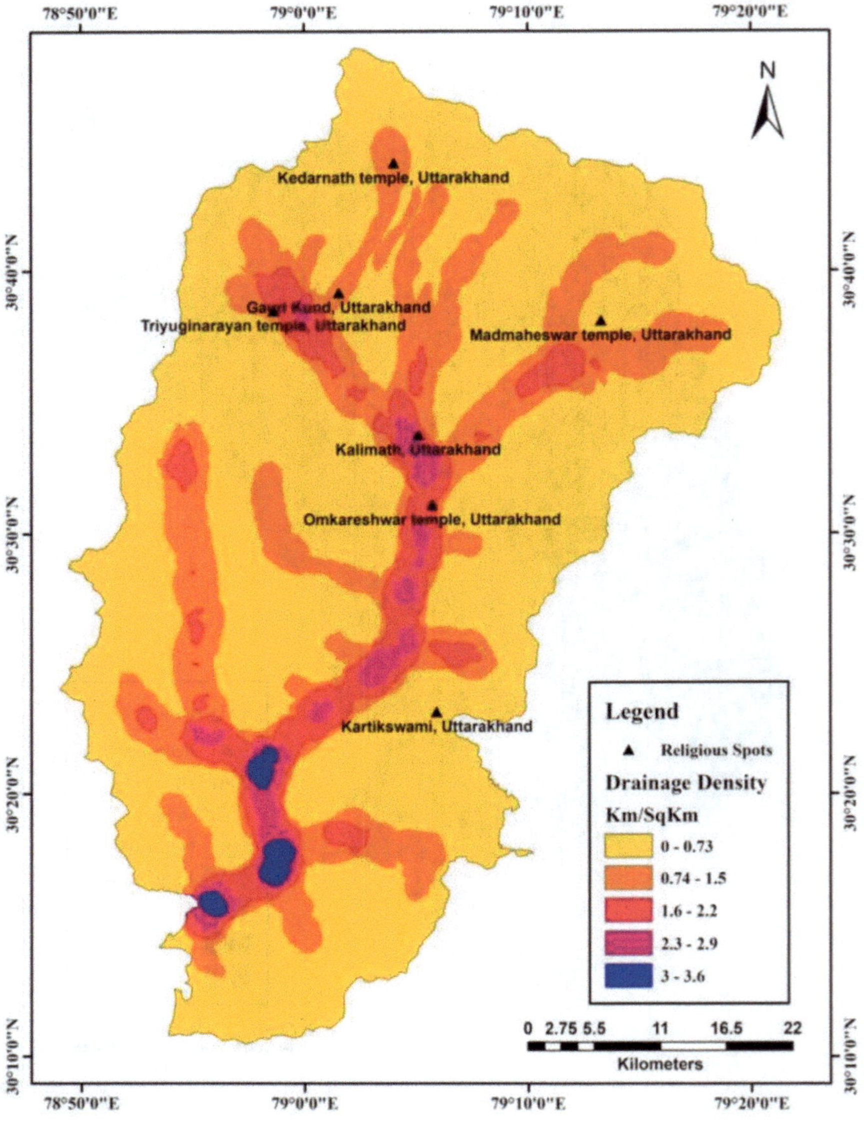

Fig. 14.7 Drainage density map of Rudraprayag

14.3.2 Landslide Susceptibility Assessment

The landslide map was prepared and was classified into five classes namely very low, low, moderate, high and very high. The area of very low and low susceptibility can be seen in the southern part of the district (Fig. 14.9) and covers a total area of about 107.69 km^2 (5.42% of total area of the district) and 395.54 km^2 (19.91% of

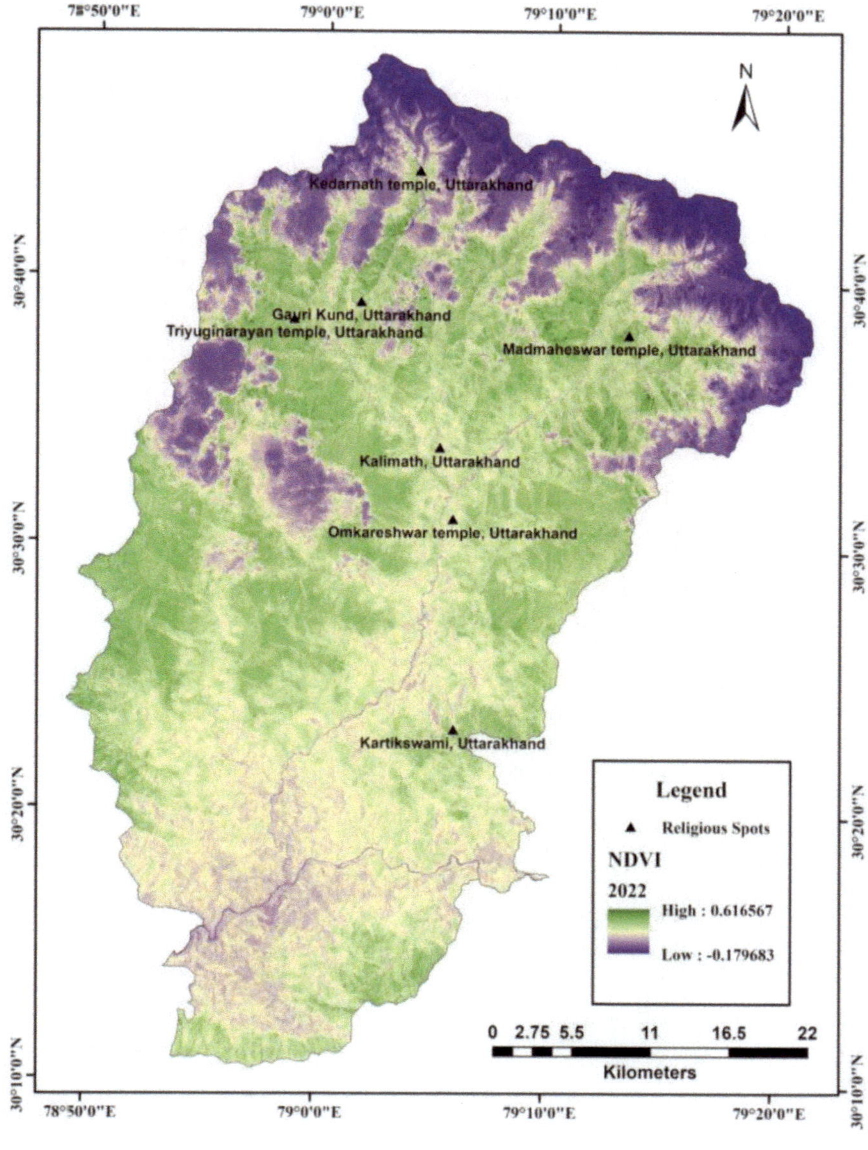

Fig. 14.8 NDVI map of Rudraprayag

total area) respectively (Table 14.2). Moderate susceptibility covers about a total area of about 444.61 km^2 (22.39% of total area) (Table 14.2) and is seen in the central part of the study area (Fig. 14.9). High and very high susceptibility zones cover about 567.48 km^2 (28.57% of the total area) and 430.88 km^2 (23.71% of total area) respectively (Table 14.2) and is seen in the western, north-western and northern part of the district (Fig. 14.9).

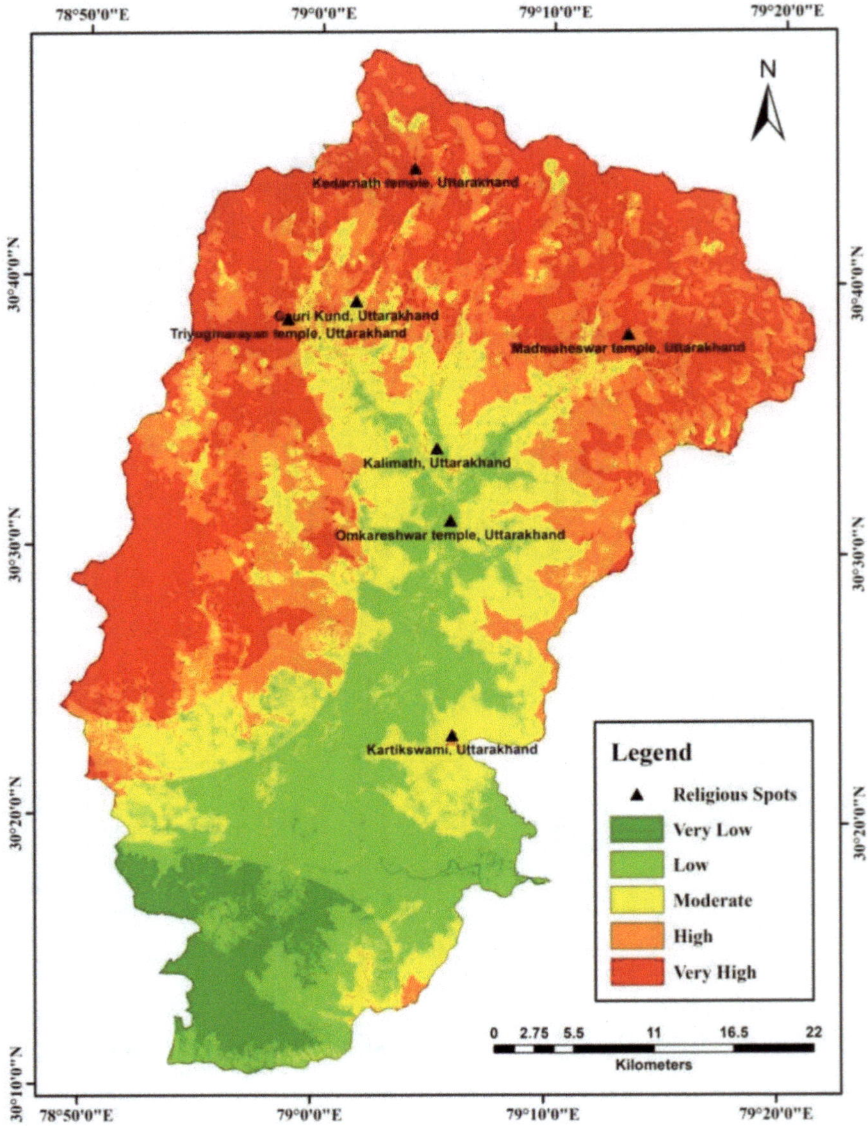

Fig. 14.9 Landslide susceptibility map of Rudraprayag

14.3.2.1 Validation of the Model

Validation is one of the essential tasks of modelling. If a model is not validated, then it becomes futile and cannot be applied in real world. In the present research work, all 327 landslide events have been used for validation. The landslide density index was computed which reveals that the highest landslide density index value of 2.24

Table 14.2 Area of different landslide susceptible zones of Rudraprayag

Class	Pixel count	Area in km^2	Area %
Very low	119,663	107.697	5.42
Low	439,489	395.540	19.91
Moderate	494,017	444.615	22.39
High	630,531	567.478	28.57
Very high	523,199	470.879	23.71
		1986.2019	100

is obtained in low landslide susceptibility zone while the lowest value of landslide density index of 0.22 is recorded in very high susceptible zone (Table 14.3) and thus gives rise to a situation of 'Geographical Paradox'. Since the landslide density values exhibit a parity with the landslide data inventory, the model is considered to be validated. It is therefore confirming that the model actually achieves its intended purpose of analyzing the landslide susceptible zones. With the help of this model it is viable to take judicious planning processes in the future. On the basis of this model high susceptible zones could be recognized and the most vulnerable areas could be refrained from intense developmental processes like high rise, urban development, road construction and others.

The study area is famous for tourism particularly religious and adventure tourism. Huge tolls of devotees visit the renowned religious tourists' spots every year along with other tourists. The economy of this region is very much dependent on tourism. So landslide affects the economy to a huge extent. Implementation of scientific mitigation techniques based on landslide susceptible zonation is the prior demand in the present state. In order to explicitly understand the landslide susceptibility, an assessment has been carried out on seven popular religious spots of Rudraprayag.

Table 14.3 Validation of the model

Parameter	Units	Range	Class	No of landslide	Percentage	Total pixels	Percentage	Landslide density index
LSM validation	Value	1–1.92	V. Low	26	7.95	119,663	5.42	1.47
		1.93–2.40	Low	146	44.65	439,489	19.91	2.24
		2.41–2.85	Mod	90	27.52	494,017	22.39	1.23
		2.86–3.23	High	48	14.68	630,531	28.57	0.51
		3.24–4.25	V. High	17	5.20	523,199	23.71	0.22

14.3.3 *Analysis of Landslide Susceptibility of Some Selected Tourist Spots of Rudraprayag*

The district of Rudraprayag has a number of tourist spots and most of them are religious spots as the district is considered to be one of the abodes of Lord Shiva. In fact, the district gets its name from the Rudra incarnation of Lord Shiva and is considered to be a pilgrim site for all the Hindus across the country as well as from other parts of the world. Every year, a lot of Hindu devotees visit the district to pay homage to Lord Shiva and mother Jagadamba. Besides, the natural beauty of the district also attracts a large number of foreign tourists from different parts of the world. Owing to this popularity, the tourism industry has turned out to be the backbone of this region's. However, the tourism industry gets severely affected when landslide takes place. The present study has selected 7 popular spots and have analyzed the susceptibility of those spots to landslides (Steps mentioned in Sect. 14.2.5). The study revealed that Omkareswar Temple and Kalimath are having susceptibility values 2.113 and 2.052 respectively and thus falls in low landslide susceptibility zone (Table 14.4). Madmaheswar temple, Kartikswami and Gauri Kund are falling in moderate susceptibility zone with susceptibility value of 2.838, 2.752 and 2.746 respectively (Table 14.4). Kedarnath temple and Triyuginarayan temple are having susceptibility values of 3.003 and 3.108 and thus is located in high landslide susceptible zone (Table 14.4).

Due to its world-wide popularity as a tourist spot, all types of developmental activities such as hotel construction, construction of roads and their expansion are done in unplanned way without taking into account the environmental consequences. Unplanned urbanization in a fragile environment is also taking place rapidly. As a result, low and gentle slope areas have become more susceptible to landslides than high elevation and steep slope areas. Areas with high altitude and steep slope have

Table 14.4 Landslide susceptibility of some tourist spots in Rudraprayag

ID	Name	Latitude	Longitude	Landslide susceptibility value	Susceptibility zone number	Susceptibility zone name
1	Kedarnath temple	30.735231	79.066895	3.003	4	High
2	Madmaheswar temple	30.635133	79.221539	2.838	3	Moderate
3	Omkareshwar temple	30.518426	79.095345	2.113	2	Low
4	Gauri Kund	30.652913	79.025658	2.746	3	Moderate
5	Triyuginarayan temple	30.641472	78.976989	3.108	4	High
6	Kartikswami	30.385782	79.098518	2.752	3	Moderate
7	Kalimath	30.562764	79.084982	2.052	2	Low

remained beyond the reach of human interference and thus have experienced less landslide events. It has been already stated that the study has revealed a situation of 'Geographical Paradox'. This is explained with the help of tourism potentiality of the district.

From the above data it is pertinent that highly susceptible regions should be of immediate concern for management process. Just by analyzing these seven spots landslide susceptibility zonation can be framed so that mitigation processes would become much easier and implementable. In broader scale if more such spots could be recognized the zonation criteria would be of much relevance.

14.4 Conclusion

There is no doubt of the fact that the Himalayan mountain belt is an abode to tourists and religious tourism is one such worth mentioning sector. As the tourism industry plays an important role in the development of the economy of the country, its continuous expansion is very much needed. However, the environmental issues are often overlooked and unscientific strategies are adopted which initially may yield fruitful results, but its long term effects may lead to several problems which in turn acts as an obstacle in the path of development. In a fragile geological and ecological environment of the Himalayas, developmental decisions should be taken keeping in mind the environmental aspects and Rudraprayag is not an exception in this regard. Expansion of roads and construction of hotels in Rudraprayag must be done keeping in mind the issue of landslide susceptibility. The landslide susceptibility map prepared will aid the decision makers and policy makers in this aspect. Landslide controlling measures should be implemented on the basis of location of a spot in the landslide zone. Conduction of Environmental Impact Assessment (EIA) must be undertaken before implementation or execution of any big project in the fragile geological structure of Rudraprayag and expert opinions is also to be taken into consideration. Moreover calculation of Tourism carrying capacity within the landslide susceptible zones is of utmost concern considering which tourist flow is to be controlled and managed likewise. Furthermore, tourism could also be restricted during the monsoon and post monsoon which is regarded to be the most vulnerable period of landslide occurrences. Several scientific mitigation techniques should be synergized with the traditional wisdom of the local people which might have the potentiality of mitigating this hazard.

It can be stated that successful implementation of the above mentioned strategies coupled with public awareness and environmental consciousness will pave the way for a well-organized tourism industry in Rudraprayag which in turn will ensure the further development of India as a whole. Lastly, the Neo deterministic approach of Griffith Taylor who postulated the 'stop and go' could help in long term sustenance, where he asks us to stop for a while after an alteration made on a natural attribute and look back to observe the repercussion and then take the next move. If this philosophy of Taylor could be adopted, it might pave sustainable future for our Earth.

References

Aditian A, Kubota T, Shinohara Y (2018) Comparison of GIS-based landslide susceptibility models using frequency ratio, logistic regression, and artificial neural network in a tertiary region of Ambon, Indonesia. Geomorphology 318:101–111. https://doi.org/10.1016/j.geomorph.2018.06.006

Akinci H (2022) Assessment of rainfall-induced landslide susceptibility in Artvin, Turkey using machine learning techniques. J Afr Earth Sci 191:104535. https://doi.org/10.1016/J.JAFREARSCI.2022.104535

Alsabhan AH, Singh K, Sharma A, Alam S, Pandey DD, Rahman SAS, Khursheed A, Munshi FM (2022) Landslide susceptibility assessment in the Himalayan range based along Kasauli–Parwanoo road corridor using weight of evidence, information value, and frequency ratio. J King Saud Univ Sci 34(2):01–12. https://doi.org/10.1016/j.jksus.2021.101759

Araújo JR, Ramos AM, Soares PMM, Melo R, Oliveira SC, Trigo RM (2022) Impact of extreme rainfall events on landslide activity in Portugal under climate change scenarios. Landslides 19(10):2279–2293. https://doi.org/10.1007/s10346-022-01895-7

Bianchini S, Solari L, Casagli N (2017) A GIS-based procedure for landslide intensity evaluation and specific risk analysis supported by Persistent Scatterers Interferometry (PSI). Remote Sens 9(11):01–20. https://doi.org/10.3390/rs9111093

Chaudhary S, Kumar A, Pramanik M, Negi MS (2022) Land evaluation and sustainable development of ecotourism in the Garhwal Himalayan region using geospatial technology and analytical hierarchy process. In: Environment, development and sustainability, vol 24, Issue 2. Springer Netherlands. https://doi.org/10.1007/s10668-021-01528-4

Crosta GB, Frattini P (2007) Rainfall-induced landslides and debris flows. Hydrol Process 22:473–477. https://doi.org/10.1002/hyp

Elmoulat M, Brahim LA (2018) Landslides susceptibility mapping using GIS and weights of evidence model in Tetouan-ras-Mazari area (Northern Morocco). Geomat Nat Haz Risk 9(1):1306–1325. https://doi.org/10.1080/19475705.2018.1505666

Fayez L, Pazhman D, Pham BT, Dholakia MB, Solanki HA, Khalid M, Prakash I (2018) Application of frequency ratio model for the development of landslide susceptibility mapping at part of Uttarakhand State, India. Int J Appl Eng Res 13(9):6846–6854

Ghorbanzadeh O, Pourmoradian S, Blaschke T, Feizizadeh B (2019) Mapping potential nature-based tourism areas by applying GIS-decision mapping potential nature-based tourism areas by applying GIS-decision making systems in East Azerbaijan Province, Iran. J Ecotour 1–23.https://doi.org/10.1080/14724049.2019.1597876

Goswami A, Sen S, Sanyal R (2022) Delineation of landslide susceptibility zone of a part of Sutlej Basin in Himachal Pradesh through frequency ratio model. In: Rana A, Rana AK, Dhawan S, Sharma S, Elngar AA (eds) Convergence of deep learning and artificial intelligence in internet of things, 1st edn. CRC Press, pp 211–230. https://doi.org/10.1201/9781003355960

Huangfu W, Wu W, Zhou X, Lin Z, Zhang G, Chen R, Song Y, Lang T, Qin Y, Ou P, Zhang Y, Xie L, Huang X, Fu X, Li J, Jiang J, Zhang M, Liu Y, Peng S et al (2021) Landslide geo-hazard risk mapping using logistic regression modeling in Guixi, Jiangxi, China. Sustainability (Switzerland) 13(9):01–20. https://doi.org/10.3390/su13094830

Khullar DR (2014) India, a comprehensive study, 3rd edn. Kalyani Publishers

Mandal B, Mandal S (2016) Assessment of mountain slope instability in the Lish River basin of Eastern Darjeeling Himalaya using frequency ratio model (FRM). Model Earth Syst Environ 2(121):01–14. https://doi.org/10.1007/s40808-016-0169-8

Mekonnen AA, Raghuvanshi TK, Suryabhagavan KV, Kassawmar T (2022) GIS-based landslide susceptibility zonation and risk assessment in complex landscape: a case of Beshilo watershed, northern Ethiopia. Environ Chall 8:100586. https://doi.org/10.1016/J.ENVC.2022.100586

Mersha T, Meten M (2020) GIS-based landslide susceptibility mapping and assessment using bivariate statistical methods in Simada area, northwestern Ethiopia. Geoenviron Disast 7(20):01–22. https://doi.org/10.1186/s40677-020-00155-x

Nohani E, Moharrami M, Sharafi S, Khosravi K, Pradhan B, Pham BT, Lee S, Melesse AM (2019) Landslide susceptibility mapping using different GIS-based bivariate models. Water (Switzerland) 11(7):1–22. https://doi.org/10.3390/w11071402

Pal R, Biswas SS, Mondal B, Pramanik MK (2016) Landslides and floods in the Tista Basin (Darjeeling and Jalpaiguri districts): historical evidence, causes and consequences. J Ind Geophys Union 20(2):66–72

Rana G, Kumar S (2016) Prospects and problems of tourism industry in Uttarakhand. In: Bansal SP, Gautam P, Walia S (eds) Tourism dimensions innovations, challenges and opportunities, 1st edn. Indu books services, pp 277–284. https://doi.org/10.2139/ssrn.3017392; https://ssrn.com/abstract=3017392

Sadiki M, Manaouch M, Aghad M, Batchi M, Al Karkouri J (2022) Identifying landslides prone-areas using GIS-based fuzzy analytical hierarchy process model in ziz upper watershed (Morocco). Ecol Eng Environ Technol 24(1):67–83. https://doi.org/10.12912/27197050/154916

Sarif MO, Gupta RD, Murayama Y (2023) Assessing local climate change by spatiotemporal seasonal LST and six land indices, and their interrelationships with SUHI and hot-spot dynamics: a case study of Prayagraj City, India (1987–2018). Remote Sens 15(1):01–40. https://doi.org/10.3390/rs15010179

Singh S (2011) Geomorphology, 5th edn. Prayag Pustak Bhawan

Sonker I, Tripathi JN, Singh AK (2021) Landslide susceptibility zonation using geospatial technique and analytical hierarchy process in Sikkim Himalaya. Quat Sci Adv 4(100039):01–17. https://doi.org/10.1016/j.qsa.2021.100039

Villuri VGK, Saha A, Bhardwaj A (2022) Development and assessment of GIS-based landslide susceptibility mapping models using ANN, Fuzzy-AHP, and MCDA in Darjeeling Himalayas, West Bengal, India. Land 11(10):01–28. https://doi.org/10.3390/land11101711

Chapter 15
Landslide Hazard Susceptibility Analysis and Modelling in the Vicinity of the Proposed Subansiri Lower HE Project, Arunachal Pradesh

P. Danuta Mohan and Shovan Lal Chattoraj

Abstract The Himalayan region of India, in particular, faces an escalating slope instability due to landslides, posing significant pressure on rapid developmental activities. In this chapter, the study focuses on generating a comprehensive landslide susceptibility map (LSM) to analyze and detect the most crucial zones susceptible to landslides for the biggest hydroelectric project undertaken in India- Subansiri lower hydroelectric project situated on the intensive slopes of the Himalaya which is under a constant threat of recurring landslides utilizing a Geographical Information System (GIS) based Analytical Hierarchy Process (AHP) model. The LSM was created by considering a range of 9 geo-environmental parameters like slope, aspect, lithology, lineament density, geomorphology, drainage density, NDVI, LULC, and Road (Euclidean distance) to identify areas susceptible to future landslides and to delineate potential hazard zones. Using the Receiver Operating Characteristic (ROC) curve model, the AHP model was validated and yielded accuracy rates of 79.2%, 79.8%, and 79.2% for the geometrical-interval, quantile, and natural breaks (Jenks) classification models, respectively. A predictive numerical rockfall model Rapid Mass Movement Simulation (RAMMS) was employed to detect the runout trajectories and spatial extent of rockfall around the vicinity of the dam. The velocity of the estimated future rockfall was detected to be 9.84 ms^{-1}, Jump height of 1 m, and kinetic energy of 132.80 kJ. The study's outcomes hold practical implications for infrastructure development and management in landslide-prone regions, as the susceptibility map can aid in the implementation of effective risk mitigation strategies.

P. Danuta Mohan (✉)
Disaster Management Studies Department, Indian Institute of Remote Sensing, Dehradun 248001, India
e-mail: pdanuta07@gmail.com

S. L. Chattoraj
Geosciences Department, Indian Institute of Remote Sensing, Dehradun 248001, India
e-mail: shovan.iitb@gmail.com

G. K. Panda et al. (eds.), *Landslide: Susceptibility, Risk Assessment and Sustainability*, Advances in Natural and Technological Hazards Research 52,
https://doi.org/10.1007/978-3-031-56591-5_15

Keywords Landslide · LSM · AHP · Rockfall · RAMMS

15.1 Introduction

Landslides, being a recurrent and profoundly destructive geological hazard, inflict significant and detrimental impacts on both natural and man-made environments (Chowdhuri et al. 2020; López et al. 2022; Ray et al. 2007). Between 1998 and 2017, this particular disaster resulted in over 1800 fatalities and affected nearly 5 million people worldwide (Huggel et al. 2012). The investigation into landslides has garnered widespread interest globally, primarily as a result of growing knowledge of the economic and societal consequences of disasters and the mounting impact of urbanization on the alpine environment (Aleotti and Chowdhury 1999). The process of initiation of landslides is extremely complex and depends on various geo-environmental factors, which may require a triggering mechanism such as rainfall or earthquakes to occur before they accelerate (Guzzetti 2002). An increased rate of slope instability of landslides, especially in the Himalayan region of India puts immense pressure on the rapid developmental activities and draws attention to the need for immediate mitigation measures to safeguard vulnerable locations, prevent casualties, and eradicate potentiall damages to the society (Onagh et al. 2012; Kumar and Devrani 2012; Sundriyal et al. 2007; Ray and Chattoraj 2014; Mandal et al. 2021; Saha et al. 2021; Das et al. 2023). The Northeastern region of India located between the Himalayan collision zone in the north and the Indo-Burma subduction zone in the east, experiences significant seismic activity and heavy rainfall. Thus, these factors contribute to frequent landslides and slope failures, causing yearly losses in lives, property, and communication disruptions (Bhusan et al. 2022).

Practically, to mitigate the entire damage due to landslide is impossible, but evaluating the hazard beforehand can help minimize its impact. Thus, scientifically estimating the likelihood of landslides becomes crucial in reducing losses, highlighting the importance of assessing potential landslide areas spatially through hazard or susceptibility evaluations (Das et al. 2023). This necessitates the need for mapping the landslide-susceptible zones (Das et al. 2022; Buah et al. 2019; Guzzetti et al. 2020).

According to Varnes et al. (1984) a landslide hazard is included in the category of natural hazards, this classification refers to the probability of a harmful event happening within a particular timeframe and location. While "hazard" pertains to the timing of a landslide, some authors have emphasized that "susceptibility" is focused on identifying the location where a landslide may occur (Guzzetti et al. 1999; Westen et al. 2006). Generally, landslide susceptibility is elucidated as the evaluation of both quantitative and qualitative characteristics, such as the size and distribution of landslides that currently exist or could arise in a particular region (López et al. 2012). Selection of any landslide susceptible model is based on the research's decision and understanding of the terrain. Thus, by establishing a landslide susceptibility map

(LSM) we can demarcate the vulnerability of an area to this hazard and take up crucial mitigation measures to sustain a future with minimised losses to landslide.

Understanding slope movements and their influencing components requires a very comprehensive understanding of landslide susceptibility mapping (Chen et al. 2016). Rockfall, a type of landslide is a significant concern in landslide fatalities even when factors at risk with a minimal level of vulnerability are incorporated (Varnes 1978; Volkwein et al. 2011). Despite typically involving smaller amounts of rock compared to other types of landslides, like rock slides and rock avalanches, they can still cause substantial damage to infrastructures, and lifelines due to their frequency, ease of occurrence, and kinetic energy (Giraud et al. 1990; Volkwein et al. 2011). Countless incidents over the past few years have shown how human activities, such as building infrastructure and residential areas in environmentally fragile regions, have exacerbated the impact of natural disasters (Leine et al. 2014). Therefore, by employing various numerical simulation models can help in elevating in the predication of the areas vulnerable to these hazards.

This study utilized a GIS-based AHP model to generate a LSM for the Subansiri lower hydroelectric (HE) project of 2000 MW by considering the ideal geo-environmental parameters to determine the regions susceptible to future landslides hazards and to demarcate the potential zones susceptible to landslides. The fundamental benefit of the AHP model is that it reduces ambiguity in subjective evaluations by using consistency ratio-based measurements, eliminating the need for any pre-existing statistical analysis in between landslides and their causal factors (Das et al. 2022). Furthermore, the novelty of this study was to implement numerical models of rockfall to safeguard the infrastructure defacements of the under-construction project, predictive 3D Rockfall modelling using RAMMS (rapid mass movement simulation) numerical modelling was initiated to have a thorough understanding of the possible future rockfall scenarios were various runout trajectories of the rockfall like the magnitude, kinetic energy, and jump height of rockfall at the most vulnerable zone prone to rockfall was analysed and assessed.

15.2 Study Area

The hydro-electric (HE) project Subansiri lower (2000 MW) of NHPC is located on the border of two Eastern states of India, Assam-Arunachal Pradesh, specifically within the geographic coordinates of 27° 29′ 26" N to 27° 49′ 07" N latitude and 94° 03′ 59" E to 94° 28′ 46" E longitude, as illustrated in Fig. 15.1. The study encompasses an estimated area of 770 km^2 with dense vegetation and the Trans-Himalayan River Subansiri flowing from the E-W direction along the Main Boundary Fault (MBT) with a drainage basin of 32,640 km^2.

The major part of the project area is dominated by the Siwaliks and Gondwanas with Mir-quartzite covering a compact extent. the lithology of the area is dominated by feldspathic sandstone, carbonaceous shale-coal, and quartzite with intercalations of the conglomerate (Goswami et al. 2018), the regional geology of the Subansiri area

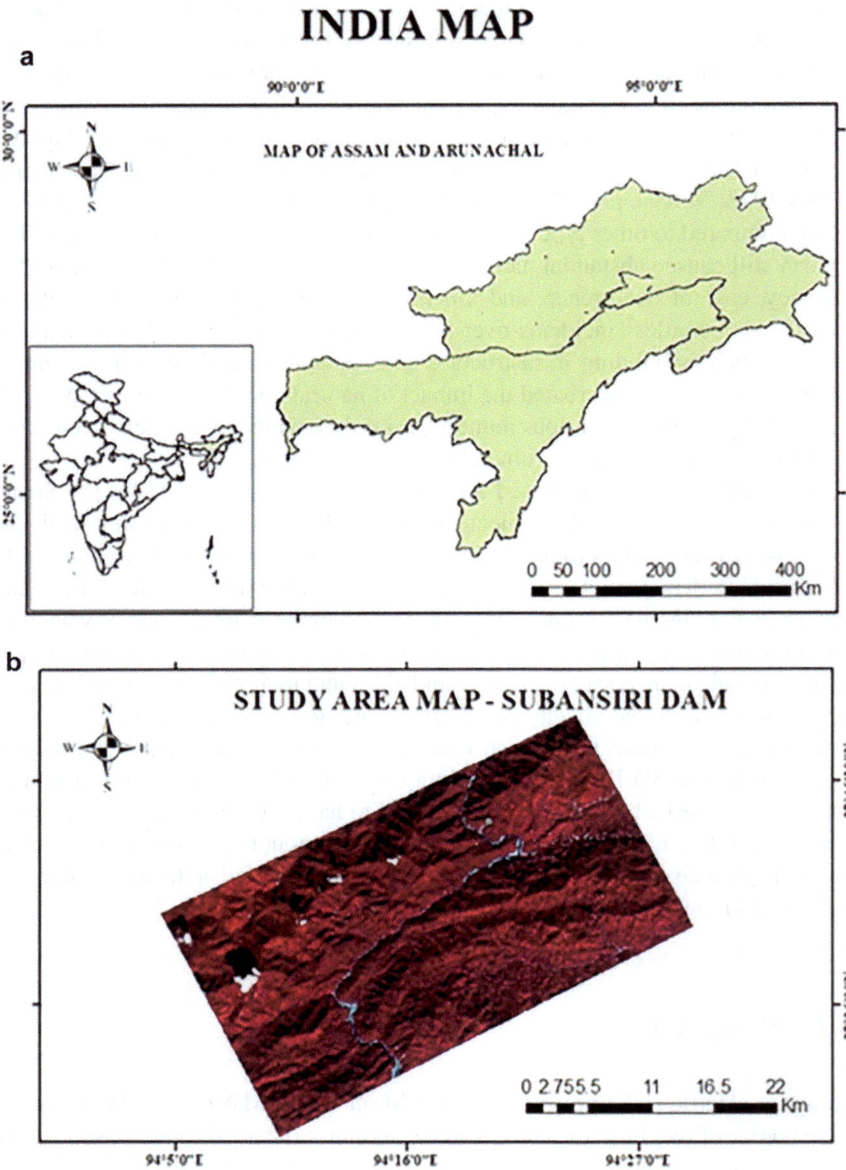

Fig. 15.1 The study area map: **a** Map of Assam and Arunachal Pradesh. **b** Subansiri dam area (Sentinel 2)

was referred from the published maps and reports of geological survey of India (GSI). The present study area receives an annual average rainfall of 400–750 cm (May to September) and is vulnerable to frequent occurrences of landslides and floods due to its underlying geology and high intensity of rainfall leading to the high discharge of water into the river respectively. Furthermore, the entire project area falls under the seismic zone V.

The three distinct geomorphic setting of Subansiri lower are namely the structural hills, the piedmont zone, and the alluvial plain, of which the structural hills represent a total area of 4.5% along the extreme northern and north-western parts of the basin. The amalgamation of various alluvial fan deposits that built up the piedmont zone makes up 7.7% of the basin's area and runs along the foothills of Arunachal Pradesh in a steeply sloping plain in a narrow band of NE-SW trending direction. The alluvial plain covers an area of 87.8% in the southern margin of the piedmont zone, and is characterized by gentle slope in the north and north-east and progressively flattens out as it merges with the flood plain of the Brahmaputra river in the south and south-east (Goswami et al. 2002).

15.3 Methodology

The categorization of landslide susceptibility zonation (LSZ) can be generally grouped into three distinct types: qualitative, semi-quantitative, and quantitative techniques. In qualitative approaches, the weights for each landslide condition factor are assigned based on the expertise knowledge. Weighted linear combinations (WLC), fuzzy logic, and AHP are examples of logical tools utilized in semi-quantitative procedures that presume weights and ranking. Quantitative approaches take into account the correlation within landslide components and a landslide inventory and are statistical methods driven by landslide inventories. Weights and ratings are derived mathematically from the landslide densities existing in the factor classes. Other classifications include bivariate and multivariate approaches. Another quantitative technique in landslide susceptibility assessment is deterministic mapping of slope instability, which relies on the specific geotechnical characteristics of a given slope. Understanding and selecting the optimal method for the LSZ will produce the best suitable result (Pardeshi et al. 2013; Anbalagan et al. 2015; Chen et al. 2017; Aleotti and Chowdhury 1999; Sarkar and Kanungo 2006).

By understanding the sensitivity and with relevant expertise and knowledge about the study region a semi-quantitative method was adopted for delineating the vulnerable slopes. The semi-quantitative approach of Landslide susceptibility mapping involves using the AHP model, which entails a matrix-based pair-wise comparison of various landslide parameters (Pourghasemi et al. 2012). A multi-criteria decision analysis (MCDA) technique—AHP is routinely used to comprehend complex challenges in a rigorous, flexible way that eliminates any decisional inconsistencies (Das et al. 2022; Saaty 1987). The AHP model deconstructs a complex and unstructured problem into its constituent factors by utilizing various selected parameters.

These factors are arranged hierarchically and assigned numerical values based on their subjective importance to determine their relative significance. Ultimately, these ratings are combined to establish priorities for the parameters in question (Saaty 1987). In this study 9 landslide controlling factors are intergraded with the AHP model to initiate the LSM.

To assess vulnerability due to rockfall, the RAMMS ROCKFALL model devised by the Swiss Federal Institute for Forest, Snow, and Landscape Research WSL, Switzerland, was employed to envisage the rockfall potential in the field of study. The RAMMS ROCKFALL model uses a hard-contact, rigid-body approach to simulate rockfall paths in three-dimensional terrain (Leine et al. 2014). The stated aim of the numerical simulation tool RAMMS is to model various types of gravitationally-induced natural disasters, including avalanches, debris flows, and hillslope debris flows, using a unified software platform (Christen et al. 2012; Zimmermann et al. 2020). The detailed workflow of this study is depicted in Fig. 15.2.

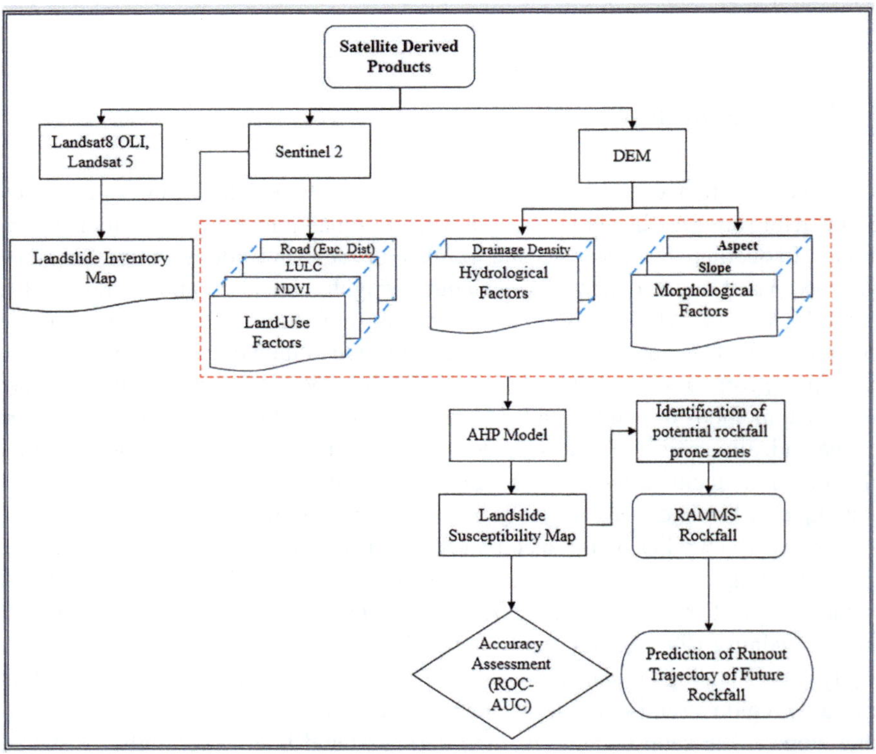

Fig. 15.2 Workflow of the methodology

15.3.1 Dataset

For performing the analysis, various medium resolution satellite datasets were used such as ALOS (Advance Land Observation Satellite) PALSAR (Phased Array L-band Synthetic Aperture Radar) DEM (12.5 m resolution), Sentinel 2 satellite images, Landsat 8 OLI, and Landsat 5 satellite images to establish and produce landslide inventory map and various thematic layers of landslide causative factors which aids in establishing the LSM and supports in the modelling of rockfall.

15.3.2 Landslide Inventory Map (LIM)

A record of landslides in a particular location is referred to as a landslide inventory. At a basic level, this inventory offers a general overview of the area where landslides occur and identify specific regions that require further investigation. Detailed inventories, on the other hand, provide a more comprehensive understanding of the various landslide processes occurring in a region, and they can be utilized to regulate or halt development in landslide-prone zones and aid in the formulation of remedial actions (Highland and Bobrowsky 2008).

Creating a landslide inventory, utilizing methods such as aerial photo interpretation, ground survey, and historical landslide databases, stands as the most direct approach to identify areas prone to landslides. The resulting output reveals the spatial distribution of mass movements, which can be visualized on a map, either as affected regions scaled to that area or represented by point symbols (Westen et al. 2006). Furthermore, a landslide inventory map for landslides illustrates their precise locations, types, and times of occurrence. The mapping of existing landslides is crucial for understanding how landslide distribution and conditioning variables are related (Westen et al. 2006; Chen et al. 2016).

The LIM as shown in Fig. 15.3 was prepared for the current study using the satellite images: Sentinel 2, Landsat 8 OLI, Landsat 5 image, and google earth pro software (the high resolution and access to different periods of data is the key benefit of this software). For the digitization of landslides, polygon-based vector analysis was used and various elements of visualization interpretation were considered (Das et al. 2022). A total of 25 landslides, which spanned across an area of 733 square meters, were detected. These landslides can be classified as a combination of debris flows, debris slides, translational slides and rockfalls.

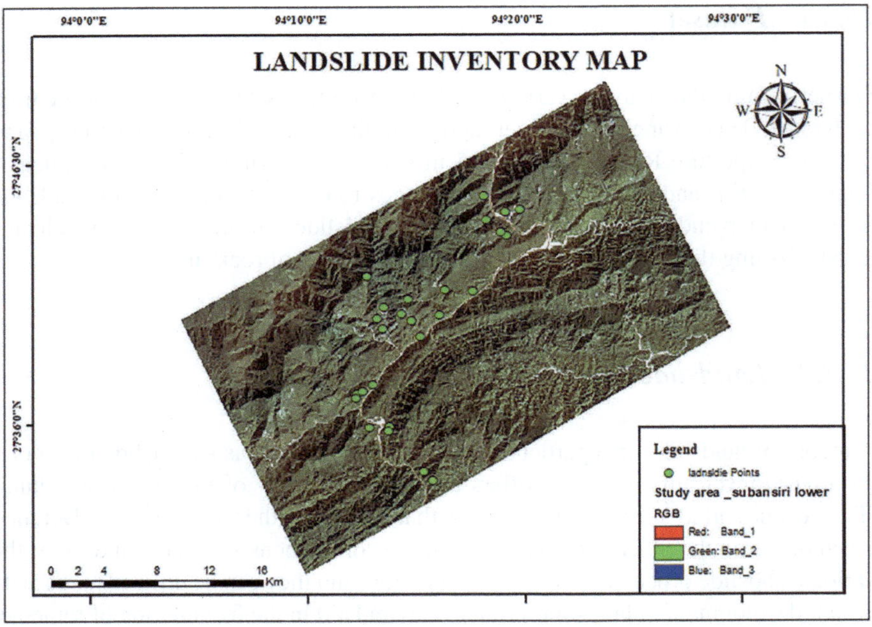

Fig. 15.3 Generated landslide inventory map for the study area

15.3.3 Landslide Condition Factors and Thematic Data Layer Preparation

The Landslide Conditioning Factor (LCF) falls under the category of preliminary elements, which are in charge of initiating landslides in the area and for which relevant data can be retrieved from available sources or the field (Pourghasemi et al. 2012). For the thematic data layer preparation 9 factors like morphological, geological, hydrological, and land use groups such as slope, aspect, drainage density, normalized difference vegetation index (NDVI), lineament density, lithology, geomorphology, land use land cover (LULC) and Euclidean distance of road were incorporated in the study. The factors that trigger events, such as rainfall and earthquakes, are highly susceptible to temporal changes. Therefore, these factors are not taken into consideration for this study.

15.3.3.1 Slope

The extent of slope inclination emerges as a paramount element influencing the initiation of slope instability. To create the slope map (Fig. 15.4a), the spatial analyst tool in Arc-GIS was utilized to process the DEM, and the degree of the slope was reclassified into five categories: <15°, 15°–25°, 25°–35°, 35°–45°, and >45°.

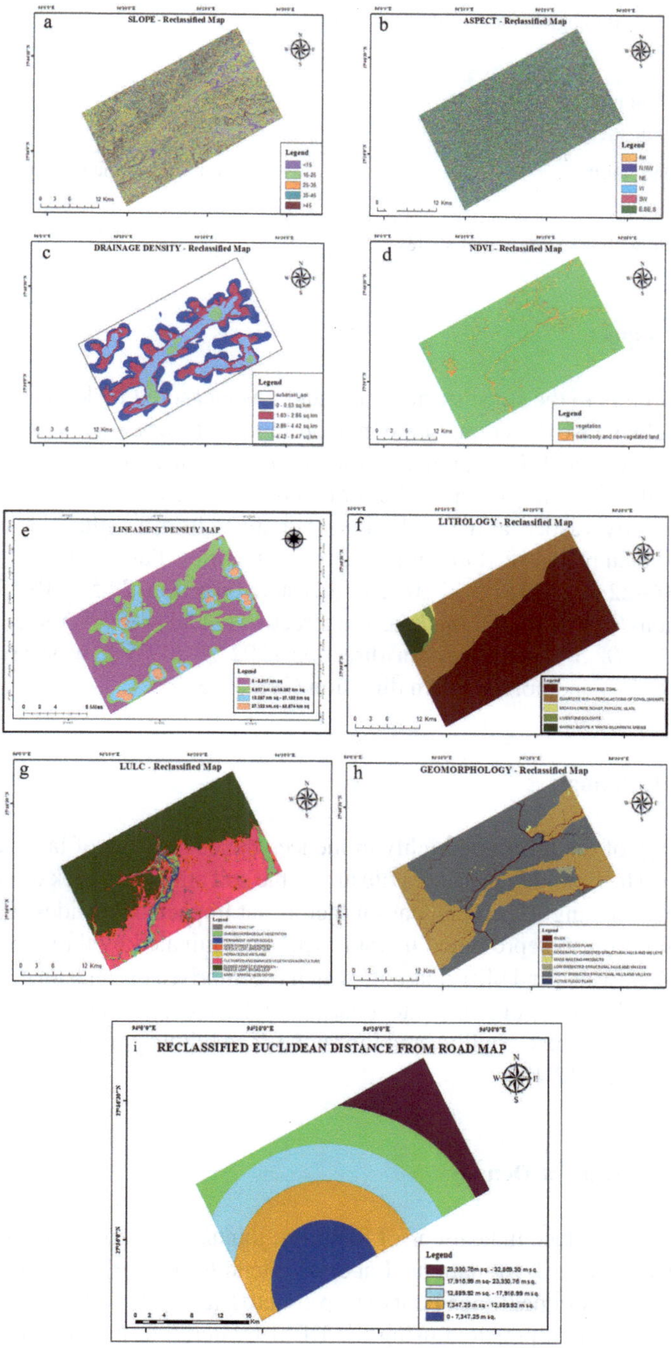

Fig. 15.4 Reclassified maps of prepared thematic layers **a** slope map **b** aspect **c** drainage density **d** NDVI **e** lineament density **f** lithology **g** LULC **h** geomorphology **i** Road (Euclidean distance)

Table 15.1 The lithological units of the study area

Sandstone	Siwaliks
Massive current bedded sandstone	
Sandstone withsiltstonee	
Main boundary fault	Lower Gondwana Group
Slate and shales	
Miri quartzite with intercalation of shales	

15.3.3.2 Aspect

The aspect refers to the slope's orientation and it is measured in degrees concerning North in a clockwise direction. Some meteorological factors, such as the amount of rainfall, sunshine, and the region's morphology are highly influenced by the slope of a region therefore it influences the probability of landslides (Pourghasemi et al. 2012). The study's aspect as in Fig. 15.4b was obtained from a digital elevation model (DEM) and then reclassified into nine distinct categories: flat terrain (−1), northern direction (0°–22.5° and 337.5°–360°), northeastern direction (22.5°- 209 67.5°), easterly direction (67.5°-112.5°), southeastern direction (112.5°–157.5°), southern direction (157.5°–202.5°), southwestern direction (202.5°–247.5°), westerly direction (247.5°–292.5°), and northwestern direction (292.5°–337.5°).

15.3.3.3 Lithology

The lithology of the hill slope highly influences the occurrence of landside (Ghosh et al. 2011). The lithology in an area illustrates the soil attributes, rock durability, and permeability, making it a crucial causal factor that impacts landslides. In the study area, the lithology as represented in Table 15.1 is dominated by three main groups: Siwaliks, Gondwana, and Miri-quartzite which is comprised of sandstone, feldspathic sandstone, shale, clay, conglomerate, quartzite with intercalations of conglomerate, coal-fossil wood, schist, phyllite, and limestone. The lithology map of the study area is depicted in Fig. 15.4f.

15.3.3.4 Lineament Density

Lineaments are surface indications of fractures, joints, shear zones, and discontinuities in subsurface rock formations. Landslides tend to be more frequent in regions where closely positioned lineaments are present (Das et al. 2022). The lineament map (Fig. 15.4e) was prepared using the DEM by deriving the hill shade for four different elevations and azimuth. Then, vector analysis using a line tool was utilized to manually digitize the lineaments and their density was found.

15.3.3.5 Geomorphology

The occurrence of landslides in a region is strongly associated with the geomorphology of that area. The study has characterized the geomorphological units in the area as comprising of highly dissected structural hills, moderately dissected structural hills, low dissected structural hills, active floodplains, older floodplains, rivers, and mass wasting products as shown in Fig. 15.4h.

15.3.3.6 Drainage Density

To demarcate the drainage network in the research region, HEC-geoHMS arc-hydro-tool was employed. The study detected stream orders varying from 1 to 4, which encompassed both major and minor channels. Additionally, a drainage density map (Fig. 15.4c) was generated using the length of streams in km^2 and Arc-GIS density tool. The established drainage density map was again reclassified into 4 classes to interpret the influence of drainage density upon landslides.

15.3.3.7 LULC

The data on LULC illustrate how the land is being utilized by humans for activities related to development and other purposes, and also shows the distribution of forests, wetlands, impermeable surfaces, agriculture, and various other land and water types within a region. The LULC map (Fig. 15.4g) of Sentinel 2 of 100 m resolution was downloaded from the Google earth engine and it was classified into 7 classes such as closed forest, open forest, wetlands, shrubs, urban/built, and sparse vegetation.

15.3.3.8 NDVI

When examining vegetation, the NDVI is utilised to assess its health, density, and growth. The index ranges from -1 to 1, where values above 0 signify vegetation and growth, while values below it indicates water bodies. Due to the natural anchoring effect of roots, vegetation commonly diminishes the likelihood of landslides and helps in preventing erosion. The NDVI of the study area was estimated using the formula:

$$NDVI = (NIR - Red)/(NIR + Red)$$

To distinguish between vegetated and non-vegetated areas, the resulting NDVI output (Fig. 15.4d) was classified into two classes.

15.3.3.9 Road Proximity (Euclidean Distance)

The current study utilized the Euclidean distance of roads to examine the correlation between road proximity and landslide occurrence. The stress condition and slope equilibrium of the road may be impacted by changes in slope (oversteepening) caused by excavation, increased loads, changes in hydrology, and drainage (Pourghasemi et al. 2012). The reclassified road map is depicted in (Fig. 15.4i).

15.3.4 AHP

The AHP model, introduced by Saaty (1980), utilizes pair-wise comparisons of relative factors to ensure consistent decision-making in situations involving multiple criteria. This methodology prioritizes three principles, which are the deconstruction, comparison, and integration of independent factors related to a dependent variable (Veerappan et al. 2017). AHP is a method of both qualitative and quantitative evaluation that is utilized to select the optimal parameter from a range of available alternatives. It is commonly applied to determine the relative importance of various components responsible for a particular issue through pair-wise comparisons. The method of determining the selective weights for different parameters is a sensitive process and it is purely subjective. By evaluating the consistency ratio of each subjectively assigned rating, the AHP model provides a key advantage in determining whether any assigned ratings require revision (Das et al. 2022).

Breaking down intricate decision-making problems into a structured hierarchy of factors and alternatives is the basis of the AHP approach. The weights for each factor and alternative are then determined through pair-wise comparisons on a nine-point ordinal scale (Fig. 15.5). Creating a matrix with an equivalent count of rows and columns to organize factors or their categories involves assigning scores on one side of the diagonal. Simultaneously, values of 1 are placed along the diagonal itself. When the factor on the vertical axis holds greater importance than the one on the horizontal axis, the assigned value will fall between 1 and 9. Conversely, in instances where the scenario is reversed, the assigned value covers the range of 1/2–1/9 (Kumar and Anbalagan 2016). The AHP model provides factors or weights represented by the eigenvector and calculates the Consistency Index (CI) and Consistency Ratio (CR), as introduced by Saaty (1980) with the following formulation:

$$CI = (\lambda\text{max} - \text{N})/(\text{N} - 1)$$

$\lambda max = maximum\ eigen\ value$
N = a number of geo-environmental factors present in rows and columns.

Intensity Importance	Description	Explanation
1	Equal importance	Two factors contribute equally to the objective
3	Moderate importance	Experience and judgment slightly to moderately favor one factor over another
5	Strong importance	Experience and judgment strongly to moderately favor one factor over another
6	Very Strong importance	A factor strongly favors over another
7	Extreme importance	Evidence of favoring one factor over another is of the highest degree possible of an affirmation
2,4,6,8	Intermediate values	Represent intermediate compromises between the preferences of 1, 3, 5, 7, and 9

Fig. 15.5 Scale of relative importance suggested by Saaty (1987)

$$CR = \frac{CI}{RI}$$

RI = Random index (Table 15.2).

The value of CI 0.1 and below is considered acceptable and the maximum threshold of consistency, value >0.1 is inconsistent. For the study 9 geo-environmental parameters were considered for deriving the landslide susceptibility map and a 9 × 9 pair-wise matrix was built to establish the AHP model. The most landslide influential parameters identified in the study were slope and lineament

Table 15.2 Random Index values for varying numbers of parameters

N	3	4	4	6	7	8	9	10	11	12	13	14	15
R	0.5	0.9	1.1	1.2	1.3	1.4	1.4	1.4	1.5	1.5	1.5	1.5	1.5
I	8	0	2	4	2	1	5	9	1	3	6	7	9

Table 15.3 Value obtained by the quantitative analysis of AHP Metrics

Nmax	9.851
CI	0.106
RI	1.45
CR	0.073

density and they were assigned higher criteria of weights than the other parameters. The CR value derived from the configured AHP model was 0.073 which is considered acceptable (Table 15.3).

15.4 Results

15.4.1 Landslide Susceptible Zonation Map

The LSZ map for Subansiri lower hydroelectric project (2000 MW) was derived from a semi-quantitative AHP model by considering a total of 9 geo-environmental factors in a pair-wise comparison matrix. The relative weights assigned to each causative factor (Table 15.4) were based on careful consideration of their influential behaviour for conditioning the slope for the occurrence of landslides. The highest priority weights were equally given to slope, lineament density, and drainage density as these factors are the major parameters that can condition slope for instability problems, and other factors were assigned relative weights accordingly. As the slope becomes more inclined the gravity influences the slope to instability mechanism and maximum landslides were recorded in slopes of angle >45°. Additionally, the proximity of slopes to tectonically active areas, such as major faults or thrusts, or lineaments, significantly impacts landslide activity. Therefore, the major occurrence of landslide frequency can be seen between lineament densities of 15.287–62.874 km^2. The drainage density influences the amount of water that infiltrates the slope, and it raises the pore water pressure and reduces shear strength, leading to slope instability. The Subansiri River flows through the study area with a maximum drainage density of 8.473 km^2. The vegetation present in an area plays a crucial role in mitigating erosion on slopes and the NDVI layer was classified into 2 classes i.e., vegetated and non-vegetated areas to derive the impact and frequency of occurrence of landslides. The lithology of an area depicts the characteristics of rocks that influence slope instability mechanisms by understanding their composition, textural properties, the extent of weathering, and physio-chemical behaviours. The study area is majorly underlain with sandstone, coal, and quartzite with intercalation of conglomerates rocks and it influences the initiation of landslides. The analysis of geomorphic factors helps in understanding the nature of past landslides. Poorly-planned construction activities associated with hill development programs, such as road cutting, infrastructure development, etc. exacerbate slope instability issues in hilly regions. The thematic layer of each factor

Table 15.4 Pairwise-comparison matrix

Pairwise comparison matrix

	SL	LD	DD (4th Order)	NDVI	LT	GM	AS	LULC	Road (Euc. dist)
SL	1	2	4	5	5	7	8	8	9
LD	1/2	1	2	4	4	6	6	7	9
DD (4th order)	1/4	1/2	1	3	3	5	7	8	9
NDVI	1/5	1/4	1/3	1	1	3	5	7	8
LT	1/5	1/2	1/3	1	1	3	5	6	7
GM	1/7	1/6	1/5	1/3	1/3	1	3	5	7
AS	1/8	1/6	1/7	1/5	1/5	1/3	1	3	5
LULC	1/8	1/7	1/8	1/7	1/6	1/5	1/3	1	3
Road (Eucl. dist)	1/9	1/9	1/9	1/8	1/7	1/7	1/5	1/3	1

was prepared and reclassified by assigning subjective weights in accordance considering their potential to cause a landslide. Additionally, the potential landslide zones within the study region were assessed using the landslide potential index (LPI) and it was achieved by multiplying the reclassified thematic layer with their respective criteria weights.

$$\begin{aligned}\text{LPI}_{\text{AHP}} = {} & (\text{Slope} * 0.307) + (\text{Lineament density} * 0.201) + (\text{Drainage density} * 0.166) \\ & + (\text{NDVI} * 0.094) + (\text{Lithology} * 0.094) + (\text{Geomorphology} * 0.059) \\ & + (\text{Aspect} * 0.036) + (\text{LULC} * 0.023) + (\text{Road} * 0.015)\end{aligned} \tag{15.4}$$

The LCZ map which was classified into five zones—very low, low, moderate, high and very high was derived by reclassifying the LPI output into 3 classes namely geometrical interval, quantile, and natural breaks (Jenks). The consistency matrix determined after deriving the criteria weights for each class is depicted in Table 15.5.

15.4.2 Model Validation—Reciever Operation Characteristics Curve

The output of LSM was validated using the ROC model and the resulted map showcased an accuracy for all three classification types: geometrical-interval, quantile, and natural breaks (Jenks) were determined to have an accuracy level of 79.2%, 79.8%, and 79.2%, respectively. The best result was yielded by the quantile classification technique as this method classify the features into equal number of classes. To validate the LSM, the landslide inventory was employed to analyze the spatial distribution of previous landslide occurrences. The outcomes are presented in Fig. 15.6a–c.

Table 15.5 Consistency matrix

Consistency matrix										
	SL	LD	DD (4th Order)	NDVI	LT	GM	AS	LULC	Road (Euc. dist)	Criteria weights
SL	1	2	4	5	5	7	8	8	9	0.307
LD	1/2	1	2	4	4	6	6	7	9	0.200
DD (4th order)	1/4	1/2	1	3	3	5	7	8	9	0.165
NDVI	1/5	1/4	1/3	1	1	3	5	7	8	0.094
LT	1/5	1/2	1/3	1	1	3	5	6	7	0.094
GM	1/7	1/6	1/5	1/3	1/3	1	3	5	7	0.059
AS	1/8	1/6	1/7	1/5	1/5	1/3	1	3	5	0.035
LULC	1/8	1/7	1/8	1/7	1/6	1/5	1/3	1	3	0.022
Road (Eucl. dist)	1/9	1/9	1/9	1/8	1/7	1/7	1/5	1/3	1	0.014

15.4.3 RAMMS: Rockfall Simulation Model

A comprehensive predictive rockfall simulation model was developed to assess potential rockfall hazards in the vicinity of the dam. The study took into account various factors such as the geological characteristics of the area, slope conditions, and the proximity to the road network. By analysing the lithology of the region, specific zones with high susceptibility to rockfall were identified, shown in Figs. 15.7 and 15.8. These zones were characterized by the presence of sandstone, a rock type known for its propensity to undergo weathering and erosion. Moreover, the chosen area had a slope angle exceeding 45°. Using advanced geospatial techniques, the simulation model incorporated data on the topography, geology, and slope conditions of the area. By integrating this information, it became possible to simulate potential rockfall scenarios and predict the area most likely to be impacted in the event of a rockfall occurrence. Therefore, In the present study, the rockfall simulation for 1 landslide occurring near the dam area was estimated to derive the velocity, kinetic energy, and jump height of the rockfall by in taking the ALOS-PALSAR digital elevation model of 12.5 m resolution in ESRI ASCII grid format in conjugation with other topographic parameters such as release area, calculation domain, forest, terrain material, and rock density. The velocity of the predicted rockfall is 9.84 ms^{-1}, the jump height of the same is initiated at a height of 1.11 m, and the kinetic rock energy of the rockfall will be 132.80 (KJ). The findings from these results are depicted in Fig. 15.9a–c. The resulting trajectory was then utilized to generate an XY plot (Fig. 15.10a–c), elucidating the correlation between altitude and key parameters including velocity (m/s), jump height, and kinetic energy.

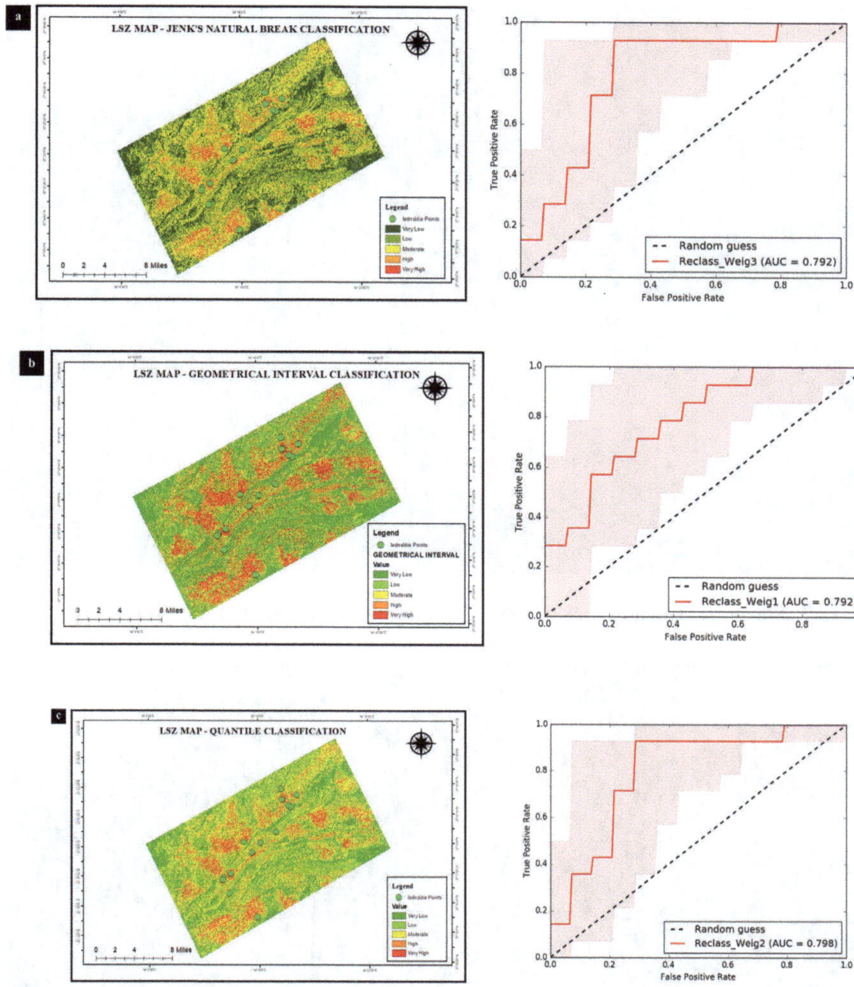

Fig. 15.6 Landslide susceptibility map and AUC ROC curve of **a** Jenk's natural break classification **b** geometrical interval classification **c** quantile classification

15.5 Discussion

Landslides present severe risks in India's hilly and plateau regions, causing fatalities and resource damage. Mapping susceptibility to landslides serves as an initial step toward developing these mountainous areas and reducing the impact of slope failures (Chowdhuri et al. 2020). The results of this chapter basically outlines the areas that are highly susceptible to landslides around the vicinity of the Subansiri Lower Hydro-electric project using AHP model. The LSM, was homogenously classified to various susceptible zones. Selecting the factors that lead to landslides is a critical step influencing the quality of landslide susceptibility analysis. Generally,

Fig. 15.7 Subansiri Dam area—28 October 2017, from Google earth pro software

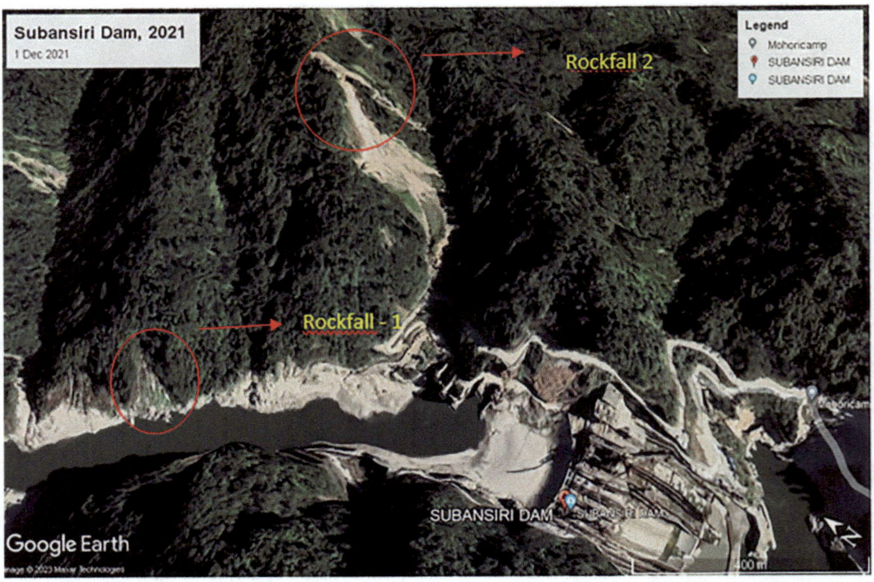

Fig. 15.8 Subansiri Dam area—1 December 2021, from Google earth pro software

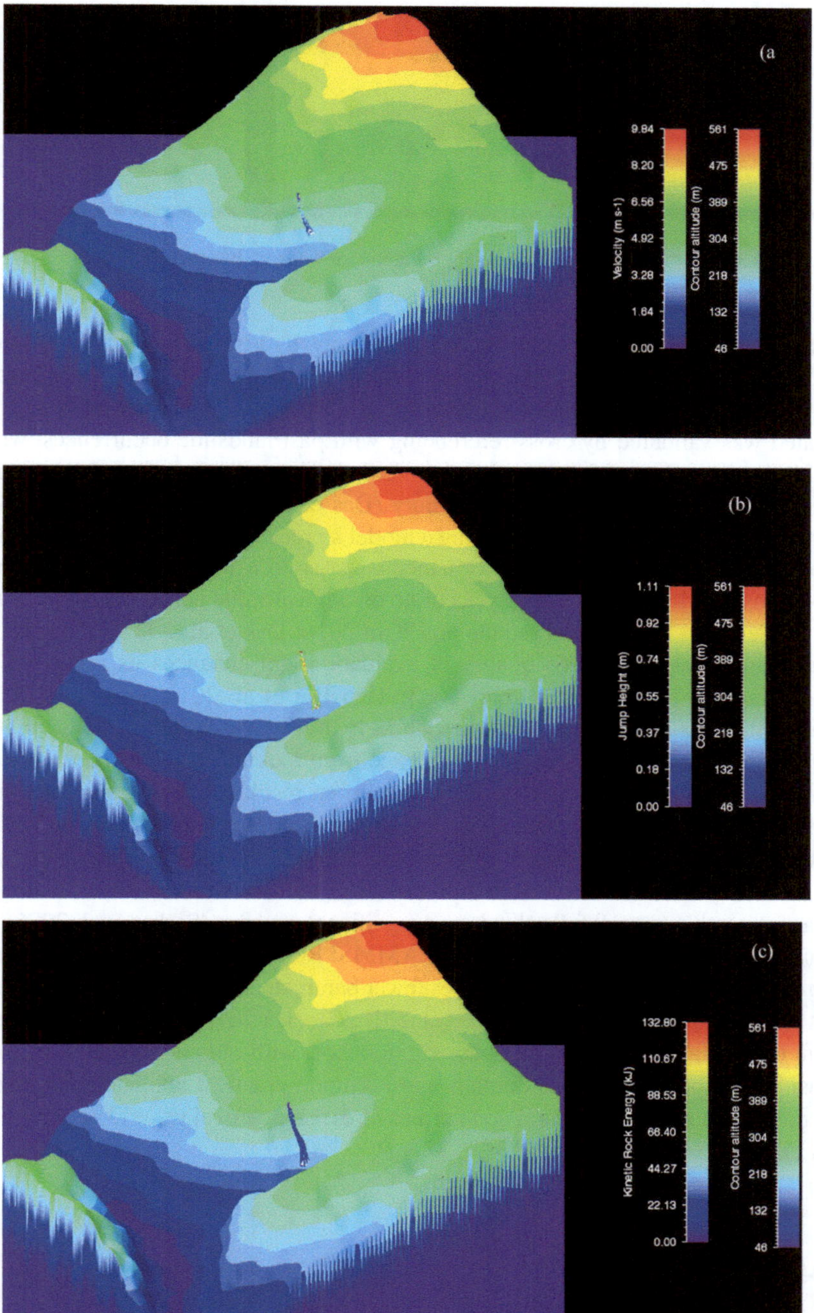

Fig. 15.9 **a** Velocity of the rockfall **b** jump height of the rockfall **c** kinetic energy of the rockfall

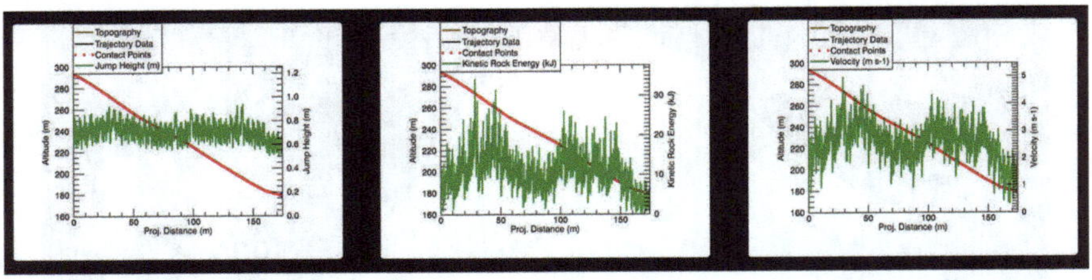

Fig. 15.10 Trajectory plot of **a** jump height **b** kinetic rock energy **c** velocity (m/s)

LSM often involves utilizing topography, geology, hydrology, and geomorphology (Aditian et al. 2018). The AHP model-derived map of predicted landslide susceptibility zonation was validated through the use of ROC–AUC method. The LSM created was validated by cross-referencing with past landslide occurrences, which highlights how the model provides a more definitive estimate of landslide probability and reduces the level of uncertainty (Das et al. 2022). The effectiveness of this method is assessed by determining a model's precision and predictive capability using the AUC. The model's performance is assessed using AUC values, where a score closer to 1 indicates a highly accurate fit, while a value below 0.5 suggests that the model lacks precision (Pradhan and Kim 2016; Das et al. 2022). This process also quantifies assessment's quality of evaluation and enhances its reliability for future analysis (Chung and Fabbri 2003; Tien Bui et al. 2012; Das et al. 2022). In mountainous areas, infrastructure such as hydroelectric facilities, transportation routes, and tourist sites are at risk of damage from rockfall incidents (Carrera et al. 2023). Rockfalls, a distinct type of failure, often result in catastrophic outcomes and pose a threat to lives nearby. The severity of these events hinges on the volume of the rock mass and the height from which it descends. Incidents involving rock falls striking vehicles have resulted in numerous injuries and fatalities. Identifying areas vulnerable to rock falls stands as a significant concern (Sarkar et al. 2006). By using the 'hard' modeling of contacts, RAMMS::ROCKFALL provides an accurate depiction of contact behaviour, requires fewer parameters, and results in a consistent mathematical formulation (Christen et al. 2012; Glocker 2013). Presently, the assessment of rockfall risks holds immense significance, with ongoing research showing considerable progress in this domain. Furthermore, practical applications utilizing uncertainty models in structural countermeasures add practical value (Volkwein et al. 2011).

15.6 Conclusion

A LIM was prepared for the study region to illustrate the spatial distribution of past landslides. The potential landslide zones were derived for the Subansiri lower hydroelectric project of 2000 MW by understanding various landslide internal causative factors that are the innate ground characteristics that render slopes vulnerable to landslides. The proper prediction of landslide hazard-prone zones is crucial for the safer construction of dams and to procure the need for stabilizing the areas before the completion of the project. The AUC values determined after configuring the AHP model showed an accuracy of 79% in all classification methods, indicating a highly predictive model. The three-dimensional rockfall RAMMS model is incredibly beneficial for planning proper mitigation measures of its state-of-the-art to estimate the velocity, kinetic energy, and run-out trajectory of the rockfall. By predicting the rockfall of the vulnerable areas, leads to have a more focused agenda of implementing robust mitigation strategies such as the construction of barriers to stop the harmful impacts of the hazard. So the implementation of various numerical models helps to achieve this. Furthermore, the current research did not incorporate factors that trigger landslides, like rainfall and seismic data, into the landslide susceptibility zonation. Hence, this research is not restricted by these factors and can be further analysed. Therefore, this chapter emphasis of assessing the most vulnerable areas to landslide by the help of LSM and then by predicting the future rockfall scenario can give diversified outlook on the mitigation measures to be implemented to safe guard the resources.

References

Aditian A, Kubota T, Shinohara Y (2018) Comparison of GIS-based landslide susceptibility models using frequency ratio, logistic regression, and artificial neural network in a tertiary region of Ambon, Indonesia. Geomorphology 318:101–111. https://doi.org/10.1016/j.geomorph.2018.06.006

Aleotti P, Chowdhury R (1999) Landslide hazard assessment: summary review and new perspectives. Bull Eng Geol Env 58(1):21–44. https://doi.org/10.1007/s100640050066

Anbalagan R, Kumar R, Lakshmanan K, Parida S, Neethu S (2015) Landslide hazard zonation mapping using frequency ratio and fuzzy logic approach, a case study of Lachung Valley, Sikkim. Geoenviron Disasters 2(1). https://doi.org/10.1186/s40677-014-0009-y

Bhusan K, Pande T, Kayal JR (2022) Landslide affected areas and challenges imposed in North Eastern Region of India: an appraisal. Aust J Eng Innov Technol 4:32–44

Bravo-López E, Fernández Del Castillo T, Sellers C, Delgado-García J (2022) Landslide susceptibility mapping of landslides with artificial neural networks: multi-approach analysis of backpropagation algorithm applying the neuralnet package in Cuenca, Ecuador. Remote Sens 14(14). https://doi.org/10.3390/rs14143495

Buah PA, Yingbin PZ, Bakah DAY, Ahiabu MK, Zhibin L (2019) Earthquake-induced landslide susceptibility analysis :the effect of DEM resolution. In: International conference on mechatronics, remote sensing, information systems, and industrial information technologies, ICMRSISIIT 2019. https://doi.org/10.1109/ICMRSISIIT46373.2020.9405915

Carrera BA, Mok C-M, Hort HM, Santi LM, Amini H (2023) Potential of rockfall impacts on critical infrastructure during earthquakes. In: ARMA US rock mechanics/geomechanics symposium, ARMA-2023

Chen W, Pourghasemi HR, Kornejady A, Zhang N (2017) Landslide spatial modeling: introducing new ensembles of ANN, MaxEnt, and SVM machine learning techniques. Geoderma 305(March):314–327. https://doi.org/10.1016/j.geoderma.2017.06.020

Chen W, Li W, Chai H, Hou E, Li X, Ding X (2016) GIS-based landslide susceptibility mapping using analytical hierarchy process (AHP) and certainty factor (CF) models for the Baozhong

region of Baoji City, China. Environ Earth Sci 75(1):1–14. https://doi.org/10.1007/s12665-015-4795-7
Chowdhuri I, Pal SC, Arabameri A, Ngo PTT, Chakrabortty R, Malik S, Das B, Roy P (2020) Ensemble approach to develop landslide susceptibility map in landslide dominated Sikkim Himalayan region, India. Environ Earth Sci 79(20). https://doi.org/10.1007/s12665-020-09227-5
Christen M, Bühler Y, Bartelt P, Leine R, Glover J, Schweizer A, Graf C, Mcardell BW, Gerber W, Deubelbeiss Y, Feistl T (2012) Integral hazard management using a unified software environment numerical simulation tool "RAMMS." In: Congress interpraevent, pp 77–86
Chung C-JF, Fabbri AG (2003) Validation of spatial prediction models for landslide hazard mapping. In: Natural hazards, vol 30
Das S, Sarkar S, Kanungo DP (2022) GIS-based landslide susceptibility zonation mapping using the analytic hierarchy process (AHP) method in parts of Kalimpong Region of Darjeeling Himalaya. Environ Monitor Assess 194(3). https://doi.org/10.1007/s10661-022-09851-7
Das S, Sarkar S, Kanungo DP (2023) A critical review on landslide susceptibility zonation: recent trends, techniques, and practices in Indian Himalaya. In: Natural hazards, vol 115, no 1. Springer Netherlands. https://doi.org/10.1007/s11069-022-05554-x
Ghosh S, Carranza EJM, van Westen CJ, Jetten VG, Bhattacharya DN (2011) Selecting and weighting spatial predictors for empirical modeling of landslide susceptibility in the Darjeeling Himalayas (India). Geomorphology 131(1–2):35–56. https://doi.org/10.1016/j.geomorph.2011.04.019
Giraud A, Rochet L, Antoine P (1990) Processes of slope failure in crystallophyllian formations. Eng Geol 29(3):241–253. https://doi.org/10.1016/0013-7952(90)90053-4
Glocker C (2013) Set-valued force laws: dynamics of non-smooth systems, vol 1. Springer Science & Business Media
Goswami U, Patgiri AD, Sarma JN (2002) Hydrogeologic and geomorphic settings of the Lower Subansiri Basin, Assam, India. J Nepal Geol Soc 27(December). https://doi.org/10.3126/jngs.v27i0.31975
Goswami TK, Bezbaruah D, Mukherjee S, Sarmah RK, Jabeed S (2018) Structures and morphotectonic evolution of the frontal fold–thrust belt, Kameng river section, Arunachal Himalaya, India. J Earth Syst Sci 127(6). https://doi.org/10.1007/s12040-018-0984-6
Guzzetti F, Carrara A, Cardinali M, Reichenbach P (1999) Landslide hazard evaluation: a review of current techniques and their application in a multi-scale study, Central Italy. Geomorphology 31(1–4):181–216
Guzzetti F, Malamud BD, Turcotte DL, Reichenbach P (2002) Power-law correlations of landslide areas in central Italy. Earth Planetary Sci Letters 195(3–4):169–183
Guzzetti F, Gariano SL, Peruccacci S, Brunetti MT, Marchesini I, Rossi M, Melillo M (2020) Geographical landslide early warning systems. Earth-Sci Rev 200(July 2019):102973. https://doi.org/10.1016/j.earscirev.2019.102973
Highland LM, Bobrowsky P (2008) The landslide handbook—a guide to understanding landslides. US Geol Surv Circular 1325:1–147
Huggel C, Clague JJ, Korup O (2012) Is climate change responsible for changing landslide activity in high mountains? Earth Surf Proc Land 37(1):77–91
KeithTurner' A, Schuster R (n.d.) IDAHO T ispomi iON DEPA1u MEN1
Kumar R, Anbalagan R (2016) Landslide susceptibility mapping using analytical hierarchy process (AHP) in Tehri reservoir rim region, Uttarakhand. J Geol Soc India 87(3):271–286. https://doi.org/10.1007/s12594-016-0395-8
Kumar K, Devrani R (2012) Micro-hazard evaluation and validation of landslide in a part of North Western Garhwal lesser Himalaya, India. Int J Geomatics Geosci 2(3):878–891. http://185.53.130.48/jggsvol1no12010/voltwo/EIJGGS3075.pdf
Leine RI, Schweizer A, Christen M, Glover J, Bartelt P, Gerber W (2014) Simulation of rockfall trajectories with consideration of rock shape. In: Multibody system dynamics, vol 32, Issue 2. https://doi.org/10.1007/s11044-013-9393-4

Mandal K, Saha S, Mandal S (2021) Applying deep learning and benchmark machine learning algorithms for landslide susceptibility modelling in Rorachu river basin of Sikkim Himalaya, India. Geosci Front 12(5):101203. https://doi.org/10.1016/j.gsf.2021.101203

Onagh M, Kumra VK, Rai PK (2012) Landslide susceptibility mapping in a part of Uttarkashi District (India) by multiple linear regression method. Int J Geol, Earth Environ Sci 2(2):102–120

Pardeshi SD, Autade SE, Pardeshi SS (2013) Landslide hazard assessment: recent trends and techniques. Springerplus 2(1):1–11. https://doi.org/10.1186/2193-1801-2-523

Pourghasemi HR, Pradhan B, Gokceoglu C (2012) Application of fuzzy logic and analytical hierarchy process (AHP) to landslide susceptibility mapping at Haraz watershed, Iran. Nat Hazards 63(2):965–996. https://doi.org/10.1007/s11069-012-0217-2

Pradhan AMS, Kim YT (2016) Evaluation of a combined spatial multi-criteria evaluation model and deterministic model for landslide susceptibility mapping. CATENA 140:125–139. https://doi.org/10.1016/j.catena.2016.01.022

Ray PKC, Chattoraj SL (2014) Sunkoshi landslide in Nepal and its possible impact in India: a remote sensing based appraisal. Int Arch Photogram, Remote Sens Spat Inform Sci—ISPRS Arch 40(8):1345–1351. https://doi.org/10.5194/isprsarchives-XL-8-1345-2014

Ray PKC, Dimri S, Lakhera RC, Sati S (2007) Fuzzy-based method for landslide hazard assessment in active seismic zone of Himalaya. Landslides 4(2):101–111. https://doi.org/10.1007/s10346-006-0068-6

Saaty RW (1987) The analytic hierarchy process—what it is and how it is used. Math Model 9(3–5):161–176. https://doi.org/10.1016/0270-0255(87)90473-8

Saha S, Arabameri A, Saha A, Blaschke T, Ngo PTT, Nhu VH, Band SS (2021) Prediction of landslide susceptibility in Rudraprayag, India using novel ensemble of conditional probability and boosted regression tree-based on cross-validation method. Sci Total Environ 764:142928. https://doi.org/10.1016/j.scitotenv.2020.142928

Sarkar S, Kanungo DP (2006) Landslide hazard zonation in India: a review. August 2017:349–354. https://doi.org/10.1142/9812701605003

Sarkar S, Kanungo DP, Patra AK (2006) Landslides in the Alaknanda valley of Garhwal Himalaya, India. Q J Eng GeolHydrogeol 39(1):79–82. https://doi.org/10.1144/1470-9236/05-020

Sundriyal YP, Tripathi JK, Sati SP, Rawat GS, Srivastava P (2007) Landslide-dammed lakes in the Alaknanda Basin, Lesser Himalaya: Causes and implications. Curr Sci 93(4):568–574

Tien Bui D, Pradhan B, Lofman O, Revhaug I (2012) Landslide susceptibility assessment in vietnam using support vector machines, decision tree, and nave bayes models. Math Prob Eng https://doi.org/10.1155/2012/974638

Van Westen CJ, Van Asch TWJ, Soeters R (2006) Landslide hazard and risk zonation—why is it still so difficult? Bulletin Eng Geology Environ 65:167–184

Varnes D (1978) Slope movement types and processes. Spec Rep 176:11–33

Varnes DJ (1984) Landslide hazard zonation: a review of principles and practice. International association of engineering geology. Commission on landslides and other mass movements on slopes, Issue 19. UNESCO. https://books.google.co.in/books?id=vT9RAAAAMAAJ

Veerappan R, Negi A, Siddan A (2017) Advancing culture of living with landslides, 2014. https://doi.org/10.1007/978-3-319-53498-5

Volkwein A, Schellenberg K, Labiouse V, Agliardi F, Berger F, Bourrier F, Dorren LKA, Gerber W, Jaboyedoff M (2011) Rockfall characterisation and structural protection—a review. Nat Hazards Earth Syst Sci 11(9):2617–2651. https://doi.org/10.5194/nhess-11-2617-2011

Wind Y, Saaty TL (1980) Marketing applications of the analytic hierarchy process. Manag Sci 26:641–658

Zimmermann F, McArdell BW, Rickli C, Scheidl C (2020) 2D runout modelling of hillslope debris flows, based on well-documented events in Switzerland. Geosciences (Switzerland) 10(2):1–17. https://doi.org/10.3390/geosciences10020070

Chapter 16
Land Use and Land Cover as a Conditioning Factor in Landslide Susceptibility: A Literature Review

Vipin Upadhyay and Himisha Dixit

Abstract Landslides have been catastrophic events in mountainous regions for ages. However, as human populations and economies expand, anthropogenic activities aimed at development have led to significant changes in land use and land cover in these regions. These changes have become crucial factors amplifying the susceptibility to landslides and escalating the threat of such disaster's multiple times over. There are two different aspects that needs to be understand to measure the susceptibility of landslide in a region. These are landslide hazard and risk to the community. Both these aspects are intricately intertwined with the spatial arrangement of land use and land cover. The occurrence of landslides is contingent upon a variety of geospatial factors, including soil texture and integrity, terrain slope, vegetation cover, drainage patterns, rainfall distribution, and soil erosion dynamics. Simultaneously, the impact of landslides on society, namely the risk posed to communities, is influenced by the availability of resources such as flora, fauna, humans, and infrastructure. Therefore, the assessment of land use and land cover holds immense significance not only in understanding susceptibility to landslide hazards but also in conducting impact assessments and devising effective mitigation strategies. This chapter provides insights into the trends of land use and land cover as conditioning factors in landslide hazards. It explores the contributions of 6,650 authors through the analysis of 3,240 articles published on landslides since 1976. Approximately 33% of these articles explicitly mention the terms "land use" and "land cover" in their abstracts, titles, and/or keywords, constituting contributions from 2,723 authors. In recent years, there has been a significant increase in studies focusing on landslides and the utilization of land use and cover as crucial terms, with the number of articles doubling in the last five years. Publications from the period 2001–2021 account for the top 20% of most cited works, contributing to over 75% of the total citations.

Keywords Landslide susceptibility · Land use land cover · Disaster · Hazard · Geography

V. Upadhyay (✉) · H. Dixit
School of Life Sciences, Central University of Himachal Pradesh, Dharamshala, HP, India
e-mail: vipinupadhaya@hpcu.ac.in

G. K. Panda et al. (eds.), *Landslide: Susceptibility, Risk Assessment and Sustainability*, Advances in Natural and Technological Hazards Research 52,
https://doi.org/10.1007/978-3-031-56591-5_16

16.1 Introduction

The rapid growth of the human population, along with human activities for development (Mukane 2014) and climate change in mountainous areas, puts significant stress on our natural environment. This makes these regions more susceptible to geological and hydrological hazards like cloud bursts, debris flows, and landslides worldwide (Hulme 2016; Mandal and Mandal 2019a, b). Human activities such as construction in hilly regions increase slope instability, making the terrain more prone to landslides (Panchal and Shrivastava 2020). Landslides, also known as mass movements or slope failures, are complex natural phenomena defined by the downward movement of soil, rocks, debris, or other materials on steep slopes under the influence of gravity (Mandal and Mondal 2019b; Nerella et al. 2019). They are significant disasters occurring in mountainous regions due to the interaction of various geological, topographical, and hydrological factors. Weak and fragile geology and surface and groundwater dynamics (Pandey et al. 2019; Coelho-Netto et al. 2007; Roslee et al. 2018), are the major factor contributing to landslide initiation (Devkota et al. 2013; Regmi et al. 2013, 2014; Gadtaula and Dhakal 2019; Bhandari and Dhakal 2020a), which are influenced by various conditionssuch as elevation, slope, aspect, curvature of the plane, soil texture, rock structure, lithology, streams, drainage pattern, land cover, and vegetation cover (Yang et al. 2003; Pour and Hashim 2017; Anucharn and Dasananda 2017). Increased vulnerability to landslides is often triggered by heavy rainfall saturating rocks and soil, reducing shear strength (Nerella et al. 2019). The soil on slopes possesses distinctive geological characteristics that significantly contribute to the initiation of landslides (Bhandari and Dhakal 2020b; Yalcin et al. 2011; Tofani et al. 2017; Bicocchi et al. 2019). Landslide events worldwide carry significant economic implications, both direct and indirect. Direct economic losses, quantifiable in monetary terms, include damages to infrastructure (Panchal and Shrivastava 2020). Indirect economic losses, such as societal dysfunction due to traffic disruptions, delays, injuries, and deaths, are not easily measurable in monetary terms (Rawat and Pant 2023; Kamal et al. 2022). These events not only result in loss of life and property but also pose serious threats to physical infrastructure, such as buildings, roads, bridges, and communication lines. They disrupt socioeconomic development and impact the natural environment, affecting land, vegetation cover, and water resources in hilly terrains globally (Nerella et al. 2019; Lee and Pradhan 2007).

Recently, there has been increased attention to landslide hazards and their studies due to a growing awareness of the socioeconomic impacts of landslides (Kumar Rai et al. 2014). In this context, the analysis of landslide susceptibility has become essential for reducing the risk of disasters, with the goal of preventing damages and casualties (Kamal et al. 2022). Landslide susceptibility maps pinpoint specific areas in the terrain that are more prone to slope failure by examining the characteristics of documented landslides (Guzzetti et al. 2012; Van Westen et al. 2012). To accomplish this, it is crucial to identify the factors that influence landslide occurrences, encompassing both triggering and conditioning factors. Various studies have categorized major conditioning factors as geological, topographical, and hydrological conditions, as well as rainfall and earthquakes as triggering factors. Among these factors, land use and land cover are regarded as significant conditioning elements in numerous studies that concentrate on hazards associated with landslides. The probability of landslides is heavily influenced by the management of land use and land cover. Land cover involves all the biological and physical features on Earth (Herold et al. 2006), whereas land use pertains to the utilization of these features. This encompasses natural resources such as rivers, lakes, ponds, mountains, forests, grasslands, rocks, and soil, as well as human-made structures like crucial infrastructure, transportation facilities, and agricultural areas. Various land features have diverse effects on both soil mechanical behavior and moisture. For instance, a robust vegetation cover serves as mechanical support for the soil, improving slope stability by preventing erosion (Masi et al. 2021; Parra et al. 2021). Conversely, actions such as deforestation, steep and improper slope cutting, irregularized drainage, uncontrolled settlement, and excessive agriculture often lead to a decrease in slope stability (Chen et al. 2019a, b, c). Therefore, the roles of land use and land cover (LULC) are pivotal as influential factors that impact landslides triggered by rainfall (Kamal et al. 2022; Panchal and Shrivastava 2020). In our research, we embark on a comprehensive investigation into how studies on landslide susceptibility integrate the multifaceted aspects of Land Use and Land Cover (LULC). Additionally, we conduct a bibliometric analysis, quantitatively examining research literature to explore how LULC data is utilized in assessing landslide susceptibility. To bring our study to a meaningful conclusion, we further scrutinize and explore prevailing trends in landslide susceptibility research. Our focus is on elucidating specific LULC data types and the various techniques used to parameterize them, paving the way for a potential system to rank the susceptibility of certain areas to landslides. The main scope of this study is to analyze patterns and trends about the role of land use and land cover as conditioning factors in landslide susceptibility within the academic literature to gain insights into the scholarly landscape, make informed decisions, and shape future research directions. The increasing availability of bibliographic databases and advanced analytical tools has further expanded the scope and impact of bibliometric analysis in the landslide hazard-related academic and research communities. It involves the statistical analysis of written publications, such as books, articles, and conference proceedings, to provide insights into various aspects of research, including authorship, collaboration, citation patterns, and publication trends. The main objectives of this study include identifying the major themes and areas in landslide hazard research and their association with land use and landcover, exploring the tempo-spatial contribution of land use and land cover in landslide research, and identifying the major research journal and author engaged in land use and land cover research in association to landslide hazard.

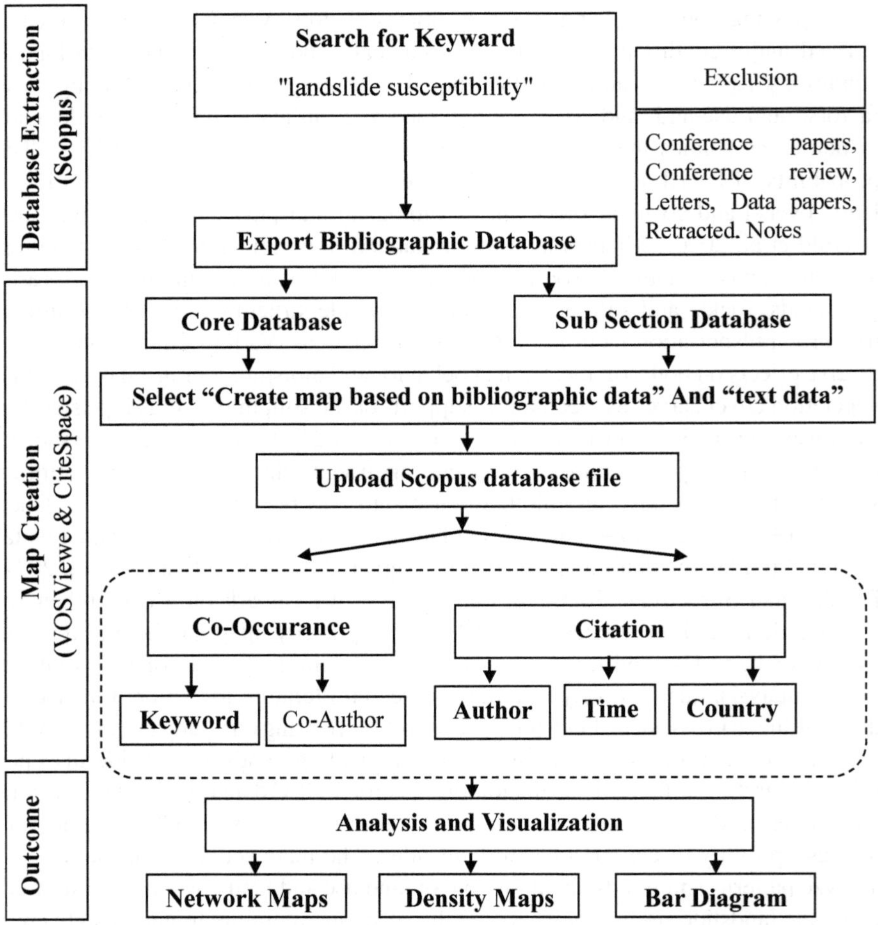

Fig. 16.1 Outline of study design used for the study

16.2 Materials and Methods

See Fig. 16.1.

16.2.1 Data Base Used: Scopus

To obtain the pertinent article, we conducted a web-based search using Elsevier's Scopus database, accessible at http://www.scopus.com. Scopus stands out as a crucial data source for bibliometric analysis, offering scholars the ability to formulate comprehensive search queries through various inquiry methods (Hirsch 2005). We were drawn to the Scopus database due to its advanced keyword search and citation analysis functions, which make it a preferred choice. Its extensive collection of articles from a diverse range of journals (Chang et al. 2015) renders it well-suited for our current study. Moreover, Scopus upholds a uniform and standardized documentation across diverse fields within the realm of literature studies (Bar-Ilan 2010). In several research contexts, it has demonstrated various advantages over Google Scholar, Medline, and even over Web of Science. These include, (1) Scopus encompasses an extensive collection of scientific literature, including content from book chapters, books, conference proceedings, research articles, conference publications, trade publications, peer-reviewed journals, and patents, etc. Its wide-ranging coverage makes it valuable for examining research trends spanning various disciplines, (2) Scopus meticulously curates its sources, ensuring the inclusion of high-quality data. The presence of content specialists within Scopus further enhances data quality, as they regularly assess and update the information to uphold accuracy, (3) Scopus facilitates sophisticated searches based on author, publication, affiliation, keyword, and citation criteria, (4) To assess research papers and authors Scopus offers various bibliometric metrics like citation metrics such as h-index, cocitation analysis, and citation counts, (5) Scopus effortlessly interfaces with data visualization and research analytics platforms, streamlining the analysis and presentation of bibliometric data, and (6) Scopus, widely regarded as the foremost and reputable research database, includes articles from influential global journals and encompasses the majority of the bibliometric repository found in Web of Science, MEDLINE, and EMBASE journals.

16.2.2 *Data Visualization Through VOSviewer*

VOSviewer, short for Visualization of Similarities Viewer, is a software tool crafted for the exploration and visualization of intricate networks (Kulkarni et al. 2009). VOSviewer enables users to analyze and visualize relationships within networks, particularly in the context of bibliometric data and scientific literature (Hood and Wilson 2001). The tool is widely used in academic and research settings for visualizing and interpreting patterns in large datasets, such as co-authorship networks, citation networks, and keyword co-occurrence networks (De Solla Price 1965). VOSviewer employs various algorithms to cluster and map nodes (representing items such as authors, documents, or keywords) based on their similarities. This allows users to identify clusters of related entities and their connections. One of the key features of VOSviewer is its ability to create interactive and customizable visualizations, providing researchers with a dynamic tool for exploring and gaining insights from their data (Bar-Ilan 2010; Soosaraei et al. 2018). The software is often used in conjunction with bibliometric analysis to uncover trends, identify influential authors, and visualize the structure of scientific fields. The visualization was used to represent the three-dimensional data i.e. (1) frequency of occurrence by the size of the node, (2) association strength by the width of interlinking lines, and (3) clustering by the color of the node in the VOSviewer map (Soosaraei et al. 2018). The size of the node is mainly used to indicate the number of publications and the number of citations per author, institute, and country; the thickness of the inter linking between the nodes indicate the intensity of the association or collaboration, and the color of each node can be used to indicates the year of publication or clustering among author, country, keywords, etc. (Falagas et al. 2008). In addition, density maps also can be produced by VOSviewer, and these density maps were used for representing the countrywide publication and citation (Boudry et al. 2018; Tao et al. 2020).

16.2.3 Search Strategy Used for Data Acquisition

There has been a growing trend in bibliometric research focused on the scientific output related to landslides (Ólafsdóttir and Tverijonaite 2018; Soosaraei et al. 2018) within the field of natural hazards. Despite this, relatively low number of studies have delved into the examination of landuse and landcover (LULC) in susceptibility analyses. In the following sections, we delineate our criteria for database searches, filtering procedures, data pre-processing steps, bibliometric analysis, and the strategies employed for the review process (refer to Fig. 16.1). The data included in this study were the core collection of Scopus database from its inception to October 2023. To optimize document retrieval, we use the keywords "Landslide susceptibility" as an entry for the "Article Abstract, Article Title, and Author Keywords" fields. To maintain the accuracy of the search scope, the search term undergoes continuous filtering, restricting the search to articles, book chapters, books, and reviews while excluding Conference papers, Conference reviews, Errata, Editorials, Notes, Retracted publications, Data papers, and Letters. The search strategy specifically targeted publications published in peer-reviewed scientific journals, including Articles, Editorial Material, Letters, and Reviews (Parish et al. 2018; Qiu et al. 2014). The language of publications is English. We excluded meeting abstracts, letters, editorial material, proceeding papers, early access, correction, book chapters, retraction, news items, reprints, or retracted publications. Therefore, to perform an overall bibliometric analysis database was shortlisted and selected using the keyword "landslide susceptibility" through the Scopus database was exported along with basic information regarding countries, institutions, publication years, publication quantity, publication types, authors, and citations. Data exported in CSV files were inspected in Microsoft Excel and further sub-sectioned for publication on landslide susceptibility mentioning landuse landcover and saved as separate CSV data files. Further, both of these files were separately uploaded to an informetric mapping software VOSViewer application version 1.6.18 (van Eck and Waltman 2014) to analyze and visualize the co-occurrence network of countries, institutions, author partnerships, keywords, and journals (Tao et al. 2020; Kulkarni 2009).

16.2.4 Bibliographic Data Analysis on Landslide Susceptibility

To conduct bibliometric analysis, we acquire the details of publications shortlisted and selected from the Scopus database using the term "landslide susceptibility". This data was further segmented into a sub-database of publications on landslide susceptibility mentioning the term "landuse landcover". Therefore, we have two data base (1) publications on landslide susceptibility and (2) publications on landslide susceptibility mentioning the term landuse landcover. Both these databases were inspected through various bibliographic analyses like (1) author and Co-authorship analysis, (2) analysis on temporal and geographic spread, (3) analysis on keywords association, and (4) analysis on landslide susceptibility focused on land use landcover. This bibliometric analysis provides important insight into (1) productivity and impact of landslide susceptibility by analyzing publication and citation counts; (2) visualize collaborations among authors and countries through network mapping; and (3) portrays the major research themes by identifying the most frequent terms and keywords (Parish et al. 2018; Boudry et al. 2018). Analyzing the citations related to countries, references, sources, and authors offers insights into the distribution and caliber of research on a specific topic. In the context of landslide susceptibility and the role of land use and land cover in it, this approach enables us to ascertain the most productive and frequently cited countries, regions, and authors. Moreover, through co-citation analysis of references, sources, and authors, we gain an understanding of the interconnections and groupings within publications, sources, and authorship within a research domain of landslide susceptibility. This analysis involves examining how frequently these elements are cited together, allowing us to pinpoint the most influential publications, sources, and authors in the realm of research on media and the digital ecosystem (Hirsch 2005; Montalván-Burbano et al. 2020). Co-word analysis furnishes insights into the prevalent keywords and their groupings within research on media and the digital ecosystem. This method entails scrutinizing the frequency of word usage across a collection of texts to identify the most significant or recurrent keywords in that set (Soosaraei et al. 2018; Tao et al. 2020). This research monitors subjects pertaining to landslide susceptibility, establishes a multidimensional relationship structure, and presents a thorough examination of the evolving trends in digital preservation of land use and land cover in relation to landslide susceptibility. The study utilizes Vosviewer to analyze the evolution of literature availability from its inception to October 2023, illustrating patterns across different time periods. The focus of the bibliometric network structure analysis lies in citation analysis, encompassing countries, references, sources, and authors (Ólafsdóttir and Tverijonaite 2018; Tao et al. 2020). The analysis of co-citation and co-word data was carried out through the use of VOSviewer software, employing the Association Strength method to gauge the strength of associations based on citation weight. Additionally, to visually depict the frequency of keyword occurrences and the connections among publications, sources, and authors, VOSviewer was utilized to generate

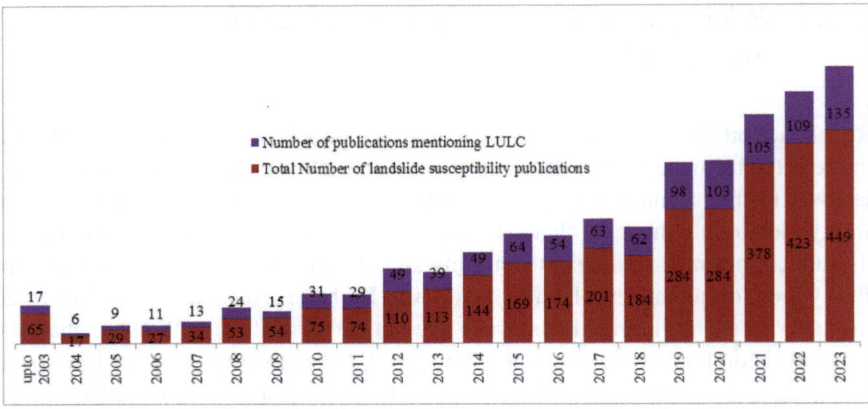

Fig. 16.2 Temporal spread and growth in the number of publications on landslide susceptibility and publications on landslide susceptibility mentioning land use land cover

network maps for co-words and co-citation. In such maps, nodes symbolize publications, sources, authors, and keywords. Larger nodes signify greater impact and influence, whereas smaller and more isolated nodes denote entities with lesser influence and lower frequency in the realm of public media and the digital ecosystem. Regarding the lines or links, they signify the co-citation relationships among nodes, and their thickness corresponds to the strength of these connections. And finally, the colors denote the clusters formed, categorized based on the most frequently recurring topics.

16.3 Results

The data incorporated into this study comprised the core collection of the Scopus database, spanning from its inception to October 2023. Initially, 4,297 articles were identified with the term "landslide susceptibility" in author keywords, titles, or abstracts, covering the period from 1976 to October 2023. This number reduced to 3,240 when our analysis was limited to articles (3,079), book chapters (186), books (6), and reviews (69), excluding Conference papers (627), Conference reviews (49), Erratum (27), Editorial (7), Note (6), Retracted (3), Data paper (2), and Letter (2). Notably, out of this subset, only 3,240 publications were composed in the English language. The core dataset for this study consisted of 3,240 publications. Within this core data, a subset of 1,085 publications was identified, specifically including the term "land use land cover" in connection with landslide susceptibility in author keywords, titles, or abstracts. This subset was extracted from the larger core database of 3,341 publications through manual searches for keywords such as "land use land cover," "land use and land cover," "landuse landcover," "land use and landcover," "land use," "land use," "land cover," or "landcover" within the previously selected 3,574 publications. Bibliographic analysis of both of these databases was performed separately. A comparative study was performed to analyze the role of land use land cover in studying the susceptibility of landslides. Research on these topics demonstrates an increasing trend from 2017–2018 to the present. It took approximately 41 years and 25 years, respectively, to reach about half of the current publication numbers for landslide susceptibility publications (1523) and publications mentioning both landslide susceptibility and land cover (535). This number doubled in the preceding five years (Fig. 16.2). The significant growth in this field could be attributed to the widespread availability of satellite imagery, topographic data, and land-use land cover data (Wulder and Coops 2014; Jun et al. 2014; Gómez et al. 2016). Additionally, there has been an increase in the number of references per article. The average reference list has approximately tripled, with an average of around 30 references per article before 2018, surpassing 80 references per article in publications from the past five years, reflecting the doubled number of publications. The surge in citations could be linked to the "Sendai Partnerships 2015–2025 for the global promotion of understanding and reducing landslide disaster risk," which was adopted by 33 countries with the aim of enhancing landslide research (Sassa 2015). The year 1991 emerged as the most impactful year in terms of publication quality, with articles published in 2003 achieving the highest average citation rate of 390 per publication for publications on landslide susceptibility. Notably, the paper with the highest citation, published in 2005, focused on the GIS-based logistic regression modeling for landslide susceptibility assessment. The top 10% of the most cited publications accounted for over ~56% of the total citations, while the top 20% of the most cited publications contributed to more than ~73% of the total citations. This analysis covered publications from 1991 to 2021 for the top 10% and from 1980 to 2022 for the top 20%. However, when evaluating quality based on citations per year, the publication from the year 2018, titled "A Review of statistically-based Landslide Susceptibility," emerged as the most impactful, with an average of ~189 citations per year. In terms of productivity, the year 2023 recorded the highest number of publications, totaling approximately 447, while publications from the year 2019 garnered the highest citation count, reaching around 11,501. Regarding publications on landslide susceptibility that mention land use and land cover, the year 2003 stood out as the most impactful year in terms of publication quality. Articles published in 2003 had the highest average citation rate, approximately ~291 per publication, with the paper published in 2008, emphasizing the role of spatial data in landslide susceptibility, hazard, and vulnerability assessment, securing the highest citation of ~912. The top 10% of most cited publications counted for more than ~50% of total citations, and the Top 20% of most cited publications counted for more than ~75% of total citations, which were published from 2001 to 2021. But if we consider citation per year as a measure of quality, the publication published in the year 2016 on a comparative assessment of Spatial prediction models for landslide monitoring proved itself as most effective with ~123 citations per year. In terms of productivity, the year 2023 has the highest number of publications, ~135, and publications published in the year 2012 have the highest citation count of ~5264. Moreover, the scrutinized articles were disseminated across approximately 544 distinct journals. Among these, the top

~10% most frequently occurring journals accounted for roughly 72% of all publications and approximately 88% of all citations. Notably, the journal "Nat. Hazards" published the highest number of articles (188), while "Geomorphology" accrued the highest number of citations (16,250). However, on considering the average publication and citations, the highlighted journal was "Environ. Earth Sci." and "Catena" with 11 publication per year and 6614 citations per year, respectively, for publication on landslide susceptibility. According to the database, the journal that demonstrated the longest activity in the field was "Geomorphology" (35 years), with publications spanning from 1989 to 2023, followed by "Earth Surf. Processes Landf." (33 years) and "Earth Surf. Processes Landf." (32 years). Regarding publications on landslide susceptibility that mention the term "land use land cover," the analyzed articles appeared in 128 different journals. Among these, the top 15 most frequent journals covered 57% of all publications and 74% of all citations. The journal "Environ. Earth Sci." published the highest number of articles (68), while "Catena" amassed the highest number of citations (6,381). However, on considering the average publication and citations, the highlighted journal was "Environ. Earth Sci." and "Environ. Geol." with 5 publication and 406 citations per year, respectively. According to the database, the journal that showed the longest activity in the field was "J. Geol. Soc. India (28 years)", with publications between 1996 and 2023, followed by "GeoJournal" (22) and the "Nat. Hazards" (21) since 2000–2021 and 2003–2023 respectively.

16.3.1 Author and Co-authorship Analysis

The core data mentioned the involvement of 6,650 authors contribution for writing the 3,240 publications on landslide susceptibility and the involvement contribution of 2,723 authors in producing 1085 publications that mentioned the term land use land cover in association with landslide susceptibility. Pradhan, B. is identified as the highest contributing author in both the case of the core (115 publications) and sub-sectioned (74 publications) database. Only 25 out of 2,723 and 177 out of 6,650 authors were identified as contributed to at least ten publications in each case, respectively, these 25 and 177 contributors contributed to ~35% of total publications in their respective areas. The articles had between a maximum of one author (~5%) to a maximum of 41, and 25 authors contributed to one article in the case of core and sub-set databases. Most contributions had reached three (672) to four (606) and two (218) to three (247) authors per publication in the case of core and subsection databases, respectively. To illustrate collaborations among co-authors, we set a minimum publication threshold of ten publications for the largest set of connected items. The resulting map reveals 177 authors for publications on landslide susceptibility and 25 authors for publications on landslide susceptibility that mention land cover (Fig. 16.3). Among these collaborations, Pradhan, B., Chen, W., and Lee, S. are the top three publishers respectively in both cases (figure). But when we go further down the list, Pham, B. T., Pourghasemi, H. R., Hong, H., Shahabi, H. and Li, Y. are the publishers placed at 4th to 8th position when we consider the publication on landslide susceptibility (Figs. 16.3a and 16.4). At the same time, the tally goes as Pourghasemi, H. R., Shahabi, H. Bui, D. T., Hong, H., and Pham, B. T. respectively for position from 4 to 8th in case of publication on landslide susceptibility mentioning landuse landcover (Figs. 16.3b and 16.4). In addition to the authorship analysis, we also explore the reach of each publisher by analyzing the number of citations and quality of publications published by the publisher by analyzing the average citation per publisher per publication. We observed that Bui, D. T. is the most effective author with the highest ratio of 218 and 234 citations per publication in case of publication on landslide susceptibility and publication on landslide susceptibility mentioning land use landcover, respectively (Table 16.1).

16.3.2 Geographic Spread Analysis

The most publications were from Asia, with China (217), India (208), Iran (150), and Malaysia (98) as the top four countries in the tally of countrywide publication list on landslide susceptibility publications mentioning landuse landcover (Fig. 16.5d). Other than Asian nations from the European region; Italy (95), Turkey (61), Norway (61), Germany (38) and Austria (31); USA (79); Australia (39); and other Asian countries; South Korea (84), Vietnam (74), and Japan (37) had more than an average of one publication per year i.e. 30 publication in last 30 years since its inception in 1993. Collaborative associations among 51 countries have a minimum of five publications each was represented as seven different clusters of country associations on publication on landslide susceptibility mentioning land use landcover (Fig. 16.5). The largest cluster shown in color red represented 19 countries most are from Europ and American regions (i.e., USA, Italy, Germany, Austria, Greece, Mexico, Brazil, etc.); second (9 countries), third (6), fourth (6), fifth (5), sixth (4) and seventh (2 countries) cluster can be represented as clusters of India (green), China (blue), Viet Nam (yellow), Malasya-South Korea (purple), Iran (cyan) and grees-south Africa (orange) with their associates in terms of collaborative efforts for studying the role of landuse landcover in landslide susceptibility (Fig. 16.5). Authors from China were the most prolific, participating in 217 publications through collaborations with over 30 countries. (Fig. 16.6c), mainly Iran (31) and Malaysia (14). India has 208 publications and most the highest publication country with also having collaboration with more than 30 countries (Fig. 16.6b), mainly with the Viet ban (27) and Iran (21). The highest collaboration of 39 publications between Iran and Malaysia placed them as the third and fourth highest publications in this field (Fig. 16.6a). These articles focused on landslide susceptibility assessment, monitoring, and modeling performance, drawing on various machine learning algorithms and statistical methods using geospatial data on geographical, topographical, and land-use landcover data. Based on the performance of each country publishing at least 30 publication, the average citation (AC) values for Norway (164), Malaysia (146), South Korea (145), and Turkey (125) was counted as excellent as they exceeded 100. Average citation (AC) values for a country like Australia (98), Vietnam (91), Iran (89), Germany (88), Italy (67), USA

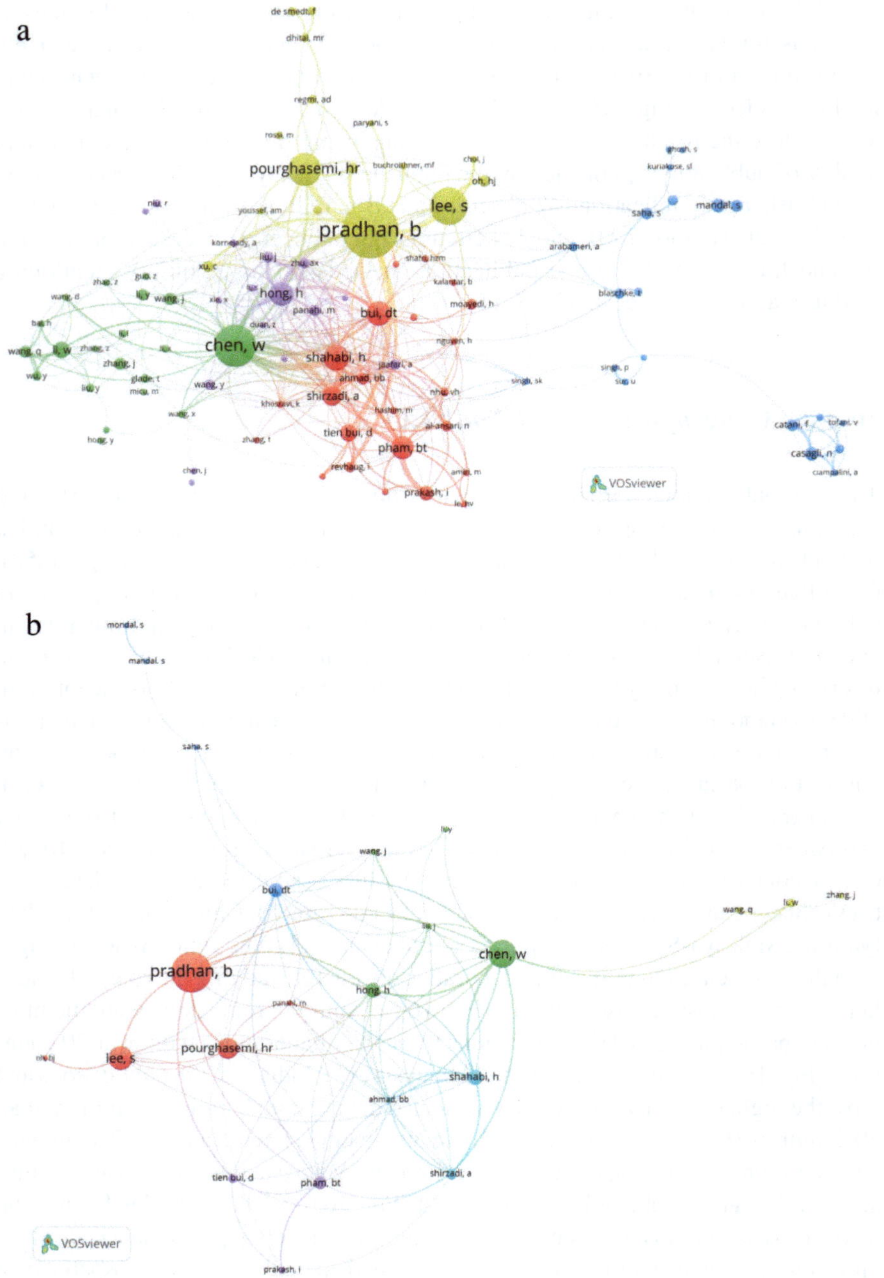

Fig. 16.3 Author–co-author (include only author having at least 10 publications) association network of publications on **a** landslide susceptibility and **b** publication on landslide susceptibility mentioning land land cover

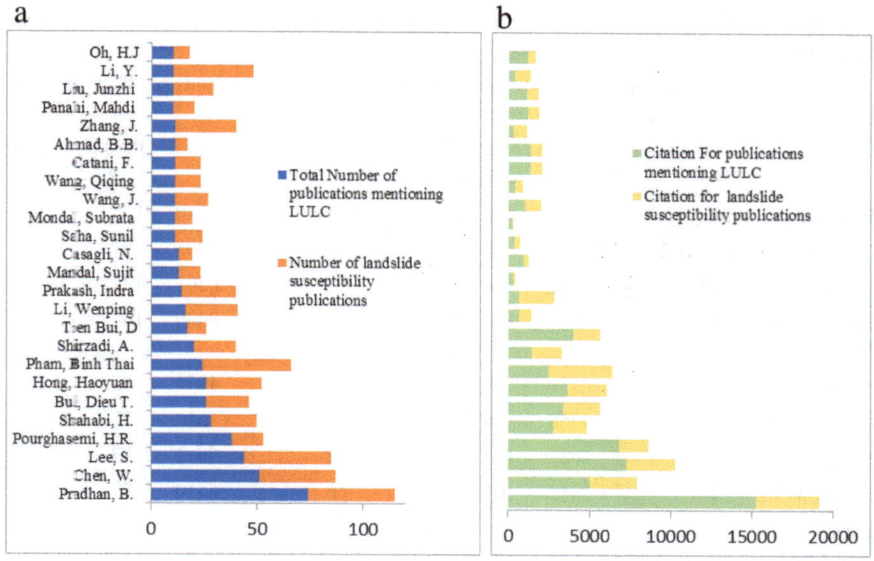

Fig. 16.4 Frequency of publication and citation per author (include only author having at least 10 publications in sub-section data) having publications on **a** landslide susceptibility and **b** publication on landslide susceptibility mentioning land use land cover

(66), and Japan (64) were counted as good as it ranged between 50–100. Countries like China (47), Australia (44), and India (28) were counted as average in terms of average citation (AC) based performance on research on the role of land use land cover in landslide susceptibility (Fig. 16.7). The geographical emphasis of each examined article, along with the national research output, supports the observation that almost all countries most severely impacted by landslides were incorporated as study areas. Additionally, countries with substantial research activity, including China, India, Iran, and Italy, have initiated national-scale Land Use and Land Cover (LULC) mapping projects (Congedo et al. 2016; Moulds et al. 2018; Zhang et al. 2019; Ghorbanian et al. 2020). Nevertheless, it is crucial to highlight that authors from these prolific countries did not represent all the regions examined in the studies. This implies that the most studied and impacted areas may not necessarily be the most significant contributors to the research. For instance, studies on Afghanistan, Colombia, Costa Rica, Ecuador, and El Salvador were conducted by researchers affiliated with countries such as China and Nepal, the Czech Republic, the USA, Germany, Canada, and Spain, respectively.

Table 16.1 Top 25 authors contributing in at least ten publications on landslide susceptibility mentioning the term "land use, landcover"

	Publication on landslide susceptibility mentioning land landcover			Publication on landslide susceptibility		
Author	Publication	Citation	Per publication citation	Publication	Citation	Per publication citation
Pradhan, B.	74	15,259	206	115	19,140	166
Chen, W.	51	5016	98	87	7905	91
Lee, S.	44	7316	166	85	10,307	121
Pourghasemi, H.R	38	6839	180	53	8609	162
Shahabi, H.	28	2770	99	50	4844	97
Bui, Dieu T.	26	3358	129	46	5653	123
Hong, H.	26	3658	141	52	6081	117
Pham, B. T.	24	2455	102	66	6419	97
Shirzadi, A.	20	1413	71	40	3288	82
Tien Bui, D.	17	3977	234	26	5655	218
Li, W.	16	662	41	41	1424	35
Prakash, I.	14	648	46	40	2820	71
Mandal, S.	13	322	25	23	440	19
Casagli, N.	13	944	73	19	1220	64
Saha, S.	11	373	34	24	702	29
Mondal, S.	11	266	24	19	293	15
Wang, J.	11	985	90	27	1980	73
Wang, Q.	11	423	38	23	884	38
Catani, F.	11	1342	122	23	2089	91
Ahmad, B.B.	11	1335	121	17	2043	120
Zhang, J.	11	289	26	40	1148	29
Panahi, M.	10	1169	117	20	1870	94
Liu, J.	10	1125	113	29	1820	63
Li, Y.	10	365	37	48	1386	29
Oh, H. J.	10	1160	116	18	1671	93

16.4 Keywords Search Analysis

The most common relevant term in the abstracts and title of all publications on landslide susceptibility and on landslide susceptibility mentioning land use land cover publications was "model", which appeared more than 9162 times in 2298 different documents and 3202 times in 774 different documents, respectively in landslide susceptibility publications and publications on landslide susceptibility mentioning land use land cover. Other common words were "area", "factor", "distance", "slope",

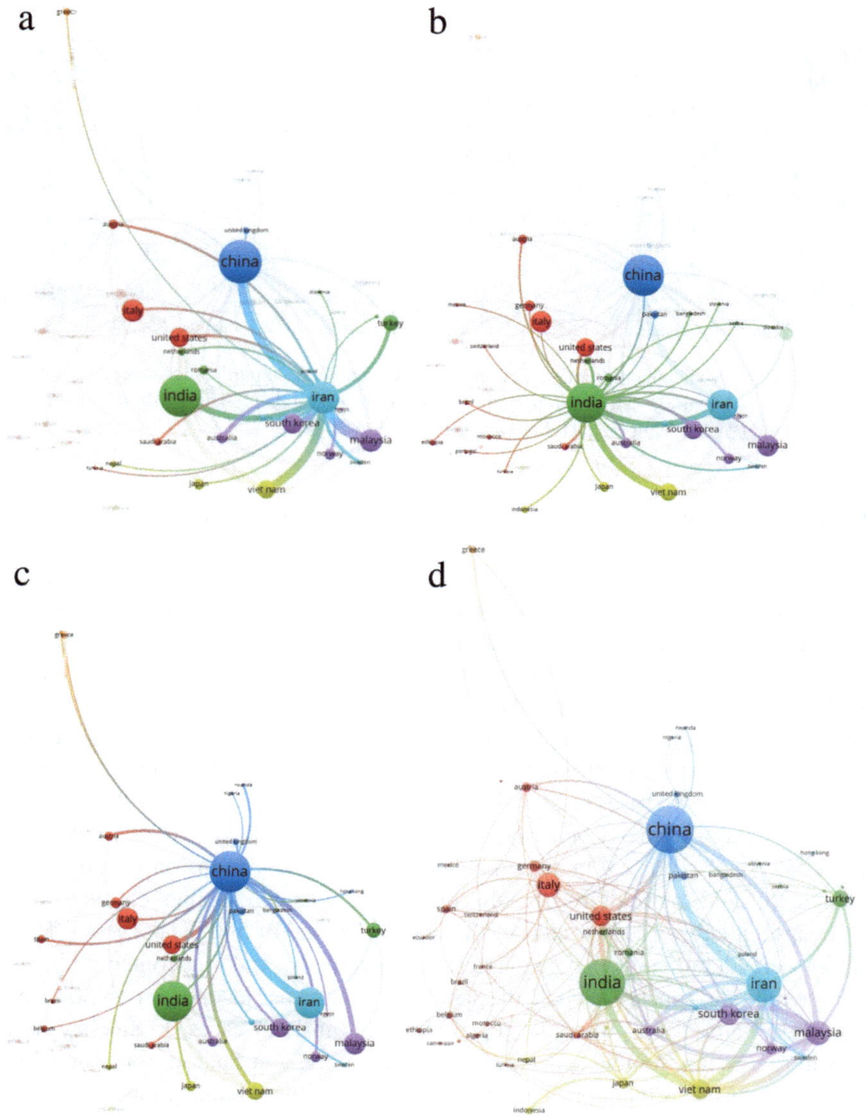

Fig. 16.6 Cooperative network visualized map of countries on publication having the association of Land use land cover mentioning in landslide susceptibility publications; **a** highlighted cooperation of Iran, **b** highlighted cooperation of India, **c** highlighted cooperation China **d** inter-country cooperation

"accuracy", "rainfall," and "lithology", which occurred more than 1000 times in more than 1000 different documents out of total 3240 documents on landslide susceptibility and more than 500 times in more than 400 different documents out of total 1085 documents on landslide susceptibility mentioning land use land cover. The term "conditioning factor" occurred around in 532 and 242 documents related to landslide

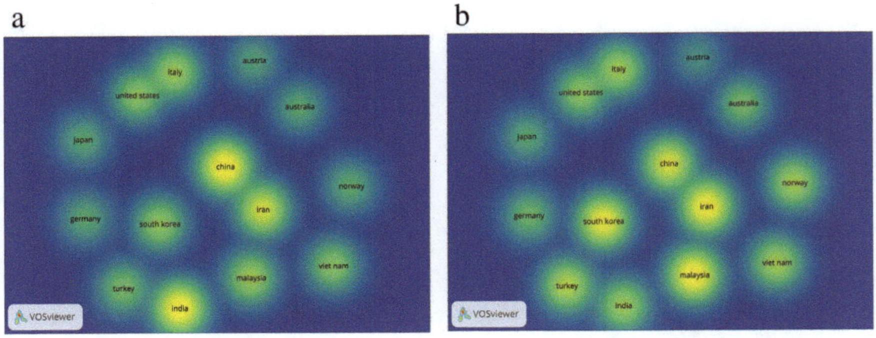

Fig. 16.7 Density visualization map of countries (having at least 30 publications) representing the number of **a** publications and **b** citations on publications on landslide susceptibility mentioning Land use land cover during the last three decades, 1993 to 2023

susceptibility (out of a total 3240) and landslide susceptibility mentioning land use land cover (out of a total 1058) respectively, about two dozen terms were related to conditioning factors in both the cases out of which 184 and 185 times in association with landuse landcover and 230 and 505 times in association with landslide susceptibility in both cases respectively (Fig. 16.8). The word land use and land cover was identified as separate words, which was corrected as a single term, "Landuse and landcover" before uploading the text file to VOSvisualizeer. Therefore, the term "landuse landcover:" was used to find associations; it shows association with more than two dozen term out of which 741 and 719 times in association with landslide susceptibility in both cases respectively. Other than the term landuse landcover, other common, though similarly generic, LULC terms such as "road", "river", "soil," and "forest" were also present in a countable amount (Fig. 16.9). In contrast, the terms "settlement" and "agriculture" remain comparably scarce in the publication records.

Regarding keywords provided by the authors, a total of 5665 and 1986 different entries were identified from both databases, respectively. And we counted for the top 50 keywords that appeared in both the databases. The terms "landslide susceptibility", "landslide," and "GIS" were the top three terms in both cases. Frequency ratio, logistic regression, remote sensing, machine learning, and random forest were the other common terms mentioned as author keywords in these publications (Fig. 16.8).

16.5 Discussion

Landslides have been catastrophic events in mountainous regions for ages. However, as human populations and economies expand, anthropogenic activities aimed at development have led to significant changes in land use and land cover in various regions. Most of the studied papers on landslide susceptibility sought better models either through statistical (Sakkas et al. 2016; Aram et al. 2022) or machine learning

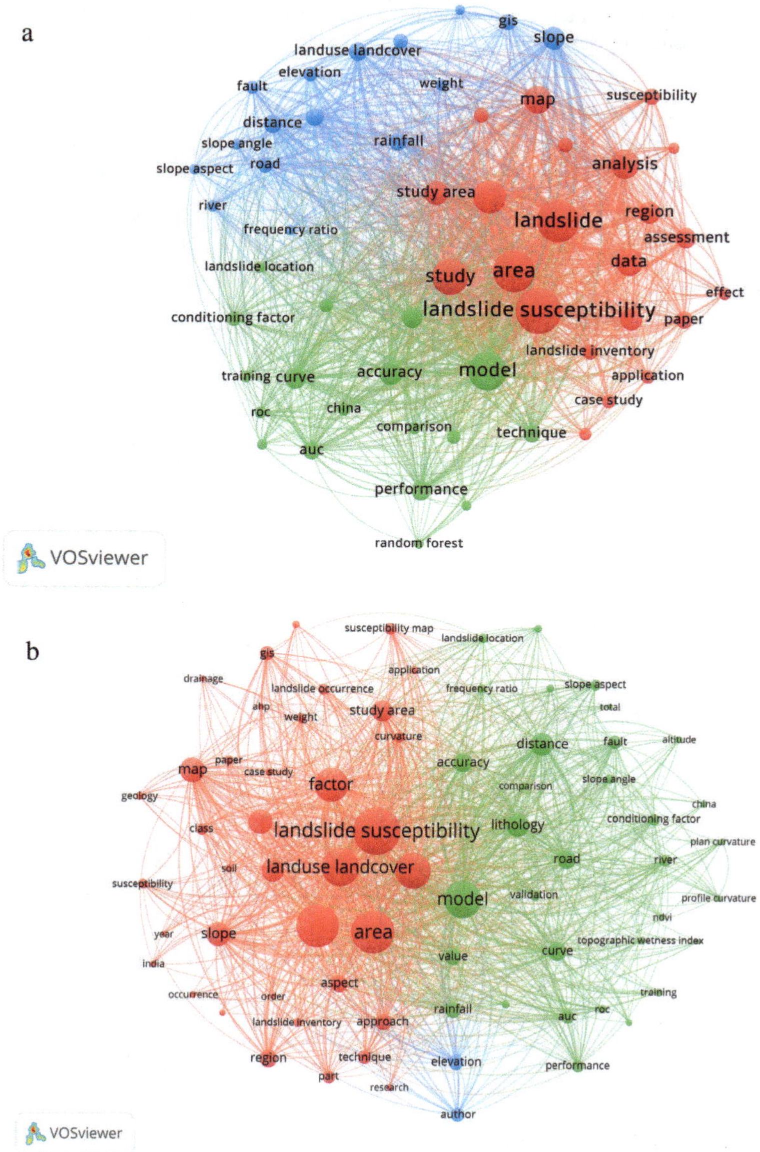

Fig. 16.8 Keywords co-occurrence network map depicting frequency and association of occurrence of major keywords that were present in the title and abstract of at least in 10% of total documents provided on **a** landslide susceptibility and **b** landslide susceptibility mentioning land use landcover. The frame represents the keywords, and lines connect the words that are provided together. The size of the frame is proportional to word frequencies, with thicker connecting lines indicating a stronger relationship between the terms

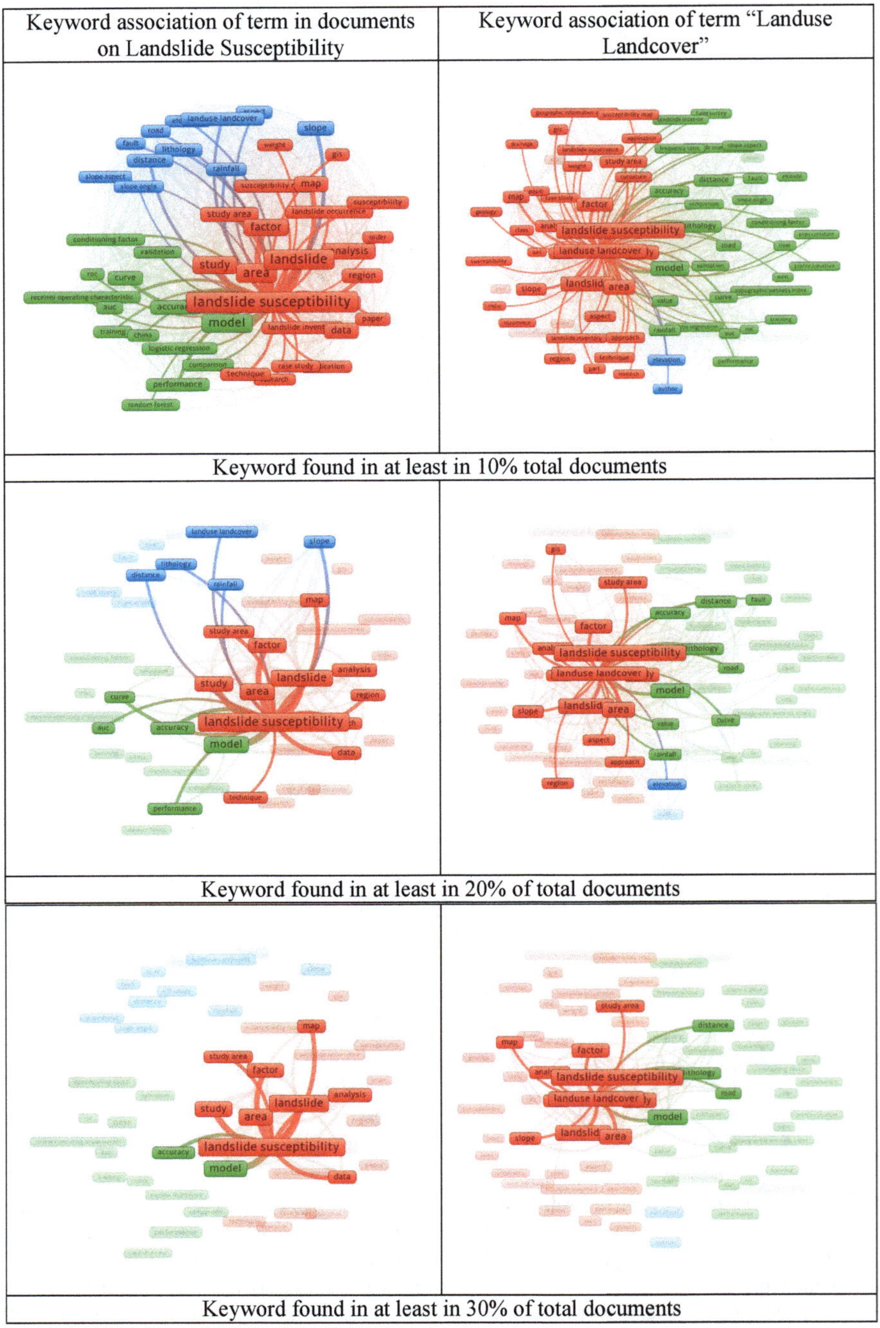

Fig. 16.9 Keywords co-occurrence network map depicting frequency and association of occurrence of keywords "landslide susceptibility" and "landuse landcover" and their association with other related terms present in title and abstract of at least 10, 20, and 30% of the total documents provided on landslide susceptibility and landslide susceptibility mentioning landuse landcover

algorithms (Shahabi et al. 2023; Sur et al. 2021) for assessment of landslide susceptibility in various parts of the globe from Asia (Singh and Veerappan 2021; Rosi et al. 2023) to Africa (Nahayo et al. 2023; Romer and Ferentinou 2016), America (Quesada-Román 2021; Palacio et al. 2020) to Australia (Palamakumbure et al. 2015; Leventhal and Kotze 2008), and even Europ (Günther et al. 2014). Numerous studies argue that alterations in land use/cover (LUCC) can increase the vulnerability to landslides (Torizin et al. 2018; Van Westen et al. 2012). Specifically, human engineering activities, such as urban development, deforestation, and extensive road expansion, have been identified as contributing factors (Van Westen et al. 2012; Liu et al. 2021; Guo et al. 2023). These studies consider the distance from roads and LULC maps as dynamic factors, assessing both short-term and long-term impacts of human activities on landslide susceptibility. Construction activities, including excavation and blasting, compromise slope stability by modifying factors such as infiltration, surface runoff, and groundwater flows (Karsli et al. 2009; Fink et al. 2017; Torizin et al. 2018; Meneses et al. 2019). Authors from all around the globe were attracted to this area of research even if the study area is not native to the author (Regmi and Poudel 2016). Landslide susceptibility assessment seems to be a multidimensional area for research, which spread over a wide range of topics from lanslide susceptibility modelling (Dandridge et al. 2023; Peng et al. 2022; Asadi et al. 2022) to lanslide susceptibility mapping (Jiaxuan et al. 2020; Tanoli et al. 2017; Ding and Hu 2014), lanslide susceptibility monitoring (Hua et al. 2021; Zhang et al. 2019), landslides susceptibility analysis (Rohan et al. 2021; Sujatha et al. 2013; Rezapour and Safaripour 2022), etc. Various using statistical or machine learning techniques such as Information Gain Ratio, frequency ratio, artificial neural networking, etc., has been used to analyze the historical landslide events in association with various landscape features like slope, aspect, drainage, landcover, lithology, geology, soil, etc. And identified urbanization and land use/land cover as key landslide drivers, alongside slope, elevation, and rainfall (Shang et al. 2023; Fang et al. 2023; Yuan et al. 2022; Ng et al. 2021; Chen et al. 2019a, b, c). Several studies emphasize that LUCC may modify landslide susceptibility (Liu et al. 2021; Chen and Huang 2013). Majorly landslide is a natural phenomenon that occurs under the influence of gravitational pull and is triggered by intensive rainfall in hilly terrain (Mandal and Mondal 2019b; Nerella et al. 2019). A large number of conditioning factors were identified that favor the occurrence of the event of a landslide. Among them, land use and land cover are some of the most influential landslide conditioning factors for susceptibility assessment (Masi et al. 2021; Parra et al. 2021). Some of the authors used the term "landuse landcover" as a single term (Naceur et al. 2022; Ozturk and Uzel-Gunini 2022; Muñoz et al. 2022) as a conditioning factor, while in some cases only land use or landcover is mentioned, similarly in some case instead using land use or landcover, similar generic term such as road, river, distance to road, distance to the river, agriculture, rangeland, forest cover, NDVI, (Dandridge et al. 2023; Rohan et al. 2021; Ozturk and Uzel-Gunini 2022) was used to calculate the landslide susceptibility assessment. Land use and land cover are two words, often used together as single terms. These are used to describe the features present on the ground. Landcover is often used to describe natural features like forests, vegetation cover, hills, rivers, etc. (Regmi and Poudel 2016; Peng et al. 2022; Kumar Rai et al. 2014). Land use is used to describe how humans are using these resources as well as modifications done by humans in this natural land cover, such as agriculture, roads, artificial drainage, dams, reservoirs, buildings, etc. Notably, research describing landuse and landcover as prominent factors in forecasting future landslide susceptibility (Pham et al. 2022) and machine learning-based predictions using modeled LULC change scenarios (Pham et al. 2022; Liu et al. 2021) exemplify the importance of LULC in predicting landslide scenarios. The significance of incorporating Land Use and Land Cover (LULC) or Land Use and Land Cover Change (LUCC) into landslide assessments stems from the influence of human activities on slopes, particularly agricultural and forestry practices, which are further influenced by global warming. However, relying solely on past landslide conditions may offer a limited perspective, as hillslope conditions undergo substantial changes in response to human interventions (Cignetti et al. 2019). Therefore, factoring in future scenarios of LULC in landslide susceptibility analysis would yield more practical outcomes, assisting public administrators in long-term land use management and the reduction of landslide disaster risks. When considering Land Use and Land Cover (LULC) as a factor influencing landslides, a complicating factor arises from the diverse impacts that different cover classes within LULC can have on soil structure (Chen et al. 2019a, b, c; Masi et al. 2021). For instance, a mature forest is likely to provide more substantial root reinforcement compared to areas with undergrowth or cultivation (Grima et al. 2020; Griffiths et al. 2020). Moreover, alterations in LULC have the potential to modify soil shear strength, leading to slope instability, and in certain instances, abrupt changes can act as triggers for landslides (Yu et al. 2023; Dandridge et al. 2023). In our examination, it was observed that the prevailing Land Use and Land Cover (LULC) classes were "road," "river," "soil," and "forest" (Fig. 16.9), likely attributed to the ease of obtaining relevant data in a more objective manner. Typically, susceptibility models often incorporate roads based on their proximity to mapped landslides (Singh et al. 2023). The construction and maintenance of roads exert a direct or indirect influence on the slope, either through slope cuts or alterations in surface water runoff (Bordbar et al. 2021; Tesfa 2022; Meusburger and Alewell 2008). The category of "soil" frequently experiences a transformation into patches of bare land or inadequately vegetated terrain rather than retaining distinct soil types. Numerous studies have highlighted that deforestation or logging exposes the soil to erosion processes and increases the risk of slope instability (Bozzolan et al. 2023; Akinci and Yavuz 2021; Wilopo et al. 2023). Additionally, forests are frequently linked with enhanced slope stability (Zhang et al. 2022; Grima et al. 2020). While a consensus exists in many studies that deforestation elevates landslide susceptibility (Alberto et al. 2022), few delve into the specific influences of forest type, structure, health, or natural disturbances (Parra et al. 2021). Therefore, although individual analyses of distinct Land Use and Land Cover (LULC) classes are possible, their interconnections can pose challenges for certain statistical models due to potential collinearity. Moreover, in the examination of Land Use and Land Cover Change

(LUCC), the alteration itself not only serves as a crucial factor in landslide occurrences but also considers the relative gains and losses within a specific Land Use and Land Cover (LULC) class (Liu et al. 2021). For instance, the conversion of forests to agriculture may exacerbate slope instability (Lehmann et al. 2019), while cultivation in previously abandoned areas could potentially reduce landslide susceptibility (Pisano et al. 2017). Approaches that consider scenarios in LULC (Jurchescu et al. 2023; Tyagi et al. 2023; Knevels et al. 2021; Arabameri et al. 2017; Reichenbach et al. 2015;) can offer valuable insights for guiding future efforts in minimizing the risk of landslide disasters.

16.6 Conclusions

This study investigated a bibliometric analysis on application of Land Use and Land Cover (LULC) data in assessing landslide susceptibility provides a comprehensive overview of the scholarly landscape, offering valuable insights into research trends, key contributors, and emerging themes. The investigation utilized the Scopus literature database, which encompassed 3,240 articles. The findings highlighted a predominant emphasis on the use of land use and land cover data as a conditioning factor in landslide susceptibility modeling. Specifically, 1,085 publications out of the total 3,240 referenced Land Use and Land Cover in their abstracts, keywords, and/or titles. Most of the scientific studies focused on evaluating model performance, ranging from statistical and index-based models to deep learning and machine learning algorithms. Notably, our results revealed that countries experiencing the greatest impact from landslides did not necessarily make substantial contributions to international authorship. This observation might suggest disparities in national investments in research and development. Significantly, Asian nations are prominently represented, whereas South American and African countries are notably underrepresented in the global author community. Our study highlights a steady rise in interest regarding the impact of Land Use and Land Cover Change (LUCC) on landslide susceptibility, notably gaining momentum since 2016. Initially, the emphasis was on understanding how LULC mapping, particularly classification methods, scale, and spatial resolution, affects landslide susceptibility. Recent studies have broadened their scope to simulate future Land Use and Land Cover (LULC) scenarios and investigate their influence on altering landslide. These prospective scenarios have the potential to provide valuable insights for the management of landslide risk and the planning of land use. In summary, the investigation into the impact of Land Use and Land Cover (LULC) on landslide susceptibility is an expanding field that offers opportunities for future scenario analyses and addressing research gaps. This understanding is crucial for researchers, policymakers, and practitioners dedicated to minimizing the impact of landslides on communities and the environment. As the field progresses, comprehending the scholarly landscape will be instrumental in guiding the future course of landslide research, enabling the development of effective mitigation strategies and furthering advancements in the field. Managing an extensive database, establishing selection criteria, choosing relevant keywords, and accessing literature and tools for data analysis pose limitations in such studies. The selection of both the scholarly database and the search keywords may introduce constraints to the scope of this research. Relying solely on Scopus might result in a skewed perspective. Furthermore, Our analysis focused on articles that specifically considered Land Use and Land Cover (LULC) as a determining factor in landslide susceptibility assessment. However, the variety of LULC classes, coupled with limitations on search keywords, might result in a constrained database of articles. By concentrating solely on terms directly linked to LULC, such as "land cover," "land use," "land use and land cover," and "LULC," there's a possibility of inadvertently excluding relevant articles that incorporate LULC-related terms like "road," "river," "forest," and so on. The omission of specific publications from the database was influenced by the selected keywords, such as choosing "landslide" instead of terms like "mass movement," "debris flow," "rockfall," or "slope destabilization." Meaningful Interpretation of trends and representation of results, that can be understandable to a wider audience is the key challenge in such studies. Moving forward, we acknowledge several challenges for future studies. These include the need to encompass additional LULC terms such as "road," "river," "soil," and "forest." Additionally, there is a call for the incorporation of multiple data sources within the same study to address the limitations associated with relying solely on a single database. With the enormously propagating literature on landslide research in recent times, this kind of study on regular time intervals become important and should be conducted independently or comparatively with previous studies.

References

Akinci H, Yavuz Ozalp A (2021) Landslide susceptibility mapping and hazard assessment in Artvin (Turkey) using frequency ratio and modified information value model. Acta Geophysica 69. https://doi.org/10.1007/s11600-021-00577-7

Anucharn T, Dasananda S (2017) A comparison of landslide susceptibility maps produced by weighted linear combination and analytical hierarchy process methods: a case study at Khao Phanom Bencha Watershed in Krabi Province. Walailak J Sci Technol 14

Arabameri A, Pourghasemi HR, Yamani M (2017) Applying different scenarios for landslide spatial modeling using computational intelligence methods. Environ Earth Sci 76. https://doi.org/10.1007/s12665-017-7177-5

Aram A, Dalalian MR, Saedi S et al (2022) An assessment of data mining and bivariate statistical methods for landslide susceptibility mapping. Scientia Iranica 29

Asadi M, Goli Mokhtari L, Shirzadi A et al (2022) A comparison study on the quantitative statistical methods for spatial prediction of shallow landslides (case study: Yozidar-Degaga Route in Kurdistan Province, Iran). Environ Earth Sci 81. https://doi.org/10.1007/s12665-021-10152-4

Bar-Ilan J (2010) Citations to the "Introduction to informetrics" indexed by WOS, Scopus and Google Scholar. Scientometrics 82. https://doi.org/10.1007/s11192-010-0185-9

Bhandari BP, Dhakal S (2020a) Spatio-temporal dynamics of landslides in the sedimentary terrain: a case of Siwalik zone of Babai watershed. Nepal. SN Appl Sci 2. https://doi.org/10.1007/s42452-020-2628-0

Bhandari BP, Dhakal S (2020b) Compositional analysis and phase relations of soil mass from the active landslides of Babai River watershed Siwalik Zone of Nepal. Eng Geol 278. https://doi.org/10.1016/j.enggeo.2020.105851
Bicocchi G, Tofani V, D'Ambrosio M et al (2019) Geotechnical and hydrological characterization of hillslope deposits for regional landslide prediction modeling. Bull Eng Geol Environ 78. https://doi.org/10.1007/s10064-018-01449-z
Bordbar M, Paryani S, Pourghasemi HR (2021) Landslide spatial modeling using a bivariate statistical method in Kermanshah Province, Iran. In: Computers in earth and environmental sciences: artificial intelligence and advanced technologies in hazards and risk management
Boudry C, Baudouin C, Mouriaux F (2018) International publication trends in dry eye disease research: a bibliometric analysis. Ocul Surf 16. https://doi.org/10.1016/j.jtos.2017.10.002
Bozzolan E, Holcombe EA, Pianosi F et al (2023) A mechanistic approach to include climate change and unplanned urban sprawl in landslide susceptibility maps. Sci Total Environ 858. https://doi.org/10.1016/j.scitotenv.2022.159412
Chang YW, Huang MH, Lin CW (2015) Evolution of research subjects in library and information science based on keyword, bibliographical coupling, and co-citation analyses. Scientometrics 105 https://doi.org/10.1007/s11192-015-1762-8
Chen L, Guo Z, Yin K et al (2019a) The influence of land use and land cover change on landslide susceptibility: a case study in Zhushan Town, Xuan'en County (Hubei, China). Nat Hazard 19. https://doi.org/10.5194/nhess-19-2207-2019
Chen W, Shahabi H, Shirzadi A et al (2019b) Novel hybrid artificial intelligence approach of bivariate statistical-methods-based kernel logistic regression classifier for landslide susceptibility modelling. Bull Eng Geol Env 78. https://doi.org/10.1007/s10064-018-1401-8
Congedo L, Sallustio L, Munafò M et al (2016) Copernicus highresolution layers for land cover classification in Italy. J Maps 12:1195–1205. https://doi.org/10.1080/17445647.2016.1145151
Chen W, Yan X, Zhao Z et al (2019c) Spatial prediction of landslide susceptibility using data mining-based kernel logistic regression, naive Bayes and RBFNetwork models for the Long County area (China). Bull Eng Geol Env 78. https://doi.org/10.1007/s10064-018-1256-z
Chen X, Chen W (2021) GIS-based landslide susceptibility assessment using optimized hybrid machine learning methods. Catena (Amst) 196. https://doi.org/10.1016/j.catena.2020.104833
Cignetti M, Godone D, Giordan D (2019) Shallow landslide susceptibility, rupinaro catchment, liguria (Northwestern Italy). J Maps 15. https://doi.org/10.1080/17445647.2019.1593252
Coelho-Netto AL, Avelar AS, Fernandes MC, Lacerda WA (2007) Landslide susceptibility in a mountainous geoecosystem, Tijuca Massif, Rio de Janeiro: the role of morphometric subdivision of the terrain. Geomorphology 87. https://doi.org/10.1016/j.geomorph.2006.03.041
Chen CY, Huang WL (2013) Land use change and landslide characteristics analysis for community-based disaster mitigation. Environ Monit Assess 185:4125–4139. https://doi.org/10.1007/s10661-012-2855-y
Dandridge C, Stanley T, Kirschbaum D et al (2023) The influence of land use and land cover change on landslide susceptibility in the Lower Mekong River Basin. Nat Hazards 115. https://doi.org/10.1007/s11069-022-05604-4
De Solla Price DJ (1965) Networks of scientific papers. Science (1979) 149. https://doi.org/10.1126/science.149.3683.510
Devkota KC, Regmi AD, Pourghasemi HR et al (2013) Landslide susceptibility mapping using certainty factor, index of entropy and logistic regression models in GIS and their comparison at Mugling-Narayanghat road section in Nepal Himalaya. Nat Hazards 65. https://doi.org/10.1007/s11069-012-0347-6
Ding M, Hu K (2014) Susceptibility mapping of landslides in Beichuan County using cluster and MLC methods. Nat Hazards 70. https://doi.org/10.1007/s11069-013-0854-0
Falagas ME, Pitsouni EI, Malietzis GA, Pappas G (2008) Comparison of PubMed, Scopus, Web of Science, and Google Scholar: strengths and weaknesses. FASEB J 22. https://doi.org/10.1096/fj.07-9492lsf

Fang H, Shao Y, Xie C et al (2023) A New Approach to Spatial landslide susceptibility prediction in karst mining areas based on explainable artificial intelligence. Sustainability (Switzerland) 15. https://doi.org/10.3390/su15043094

Fink M, Fischer C, Laux P, et al (2017) Distributed assessment of sediment Dynamics in Central Vietnam. In: water resources development and management

Gadtaula A, Dhakal S (2019) Landslide susceptibility mapping using weight of evidence method in Haku, Rasuwa District, Nepal. J Nepal Geol Soc 58. https://doi.org/10.3126/jngs.v58i0.24601

Gómez C, White JC, Wulder MA (2016) Optical remotely sensed time series data for land cover classification: a review. ISPRS J Photogramm Remote Sens 116:55–72. https://doi.org/10.1016/j.isprsjprs.2016.03.008

Ghorbanian A, Kakooei M, Amani M (2020) Improved land cover map of Iran using Sentinel imagery within Google Earth Engine and a novel automatic workflow for land cover classification using migrated training samples. ISPRS J Photogramm Remote Sens 167:276–288. https://doi.org/10.1016/j.isprsjprs.2020.07.013

Griffiths JW, Lukens CE, May R (2020) Increased forest cover and limits on clear-felling could substantially reduce landslide occurrence in tasman, new zealand. N Z J For Sci 50. https://doi.org/10.33494/nzjfs502020x94x

Grima N, Edwards D, Edwards F et al (2020) Landslides in the Andes: forests can provide cost-effective landslide regulation services. Sci Total Environ 745. https://doi.org/10.1016/j.scitotenv.2020.141128

Günther A, Hervás J, Van Den Eeckhaut M, et al (2014) Synoptic pan-European landslide susceptibility assessment: the ELSUS 1000 v1 map. In: Landslide science for a safer geoenvironment

Guo Z, Ferrer JV, Hürlimann M et al (2023) Shallow landslide susceptibility assessment under future climate and land cover changes: a case study from southwest China. Geosci Front 14. https://doi.org/10.1016/j.gsf.2023.101542

Guzzetti F, Mondini AC, Cardinali M et al (2012) Landslide inventory maps: New tools for an old problem. Earth Sci Rev 112

Herold M, Latham JS, Gregorio Di A, Schmullius CC (2006) Evolving standards in land cover characterization. J Land Use Sci 1. https://doi.org/10.1080/17474230601079316

Hirsch JE (2005) An index to quantify an individual's scientific research output. Proc Natl Acad Sci U S A 102. https://doi.org/10.1073/pnas.0507655102

Hood WW, Wilson CS (2001) The literature of bibliometrics, scientometrics, and informetrics. Scientometrics 52. https://doi.org/10.1023/A:1017919924342

Hua Y, Wang X, Li Y et al (2021) Dynamic development of landslide susceptibility based on slope unit and deep neural networks. Landslides 18. https://doi.org/10.1007/s10346-020-01444-0

Hulme M (2016) 1.5 °C and climate research after the Paris Agreement. Nat Clim Chang 6

Jiaxuan H, Mowen X, Atkinson PM (2020) Dynamic susceptibility mapping of slow-moving landslides using PSInSAR. Int J Remote Sens 41. https://doi.org/10.1080/01431161.2020.1760398

Jun C, Ban Y, Li S (2014) Open access to Earth land-cover map. Nature 514:434–434. https://doi.org/10.1038/514434c

Jurchescu M, Kucsicsa G, Micu M et al (2023) Implications of future land-use/cover pattern change on landslide susceptibility at a national level: a scenario-based analysis in Romania. Catena (Amst) 231. https://doi.org/10.1016/j.catena.2023.107330

Kamal ASMM, Ahmed B, Tasnim S, Sammonds P (2022) Assessing rainfall-induced landslide risk in a humanitarian context: The Kutupalong Rohingya Camp in Cox's Bazar Bangladesh. Nat Hazards Res 2. https://doi.org/10.1016/j.nhres.2022.08.006

Karsli F, Atasoy M, Yalcin A et al (2009) Effects of land-use changes on landslides in a landslide-prone area (Ardesen, Rize, NE Turkey). Environ Monit Assess 156. https://doi.org/10.1007/s10661-008-0481-5

Knevels R, Brenning A, Gingrich S et al (2021) Towards the use of land use legacies in landslide modeling: current challenges and future perspectives in an austrian case study. Land (Basel) 10. https://doi.org/10.3390/land10090954

Kulkarni AV, Aziz B, Shams I, Busse JW (2009) Comparisons of citations in web of science, Scopus, and Google Scholar for articles published in general medical journals. JAMA 302. https://doi.org/10.1001/jama.2009.1307

Kumar Rai P, Mohan K, Kumra VK (2014) Landslide hazard and its mapping using remote sensing and gis. J Sci Res 58

Lehmann P, von Ruette J, Or D (2019) Deforestation effects on rainfall-induced shallow landslides: remote sensing and physically-based modeling. Water Resour Res 55:9962–9976. https://doi.org/10.1029/2019WR025233

Lee S, Pradhan B (2007) Landslide hazard mapping at Selangor, Malaysia using frequency ratio and logistic regression models. In: Landslides

Leventhal AR, Kotze GP (2008) Landslide susceptibility and hazard mapping in Australia for land-use planning—with reference to challenges in metropolitan suburbia. Eng Geol 102. https://doi.org/10.1016/j.enggeo.2008.03.021

Liu J, Wu Z, Zhang H (2021) Analysis of changes in landslide susceptibility according to land use over 38 years in lixian county, China. Sustainability (Switzerland) 13. https://doi.org/10.3390/su131910858

Mandal S, Mondal S (2019a) Prediction of landslide susceptibility using bivariate models. In: Environmental science and engineering

Mandal S, Mondal S (2019b) Geomorphic, geo-tectonic and hydrologic attributes and landslide susceptibility. In: Environmental science and engineering

Masi EB, Segoni S, Tofani V (2021) Root reinforcement in slope stability models: a review. Geosciences (Switzerland) 11

Meneses BM, Pereira S, Reis E (2019) Effects of different land use and land cover data on the landslide susceptibility zonation of road networks. Nat Hazards Earth Syst Sci 19. https://doi.org/10.5194/nhess-19-471-2019

Meusburger K, Alewell C (2008) Impacts of anthropogenic and environmental factors on the occurrence of shallow landslides in an alpine catchment (Urseren Valley, Switzerland). Nat Hazards Earth Syst Sci 8. https://doi.org/10.5194/nhess-8-509-2008

Moulds S, Buytaert W, Mijic A (2018) A spatio-temporal land use and land cover reconstruction for India from 1960–2010. Sci Data 5:180159. https://doi.org/10.1038/sdata.2018.159

Montalván-Burbano N, Pérez-Valls M, Plaza-Úbeda J (2020) Analysis of scientific production on organizational innovation. Cogent Bus Manag 7.https://doi.org/10.1080/23311975.2020.1745043

Mukane P (2014) India's worst landslides, and why these might not be the last ones. http://www dnaindiacom/india/

Muñoz-Torrero Manchado A, Allen S, Ballesteros-Canovas JA, Dhakal A, Dhital MR, Stoffel M (2022) Deforestation controls landslide susceptibility in Far-Western Nepal. CATENA 219:106627. https://doi.org/10.1016/j.catena.2022.106627

Naceur HA, Abdo HG, Igmoullan B et al (2022) Performance assessment of the landslide susceptibility modelling using the support vector machine, radial basis function network, and weight of evidence models in the N'fis river basin. Morocco Geosci Lett 9. https://doi.org/10.1186/s40562-022-00249-4

Nahayo L, Peng C, Lei Y, Tan R (2023) Spatial understanding of historical and future landslide variation in Africa. Nat Hazards 119. https://doi.org/10.1007/s11069-023-06126-3

Nerella SP, Alajangi S, Dhakal D (2019) Landslide susceptibility mapping using GIS-based lkelihood frequency ratio model: a case study of Pakyong—Pacheykhani Area, Sikkim Himalaya. In: Rao P, Rao K, Kubo S (eds) Proceedings of International Conference on Remote Sensing for Disaster Management. Springer Series in Geomechanics and Geoengineering. Springer, Cham. https://doi.org/10.1007/978-3-319-77276-9_51

Ng CWW, Yang B, Liu ZQ et al (2021) Spatiotemporal modelling of rainfall-induced landslides using machine learning. Landslides 18. https://doi.org/10.1007/s10346-021-01662-0

Ólafsdóttir R, Tverijonaite E (2018) Geotourism: a systematic literature review. Geosciences (Switzerland) 8

Ozturk D, Uzel-Gunini N (2022) Investigation of the effects of hybrid modeling approaches, factor standardization, and categorical mapping on the performance of landslide susceptibility mapping in Van Turkey. Nat Hazards 114. https://doi.org/10.1007/s11069-022-05480-y

Pal I, Shaw R, Djalante R, Shrestha S (2021) Disaster resilience and sustainability: adaptation for sustainable development

Palacio Cordoba J, Mergili M, Aristizábal E (2020) Probabilistic landslide susceptibility analysis in tropical mountainous terrain using the physically based r.slope.stability model. Nat Hazards Earth Syst Sci 20. https://doi.org/10.5194/nhess-20-815-2020

Palamakumbure D, Flentje P, Stirling D (2015) Consideration of optimal pixel resolution in deriving landslide susceptibility zoning within the Sydney Basin, New South Wales, Australia. Comput Geosci 82. https://doi.org/10.1016/j.cageo.2015.05.002

Panchal S, Shrivastava AK (2020) Landslide Susceptibility Mapping Along Highway Corridors in GIS Environment. In: Lecture notes in civil engineering

Pandey VK, Sharma KK, Pourghasemi HR, Bandooni SK (2019) Sedimentological characteristics and application of machine learning techniques for landslide susceptibility modelling along the highway corridor Nahan to Rajgarh (Himachal Pradesh) India. Catena (Amst) 182. https://doi.org/10.1016/j.catena.2019.104150

Parish AJ, Boyack KW, Ioannidis JPA (2018) Dynamics of co-authorship and productivity across different fields of scientific research. PLoS ONE 13. https://doi.org/10.1371/journal.pone.0189742

Parra E, Mohr CH, Korup O (2021) Predicting patagonian landslides: roles of forest cover and wind speed. Geophys Res Lett 48

Peng T, Chen Y, Chen W (2022) Landslide susceptibility modeling using remote sensing data and random subspace-based functional tree classifier Remote Sens (Basel) 14. https://doi.org/10.3390/rs14194803

Pham BT, Van PT, Nguyen-Thoi T et al (2022) Ensemble modeling of landslide susceptibility using random subspace learner and different decision tree classifiers. Geocarto Int 37. https://doi.org/10.1080/10106049.2020.1737972

Pisano L, Zumpano V, Malek Ž (2017) Variations in the susceptibility to landslides, as a consequence of land cover changes: a look to the past, and another towards the future. Sci Total Environ 601–602:1147–1159. https://doi.org/10.1016/j.scitotenv.2017.05.231

Pour AB, Hashim M (2017) Application of Landsat-8 and ALOS-2 data for structural and landslide hazard mapping in Kelantan, Malaysia. Nat Hazard 17. https://doi.org/10.5194/nhess-17-1285-2017

Qiu JP, Dong K, Yu HQ (2014) Comparative study on structure and correlation among author co-occurrence networks in bibliometrics. Scientometrics 101. https://doi.org/10.1007/s11192-014-1315-6

Quesada-Román A (2021) Landslides and floods zonation using geomorphological analyses in a dynamic catchment of Costa Rica. Rev Cartogr 2021. https://doi.org/10.35424/rcarto.i102.901

Rawat PK, Pant B (2023) Geoenvironmental GIS development to investigate landslides and slope instability along frontal zone of Central Himalaya. Nat Hazards Res 3. https://doi.org/10.1016/j.nhres.2023.03.005

Regmi AD, Devkota KC, Yoshida K et al (2014) Application of frequency ratio, statistical index, and weights-of-evidence models and their comparison in landslide susceptibility mapping in Central Nepal Himalaya. Arab J Geosci 7. https://doi.org/10.1007/s12517-012-0807-z

Regmi AD, Poudel K (2016) Assessment of landslide susceptibility using GIS-based evidential belief function in PatuKhola watershed, Dang Nepal. Environ Earth Sci 75. https://doi.org/10.1007/s12665-016-5562-0

Regmi AD, Yoshida K, Dhital MR, Devkota K (2013) Effect of rock weathering, clay mineralogy, and geological structures in the formation of large landslide, a case study from DumreBesei landslide Lesser Himalaya Nepal. Landslides 10. https://doi.org/10.1007/s10346-011-0311-7

Reichenbach P, Mondini AC, Rossi M (2015) Land use change scenarios and landslide susceptibility zonation: The briga catchment test area (Messina, Italy). In: Engineering geology for society and territory - volume 1: climate change and engineering geology

Rezapour Andabili N, Safaripour M (2022) Identification of precipitation trend and landslide susceptibility analysis in Miandoab County using MATLAB. Environ Monit Assess 194. https://doi.org/10.1007/s10661-022-10069-w

Rohan TJ, Wondolowski N, Shelef E (2021) Landslide susceptibility analysis based on citizen reports. Earth Surf Process Landf 46. https://doi.org/10.1002/esp.5064

Romer C, Ferentinou M (2016) Shallow landslide susceptibility assessment in a semiarid environment—a Quaternary catchment of KwaZulu-Natal South Africa. Eng Geol 201.https://doi.org/10.1016/j.enggeo.2015.12.013

Rosi A, Frodella W, Nocentini N et al (2023) Comprehensive landslide susceptibility map of Central Asia. Nat Hazard 23. https://doi.org/10.5194/nhess-23-2229-2023

Roslee R, Tongkul F, Simon N (2018) A comparative evaluation of Normal Polygon Geotechnical Deterministic Analysis (NPGDA) and GEOStatisticalINterpolation Techniques (Kriging) (GEOSTAINT-K): a case study from Kota Kinabalu area, Sabah, Malaysia. ASM Sci J 11

Sakkas G, Misailidis I, Sakellariou N et al (2016) Modeling landslide susceptibility in Greece: a weighted linear combination approach using analytic hierarchical process, validated with spatial and statistical analysis. Nat Hazards 84 .https://doi.org/10.1007/s11069-016-2523-6

Sassa K (2015) ISDR-ICL Sendai Partnerships 2015–2025 for global promotion of understanding and reducing landslide disaster risk. Landslides 12:631–640. https://doi.org/10.1007/s10346-015-0586-1

Shahabi H, Ahmadi R, Alizadeh M et al (2023) Landslide susceptibility mapping in a mountainous area using machine learning algorithms. Remote Sens (Basel) 15. https://doi.org/10.3390/rs15123112

Shang H, Su L, Chen W, Tsangaratos P, Ilia I, Liu S, Cui S, Duan Z (2023) Spatial prediction of landslide susceptibility using logistic regression (LR), Functional Trees (FTs), and Random Subspace Functional Trees (RSFTs) for Pengyang County, China. Remote Sens 15(20):4952. https://doi.org/10.3390/rs15204952

Singh NB, Veerappan R (2021) GIS-based landslide susceptibility zonation mapping using fuzzy gamma operator model in part of Trans-Asian Highway (Mao-Kangpokpi), Manipur, India. In: Disaster resilience and sustainability: adaptation for sustainable development

Singh P, Sur U, Rai PK, Singh SK (2023) Landslide susceptibility prediction using frequency ratio model: a case study of Uttarakhand, Himalaya (India). Proc Indian Natl Sci Acad 89. https://doi.org/10.1007/s43538-023-00171-z

Soosaraei M, Khasseh AA, Fakhar M, Hezarjaribi HZ (2018) A decade bibliometric analysis of global research on leishmaniasis in Web of Science database. Ann Med Surg 26

Sujatha ER, Rajamanickam V, Kumaravel P, Saranathan E (2013) Landslide susceptibility analysis using probabilistic likelihood ratio model-a geospatial-based study. Arab J Geosci 6. https://doi.org/10.1007/s12517-011-0356-x

Sur U, Singh P, Rai PK, Thakur JK (2021) Landslide probability mapping by considering fuzzy numerical risk factor (FNRF) and landscape change for road corridor of Uttarakhand India. Environ Dev Sustain 23. https://doi.org/10.1007/s10668-021-01226-1

Tanoli JI, Ningsheng C, Regmi AD, Jun L (2017) Spatial distribution analysis and susceptibility mapping of landslides triggered before and after Mw7.8 Gorkha earthquake along Upper Bhote Koshi, Nepal. Arab J Geosci 10. https://doi.org/10.1007/s12517-017-3026-9

Tao J, Qiu D, Yang F, Duan Z (2020) A bibliometric analysis of human reliability research. J Clean Prod 260

Tesfa C (2022) GIS-Based AHP and FR methods for landslide susceptibility mapping in the Abay Gorge, Dejen-Renaissance Bridge, Central Ethiopia Geotech. Geol Eng 40. https://doi.org/10.1007/s10706-022-02197-4

Tofani V, Bicocchi G, Rossi G et al (2017) Soil characterization for shallow landslides modeling: a case study in the Northern Apennines (Central Italy). Landslides 14. https://doi.org/10.1007/s10346-017-0809-8

Torizin J, Wang L chao, Fuchs M, et al (2018) Statistical landslide susceptibility assessment in a dynamic environment: A case study for Lanzhou City, Gansu Province, NW China. J Mt Sci 15. https://doi.org/10.1007/s11629-017-4717-0

Tyagi A, Tiwari RK, James N (2023) Prediction of the future landslide susceptibility scenario based on LULC and climate projections. Landslides 20. https://doi.org/10.1007/s10346-023-02088-6

Van Ech NL, Waltman L (2014) Visualizing bibliometric networks. In: Ding Y, Rousseau R, Wolfram D (eds) Measuring scholarly impact: methods and practice

Van Westen CJ, Jaiswal P, Ghosh S, et al (2012) Landslide inventory, hazard and risk assessment in India. In: Terrigenous mass movements: detection, modelling, early warning and mitigation using geoinformation technology

Wulder M, Coops N (2014) Satellites: make earth observations open access. Nature 513:30–31. https://doi.org/10.1038/513030a

Wilopo W, Putra DPE, Fathani TF, et al (2023) Comparison of two landslide hazard zonation methods in the volcanic terrain of Temanggung Regency, Central Java, Indonesia. J Degrad Min Lands Manag 10. https://doi.org/10.15243/jdmlm.2023.103.4537

Yalcin A, Reis S, Aydinoglu AC, Yomralioglu T (2011) A GIS-based comparative study of frequency ratio, analytical hierarchy process, bivariate statistics and logistics regression methods for landslide susceptibility mapping in Trabzon NE Turkey. Catena (Amst) 85. https://doi.org/10.1016/j.catena.2011.01.014

Yang F, Kumar A, Schlesinger ME, Wang W (2003) Intensity of hydrological cycles in warmer climates. J Clim 16. https://doi.org/10.1175/2779.1

Yu X, Xia Y, Zhou J, Jiang W (2023) Landslide susceptibility mapping based on multitemporal remote sensing image change detection and multiexponential band math. Sustainability (Switzerland) 15. https://doi.org/10.3390/su15032226

Yuan X, Liu C, Nie R et al (2022) A comparative analysis of certainty factor-based machine learning methods for collapse and landslide susceptibility mapping in Wenchuan County China. Remote Sens (Basel) 14. https://doi.org/10.3390/rs14143259

Zhang H, Zhang G, Jia Q (2019) Integration of analytical hierarchy process and landslide susceptibility index based landslide susceptibility assessment of the Pearl River Delta Area, China. IEEE J Sel Top Appl Earth Obs Remote Sens 12. https://doi.org/10.1109/JSTARS.2019.2938554

Zhang Y, Shen C, Zhou S, Luo X (2022) Analysis of the influence of forests on landslides in the Bijie Area of Guizhou. Forests 13. https://doi.org/10.3390/f13071136

Chapter 17
A Geospatial Review Analysis of Increased Frequency of Large Glacier-Related Landslides in Mountainous Regions

Naima Umar

Abstract As the world warms, many glaciers throughout the world are melting. Glacier melting has been linked to climate change as a major cause of landslides in mountainous regions. The whole effect of glacier growth and sublimation on landslide-induced denudation cannot be understood from an evaluation of the triggering mechanism alone. The proposed chapter constructed landslide and glacial statistics to represent the HMA employing a Landsat time-series spanning the last years to evaluate recent tendencies. Large earthquake-related landslides were isolated and deleted from the database, leaving a resource for investigating potential climate-change-related shifts. Mountainous locations are particularly susceptible to the impacts of climate change. The intricate web of connections among climatic, ecological, and sociocultural systems in these areas is being modified by the shifting climate. This chapter aims to investigate the forthcoming obstacles that these distinctive locations will face. Analysis using satellite images and landslide inventories indicates a substantial increase in the frequency of significant landslides associated with glaciers throughout recent decades. Our investigation is based on the research conducted in 2021 on the High Mountain Asia (HMA) region revealed that the extent of land impacted by landslides almost quadrupled during the time periods of 1999–2008 and 2009–2018. Additionally, the frequency of big landslides (>1 km^2) also saw an increase throughout this timeframe. Present chapter examine the significance of these factors and identify the challenges that arise in the current environment and political landscape. The expansion of landslides is linked to the shrinking of glaciers. Proposed work based on the secondary sources of data, given dataset is very useful for risk assessment, is an inventory of landslides connected to glaciers.

Keywords Glacier melting · Landslide · Climate change · Geospatial analysis

N. Umar (✉)
Assistant Professor, Department of Geography, School of Liberal Arts, Noida International University, Greater Noida, Uttar Pradesh, India
e-mail: naima.umar@niu.edu.in

G. K. Panda et al. (eds.), *Landslide: Susceptibility, Risk Assessment and Sustainability*, Advances in Natural and Technological Hazards Research 52,
https://doi.org/10.1007/978-3-031-56591-5_17

17.1 Introduction

The rate at which glaciers are losing mass is likewise increasing (Hugonnet et al. 2021). Global projections based on best-case and worst-case scenarios, which depend on the quantity of CO_2 in the atmosphere, indicate that polar and mountain glaciers will experience a loss of approximately 18–36% of their masses by the end of the twenty-first century, compared to their mass in 2015 (IPCC 2019). The reduction of ice and snow in high-altitude regions is a consequence of the rise in the Earth's mean temperature caused by the ongoing release of greenhouse gases by human activities since the beginning of the Industrial Revolution.

Climate change impacts in the Pyrenees have been seen during the last century. Undoubtedly, the average temperature in this area has significantly risen during the last 50 years. Furthermore, there is a decline in precipitation at a rate of 2.5% every decade, as reported by Amblar-Francés et al. (2020). The highest temperature experienced annually in the Pyrenees by the year 2030 would rise by 1.0–2.7 °C compared to the temperature levels recorded between 1961 and 1990. These predictions are based on the RCP8.5 scenario, as stated by Amblar-Francés et al. (2020). In addition, according to Amblar-Francés et al. (2020), the increase in temperature by 2050 would range from 2.0 to 4.0 °C. Glaciers are also under jeopardy. Over the past 150 years, this Monte Perdido Glacier in the Huesca region of the Spanish Pyrenees has seen more significant melting compared to the previous 2000 years. Data indicates that, under the current climate conditions, the glacier may ultimately vanish (Moreno et al. 2021).

When evaluating the effects of climate change, it uses a vector form to show how quickly the climate is changing. It is necessary to interpret velocity on the supposition that exposure to climate change is correlated with the length of the climatic trajectory; longer pathways indicate more exposure. Distance, however, is not a reliable indicator of exposure since it does not account for the amount of time individuals spend in climatically different places. Here, we compute the minimal cumulative exposure (MCE) and velocity along North American climatic paths in degrees Celsius. We discover that velocity and MCE have a poor relationship, with each measure indicating different regions of climate change susceptibility. Velocity, in particular, undervalues exposure in mountainous areas where climatic trajectories cross several climates, leading to elevated MCE. In contrast, since there is less environmental barrier to movement in plain places, velocity becomes high when minimum cumulative exposure is low. According to our findings, mountainous regions are climatically more detached comparatively others (Dobrowski and Parks 2016).

It is abundantly clear that the rapid retreat of glaciers all across the globe is a direct result of the enormous effect that climate change is having. A serious consequence has been prompted as a result of this change in glacial dynamics, and that is an increase in the number of big glacier-related landslides that occur in mountainous places. These landslides provide enormous dangers to the ecosystems, infrastructure, and populations that are located in the surrounding area. In order to create efficient ways for reducing this surge, it is crucial to have a good grasp of the underlying patterns and processes.

Present chapter, carefully investigate occurrences of landslides in the HMA region and based on Landsat data spanning the period of 1999 to 2018, including a total of two decades. Additionally, glacier area maps were made for each of the 20 consecutive years. However, the maps of landslides and glacial areas provide comprehensive and precise data on their location and time. Our study conducted a comprehensive analysis of the relationship between landslides and glaciers on a worldwide scale. Global warming is causing a significant number of glaciers to recede globally. Glacier melting has been shown to be main reason of landslide in higher elevation of alpine regions, along with climate change. According to our findings, there has been an increase in the frequency-area distribution of significant landslides in the HMA during the last ten years. An increase in landslide area is correlated with a decrease in glacier area.

17.2 Objectives

The present chapter has the following objectives:

1. To analyse the geographical and temporal distribution of landslides in mountainous regions.
2. To investigate the forthcoming obstacles that these distinctive locations will face.

17.3 Discussion

In mountainous places, landslides are common and may be caused by several occurrences including rainfall, earthquakes, and freeze–thaw cycles (Balzano et al. 2019; Iverson 2000; Serey et al. 2019; Laimer 2017). Risks such as glacier flee, frozen deterioration, and lake reduction are becoming increasingly common in high-altitude regions due to climate change. These occurrences often lead to unstable slopes and an increase in the occurrence of landslides with deeply embedded roots.

By weakening permafrost (Noetzli and Gruber 2009) and melting glaciers (Fischer et al. 2006), climate change indirectly affects landslides that happen at high elevations and may increase their frequency and magnitude (Haeberli et al. 1997). In addition to altering precipitation patterns, glacial melting may also cause severe rainfall, which exacerbates slope instability. These theories on slope stability and climate change are supported by observations of recent large-scale landslides in certain mountainous areas, although the data is usually inconclusive. This is partially because, in many places around the world's alpine and glacier ecosystems, it is not possible to compare current landslide statistics with historical records due to the lack of systematic documentation of landslides (Coe et al. 2018; Schneider et al. 2011; Coe 2020).

Rain, earthquakes, and cycles of freezing and thawing are common causes of landslides, which happen often in mountainous areas (Balzano et al. 2019; Iverson 2000; Serey et al. 2019; Laimer 2017). As a result of global warming, high-altitude areas are more vulnerable to risks such glacier melting, permafrost deterioration, and lake shrinkage. The instability of slopes and the occurrence of deep-seated landslides are common outcomes of these dangers.

As a result of climate change, permafrost is eroding and glaciers are melting, which may result in landslides occurring more often and with greater severity. This has an indirect influence on landslides that occur at high altitudes. There is a propensity for sediments to accumulate on the edge of a glacier that is receding, and as the glacier continues to recede, the sediments may become unstable. It is possible for the melting of glaciers to change the patterns of precipitation, particularly in relation to heavy rainfall, which may further worsen the instability of slopes. The assumptions about climate change and slope stability are somewhat supported by observations of recent significant landslides in certain mountain locations. However, the data is sometimes inconclusive. This issue arises in alpine and glacier regions worldwide owing to the absence of organized documentation of past landslides that can be used for comparison with current landslide data.

The repercussions of climate change are dramatically showing themselves over the wide alpine landscapes of Asia, namely in the rapid retreat of glaciers to lower elevations. This occurrence has resulted in a trend that is cause for concern: a significant rise in the number of big landslides that are caused by glaciers. Not only do these occurrences pose a risk to human populations and infrastructure, but they also upset the delicate biological balance that exists in these areas. It is possible to shed light on the patterns and reasons that are driving this surge via the use of a careful geospatial analysis, which may help in the creation of targeted mitigation solutions. When it comes to studying the dynamics of glacier-related landslides using geospatial analysis, Asia, which is home to a variety of mountain ranges such as the Himalayas, Karakoram, and Pamirs, provides an environment that is unlike any other for this purpose.

It has been discovered via geospatial analysis that there is a significant connection between the retreat of glaciers and the incidence of big landslides in mountainous parts of Asia. During the process of glaciers receding, unstable slopes that were formerly supported by ice are revealed. It is because of this exposure that slope instability is enhanced, and the likelihood of landslides occurring is raised as well. The monitoring of glacier retreat has been made possible by satellite photography and geographic information system mapping, which has made it easier to identify regions that are at a higher danger.

With geological movements contributing to the susceptibility of mountainous regions to landslides, Asia is a region that is prone to seismic activity or earthquakes. The intricate relationship that exists between tectonic pressures and the stability of slopes may be better comprehended with the use of geospatial analysis. When fault lines, seismic activity, and landslide occurrences are mapped, it is possible to get valuable insights into places where geological variables increase the likelihood of landslides occurring.

The different alpine landscapes of Europe, ranging from the Alps to the Scandinavian ranges, are undergoing dramatic modifications as a result of climate change. The increasing incidence of big landslides that are caused by glaciers is one of the results that gives cause for concern. Communities, infrastructure, and the environment are all exposed to significant dangers as a result of these occurrences. It is possible to decipher the underlying patterns and causes of this surge via the utilization of geospatial analysis, which provides insights that are essential for the development of effective mitigation solutions.

The relationship between the retreat of glaciers and the increased incidence of big landslides in mountainous parts of Europe has been brought to light using geospatial analysis. As glaciers retreat, they expose slopes that were formerly supported, which results in an increase in instability and a higher likelihood of landslides having an adverse impact. Geospatial information system (GIS) mapping and satellite imaging make it possible to monitor glacier retreat, which helps in locating regions that are prone to experiencing such occurrences.

The vulnerability of mountainous regions in Europe to landslides is further exacerbated by environmental factors such as shifting patterns of precipitation and fluctuations in temperature brought on by climate change. The monitoring of climatic fluctuations and the influence such shifts have on slope stability is made easier with the use of geospatial analysis, which contributes to a more thorough knowledge of the dangers that are presented by environmental changes.

More research is concentrating on simulating the huge cover of ice in high elevated areas that extended duration in order to recreate paleoclimate or investigate the impact of glacial isostatic forces on the Earth's lithosphere. It is difficult to strike a balance between processing costs and the geographical breadth and model complexity. 1–2 km spatial resolution was often used in earlier large-scale ice cover modeling experiments conducted in mountainous regions including the Tibetan Plateau, New Zealand, and the European Alps. On the other hand, high elevated areas typified with its accelerated elevation which has geographic characteristics, while poorly defined in coarse resolution topography, are essential for glacier mass balance and dynamics.

An Instructed Glacier Model (IGM) represents a distinctive 3D ice model that incorporates a neural network based on convolution trained using advanced ice models to accurately simulate ice movement. As a consequence, run times are significantly accelerated, making better spatial resolution possible to achieve. We simulate the whole European Alps (480, 240 km^2) using IGM and compare models with a resolution of 200 and 2000 m. To replicate glacial conditions and mimic the expansion of ice over 70,000 years, we use an exponential cooling rate to gradually decrease the temperature of the existing environment by 6 °C.

Initial findings point to systematic variations linked to resolution: When ice builds up at high altitudes at the start of cooling, the reduced resolution produces somewhat higher ice volume. But after about 41,000 years, this pattern reverses, just before the huge valleys are completely covered in thick ice. In the higher resolution model, we detect up to 15% greater ice volume when the Alps are completely covered in ice. The variations in ice volume are not dispersed equally over space. In certain sections of the Alps, mainly huge valley systems, of up to 12,000 km^2, the greater precision model generates ice that is thicker and covers a larger area, whereas in other somewhat smaller areas, the frozen surface is thinner and covers a smaller area. Currently, we are investigating the extent to which the flow of glaciers from steep slopes to wider, less steep valleys is evident at different levels of detail (Werner et al. 2023).

Large landslides are a common hazard in high mountains and shows a significant characteristic in the growth of environment throughout Quaternary timescales (Balzano et al. 2019). The stability of valley slopes may weaken and they may fall as mountain glaciers recede and permafrost thaws (Iverson 2000). Paraglacial landslides physical effects may trigger a series of secondary risks, including tsunamis, floods caused by glacier lake, or any permutation of these (Coe and Godt 2012). Deglaciated and continued melting landscapes provide a hazard for tens of kilometres downstream from their genesis (Geertsema et al. 2013). Over the last several decades, glacier retreat has triggered many devastating landslides all over the world (Balzano et al. 2019; Kaab et al. 2005; Noetzli and Gruber 2009), some of which have claimed hundreds of lives. Because of the anticipated ongoing shrinkage of bigger ice sheets, landslides are expected to occur more often both locally, near mountain glaciers, and across wider regions. Therefore, in order to reduce the hazards linked to an anticipated rise in mass movements, we must comprehend how landslide spatial distribution is regulated at both the local and regional scales.

The distribution and factors influencing paraglacial landslides throughout whole mountain ranges or ice sheets have not been extensively studied, but several have examined the spatial distribution of landslides (Fischer et al. 2006; Haeberli et al. 1997; Ren et al. 2020; Coe et al. 2018; Schneider et al. 2011). As a result, it is still unknown what complicated factors contribute to slope instability in these deglaciated places. Large bedrock landslides are thought to be primarily influenced based on thickness of the icesheet at the time of the glacial retreat and the rate of glacial decompression (Fischer et al. 2006; Coe 2020), post-glacial uplift linked to increased seismic activity (Ren et al. 2020; Gul et al. 2017), the scattering of fragile rock, and the Spatial and weather-related elements. Because of this, the steepest and wettest areas of deglaciated mountain belts (Hewitt 1988) are home to a large number of paraglacial landslides, which have a propensity to group together along faults that are prone to earthquake activity (Kyriou and Nikolakopoulos 2019).

Using data on the distribution of large landslides over the almost 305,000 km^2 of Patagonia that have been deglaciated, we find a clear trend. Between about 38 and 56°S, our mapped region extends for approximately 2000 km across the southernmost region of the Andes, namely the Patagonian and Fuegian Andes. Along with the Patagonian Ice Sheet (PIS), it encompasses the area where the maximum glacial activities occurred. According to research, the mountain belts in Southern Patagonia are more prone to large landslides in lower, drier, and less seismically active areas. We contend that this results from the fact that conceivably dangerous materials have not been completely eliminated by glacier throughout Quaternary, and periphery of the mountain ranges was not extensively glaciated as their centre massifs.

17.4 Local Environments

The plotted region extends over around 2000 km, from approximately 38–56°S, and includes the southernmost ranges of the Andes, the Patagonian and Fuegian Andes (Fig. 17.1). This region is characterized by the Patagonian Glacier Sheet (PIS) borders that Davies et al.22 reconstructed from the Last Glacial Maximum (LGM). The Andes Mountains in the west, with elevations between 3000 and 4000 m a.s.l., the piedmont zone in the east, and a rough fjord landscape with groups of islands like Tierra del Fuego and Chiloé in the east make up the three distinct regions that make up this area. Figure 17.1 shows that the region is home to several mountain glaciers and smaller ice caps in addition to several huge extra polar snow areas, like the North Patagonian Snow area and the Southern Patagonian Snow area (Strom and Abdrakhmatov 2018). The area climate is impacted by the Southern Westerly Wind (SWW), which deliver about 5000 mm of rainfall annually through the Pacific Ocean to the western portion of the mountain range. However, due to their location in a rain shadow, the eastern foothills get a maximum precipitation every year (Moragues et al. 2019).

The topography and geology have a significant impact in influencing the extent of landslides throughout vast regions of glaciated and non-glaciated mountain ranges. Patagonia prone to experiencing significant landslides, particularly on broad ridges characterized by steep slopes and substantial elevation differences resulting from volcanic and sedimentary materials. Long-term glacial erosion, which exposes a stable crystalline base and forms U-shaped valleys interspersed with small hills decreases the likelihood of significant landslides happening over a distance of one kilometre. The Patagonian and Fuegian Andes include four distinct characteristics that explain the distinct geographic distribution of landslides in the subject matter region, unlike other mountain ranges, where substantial landslides often occur on the highest and most precipitous slopes (Blondeau et al. 2021). The PIS is characterized by an extensive glaciation period, a massive ice sheet during the Quaternary period, a resilient granite massif dominating the alpine region, and the creation of vast glacial lakes that submerged a substantial portion of the PIS's eastern border following maximum glaciation. The eastern boundary consists of geologically unstable volcanic and sedimentary strata. Additional research is required to examine the relationship between the duration of glaciation, the extent of glacial landform formation, and the incidence of significant landslides in various mountain ranges with diverse climatic and geological attributes.

Jia Liu, Yuming Wu & Xing Gao highlighted the fact in the light of "Increase in occurrence of large glacier-related landslides in the high mountains of Asia "Scientific Reports, this chapter analysed significant landslides caused by glaciers in high mountain regions using a unique dataset specifically focused on glacier-related landslides. The research investigates the spatial and temporal patterns of landslide in the high mountain's areas across a 20-year timeframe from 1999 to 2018. We want to determine whether there have been any changes in the frequency and intensity of landslides in this area as a result of warming. The study area covers an expansive territory of over 600,000 square kilometres. It includes part of Pamir, western

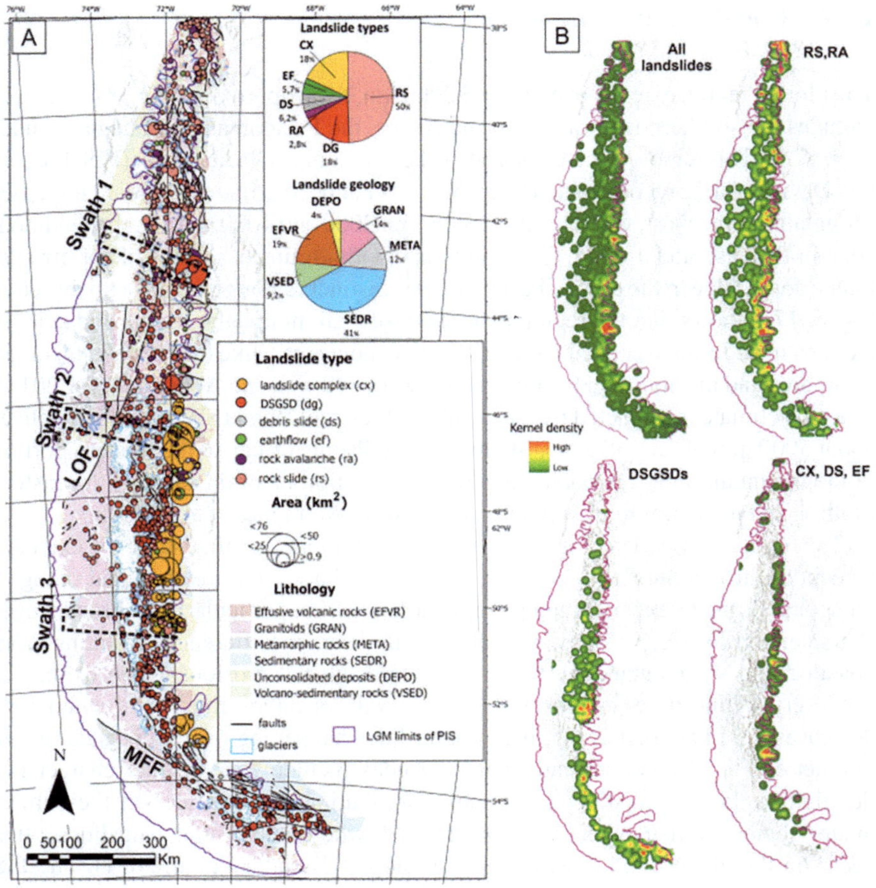

Fig. 17.1 (**A**) The spatial distribution of significant landslides (>1 km^2) and the geological characteristics inside the Last Glacial Maximum (LGM) boundaries of the PIS are being analyzed. The landslides are shown as centroids graded by size, accompanied by pie diagrams in the inset that indicate the percentage of different kinds of landslides and their lithology. (**B**) Kernel density maps depicting the distribution of all landslide and their particular categories. Primary faults: The Liquiñe-Ofqui fault zone (LOF) and the Magallanes-Fagnano fault (MFF)

Himalayas, Hindu Kush, Karakoram, and western Kunlun Mountains, located within the geographical coordinates of 33.8°–39.5°N and 70.5°–78.7°E. Using Landsat images, a list of glacier extent and landslide maps over the last 20 years was created. Because of their close proximity to large earthquakes that have been documented in both space and time, landslides that may have seismic origins were not included in the catalog.

Numerous high-altitude zones are similar to the HMA region. Due to their widespread distribution and abundance as a water resource, glaciers play a significant role in providing millions of people in the surrounding regions with clean water. The HMA's glaciated regions are more temperature-sensitive than the plains lower elevations (Kraaijenbrink et al. 2017). Predictions for climate change during this age include an increase in the frequency of severe weather occurrences, glacier retreat, and changes to the water levels in glacial lakes (Gul et al. 2017). The HMA is prone to geologic hazards, and landslides might have an impact on vital transportation routes that link adjacent Asian nations and pass these mountains (Hewitt 1988; Clague and Evans 2000).

Variables affecting landslide identification. In alpine and glacial settings, landslides may be distinguished by three characteristics: (1) a stark contrast with the encompassing frozen landscape; (2) The effects of the landslides on the river and the lobate patterns, which are characteristic of deposits resulting from rock avalanches, are examined. Gradient imbalances may also be deduced from regional variations in attitude form, or patterning, as well as the presence of lineaments (Soeters and Van Westen 1996). Nonetheless, there are a few things that influence our ability to recognize a specific landslide throughout the identification phase. The qualities of remote sensing images, snow cover, and landslide features all have an impact on landslide detection.

A complicated process of movement is a landslide. Failure, transportation, and accumulation areas are all included in landslides. It is challenging to identify some slides that did not result in optical distinctive change in the picture due to the effect of alpine geological features (such as glacier development and absence of flora) and image quality factors (such as resolution, clouds, etc.). They consist of a few minor debris flows, small-scale moraine collapses, and insignificantly accumulated rock avalanches. We discovered throughout the identification procedure that the impact of landslides on rivers helps with the identifying process.

Rock-type landslides are common across the Hindu Kush and the western Himalayas, although they are most prevalent in the western Kunlun Mountains of the HMA. The West Kunlun Mountains and the Hindu Kush are the primary locations for debris land-slides, while there are also occasional occurrences in the Pamirs and the western Himalayas. At an average height of 4139 m, moraine landslides are common at higher altitudes, such as the Karakorum Mountains, the western Himalayas, the Kunlun Mountains, and the Pamirs. Additionally, we examined the distribution of major landslides, which are those that impact regions larger than.

This present chapter examining data of Landsat from 1999 to 2018 and glacier area maps were made on a yearly basis for the same 20-year period. The presence

Table 17.1 Earthquakes that potentially triggered landslides

Date	Latitude (°)	Longitude (°)	Epicenter	Magnitude
7 December 2015	38.21	72.77	104 km W of Murghob, Tajikistan	7.2
3 July 2015	37.45	78.15	97 km SE of Yilkiqi, China	6.4
8 October 2005	34.53	73.58	Pakistan	7.6
20 November 2002	35.41	74.51	Northwestern Kashmir	6.3

of rock surface and ice cover in the HMA complicates the detection of landslides, in addition to the influence of picture resolution. This may result in an underestimation of the frequency of landslides. However, the maps of landslides and glacial areas provide precise and comprehensive data on their location and time of occurrence. Using our customized inventory, study conducted a comprehensive analysis of the correlation between landslide and snow sheet designated in research region. There are given significant deductions may be derived with the help of study.

Plot illustrating the extent of land affected by landslides and the extent of glacier coverage. GA denotes the extent of glacier coverage, whereas LA denotes the extent of landslide coverage. It indicates the natural logarithm. Study is showing trend line with the help of dots in black color.

1. Landslides are prevalent in the HMA. The extent of landslide catastrophes has grown during the previous 20 years. The frequency of landslides and the extent of their impact were greater in the most recent 10-year period (2009–2018) compared to the preceding 10-year period (1999–2008). The occurrence of landslides was exceptionally elevated during 3 of the 4 El Niño years.
2. Landslides in the HMA exhibit power-law (fractal) frequency characteristics. The attenuation coefficient for the period from 2009 onwards demonstrated a reduction in comparison to the preceding 10 years, indicating an increased likelihood of significant landslides.
3. The glaciated area in the High Mountain Asia (HMA) region has seen a consistent decrease over the previous two decades. With the exception of the Karakoram Mountains, the majority of glaciers in the studied region have undergone retreat. The reduction of glacial area has correspondingly led to an increase in the occurrence of yearly landslides.

The melting of glaciation in the High Mountain Asia region over a period of 20 years seems to be linked to an increase in the occurrence of landslides, as well as an increase in the size of these landslides. There is a discernible weak inverse relationship between the occurrence of landslides on an annual basis and the extent of glacial regions. While local factors may have initiated landslides in the HMA, the retreat of glaciers might serve as a valuable indicator for the climate-related factors that affect the stability of slopes. Further global warming will result in a continued decrease in the extent of glaciers. Therefore, the probability of substantial landslides within the HMA is anticipated to continue increasing.

It is necessary to take a comprehensive strategy to mitigation in order to address the growing number of big landslides that are caused by glaciers in mountainous parts of Europe. The use of geospatial analysis is an essential instrument for elucidating the intricate relationships that exist between various environmental parameters, which in turn helps to direct the development of specific strategies. Through the incorporation of scientific findings, proactive policies, and community involvement, Europe has the ability to lessen the effects of these natural catastrophes and strengthen its resilience in the face of climate challenges that are constantly altering (Jia Liu, Yuming Wu & Xing Gao "Increase in occurrence of large glacier-related landslides in the high mountains of Asia "Scientific Reports).

Velocity and MCE are two aspects that complement each other in describing exposure. Velocity refers to the prospective migration pathways' length, while MCE represents the impediment to movement caused by climatic conditions along these corridors During the course of time. Estimations of MED and velocity ED align with previous research indicating that mountainous areas exhibit a relatively low susceptibility to climate change (longer durations of climatic residence) (Serey et al. 2019; Pánek 2019; Haeberli et al. 1997; Ren et al. 2020). Nevertheless, this explanation presupposes that there is no spatial variation in climate resistance to mobility. We demonstrate that in regions with intricate topography (such as the mountain ranges of North America), the magnitude of Maximum Climate Envelope (MCE) may be substantial even for brief climate trajectories due to their frequent traversal of landscapes with diverse climatic. Within this particular context, the presence of diverse climate patterns throughout space serves to enhance the level of barrier to migration, hence decreasing the degree of climatic interconnectedness between a starting point and a destination point. Conversely, extended climate patterns throughout much of the continent go across climatically uniform regions with little or no hindrance. Velocity is significant whereas MCE is absent in certain circumstances, namely in the Great Plains areas and the Canadian Boreal area.

The climate trends predicted by ED and MED represent two extreme possibilities among a range of conceivable trajectories. These possibilities depend on the physical characteristics of the place and the characteristics of the species that inhabit there. In places with a flat topography, such as the one shown in Fig. 17.2, There is limited disparity among both of these end-members. The values of both MED and ED velocities are similar. Nevertheless, there are notable distinctions, especially in areas characterized by complex terrain. In addition, MCE denotes the least amount of exposure to different climatic conditions along a prospective route of migration (Fig. 17.3).

The specific vulnerability of particular creatures to climate change will also be influenced by the characteristics of the species that inhabit the area (Clague and Evans 2000; Kyriou and Nikolakopoulos 2019). Vagile creatures have the ability to match the speed of huge climatic velocities, but sessile species may lag behind as the climate changes. This may lead to greater vulnerability when climate change surpasses their ability to move. Furthermore, the characteristics related to the distribution of organisms are likely to influence the paths taken during migration. For instance, avian species like birds may exhibit linear trajectories that may be accurately represented by velocity vectors. Temperature sensitivities may also influence an organism's preference for either linear paths or ones that minimize exposure to varying climates. Thermal experts, such as ectotherms found in tropical regions, are more likely to limit their exposure to various climates, as measured by speed MED, compared to temperature specialists, as endotherms found in high latitudes. Similarly, plant species with a restricted range of reproductive circumstances may likely need longer migratory routes than those calculated using velocity ED.

When we discuss about MED and MCE regarding velocity it offers a systematic overview for evaluating exposure based on the ability of species to disperse and their

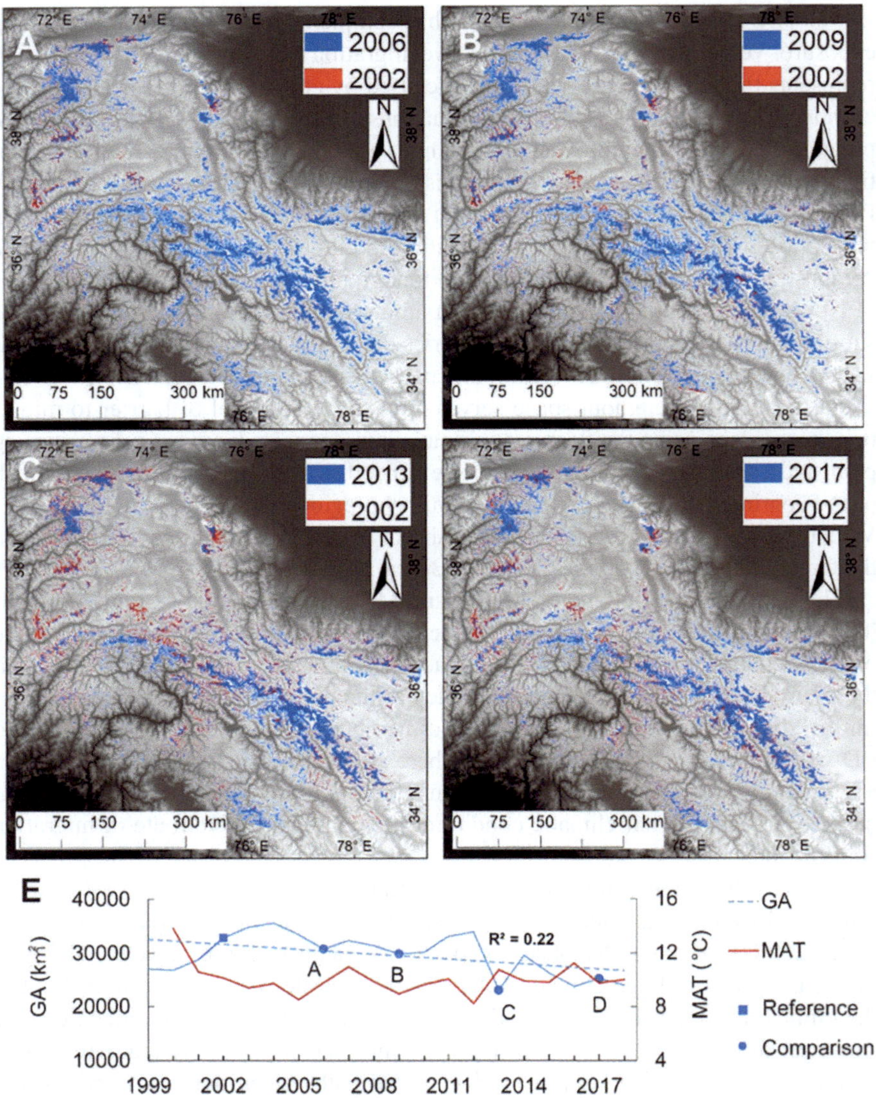

Fig. 17.2 An analysis of the extent of glacial coverage in the research area for certain years. Panels A–D depict the spatial coverage of glaciers in the years 2002, 2006, 2009, 2013, and 2017, accordingly. Red represents the amount of glacial extent that disappeared over the specified period, whereas blue denotes the presence of glaciers throughout both periods. Panel E provides a concise overview of the general patterns seen in the size of glaciers (glacial area) and the average temperature over 20 years. These trends were analysed and shown using Arcgis 10.2 software. https://www.esri.com/en-us/arcgis/products/arcgis-desktop

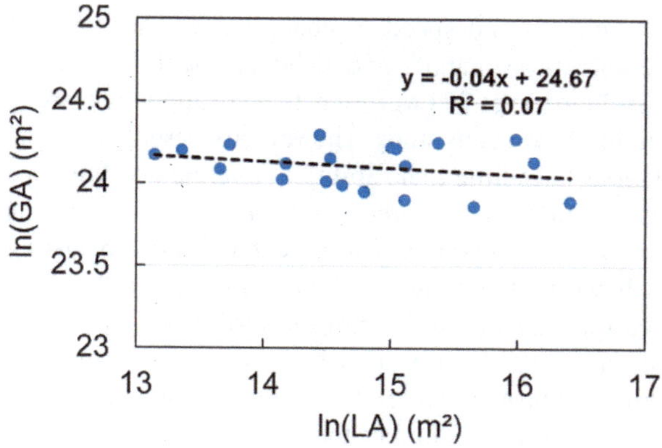

Fig. 17.3 Area of glaciers and landslides graphed. The glacier area is represented by GA, the landslide area by LA, and the natural log is represented by ln. The linear trend line is shown by the black dot line

tolerance to temperature (Fig. 17.4a). Study highlights a straightforward categorization that is (Fig. 17.4b): low level of velocity, low level of MCE, high velocity of MCE that is high level of velocity, MCE low level velocity with high MCE velocity. These words are used to categorize the quality and quantity of data, without implying any biological significance to the specific thresholds we choose (refer to the methodology section). However, regions with high velocity provide a greater danger of exposure to species that do not disperse well. High MCE areas provide a greater danger for thermal experts.

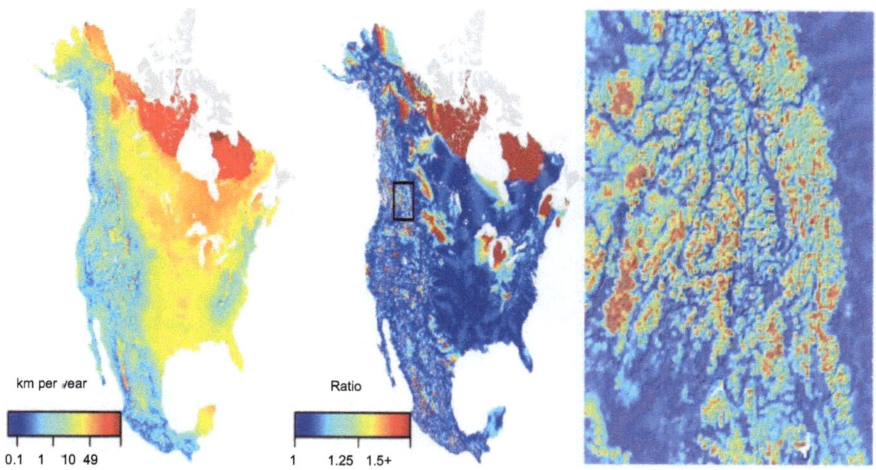

Fig. 17.4 Climate velocity in North America from 1995 to 2085. The concept of climate velocity, as measured by the minimum exposure distance (velocity MED), is introduced in (**a**). The ratio of velocity MED to velocity determined by the Euclidean distance (velocity ED) is shown as (**b**). A downscaled version (black box) of (**b**), recreated to a level of 1 km to enhance visual perception, is shown in (**c**). The grey areas indicate pixels where either the source or source pixel is situated on the island

Velocity and speed of changing weather and MCE should be utilized, which require more quantity of data but consider the unique responses of particular species in relation to global warming, that are applied regardless of specific data, like the data highlighted in this study. The velocity based on distance presupposes that local populations have little to no ability to tolerate or adjust to changes in temperature (Korup et al. 2007). Therefore, it indicates the maximum need for migration of climate-sensitive organisms. On the other hand, sites consisting of species with wider climatic behaviour shown mean climate envelope with values showing velocity. The main reason is a larger portion of the surrounding areas shows favourable for that kind of species. Additionally, there have been advancements in fine-filter methodologies that simulate probable migratory routes for specific species. For instance, circuit theory has been used to simulate the migratory paths of animals that follow climatic changes (Richter et al. 2016). Furthermore, both distance- and gradient-based methods have been modified to include the climate tolerances of specific species. This adaptation calculates locational extent and velocity of a particular organism as they have to migrate in order to keep up with its climatic niche (Ojala et al. 2017). The limits of biotic velocity are similar to the ones we have identified in particular areas. This will not accurately indicate degree that is prospective migratory areas which highlights particular organism with various climatic conditions (Fig. 17.5).

The quality of the input information and the parameters needed to calculate speed MED, movement ED, or MCE are also affected. Reducing the data quality increases

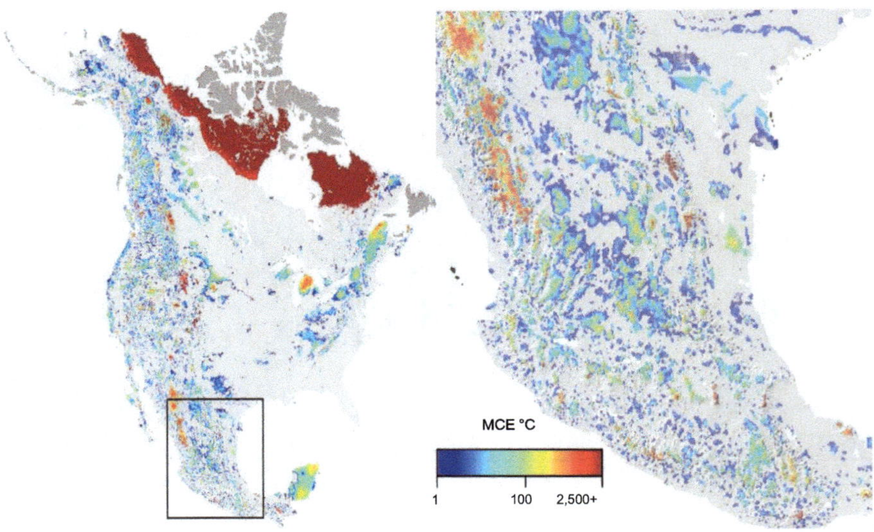

Fig. 17.5 Minimal aggregate exposure. The Mean Cumulative Emissions (MCE) for North America from 1995 to 2085 are shown in a. Figure b displays an inset, represented by a black box, that has been recreated to a level of 1 km in order to enhance visual perception. The MCE values of 40 are presented in colour, whereas an MCE value of ¼ 0 is displayed in light grey. The MCE values of pixels located on islands and images that have future climatic analogues on islands are shown in a dark grey colour (see to the Methods section for further details)

movement MED, speed ED, and MCE, whereas expanding the bin width required to establish a climatic analogue decreases them. These findings align with prior research that has investigated the sensitivity of velocity calculations to scale. They indicate that when comparing studies or assessing species dispersal rates in relation to climate velocities, it is crucial to take into account the specific scale of analysis.

Mountains sustain about 25% of the Earth's land-based biodiversity, include 32% of designated protected areas, and provide shelter more than 50% of the world's biodiversity hotspots (Turcotte et al. 2002). Given these circumstances, our findings prompt a logical inquiry into how to reconcile the concept of mountains serving as refuges in the face of climate change, as opposed to mountains being regions with strong resistance to climatic shifts. Climate change adaptation solutions frequently rely on the assumption that regional variability in climate may mitigate the consequences of climate change via short-distance dispersion. This paradigm forms the basis for strategies used in reserve design (Malamud et al. 2004; Bessette-Kirton and Coe 2016) and techniques used in the identification of microrefugia (Noetzli and Gruber 2009; Schneider et al. 2011; Soeters and Van Westen 1996; Lei et al. 2019). However, a crucial and sometimes disregarded factor to examine is the accessibility of places to migratory organisms (Lei et al. 2019). Accessibility is influenced by the degree of climatic connectedness to a certain location and is influenced by several variables, including physical obstacles to mobility, habitat fragmentation, and changes in land use and land cover. When attempting to reconcile these two positions, it is crucial to take into account the temporal scale. Velocity MED and MCE are influenced by the size of climatic changes over time compared to the local geographical difference in climate. During brief time periods or in areas with distinct climate variations, the process of local dispersal can enable an organism to adapt to climate change without being exposed to different climates. This can be achieved through simple movement towards higher elevations along a spatial climate gradient. Nevertheless, if the extent of environmental change surpasses the localized geographic difference in temperature, the magnitude of climate exposure (MCE) and possibly the magnitude of ecological disruption (MED) would significantly rise. This is because organisms will have to navigate through inappropriate climates in order to reach a future climate that is similar to their current habitat. Outside of these specific spatial and temporal areas, these places will only provide temporary refuge for organisms.

Study demonstrate that distance (as well as velocity) is an inadequate indicator of climate connection due to its failure to include the geographical variations in climatic resistance to movement, often known as MCE (Movement Cost Equivalent). The presence of high MCE (Mountain Climate Exclusivity) in mountainous regions indicates that these locations are more geographically separated in terms of climate than previously documented. In contrast, the flat areas of the continent exhibit significant velocities while offering less impediment to movement due to climate. The climatic connectedness between these locations and their future climate equivalents may be larger than presently acknowledged. Therefore, it is necessary to conduct a more detailed evaluation of climate exposure, specifically considering the resistance of climate to movement. Therefore, we support the need for more collaboration

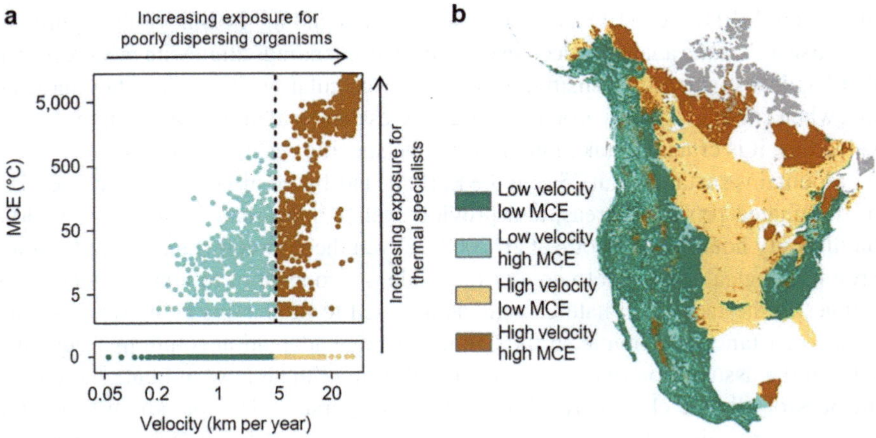

Fig. 17.6 (**a**) Acceleration and minimum total exposure classification. The provided data illustrates the correlation with velocity MED and minimal total exposure (MCE) for a sample chosen at random of 5000 images from North America. The acceleration MED and MCE measurements of 40 are presented on a logarithmic scale. The bottom panel displays MCE values of 0, which account for 71% of the pixels, and are classified as low. MCE40 values, on the other hand, are classified as high. The median velocity, denoted as MED, is shown by a dashed vertical line. This line serves to distinguish among the high and low velocity classes. A map displaying the spatial arrangement of the four categories, determined by the multivariate distributions of the MCE and velocity MED, is shown in (**b**). The color of dots in A correlate to the color of regions on map B. The gray areas indicate pixels where either the original or target pixel is situated on an island, as explained in the Methods section

between connectivity modeling methodologies that take climate into account and the creation of measures that measure susceptibility to climate change. This integration will provide a more accurate and comprehensive description of the many aspects of biota exposure resulting from continuing climate change (Figs. 17.6 and 17.7).

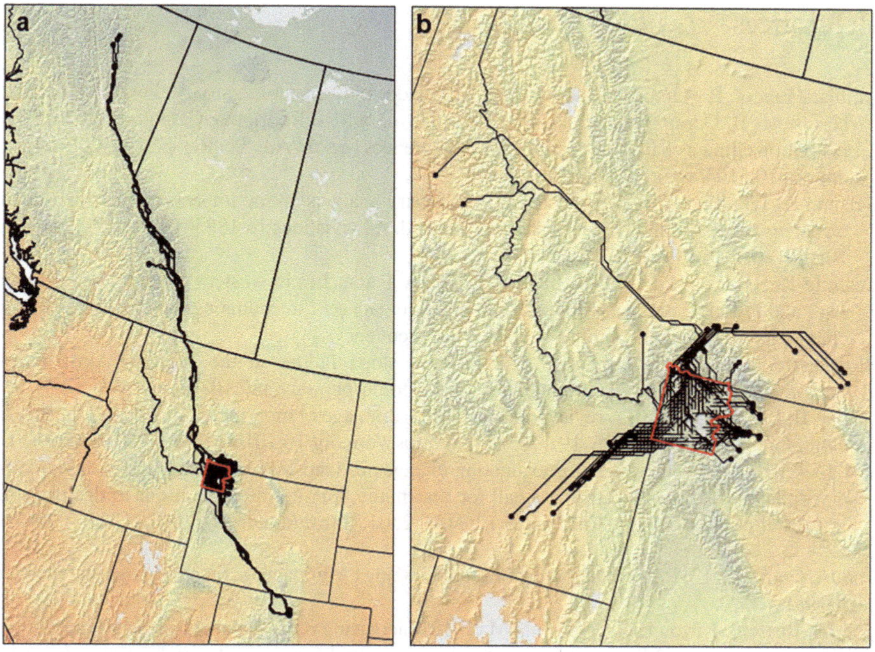

Fig. 17.7 Climate projections of Yellowstone National Park in the United States. Climatic predictions of Yellowstone National Park (red colour polygon) from 1995 to 2085, both in the forward (**a**) and backward (**b**) directions. Future patterns illustrate the approaches that minimize the total exposure to different temperature levels between a starting point and its corresponding future climatic location. Forward trajectories, as defined by the concept of 'forward velocity', describe the possible impact of climate change on organisms. Reverse climate trajectories illustrate the pathways that have the least Minimum Climate Error (MCE) across a hypothetical cell and an analogous cell under present climate circumstances. Reverse directions, as defined by the concept of 'reverse velocity', refer to the measure of how easily a location may be colonized by other places in terms of climate accessibility

17.5 Conclusion

In conclusion, the retreat of glaciers in the High Mountain Asia region over a period of 20 years seems to be linked to an increase in the occurrence of land- slides, as well as an increase in the size of these landslides. There is a discernible weak inverse relationship between the occurrence of landslides on an annual basis and the extent of glacial regions. While local factors may have first caused landslides in the HMA, the retreat of glaciers might serve as a valuable indicator for the climate-related factors that affect the stability of slopes. In the event that climatic warming persists, the extent of glaciers will continue to diminish. Consequently, the likelihood of significant landslides in the HMA is expected to keep rising.

References

Amblar-Francés P, Ramos-Calzado P, Sanchis-Lladó J, Hernanz-Lázaro A, Peral-García MC, Navascués B, Dominguez-Alonso M, Pastor-Saavedra MA, Rodríguez-Camino E (2020) High resolution climate change projections for the Pyrenees region. Adv Sci Res 17:191–208. https://doi.org/10.5194/asr-17-191-2020

Balzano B, Tarantino A, Ridley A (2019) Preliminary analysis on the impacts of the rhizosphere on occurrence of rainfall-induced shallow landslides. Landslides 16:1885–1901. https://doi.org/10.1007/s10346-019-01197-5

Bessette-Kirton E, Coe JA (2016) Inventory of rock avalanches in western Glacier Bay National Park and Preserve, Alaska, 1984–2016: a baseline data set for evaluating the impact of climate change on avalanche magnitude, mobility, and frequency

Blondeau S, Gunnell Y, Jarman D (2021) Rock slope failure in the Western Alps: a first comprehensive inventory and spatial analysis. Geomorphology 380:107622

Clague JJ, Evans SG (2000) A review of catastrophic drainage of moraine-dammed lakes in British Columbia. Qunatum Sci Rev 19:1763–1783. https://doi.org/10.1016/s0277-3791(00)00090-1

Coe JA (2020) Bellwether sites for evaluating changes in landslide frequency and magnitude in cryospheric mountainous terrain: a call for systematic, long-term observations to decipher the impact of climate change. Landslides 17:2483–2501. https://doi.org/10.1007/s10346-020-01462-y

Coe JA, Godt J (2012) Review of approaches for assessing the impact of climate change on landslide hazards

Coe JA, Bessette-Kirton EK, Geertsema M (2018) Increasing rock-avalanche size and mobility in Glacier Bay National Park and Preserve, Alaska detected from 1984 to 2016 Landsat imagery. Landslides 15:393–407. https://doi.org/10.1007/s10346-017-0879-7

Dobrowski SZ, Parks SA (2016) Climate change velocity underestimates climate change exposure in mountainous regions, August 2016. Nat Commun. https://www.nature.com/articles/ncomms12349

Fischer L, Kaab A, Huggel C, Noetzli J (2006) Geology, glacier retreat and permafrost degradation as controlling factors of slope instabilities in a high-mountain rock wall: the Monte Rosa east face. Nat Hazards Earth Syst Sci 6:761–772. https://doi.org/10.5194/nhess-6-761-2006

Geertsema M, Clague JJ, Hasler A (2013) Influence of climate change on geohazards in Alaskan parks. Alaska Park Sci 12:81–85

Gul C et al (2017) Using Landsat images to monitor changes in the snow-covered area of selected glaciers in northern Pakistan. J Mt Sci 14:2013–2027. https://doi.org/10.1007/s11629-016-4097-x

Haeberli W, Wegmann M, Vonder Muhll D (1997) Slope stability problems related to glacier shrinkage and permafrost degradation in the Alps. Eclogae Geol Helv 90:407–414

Hewitt K (1988) Catastrophic landslide deposits in the Karakoram Himalaya. Science 242:64–67. https://doi.org/10.1126/science.242.4875.64

Hugonnet R, McNabb R, Berthier E, Menounos B, Nuth C, Girod L, Farinotti D, Huss M, Dussaillant I, Brun F, Kääb A (2021) Accelerated global glacier mass loss in the early twenty-first century. Nature 592(7856):726–731. https://doi.org/10.1038/s41586-021-03436-z

IPCC (2019) IPCC special report on the ocean and cryosphere in a changing climate (Pörtner H-O, Roberts DC, Masson-Delmotte V, Zhai P, Tignor M, Poloczanska E, Mintenbeck K, Alegría A, Nicolai M, Okem A, Petzold J, Rama B, Weyer NM, eds.) (In Press)

Iverson RM (2000) Landslide triggering by rain infiltration. Water Resour Res 36:1897–1910. https://doi.org/10.1029/2000wr900090

Kaab A et al (2005) Remote sensing of glacier- and permafrost-related hazards in high mountains: an overview. Nat Hazards Earth Syst Sci 5:527–554

Korup O, Clague JJ, Hermanns RL, Hewitt K, Strom AL, Weidinger JT (2007) Giant landslides, topography, and erosion. Earth Planet Sci Lett 261:578–589

Kraaijenbrink PDA, Bierkens MFP, Lutz AF, Immerzeel WW (2017) Impact of a global temperature rise of 1.5 °C on Asia's glaciers. Nature 549:257. https://doi.org/10.1038/nature23878

Kyriou A, Nikolakopoulos K (2019) Landslide mapping using optical and radar data: a case study from Aminteo, Western Macedonia Greece. Eur J Remote Sens. https://doi.org/10.1080/22797254.2019.1681905

Laimer HJ (2017) Anthropogenically induced landslides—a challenge for railway infrastructure in mountainous regions. Eng Geol 222:92–101. https://doi.org/10.1016/j.enggeo.2017.03.015

Lei YB et al (2019) Extreme Lake level changes on the Tibetan Plateau associated with the 2015/2016 El Nino. Geophys Res Lett 46:5889–5898. https://doi.org/10.1029/2019gl081946

Malamud BD, Turcotte DL, Guzzetti F, Reichenbach P (2004) Landslide inventories and their statistical properties. Earth Surf Process Landf 29:687–711

Moragues S et al (2019) Slope instability analysis in South Patagonia applying multivariate and bivariate techniques on Landsat images during 2001–2015 period. CATENA 174:339–352. https://doi.org/10.1016/j.catena.2018.11.024

Moreno A, Bartolomé M, López-Moreno JI, Pey J, Corella JP, García-Orellana J, Sancho C, Leunda M, Gil-Romera G, González-Sampériz P, Pérez-Mejías C, Navarro F, Otero-García J, Lapazaran J, Alonso-González E, Cid C, López-Martínez J, Oliva-Urcia B, Faria SH, … García-Ruíz JM (2021) The case of a southern European glacier disappearing under recent warming that survived Roman and Medieval warm periods. The Cryosphere 15:1157–1172. https://doi.org/10.5194/tc-15-1157-2021

Noetzli J, Gruber S (2009) Transient thermal effects in Alpine permafrost. Cryosphere 3:85–99. https://doi.org/10.5194/tc-3-85-2009

Ojala AEK, Matilla J, Markovaara-Koivisto M, Ruskeeniemi T, Palmu JP, Sutinen R (2017) Distribution and morphology of landslides in northern Finland: an analysis of postglacial seismic activity. Geomorphology 326:190–201

Pánek T (2019) Landslides and quaternary climate changes—the state of the art. Earth-Sci Rev 196:22. https://doi.org/10.1016/j.earscirev.2019.05.015

Ren LY, Duan KQ, Xin R (2020) Impact of future loss of glaciers on precipitation pattern: a case study from south-eastern Tibetan Plateau. Atmos Res 242:7. https://doi.org/10.1016/j.atmosres.2020.104984

Richter A, Ivins E, Lange H, Mendoza L, Schröder L, Hormaechea JL, Cacassa G, Marderwald E, Fritsche M, Perdomo R, Horwath M, Dietrich R (2016) Crustal deformation across the Southern Patagonian Icefield observed by GNSS. Earth Planet Sci Lett 452:206–215

Schneider D, Huggel C, Haeberli W, Kaitna R (2011) Unraveling driving factors for large rock–ice avalanche mobility. Earth Surf Process Landf 36:1948–1966

Serey A et al (2019) Landslides induced by the 2010 Chile megathrust earthquake: a comprehensive inventory and correlations with geological and seismic factors. Landslides 16:1153–1165. https://doi.org/10.1007/s10346-019-01150-6

Soeters R, Van Westen C (1996) Slope instability recognition, analysis and zonation. Landslides Investig Mitig 247:129–177

Strom A, Abdrakhmatov K (2018) Rockslides and rock avalanches of Central Asia: distribution, morphology, and internal structure. Elsevier, Amsterdam

Thomson SN, Brandon MT, Reiners PW, Tomkin JH, Vasquez C, Wilson NJ (2010) Glaciation as both destructive and constructive long-term control on mountain building. Nature 467:313–317

Turcotte DL, Malamud BD, Guzzetti F, Reichenbach P (2002) Self-organization, the cascade model, and natural hazards. Proc Natl Acad Sci USA 99:2530–2537. https://doi.org/10.1073/pnas.012582199

Werner H, Scherler D, Winkelmann R, Jouvet G (2023) Impact of spatial resolution on large-scale ice cover modeling of mountainous regions, EGU General Assembly 2023, Vienna, Austria, 24–28 April 2023, EGU23-4200. https://doi.org/10.5194/egusphere-egu23-4200

Chapter 18
Landslide Detection Using DInSAR Technique: A Case Study

Swati Sharma, Rohan Kumar, and Nandakishore

Abstract This work aimed to utilize the capabilities of the Interferometric Synthetic Aperture Radar (InSAR) technique to assess the slope instability condition in part of Himachal Pradesh, India. India is one of the countries to report the largest number of landslide events, especially in the monsoon season. The Himalayan regions in India are prone to heavy precipitation (rainfall) which loosens the soil and causes slope displacement. This study involves the rain-induced (pre and post-event) landslide study to detect the magnitude of ground deformation in Bariara village in district Kangra, Himachal Pradesh, India. In this study, the microwave repeat pass Interferometric Synthetic Aperture Radar (InSAR) technique was used for deducing the slope deformations. Two SAR imageries were used to detect the phase difference for developing the deformation map showing a downslope movement of almost 1.5 m. The results from the interferometric method have been cross-validated by data collected from the field survey. This study aimed to find the absolute rate of movement of a Himalayan landslide using the DInSAR method and has effectively added to the landslide literature for extensive future research in this study area.

Keywords Himalayan landslide · Rain-induced slope failure · Interferometry · Synthetic aperture radar

---

S. Sharma (✉)
Amity Institute of Geoinformatics and Remote Sensing, Amity University, Noida, India
e-mail: ssharma27@amity.edu

R. Kumar
Department of Chemical and Petroleum Engineering, Lovely Professional University, Punjab, India
e-mail: rohan.25322@lpu.co.in

Nandakishore
Amity Institute of Geoinformatics and Remote Sensing, Amity University, Noida, India

G. K. Panda et al. (eds.), *Landslide: Susceptibility, Risk Assessment and Sustainability*, Advances in Natural and Technological Hazards Research 52,
https://doi.org/10.1007/978-3-031-56591-5_18

18.1 Introduction

In the last few years, the number of disaster events has sprucely increased. Roughly, 15% of the Indian mainland is prone to landslides (GSI, http://www.portal.gsi.gov.in) inclusive of the vast terrain of the Himalayas, Nilgiris and Western Ghat. Factors affecting the landslides in the Himalayan region are steep slopes, weak slope material, structural discontinuities, large-scale anthropogenic interference, natural calamities, etc. In recent history, dams, tunnels, roads etc. have developed dramatically in the fragile Himalayan terrain (Kumar and Anbalagan 2019). The positive aspects of the development work also bring along some major challenges related to the viability of natural and physical elements. This work deals with landslide assessment of the slopes surrounding the Kangra Valley, Himachal Pradesh using microwave remote sensing techniques.

Over the past 10 years, InSAR is a general method widely accepted for landslide mapping and observation. This is basically due to the emergence of differential interferometry techniques known as DInSAR that utilize radar images from synthetic aperture radar (SAR) sensors. The main objective of this study is to generate results in the form of an interferogram which shows the phase difference for all image pixels of two passes of the satellite over the same area. Using several SAR images and the resulting interferograms, it is then possible to obtain the temporal evolution of ground displacement over time. Time series of deformation can be obtained from those pixels of the SAR image that maintain an optimal level of consistency (i.e. the stability of the signal) over the entire monitoring period. These pixels can be selected using different algorithms (Crosetto et al. 2016), typically grouped in point-like (or persistent) scatterers interferometry (Casagli et al. 2023) and distributed scatterer interferometry. The final result of this study is a deformation map inclusive of points with a value of slope movement velocity. DInSAR is extremely useful for landslide studies as it is a medium for high temporal and spatial resolution studies (Casagli et al. 2023). Many researchers have used the application of remote sensing data for landslide studies using InSAR analysis: Mondini et al. (2021), Bhasin et al. (2023), Liu et al. (2021), and Roy et al. (2022) are all valuable examples of literature describing the exploitation of remote sensing data in landslide best practices. Devaraj et al. (2022) studied the landslides in the mountainous areas of Western Ghat using the DInSAR analysis and correlated the land use land cover pattern with the slope failures. In their study, the loss of DInSAR information due to dense vegetation cover was taken into consideration. Niraj et al. (2023) have used the DInSAR-based time series analysis using scenes (SAR data) from consecutive years to predict the landslide probability at the Kotrupi region in Himachal Pradesh. This study involves slope movement map preparation using the DInSAR technique by taking the landslide scenario at the Bariara Village in Kangra district, Himachal Pradesh, India. Because of the complexity of physiography and growing anthropological activities, the state of Himachal Pradesh in India is recognized for frequent landslides, especially during monsoon season. The landslide caused by rain in the monsoon season of 2023; was such a devastating case that hundreds of people lost their lives in addition to unprecendented damage to property. According to research done in the past, the occurrence of landslides in Himachal Pradesh is extensive (Sharma et al. 2019; Kumar et al. 2020; Paswan and Shrivastava 2022). The effect of landslides has increased significantly recently because of the growth of population and expansion of urban centres. For the proper mitigation of landslide hazards, prior information on the slope instability condition is important, thus many methods are used for establishing the slope instability condition. The difference InSAR (DInSAR) method has been reported to work well in identifying slope movement conditions thus here we have used the DInSAR method on the slopes of Bariara Village, Kangra district in Himachal Pradesh. The objective of this study was to leverage the interferometry technique in detecting the ground movement magnitude due to landslide activity and to document this event for detailed investigation in future using field measures to mitigate the studied slope.

18.2 Study Area

This study involves the slope movement map preparation using the DInSAR technique by taking the landslide scenario at the Bariara Village, Kangra district, Himachal Pradesh (32°17′50.98″ N, 75°59′44.34″ E) (Fig. 18.1). According to Sharma and Mahajan (2018), this region is part of the Bhed Khad in the Kangra district, where the local tributaries empty into the Beas River basin. The Kangra region is sub-humid, with the monsoon season (July to September) seeing the most rainfall, and an annual precipitation of up to 1751 mm, according to the district's groundwater information booklet (http://cgwb.gov.in/Regions/NHR/Reports/Year%20Book%202016-17%20NHR.pdf). A local seasonal stream drains the right edge of the landslide zone, which is a tributary of the Beas River. Slightly south of the toe portion of this landslide, approximately 9 settlements were completely damaged from a destructive landslide activity in the year 2022 and further south, the main road and more settlements are vulnerable to any future slope displacement. The inset study area polygon in Fig. 18.1 shows the Bariara landslide covering a total area of 20,725 m^2 which was observed in this study. Due to their primary orientation toward the southwest and the maximum amount of sunlight, the slopes in this landslide zone may have a healthy forest cover. Mostly agricultural terraces have been seen to the south of the landslide zone, providing a means of subsistence for the nearby locals. There are other underlying causes for this landslide zone, which have moved nearly a large part of the slope presented in this study.

On the regional scale, this area contains several significant thrust boundaries, including the Main Boundary Thrust, which divides the Lesser Himalayan and the Outer Himalayan rocks. Local thrusts namely Jwalamukhi Thrust, and the Murree Thrust traverse across the study area toward the southern side. This aligns with the region's tectonic significance and the seismic activity that has shaken the study area in past (Mahajan et al. 2022). This landslide site of Village Bariara is at a distance of 3.5 km from the National Highway connecting District Mandi and Pathankot City

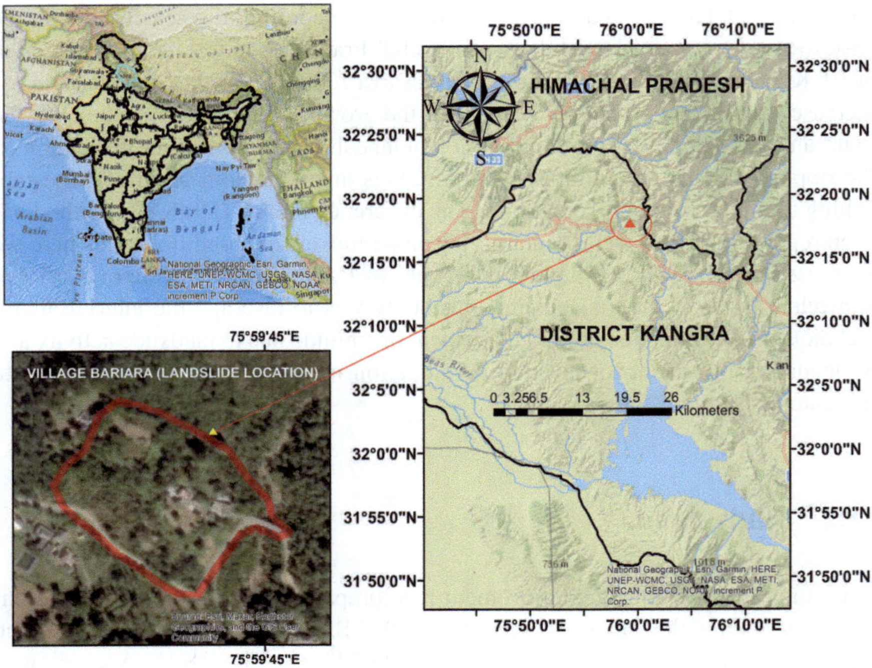

Fig. 18.1 Location map of study

which is an important economic route for the tourism industry in Himachal Pradesh. Due to the recent expansion of this highway such slope failures have become very active in monsoon season which directly hints toward the weakening of lithology on a few km scale in the vicinity of the presently studied landslide.

Field Evidence

This section shows the field evidence of damage caused by the landslide in the study area. The toe part of the damaged slope depicted the impact of slope draining due to seasonal rainfall (Fig. 18.2) since the landslide activity was reported in the month of August 2022. The length of lateral cracks on the ground surface (Fig. 18.3) shows the intensity of slow-moving but vandalizing slope damage. The houses developed cracks within a period of 3–4 days and no casualties were reported. The cracks mapped on the ground were 8 to 10 m in length in one stretch with orientation in the NW–SE direction. The intensity of damage is well pictured through the field photographs showing the ground surface undulating in the form of rise and fall due to downslope movement (Fig. 18.4). There were around 9–10 houses which were completely destructed to the extent that the resettlement is impossible on the same slope.

Fig. 18.2 Field photographs showing water logging at the toe of the damaged slope in the study area

Fig. 18.3 Field photographs showing lateral cracks across the deformed slope

Fig. 18.4 Field photographs A and B showing the damage effects of the landslide on settlements

18.3 Materials and Method

The DInSAR approach has been applied to gauge the surface movements by utilizing two passes in the interferometric methods. The DInSAR was utilized to track the phase difference of the complex SAR signals between the acquisitions from two unique passes. The dataset has been picked such that the distinction in orbital position would be insignificant. An interferogram was created by the superimposing of two SAR pictures by deduction of phase values in every pixel. The interferogram contains a few phase data, such as—the phase difference between the spatial pattern of the data pair, and the phase loss due to the impact of moving target objects which are components of dynamic physical elements (Figs. 18.5 and 18.6). For the soundness of this study, Sentinel-1A—SAR data was gathered from the Alaska Space Facility site (https://vertex.daac.asf.alaska.edu).

18.4 Result and Discussion

D-InSAR technique with two pairs of Sentinel-1A SLC images dated 15th August 2022 and 27th August 2022, respectively, were used to deduce the temporal and spatial evolution of land surface deformation in the study area (Bariara Village, District Kangra). Using this pair of Sentinel-1A data for August 2022, the generated D-InSAR interferogram is shown in Fig. 18.7.

The fringes are shown in Fig. 18.8 and then the phase image was deburst and unwrapped with the help of the Snaphu plugin in SNAP software. The phase difference between the two images was then used to extract the displacement map (Fig. 18.9) to know how much the slope has moved downwards.

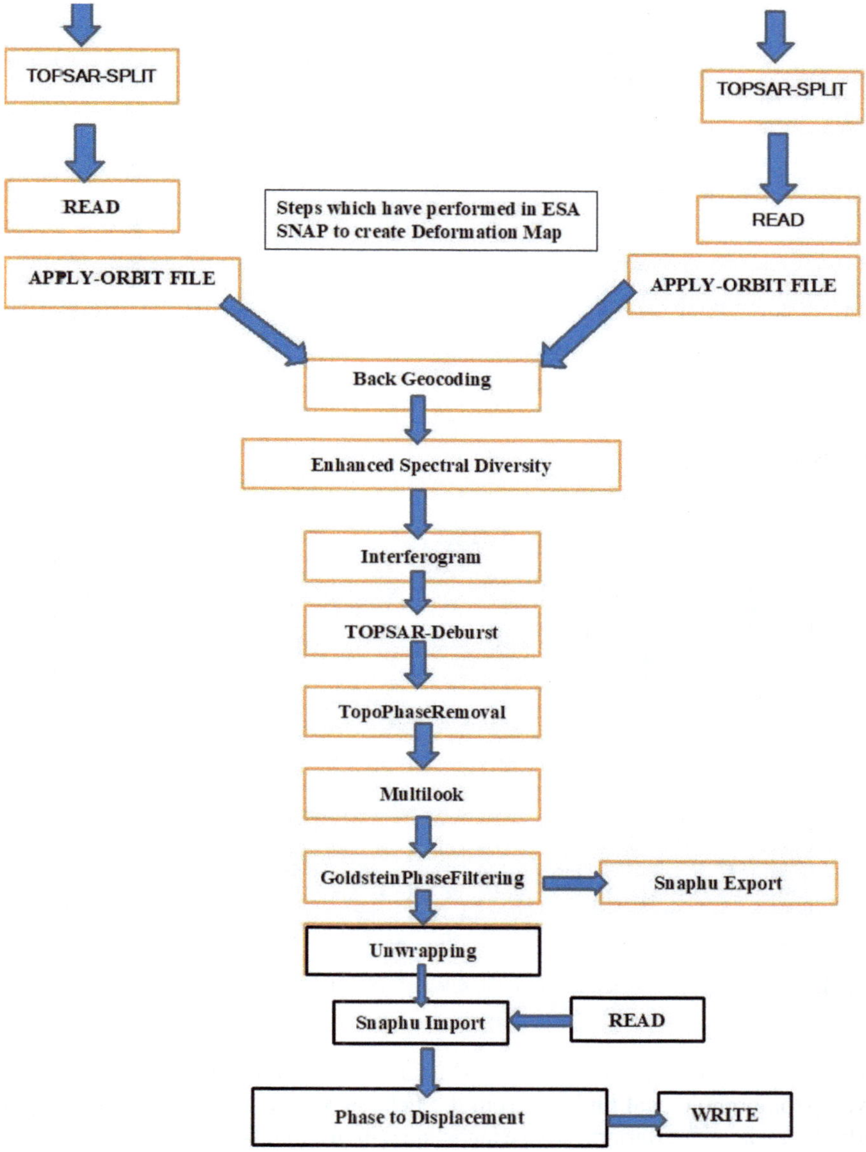

Fig. 18.5 Flowchart depicting the methodology

Sentinel-1A information was adequately used in this study for further analysis of the amount of slope movement. The study has inferred a total downward slope displacement of approximately 1.50 m with a total area of 642.12 m^2 under the landslide debris near the lower middle and toe part of the slope. The pixel displacement between the two imageries (Table 18.1) was calculated by extracting the index values

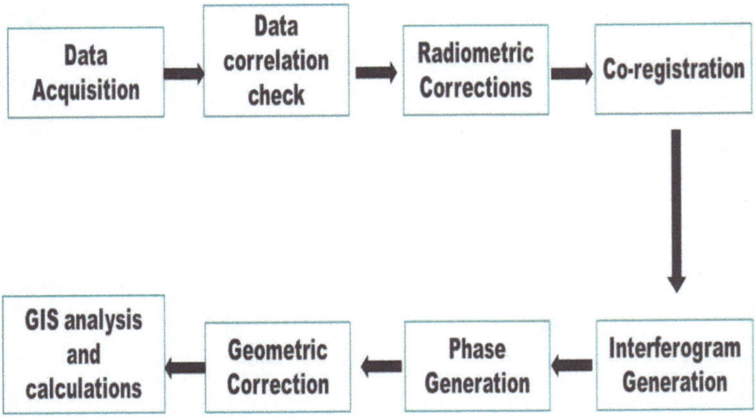

Fig. 18.6 Overall study flow

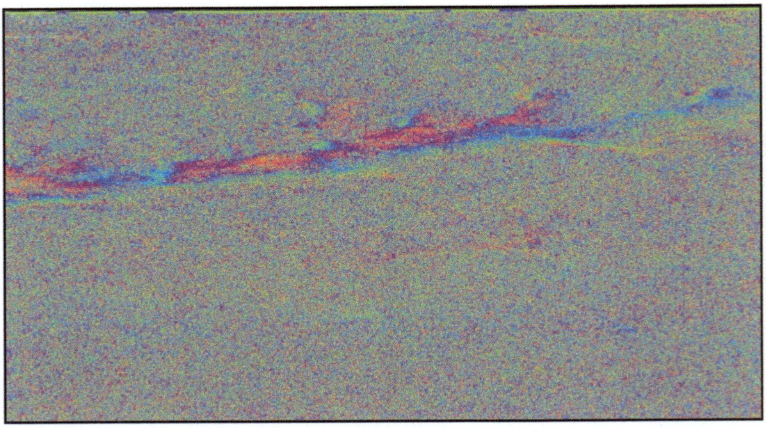

Fig. 18.7 Interferogram generated from the selected dataset

of pixels from the displacement map (Fig. 18.9). The integration of DInSAR data with geomorphological findings from the field survey helped in understanding the advancement of the Bariara Village landslide within a month in the monsoon season. The landslide area falls under District Kangra where the average monthly rainfall recorded for August 2022 was 614.7 mm as per the Indian Meteorological Division (IMD) website mausam.imd.gov.in which was a record high rainfall throughout the year. In the same year, District Kangra witnessed many other landslide episodes in nearby areas of the present landslide zone.

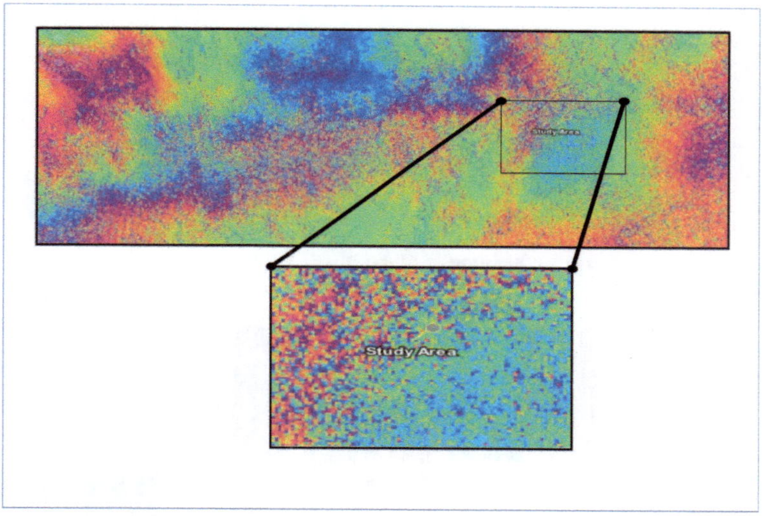

Fig. 18.3 Phase image extracted from the interferogram

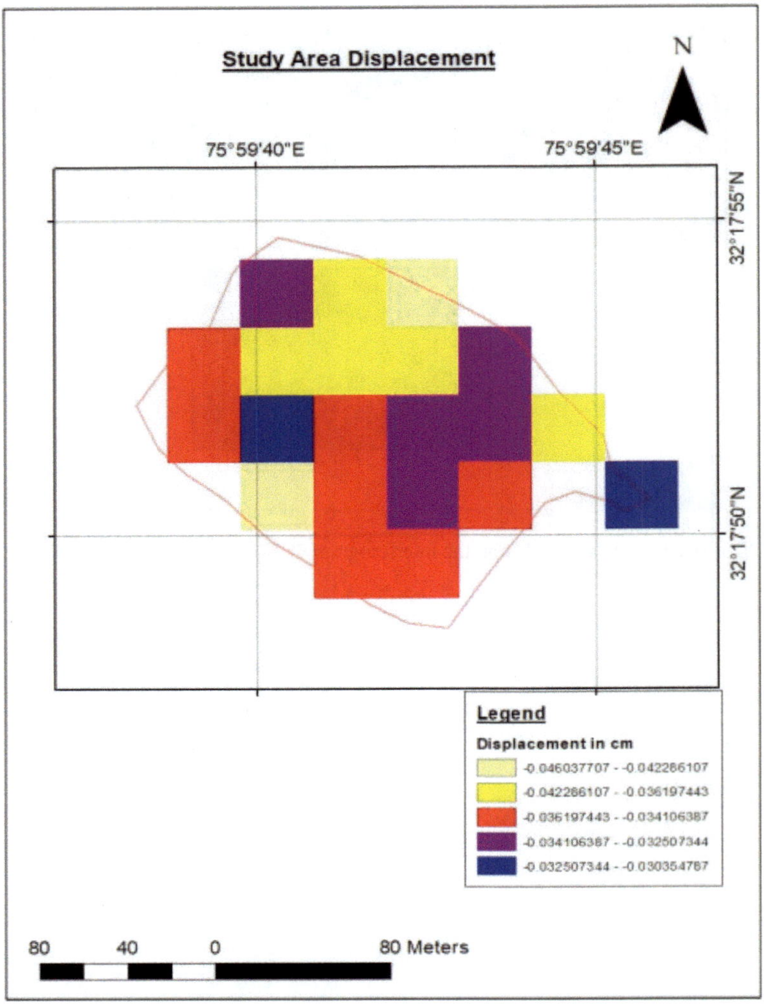

Fig. 18.9 The displacement map of the study area is illustrated as a thematic pixel map

Table 18.1 Spatio-temporal characteristics of Sentinel-1A—SAR Data used

Acquisition date	Master/slave	Direction	Perpendicular baseline [m]	Acquisition mode	Product type
15th August 2022	Master	Descending	0	Interferometric wide (IW) swath	Single looked complex (SLC)
27th August 2022	Slave	Descending	12.5	Interferometric wide (IW) swath	Single looked complex (SLC)

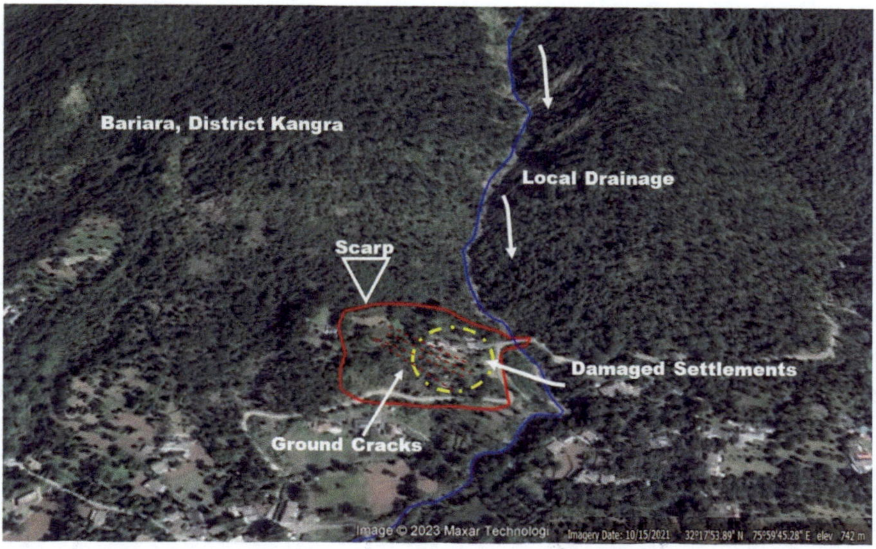

Fig. 18.10 Picture showing the damage extent from landslide event in the study area

18.5 Conclusion

Slope instabilities in valley slopes can be effectively detected and tracked with the use of DInSAR techniques. The kinematics and evolution of the landslide in the study area were understood with the aid of the integration of DInSAR data with traditional geomorphological research in the field. The displacement time series obtained from the DInSAR analysis has illustrated the landslide behaviour and its damage potential which is also depicted from the overall field scenario (Fig. 18.10). This displacement has pointed to more extensive relationships between the movement of the global slope in this region.

Figures 18.11 and 18.12 resulted from the NDVI change analysis using natural colour and infrared band of sentinel 2 imagery respectively to validate the changes due to rain-induced landslide activity where the landslide scar and change in the vegetation cover is evident. The pre-landslide image is from the first week of August 2021 and the post-landslide image is from October 2021. The NDVI change in two months validates the landslide occurrence and also depicts the vulnerability of the surrounding areas to a global slope movement activity due to rain as a triggering agent. The continuous application of DInSAR in conjunction with other monitoring techniques is necessary to identify any pre-failure precursor and critical deformation along road segments to prevent irreversible damage and to support appropriate management of the area. Using the DInSAR techniques, we can study these kinds of disasters and prompt the study area's administration to act right away to prevent serious harm to social life.

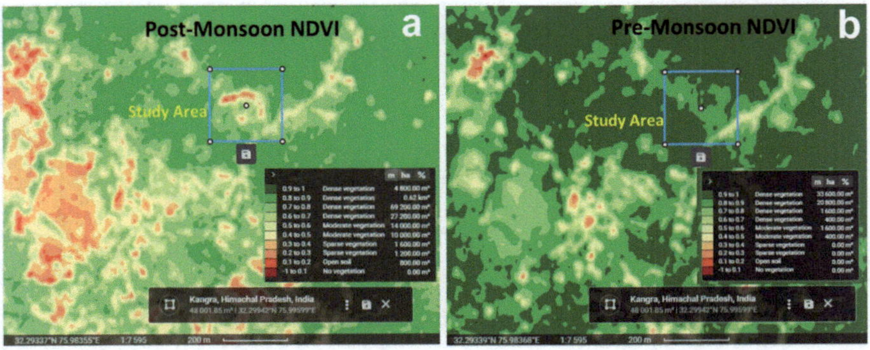

Fig. 18.11 Picture showing the change in NDVI pattern Post-monsoon and Pre-monsoon in the study area (Sentinel 2 image)

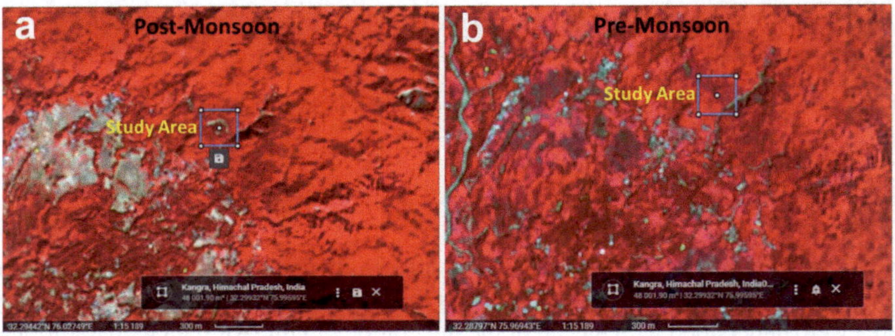

Fig. 18.12 Picture showing the landslide signature (scar) in the study area pre and post-monsoon (*Source* Sentinel 2 Infrared band)

Acknowledgements The authors would like to thank the Amity Institute of Remote Sensing and GIS for administrative and logistic support during the research work. The authors also thank the editors for the smooth submission and review process.

References

Bhasin R, Aslan G, Dehls J (2023) Ground Investigations and detection and monitoring of landslides using Sar interferometry in Gangtok Sikkim Himalaya. Geohazards 4(1):25–39

Casagli N, Intrieri E, Tofani V, Gigli G, Raspini F (2023) Landslide detection, monitoring and prediction with remote-sensing techniques. Nat Rev Earth Environ 4(1):51–64

Crosetto M, Monserrat O, Devanthéry N, Cuevas González M, Barra A, Crippa B (2016) Persistent scatterer interferometry using Sentinel-1 data. Int Arch Photogramm, Remote Sens Spat Inf Sci 41(B7):835–839

Devaraj S, Yarrakula K, Martha TR, Murugesan GP, Vaka DS, Surampudi S, … , Budamala V (2022) Time series SAR interferometry approach for landslide identification in mountainous areas of Western Ghats, India. Journal of Earth System Science 131(2):133

Kumar A, Sharma RK, Mehta BS (2020) Slope stability analysis and mitigation measures for selected landslide sites along NH-205 in Himachal Pradesh, India. J Earth Syst Sci 129:1–14

Kumar R, Anbalagan R (2019) Landslide susceptibility mapping of the Tehri reservoir rim area using the weights of evidence method. J Earth Syst Sci 128:1–18

Liu Z, Qiu H, Ma S, Yang D, Pei Y, Du C, ... , Zhu Y (2021) Surface displacement and topographic change analysis of the Change the landslide on September 14, 2019, China. Landslides 18:1471–1483

Mahajan AK, Sharma S, Patial S et al (2022) A brief address of the causal factors, mechanisms, and the effects of a major landslide in Kangra valley, North-Western Himalaya, India. Arab J Geosci 15:925. https://doi.org/10.1007/s12517-022-10163-w

Mondini AC, Guzzetti F, Chang KT, Monserrat O, Martha TR, Manconi A (2021) Landslide failures detection and mapping using synthetic aperture radar: past, present and future. Earth Sci Rev 216:103574

Niraj KC, Chatterjee RS, Shukla PD (2023) Estimating the period of probable landslide event using advanced D-InSAR technique for time-series deformation study of Kotrupi region. Geomat Nat Haz Risk 14 1 2281245

Paswan AP, Shrivastava AK (2022) Modelling of rainfall-induced landslide: a threshold-based approach. Arab J Geosci 15(8):795

Roy P, Martha TR, Khanna K, Jain N, Kumar KV (2022) Time and path prediction of landslides using InSAR and flow model. Remote Sens Environ 271:112899

Sharma P, Rawat S, Gupta AK (2019) Study and remedy of kotropi landslide in Himachal Pradesh, India. Indian Geotech J 49:603–619

Sharma S, Mahajan AK (2018) Comparative evaluation of GIS-based landslide susceptibility mapping using statistical and heuristic approach for Dharamshala region of Kangra Valley, India. Geoenvironmental Disasters 5:1–16

Chapter 19
Landslide Hazard Risk Assessment Using GIS and Analytical Hierarchy Process (AHP) Approach: Evidence from 2017 Rangamati Hill Tracts Landslide Event, Bangladesh

Biswajit Nath and Anjuman Ara

Abstract A large number of local people and property losses have been reported in the Rangamati Hill Tracts (RHT) of Bangladesh due to the 17 June 2017 Landslide event. This chapter investigated the actual causes of this landslide event which focuses on the coupling effects of earthquake-rainfall-induced landslide activities that occurred in this area. To know this scenario better, the study considered freely available earth observation multi-temporal Landsat satellite scenes for landslide-induced change assessment. In addition, thirteen associated conditioning factors were integrated within the GIS environment by adopting a multicriteria analytical hierarchy process (AHP) approach to prepare a landslide hazard risk (LHR) map of the study area. The study results quantified 3.81, 3.00, and 8.90% of areas as very high-risk, high, and moderate LHR categories. Moreover, the results also revealed that multiple issues (i.e., earthquakes, heavy rainfall, hill cutting, deforestation, and haphazard human settlement construction on the hill slope) are the main catalysts behind this landslide event. Finally, the study recommends integrated Remote sensing, GIS, and AHP techniques for future landslide risk assessment across the globe.

Keywords 2017 Rangamati Landslide · Risk · GIS · Analytical hierarchy process · Earthquake · Rainfall

B. Nath (✉) · A. Ara
Department of Geography and Environmental Studies, Faculty of Biological Sciences, University of Chittagong, Chittagong 4331, Bangladesh
e-mail: nath.gis79@cu.ac.bd

A. Ara
e-mail: aara23@student.aau.dk

A. Ara
Department of the Built Environment, Faculty of Engineering and Science, Aalborg University, 9220 Aalborg East, Denmark

G. K. Panda et al. (eds.), *Landslide: Susceptibility, Risk Assessment and Sustainability*, Advances in Natural and Technological Hazards Research 52,
https://doi.org/10.1007/978-3-031-56591-5_19

19.1 Introduction

In world wide natural disasters research, it is clearly visible that landslide activity in recent time is increasing and due to its intensification it is identified as the third most common category (Chen et al. 2017). This phenomenon is triggering due to extensive development activities in the hilly areas, tourism activities, unplanned construction, and high population growth (Guzzetti et al. 2012). From the perspective of Bangladesh, landslide activity has also been observed in the Chittagong City Corporation (Ahmed and Dewan 2017), and the Chittagong Hill Tracts regions (Rabby and Li 2019; Abedin et al. 2020) in recent times. Therefore, it is considered the most common and devastating disaster in Bangladesh. The south-eastern hilly region, especially Rangamati Hill Tracts (RHT) (commonly known as the indigenous hilly district) experienced a chronological landslide event and moderate magnitude earthquake events in recent history. The most recent devastating landslide occurred on 12 June 2017 which claimed the lives of 152 people and severely destroyed 6,000 dwellings, roads, telecommunication systems, and power supply, with an estimated economic loss of about USD 223 million (UNRC 2017; Bajracharya and Maharjan 2018; Abedin et al. 2020). Among the entire RHT, the Vedvedi neighborhood of Rangamati Sadar occupies 1130 families which were affected due to its direct consequences (Chowdhury 2017).

Landslide identification involves the determination of the spatial extent of a particular type of landslide, its volume, and the probability of its occurrences over some time and space (Cruden and Varnes 1996). After the recent landslide occurrence, landslide susceptibility mapping of the RHT has become indispensable to identify the reason and changing conditions of landslide hazard-prone areas. The recent multiple landslide events in RHT are identified as a major environmental problem that hinders to achievement of sustainable development of hilly areas, and if this continues, it will be a major threat of landslide occurrence in the future also. Due to these uncertainties, this chapter went through the past literature about landslide activity in RHT and observed a lack of research work on the present topic.

To minimize the research gap, risk areas delineation in the entire RHT and existing landslide-induced change measurements are the prime focuses of the present investigation. This chapter shortlisted before and after landslide scenarios to know the environmental changes in the study areas. The number of landslides on 12 June 2017 in Rangamati was staggering which has motivated us to focus on this event with the unique idea. Therefore, the specific objectives of the present study were, (i) to identify the landslides by considering the associated risk factors, (ii) to map hazard risk areas through a multicriteria-based analytical hierarchy process (AHP) approach, and (iii) to assess landslide induced environmental changes of the RHT.

This chapter focuses in-depth study of the 2017 RHT landslide event by considering multi-temporal earth observation Landsat satellite imageries and available secondary data sources. The study result was later checked with the immediate emergency GPS survey data conducted by ICIMOD, Nepal (Bajracharya and Maharjan

2018) as the field evidence of the occurred landslides events. The results generated through the geo-integrated techniques (GIS, and AHP approach) along with the observed linkage of landslide event occurrences with immediate past earthquakes make this study unique. Therefore, the present findings can be beneficial to planners and decision-makers to mitigate landslide activity in hilly regions in a more sustainable way. In addition, this chapter finds the method suitable for detecting the landslide in advance before the occurrences of landslide events.

19.1.1 Basics of Landslide Inventories

A landslide is a geomorphic process influenced by many geomorphic features such as slope, drainage, weak geology, faults, folds, lithology, relief, seismic and tectonic activities, etc. Besides natural causes, intense and unplanned human activities also trigger landslide incidents. Wang et al. (2010) identified that landslide activities are down-slope movements of masses of rock debris because of the losses in their shearing strengths. It may occur suddenly in hilly terrain during earthquakes or in monsoon season. According to Crozier (1986), landslides are categorized into two basic types, where the first category includes slides that occurred due to mechanical, hydrostatic pressure and erosion, and the second category considers physical and/or chemical properties changes of the soil.

Swanston and Swanson (1976) stated that forest destruction and road construction initiate landslide activity. However, Bhandari and Gupta (1985) blame the man for his interference in the hill ecosystem that leads to a landslide. Further, Valdiya's (1987) study shows that unplanned road construction causes destructive landslides. Cruden (2003) and Varnes (1978) observed that landslide activity occurred due to slope failure, and geological, morphological, physical, and human causes. The problems of landslide, subsidence, and erosion are quite common in the hilly terrain due to the combination of several factors like the geological movements in fragile geology and high steep relief, lithology, fracture patterns, increased anthropogenic activities, water seepage, vegetation cover, weather and climatic changes (Basu and Pal 2020). Moreover, several authors across the globe observed that landslide activity is common in hilly and mountainous terrain especially during and immediately after the rainy season (Bartarya et al. 1989; Bartarya and Valdiya 1989; Bartarya 1991; Gupta and Bist 2004).

Following the above pieces of literature, we have later moved directly to the RHT to search past major landslide events. Due to the sheer shortfall of this kind of study in this RHT, we have relied on the event background information, where the survey was carried out by the ICIMOD, Nepal. During our field investigation time in the month of April–May 2019 in RHT, there was no landslide event was observed, therefore the pictures of the landslide event were taken from published reports and other internet sources (see Fig. 19.1).

Fig. 19.1 Landslides event of RHT that occurred on 12th June 2017 (*Source* Dhaka Tribune, photo captured on 15 June 2017)

19.1.2 Application of GIS and Remote Sensing on Landslide Hazard Mapping

For nearly four decades, several researchers and scientists attempted to map landslide hazard zones through cartographic methods. In this regard, remote sensing and modern GIS techniques help to assess landslide activity. GIS gives multiple abilities to users to handle and analyze the spatial data more efficiently and accurately as well as new data integration with the existing ones. GIS is defined as a computer-based system with excellent geospatial data processing capacity (qualitative and quantitative) as well as its practical applicability in natural disaster assessment (Carrara et al. 1982, 1991, 1992, 1995).

Further, the application of GPS is also important for locating landslide sites. Crozier and Glade (2005) observed that GIS plays an important role in geospatial modeling, hazards, risk mapping, and planning. Several authors have conducted research in studying rainfall-based slope failure (Bhandary et al. 2013), landslide susceptibility mapping (Sarkar and Kanungo 2004; Van Westen 1993; Van Westen et al. 2006, 2008), spatial landslide prediction by employing statistical models (Basu and Pal 2020) and landslide hazard assessment using deterministic modeling (Terlien et al. 1995; Zêzere 2002). Therefore, GIS and Remote sensing integrated techniques emerged as a powerful tool in mapping and monitoring the earth's surface.

In the perspective of qualitative analysis, Saha et al. (2002) considered subjective decision principles based on the weights and the expert ratings applied to prepare thematic maps and overlay analysis. In this chapter, the fundamental objective of the study is to identify the risk areas and change assessment. Landslide inventory, susceptibility, and hazard risk map generation are helpful for different policymakers for proper sustainable LULC planning, and LHR mitigation that help them to maintain and achieve particular Sustainable Development Goals (SDG) within the specific period set by the UN 2030 SDGs.

19.1.3 Landslide-Inducing Factors

Crozier (1986) classified three responsible factors (i.e. the preparatory factors, triggering factors, and controlling factors) that caused the landslide activity in any region. It involves lithology of slope material, geomorphic and structural features, geology, hydrogeologic condition, and vegetation of the region. Landslide is identified as one of the most vulnerable environmental issues at present (Lee and Pradhan 2006). However, other researchers (Van Westen 2000; Horelli 2005) observed that anthropogenic activities (i.e., unplanned urbanization and deforestation) in combination with geological and geomorphological aspects contributed a lot to the occurrence of the landslide. In the recent past, Sarwar (2008) identified landslides as a hazard as it has widely impacted in several years (i.e., 2007, 2008, 2010, 2012, and 2015) in different south-eastern regions of Bangladesh. The recent landslide disaster that took place in RHT on 12 May 2017 due to torrential rainfall (UNPO 2017) has many miseries with the loss of lives and properties (UNRC 2017). Moreover, the recent study by Khatun et al. (2023) identified climatic variabilities such as temperature increase and extreme rainfalls on consecutive days are the major factors to triggered the landslide event in RHT.

19.2 Materials and Methods

19.2.1 Information on the Study Area

Rangamati is considered as a hill district in the south-eastern part of the Chittagong Division, Bangladesh and the town of Rangamati serves as the headquarters of the district. The district is bordered by the Tripura state of India to the north, Bandarban District to the south, Mizoram State of India and Chin State of Myanmar to the east, and Khagrachari and Chittagong Districts to the west. The geographical extension of RHT lies in between 22° 27′ and 23° 44′ north latitudes and 91° 56′ and 92° 33′ east longitudes. The location of the study area is shown in Fig. 19.2.

The area is hilly in origin and covered by thick vegetation. The RHT is physically constituted of earthen and rocky hills, forests, waterfalls, rivulets (chara), and river valleys. The area of the district is 5759.16 km^2 of which 1292 km^2 is riverine and 4825 km^2 is under forest vegetation. The five rivers, namely Chengi, Maini, Karnaphuli, and Kasalong located in the north and Raingkhiong in the south. The study area belongs to the tropical climate and monsoon rainfall that occurs in most months of the year. The average temperature in RHT is 26.2 °C and the rainfall is around 2609 mm per year.

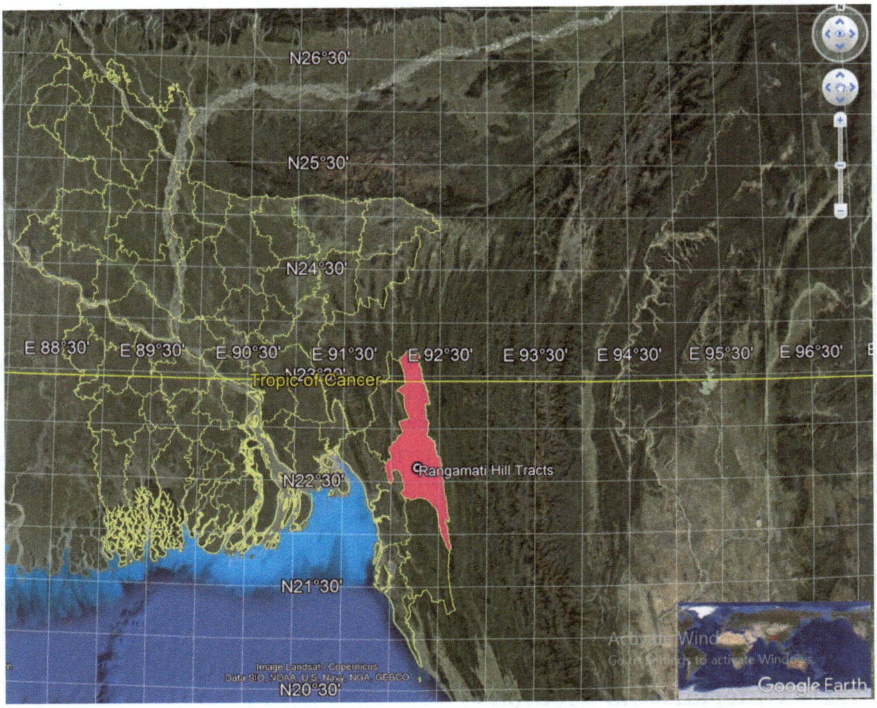

Fig. 19.2 Location of RHT in Bangladesh and the study area (RHT district boundary as Google Earth kml file) highlighted in pink color

19.2.2 Data Collection and Processing Methods

The study includes secondary data sources. The study used various data methods and the landslide was identified by Google Earth through visual inspection including landslide survey data carried out by ICIMOD, Nepal and temporal data of before, after, and recent Landsat Satellite imageries were used for event investigation. Different types of data have been used in this study such as *rainfall*, *land use patterns*, *Digital Elevation Model (DEM)*, etc. The DEM data was acquired from the USGS Earth Explorer archive as an ASTER image with 1-arc second (30-m resolution), which was later used to derive the *slope*, *aspect*, and *elevation* of the area.

Digital Geologic and geophysical data of Bangladesh were downloaded from Cat alog.data.gov/Department of the Interior. The geological faults and regional geology were subsets based on the study area boundary shape file. The soil information of the study area was clipped from the Digital Soil Map of the World (DSMW)-ESRI shape file format (updated on 2011-03-15), available in the FAO/GeoNetwork domain. The entire analyses have undergone many processes which were performed through multi-software integration techniques such as ENVI 5.3, Catalyst Earth (former PCI Geomatica), and ArcGIS 10.8, respectively. The flowchart of the methodology

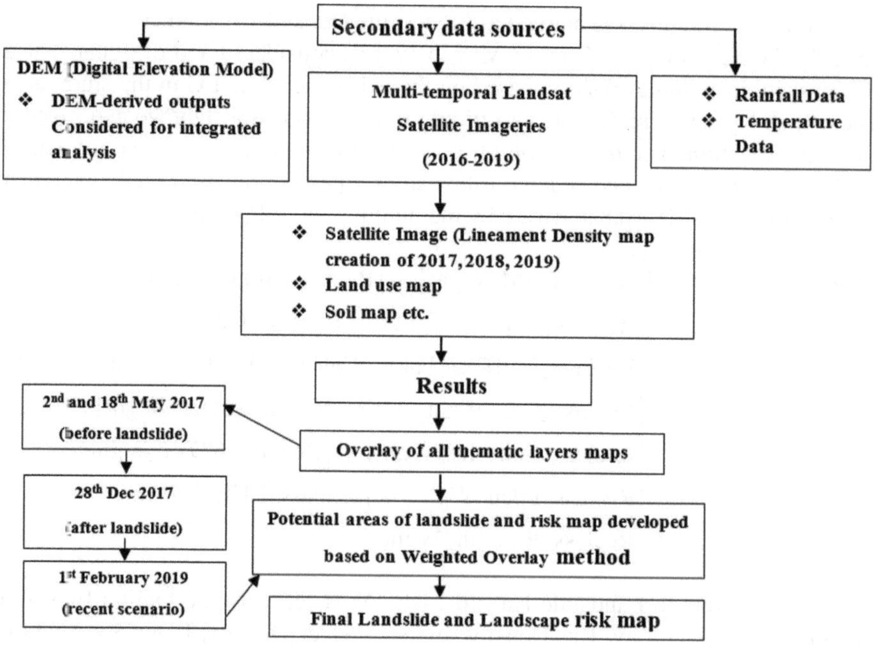

Fig. 19.3 Flowchart of the methodology used in this study

followed in this study is shown in Fig. 19.3, and next the details processes of each method are presented systematically.

For structural analysis, *lineaments* were retrieved digitally by using a user-defined value of LINE algorithm parameters of CATALYST Earth (former PCI Geomatics) software. Various computer-aided methods, such as edge detection, thresholding, and curve extraction steps, were carried out over Principal Component image analysis to derive an image referred to as PCA image (i.e., best band selection) of multi-temporal Landsat 8 OLI scenes. In connection to this, six parameters with user-defined values of LINE Algorithm such as RADI-(10), GTHR-(15), LTHR-(20), FTHR-(3), ATHR-(15), and DTHR-(10) were considered for lineament feature extraction that saved into a shape file, later imported in ArcGIS 10.8 software for line splitting at vertices, where the compound line split into simple lines, and final lineaments considered in the Lineament density (LD) mapping. This has been carried out in the model builder of ArcGIS 10.8 software.

Furthermore, to detect the vegetation of the area, land use patterns are considered and the Normalized Difference Vegetation Index (NDVI) method is used to detect the vegetation status at different time frames. The MCE-AHP is a semi-quantitative method in which decisions are taken using weights through pair-wise relative comparisons (Saaty 1980). At this stage, calculations of AHP were done in a spreadsheet environment by assigning the relative weights (5, 4, 3, 2, 1) for individual layers.

In the relative weights, 5 is marked as high influence, and 1 is marked as low influence. In the next, the LULC of 18 May 2017 was considered and designated with risk priority in a similar fashion based on the existence of LULC in the study area. Then, each layer was reclassified according to the assigned weightage and saved in a specific location. The integration of the various factors that may have influenced the landslide events was considered followed by a procedure based on the weighted linear sum in ArcGIS 10.8 software by maintaining the following equation:

$$\begin{aligned} \text{LHR}[\text{T\_Weight}] = {} & [\text{Reclass\_DEM\_Wght}] + [\text{Reclass\_Slope\_Wght}] \\ & + [\text{Reclass\_Aspect\_Wght}] + [\text{Reclass\_Geol\_Wght}] \\ & + [\text{Reclass\_BuffDist\_Fault\_Wght}] + [\text{Reclass\_Rivdensity\_Wght}] \\ & + [\text{Reclass\_Soil\_Wght}] + [\text{Reclass\_RoadDen\_Wght}] \\ & + [\text{Reclass\_Ldensity\_Wght}] + [\text{Reclass\_NDVI\_Wght}] \\ & + [\text{Reclass\_Fden\_Wght}] + [\text{Reclass\_LULC\_Wght}] \\ & + [\text{Reclass\_Rainfall\_Wght}] \end{aligned}$$

where LHR refers to Landslide Hazard Risk, DEM refers to the Digital Elevation Model, Geol is Geology, Riv Density refers to River Density, LDensity refers to Lineament Density, NDVI refers to Normalized Difference Vegetation Index, Fden represents Fault Density, LULC refers to Land use and Land cover, and others are mentioned in the above equation.

Finally, all the layers are integrated into the GIS environment using relative weights. The landslide hazard risk (LHR) is calculated and maps of LHR and environmental changes are prepared based on the assigned weighted factor of individual layers with priority ranking related to landslide importance, generated based on the weighted overlay method of the Multiple Criteria-Analytical Hierarchy Process (MC-AHP) approach.

Based on the layer integration technique, the total weighted score was calculated and for quality visualization, the risk-weighted score was further classified into five class categories i.e., *very high risk*, *high risk*, *medium risk*, *low risk*, and *very low* category. This has been performed in the ArcGIS software following the standard deviation method (1 Std Dev). Based on the raster image, pixel-based vectorization was performed and the area was calculated in square kilometers. Deep green to Red color intensification was selected to represent the five risk class categories of the LHR map of the RHT.

19.3 Results and Discussion

Various multiple factors are considered in LHR mapping and change assessment. According to the Geological Survey of India (GSI) scientists, hazard zone mapping has been done based on the past occurrence of landslide events. The study considered 13 important factors that may influence the LHR area mapping by adopting the multiple criteria evaluation technique of the Analytical Hierarchy Process (AHP) approach and this has been done using ArcGIS 10.8 software.

19.3.1 Multiple Geo-environmental Factors Analysis

19.3.1.1 DEM, Slope, and Aspect Detection

To monitor landslide activity in a specific region, DEM (30 m resolution) data from USGS Earth Explorer (available at: https://pubs.er.usgs.gov) is considered an excellent source that helps to derive topographic attributes. In RHT, DEM shows very high topographic attributes area ranges (473–1003 m) and very low areas (0–82 m) which are considered as very susceptible or landslide occurrence areas. Slope and aspect are derived from DEM which are shown systematically in Fig. 19.4a–c.

For slope, the very highly susceptible to landslide areas of RHT are extracted between 26.05–67.79 and the very low areas are extracted between 0–5.58. On the other hand, the highest aspect is recorded in the north which scores between 337.5–360. Therefore, these aspects are very susceptible to landslide activity.

19.3.1.2 Geology

Geology affects landslides in some ways such as lithology, seismicity, and fault. The existing geology of RHT is Neogene covered by the whole water bodies referred to as Kaptai Lake (a prominent tourist spot). The geology of RHT is subtracted from the shapefile of the Geologic map of South Asia. The Geology of the RHT is shown in Fig. 19.5.

19.3.1.3 Fault Based Buffer and Soil Type

Due to high influence of closeness of the two major faults and two recent earthquakes of M_W 4.7 struck in the vicinity of the area, we have created fault-based buffer distance to visualize its impacted areas. Here in this case, the area close to faults is considered a very high-risk category and the distance far from itis considered as less risk in category.

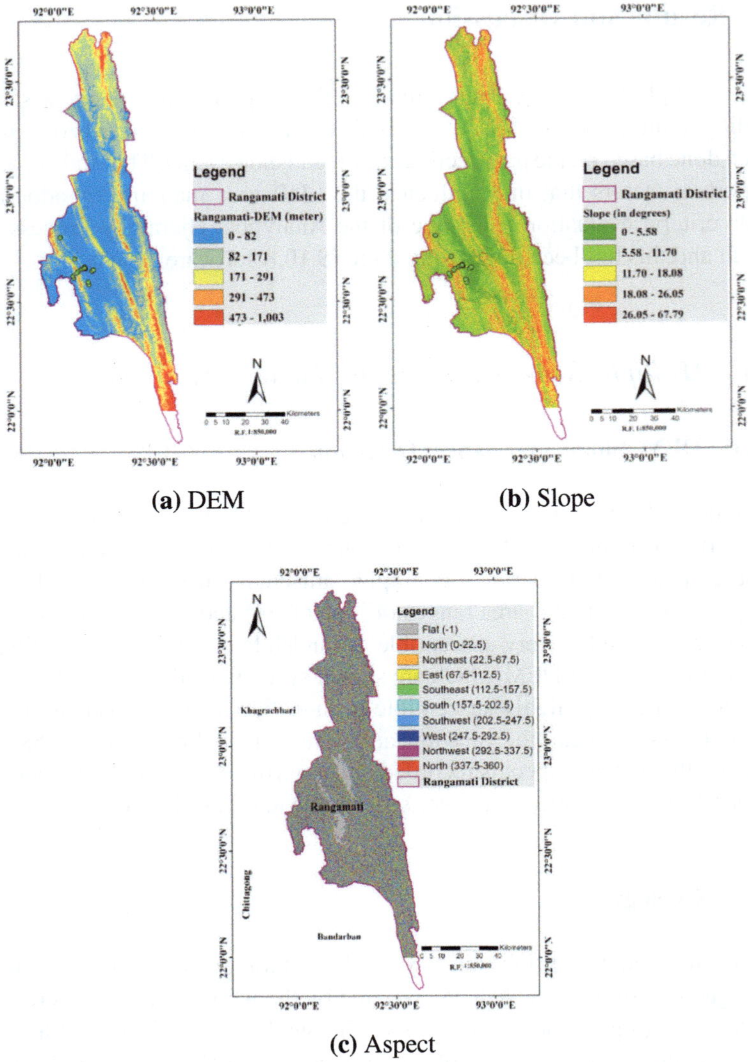

Fig. 19.4 DEM, slope, aspect mapping of RHT

On the other hand, an investigation of the soil type of the area has been taken from Soil Grids V0. 5.3 (*Source: ISRIC World Soil information.org.*). Buffer fault and Soil type are shown in Fig. 19.6a, b. In this study, the buffer fault distance is considered 25 km at the very high-risk areas and 5 km at the very low-risk areas (Fig. 19.6a). The soil type investigation shows that Bd61-2c soils are very highly susceptible to landslide risk at RHT (Fig. 19.6b).

Fig. 19.5 Geology of RHT

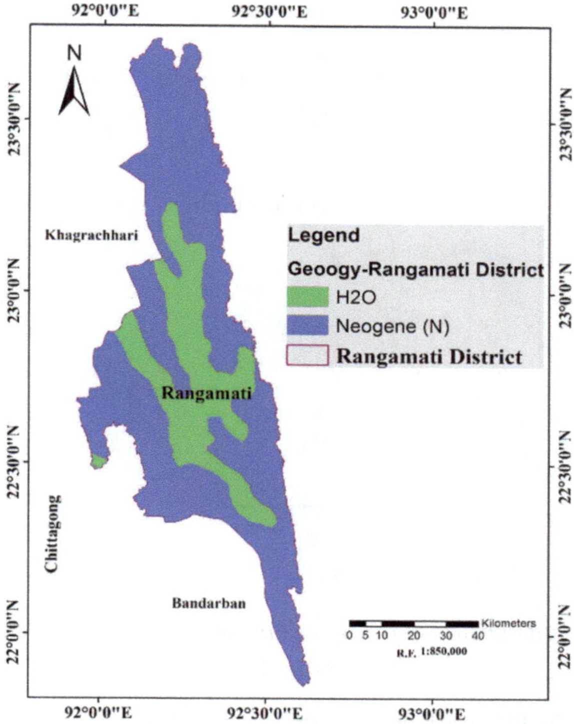

19.3.1.4 Road and Water Bodies Density

In the case of road distance, the highest density of road in RHT is 104.21–187.13 km/km^2. The study observed its direct relation to the landslide event. On the other hand, when water bodies' density reaches 7.85–13.09 km/km^2 are considered at a greater risk of landslide occurrence. The road density and water bodies density maps are shown in Fig. 19.7a, b.

19.3.1.5 Lineament Density (LD) Map

The lineaments are linear features on the earth's surface that reflect general surface manifestations of underground fractures. The LD map is created after removing artifacts in the ArcGIS 10.8 model builder environment based on the derived split final lineament data layers. This has been derived from PCI images generated from original Landsat atmospheric corrected reflectance satellite images (2017–2019) (Fig. 19.8a–d).

In this case, LD represents the total length of the linements in a unit area. During this study, a high fracturization area has been identified at RHT through LD analysis. The results found that *before the landslide* event on 2 May and 18 May 2017, high

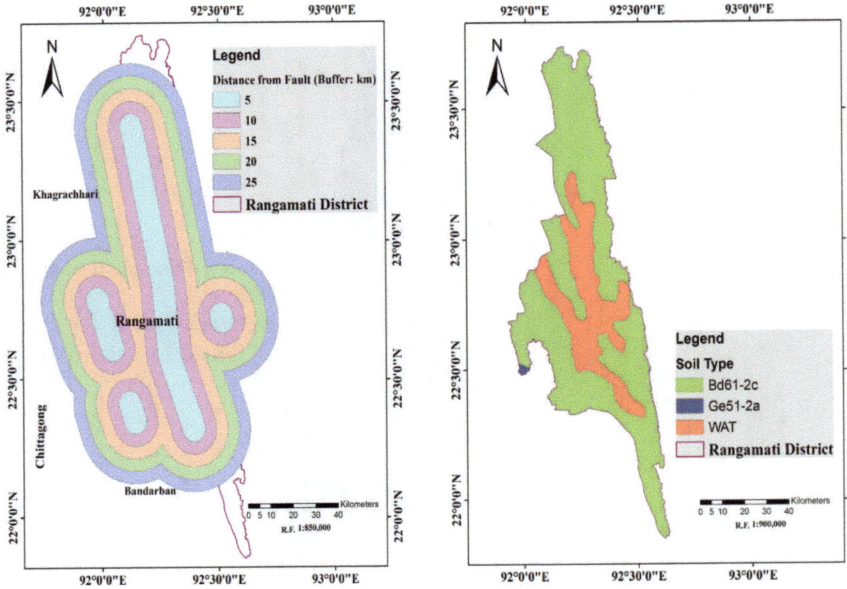

***(a)** Distance from fault based on buffer creation*

***(b)** Soil Type*

Fig. 19.6 **a, b** Buffer fault and soil type of RHT

fracturization areas were observed where LD was observed at 1.00–1.79 km/km^2 (Fig. 19.8a), and 0.82–1.59 km/km^2 (Fig. 19.8b). However, after the landslide event on 28 Dec 2017, a high fracturization area was also observed in the range of LD 1.07–1.84 km/km^2 (Fig. 19.8c). On the other hand, the LD result on 1st Feb 2019 shows a high rock fracturization area in the range of 0.83–1.13 km/km^2 (Fig. 19.8d).

19.3.1.6 NDVI

The NDVI map consists of five classes, which are derived from the Landsat-8 Operation Land Imager (OLI) satellite images using ENVI 5.3 software. These images were acquired at different times referring to *before the landslide* on 2 May and 18 May 2017 and *after the landslide* on 28 Dec 2017 (Fig. 19.9a, b). The calculation has been done using the Eq. 19.1:

$$\text{NDVI} = (\text{NIR} - \text{RED})/(\text{NIR} + \text{RED}) \qquad (19.1)$$

The results of NDVI outputs suggest that *before the landslide* (2 May 2017), the lowest NDVI value was −0.41 to −0.05, and the highest value was observed 0.62–0.79 (Fig. 19.9c), whereas on 18 May 2017, the lowest NDVI value was −0.35 to

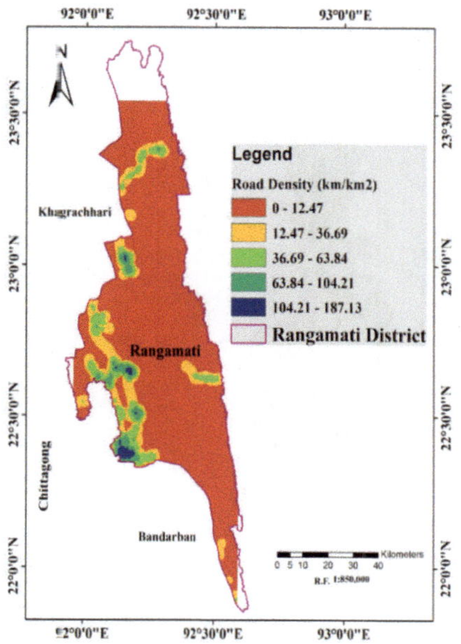

(a) Road Density

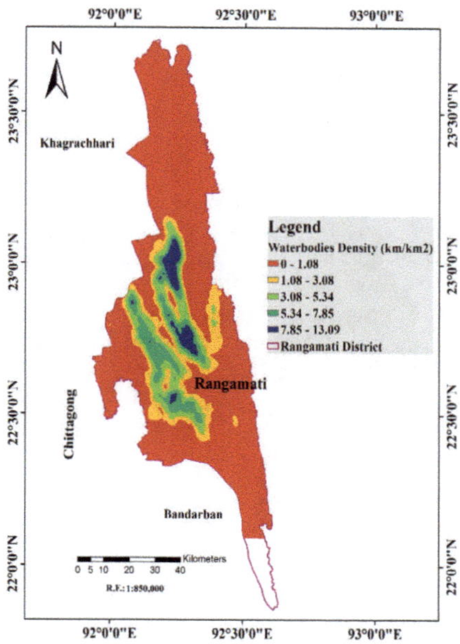

(b) Water bodies Density

Fig. 19.7 **a**, **b** Road density and water bodies density map of RHT

0.59 and the highest value was observed 0.72–0.80 (Fig. 19.9a, d). However, after the landslide event on 28th Dec 2017, the lowest NDVI value was observed from −1 to −0.17 (Fig. 19.9b) and the highest value was observed from 0.69 to 0.86 which is shown sequentially in Fig. 19.9e.

19.3.1.7 Land Use

Land use is also identified as a key factor related to landslide occurrences. The landslide occurrence is inversely related to the vegetation density. Barren land is more prone to landslide activity as compared to the forest area (Kanungo et al. 2009; Saha et al. 2005). The LU has been generated based on a single Landsat image (17 May 2017) in different categories such as *water bodies, agricultural land, forests with dense vegetation, barren land, settlements, and hills with less vegetation* (Fig. 19.10). According to the investigation, the LU area consists of a mixture type where a very high-risk area is consistent with less vegetation which is very much susceptible to landslides.

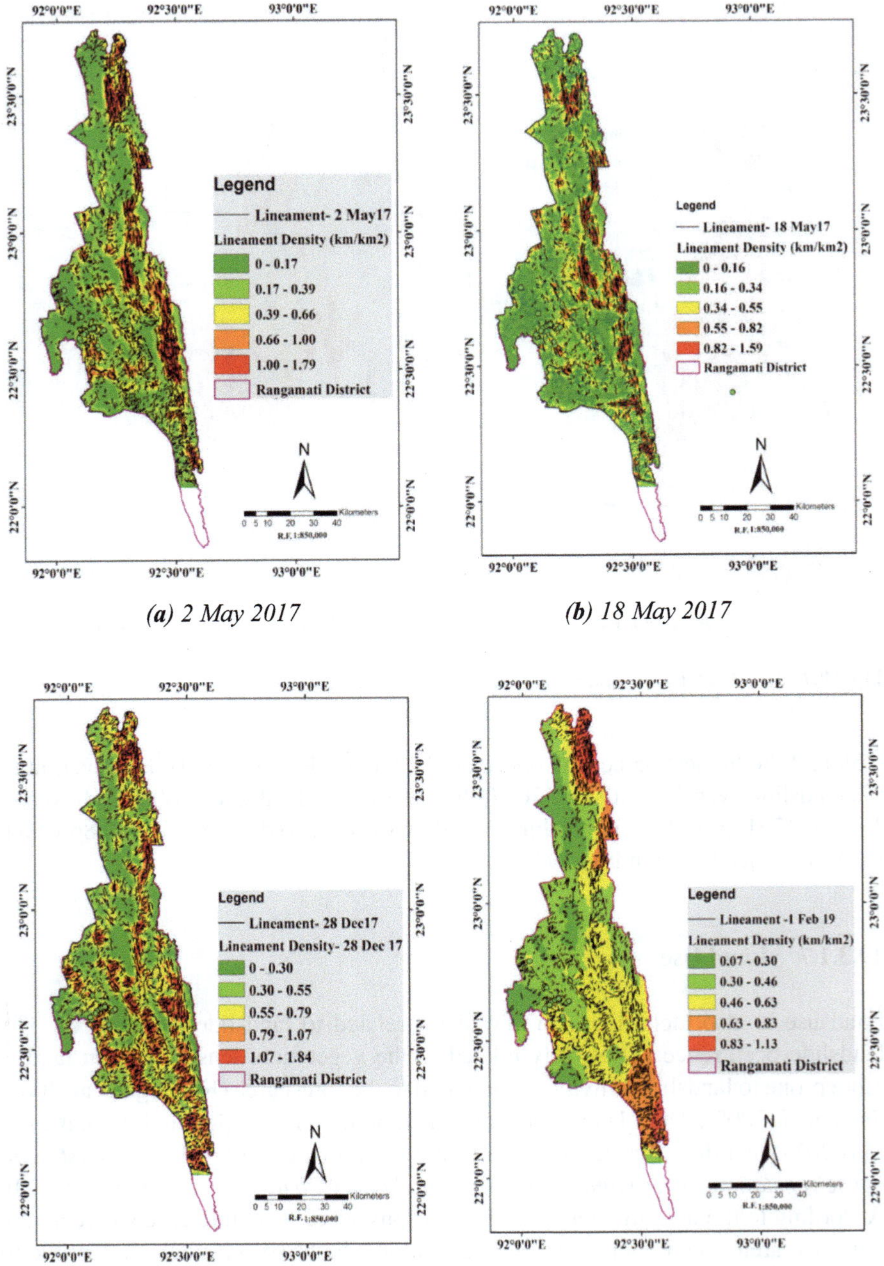

(a) 2 May 2017

(b) 18 May 2017

(c) 28 Dec 2017

(d) 1 Feb 2019

Fig. 19.8 a, b, c, d Lineament and density map creation based on before, immediate before, after, and recent scenario observations as evident through satellite images (2017–2019)

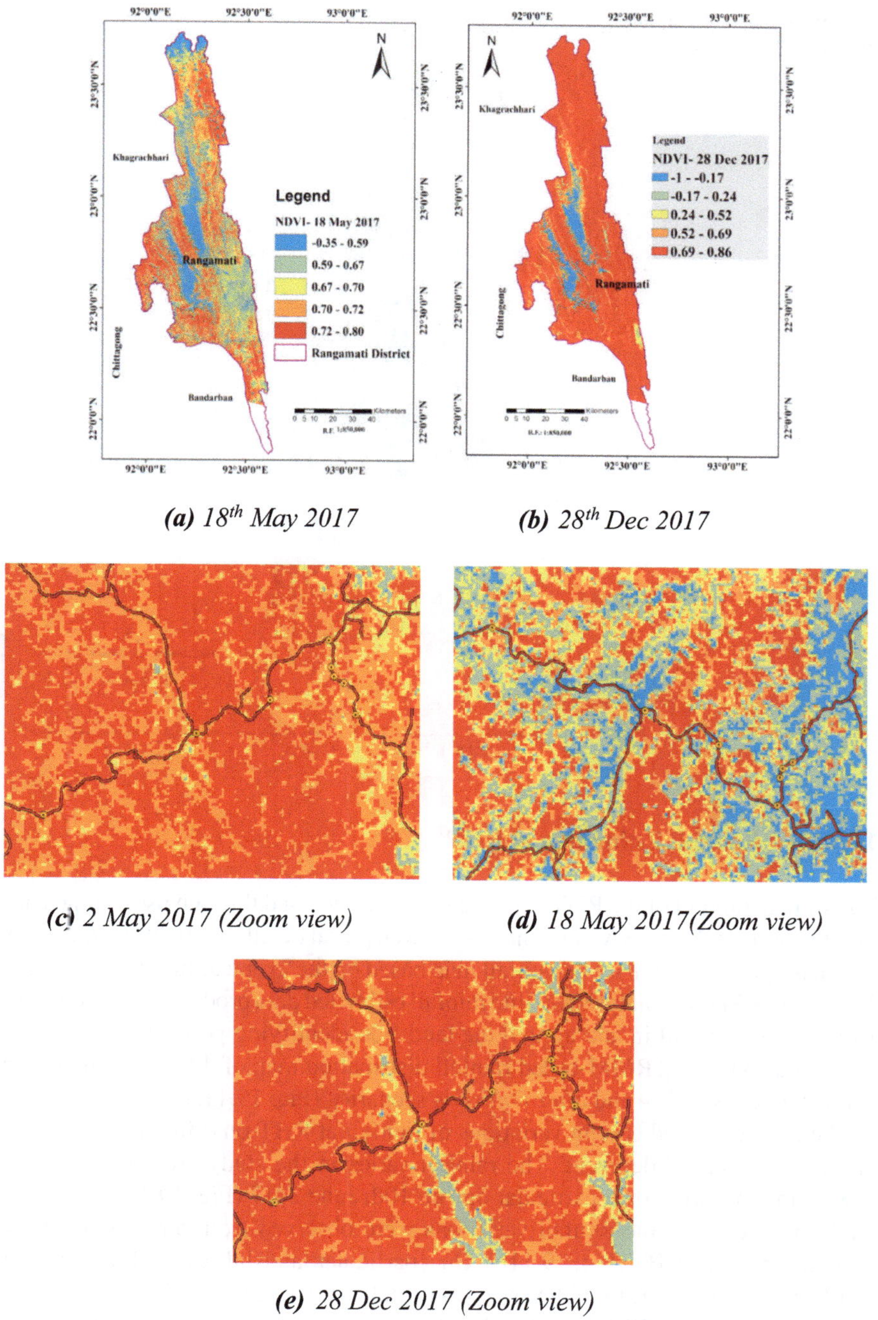

(a) 18th May 2017 *(b) 28th Dec 2017*

(c) 2 May 2017 (Zoom view) *(d) 18 May 2017(Zoom view)*

(e) 28 Dec 2017 (Zoom view)

Fig. 19.9 NDVI-based vegetation change detection of RHT; **a** 18th May 2017 (before landslide), **b** 28th Dec 2017 (after landside), **c** zooming view of NDVI on 2 May 2017, **d** zooming view of NDVI on 18 May 2017, and **e** zooming view of NDVI on 28 Dec 2017

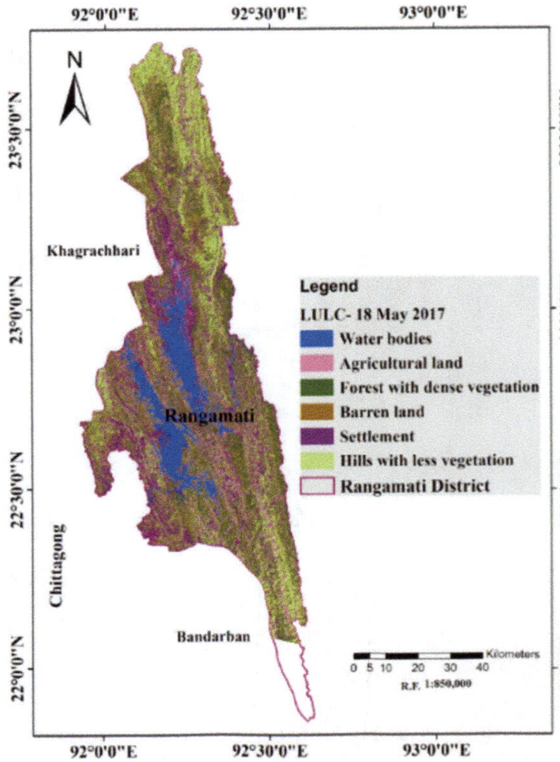

Fig. 19.10 Land use of the study area 18 May 2017

19.3.1.8 Rainfall Estimation at RHT

To estimate rainfall of the RHT, the data has been accessed through user defined area of interest from the NASA-Giovanni v 4.36 web portal (available at: https://giovanni.gsfc.nasa.gov/giovanni/). The rainfall map based on the Tropical Rainfall Measuring Mission (TRMM) reanalysis satellite rainfall estimated data products observed subtle variations of rainfall in and around the landslide identifying points (see green dots in all images). The TRMM-based rainfall distribution map of different time frames is referred to as *before* and *after landslide* scenarios (Fig. 19.11a–d).

We have observed certain variations of rainfall at RHT on different days during and after the landslide event. The rainfall varies in the study area from 36.36 to 178.14 mm occurred on a particular day on 10th June 2017 (Fig. 19.11a). Whereas, on 11th June 2017, the rainfall variations were observed from 5.82 to 48.45 mm rainfall across the RHT (Fig. 19.11b). On a particular landslide event day (12th June 2017) (Fig. 19.11c), the rainfall varied from 76.38 to 264.33 mm in the study area. On 13th June 2017 (Fig. 19.11d), immediately after the landslide event, the rainfall was observed from 41.20 to 134.52 mm which is half of the landslide event day. Due to excessive rainfall on consecutive days, a series of landslide events have been observed

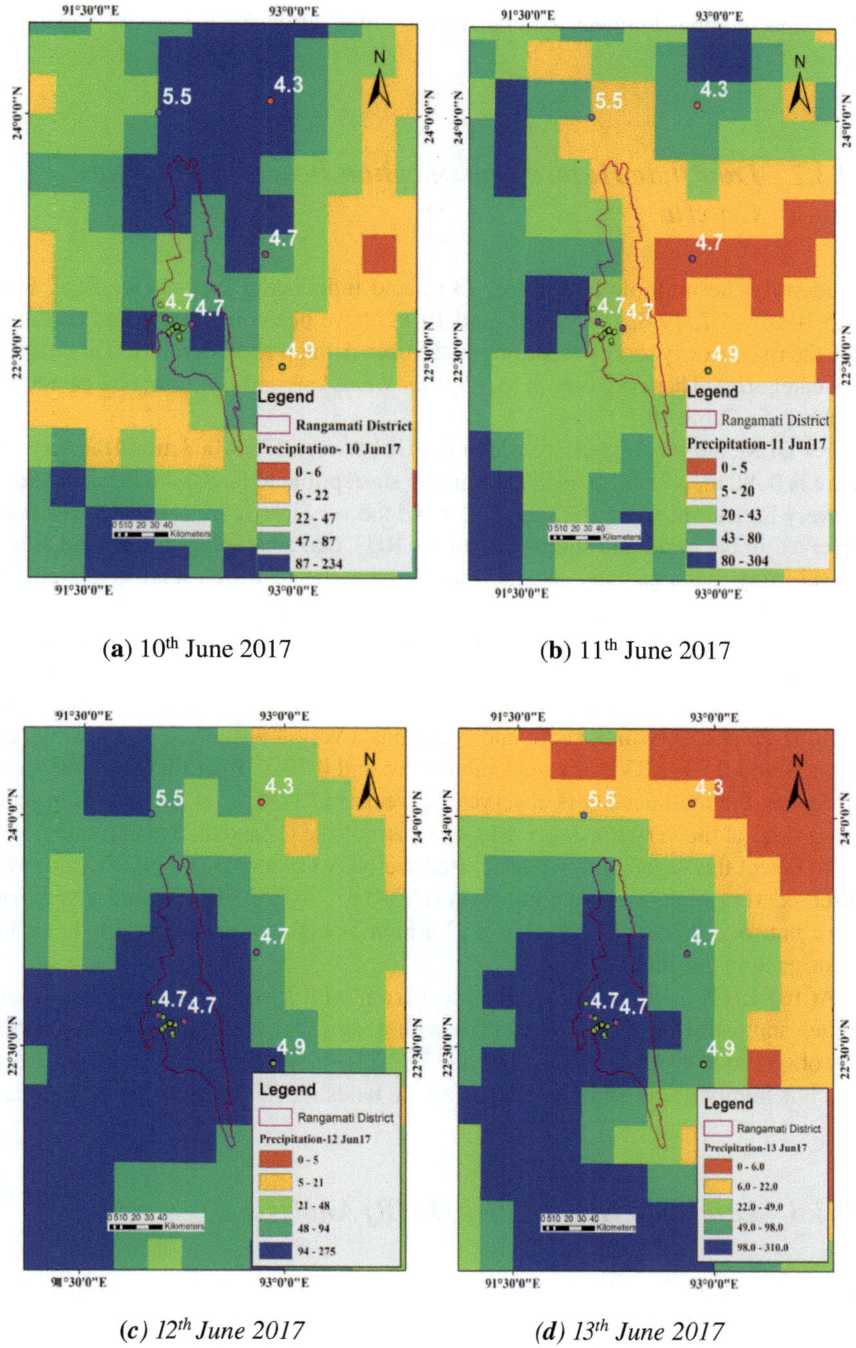

(**a**) 10th June 2017 (**b**) 11th June 2017

(*c*) *12th June 2017* (*d*) *13th June 2017*

Fig. 19.11 TRMM satellite-based rainfall measurement at RHT and surrounding areas; **a** 10th June 2017, **b** 11th June 2017, **c** 12th June 2017, and **d** 13th June 2017 along the national highway number (*N106*) of *Chittagong-Hathazari-Rangamati Road*.

19.3.2 Landslide Hazard Classification Based on Multiple Criteria

To identify the multiple landslides, 13 related influencing factors (see Figs. 19.4, 19.5, 19.6, 19.7, 19.8, 19.9, 19.10 and 19.11) have been considered and classified into the risk priority scores according to the user-defined choices referred to different risk categories. These are *very high, high, medium, low,* and *very low-risk* categories as shown in Table 19.1.

For DEM, the very high score is identified in between 473–1003 m and the very low score is 0–82 m, which is considered highly susceptible to landslide. For landslides, the very high slope score is 23.95–54.53 and the very low score is 0–3.63 which are observed in the north and south parts of the RHT that are susceptible to landslides.

Referring to the geology category, we find that the very high score is in Neogene type and the low score is in H_2O. On the other hand, the fault-based buffer is at a very high score which is responsible for landslide occurrence. In the case of river density estimation, a very high score was observed between 104.21–187.13, and a very low score area was observed in the range of 0–12.47. Whereas the soil is categorized as Bd61-2c at a very high score and water has a very low score. For landslide, the LD is found 0.83–1.83 at the very high score and 0.70–0.30 at the very low score. Moreover, the fault density is observed between 30.53–35.72 at the very high score and 0–5.45 at the very low score which is susceptible to landslide occurrence.

To detect the vegetation, we have used the NDVI method. The NDVI value was observed very high score ranging from 0.59–0.67, −0352 to 0.59, and a very low score between 0.72–0.80 and 0.7–0.72 which is highly responsible for landslide occurrence in the study area.

At the RHT, land use is identified as barren with a very high score, and water bodies and forests with less vegetation with a very low score, whereas the rainfall was observed at 483.93–757.53 mm, and 757.53–992.05 mm at the very high score, which is highly responsible for triggering the landslide activities in the study area.

19.3.3 Landslide Hazard Risk (LHR) Analysis

By overlaying the layers and using relative weights to the specific factors maps (see details in the method section), we have calculated the final weighted score of LHR. Based on the weighted score result, the LHR map is finally prepared and classified into five (5) risk hazard class categories range and risk types including Very High Risk,

Table 19.1 Classification of landslide based on multiple criteria evaluation (MCE) score

Multiple criteria	Very high	High	Medium	Low	Very low
1. DEM (meter) (Digital elevation model)	473–1003	291–473	171–291	82–171	0–82
2. Slope (degree)	23.95–54.53	16.04–23.95	9.62–16.04	3.63–9.62	0–3.63
3. Aspect	North	Northeast and northwest	West and east	Southeast and southwest	South
4. Geology	Neogene	–	–	–	H_2O
5. Distance from fault according to buffer (km)	5, 10	15	–	20, 25	–
6. River density	104.21–187.13	63.84–104.21	36.69–63.84	12.47–36.69	0–12.47
7. Soil	Bd61-2c	Ge51-2a	–	–	Water
8. Road density	104.21–187.13	63.84–104.21	36.69–63.84	12.47–36.69	0–12.47
9. Lineament density	0.83–1.13	0.63–0.83	0.46–0.63	0.30–0.46	0.07–0.30
10. NDVI	0.59–0.67, –0.352 to 0.59	–	–	0.67–0.70, 0.58–0.67	0.72–0.80, 0.70–0.72
11. Fault density	30.53–35.72	23.53–30.53	15.27–23.53	5.46–15.27	0–5.46
12. Land use map	Barren land	Hills with less vegetation, settlement, agriculture land			Water bodies, forest with less vegetation
13. Rainfall (mm)	483.93–757.53, 757.53–992.05	992.05–1178.79	1178.79–1326.45	1362.45–1591.37	–

High Risk, Medium Risk, Low Risk, and Very Low-Risk categories, respectively. The detail of these statistics is shown in Table 19.2.

According to the result (Table 19.2), about 219.33 km^2 of the northern part of RHT is designated as a very high risk of landslide, whereas 173.02 km^2 of the northeast and northwest parts are identified as high-risk areas. On the other hand, 512.56 km^2 of the west and east area are at medium risk, 3757.42 km^2 of southeast and southwest are at low and 1096.83 km^2 areas are identified as very low risky categories. The true representation of the LHR map and the overall risk area was measured based on the percentage share of RHT is reflected in Fig. 19.12a and b, respectively.

Table 19.2 Landslide hazard risk (LHR) zone and corresponding area statistics (*Source* Author's analysis, 2023)

Landslide hazard risk class range	LHR category	Risk directions	Risk area (km²)	Risk area (%)
23–34.92	Very low	South	1096.83	19.05
34.92–58.03	Low	Southeast and southwest	3757.42	65.24
58.03–81.14	Medium	West and east	512.56	8.90
81.14–104.25	High	Northeast and northwest	173.02	3.00
104.25–196.12	Very high	North	219.33	3.81
Total			5759.16	100.00

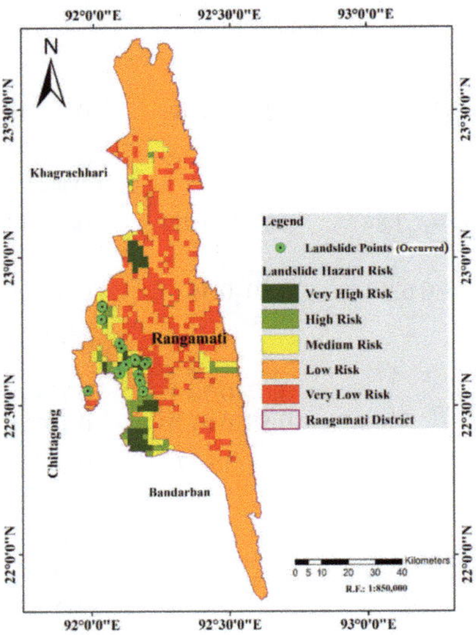

(a) LHR zone of RHT

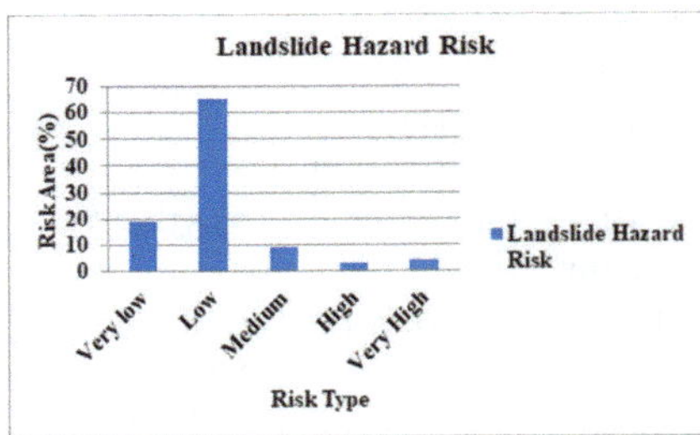

(b) LHR area in percentage share of RHT

Fig. 19.12 **a** LHR zone of RHT with occurred landslides, **b** landslide hazard risk classes and their percentage of sharing in the RHT

Based on the risk statistics, about 3.81% of areas of RHT are very high-risk and identified as landslide hazard-prone areas. The reported landslides surveyed by the ICIMOD team (Bajracharya and Majarjan 2018) plotted in the RHT map have fallen within the medium, high, and very high landslide risk zones. Moreover, the area also has shown to be low to very low-risk zones where a small number of landslide points are observed.

19.3.4 Environmental Changes Analysis

Landslide causes a great environmental change in the RHT. Based on the different landslide determining factors like LD, and NDVI, we have observed and analyzed the changes at zooming view that could identify the environmental changes at pixel level. For LD, before the landslide event on 2 May 2017 (42 days before) (Fig. 19.13a) and 18 May 2017 (26 days before) (Fig. 19.13b), we found rock-fractured areas in the study sites between 1.00–1.79 and 0.82–1.59. The changes of LD after landslide on 28 December 2017 and then recent progress on 1 February 2019 are observed in details in Fig. 19.13c and d, respectively.

With the circumstances of the landslide event, the NDVI changes are shown in Fig. 19.14a–d.

Landslides have played a major role in decreasing forest coverage in the affected areas as noticed in the NDVI values as well as the corresponding area's lineament values which were confirmed through LD images. According to the present study, the recent past earthquake events within or the close vicinity of the study area were

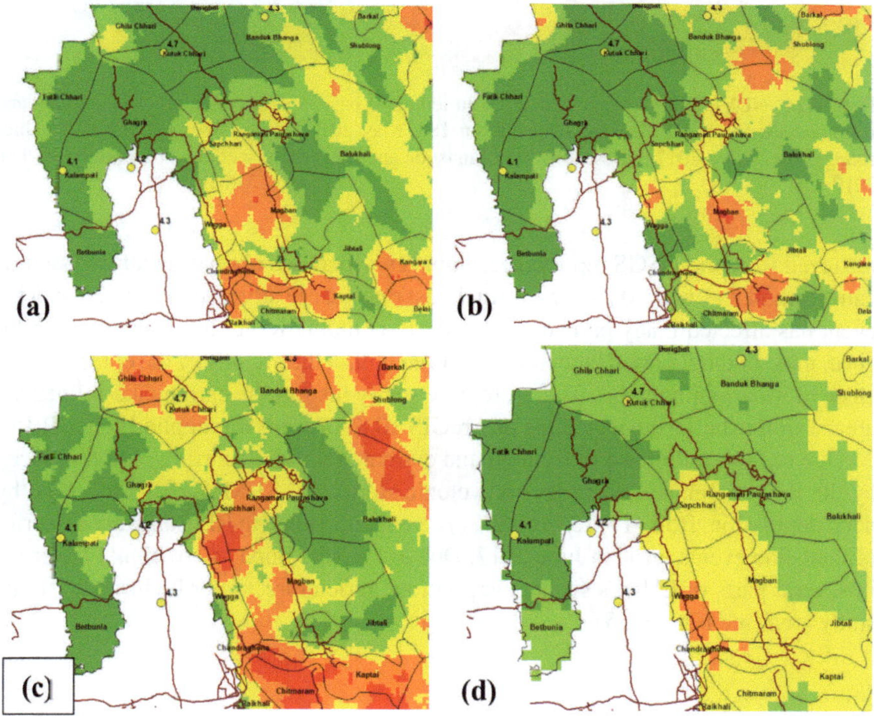

Fig. 19.13 LD changes before and after landslide activity **a** 42 days before as observed on 2nd May 2017, **b** immediately before landslide on 18th May 2017 (26 days before), **c** after landslide on 28 Dec 2017 and **d** recent changes in LD on 1st Feb 2019

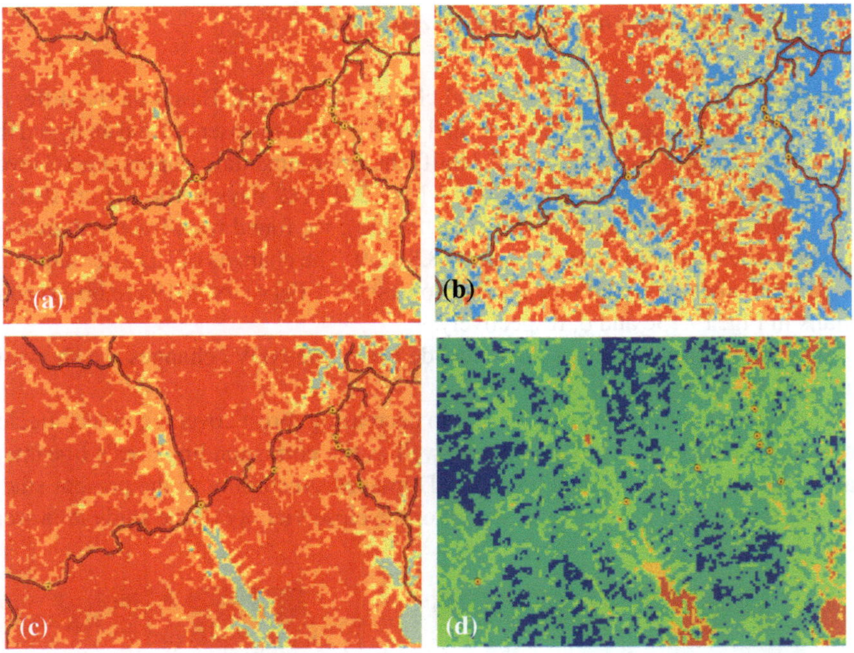

Fig. 19.14 NDVI values changes observed in temporal basis **a** 42 days before on 2nd May and **b** immediate changes before the landslide on 18th May 2017 (26 days before), **c** NDVI values changes on 28 Dec 2017 representing after landslide and **d** as recent scenario observed on 1st Feb 2019

retrieved from the USGS reported earthquakes, where we observed that two significant earthquakes with M_w 4.7 had struck just one year in advance in June 2016. This event has affected many parts of Bangladesh including RHT. To observe this influence, most of the reported landslides occurred within the vicinity of the epicenter and close to its adjoining areas. Therefore, to visually inspect the earthquake-influenced area, a simple buffering technique of ArcGIS 10.8 software was applied (Fig. 19.15).

Due to this earthquake movement and extreme rainfall occurring period induced by the Bay of Bengal (BoB) severe cyclonic storm (28–31 May 2017) followed by deep depression, BoB (11–13 June 2017) heavily impacted in RHT and faced a series of landslide events on 12th June 2017. Due to this, the impacts are highly observed in several parts of LULCs such as vegetation, agriculture, water bodies, etc. in the study area (see Fig. 19.15).

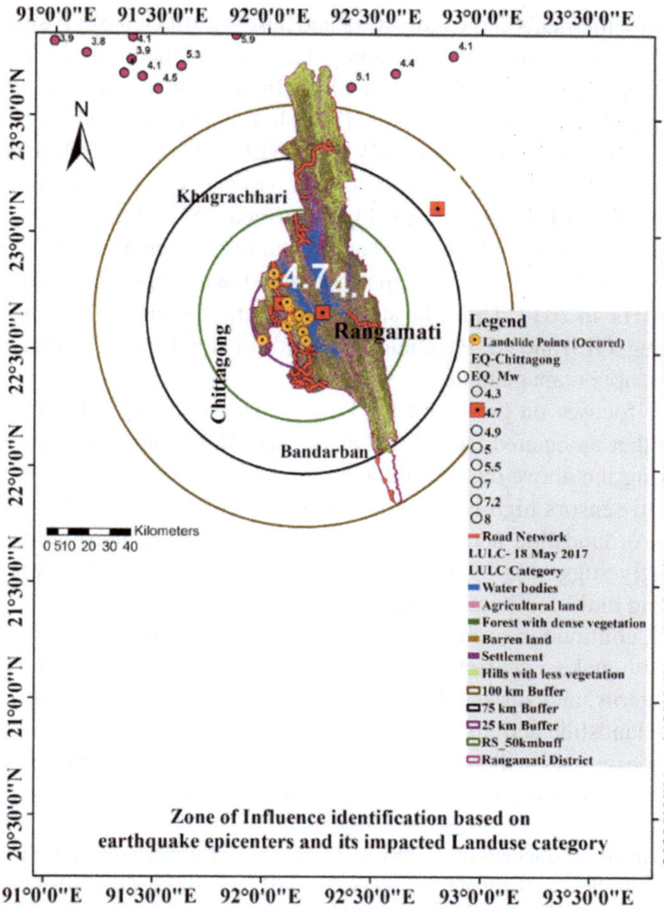

Fig. 19.15 Zone of influence based on earthquake epicenters and its impacted land use category

19.4 Conclusion

Landslide is a very important content considered for LHR assessment. Determination of landslide-prone areas early can help managers, local planners, district administrators, and decision-makers to know where it will occur before the event happens. This study shortlisted the earlier scenario which helps to detect the landslides in advance which was unnoticed till this research activity. Moreover, the study helps to know the environmental changes after the landslide including the recent observations. The number of landslides on 12 June 2017 in the RHT was staggering, which has motivated us to conduct the present study using the combination of GIS, Remote sensing, and AHP techniques. The topsoil in this area was saturated by heavy rainfall and one year before earthquake events severely affected the area to become risky to landslide.

The roads and settlements at the gully head and base are affected by erosion and landslide. Therefore, it is confirmed in this study that high and medium LD concentration areas are susceptible to high landslide activity. In addition, unplanned land use and barren land are also liable for increasing the risk of landslide activity.

To illustrate the details reflections of RHT, the LD, DEM, and its derived products (i.e., slope, aspect), LULC, geology, and other associated factors are found helpful to delineate various LHR areas. The study quantified 3.81, 3.00, and 8.90% areas as very high-risk, high, and moderate LHR areas. Our results generated in this study also confirmed that LHR areas are better matched with the multiple landslide events that occurred in RHT in 2017. Thus, the study further emphasizes the best utilization of remote sensing, GIS, and AHP techniques found suitable for landslide risk mapping and its sustainable management.

Our study focuses on the entire RHT by considering only 2017 LULC, lineaments, and other associated factors that are subjected to change as time progresses. By considering the above datasets changing nature, the study further recommends utilizing multi-sensors high-resolution satellite imageries that have 10–20 m pixel resolutions. As landslide activities occurred at the local scale, micro-level analysis is strongly suggested to figure out the precise changes in the affected areas by considering multi-temporal satellite datasets.

Moreover, continuous surface manifestations through lineament changes, monitoring of earthquake occurrence areas along the geological fault lines, rainfall changing patterns, and surface deformation change analyses are essential to monitoring future landslide activity in the RHT areas. Therefore, continuous monitoring and mapping exercises will be required to better understand the complex nature of the resident's living place to be safe, sound, and sustainable.

Acknowledgements The authors of this chapter highly showed heartfelt sorrow to the local people who lost their valuable lives due to the 2017 June landslide event. The authors would like to express sincere gratitude to the editors of this book for inviting us to write a chapter and special thanks to the reviewers for their critical comments on our chapter and the production team for the final proof and its production which helped us to improve our previous version in current form.

References

Abedin J, Rabby YW, Hasan I et al (2020) An investigation of the characteristics, causes, and consequences of June 13, 2017, landslides in Rangamati District Bangladesh. Geoenviron Disasters 7:23

Ahmed B, Dewan A (2017) Application of bivariate and multivariate statistical techniques in landslide susceptibility modeling in Chittagong City Corporation, Bangladesh. Remote Sens 9(4):304. https://doi.org/10.3390/rs9040304

Bajracharya SR, Majarjan SB (2018) Landslides induced by June 2017 rainfall in Chittagong Hill Tracts, Bangladesh: causes and prevention—field report, International Centre for Integrated Mountain Development (ICIMOD), Kathmandu, Nepal, p 26. https://doi.org/10.53055/ICIMOD.726. https://lib.icimod.org/record/34041. Accessed 20 Jan 2019

Bartarya SK (1991) Watershed management strategies in central Himalaya: the Gaula River Basin, Kumaun, India. Land Use Policy 8(3):177–184

Bartarya SK, Valdiya KS (1989) Landslides and erosion in the catchment of the Gaula River, Kumaun Lesser Himalaya, India. Mt Res Dev 9(4):405–419. https://doi.org/10.2307/3673588

Basu T, Pal S (2020) A GIS-based factor clustering and landslide susceptibility analysis using AHP for Gish River Basin, India. Environ Dev Sustain 22(5)

Bhandari RK, Gupta C (1985) Problems of landslides in the Himalaya and future directions. In: Singh JS (ed) Environmental regeneration in the Himalaya. Gynanodaya Prakashan, Nainital, pp 17–37

Bhandary NP, Daha RK, Timilsina M, Yatabe R (2013) Rainfall event-based landslide susceptibility zonation mapping. Nat Hazards 69:365–388

Carrara A, Sorriso-Valvo M, Reali C (1982) Analysis of landslide form and incidence in southern Italy. CATENA 9:35–62

Carrara A, Cardinali M, Detti R, Guzzetti F, Pasqui V, Reichenbach P (1991) GIS techniques and statistical models in evaluating landslide hazard. Earth Surf Process Land 16(5):427–445

Carrara A, Cardinali M, Guzzetti F (1992) Uncertainty in assessing landslide hazard and risk. ITC J 2:172–183

Carrara A, Cardinali M, Guzzetti F, Reichenbach P, Carrara A, Guzzetti F (1995) GIS technology in mapping landslide hazard geographical information system in assessing natural hazard. Kluwer Academic Publisher, Netherlands, pp 135–175

Chen CW, Chen H, Wei LW et al (2017) Evaluating the susceptibility of landslide landforms in Japan using slope stability analysis: a case study of the 2016 Kumamoto earthquake. Landslides 14:1793–1801. https://doi.org/10.1007/s10346-017-0872-1

Chowdhury M (2017) Rangamati landslides batter 1,700 families. bdnews24.com

Crozier MJ (1986) Landslides: causes, consequences and environment. Croom Helm Australia Pty. Ltd., London

Crozier MJ, Glade T (2005) Landslide hazard and risk: issues, concepts, and approach. In: Glade T, Anderson M, Crozier M (eds) Landslide hazard and risk. Wiley, Chichester, pp 1–40

Cruden DM (2003) The first classification of landslides? Environ Eng Geosci 9(3):197–200. https://doi.org/10.2113/9.3.197

Cruden DM, Varnes DJ (1996) Landslide types and processes. In: Turner AT, Schuster RL (eds) Landslides—investigation and mitigation. Transportation Research Board special report no, vol 247. National Academy Press, Washington, DC, pp 36–75

Gupta V, Bist KS (2004) The 23 September 2003 Varunavat Parvat landslide in Uttaranchal township, Uttaranchal. Curr Sci 87(11):1600–1605

Guzzetti F, Cesare Mondini A, Cardinali M, Fiorucci F, Santangelo M, Chang K-T (2012) Landslide inventory maps: new tools for an old problem. Earth-Sci Rev 112(1–2):42–66. ISSN 0012-8252. https://doi.org/10.1016/j.earscirev.2012.02.001

Horelli JA (2005) Landslides in Hong Kong. University of Helsinki

Kanungo DP, Arora MK, Sarkar S, Gupta RP (2009) Landslide susceptibility zonation (LSZ) mapping—a review. J South Asia Disaster Stud 2(1):81–106

Khatun M, Hossain A, Sayem H (2023) Climate variability & establishment of rainfall threshold line for landslide hazards in Rangamati, Bangladesh. Open J Geol 13:959–979. https://doi.org/10.4236/ojg.2023.139041

Lee S, Pradhan B (2006) Landslide hazard assessment at Cameron Highland Malaysia using frequency ratio and logistic regression models. Geophy Res Abstr 8. SRef-ID: 1607–7962/gra/EGU06-A-03241

NASA: Giovanni v 4.36. https://giovanni.gsfc.nasa.gov/giovanni/. Last accessed 15 Nov 2023

Rabby YW, Li Y (2019) An integrated approach to map landslides in Chittagong hilly areas, Bangladesh, using Google earth and field mapping. Landslides 16:633–645. https://doi.org/10.1007/s10346-018-1107-9

Saaty LT (1980) A scaling method for priorities in hierarchical structures. J Math Psychol 15:234–281

Saha AK, Gupta RP, Arora MK (2002) GIS-based landslide hazard zonation in the Bhagirathi (Ganga) valley, Himalayas. Int J Remote Sens 23(2):357–369
Saha AK, Gupta RP, Sarkar I et al (2005) An approach for GIS-based statistical landslide susceptibility zonation—with a case study in the Himalayas. Landslides 2:61–69. https://doi.org/10.1007/s10346-004-0039-8
Sarkar S, Kanungo DP (2004) An integrated approach for landslide susceptibility mapping using remote sensing and GIS. Photogram Eng Remote Sens 70(5):617–625
Sarwar MGM (2008) Landslide tragedy of Bangladesh. In: The first world landslide forum. Tokyo. http://www.academia.edu/3588839/Landslide_Tragedy_of_Bangladesh
Swanston DN, Swanson FJ (1976) Timber harvesting, mass erosion and steepland forest geomorphology in the Pacific Northwest. In: Coates DR (ed) Geomorphology and engineering. Dowden, Hutchinson and Ross, Stroudsburg, PA, pp 199–221
Terlien MTJ, Van Westen CJ, Van Asch TWJ (1995) Deterministic modelling in GIs-based landslide hazard assessment. In: Carrara A, Guzzetti F (eds) Geographical information systems in assessing natural hazards. Kluwer Academic Publishers, Dordrecht, pp 57–78
UNPO (2017) Chittagong Hill tracts: torrential rainstorms and wide-scale landslides leave thousands homeless. https://unpo.org/article/20199?id=20199. Accessed 22 Jan 2019
UNRC (2017) Bangladesh: HCTT response plan (June–December 2017)—Bangladesh | Relief Web. https://reliefweb.int/report/bangladesh/bangladesh-hctt-response-plan-june-december-2017. Accessed 01 Jan 2019
Valdiya KS (1987) Environmental geology: the Indian context. Tata McGraw-Hill, New Delhi, p 537
Van Westen CJ (1993) Application of geographic information system to landslide hazard zonation, ITC publication number 15, Part 1. ITC, Enschede, 245 pp
Van Westen CJ, Van Asch TW, Soeters R (2006) Landslide hazard and risk zonation—why is it still so difficult? Bull Eng Geol Environ 65:167–184
Van Westen CJ, Castellanos-Abella E, Sekhar LK (2008) Spatial data for landslide susceptibility, hazards and vulnerability assessment: an overview. Eng Geol 102(3–4):112–131
Van Westen CJ (2000) The modelling of landslide hazards using GIS. Geophysics 21:241–255
Varnes DJ (1978) Slope movement types and processes. Transp Res Board Spec Rep (176)
Wang G, Suemine A, Schulz WH (2010) Shear-rate-dependent strength control on the dynamics of rainfall-triggered landslides, Tokushima Prefecture, Japan. Earth Surf Proc Land 35:407–416
Zêzere JL (2002) Landslide susceptibility assessment considering landslide typology. A case study in the area north of Lisbon (Portugal). Nat Hazards Earth Syst Sci 2:73–82

Part II
Sustainability

Chapter 20
Landslide Risk Assessment, Awareness, and Risk Mitigation: Case Studies and Major Insights

Md. Humayain Kabir

Abstract Landslide is one of the major environmental disasters which affects both our natural ecosystems and human well-being. Therefore, a comprehensive understanding of landslide risk management is crucial. A total of 444 mass movement-related events occurred from 2000 to 2022. Among all these events, landslides were 346 and China (56), Indonesia (52), India (28), Colombia were the most landslide-affected countries. In this chapter, landslide risk assessment, awareness, and risk mitigation were discussed. After a thorough analysis of more than 50 case studies globally, this chapter showcased the major insights. Results found that both quantitative and qualitative methods were used for landslide risk assessment and mapping. More widely used methods include multi-criteria analysis, cluster analysis, multivariate analysis, Fuzzy modelling, analytical hierarchy approach, machine learning and artificial intelligence. Landslide risk mitigation encompasses both structural and non-structural measures. Particularly, nature-based solutions such as vegetation coverage are cost-effective and environment-friendly measures. It is also found that community-based landslide risk reduction and early warning based landslide risk mitigation are more successful. In addition to risk mitigation, awareness programs play a vital role in educating the public and children about landslide risk management Existing scientific knowledge from case studies explains the different forms of such awareness campaign. However, there is a gap between the affected communities and relevant stakeholders. To bridge this gap, a strong collaboration and cooperation between all the actors and stakeholders needs to be ensured towards an integrated management of landslide risks.

Md. Humayain Kabir (✉)
Institute of Forestry and Environmental Sciences, University of Chittagong, Chittagong, Bangladesh
e-mail: mh.kabir@uni-graz.at

Wegener Center for Climate and Global Change, University of Graz, Graz, Austria

Department for Knowledge and Communication Management, University for Continuing Education Krems, Krems an der Donau, Austria

G. K. Panda et al. (eds.), *Landslide: Susceptibility, Risk Assessment and Sustainability*, Advances in Natural and Technological Hazards Research 52,
https://doi.org/10.1007/978-3-031-56591-5_20

Keywords Landslide · Risk management · Awareness · Mitigation measures · Integrated approach

20.1 Introduction

A landslide is defined as movement of rock or soil mass, or both caused by the effects of gravitational force (Highland and Bobrowsky 2008). It is a common and quite complex natural phenomenon in mountainous area causes by geological, hydrological, anthropological, environmental factors (Ahmed et al. 2014; Marana 2017). In other words, it is a mass displacement of surface of the earth's slope materials along discrete shear under the change of natural gravity (Cruden 1993). Landslide shows an evidence of slope instability (Crozier and Glade 2005) in the forms of rockfalls, rockslides, massive debris flow, soil slips, rock avalanches, and mud-flows (Huang 2014).

Geologically, long-term landscape changes and the evolution of the hillslope are significantly influenced by landslides. This natural process has negative socioeconomic effects and causes significant financial losses, deaths, disruptions to the natural environment, ecosystems, and infrastructure (Tyoda 2013). Landslides might be shallow or deep-seated and are mostly sporadic, localized and dispersed (Ahmed et al. 2014).

In mountainous regions, earthquakes, extreme precipitation, and anthropogenic activities are the main causes of landslides. However, cut and fill failures, river bluff failures, the collapse of mine waste piles, quarries, open pit mines, and lateral spreading can also cause landslides in low-lying areas (Tyoda 2013). Underwater landslides or in an offshore marine setting landslide also occur in low relief and small slope gradients area.

Natural catastrophes, such as landslides, are becoming more frequent and severe worldwide, seriously affecting infrastructure and people. The amount of money lost to landslides has tripled over the past ten years, from around USD 50 billion in the 1980s to over USD 200 billion. Millions of people have been impacted by and lost a great deal of life due to landslides globally (Hidalgo and Vega 2021). All around the world, in all types of terrain and climate, landslides cause billions of dollars' worth of financial damage and hundreds of fatalities and injuries annually. They frequently result in long-term economic upheaval, population displacement, and adverse environmental repercussions. According to global records, landslides frequently occur in China, the Philippines, India, Indonesia, Nepal, Japan, and the United States. In global south, Bangladesh is one of the most populous countries, and due to its physiographic characteristics, it is particularly susceptible to landslides (Islam et al. 2017). People are compelled to dwell along the foothills of hills because to land shortages, natural catastrophe porn areas, and extreme population pressure (Islam et al. 2017). However, the unconsolidated sedimentary rocks that make up the hill region's soil, including sandstone, siltstone, shale, and conglomerate, increase the area's susceptibility to landslides.

Although landslides are one of the disastrous events, there is knowledge gap on stocktaking the comprehensive landslide risk management globally. Considering this background, this chapter aims to showcase the global overview of landslide, the-state-of-art of landslide risk assessment and vulnerability mapping, widely used mitigation measures with their scientific innovations and societal implications, and necessary awareness campaigns for landslide risk reduction and management.

20.2 Global Landslide: An Overview

Landslides are a major and common geologic hazard globally which causes a number of loss and damages to the property and human life. A total of 444 mass movement-related events occurred from 2000 to 2022. Among the landslide events were 346 (Fig. 20.1) whereas avalanche was 52. Most of the landslide events were in Asia (274) followed by Americas (86) and Africa (54), respectively (Fig. 20.1). The most landslide-prone countries were China (56), Indonesia (52), India (28), Colombia, Afghanistan, Pakistan, Philippines, Nepal, Peru, and Guatemala, respectively (Fig. 20.1). In the last decades, a sizable amount of people died due to adverse effects of landslides. The highest number of 3427 people died in 2010 globally followed by 2017 (total 2312).

20.3 Landslide Hazard Zonation and Susceptibility Mapping

Landslide susceptibility is the ability of soil or rock materials to produce a variety of landslides as a result of various circumstances or actions (Guzzetti et al. 1999). "Landslide susceptibility mapping" is the division of the terrain into zones of varying stability according to an assessed importance of causative factors (Φουμέλης et al. 2004). Planning for disaster management, development initiatives, early warning systems, and mitigation processes all heavily rely on landslide susceptibility maps (Chalkias et al. 2014; Dahal et al. 2008). It also presents the potential land sliding area in the future based on past and future occurrence (Crozier and Glade 2005).

"Zonation" refers to the process of classifying land surface into zones based on the level of current or anticipated landslide threat (Varnes 1984). "The division of land into somewhat homogeneous areas or domain and the ranking according to the degrees of actual or potential landslide susceptibility, hazard, risk, or applicability of certain landslide related regulation," is the definition of landslide hazard given by Couture et al. (2013). It used to be believed that mapping and assessing a region's vulnerability to landslides were laborious and time-consuming processes. However, modern advances in computer applications, geographic information systems, and remote sensing are crucial to the mapping and evaluation of landslide vulnerability.

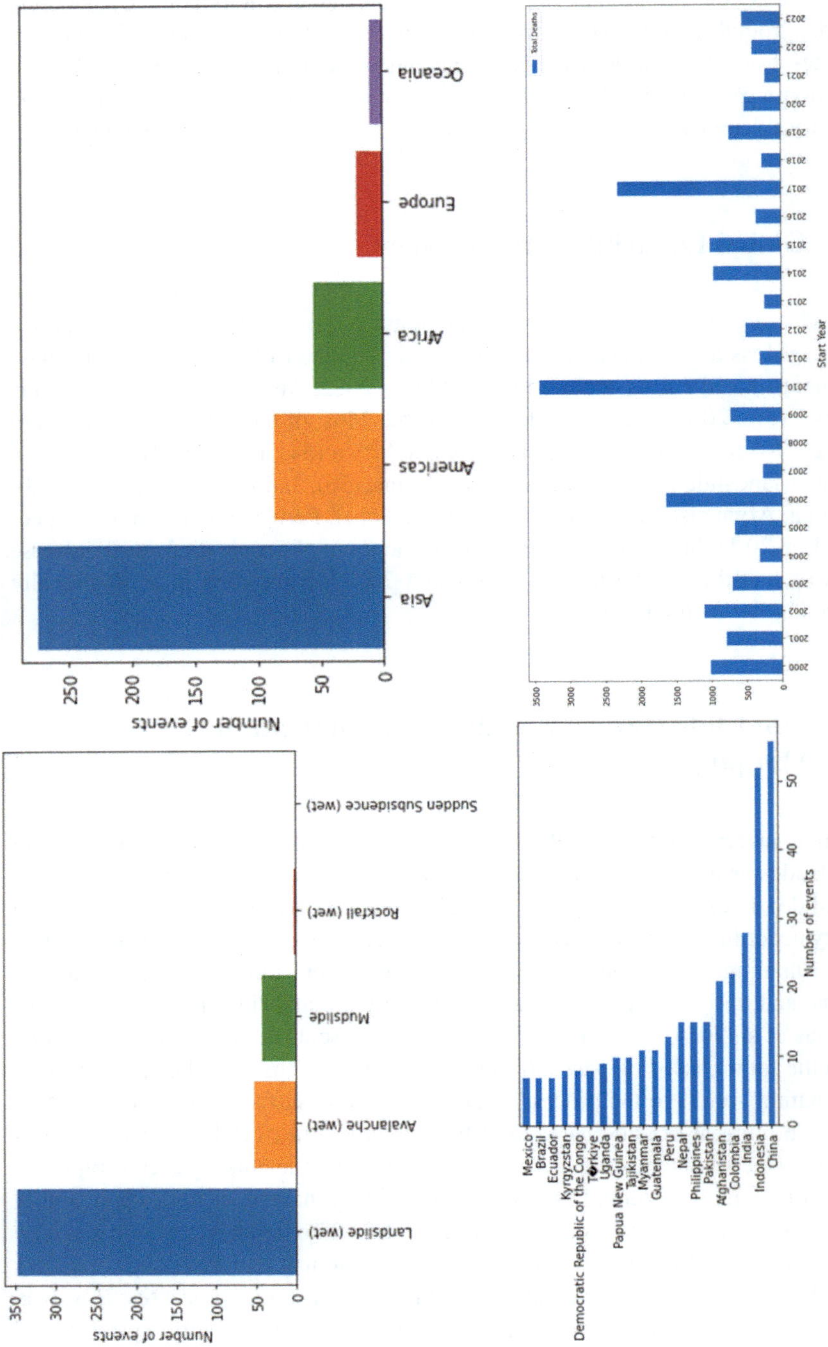

Fig. 20.1 Number of landslide events, their distribution, and total number of deaths from 2000 to 2022 (*Data Source* https://public.emdat.be/data)

20.4 Methods Used in Landslide Hazard Mapping Susceptibility Mapping

Assessing the risk and susceptibility of landslides is a crucial first step in risk and hazard management (Pardeshi et al. 2013). Numerous techniques are employed for this, but no one technique is acknowledged as being generally beneficial for evaluation (Pardeshi et al. 2013; Tyoda 2013).

Many landslide hazard and susceptibility mapping techniques have been applied for various reasons across the world throughout the past three decades in various locations. These consist of inventory-based mapping, heuristic, probabilistic, and deterministic techniques, statistical analysis, and multi-criteria decision making (Pardeshi et al. 2013). There are five general categories into which landslide susceptibility mapping approaches may be divided (Guzzetti et al. 2005). These techniques include: 1. Direct geomorphological mapping; 2. Landslide inventory analysis; 3. Heuristic or index-based approaches; 4. Statistical methodology; and 5. Process-based conceptual models. Two primary categories, such as the qualitative and quantitative approaches, can also be used to group it (Ayalew and Yamagishi 2005; Glade et al. 2006). Different methods used for landslide hazard assessment, zonation, and susceptibility mapping are summarized in Table 20.1 (Pardeshi et al. 2013; Tyoda 2013).

20.5 Remote Sensing and Geographic Information System (GIS)

In order to monitor the causal variables impacting landslide susceptibility and to create thematic information for landslide susceptibility mapping, remotely-sensed satellite information can be employed (Lee et al. 2004). Digital image processing is an effective tool for data merging, enhancement, classification, and accuracy assessment. Synthetic Aperture Rader (SAR) data and LiDAR data used for delineation of landslide morphological features and landslide classification (Joyce et al. 2014; Tyoda 2013). Use of Digital Elevation Model (DEM) for landslide assessment is more useable at present time. Thematic data layers such as slope aspect, slope angel, drainage, elevation, curvature etc. are extracted from DEM layer for landslide assessment (Barla et al. 2010; Pardeshi et al. 2013; Rotigliano et al. 2011; Saraf et al. 2009).

GIS is widely used for landslide assessment. It helps to generate thematic data layers, assign weight of causative factors, integration of data and prepare susceptibility map of landslide. Using GIS, several methods such as weight overlay, AHP, MCDA, IVM, logistic regression, frequency distribution, ANN models used for landslide susceptibility mapping and hazard zonation.

Table 20.1 Review of different approaches used for landslide Susceptibility Mapping

Methods	Description
Qualitative process	
Distribution analysis	Distribution analysis, sometimes referred to as landslide inventory, gives a map's geographic distribution of landslides that are currently occurring, shown as either polygons or points (Wieczorek 1984). Based on the distribution of landslides, movement kinds, movement rates, and types of displaced materials, inventory maps are generated. Landslide data were collected by intensive field inventory, historical records and documents, satellite image interpretation, and aerial photogrammetry (Pardeshi et al. 2013). This method is economically feasible and cover a large area but does not relate landslide with its causative factors (Wieczorek 1984)
Map combination	A straightforward process known as the "map combination approach" merges several thematic map layers according to causal factors. This approach relies on specialized expertise. It is a simple procedure that entails the several phases (Soeters and van Westen 1996): 1. Identifying the elements that may cause a landslide. 2. Creation of a theme layer with pertinent categories. 3. The distribution of rankings and weights, correspondingly. 4. Thematic data layers combined or integrated. 5. Making a map of the vulnerability to landslides
Quantitative process	
Statistical analysis	Statistical methods have been more popular in recent years for mapping landslide vulnerability. Knowledge-based techniques were once the most popular, although they required more time and effort. This method has been replaced with a data-driven or statistical approach for reducing subjectivity in weightage assignment and ensure more objective and reproducible results (Kanungo et al. 2008). These methods can be grouped into two categorizes (Tyoda 2013): bivariate statistical analysis, and multivariate statistical analysis
Bivariate statistical analysis	In the bivariate statistical analysis for landslide susceptibility mapping, each data layer of the causative components is compared to the present landslide distribution. (Kanungo et al. 2008; Tyoda 2013). The landslide density determines the weight values of the causal components (Pardeshi et al. 2013; Tyoda 2013). Bivariate statistical methods for mapping the susceptibility to landslides include weights of evidence modeling, information value modeling, frequency analysis, and weight overlay modeling. (Pardeshi et al. 2013)

(continued)

Table 20.1 (continued)

Methods	Description
Frequency analysis approach	The frequency ratio is derived from the correlation that has been shown to exist between the distribution of landslides and each of their contributing causes (Pardeshi et al. 2013). This method is helpful for establishing a physical connection between the location of landslides and the factors that cause them. Using this method, the relationship between each subclass of causative factors and the incidence of landslides is calculated. By applying this approach, the landslide susceptibility index, or LSI, is calculated by adding the frequency ratio values of each element (Lee 2005; Pardeshi et al. 2013)
Information Value Model (IVM)	Based on correlations between the occurrence of landslides and associated characteristics, the Information Value Model (IVM) is a bivariate statistical technique for landslide spatial information prediction (Sarkar et al., n.d.). According to Pardeshi et al. (2013), the information values are established for every subclass of landslide-related parameter based on the existence of landslides in a certain mapping unit. IVM has shown to be an effective technique for assessing the relative importance of each causal element that contributes to the occurrence of landslides (Champati Ray et al. 2006; Kanungo et al. 2008)
Weight of evidence model	Pardeshi et al. (2013) defined a Weight of Evidence is a log linear based Bayesian probability model for landslide susceptibility assessment that uses the incidence of landslides as training data for prediction. It computes the likelihood of landslide hazards, both unconditionally and conditionally. This technique uses the degree of geographical correlation between the incidence of landslides and each class of explanatory factors by calculating positive and negative weights
Weight overlay method	One basic bi-variate statistical technique is weighted overlay. Based on the correlation between the frequency of landslides and the causes that cause them, weights are allocated. The causal variables are assigned numerical weights based on how frequently they cause landslides. Additionally, an overlay of the data layers creates an LHZ map (Pardeshi et al. 2013)
Multivariate statistical analysis	The proportional contribution of each theme data layer to the overall landslide susceptibility is taken into account by this technique for landslide hazard zonation (LHZ) (Kanungo et al. 2012). Commonly used techniques for LHZ mapping include conditional analysis, deep learning technique such as artificial neural networks (ANN), discriminant analysis, multiple regression models, logistic regression models, and others (Pardeshi et al. 2013). These techniques determine the proportion of landslide incidence area for every pixel and landslide event

(continued)

Table 20.1 (continued)

Methods	Description
Logistic regression model	The foundation of the logistic regression statistical model is the idea that the same factors that led to slope collapse in a particular area would also likely produce landslides in the future (Guzzetti et al. 1999). Based on the values of the variables, it may be used to forecast whether a result will occur or not. These models can be either binary or multinomial logistic regression (Pardeshi et al. 2013)
Discriminant analysis	The largest difference between the landslide occurrence group and the non-landslide group for each independent landslide causative factor is found using discriminant analysis. It also establishes these factors' weights (Lee et al. 2008). By dividing the slope unit into landslide impacted and landslide free classes, the Standardized Discriminant Function Coefficient (SDFC) may be calculated (Pardeshi et al. 2013)
Probabilistic analysis	This technique contrasts the geological distribution of landslides with various causal components and aids in determining the spatial, temporal, and magnitude likelihood of landslide events (Guzzetti et al. 2005). It consists of the favorability function, the certainty factor, and the Bayesian probability weight of evidence techniques (Pardeshi et al. 2013; Tyoda 2013). According to Kanungo et al. (2012), the degree of correlation between each theme data layer and the landslide distribution is converted to a number based on the probability distribution function
Fuzzy set based approach	By using this method, fuzzy membership values are assigned to various categories of thematic data layers, and the input neurons of a neural network model receive data from these membership values. Based on landslide inventory data, the magnitude of landslide susceptibility is represented by values ranging from 0 to 1 (Pardeshi et al. 2013; Tyoda 2013)
Artificial neural network-based approach	An artificial neural network (ANN) is a system that can learn a certain phenomenon that is comparable to human behavior (Pardeshi et al. 2013). Over three layers of neurons coupled by weights make up an ANN. The "Back propagation learning algorithm" used by this model establishes guidelines for weight assignment. Then, each variable's weight is changed to reduce errors. According to Catani et al. (2005), the Artificial Neural Network (ANN), a non-linear model, has shown to be more successful in landslide zonation and hazard risk assessment

(continued)

Table 20.1 (continued)

Methods	Description
Analytical hierarchy process approach	According to Saaty (2008), the Analytical Hierarchy approach (AHP) is a multi-criteria decision-making approach that measures outcomes through pairwise comparisons and depends on expert opinions to determine priority scales. AHP functions on four levels: problem definition, goal and option identification, pairwise comparison matrix generation, weight determination, and overall priority decision. Various elements that cause landslides are taken into consideration as possibilities in LHZ. Each landslide-related statistic is given an absolute number (ranging from 1 to 9) according to its relative relevance, and comparison matrices are created to calculate the consistency ratio (CR) and consistency index (CI) (Crozier and Glade 2005). There have been several attempts recently to map the vulnerability of landslides worldwide using GIS-based AHP (Mondal and Maiti 2012a)
Rainfall threshold	This method refers to the lowest amount of rainfall or duration required to produce a landslide (Varnes 1984). The most often utilized characteristics to define rainfall threshold are cumulative rainfall, rainfall intensity, antecedent rainfall, and rainfall duration (Pardeshi et al. 2013). The wet soil bulk density, slope angle, upslope drainage, and water density all contribute to the critical rainfall threshold model
Physically based landslide susceptibility model	This method, which is based on mechanical processes, explains the physical mechanism that causes landslides to occur (Pardeshi et al. 2013). Based on basic mechanical rules, physically based models for assessing the risk of landslides describe the physical processes that precede the landslide occurrence. According to Kuriakose (2010), these models take into consideration the slope's transitory ground water reaction to rainfall. These models may be used in locations with inadequate landslide inventory since they do not require time series landslide data (Kuriakose 2010)

20.6 Landslide Risk Assessment: Selected Case Studies

For better preparedness against landslide, it is highly needed to assess the landslide risks more accurately. A wide range of methods are used for landslide risk assessment. The major findings and insights from the case studies are presented in the following Fig. 20.2 and Table 20.2.

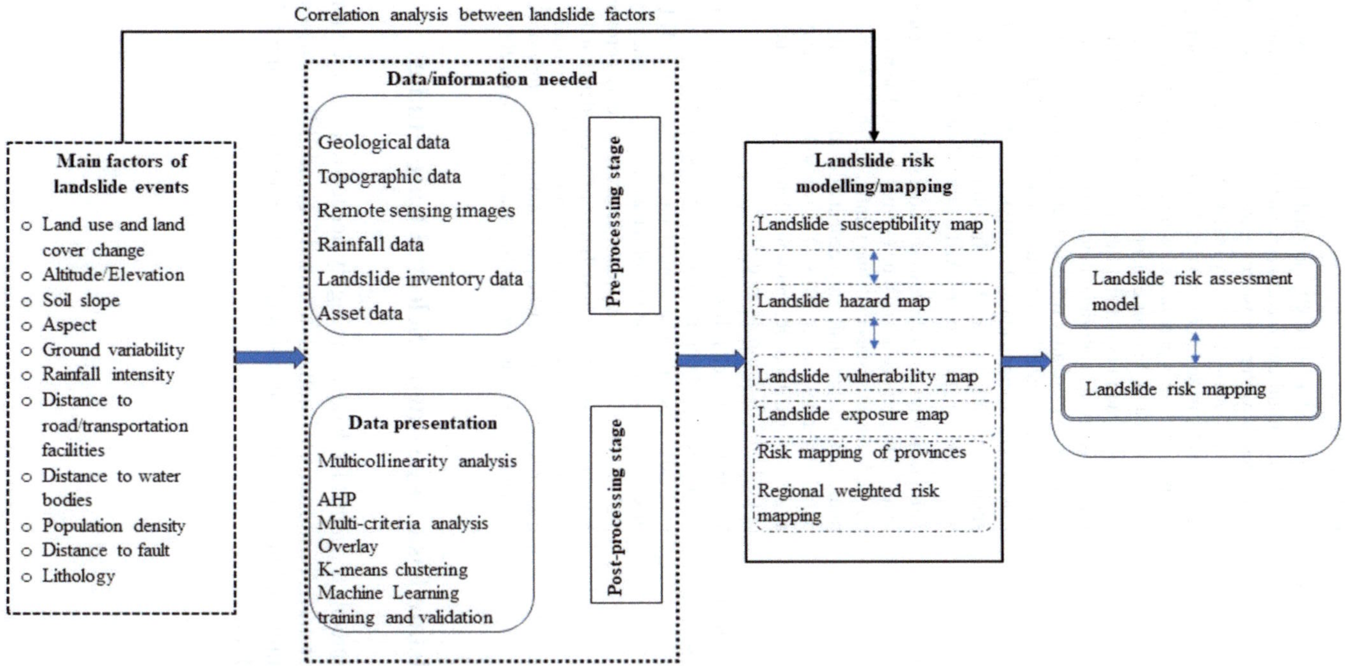

Fig. 20.2 Steps of landslide risk assessment and mapping (adopted from Akgul et al. 2022; Chang et al. 2021; Zhao et al. 2021)

Table 20.2 Major finding and insights from the selected landslides case studies

Case study	Major insights	References
– Landslides in Puerto Rico – Vegetation damage was closely linked to landslides and slope aspect	– Predominant landslide locations in Puerto Rico were associated with certain slope aspects – Vegetation damage was highly correlated with landslide occurrence and slope aspect of area – Errors in landslide locations can introduce bias in data analysis – High-resolution LiDAR data are more accurate but influenced by artificial surfaces	Gorokhovich and Vustianiuk (2021)
– Developed a spatial case-based reasoning method for assessing regional landslide risks – Integrated spatial proximity and topological interlinkages as spatial characteristics	– Optimized support vector machine (SVM) and convolutional neural network (CNN) methods were selected – They used attribute similarity and spatial similarity for joint reasoning of landslide events – Highly effective for regional landslide risk assessment and mapping – Limitations include case base richness, regional differences, and attribute calculations	Zhao et al. (2021)
– Quantitative landslide risk assessment in the China–Pakistan Economic Corridor (CPEC) – An assessment index method for sensitivity, risk, and susceptibility to landslides	– Evaluation of landslide vulnerability based on drainage, transportation facilities, population density, and physical property factors – Helps with transportation and communication planning, rehabilitation, and mitigation measures in the highly landslide-prone areas	Chang et al. (2021)

(continued)

Table 20.2 (continued)

Case study	Major insights	References
– Proposed a quantitative landslide risk assessment method – Monte Carlo simulation and B-spline Material Point Method are combined	– Numerical modelling was deployed to test and demonstrate the newly developed method – The variation of soil's shear strength has significant effect on the dynamic motion of landslides – The proposed method provides a more detailed and clear understanding of landslide risk – This new method is useful for landslide related hazard identification, assessment, and risk factor analysis	Zhou and Sun (2020)
– New approach for time series analysis of natural disasters in Turkey – Cluster analysis used to determine landslide risks in the provinces	– Trabzon is the riskiest province in Turkey for all four disaster types – The new approach minimizes errors caused by surface area differences – The results will help prioritize disasters and establish protective policies	Akgul et al. (2022)
– Identified risk factors for landslide disasters in China – Distance to roads, extreme precipitation, and land use are main risk drivers	– Landslide susceptibility map was prepared using the maximum entropy model – Potential landslide risk was higher in eastern and southern parts of China – MaxEnt model used to estimate risk factors and landslide distribution in China – Limitations include lack of data on anthropogenic factors and earthquakes	Wang et al. (2021)
– Studied location was in Shafe and Baso catchments in Gamo highland, Ethiopia – They applied fuzzy set theory and danger pixel concept for landslide risk assessment	– The most applicable biological and mechanical remedial measures are required to landslide risk mitigation – Fuzzy set linguistic approach provides more accurate and realistic results for landslide risk assessment	Shano et al. (2022)

(continued)

Table 20.2 (continued)

Case study	Major insights	References
– Integrated method for landslide risk assessment in Hindu Kush Mountain ranges – Utilizes geospatial data and semi-quantitative techniques	– Integrated approach for landslide hazard, vulnerability, and risk assessment – Landslide risk assessment map classified into five risk classes – Assists in highlighting landslide risk hotspots for landslide risk reduction and mitigation	Shah et al. (2023)
– Study conducted on landslide hazard, vulnerability, and risk assessment in Mussoorie, India – Four statistical approaches used for landslide hazard assessment with 75–80% success rate	– Four statistical approaches were used for landslide hazard assessment – Risk map shows certain wards with high and very high-risk zones – Certain wards are under high and very high landslide prone areas	Hader et al. (2022)
– Methodology to integrate spatial information for landslide risk assessment in Cubatão municipality, São Paulo, Brazil – Applied susceptibility, vulnerability, and rainfall thresholds to produce risk map – Random Forest algorithm used for susceptibility assessment – Social vulnerability assessment based on socioeconomic and demographic variables	– Three approaches used to calculate rainfall thresholds – Useful tool for risk mitigation and early warning system development – Potential for replication in other areas – High potentiality for real-time based forecast and dynamic risk map – Validation shows promising performance in predicting landslides	
– Risk assessment for landslide-prone areas in eastern Kentucky, USA – Broad range of risk assessment approaches demonstrated	– Little changes in data quality can change landslide risk assessment results – Different hazard inputs and exposure data showed significant role in risk maps – Coarse slope-based landslide risk map is inadequate for a broader area	Crawford et al. (2022)

(continued)

Table 20.2 (continued)

Case study	Major insights	References
– Landslide inventory mapping for Chittagong Hilly Areas using Google Earth Engine and field inventory – Combination of both approaches provides comprehensive landslide inventory for the area	– Field mapping identified higher number of landslide events than Google Earth mapping – Google Earth-based landslide mapping is ineffective in urban areas – Integration of field mapping and Google Earth improves landslide inventory mapping – Google Earth may not able to identify landslides due to fast vegetation recovery – Field mapping provides more detailed information on landslide cause and type	Rabby and Li (2019)
– Assessed landslide risk in water supply basins in La Liboriana River basin, Salgar, Colombia – Methodology includes probabilistic methods considering rainfall and earthquakes	– Results suggest method accurately assesses landslide hazard and expected losses – The risk analysis model is useful in data scarce areas – Simple and less complicated method is an alternative approach for a data-scared region	Hidalgo and Vega (2021)

20.7 Landslide Mitigation: Insights from Selected Case Studies

For taking better mitigation measures against landslides, it is highly needed to understand the landslide risks more accurately. Different methods are used for landslide risk assessment and mitigation. In the following Table 20.3, the major findings and insights from the case studies are summarized and presented.

20.8 Landslide Awareness: Insights from Selected Case Studies

Landslide awareness is important not for only mitigating and adapting risks, but also for the protection of properties and educating new generations and children for a disaster resilient future. Therefore, it is highly needed to get involved with the landslide awareness campaign. A wide range of tools are used for landslide awareness. In the following Table 20.4, the major findings and insights from the case studies are presented.

Table 20.3 Major finding and insights from the selected landslide mitigation case studies

Case study description	Major insights/practical implications/ conclusion
Use of vegetation for landslide risk mitigation (Kalsnes and Capobianco 2022*)*	
– Modelling of climate-induced landslides is important for mapping and risk reduction – Vegetation coverage has great role in erosion protection and landslide mitigation – Landslide risk management requires modelling for mapping and mitigation – Use of vegetation as Nature-Based Solutions (NBS) for reducing soil and landslide risk reduction – Vegetation affects soil-atmosphere interaction and soil hydro-mechanical properties – More efforts needed to quantify the effectiveness of vegetation in landslide mitigation	– Modelling slope stability with vegetation is complex due to various factors – There is a high necessity for advancement of modelling tools for landslide risk management – Vegetation as a cost-effective measure in landslide risk mitigation and reduction – Need to study spatio-temporal variation of vegetation effects on a slope – A joint initiatives in local instrumentation and risk modelling advancement is needed – The nexus between soil stability and vegetation coverage is complex and significantly varies with time, soil type, and local climatic conditions – Modelling the vegetation effects on slope stability requires understanding of vegetation features and growth effects
Toward negotiated mitigation of landslide risks in information settlement (Smith et al. 2020*)*	
– Strategies for reducing the potential risks from landslide events in informal settlements – Focused on a pilot project in Medellín, Colombia – Aims to find viable approaches to risk reduction strategies – They used a two-layered community-local government spaces for effective negotiation – Highlights the importance of tackling challenges from the state-local community trade-off – The approach involves a shift towards short and medium-term safety measures – Explored community and state perspectives on risk-reducing strategies – Identified barriers and viable approaches to risk reduction in informal settlements	– Local community is an integral part of participatory landslide risk monitoring and mitigation – Trust and a strong collaboration between community and scientific experts were recognized – Community-based monitoring system of landslide and pilot cases of mitigation measures pilots were employed – Valuing community knowledge and incorporating it into decision-making processes – Addressing limitations and potential adverse community dynamics in project implementation – Showcased the importance of beating state-community stand-off over land rights – Reorienting the associated safety related problems arising in in the short and medium term – Using an incremental strategy to provide negotiated landslide risk mitigation

(continued)

Table 20.3 (continued)

Case study description	Major insights/practical implications/ conclusion
Root reinforcement as a measure for shallow landslides risk mitigation (Cardile et al. 2019*)*	
– Vegetation can be used as a sustainable solution for slope reinforcement – Use of vegetation as risk mitigation and erosion control method	– Reinforcing slopes with roots provides sustainable bioengineering solutions – Root system improve soil shear strength by means of hydrological and mechanical processes – Reinforced soils have higher shear strength compared to non-reinforced soils – Mechanical contribution depends on root tensile strength, density, and distribution – Roots act as anchors, contributing to increased shear resistance until failure conditions are reached
Mitigation measures of debris flow and landslide risk (Genevois et al. 2022*)*	
– Case studies describe specific mitigation measures implemented in North-Eastern Italy – Developing of effective mitigation measures for debris flow and landslides – Remedial measures include slope stabilization, improved drainage system, retaining walls, and reinforcement – Debris flow site needs building of debris basin, blocks, and breakers	– Remedial measures for landslides and debris flow control are necessary – Effectiveness of remedial works should be evaluated over time – Mitigation procedures involve knowledge acquisition, risk assessment, planning, and monitoring – Naturalistic engineering operations should be considered when possible – Complex landslides may require urban planning and community preparedness strategies – Active landslide site required deep ground improvement and hard engineering works
Landslide risk mitigation of in Lisbon Historical Center (Fartaria et al. 2019*)*	
– The monitoring and stabilization of a slope in Lisbon – Horizontal cumulative movements and cracks were detected on the slope – Structural and geotechnical reinforcement measures were implemented to mitigate landslide risk – Monitoring system installed after stabilization works, including inclinometers and piezometers	– Solutions included concrete buttress, reinforced bored piles, and drainage system refurbishment – Importance of geological and geotechnical information for managing geotechnical risk – Surveying and monitoring were essential to planning and carrying out stabilizing projects – The bottom masonry wall received additional restoration work – Stabilization works included bored piles and concrete buttress

(continued)

Table 20.3 (continued)

Case study description	Major insights/practical implications/ conclusion
Geomorphological analysis and early warning systems for landslide risk mitigation (Thapa et al. 2023)	
– Landslides in Nepal caused by seismic activity, excessive rainfall, and unplanned land-use changes – Low-cost bioengineering and early warning systems used for landslide mitigation – Case study of Methum landslide in Lalitpur district, central Nepal – Analysis of landslide geomorphology and identification of triggering factors	– Effectiveness of landslide early warning system (LEWS) was evaluated – LEWS assesses rainfall intensity, soil moisture, and displacement activity – LEWS reduces the observed landslide risk at the local level – Thresholds customized based on previous studies and experiences – Alert messages sent to designated persons and authorities during landslide events – Anthropogenic factors can be managed to reduce landslide risk – Strategies include replacing terraced farming with agroforestry or seed orchards – LEWS should be customized for specific landslide conditions and parameters – Government authorities should promote and support the use of LEWS – LEWS should incorporate disaster response actions and encourage evacuations
Participatory process for landslide risk mitigation (Scolobig et al. 2016)	
– Investigated the participatory-based landslide risk mitigation in Nocera Inferiore, Italy – Multiple worldviews considered, compromise reached on risk reduction measures – Experts provided policy options corresponding to participants' perspectives	– They used analytical-deliberative process for identifying the best landslide risk mitigation measures – Participation of stakeholders mandated by the European Union – Multiple worldviews among participants structured to achieve compromise recommendation – Active participation of citizens and experts in landslide risk management process is highly feasible and valuable input – Mutual learning process between network building with experts, and improved public perceptions of landslide risks – Raised awareness and broke existing deadlock in decision-making process

(continued)

Table 20.3 (continued)

Case study description	Major insights/practical implications/ conclusion
Effectiveness of landslide slope stability (Kamal et al. 2023)	
– The effectiveness of landslide mitigation measures and community risk perception inKutupalong Rohingya Camp in Bangladesh – Existing structural mitigation measures ensures partial protection against landslide risks – Study showed that concrete retaining walls were perceived as more effective – Cohesion and friction angle of lower sandstone are important factors for slope protection – Several soft and hard mitigation measures implemented but many failed	– Highlights the ineffectiveness of structural mitigation measures in preventing landslides – Community perception plays a crucial role in evaluating the effectiveness of interventions – Concrete retaining walls are perceived as the most effective measure by the community – Non-structural measures and nature-based solutions are recommended for landslide risk reduction and mitigation – The interdisciplinary approach used in the study enhances public trust – Blending physical and social science components enhances the trustworthiness of the findings
Hydrological alert systems and landslide risk mitigation (Floris et al. 2012)	
– Used process-based physical model for hydrological alert system in landslide risk mitigation in Val di Maso in northeastern Italian Alps – Identification of rainfall pattern and minimum triggering threshold for landslides – Effective and easy to handle alert system for landslide early warning	– Empirical models identify rainfall patterns and evaluate recurrence of critical conditions – Physically-based models evaluate effects on slope stability and set minimum alert threshold – Newly developed hydrological alert system can be applied to different scales and geographic contexts – Alert system components: rain gauge station, software, transmission system

20.9 Conclusion

The severity and intensity of landslide events has increased over the last few decades. Therefore, a holistic understanding and an integrated management of landslide risks is the necessity of this century. Considering this context, this chapter focuses on landslide risk assessment, mitigation, and awareness. For stock staking the existing knowledge on landslide risk management, the most relevant case studies were reviewed thoroughly and identified the most important insights and potential gaps. Over the last two decades, more than 400 mass movement-related events have occurred globally. Current knowledge bulk highlighted multi-criteria analysis, cluster analysis, multi-variate analysis, Fuzzy modelling, analytical hierarchy approach, geo-information-based modelling, community-based mapping, use of machine learning and artificial intelligence are the most popular methods for landslide risk assessment. Landslide risk mitigation encompasses both structural and non-structural measures. Particularly, nature-based solutions such as afforestation and reforestation are cost-effective and environment-friendly measures. In addition to risk mitigation, awareness programs play a vital role in educating the public

Table 20.4 Major finding and insights from the selected landslides awareness case studies

Case study description	Major insights/implication/conclusion
Public awareness of landslide hazards (Solana and Kilburn 2003*)*	
– Survey conducted in landslide prone areas in the Barranco de Tirajana (BdT) Basin, Gran Canaria – Communities are generally aware of landslides but not of their own vulnerability – Limited pressure on local authorities to prepare landslide contingency plans or enforce building codes – Survey conducted to assess landslide awareness among the most vulnerable communities – Communities aware of landslides but not of potential hazards to themselves	– The survey revealed a satisfactory level of general awareness of landslide hazards – Rainfall and earthquakes were identified as main triggers of landslides – At least 50% of the surveyed communities usually not react timely and effectively during landslide emergency – A public awareness campaign for landslide response is warranted – Public information pamphlet is a powerful instrument to raise landslide mitigation awareness – A follow-up study is needed to investigate the impact of the pamphlet
Improving the awareness index for landslide cases (Susanto et al. 2018*)*	
– Analyzed the readiness of government and non-government authorities for landslide disaster management in Semarang City – Calculation of awareness index using UNESCO formulation – Identification of poor awareness indexed areas in Semarang city – Parameters for disaster preparedness index were analyzed using Analytic Hierarchy Process – Preparedness policy had the highest weight in the AHP results	– Preparedness index for disaster-related stakeholders helps for better landslide risk management – The preparedness index for Kembang Arum was 68.3 (Ready) whereas Manyaran was 57.6 (Almost Ready) – The preparedness index for BPBD Semarang was 94.0 (Very Ready) – The highest weighted parameter is the preparedness policy – Recommendations included detailed action plans, improved warning systems, better resource mobilization, and high-quality community education – Formulation of an integrated management system to implement recommendations – Importance of coordination among stakeholders for disaster preparedness and management

(continued)

Table 20.4 (continued)

Case study description	Major insights/implication/conclusion
Satellite-based assessment of landslide hazard and situational awareness (Kirschbaum and Stanley 2018*)*	
– Landslide Hazard Assessment for Situational Awareness (LHASA) model encompasses of satellite-based precipitation estimation with landslide susceptibility map – It provides near real-time assessment of potential landslide events – LHASA nowcasts have a wider range of probability of landslide detection – LHASA is an open access framework that can be adapted to different scales	– LHASA has two categories for potential landslide activity – The highest hazard level locations tailor with the factors such as steep slopes, deforestation rate, seismicity, and physical infrastructure – LHASA provides near real-time information related to landslide events for situational awareness – LHASA facilitates cross-regional comparison and encourages more study – Potential landslide activity may be examined using LHASA at seasonal, yearly, and decadal stages – Organizations needing situational knowledge of landslide dangers at regional to global sizes are intended to use LHASA
Global landslide forecasting system and situational awareness (Khan et al. 2022*)*	
– Near real-time estimations of possible landslide hazards and exposure are provided globally by the Landslide Hazard Assessment for Situational Awareness (LHASA) – LHASA now has a precipitation forecast module that allows it to evaluate the risk of landslides up to three days ahead of time – The model uses assimilation products and precipitation, soil moisture, and snow depth data from the GEOS-FP prediction – The LHASA-F system shows promise in resolving extreme precipitation induced landslide events – This system provides useful information for landslide advisories and has the potential for predicting landslide globally	– LHASA-F provides landslide forecast outputs at global, regional, and local scales – Results are evaluated through mean absolute difference (MAD) and mean difference (MD) – The forecasted results can be mapped differently depending on the type of applications – Stakeholder engagement is important for the development and implementation of the system – Periodic assessment and regular user feedback are vital for operational success – Availability of diverse regional landslide inventories are crucial for system improvement – Rescaling of precipitation data is critical for incorporating forecasted data – Combining different forecasted data streams shows promise for larger events

(continued)

Table 20.4 (continued)

Case study description	Major insights/implication/conclusion
Community awareness on landslide disaster (Jamir and Khan 2018*)*	
– Nagaland is prone to natural disasters like landslides, earthquakes, and floods – Landslide disasters are frequent and increasing during the monsoon season – The government needs to involve the community and enforce strict rules – Losses due to disasters depend on the type and location of the event	– Emphasizes the need for community involvement in landslide disaster awareness – The State Government does not provide counseling or rehabilitation facilities to disaster-affected people – Most casualties from landslide disasters in Nagaland occurred in Mokokchung town – Local communities should be involved in awareness programs and training – NGOs and CBOs should work together with the government on disaster-related issues – Counselling and rehabilitation facilities should be provided to disaster-affected people – Strict rules should be implemented for construction to tackle disasters
Landslide public awareness and education programs (Motoyama and Abdullah 2013*)*	
– Malaysia has experienced significant landslides and floods due to intense rainfall – The National Slope Master Plan aims to create public awareness and education on landslide started in 2008 – The program focuses on identifying target audiences, building capacity, and effective communication	– Establishment of slope oversight committees at national level – Reporters published slope issues in newspapers regularly – Setup of SlopeWatch, a community-based landslide awareness building organization – Roadshows and exhibit display for general public awareness – Use of media to shape public opinion on slope safety – Key messages include "Learn, Monitor, Maintain, and Report" – A community-based organization and slope units have been formed in at-risk areas – Dramatic increase of public awareness campaigns since the major landslide in Bukit Antarabangsa

(continued)

Table 20.4 (continued)

Case study description	Major insights/implication/conclusion
Landslide awareness system (Laws) to increase the resilience and safety of transport infrastructure (Miele et al. 2021*)*	
– The Pan-American Highway in Ecuador was used as a case study – A mixed mothed was used for infrastructure monitoring – DInSAR technique, field surveys, and QPROTO tool were utilized – A cognitive map was created to characterize 22 km of the highway – Aims to increase awareness and manage landslide risks – About 7 km of the highway was found to be exposed to landslides – Different landslide types were identified and analyzed	– Implementation of a near real-time monitoring system – Identification of clusters for mass movements and development of a cognitive tool for better understanding of slope dynamics – The tool increases transportation safety and ensures resilient development – The suggested procedure provides strong support to disaster risk reduction and management – The inventory identified 141 landslides affecting or potentially affecting road infrastructure – Rockfall analysis indicates critical sectors with high propagation potential
Landslide risk perception, public awareness and knowledge (Hernández-Moreno and Alcántara-Ayala 2017*)*	
– Analyzes landslide risk perception in Teziutlán, Puebla—an area historically affected by landslides—Mexico – The questionnaire revealed that all participants were aware of landslide risk – The city center was considered the safest area due to its geological properties – The respondents expressed the need for disaster risk reduction initiatives – The Mexican Army was seen as the most accountable for landslide response	– Need for disaster risk reduction initiatives at the community level – Mexican Army has highest accountability for landslide disaster response – Municipal government and health institutions perceived as providing insufficient response – Disaster risk reduction efforts in the area have been limited – High level of confidence in Red Cross and scientists for information – There is a need to reduce vulnerability and enhance awareness and knowledge – Communication strategies should focus on hazard understanding, vulnerability, and integrated approaches – Resilience building provides positive externalities related to overall development process – Understanding landslide perception is crucial for enhancing risk awareness and knowledge

(continued)

Table 20.4 (continued)

Case study description	Major insights/implication/conclusion
Landslide exposure awareness and engagement of children (Ruiz-Cortés and Alcántara-Ayala 2020)	
– Focuses on raising awareness of landslide risks in peri-urban mountain areas of Mexico – Community-based mapping involving children were conducted to enhance disaster risk knowledge – Community-based landslide mapping is crucial for increasing landslide disaster risk awareness – Maps used to depict historical and current processes of landslide risks	– Community-based landslide exposure map addressed knowledge of local people including children – Map integrated elements on natural landscape and human impact contributing to landslide exposure – Challenges include inadequate disaster risk reduction (DRR) policies and associated social issues – Capacity building of children is necessary for effective DRR – Additional investment in community-centric DRR and engagement of authorities is needed – Decision-making structures should include local stakeholders for IDRiM implementation – DRR education is crucial for societal transformation and wise policymaking

and children about landslide risk management through active engagement and role playing. However, in this chapter, selected case studies were considered. Historical evaluation of landslide risk assessment and management, possible mitigation measures, and global awareness campaigns on landslide risk reduction was out of the scope of this chapter. Therefore, an-depth analysis is needed to better understand landslide risk management. Particularly, anthropogenic impacts on landslide events and future landslide impacts and mitigation scenarios under climate change and societal change, integrated management of landslide risks are the demand of time. Considering these future research directions, this chapter suggests that for long-term sustainable and integrated management of landslide risks, a joint effort involving all the relevant stakeholders is the key.

References

Ahmed B, Rahman MS, Rahman S, Huq FF, Ara S (2014) Landslide inventory report of Chittagong Metropolitan Area, Bangladesh

Akgul M, Akay AO, Ozocak M, Esin Aİ, Şenturk N (2022) A new approach to spatial risk analysis in the long-term (1950–2020) assessment of natural disasters (avalanche, landslide, rockfall, flood) in Turkey. Nat Hazards 114(3):3471–3508

ArcGIS for Desktop, Help (n.d.) NDVI function [WWW Document]. http://desktop.arcgis.com/en/arcmap/10.3/manage-data/raster-and-images/ndvi-function.htm. Accessed 23 Feb 19

Ayalew L, Yamagishi H (2005) The application of GIS-based logistic regression for landslide susceptibility mapping in the Kakuda-Yahiko Mountains, Central Japan. Geomorphology 65:15–31. https://doi.org/10.1016/j.geomorph.2004.06.010

Barla G, Antolini F, Barla M, Mensi E, Piovano G (2010) Monitoring of the Beauregard landslide (Aosta Valley, Italy) using advanced and conventional techniques. Eng Geol 116:218–235

Cardile G, Pisano M, Moraci N, Cazzuffi D (2019) Root reinforcement as a measure for shallow landslides risk mitigation. ArcHistoR EXTRA 6

Catani F, Casagli N, Ermini L, Righini G, Menduni G (2005) Landslide hazard and risk mapping at catchment scale in the Arno River basin. Landslides 2:329–342

Chalkias C, Ferentinou M, Polykretis C (2014) GIS-based landslide susceptibility mapping on the Peloponnese Peninsula, Greece. Geosciences 4:176–190

Champati ray PK, Dimri S, Lakhera RC, Sati S (2006) Fuzzy-based method for landslide hazard assessment in active seismic zone of Himalaya. Landslides 4:101. https://doi.org/10.1007/s10346-006-0068-6

Chang M, Cui P, Dou X, Su F (2021) Quantitative risk assessment of landslides over the China-Pakistan economic corridor. Int J Disaster Risk Reduction 63:102441

Chittagong Hill Tracts Development Facility (CHTDF) (n.d.) Earthquake risk reduction and recovery preparedness programme in Chittagong Hill Tracts Area

Couture R, Blais-Stevens A, Bobrowsky P, Wang B, VanDine D (2013) Canadian technical guidelines and best practices related to landslides: a national initiative for loss reduction. In: Sassa K, Rouhban B, Briceño S, McSaveney M, He B (eds) Landslides: global risk preparedness. Springer Berlin Heidelberg, Berlin, Heidelberg, pp 315–322. https://doi.org/10.1007/978-3-642-22087-6_22

Crawford MM, Dortch JM, Koch HJ, Zhu Y, Haneberg WC, Wang Z, Bryson LS (2022) Landslide risk assessment in Eastern Kentucky, USA: developing a regional scale limited resource approach. Remote Sens 14(24):6246

Crozier MJ, Glade T (2005) Landslide hazard and risk: issues, concepts and approach. Landslide Hazard Risk 1–40

Cruden DM (1993). Cruden DM, Varnes DJ (1996) Landslide types and processes. Special Report, Transportation Research Board, National Academy of Sciences, vol 247, pp 36–75

Dahal RK, Hasegawa S, Nonomura A, Yamanaka M, Masuda T, Nishino K (2008) GIS-based weights-of-evidence modelling of rainfall-induced landslides in small catchments for landslide susceptibility mapping. Environ Geol 54:311–324

Fartaria C, Pinto A, Cristovao A, Godinho L (2019) Landslide risk mitigation of "São Pedro de Alcântara Viewpoint Slope" in Lisbon Historical Center. In: 17th European conference on soil mechanics and geotechnical engineering

Floris M, D'Alpaos A, De Agostini A, Stevan G, Tessari G, Genevois R (2012) A process-based model for the definition of hydrological alert systems in landslide risk mitigation. Nat Hazard 12(11):3343–3357

Genevois R, Tecca PR, Genevois C (2022) Mitigation measures of debris flow and landslide risk carried out in two mountain areas of North-Eastern Italy. J Mt Sci 19(6):1808–1822

Glade T, Anderson MG, Crozier MJ (2006) Landslide hazard and risk. Wiley

Gorokhovich Y, Vustianiuk A (2021) Implications of slope aspect for landslide risk assessment: a case study of Hurricane Maria in Puerto Rico in 2017. Geomorphology 391:107874

Guzzetti F, Carrara A, Cardinali M, Reichenbach P (1999) Landslide hazard evaluation: a review of current techniques and their application in a multi-scale study, Central Italy. Geomorphology 31:181–216

Guzzetti F, Reichenbach P, Cardinali M, Galli M, Ardizzone F (2005) Probabilistic landslide hazard assessment at the basin scale. Geomorphology 72:272–299

Hader PRP, Reis FAGV, Peixoto ASP (2022) Landslide risk assessment considering socionatural factors: methodology and application to Cubatão municipality, São Paulo, Brazil. Nat Hazards 110(2):1273–1304

Hernández-Moreno G, Alcántara-Ayala I (2017) Landslide risk perception in Mexico: a research gate into public awareness and knowledge. Landslides 14:351–371
Hidalgo CA, Vega JA (2021) Probabilistic landslide risk assessment in water supply basins: La Liboriana River Basin (Salgar-Colombia). Nat Hazards 109(1):273–301
Highland L, Bobrowsky PT (2008) The landslide handbook: a guide to understanding landslides. US Geological Survey Reston
Huang J (2014) Investigation on landslide susceptibility using remote sensing and GIS methods
Islam MA, Islam MS, Islam T (2017) Landslides in Chittagong hill tracts and possible measures. In: Proceedings, international conference on disaster risk mitigation, Dhaka, Bangladesh
Jamir S, Khan NU (2018) Community awareness on landslide disaster: 'Experience of Nagaland'. In: Development and disaster management: a study of the Northeastern States of India, pp 361–371
Joyce KE, Samsonov SV, Levick SR, Engelbrecht J, Belliss S (2014) Mapping and monitoring geological hazards using optical, LiDAR, and synthetic aperture RADAR image data. Nat Hazards 73:137–163
Kalsnes B, Capobianco V (2022) Use of vegetation for landslide risk mitigation. Climate adaptation modelling. Springer International Publishing, Cham, pp 77–85
Kamal AM, Hossain F, Ahmed B, Rahman MZ, Sammonds P (2023) Assessing the effectiveness of landslide slope stability by analysing structural mitigation measures and community risk perception. Nat Hazards 117(3):2393–2418
Kanungo DP, Arora MK, Sarkar S, Gupta RP (2012) Landslide susceptibility zonation (LSZ) mapping—a review
Kanungo DP, Arora MK, Gupta RP, Sarkar S (2008) Landslide risk assessment using concepts of danger pixels and fuzzy set theory in Darjeeling Himalayas, Landslides 5:407–416. https://link.springer.com/article/10.1007/s10346-008-0134-3
Khan S, Kirschbaum DB, Stanley TA, Amatya PM, Emberson RA (2022) Global landslide forecasting system for hazard assessment and situational awareness. Front Earth Sci 10:878996
Kirschbaum D, Stanley T (2018) Satellite-based assessment of rainfall-triggered landslide hazard for situational awareness. Earth's Fut 6(3):505–523
Kuriakose SL (2010) Physically based dynamic modelling of the effects of land use changes on shallow landslide initiation in the Western Ghats, Kerala. ITC, India
Lee S (2005) Application of logistic regression model and its validation for landslide susceptibility mapping using GIS and remote sensing data. Int J Remote Sens 26:1477–1491. https://doi.org/10.1080/01431160412331331012
Lee S, Choi J, Min K (2004) Probabilistic landslide hazard mapping using GIS and remote sensing data at Boun, Korea. Int J Remote Sens 25:2037–2052
Lee C-T, Huang CC, Lee JF, Pan KL, Lin ML, Dong JJ (2008) Statistical approach to storm event-induced landslides susceptibility. Nat Hazards Earth Syst Sci 8:941–960
Marana B (2017) A simplified ArcGIS approach for landslides risk assessment in the Province of Bergamo. J Geogr Inf Syst 9:699–716
Miele P, Di Napoli M, Guerriero L, Ramondini M, Sellers C, Annibali Corona M, Di Martire D (2021) Landslide awareness system (Laws) to increase the resilience and safety of transport infrastructure: The case study of pan-American highway (Cuenca–Ecuador). Remote Sens 13(8):1564
Mondal S, Maiti R (2012) Landslide susceptibility analysis of Shiv-Khola watershed, Darjiling: a remote sensing & GIS based Analytical Hierarchy Process (AHP). J Indian Soc Remote Sens 40:483–496
Motoyama E, Abdullah CH (2013) Landslide public awareness and education programs in Malaysia. In: Landslide science and practice: Volume 7: Social and economic impact and policies, pp 291–296
Pardeshi SD, Autade SE, Pardeshi SS (2013) Landslide hazard assessment: recent trends and techniques. SpringerPlus, p 2. https://doi.org/10.1186/2193-1801-2-523

Rabby YW, Li Y (2019) An integrated approach to map landslides in Chittagong Hilly areas, Bangladesh, using Google Earth and field mapping. Landslides 16(3):633–645

Rotigliano E, Agnesi V, Cappadonia C, Conoscenti C (2011) The role of the diagnostic areas in the assessment of landslide susceptibility models: a test in the Sicilian chain. Nat Hazards 58:981–999

Ruiz-Cortés NS, Alcántara-Ayala I (2020) Landslide exposure awareness: a community-based approach towards the engagement of children. Landslides 17:1501–1514

Saaty TL (2008) Decision making with the analytic hierarchy process. Int J Serv Sci 1:83–98

Saraf A, Das J, Rawat V (2009) Satellite based detection of early occurring of co-seismic landslides. J South Asia Disaster Stud J. SAARC Disaster Manag Cent 2:47–55

Sarkar S, Kanungo DP, Patra AK, Kumar P (n.d.) GIS based landslide susceptibility mapping—a case study in Indian Himalaya, pp 9

Scolobig A, Thompson M, Linnerooth-Bayer J (2016) Compromise not consensus: designing a participatory process for landslide risk mitigation. Nat Hazards 81:45–68

Semlali I, Ouadif L, Bahi L (n.d.) Landslide susceptibility mapping using the analytical hierarchy process (AHP) and GIS: a study from Morocco

Shah NA, Shafique M, Ishfaq M, Faisal K, Van der Meijde M (2023) Integrated approach for landslide risk assessment using geoinformation tools and field data in Hindukush Mountain ranges, Northern Pakistan. Sustainability 15(4):3102

Shano L, Raghuvanshi TK, Meten M (2022) Fuzzy set theory and pixel-based landslide risk assessment: the case of Shafe and Baso catchments, Gamo highland, Ethiopia. Earth Sci Inform 15(2):993–1006

Smith H, Coupé F, Ferrari SG, Rivera H, Mera WEC (2020) Toward negotiated mitigation of landslide risks in informal settlements: reflections from a pilot experience in Medellín, Colombia. Ecol Soc 25(1):19

Soeters R, van Westen CJ (1996) Landslides: investigation and mitigation. Chapter 8—Slope instability recognition, analysis, and zonation. Transportation Research Board Special Report

Solana MC, Kilburn CR (2003) Public awareness of landslide hazards: the Barranco de Tirajana, Gran Canaria, Spain. Geomorphology 54(1–2):39–48

Susanto N, Prastawa H, Putranto TT, Zakina OA (2018) Improving the awareness index of government and non-departmental government institutions for landslide cases in Semarang city. In: AIP conference proceedings, vol 2019, no 1. AIP Publishing

Thapa PS, Adhikari BR, Shaw R, Bhattarai D, Yanai S (2023) Geomorphological analysis and early warning systems for landslide risk mitigation in Nepalese mid-hills. Nat Hazards 1–20

Tyoda Z (2013) Landslide susceptibility mapping: remote sensing and GIS approach (Thesis). Stellenbosch University, Stellenbosch

Varnes DJ (1984) Landslide hazard zonation: a review of principles and practice. Nat Hazards (3)

Wang D, Hao M, Chen S, Meng Z, Jiang D, Ding F (2021) Assessment of landslide susceptibility and risk factors in China. Nat Hazards 108:3045–3059

Wieczorek GF (1984) Preparing a detailed landslide-inventory map for hazard evaluation and reduction. Environ Eng Geosci Xxi:337–342. https://doi.org/10.2113/gseegeosci.xxi.3.337

Zhao Z, Chen J, Xu K, Xie H, Gan X, Xu H (2021) A spatial case-based reasoning method for regional landslide risk assessment. Int J Appl Earth Obs Geoinf 102:102381

Zhou X, Sun Z (2020) Quantitative assessment of landslide risk using Monte Carlo material point method. Eng Comput 37(5):1577–1596

Φουμέλης M, Λέκκας ΕΛ, Παρχαρίδης ΙΑ (2004) Landslide susceptibility mapping by GIS-based qualitative weighting procedure in Corinth area. Δελτίον Της Ελληνικής Γεωλογικής Εταιρίας 36:904–912

Chapter 21
Community-Based Landslide Disaster Mitigation on the Northern Slope of "Telaga Lele" Hill, Banjarnegara Regency, Indonesia

Hari Harjanto Setiawan and Mahdi Ibrahim Tanjung

Abstract This chapter will present community-based landslide disaster mitigation on the northern slopes of the "Telaga Lele" hill, Banjarnegara Regency, Indonesia. The landslide that occurred in December 2014 caused dozens of houses to be damaged, and hundreds of people were buried by landslide material. A total of 2,038 refugees and a total of 95 victims were found dead, and 13 people were declared missing. The method used to obtain data is qualitative, namely by interviewing people who know about the disaster. The limitation of this research is that observations were not carried out when the disaster occurred. Data is strengthened with GIS before and after the disaster. Disaster management reports are also used in this paper. The research results show that the Telaga Lele Hill Landslide event was caused by geological conditions with a slope of around 45°, heavy rainfall for three days, and a height of more than 1000 m. The community's lack of strategy in preventing disasters makes them vulnerable to landslides. Community-based disaster mitigation can be carried out in two ways: strengthening understanding of anticipating landslide hazards and biosystems by paying attention to the balance of slopes formed by plants.

Keywords Landslide · Natural disaster · Community-based · Adaptation · Mitigation

H. H. Setiawan (✉)
National Research and Innovation Agency (BRIN), Jakarta, Indonesia
e-mail: hari.harjanto.setiawan@brin.go.id

M. I. Tanjung
Ministry of Public Works and Housing (PUPR), Bandung, Indonesia

G. K. Panda et al. (eds.), *Landslide: Susceptibility, Risk Assessment and Sustainability*, Advances in Natural and Technological Hazards Research 52,
https://doi.org/10.1007/978-3-031-56591-5_21

21.1 Introduction

Indonesia often experiences landslides because of the many mountains or hills, including Banjar Negara Regency. Landslides have caused significant damage to public facilities and agricultural land and have resulted in casualties, affecting the economic activities of the affected population (Safriani and Wibowo 2022). Based on 2014 data, Indonesia's landslides have the highest incidents and death rates. As a result of the high frequency and devastating impact of landslides, it is evident that effective and targeted mitigation strategies need to be implemented in landslide-prone areas, such as the northern slope of "Telaga Lele" hill in Banjarnegara district. Before determining a strategy to overcome landslides, a study must first be conducted on the affected communities' physical and social conditions so that the strategy implemented can overcome them comprehensively.

Previous studies have focused on the social sciences of landslide disaster mitigation in Indonesia (Zamroni et al. 2020). These studies have explored various approaches and aspects of landslide risk reduction, including community-based risk reduction efforts, sociotechnical approaches, disaster mitigation programs, driving factors of social vulnerability, and public perception of disaster landslides. Understanding the weaknesses in landslide disaster mitigation in Indonesia is crucial, considering the actions taken by governments, researchers, and local communities. One area requiring specific attention is the northern slope of "Telaga Lele" hill in Banjarnegara district. The Banjarnegara district in Central Java has a history of landslide events (Muhammad et al. 2022).

Several studies have been conducted on evaluating, managing, and mitigating slide risk in various countries (Cantarino et al. 2021). Tarraguel et al. conducted a landslide risk assessment using a spatial analysis approach. Radosavljevic et al. analyze the dangers of flooding and erosion caused by global warming and rising sea levels (Zhao et al. 2020). Little research still focuses on community-based landslide disaster mitigation on the northern slopes of the "Telaga Lele" hill, Banjarnegara Regency, Indonesia. Community-based landslide disaster mitigation is crucial on the northern slopes of "Telaga Lele" hill, Banjarnegara district, Indonesia.

The northern slope of the "Telaga Lele" hill in Banjarnegara, Indonesia, is prone to landslides. Therefore, it is essential to implement community-based landslide disaster mitigation strategies in this region to minimize the risk and impact of landslides. Community-based landslide disaster mitigation is essential in reducing landslide risk (Hostettler et al. 2019) on the northern slopes of the "Telaga Lele" hill, Banjarnegara district, Indonesia. Implementing the Landslide Disaster Mitigation Strategy on "Telaga Lele" Hill can help strengthen community resilience and reduce the potential for damage due to landslides.

Disaster risk reduction efforts involving the community are very effective for disaster mitigation. The forms of activity are public awareness campaigns, evacuation drills, and developing an early warning system (Klimeš et al. 2019). This approach recognizes community members' local knowledge and capabilities, making them active participants in the mitigation process. Community-based landslide disaster mitigation can also utilize the community's traditional knowledge and oral traditions (Wibowo et al. 2018). Incorporating traditional community knowledge and practices can further increase the effectiveness of landslide disaster mitigation efforts on the northern slopes of "Telaga Lele" hill, Banjarnegara district, Indonesia. Implementing community-based landslide disaster mitigation strategies on the northern slopes of the "Telaga Lele" hill is essential to reduce the impact of landslides in vulnerable areas.

This landslide disaster shows how important community-based mitigation and preparedness is, which can be used to measure community recovery capacity. This is an important lesson because Indonesia is in a disaster-prone area, so the risk can be reduced. Massive human intervention in ecology and low mitigation readiness means disaster management needs improvement. This must be a priority for all policymakers, including society. When facing disaster risks, community-based disaster mitigation is key to maintaining security, welfare, and community resilience. Disaster mitigation requires proactive actions to reduce the negative impacts of disasters. Prioritizing community-based disaster mitigation is a sustainable and security investment that protects future generations from various threats of landslides.

This book chapter will answer three important questions: (1) How did the "Telaga Lele" Hill landslide occur? (2) What is the condition of vulnerable communities affected by landslides and (3) How is community-based disaster mitigation? So, it will contribute to developing community-based disaster mitigation in handling landslide disasters. Through these disaster risk reduction efforts, communities on the northern slopes of the "Telaga Lele" hill, Banjarnegara Regency, Indonesia, can be better prepared to face disasters. This chapter's novelty is that involving the community in every stage of disaster mitigation will reduce the impact of landslides. These three questions must be answered with data obtained using appropriate methods. So, interviews with people who experienced disasters and historical maps of landslides based on remote imagery are necessary.

21.2 The Rationale of the Study

21.2.1 Landslide Disaster Risk Analysis

To effectively implement community-based landslide disaster mitigation strategies, conducting a thorough risk analysis of the area is essential (Sultana and Tan 2021). This analysis should include an assessment of the geomorphology (Morino et al. 2021) of the northern slope of "Telaga Lele" hill and factors such as slope gradient, soil composition, and rainfall patterns. By analyzing the area's geomorphology, researchers can identify areas more prone to landslides and develop appropriate mitigation measures. Additionally, a thorough understanding of the hydrological characteristics of the area, such as surface water drainage patterns and groundwater conditions, is necessary to assess the potential impact of water on slope stability and landslide occurrence.

Furthermore, a comprehensive analysis of historical landslide events and their characteristics can provide valuable insights into the factors contributing to landslides in the area (Alfarabi et al. 2019). Conducting a risk analysis of the northern slope of "Telaga Lele" hill in Banjarnegara district, Indonesia, is crucial for implementing community-based landslide disaster mitigation strategies. Researchers can identify high-risk areas and prioritize mitigation efforts by analyzing the area's geomorphology, slope gradient, soil composition, rainfall patterns, and historical landslide events.

The results of this risk analysis will contribute to determining effective and targeted landslide disaster mitigation strategies in the community (Lin and Chang 2020). These strategies can include measures such as slope stabilization, reforestation, and establishing early warning systems. By involving the local community in these efforts and integrating disaster knowledge into education programs, residents can be empowered to participate actively in landslide risk reduction activities. This understanding will inform decision-making processes regarding spatial planning, disaster management, and implementing community-based landslide disaster mitigation strategies. Detailed mapping and analysis are crucial for spatial planning and decision-making regarding landslide-prone areas (Vojteková and Vojtek 2020).

The level of vulnerability is a factor that influences the risk of a disaster occurring (Di Napoli et al. 2023). Risk is a combination of danger and vulnerability in a landslide disaster. The level of physical, economic, social, and environmental hazards influences vulnerability. The level of physical vulnerability is calculated based on the percentage of built-up land area and the percentage of damaged road length (Cardoso et al. 2023). The larger the land area, the greater the building damage when a disaster occurs. Economic vulnerability is calculated based on the percentage of the population working in vulnerable sectors, the percentage of poor households, and the percentage of agricultural land and plantations (Jiang Li et al. 2023). Environmental vulnerability is calculated based on the percentage of forest area and bush area (Chang et al. 2023).

21.2.2 Community-Based Mitigation Measures

Based on landslide risk analysis, community-based mitigation measures can be implemented on the northern slopes of "Telaga Lele" hill, Banjarnegara district, Indonesia. Development and enforcement of land use regulations (Cheung 2021). Establishing regulations that guide appropriate land use in landslide-prone areas is essential. This may include zoning restrictions, building codes, and construction practice codes to ensure that structures are built in safe locations and designed to withstand potential landslide hazards. Education and awareness programs for communities in landslide-prone areas (Islam et al. 2021; Pollock and Wartman 2020). Community

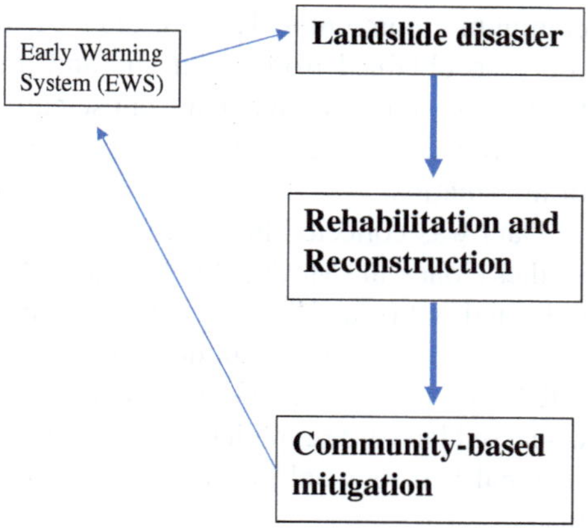

Fig. 21.1 Landslide disaster mitigation flow

members must be educated about the risks and impacts of landslides and the importance of implementing mitigation measures. This can be achieved through workshops, training sessions, and public awareness campaigns (Ahmed et al. 2023).

Implementing early warnings can alert communities to the danger of impending landslides, giving them time to evacuate or take necessary precautions. (Calvello et al. 2020). This may include installing monitoring equipment such as rain gauges, seismometers, and inclinometers to detect potential signs of slope instability. Partnerships with local communities and cross-ministerial initiatives in overcoming landslide disasters (Abdeen et al. 2021). Collaboration between government agencies, academics, non-governmental organizations, and local communities is critical to the success of landslide mitigation efforts. This can involve joint planning, resource sharing, and capacity building to implement mitigation measures effectively (Sassa et al. 2023). Reforestation and slope stabilization are essential in maintaining slope stability because they help absorb excess rainfall, reduce surface runoff, and bind the soil. Reforestation efforts can be carried out to restore conditions where the landslide occurred (Sirimorok et al. 2023).

These steps aim to reduce the risk to communities and buildings from landslides. This action will also reduce casualties, damage, and economic losses. Additionally, ongoing research and data collection are essential to enhance our understanding of landslide hazards and vulnerability in the area. Policymakers can prioritize funding for landslide risk mitigation (Fig. 21.1).

21.3 Materials and Methods

This study describes community-based landslide disaster mitigation on the northern slopes of the "Telaga Lele" hill, Banjarnegara district, Indonesia. Next, it describes the adaptation of local communities (Pittock and Jones 2000) to facing landslide disasters. Qualitative data was obtained through interviews and observations of the community, community leaders, and local government. Information about landslides was also obtained from electronic and print news analysis. Literature studies on landslides were also obtained from scientific journals in various countries. Reports, news from electronic media, and reputable scientific journals are also used as sources of information.

Data was collected through in-depth interviews (Prentice 2017, documentation studies (Jones and McCoy 2019), and observation (Greatorex 2014). The information collected is (a) landslide incidents, (b) how to survive in the atmosphere, and (c) related stakeholders. The in-depth interview process will be carried out in December 2022. The in-depth interview lasted approximately 60 to 90 min. The community discussed landslide experiences and how to overcome them. Community leaders and regional and regional governments discussed disaster management programs and plans.

The collected data was analyzed using thematic analysis (Sundler et al. 2019). Before data collection, the desired theme is first determined. The theme chosen was Community-based landslide disaster mitigation on the northern slope of "Telaga Lele" hill, Banjarnegara district, Indonesia. Writing this book chapter has limited time; namely, the time to observe disasters is done when a disaster occurs. The conditions described above do not fully describe the condition of Indonesia as a whole because the characteristics of landslides in each region are different. Information regarding the landslide disaster on Telaga Lele Hill uses news from electronic media. Apart from electronic media, it also uses Google Earth Satellite Photos regarding the history of landslides.

The historical landslide inventory map combines the results of field investigations and the interpretation of remote sensing data (Confuorto et al. 2023). Two-time intervals from historical landslide inventory maps were used for landslide frequency assessment to evaluate the impact of data inhomogeneity. The map consists of two periods, namely 2014 before the landslide occurred and 2015, 2016, and 2023 after the landslide occurred. A historical landslide inventory map was used for vulnerability analysis.

21.4 Results and Discussion

21.4.1 The "Telaga Lele" Hill Landslide Incident

The landslide disaster on Telaga Lele Hill occurred in 2014. This area is prone to landslides with moderate to high intensity. The surrounding area received heavy rain on December 10 and 11, 2014. The soil at that location is old volcanic deposits, so the soil layer is thick. The rain that fell for two days caused it to fill with water, causing the Telaga Lele Hill to experience weathering. The slope of the hill is less than 60%, and the plants around the hill cannot hold water. The cause of the landslide was the condition of the area, which did not allow it to be used as a residential area.

Fig. 21.2 "Telaga Lele" Hill Landslide Incident

The hilly topography and loose soil texture make this area vulnerable to landslides (Biswas et al. 2017).

The role of the state and society is very necessary in dealing with landslides. The state's role is represented by the existence of the National Agency for Disaster Countermeasures, which immediately takes action when a disaster occurs. Community participation is necessary to reduce and avoid disaster risks (Marston et al. 2020). Increasing community awareness and capacity is a form of community participation in landslide disaster mitigation. The community is the party that has direct experience of disaster events, so their understanding becomes capital for reducing disaster risk. Community response to disasters is very important to reduce disaster risk (Fig. 21.2).

21.4.2 Geological Condition

This area has relatively wavy topography, and the slopes are not too steep. Telaga Lele Hill is located in these mountains, with hills with a slope of 35–45° and a height of 900–1100 m above sea level. The landslide point is located on the upper slope of Mount Telaga Lele with a slope of around 45°, including a zone prone to high-ground movement (Sobarna and Hartanto 1992); see Fig. 21.3. Meanwhile, residential areas affected by landslides are in the valley around the Petir River.

Telaga Lele Hill is characterized by its hilly terrain. This hill has steep and undulating slopes, with dense vegetation covering its surface. This hill is famous for its beautiful views and lush greenery. This hill is also intersected by several streams and rivers, which add to its overall beauty and provide water for the surrounding area. However, based on the available details, it can be concluded that Telaga Lele Hill

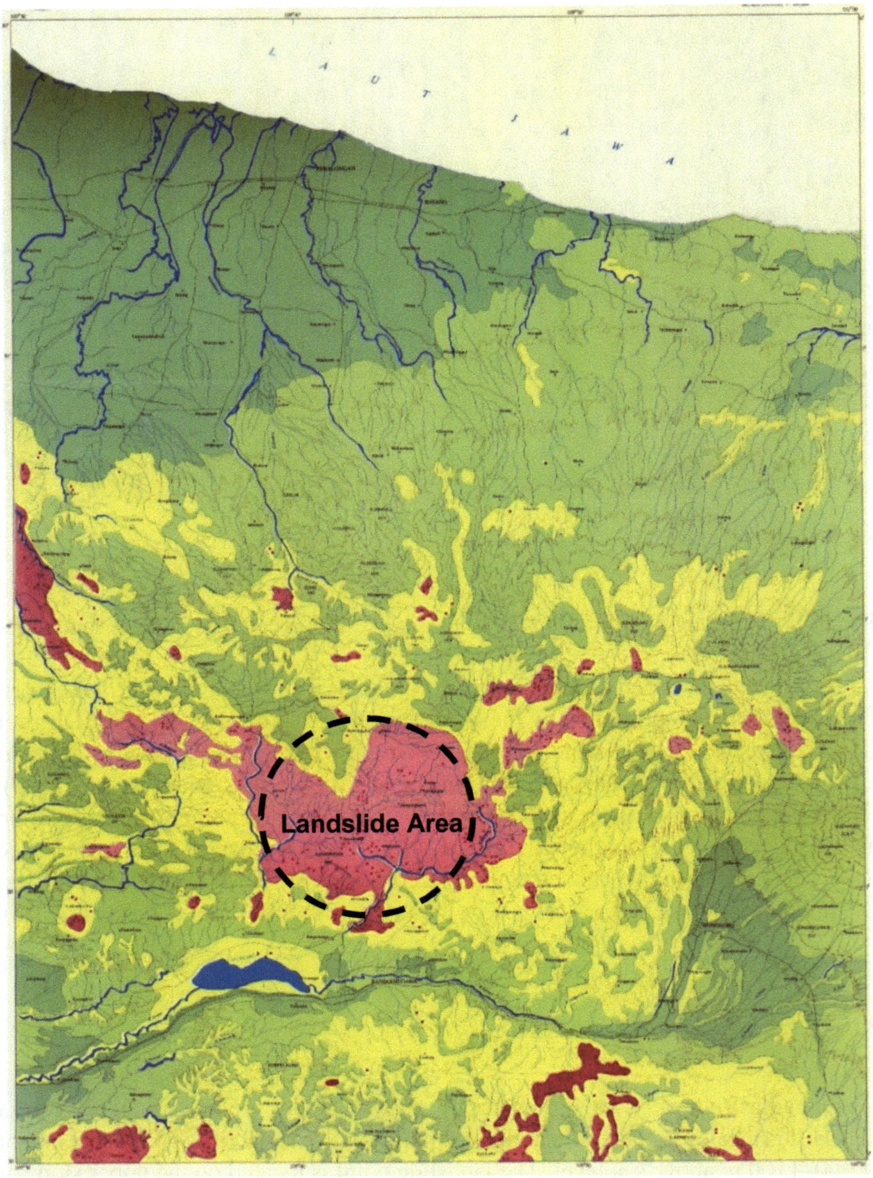

Fig. 21.3 Map of Land Movement Hazard Zones in the Banjarnegara and Surrounding Areas. (*Source* Sobarna and Hartanto 1992)

is a prominent natural feature in the area, offering beautiful views and opportunities for outdoor activities such as hiking and nature exploration.

Hills dominate the topography of Telaga Lele Hill compared to the lowlands. Landslide disasters are nothing new for society. The landslide disaster was caused by rain that fell for two consecutive days in this area. The landslide just happened, flattening all the buildings in front of it. Telaga Lele Hill, 400 m high, experienced a landslide that hit Jemblung Hamlet. The direction of the landslide from the starting point was towards the northwest, then hit the river wall and turned southwest towards dense settlements. The type of landslide was a slumping (Auger et al. 2023) (rotational slide) type. Rotational landslides have upward curves and often occur in soil masses that move in one unit.

The characteristic of this type of landslide is movement along curved fracture surfaces, including unconsolidated or weakly consolidated material. This is caused by erosion along the base of the slope. Material from this type of slump failure moves down along the curved plane of the crack so that the block experiences downward rotation. The landslide disaster caused the community to experience losses, which can be grouped into physical and non-physical aspects (Li and Mind'je 2023). Losses from the physical aspect include damage to residential buildings, closure of primary road access, damage to residents' rice fields and gardens, and the closure of rivers due to landslides. Landslide disaster losses, when viewed from a non-physical aspect, can be losses when viewed from the psychological aspect of society. The losses incurred are in the form of trauma experienced by the community (Kabunga et al. 2022) (Fig. 21.4).

The land use around the disaster location is a 15-hectare village, rice fields, mixed gardens, and Telagalele Hill, a steep hill covered with hardy plants for seasonal vegetable crops such as mustard greens and cabbage. The rocks that make up the Telagalele hill that collapsed are debris from volcanic eruptions in the form of lava and colluvial deposits that have experienced high levels of weathering and turned into loose, brownish-red soil with a thickness of around 2–3 m. The geological structure developed around the landslide location is a downward fault trending west–east. From the results of field tests at the Bukit Telaglele village road using hand vane shear, the peak undrained shear strength values varied between 21 and 68 kPa, and the remolded undrained shear strength varied between 5 and 21 kPa. Thus, when this soil experiences large deformation, this land is included in the medium-sensitive category.

21.4.2.1 Rainfall

The Class III Geophysical Station of the BMKG (Meteorology, Climatology, and Geophysics Agency) Banjarnegara recorded rainfall throughout Thursday, December 11, 2014, reaching 112.7 mm, and on Friday the rainfall was still 101.8 mm. In two days, the intensity of rain that fell in Banjarnegara reached 214.5 mm. This heavy rainfall is believed to be the cause of the landslide that occurred on Telaga Lele Hill. The landslide on Telaga Lele Hill is thought to have been caused by heavy rainfall

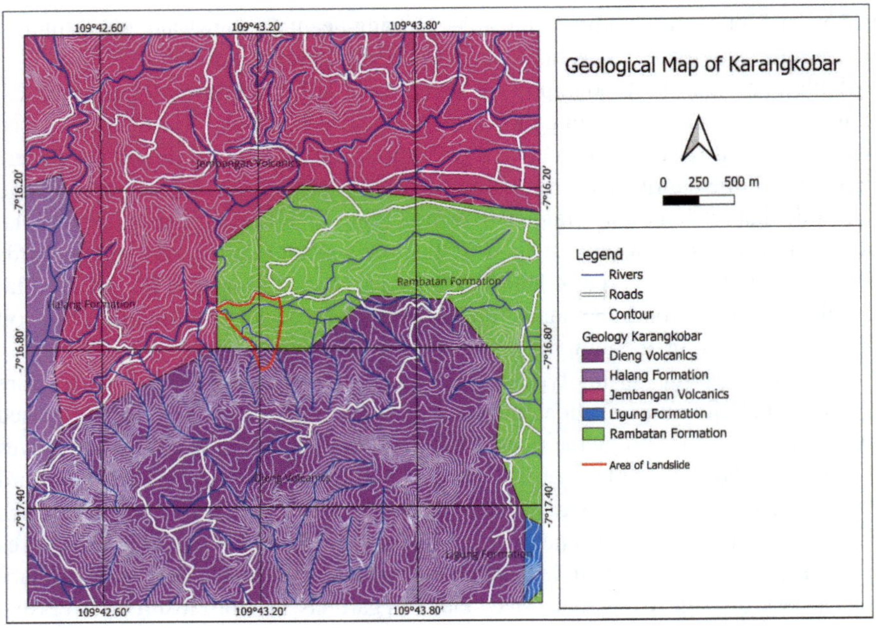

Fig. 21.4 Geological map of the Karangkobar region and surrounding areas (*Source* Karangkobar District)

(Dikshit et al. 2019). This heavy rainfall makes the rocks on the hill increasingly unstable, leading to landslides. The intensity of rainfall in Bukit Telaga Lele on 11 and 12 December 2014 exceeded the threshold. The threshold value for rainfall in Banjarmangu District, Banjarnegara Regency, cumulative for three days, is 40 mm. This shows that rainfall on Telaga Lele Hill exceeds the threshold value, thereby increasing the possibility of landslides.

The high rainfall caused Telaga Lele Hill to become increasingly unstable, resulting in landslides. The increasing frequency and intensity of extreme weather and weighty rainfall also contribute to the vulnerability of hilly areas and steep slopes to landslides. This vulnerability was seen in the landslide incident on Telaga Lele Hill, where high rainfall exceeding the threshold value for cumulative rainfall for three days made rocks unstable and caused landslides.

21.4.2.2 Causes and Mechanisms of Land Slide

Landslides are the movement of masses of rock, soil, or debris, a mixture of soil and rock, by gravity towards the base of a slope (Sun et al. 2023). Geologically, a landslide is a geological event with ground movement, such as the fall of rocks or large lumps of soil. Banjarnegara Regency is an area prone to natural disasters, especially landslides. Most of the topography of Banjarnegara Regency (65%) is at an altitude of between 100 and 1000 m above sea level. The highest potential for landslides in this area is in Karangkobar District. This is because Karangkobar District has an altitude of more than 1000 m above sea level. Therefore, Karangkobar District is classified as an area with a high potential for a landslide. Banjarnegara is one of the Central Java districts with relatively high landslide-prone areas.

Data from the Regional Disaster Management Agency Banjarnegara shows 57 landslides occurred in 2007. The number of incidents increased in 2008 to 76. In 2009, it increased by 126 times; in 2010, it increased sharply by 200. Jemblung Hamlet, located in Sampang Village, Karangkobar District, Banjarnegara Regency, a landslide occurred on Friday, December 12, 2014. The number of residents in Jeblung Hamlet was 308 people. Two hundred people survived, 51 victims died, and 57 other victims are still missing. Jemblung Hamlet is an area prone to landslides with medium–high intensity. The factor that causes landslides is that many fault lines pass through the area where the landslide occurs, a hilly land texture with steep and upright slopes (Meng et al. 2021). The landslide was preceded by the area around Jemblung Hamlet, Banjarnegara, being hit by heavy rain for two days.

Several factors caused the landslide disaster at this location: (1) High rainfall on December 11, 2014, reached 112.7 mm. (2) The rocks that made up the Telagalele hill that collapsed were the remains of volcanic eruptions in the form of lava and colloquial deposits that had undergone high levels of weathering and turned into loose soil. (3) There is a downward fault on Telagalele Hill's slopes, which is considered a landslide area. (4) The slope of the Telagalele hill that experienced the landslide was $\pm 45^\circ$.

21.4.3 *Communities Are Vulnerable to Being Affected by Landslides.*

Almost 70–75% of the Banjarnegara area is prone to landslides because it is located on the slopes of the Dieng plateau. At least 12 sub-districts out of 20 in Banjarnegara Regency, Central Java, are included in the red zone or prone to landslides because the area is hilly and the ground moves quickly. Landslides often occur in Banjarnegara. The local community should understand and anticipate them so they can live in harmony with nature (Ali et al. 2021). This is important because, so far, the concept of self-help disaster mitigation for communities living in disaster-prone areas is still weak. Until now, people generally still rely on government initiatives and assistance. As a result, the risk of falling victim when a disaster occurs cannot be reduced. Many residents who live in the hilly area of Banjarnegara admit that they do not have insight into disaster mitigation, which relies only on natural phenomena. Communities are vulnerable to being affected by landslides. We can see a comparison of conditions before the landslide, after the landslide, and in the current conditions.

The mechanism of soil movement is as follows: due to high rainfall, the soil becomes saturated with water. The density of loose, weathered soil decreases, and its shear strength decreases. The fault structure cuts the hillside into a weak, landslide shear plane. Saturated weathering soil adds weight to the land mass on a steep slope so that the slope moves to find a new balance, and soil movement occurs. To reduce the risk of victims of natural disasters due to land movement, settlements that are not directly affected by landslides must be relocated to places that are safe from the possibility of landslides. Outreach to residents in areas with the potential for landslides is very necessary. There is a need for an early warning system to detect landslides (Hidayat et al. 2019). Carry out greening on the hills to prevent land erosion.

The length of the landslide is ±530 m and the width at the foot is ±430 m. Based on Google Earth satellite photos (Fig. 21.5) taken in 2014 before the landslide, it shows that most of the land was on plantations. Most of the residential areas were only at the foot of the landslide. Landslides hit residential areas and vehicles passing on the road at the foot of the landslide (Fig. 21.6).

After one year, namely 2015, the landslide area was still not covered by plants. This shows that the land is still unstable and vulnerable to further landslides. Therefore, local communities must develop community-based landslide disaster mitigation strategies to reduce vulnerability and increase resilience in facing future landslide events (Raška 2019). This strategy should include identifying disaster-prone areas, implementing control techniques, and establishing an early warning system.

Two years later, in 2016, plants began to grow. This shows the positive impact of vegetation cover in stabilizing landslide-prone areas Fig. 21.7 and reducing the risk

Fig. 21.5 "Telaga Lele" Hill before the landslide occurred

Fig. 21.6 The condition of the landslide area after one year was still not covered by plants

of further landslides. Implementing community-based landslide disaster mitigation efforts, such as establishing early warning systems and promoting vegetation cover, has shown positive results in stabilizing landslide-prone areas and reducing the potential for future disasters. By engaging the local community in disaster preparedness and mitigation, the reliance on natural phenomena alone can be overcome, creating a safer living environment for those in the area.

Actively involving local communities in community-based landslide disaster mitigation can foster a sense of responsibility for their safety and welfare. Apart from that, swales or drainage channels need to be made on land planted with vegetables (seasonal crops). Roads that experience landslides need to be strengthened to withstand further landslides. Geotechnical investigations and topographic measurements are required to determine the type of reinforcement (Fig. 21.8).

Four years later, namely in 2018, the landslide area was re-grown with plants, and gradually, until 2023, the landslide area returned to plantations. In 2023, ten years after the landslide occurred, the affected area still shows signs of significant damage and instability. This shows that long-term and sustainable efforts are needed to rehabilitate areas affected by landslides and ensure their stability. Usually, they rely on farming for their economy, so living in a vulnerable hilly area like this seems a rational choice. In general, rural areas prone to landslides are located under large hills, some of which are covered with seasonal plants. As a result, landslides are very easily triggered during the rainy season. So far, residents have relied more on disaster outreach from the government.

Fig. 21.7 Landslide conditions after two years of planting

Fig. 21.8 Google Earth Satellite Photo of the landslide area in 2023 after the landslide

Residents have initiated no self-help movement to raise awareness of disasters (Carrasco & O'Brien 2022). The behavior of residents around disaster-prone areas is no longer by conservation principles. An example is Jemblung Hamlet, which experienced a landslide at the end of 2014. Of the around 150 families in the hamlet, 75% of them have fish ponds around their houses. These fish ponds are vulnerable to triggering landslides because they will make the soil soft. The first step that must be taken by the government and the community is to recognize the characteristics of Jemblung Hamlet and its surroundings, for example, regarding the landform in the form of hills with moderate to steep slopes, volcanic rocks, varying slope shapes, soil types, rainfall conditions, and the intensity of rivers in the hamlet.

In general, the geological condition of this region has unstable soil. When it rains, the soil layer is vulnerable to sliding or landslides. Another cause that is no less important is the inappropriate land use system by the community. For almost ten years, the hillside, previously covered with trees and perennials, has now been replaced by salak plants (Salacca zalacca). This plant with thorny stems, developed in monoculture by residents, has a drawback because it does not have a firm rooting texture in the soil (Kool et al. 2019). The short root fibers do not go deep into the soil, making the texture of the soil layer unstable.

The root fibers of the salak tree are short in the soil and make the soil looser and brittle. Because of this, the large number of snake fruit trees planted by residents in Karangkobar is one of the factors that makes the land in the area accessible to move. Therefore, an alternative policy that can be implemented is restructuring the area's residential areas. Areas designated as prohibited areas for residential areas should be enforced so that they do not become disasters in the future. Naturally, the Karangkobar area is vulnerable but not necessarily potentially dangerous. However, dangers can quickly arise, either through rain or incorrect land use.

Based on the author's observations in the field, many people in Karangkobar District know that their land is prone to landslides. However, they are determined to remain in the area for various reasons, generally because of their socio-economic background. Therefore, for the Banjarnegara Regency Government, the issue of land use in this disaster-prone area needs to be prioritized. Relatively uncontrolled land use can cause landslides now and in the future. One of the things that needs our attention in the future is the management of donations for refugees. So far, the average donor wants donations directly from the victim or the victim's family. As a result, donations pile up. If victims can manage aid funds well, these funds can be used as well as possible in the aftermath of a disaster. On the other hand, if the victim community cannot manage aid funds well, then the aid funds will quickly run out, and nothing will be left.

According to Musobihin, a donation management volunteer, when the landslide disaster occurred in Banjarnegara, there was a huge flow of aid funds. The institution he founded manages donations of almost IDR 2 billion. Not to mention other donation-recipient institutions that existed before the landslide occurred. This is where efforts to ensure transparency in the flow of funds must be prioritized. Refugees or relocated communities are the part that must be prioritized so that they can survive after a disaster occurs. Relocation is an effort to move part or all of an activity, along with facilities and infrastructure supporting the activity, from one place to another to increase the safety, feasibility, and legality of use while still paying attention to the relationship between what is being moved and the natural and built environment at the destination.

The relocation destination areas for refugees in Banjarnegara are only different villages. The location of the relocation village is only next to each other so that it does not entirely uproot the socio-cultural roots of the previous area. Relocation land is also proprietary. This differs from Sinabung refugees, whose land status is still on lease. With the relocation location only being a different village, the cultural roots of the Banjarnegara refugees have not been completely uprooted. They can still easily access the location of their destroyed house, even though they are prohibited from using it because the area has been declared a "red area or cannot be used anymore."

21.4.4 Community-Based Disaster Mitigation

The adaptation process is essential for society's adaptation to the existing environment (Rotaru et al. 2021). The adaptation process for people is essential because they have to readjust to their new environment. This adaptation process is carried out through the initial stages until an outcome of the adaptation process carried out by the community is achieved. The adaptation strategies (Pennington et al. 2022)used by society are adaptation by reaction, adaptation by reaction, and adaptation by withdrawal.

The first adaptation strategy used by society is adaptation by reaction (Setyawan and Mukminan 2021). This adaptation was carried out by planting several trees in locations affected by landslides. The second strategy used by society is adaptation by adjustment. This strategy takes the form of changing people's behavior towards the environment to change the environment for the better. The primary adaptation strategy chosen in dealing with landslides is adaptation by withdrawal. This adaptation involves moving out of the living environment to a safer place. This is because it is prone to landslides again. Conditions after the landslide disaster made it impossible to rebuild their homes, forcing all residents to move.

People adapt by moving to new environments from old environments (Ahmed 2021). The landslide location is not recommended as a place to live again. The government provides options for residents to move to safer places. The location used as a new residence is chosen based on the community's choice. The community was moved to a new location. The land area provided to build new settlements for the community is the community can take 6,100 m^2. 4 action options to live with the risk of disaster, including (1) Do nothing. (2) Protect the society from danger (3) Avoid disasters (4) Living with the dangers and risks.

So, from these four options, how society adapts and what factors influence this adaptation strategy can be drawn. Three factors determine a community adaptation strategy to face landslides: environmental conditions, community understanding of disasters, and the assistance provided to the community. The environmental conditions are those that can no longer be used as residential land due to landslides (Pereira et al. 2020). Jemblung Hamlet has been designated as an area prone to landslides. Therefore, the government has prohibited the construction of buildings in the landslide area. The hill above is prone to landslides because its condition is very worrying, which is one of the factors why it is no longer allowed to be used again.

Landslide locations surrounded by hills should not be used as residential land. Its location on a fault line makes this hamlet prone to landslides. Community understanding of landslide disasters can be known through community interpretation of the level of landslide vulnerability in their area. An area that has a high level of landslide vulnerability but the public's understanding of the level of landslides is low will endanger the people living in that area. The low level of public understanding of areas with a high level of vulnerability to landslides will cause people to be less aware of the potential for landslides to occur, which will cause losses to the community and environmental damage (Antronico et al. 2020). The public's understanding and knowledge about landslides is that landslides are a collection of soil and rocks that fall from the top of a hill onto all objects below it.

The public's understanding of landslides is still insufficient to say that the public understands (Ruiz-Cortés and Alcántara-Ayala 2020). People do not yet have insight into disaster mitigation, so they only rely on natural phenomena that occur in the environment around which they live. After conducting interviews regarding the community's understanding of the landslide disaster, the community did not understand what caused the landslide in their village to occur. It can be concluded that public understanding of landslide disasters and risks is still low. The people of Jemblung Hamlet generally do not know the cause of landslides in their hamlet because their behavior does not pay attention to the environment. People do not understand how to preserve the environment they live in. The landslide disaster attracted a lot of sympathy from many people who saw it.

Landslide incidents are not only borne by the community itself; in this case, the government and other institutions also take part in overcoming the landslide disaster (Nishikawa 2023). Many parties donated aid to the victims of the Jemblung Hamlet landslide. Assistance for victims of the landslide disaster in Jemblung Hamlet does not only come from Banjarnegara residents but also from residents of other districts or individuals, NGOs, schools, banks, cooperatives, foundations, villages, and agencies from other districts. Assistance provided by the government is one of the factors that influences people to move to a new environment. The government provided permanent housing to the people of Jemblung Hamlet to replace their houses, which landslides had destroyed. This permanent residence was given to 27 heads of families, which means there were 27 types of 36 permanent residences built by the government. The permanent residential complex built by the government not only contains houses for landslide victims, but the government is also building facilities for the community.

Community understanding is in the form of knowledge and perceptions, which are actualized in attitudes and actions in dealing with disasters (Bukhari et al. 2023). The result of people's attitudes and actions in facing disasters is an adaptation strategy, which means adjustments made due to threats to the surrounding environment. Considering the importance of adaptation and social resilience in communities experiencing disasters (Peng et al. 2019), in the future, there is a need to increase studies on adaptation strategies carried out by communities, both at the individual and group levels, in facing disasters. Adaptation is the final result of people's attitudes that emerge based on their perception and knowledge of potential disasters. Studies on adaptation strategies and social resilience are rarely carried out in Indonesia.

Understanding the anticipation of landslide danger should be known to the entire community so that it is hoped that the community will be able to overcome the conditions in their area. The central and regional governments should focus on efforts to raise awareness among the people of Banjarnegara that they live in a disaster-prone area. They must be properly educated about disaster mitigation (Alam and Ray-Bennett 2021). As in the Jemblung Hamlet area, the most appropriate treatment is a bio-engineering system. This bio-engineering method handles landslides by paying attention to the balance of the slopes formed by plants.

On Telaga Lele Hill and its surroundings, plant arrangements should pay attention to the slope angle and position of the slope (Huang et al. 2021). Secondary crops can be planted on lower slopes, and the slope angle is not steep, but there must be plants that strengthen the slope, such as calliandra and other plants that have strong roots, even though there are not many of them. Then, further up, with a higher slope angle, it is still permissible to plant secondary crops, but the number of plants supporting the slope must be increased because the slope condition is quite steep and has the potential for landslides. Furthermore, on very steep slopes, only slope-strengthening plants may be permitted.

Perennial plants can also function as water absorbers, thereby reducing the water concentration in the soil, which can naturally prevent landslides. But the most important thing is that this plant is a soil support plant in slope areas, which will function to defend the soil from landslides. The community around landslide-prone areas must truly understand this. Non-structural activities such as outreach on landslide prevention are essential in handling landslides and must be carried out immediately.

The government should collaborate with all elements of society for this activity (Matsuoka & Gonzales Rocha 2020), for example, by involving universities, NGOs, and community organizations in the community so that the frequency of landslides can be reduced and the community can prevent landslides. Landslide disasters are nothing new for the people of Karangkobar District, Banjarnegara, Central Java. Almost every year, similar disasters occur in this sub-district, which has 13 villages. Almost every year, landslides claim the lives and property of residents in this area. Naturally, the condition of the area in Karangkobar District, where Jemblung Hamlet is located, does not allow it to be used as a settlement.

21.5 Limitations of the Study

This research is only limited to answering research problems. This study cannot be generalized, considering that the characteristics of each landslide area are always different. However, the results of this research can be used to strengthen community mitigation in dealing with landslides. Therefore, local communities with other characteristics must prepare an assessment of community conditions. This research has limited observation time, which impacts the research results. Researchers did not observe directly when the event occurred. Researcher events from the media recorded by the community when a landslide occurred. So, research results cannot be used at different times. The reliability of this research cannot be used as a measurement standard because the social situation of society is unique and unstable.

21.6 Recommendations

Based on the data above regarding landslide conditions, vulnerability conditions, and community-based landslide disaster mitigation, the following can be recommended:

a. Communities in landslide-prone areas should prioritize flood preparedness and adaptation strategies to reduce disaster risk.
b. Disaster management agencies should develop practical tools and strategies that consider cultural perceptions and attitudes to achieve disaster risk reduction objectives.
c. Local chiefs and community leaders should be actively engaged as critical partners in all stages of disaster preparedness and capacity-building programs.
d. Local communities should be consulted and involved in decision-making processes related to adaptation and capacity-building activities.
e. Disaster resilience efforts should shift from short-term relief to long-term support for affected communities.
f. More studies and research on adaptation strategies and social resilience in the face of disasters should be conducted in Indonesia.
g. Community-based bio-engineering systems should be implemented in landslide-prone areas as an effective slope stabilization and mitigation method.

21.7 Conclusions

Community-based landslide disaster mitigation is an effort to reduce disaster risk. The government's strategy is to build public awareness about reducing disaster risk by holding community meetings to remind them when a disaster occurs. Community participation and empowerment are crucial in building resilience and reducing disaster risk. Community awareness, risk perception, and adaptation are important factors that governments must consider when planning and implementing community-based disaster management strategies.

These factors are important because communities are often the first to respond to disasters and can play an important role in disaster prevention and response efforts. Public awareness, risk perception, and adaptation in disaster management cannot be underestimated. Another recommendation is to foster community participation to carry out community-based disaster management efforts. To improve disaster mitigation efforts, the government must proactively encourage effective communication between scientists, policymakers, and local communities. Furthermore, programs that have been implemented are routinely evaluated. This evaluation is regularly carried out from the lowest to the highest level of government.

Acknowledgements We thank the National Research and Innovation Agency (BRIN) and the Ministry of Public Works and Housing (PUPR) for supporting this research in 2022.

References

Abdeen FN, Fernando T, Kulatunga U, Hettige S, Ranasinghe KDA (2021) Challenges in multi-agency collaboration in disaster management: a Sri Lankan perspective. Int J Disaster Risk Reduct 62:102399. https://doi.org/10.1016/J.IJDRR.2021.102399

Ahmed B (2021) The root causes of landslide vulnerability in Bangladesh. Landslides 18(5):1707–1720. https://doi.org/10.1007/S10346-020-01606-0/FIGURES/9

Ahmed B, Arfanul Alam SMR, Ahmed I, Sammonds P (2023) The anthropogenic aggravation of landslide disasters in Bangladesh: key informants' perspectives, pp 385–401. https://doi.org/10.1007/978-3-031-18471-0_28

Alam E, Ray-Bennett NS (2021) Disaster risk governance for district-level landslide risk management in Bangladesh. Int J Disaster Risk Reduct 59:102220. https://doi.org/10.1016/J.IJDRR.2021.102220

Alfarabi MS, Manessa MDM, Rustanto A, Ristya Y (2019) Geomorphology and Landslide-Prone Area in Cisolok District, Sukabumi Regency. In: E3S web of conferences 125:01005. https://doi.org/10.1051/E3SCONF/201912501005

Ali T, Paton D, Buergelt PT, Smith JA, Jehan N, Siddique A (2021) Integrating Indigenous perspectives and community-based disaster risk reduction: a pathway for sustainable Indigenous development in Northern Pakistan. Int J Disaster Risk Reduct 59:102263. https://doi.org/10.1016/J.IJDRR.2021.102263

Antronico L, De Pascale F, Coscarelli R, Gullà G (2020) Landslide risk perception, social vulnerability and community resilience: the case study of Maierato (Calabria, southern Italy). Int J Disaster Risk Reduct 46:101529. https://doi.org/10.1016/J.IJDRR.2020.101529

Auger B, Nevalainen L, Blais JM, Thienpont JR, Korosi JB (2023) A comparative paleolimnological analysis of Chydorus exposure to ultraviolet radiation associated with shoreline retrogressive thaw slumping in lakes of the Mackenzie Delta uplands (Northwest Territories, Canada). J Paleolimnol 70(3):193–204. https://doi.org/10.1007/S10933-023-00290-Z/FIGURES/4

Biswas RN, Islam MN, Islam MN (2017) Modeling on management strategies of slope stability and susceptibility to landslides catastrophe at hilly region in Bangladesh. Model Earth Syst Environ 3(3):977–998. https://doi.org/10.1007/S40808-017-0346-4/TABLES/10

Bukhari MH, da Silva PF, Pilz J, Istanbulluoglu E, Görüm T, Lee J, Karamehic-Muratovic A, Urmi T, Soltani A, Wilopo W, Qureshi JA, Zekan S, Koonisetty KS, Sheishenaly U, Khan L, Espinoza

J, Mendoza EP, Haque U (2023) Community perceptions of landslide risk and susceptibility: a multi-country study. Landslides 20(6):1321–1334. https://doi.org/10.1007/S10346-023-02027-5/TABLES/7

Calvello M, Devoli IG, Freeborough IK, Gariano ISL, Guzzetti IF, Kirschbaum ID, Nakaya IH, Robbins IJ, Stähli IM (2020) LandAware: a new international network on Landslide Early Warning Systems. Landslides 17(11):2699–2702. https://doi.org/10.1007/S10346-020-01548-7

Cantarino I, Carrion MA, Palencia-Jimenez JS, Martínez-Ibáñez V (2021) Landslide risk management analysis on expansive residential areas-case study of la Marina (Alicante, Spain). Nat Hazard 21(6):1847–1866. https://doi.org/10.5194/NHESS-21-1847-2021

Cardoso A, Pereira S, Ferreira TM, Zêzere JL, Melo R, Vaz T, Oliveira SC, Garcia RAC, Santos PP, Reis E (2023) On the physical vulnerability of buildings exposed to landslide hazards in the Lisbon Metropolitan Area, pp 117–144. https://doi.org/10.1007/978-981-99-0745-8_7

Carrasco S, O'Brien D (2022) Incremental pathways of post-disaster housing self-recovery in villa verde, chile. Architecture 2(3):544–561. https://doi.org/10.3390/ARCHITECTURE2030030

Chang M, Dou X, Su F, Yu B (2023) Remote sensing and optimized neural networks for landslide risk assessment: Paving the way for mitigating Afghanistan landslide damage. Ecol Ind 156:111179. https://doi.org/10.1016/J.ECOLIND.2023.111179

Cheung RWM (2021) Landslide risk management in Hong Kong. Landslides 18(10):3457–3473. https://doi.org/10.1007/S10346-020-01587-0/FIGURES/16

Confuorto P, Casagli N, Casu F, De Luca C, Del Soldato M, Festa D, Lanari R, Manzo M, Onorato G, Raspini F (2023) Sentinel-1 P-SBAS data for the update of the state of activity of national landslide inventory maps. Landslides 20(5):1083–1097. https://doi.org/10.1007/S10346-022-02024-0/FIGURES/10

Di Napoli M, Miele P, Guerriero L, Annibali Corona M, Calcaterra D, Ramondini M, Sellers C, Di Martire D (2023) Multitemporal relative landslide exposure and risk analysis for the sustainable development of rapidly growing cities. Landslides 20(9):1781–1795. https://doi.org/10.1007/S10346-023-02065-Z/FIGURES/1

Dikshit A, Satyam N, Pradhan B (2019) Estimation of rainfall-induced landslides using the TRIGRS model. Earth Syst Environ 3(3):575–584. https://doi.org/10.1007/S41748-019-00125-W/FIGURES/9

Greatorex M (2014) Observation. Wiley encyclopedia of management, edited by Professor Sir Cary L Cooper.https://doi.org/10.1002/9781118785317.weom090543

Hidayat R, Sutanto SJ, Hidayah A, Ridwan B, Mulyana A (2019) Development of a landslide early warning system in Indonesia. Geosciences 9(10):451.https://doi.org/10.3390/GEOSCIENCES9100451

Hostettler S, Jöhr A, Montes C, D'Acunzi A (2019) Community-based landslide risk reduction: a review of a Red Cross soil bioengineering for resilience program in Honduras. Landslides 16(9):1779–1791. https://doi.org/10.1007/S10346-019-01161-3/FIGURES/7

Huang F, Tao S, Chang Z, Huang J, Fan X, Jiang SH, Li W (2021) Efficient and automatic extraction of slope units based on multi-scale segmentation method for landslide assessments. Landslides 18(11):3715–3731. https://doi.org/10.1007/S10346-021-01756-9/FIGURES/14

Islam MM, Amir AA, Begum RA (2021) Community awareness towards coastal hazard and adaptation strategies in Pahang coast of Malaysia. Nat Hazards 107(2):1593–1620. https://doi.org/10.1007/S11069-021-04648-2/FIGURES/11

Jones KML, McCoy C (2019) Reconsidering data in learning analytics: opportunities for critical research using a documentation studies framework. Learn Media Technol 44(1):52–63. https://doi.org/10.1080/17439884.2018.1556216

Kabunga A, Okalo P, Nalwoga V, Apili B (2022) Landslide disasters in eastern Uganda: post-traumatic stress disorder and its correlates among survivors in Bududa district. BMC Psychol 10(1):1–7. https://doi.org/10.1186/S40359-022-01001-5/TABLES/4

Klimeš J, Rosario AM, Vargas R, Raška P, Vicuña L, Jurt C (2019) Community participation in landslide risk reduction: a case history from Central Andes Peru. Landslides 16(9):1763–1777. https://doi.org/10.1007/S10346-019-01203-W/FIGURES/6

Kool D, Tong B, Tian Z, Heitman JL, Sauer TJ, Horton R (2019) Soil water retention and hydraulic conductivity dynamics following tillage. Soil Tillage Res 193:95–100. https://doi.org/10.1016/J.STILL.2019.05.020

Li CJ, Hu YX, Jiang N, Li HB, Zhou JW (2023) Risk assessment and landslide prevention design using numerical modelling—a case study in Qingliu China. J Mt Sci 20(4):943–961.https://doi.org/10.1007/S11629-022-7814-7/METRICS

Li L, Mind'je R (2023) Basic information on hydrogeological hazards (flood and landslide). In: Hydrogeological hazard susceptibility and community risk perception in Rwanda, pp 7–22. https://doi.org/10.1007/978-981-99-1751-8_2

Lin PSS, Chang KM (2020) Metamorphosis from local knowledge to involuted disaster knowledge for disaster governance in a landslide-prone tribal community in Taiwan. Int J Disaster Risk Reduct 42:101339. https://doi.org/10.1016/J.IJDRR.2019.101339

Marston C, Renedo A, Miles S (2020) Community participation is crucial in a pandemic. Lancet 395(10238):1676–1678. https://doi.org/10.1016/S0140-6736(20)31054-0

Matsuoka Y, Gonzales Rocha E (2020) Sendai voluntary commitments: landslide stakeholders and the all-of-society approach enhanced by UNDRR. Landslides 17(10):2253–2269. https://doi.org/10.1007/S10346-020-01519-Y/TABLES/2

Meng ZJ, Ma PH, Peng, JB (2021) Characteristics of loess landslides triggered by different factors in the Chinese Loess Plateau. J Mt Sci 18(12):3218–3229. https://doi.org/10.1007/S11629-021-6880-6/METRICS

Morino C, Conway SJ, Balme MR, Helgason JK, Sæmundsson Þ, Jordan C, Hillier J, Argles T (2021) The impact of ground-ice thaw on landslide geomorphology and dynamics: two case studies in northern Iceland. Landslides 2021 18:8, 18(8), 2785–2812. https://doi.org/10.1007/S10346-021-01661-1

Muhammad I, Aditya I, Priyono R, Rahmania O, Maulana MR, Silitonga M, Taufiqurrahman MR (2022) Mapping of landslide potential zone with frequency ratio method and mitigation efforts in Banjarnegara Regency, Central Java, Indonesia. J Soc Sci 3(3):633–640. https://doi.org/10.46799/JSS.V3I3.354

Nishikawa S (2023) Application of Ichi-Nichi-Mae (The day before the disaster) project for landslide awareness and risk communication, pp 317–321.https://doi.org/10.1007/978-3-031-18471-0_24

Peng L, Xu D, Wang X (2019) Vulnerability of rural household livelihood to climate variability and adaptive strategies in landslide-threatened western mountainous regions of the Three Gorges Reservoir Area China. Clim Dev 11(6):469–484. https://doi.org/10.1080/17565529.2018.1445613

Pennington CVL, Bossu R, Ofli F, Imran M, Qazi U, Roch J, Banks VJ (2022) A near-real-time global landslide incident reporting tool demonstrator using social media and artificial intelligence. Int J Disaster Risk Reduct 77:103089. https://doi.org/10.1016/J.IJDRR.2022.103089

Pereira S, Santos PP, Zêzere JL, Tavares AO, Garcia RAC, Oliveira SC (2020) A landslide risk index for municipal land use planning in Portugal. Sci Total Environ 735:139463. https://doi.org/10.1016/J.SCITOTENV.2020.139463

Pittock AB, Jones RN (2000) Adaptation to what and why? Environ Monit Assess 61(1):9–35. https://doi.org/10.1023/A:1006393415542/METRICS

Pollock W, Wartman J (2020) Human vulnerability to landslides. GeoHealth 4(10):e2020GH000287. https://doi.org/10.1029/2020GH000287

Prentice CM (2017) Interviews, qualitative. Int Encycl Intercult Commun:1–8. https://doi.org/10.1002/9781118783665.ieicc0108

Raška P (2019) Contextualizing community-based landslide risk reduction: an evolutionary perspective. Landslides 16(9):1747–1762. https://doi.org/10.1007/S10346-018-1099-5/TABLES/2

Rotaru A, Pescaru RA, Olteanu-Donţov I, Nicuţă AM, Mihai P, Ciocan V, Balan MC (2021) Hazard risk mitigation for a sustainable built environment. In: Springer series in geomechanics and geoengineering, pp 3–34.https://doi.org/10.1007/978-3-030-61118-7_1/COVER

Ruiz-Cortés NS, Alcántara-Ayala I (2020) Landslide exposure awareness: a community-based approach towards the engagement of children. Landslides 17(6):1501–1514. https://doi.org/10.1007/S10346-020-01391-W/TABLES/3

Safriani EW, Wibowo YA (2022) Preparedness and adaptive capacity of students for landslide disasters in Karangkobar, Central Java, Indonesia. KnE Soc Sci:1–19. https://doi.org/10.18502/KSS.V7I5.10536

Sassa K, Canuti P, Bobrowsky P, Casagli N (2023) International consortium on landslides: from IDNDR, IGCP, UNITWIN, WCDRR 2 & 3 to Kyoto Landslide Commitment 2020, pp 11–43.https://doi.org/10.1007/978-3-031-16898-7_2

Setyawan HH, Mukminan (2021) The community adaptation strategy in dealing with landslide disaster in Banaran Village Ponorogo Regency. IOP Conf Ser: Earth Environ Sci 884(1):012003. https://doi.org/10.1088/1755-1315/884/1/012003

Sirimorok N, Batiran KB, Fisher MR, Verheijen B, Nursaputra M, Sahide MAK (2023) Resettling at the precipice: deepening vulnerabilities in disaster recovery from Upland Sulawesi. Hum Ecol 51(4):737–751. https://doi.org/10.1007/S10745-023-00422-Z/FIGURES/2

Sobarna R, Hartanto (1992) Susceptibility to landslide zone map of Banjarnegara—Pekalongan Quadrangle, Java. Directorate of Environmental Geology

Sultana N, Tan S (2021) Landslide mitigation strategies in southeast Bangladesh: lessons learned from the institutional responses. Int J Disaster Risk Reduct 62:102402. https://doi.org/10.1016/J.IJDRR.2021.102402

Sundler AJ, Lindberg E, Nilsson C, Palmér L (2019) Qualitative thematic analysis based on descriptive phenomenology. Nursing Open 6(3):733–739

Sun H, Li T, Pei X, Guo J, Tian J, Wang S, Pu M (2023) Gravitational deformation and reactivation mechanism of a fault-bounded slope, Eastern Yanshan Mountains, China. Forests 14(3):495.https://doi.org/10.3390/F14030495

Vojteková J, Vojtek M (2020) Assessment of landslide susceptibility at a local spatial scale applying the multi-criteria analysis and GIS: a case study from Slovakia. Geomat Nat Haz Risk 11(1):131–148. https://doi.org/10.1080/19475705.2020.1713233

Wibowo B, Wasino M, Joebagio H (2018) The Educational values of oral tradition as an attempt to preserve customary land in Ketapang District West Kalimantan, pp 80–85.https://doi.org/10.2991/ISET-18.2018.17

Zamroni A, Kurniati AC, Nur H, Prasetya E (2020) The assessment of landslides disaster mitigation in Java Island, Indonesia: a review. J Geosci, Eng, Environ, Technol 5(3):124–128. https://doi.org/10.25299/JGEET.2020.5.3.4676

Zhao W, You Y, Chen X, Liu J, Chen J (2020) Case study on debris-flow hazard mitigation at a world natural heritage site, Jiuzhaigou Valley, Western China. Geomat Nat Haz Risk 11(1):1782–1804. https://doi.org/10.1080/19475705.2020.1810784

Chapter 22
Evaluation of Double Fusion Satellite Rainfall Dataset in Establish Rainfall Thresholds for Landslide Occurrences Over Badung Regency-Bali

Putu Aryastana, Listya Dewi, Putu Ika Wahyuni, I. Nengah Sinarta, Jason Pajimola Punay, and Jackson Chang Hian Wui

Abstract Rainfall stations provide reliable rainfall data, but their availability is limited in mountainous areas, complex terrain, and remote areas. Satellite rainfall datasets (SRDs) provide high-resolution worldwide rainfall estimation, which has the potential to be used in identifying rainfall conditions that trigger landslides. Landslides can be predicted through rainfall threshold modeling, serving as an early warning system. It is essential to validate the chosen threshold model to assess the accuracy of forecasting landslide occurrences triggered by rainfall events. This study aims to evaluate the effectiveness of a dual fusion approach, utilizing two SRDs, in establishing rainfall thresholds for landslide prediction in the Badung regency over the period from 2015 to 2022. Rainfall threshold analysis in this investigation focuses on cumulative rainfall events occurring 5, 7, 10, and 15 days prior to the onset of landslides. The first fusion was established through the application of the cumulative

P. Aryastana (✉)
Department of Civil Engineering, Warmadewa University, Denpasar, Indonesia
e-mail: aryastanaputu@warmadewa.ac.id

L. Dewi · P. I. Wahyuni · I. N. Sinarta
Master Program of Infrastructure and Environmental Engineering, Postgraduate Program, Warmadewa University, Denpasar, Indonesia
e-mail: putuikawahyuni@warmadewa.ac.id

I. N. Sinarta
e-mail: inengahsinarta@warmadewa.ac.id

J. P. Punay
Department of Physics, College of Science, Bicol University, Legazpi City, Philippines
e-mail: jppunay@bicol-u.edu.ph

J. C. H. Wui
Preparatory Center for Science and Technology, University Malaysia Sabah (UMS), Sabah, Malaysia
e-mail: jacksonchw@ums.edu.my

G. K. Panda et al. (eds.), *Landslide: Susceptibility, Risk Assessment and Sustainability*, Advances in Natural and Technological Hazards Research 52,
https://doi.org/10.1007/978-3-031-56591-5_22

distribution functions method, involving a comparison between the SRDs and the datasets from rain gauges. Subsequently, the analysis transitioned to the second fusion, where a weighted correlation coefficient function was employed to assess the connections between rain gauges and individual SRDs. The results illustrate that the second fusion of SRDs yields an area under the curve value of 0.82 for a 15-day cumulative rainfall, surpassing the performance of the first fusion. The first quartile approach demonstrates the highest accuracy compared to alternative methods, providing a reliable estimation of landslide occurrences with minimal error.

Keywords Double fusion · Landslides · Rainfall · SRDs · Thresholds

22.1 Introduction

Landslides induced by intense precipitation represent a frequent natural disaster in regions characterized by intricate topographical features (Liao et al. 2010; Ferardi et al. 2018; Park et al. 2020; Muntohar et al. 2021). These occurrences can exert a profound influence on various facets of society, including human lives, infrastructure integrity, and environmental well-being (Pradhan et al. 2019; Katiraie-Boroujerdy et al. 2020). To mitigate the adverse effects of landslides, two principal strategies have been employed: structural and non-structural mitigation measures (Sassa et al. 2022). Structural interventions typically necessitate substantial financial investments and protracted planning and development periods. Consequently, non-structural mitigation approaches offer a viable alternative to minimize casualties in the event of a disaster (Iskandar et al. 2021) (Department of Regional Development and Environment Organization of American States. 1990). One commonly applied preventive measure in various academic fields involves the utilization of landslide early warning systems that integrate a rainfall threshold framework (Popescu and Trandafir 2014; Segoni et al. 2018). The concept of rainfall threshold encompasses two distinct methodologies: physical method and empirical method (Guzzetti et al. 2007; He et al. 2020). The physical model demands an extensive array of ancillary data, as it integrates spatial analysis, soil composition, lithological characteristics, topography, and slope information (Guzzetti et al. 2007). The accurate collection of such data presents a formidable challenge and often necessitates significant resources (He et al. 2020). Conversely, the empirical method relies on historical analyses of rainfall events that have historically triggered landslides. These empirically derived thresholds can be established on a world-wide, region-wide, or locally-based scales and are independent of variables such as soil composition, the geology, topography, and land use patterns (Nanda Pratama et al. 2017; Gariano et al. 2020; He et al. 2020). Consequently, the adoption of the empirical method for determining rainfall thresholds is favored due to its ease of access to required data, namely, landslide occurrence records and rainfall data set.

Rainfall data plays a pivotal role in the determination of rainfall thresholds (Levizzani and Cattani 2019). Nevertheless, the utilization of rain gauges for rainfall measurements is constrained by their limited coverage and the irregular spatial distribution inherent to such instruments, particularly in regions characterized by intricate topographical features (Zhao et al. 2023). Recent advancements have ushered in the era of satellite rainfall dataset (SRDs), a classification that can be broadly categorized into three distinct subtypes, each characterized by variations in dataset sources, and dataset collection methodologies. These subdivisions encompass interpolated rainfall data generated from ground-based weather stations (Yatagai et al. 2012), satellite-dependent rainfall data derived from microwave and infrared rainfall estimation conducted on a world-wide (Zhao et al. 2023), and reanalysis-derived rainfall data, which are products of predictive modeling and data integration processes (Agou et al. 2022). These methodologies establish a nexus between computational models and observational data originating from diverse sources, encompassing satellite-derived datasets and ground-based observations (Bengtsson et al. 2004). Previous investigations have offered substantial evidence attesting to the efficacy of rainfall data within the context of these three classification categories. However, it is noteworthy that the satellite-based rainfall dataset consistently yields superior results (Yu et al. 2020). Consequently, satellite data emerges as a compelling alternative capable of furnishing more precise and contextually relevant information (Nomnafa et al. 2022).

The utilization of satellite rainfall data in the development of early warning systems reliant upon rain threshold criteria has garnered widespread adoption, primarily attributable to the availability of pertinent components for rainfall forecasting (Chikalamo et al. 2020). Rainfall events are the primary trigger for the majority of slope collapses and landslides. Consequently, several researchers have endeavored to establish rainfall thresholds that can facilitate accurate predictions of slope collapses and landslides. These thresholds are constructed based on a range of parameters, encompassing average rainfall, rainfall event duration, the ratio of rainfall to daily precipitation, past rainfall in relation to annual average precipitation, and the ratio of daily rainfall to maximum past rainfall (Guzzetti et al. 2007; Allo 2010; Alvioli et al. 2014; Gariano et al. 2015; Nikolopoulos et al. 2017; Ferardi et al. 2018; He et al. 2020). In their research endeavors, these scholars have harnessed satellite rainfall datasets, including The Tropical Rainfall Measuring Mission (TRMM) researched by (Nanda Pratama et al. 2017; Rossi et al. 2017; Hidayat et al. 2019; Chikalamo et al. 2020), Global Satellite Mapping of Precipitation (GSMaP) researched by (Arrisaldi et al. 2021; Aryastana 2023), Integrated Multi-satellitE Retrievals for GPM (IMERG) researched by (Nanda Pratama et al. 2017; Wang et al. 2021; Brunetti et al. 2021; Yuniawan et al. 2022; Aryastana 2023), Precipitation Estimation from Remotely Sensed Information using Artifial Neural Networks (PERSIANN) researched by (Brunetti et al. 2018; Aryastana 2023) and Climate Prediction Center Morphing Algorithm (CMOPRH) researched by (He et al. 2020). Earlier studies have shed light on the disparities in the performance of satellite-derived rainfall datasets, which depend on the distinct climatic conditions present in each region. Notably, a discernible margin of error persists when comparing the utility of these datasets with that of rainfall data collected from ground-based weather stations. Moreover, it is noteworthy that the studies mentioned have solely relied upon a single satellite data source in their analyses aimed at establishing thresholds for rainfall-induced landslides. Consequently, there arises a compelling necessity for an alternative approach that integrates multiple rainfall satellite images. This integration is poised to mitigate the inherent uncertainty encountered when delineating thresholds indicative of landslide triggering rainfall events.

Several techniques for bias correction have been developed to improve the precision of satellite rainfall datasets when compared to measurements acquired from rain gauge stations (Katiraie-Boroujerdy et al. 2020; Guo et al. 2020; Vaittinada Ayar et al. 2021). These correction methods are particularly valuable in areas where there is a shortage of rain gauge stations and where complex topographical features, such as high mountainous regions and vast desert expanses, exist. A preceding investigation reported that the correction of satellite data yielded more precise and superior results, particularly in cases of high-intensity rainfall events (Katiraie-Boroujerdy et al. 2020). In addition to rectifying satellite-derived data, there arises a need for an alternative approach that involves the integration of rainfall satellite images. Such integration is indispensable for mitigating the inherent uncertainty associated with the establishment of rainfall thresholds conducive to landslide occurrences. A promising avenue for improving the accuracy of early rainfall identification, aimed at curtailing landslide impacts, is the prediction of rainfall through the amalgamation of ground-based field data and satellite-derived weather data. This synergy results in composite rainfall datasets that exhibit greater spatial coverage in comparison to single-source satellite data (Zhao et al. 2023). The current application of integrating two distinct satellite rainfall datasets to establish landslide-triggering rainfall thresholds faces notable limitations. Earlier investigations explored the merging of SRDs, exemplified by combining SM2RAIN and IMERG datasets. Through analysis, this fusion aimed to formulate a rainfall threshold model tailored to the Indian context. Integrating diverse satellite data products enables harnessing each product's unique advantages while mitigating SM2RAIN's tendency to underestimate rainfall or IMERG's inclination to overestimate, particularly in instances of low-intensity rainfall events (Brunetti et al. 2021). Prior research pinpointed limitations in using daily precipitation data for landslide-triggering thresholds, prompting this study to incorporate hourly rainfall data. The effective establishment of a landslide early warning system hinges on accurately applying rainfall thresholds derived from hourly data integration (Gariano et al. 2020).

The application of SRDs in establishing rainfall thresholds for landslide occurrences is still limited, particularly in Bali Province. Previous researchers examined rainfall events that trigger landslides to determine rain threshold values using daily, basic, and monthly rainfall data, relying on a single rain satellite (Guzzetti et al. 2007; Allo 2010; Alvioli et al. 2014; Gariano et al. 2015; Nikolopoulos et al. 2017; Ferardi et al. 2018; He et al. 2020; Hidayat and Zahro 2020). Scholars studying the application of SRDs like IMERG in rain threshold development have found superior performance for hourly rainfall data (Brunetti et al. 2021; Aryastana 2023). Furthermore, past studies have not investigated rainfall thresholds based on the integration of high temporal-spatial resolution of SRDs. Hence, the innovation in this study lies in determining the rainfall threshold by merging high temporal-spatial resolution datasets, namely IMERG and GSMaP. Conversely, no research has been conducted on the analysis of rainfall thresholds triggering landslides in Badung Regency. Therefore, the objective of this recent study is to evaluate the effectiveness of SRDs in constructing a rainfall threshold for predicting landslides through a dual fusion approach in Badung Regency. The anticipation is that this study will enhance the effectiveness of SRDs, making them a viable alternative for determining rainfall thresholds that lead to landslide events.

22.2 Rationale of the Study

This research is motivated by the need for more accurate and reliable rainfall data for effective landslide prediction and early warning systems. Recent advancements in satellite technology have introduced the double fusion SRDs (IMERG and GSMaP), presenting an opportunity to overcome the limitations of traditional monitoring systems (Salahi et al. 2023). This study aims to evaluate the efficacy of this dataset in providing high-resolution and real-time rainfall information. Furthermore, the focus on the Badung Regency in Bali is justified by the region's vulnerability to landslides and the potential impact on local communities. The outcomes of this research have direct implications for local authorities and disaster management agencies in implementing targeted mitigation measures (Raška et al. 2023). Moreover, this research contributes to the scientific community's understanding of satellite-based rainfall datasets' effectiveness in landslide prediction. By offering insights into the applicability of the double fusion SRDs (IMERG and GSMaP), the study aims to advance the field of geo-hazard studies. In summary, this book chapter seeks to provide a comprehensive evaluation of the double fusion SRDs utility in establishing accurate rainfall thresholds for landslide occurrences in the Badung Regency. The study's unique contribution lies in its regional focus, technological integration, and potential impact on community resilience.

22.3 Materials and Method

22.3.1 Study Area

The study was carried out within the Badung regency located in Bali (Fig. 22.1). Badung regency is located between 8°14′01″–8°50′52″S and 115°05′03″–115°26′51″E, Badung regency has an area of 418.52 km^2, or about 7.43% of the area of Bali Province. In terms of topography, the Badung regency displays a diverse range of terrain contours. Elevation-wise, the South Kuta, North Kuta, and Kuta districts are situated at altitudes ranging from 0 to 65 MASL. Meanwhile, the Mengwi district covers elevations from 0 to 350 MASL, while

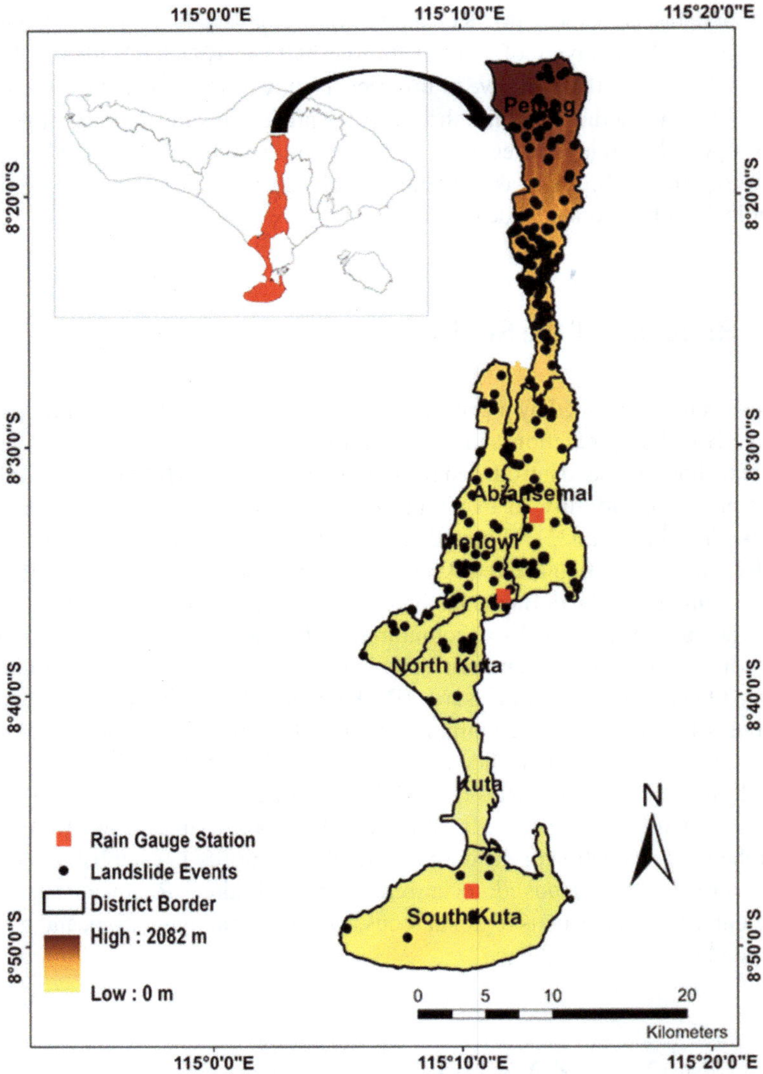

Fig. 22.1 The Badung regency map displays the locations of landslide incidents, the positions of rain gauge stations, and elevation data

the Abiansemal district encompasses altitudes between 75 and 350 MASL. The Petang district extends from 250 to 2075 MASL. Similar to other regions in southern Indonesia, the air temperature in the Badung area varies with changes in land surface altitude. Overall, the air temperature in the Badung regency ranges from 22 to 34 °C, except for hilly and mountainous areas where the average temperature typically remains below 26 °C. Relative humidity levels in the Badung regency typically range from 50 to 90% (Pratiwi 2022).

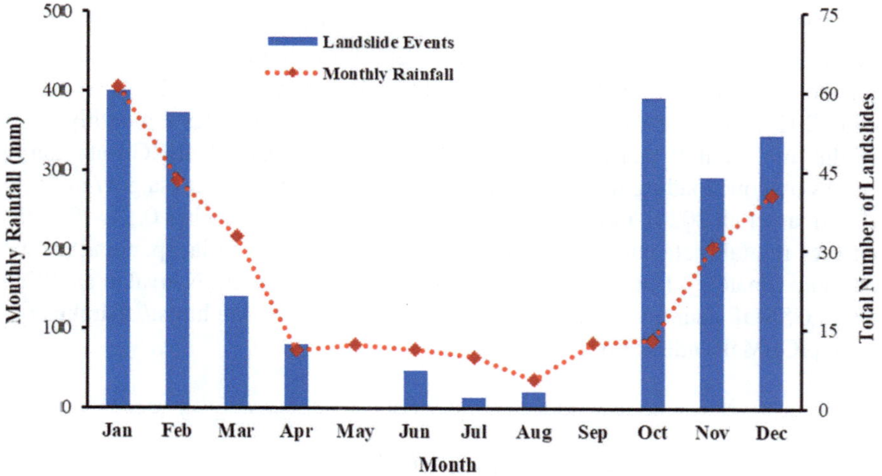

Fig. 22.2 Landslide events and mean of monthly rainfall from 2015 to 2022

22.3.2 Landslide Events

The research employs landslide event data spanning the years 2015 to 2022 obtained from Regional Disaster Management Agency of Badung regency. This dataset encompasses crucial information concerning landslide events, including event location specifics, event dates, event location coordinates, the affected area, and the extent of resulting losses. The Regional Disaster Management Agency of Badung regency compiled the report from which the landslide data originates. Landslide data is needed to analyze the rainfall threshold.

Badung Regency has a tropical climate with a monsoon climate type with two different seasons, namely the rainy season and the dry season. The rainy season in the Badung region occurs due to westerly monsoon gusts that are wet, humid, and bring a lot of water vapor producing rain clouds and the rainy season usually occurs between November to April with the peak usually occurring between January and February (Liu et al. 2020). This is evidenced by the occurrence of landslides that occur in the rainy season more (Fig. 22.2). Meanwhile, the dry season in the Badung region lasts from May to October caused by the dry and cold easterly monsoon.

22.3.3 Rainfall Dataset

This research uses 2 (two) rainfall data, namely rainfall data consisting of hourly rainfall measurements obtained from the River Basin Development Agency Bali-Penida with selected rain gauge stations namely Mambal, Sading, and Unud. The second set of rainfall data originated from satellite-based rainfall datasets, specifically

IMERG and GSMaP. IMERG data has a temporal-spatial resolution of 0.1° × 0.1° every 30 min (Wei et al. 2018). IMERG Satellite Rainfall is the latest replacement for the TRMM satellite (Huffman et al. 2020; Rahmawati et al. 2021; Aryastana et al. 2022). As an earth-orbiting satellite, IMERG data provides a monthly report of the total rainfall that falls in an area (Yuda et al. 2020). IMERG data can be access by downloading it from the website https://giovanni.gsfc.nasa.gov/giovanni/ (Huffman et al. 2020). GSMaP has a spatial resolution that reaches 0.1° × 0.1° this ensures rainfall data can be acquired for every area in Indonesia, spanning an area of approximately 11.06 × 11.06 km (Kubota et al. 2006, 2020; Nauval et al. 2021). Daily GSMaP data by downloading it from the JAXA website https://sharaku.eorc.jaxa.jp/GSMaP/index.htm.

22.3.4 *Determinination of Rainfall Thresholds*

Thresholds play a crucial role in ascertaining both the boundary for landslide incidents and the minimal duration required to initiate a landslide, as discussed in (Allo 2010). The cumulative rainfall associated with rainfall events was computed across different time intervals, specifically 5, 7, 10, and 15 days preceding the occurrence of landslides (Fig. 22.3). In order to establish the rainfall threshold, this investigation employed statistical location parameters, specifically the first quartile (Q1), median (Q2), and third quartile (Q3).

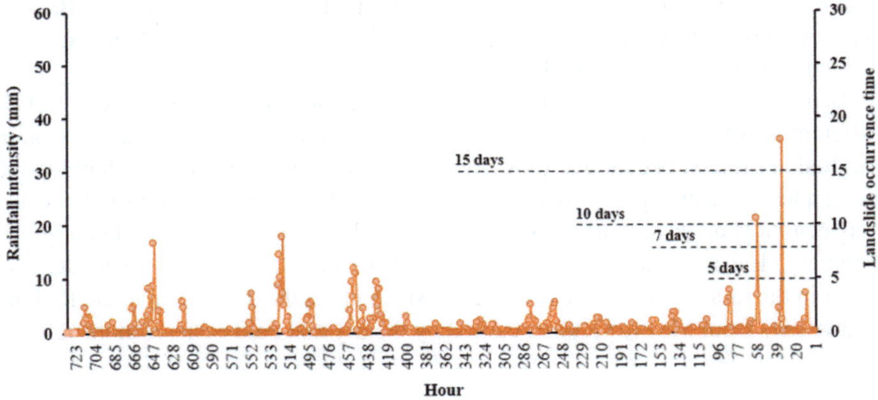

Fig. 22.3 Definition of rainfall events before landslide events

22.3.5 Fusion Rainfall Dataset

In this research, integrated satellite rainfall dataset, resulting from the fusion of rainfall estimates obtained through IMERG and GSMaP, was formulated and employed as an input for determining rainfall thresholds to predict landslides. This fusion of satellite rainfall datasets serves the purpose of mitigating errors associated with individual satellite data sources, such as the potential underestimation of rainfall by GSMaP or overestimation by IMERG, especially during low-intensity rainfall events. The process of merging these satellite datasets is accomplished through the application of the following Eq. (22.1) (Brunetti et al. 2021).

$$S_{fussion} = S_{GSMaP} + w_i (S_{IMERG} - S_{GSMaP}) \tag{22.1}$$

In this context, "w_i" represents the integration weight which varies from 0 to 1, and is calculated individually for every pixel using the following Eq. (22.2) (Kim et al. 2015).

$$w_i = \frac{\rho_{GSMaP.R} - (\rho_{GSMaP.IMERG} \cdot \rho_{IMERG.R})}{\rho_{IMERG.R} - (\rho_{GSMaP.IMERG} \cdot \rho_{GSMaP.R}) + \rho_{GSMaP.R} - (\rho_{GSMaP.IMERG} \cdot \rho_{IMERG.R.})} \tag{22.2}$$

22.3.6 Performance Analysis of Rainfall Thresholds

Threshold performance can be analyzed using a confusion matrix that encompasses the occurrences and non-occurrences of landslides, resulting in four distinct conditions. A "True Positive" occurs when rainfall surpasses a specified threshold, and landslides indeed transpire. Conversely, a "True Negative" is recorded when rainfall remains below the threshold, and no landslides transpire. Conversely, a "False Positive" scenario occurs when rainfall surpasses the threshold, yet no landslides transpire. Similarly, a "False Negative" situation is identified when rainfall falls below the threshold, but a landslide does occur (1.0) (Table 22.1) (Muntohar et al. 2021; Yuniawan et al. 2022).

In this study, ROC (Receiver Operating Characteristic) analysis is employed to assess the accuracy of rainfall threshold models in predicting rainfall events that trigger landslides. The ROC curve delineates the connection between True Positive

Table 22.1 Contingency table

Model predictions	Landslide events	
	Landslide	No landslide
> Threshold	True positive (TP)	False positive (FP)
≤ Threshold	False negative (FN)	True negative (TN)

Table 22.2 Statistical metrics employed for assessing the effectiveness of the thresholds

Index	Equation	Range	Best value
True positive rate (TPR)	$TPR = \frac{TP}{TP+FN}$	(0; 1)	1
False positive rate (FPR)	$FPR = \frac{FP}{FP+TN}$	(0; 1)	0
Accuracy	$Acc = \frac{TP+TN}{TP+TN+FP+FN}$	(0; 1)	1

Rate (TPR) and False Positive Rate (FPR) metrics at different thresholds, offering insights into the effectiveness of rainfall thresholds in predicting landslides. When a threshold exhibits a high degree of accuracy, the ROC curve converges towards the point of perfect classification, exemplified by TPR = 1 and FPR = 0 (refer to Table 22.2) (Fawcett 2006). The area under the curve (AUC) quantifies the accuracy of the empirical model, with values ranging from 0 to 1. Higher AUC values signify greater accuracy (Fawcett 2006). If the AUC is close to 1, the accuracy is very high.

In the context of this investigation, Fig. 22.4 presents a detailed flow chart outlining the research methodology employed in this study. This visual representation serves as a guide to elucidate the sequential steps undertaken throughout the research process. The diagram encapsulates key stages, offering a visual roadmap from the initial conceptualization of the research question to the ultimate interpretation of findings.

22.4 Results and Discussion

22.4.1 Result

22.4.1.1 Rainfall Thresholds of First Fusion

In the current study, an analysis of rainfall thresholds encompassing cumulative rainfall durations of 5, 7, 10, and 15 days. This analysis utilized two distinct sources of satellite rainfall data: IMERG and GSMaP, as well as the integrated IMERG-GSMaP dataset, each of which offers varying values. The objective was to demonstrate the correlation between specific landslide occurrences and both the intensity and duration of rainfall events. In addition, landslides generally occur after heavy rainfall and last for several days. The highest threshold value was derived from the third approach (Q3), with the second method (Q2) coming next, and lastly, the first method (Q1) yielded the lowest threshold value. The largest thresholds occurred in the third method in the 15-day cumulative rainfall which had values of 240.40, and 237.20 mm. Followed by the second method thresholds of 159.18, and 160.03 mm. Last is the lowest threshold of the first method, with threshold values of 112.12, and 126.34 mm. Then for the smallest threshold value obtained from the 5-day cumulative rainfall. The threshold of the third method is 90.06 and 90.23 mm. Then for the second

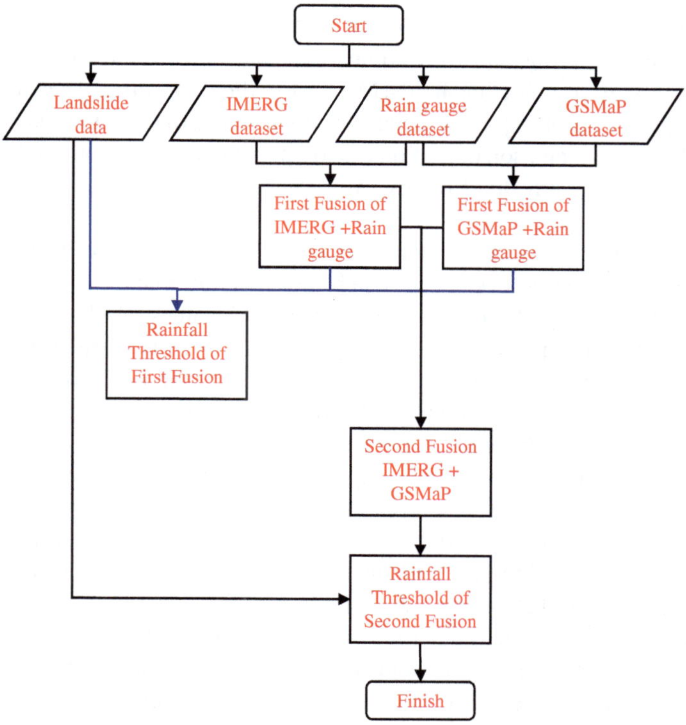

Fig. 22.4 Research flowchart

method of 60.91 and 61.77 mm. As for the values of 26.42, and 27.87 mm for the first method (Table 22.3).

Landslide thresholds derived from cumulative rainfall exhibit a wide range of values, spanning from 20 mm to over 200 mm. This variability underscores the strong dependence of landslide thresholds on factors such as location, climate, and the methodology used to establish the threshold line (Guzzetti et al. 2007). Areas

Table 22.3 Rainfall threshold of first fusion

Threshold line	SRDs	Duration (days)			
		5	7	10	15
Q1	GSMaP	26.42	39.27	72.03	112.12
	IMERG	27.87	42.79	78.98	126.34
Q2	GSMaP	60.91	76.67	113.47	159.18
	IMERG	61.77	77.47	113.47	160.03
Q3	GSMaP	90.06	118.50	171.48	240.40
	IMERG	90.23	120.30	165.69	237.20

characterized by highland terrain featuring steep slopes will have different rainfall intensities triggering landslides compared to lowlands with relatively flat slopes, resulting in distinct rainfall thresholds. Moreover, when determining landslide thresholds in a specific location, it is crucial to account for variations in seasons, climate patterns, land cover, and soil conditions in comparison to other areas, as these factors can lead to differing threshold values, even when examining the same geographical locations.

To validate the accuracy of each threshold value, their ability to predict landslide occurrence and non-occurrence can be assessed through the evaluation of True Positive Rate (TPR) and False Positive Rate (FPR) values. Higher TPR and FPR values indicate better performance of the threshold. Accuracy, which measures the percentage of predictions made by the threshold regarding landslide occurrence or non-occurrence relative to the total number of predictions generated by the threshold, also provides valuable insights. A higher accuracy value signifies a more effective threshold. Among the three methods employed in this study, the first method outperforms the other two. It boasts the highest TPR value at 0.75, indicating its superior ability to predict landslide occurrences. In contrast, the second method yields a lower TPR value of 0.53, and the third method exhibits the lowest TPR value among all methods, standing at 0.25, as evident in Table 22.4.

The performance of satellite rainfall data was assessed based on the results of the confusion matrix. Among the 316 landslide events recorded across Badung Regency, the counts of rainfall events causing landslides (TP), not causing landslides (TN), and the Accuracy (ACC) were determined. The analysis of the confusion matrix revealed that, among the three methods employed to establish rainfall thresholds, the first method consistently provided the most accurate predictions for both landslide and non-landslide events for each satellite rainfall dataset. Additionally, this method exhibited a lower prediction error rate compared to actual landslide events.

Table 22.4 Confusion matrix of first fusion

Threshold line	Duration	IMERG			GSMaP		
Q_1	5	0.75	0.30	0.74	0.74	0.30	0.73
	7	0.75	0.28	0.75	0.75	0.28	0.75
	10	0.75	0.28	0.75	0.73	0.28	0.73
	15	0.75	0.26	0.75	0.72	0.26	0.72
Q_2	5	0.50	0.22	0.54	0.50	0.22	0.53
	7	0.50	0.22	0.54	0.50	0.22	0.53
	10	0.53	0.16	0.58	0.50	0.16	0.55
	15	0.50	0.16	0.55	0.48	0.16	0.53
Q_3	5	0.25	0.22	0.32	0.23	0.18	0.31
	7	0.25	0.12	0.34	0.24	0.12	0.33
	10	0.25	0.12	0.34	0.25	0.12	0.34
	15	0.25	0.04	0.35	0.25	0.04	0.35

The AUC analysis was conducted to evaluate the accuracy of detecting rainfall events triggering landslides and those that do not. When considering cumulative rainfall over 5, 7, 10, and 15 days for both IMERG and GSMaP satellite data, the 15-day rainfall data demonstrated superior performance. The Area Under the Curve (AUC) obtained from the ROC curve indicated that the rainfall thresholds achieved reasonably good accuracy. Specifically, the results for each satellite rainfall product were AUC = 0.76 (76%) for GSMaP and AUC = 0.75 (75%) for IMERG (Fig. 22.5 and Table 22.5).

However, it is essential to optimize rainfall data by addressing both the correction of rain gauge station data and satellite rainfall data. This is necessary because previous research has demonstrated that the 15-day rainfall threshold, when applied before landslide events, yields the highest accuracy of 86% (Chikalamo et al. 2020). Derived from the AUC acquired from the ROC curve, the rainfall threshold exhibits commendable accuracy and can serve as an effective early warning system for landslides. Nevertheless, in the context of this study, it is imperative to enhance the accuracy of rainfall data by optimizing the correction of rain gauge station data.

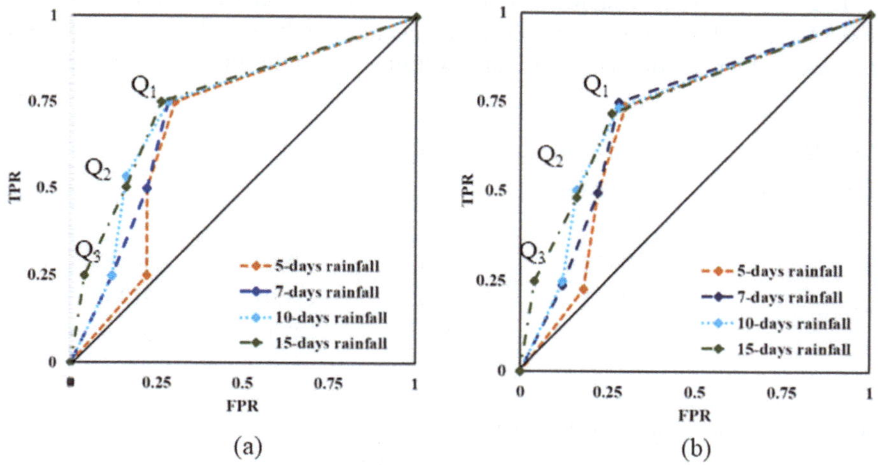

Fig. 22.5 ROC of first fusion: **a** IMERG; **b** GSMaP

Table 22.5 AUC value of first fusion

Dataset satellite	Cumulative rainfall			
	5	7	10	15
IMERG	0.69	0.72	0.74	0.76
GSMaP	0.69	0.72	0.73	0.75

22.4.1.2 Rainfall Threshold of Second Fusion

Previous analyses have shown variations in the performance of satellite rainfall data and there is still a degree of error when compared to the use of field station rainfall. Therefore, an alternative integration between satellite rainfall data is needed, to reduce the uncertainty that occurs when building rainfall thresholds due to landslides. The threshold value resulting from the integration of IMERG-GSMaP produces the highest value compared to without integration (Table 22.6). The highest threshold value was obtained from the 15-day cumulative rainfall of 302.62 mm for the third method; 212.13 mm for the second method; and 162.81 mm for the first method.

The evaluation of accuracy for the integration of IMERG-GSMaP produced the most favorable result, registering a value of 0.78 in the initial approach. Following that, the second approach yielded a value of 0.60, whereas the third approach yielded a value of 0.43. Importantly, it should be emphasized that the first approach exhibited a higher percentage of True Positive Rate (TPR) values in comparison to both the second and third approaches for identifying rainfall events that lead to landslides, as depicted in Table 22.7 below.

The integration between IMERG-GSMaP has an AUC value at 15 days cumulative of 0.82 (82%) (Fig. 22.6) which states that the rainfall threshold is good to be used for the development of a landslide early warning system in Badung regency.

Table 22.6 Rainfall threshold of the second fusion

Threshold line	Duration (days)			
	5	7	10	15
Q1	36.78	56.44	98.09	162.81
Q2	78.43	100.59	151.05	212.13
Q3	113.03	150.91	207.10	302.62

Table 22.7 Confusion matrix of the second fusion

Threshold line	Statistical indices	Duration (days)			
		5	7	10	15
Q_1	TPR	0.75	0.75	0.75	0.75
	FPR	0.19	0.17	0.2	0.19
	ACC	0.77	0.77	0.77	0.78
Q_2	TPR	0.5	0.5	0.45	0.48
	FPR	0.15	0.14	0.14	0.13
	ACC	0.58	0.6	0.56	0.59
Q_3	TPR	0.25	0.25	0.25	0.25
	FPR	0.12	0.09	0.1	0.03
	ACC	0.41	0.42	0.42	0.43

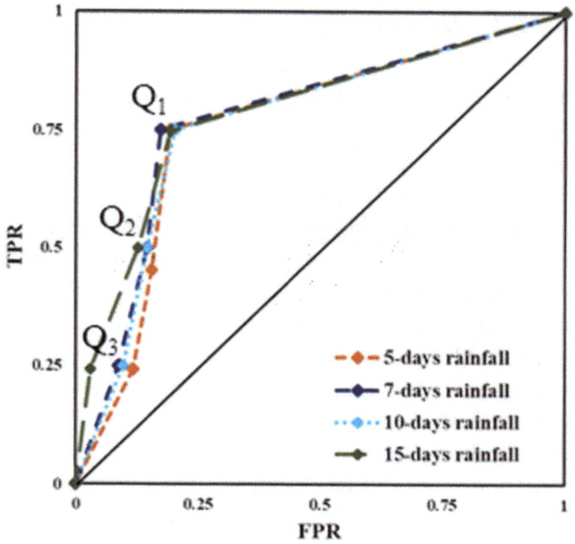

Fig. 22.6 ROC of the second fusion

This study's findings emphasize the effectiveness of the satellite rainfall data fusion method in determining rainfall thresholds for landslides, surpassing the results achieved without fusion. The established threshold holds potential significance for the development of a landslide early warning system in Badung regency. Table 22.8 provides a summary of the AUC values for rainfall thresholds using the fusion method.

It is worth noting that this study focused exclusively on evaluating thresholds for the entire Badung regency, without accounting for localized variations in factors like seasonality, land cover, and soil conditions. Multiple factors contribute to this outcome, including the region's intricate topography, variable climatic patterns, elevated terrain in mountainous areas, and the uneven distribution of rain gauge stations, especially in sparsely populated zones. It's noteworthy that previous research has also highlighted how intricate topography can influence the precision of satellite data (AghaKouchak et al. 2009; Dinku et al. 2010). Consequently, it is of paramount importance to develop validation methods and assess the performance of satellite data. In recent years, several endeavors have been undertaken to create various validation methods for satellite-based rainfall estimation (Fatkhuroyan et al. 2018).

Table 22.8 AUC value for second fusion

Fusion	Cumulative rainfall			
	5	7	10	15
IMERG-GSMaP	0.77	0.79	0.79	0.82

22.4.2 Discussion

The rainfall thresholds established in this research are applicable to the entire Badung Regency in cases of rainfall-induced landslides. In comparison to previously assessed thresholds, those derived through the dual fusion method exhibit superior performance. Specifically, the initial fusion of IMERG and rain gauge data demonstrates enhanced efficacy in determining rainfall thresholds for landslides across Badung Regency when juxtaposed with the fusion of GSMaP and rain gauge data. This outcome is attributed to IMERG's heightened proficiency in identifying light and moderate rainfall, as corroborated by existing studies (Sunilkumar et al. 2019; Shi et al. 2020; Rahmawati et al. 2021). Moreover, IMERG boasts a shorter temporal resolution of 30 min, proving advantageous in capturing brief yet impactful rainfall events (Huffman et al. 2020). This aligns with prior research utilizing IMERG satellite data for the analysis of rainfall thresholds in landslide early warning system development, affirming IMERG's superior performance in hourly rainfall data (Brunetti et al. 2021). Furthermore, IMERG's precipitation products demonstrate significant potential as a reliable source of rainfall data for remote or inaccessible areas lacking rain observation infrastructure (Yuda et al. 2020). Additionally, the precipitation products from IMERG exhibit notable accuracy on both daily and hourly time scales (Tan and Duan 2017; Brunetti et al. 2021). The findings underscore the viability of IMERG as a robust and versatile tool for rainfall data acquisition, particularly in challenging terrains or regions without established rain observation networks.

The discernible increase in the AUC value observed in the second fusion, as compared to the initial fusion, suggests a notable enhancement in the predictive accuracy of the integrated model. This improvement can be ascribed to the corrective role played by the second fusion, specifically in addressing inherent errors associated with individual satellite data sources. Notably, it effectively tackles challenges such as the potential underestimation of rainfall by GSMaP or overestimation by IMERG, particularly during instances of low-intensity rainfall events. This corrective function is crucial in refining the model's ability to provide more accurate and reliable predictions. Moreover, the findings align with the observations made by (Gariano et al. 2020) regarding the superior efficacy of utilizing hourly rainfall data compared to daily rainfall data. Their insights emphasize the importance of temporal granularity, wherein hourly data proves to be more responsive and reflective of dynamic weather patterns. The preference for hourly data contributes significantly to the model's enhanced precision, reducing the overall predictive limitations associated with relying on daily rainfall data. This nuanced understanding of temporal resolution further substantiates the robustness of the second fusion approach and reinforces its potential for improved landslide prediction capabilities, especially in scenarios involving varying rainfall intensities.

22.5 Limitations of the Study

The recent study had constraints, focusing solely on the analysis and evaluation of the double fusion SRDs' capacity to determine the cumulative rainfall threshold under general conditions. This study exhibits several constraints. Firstly, it omits an analysis of the community's ability to manage landslide disasters. Secondly, there is a dearth of planning for landslide disaster control in the designated study area. Thirdly, the investigation neglects the assessment of slope stability. Moreover, the study abstains from scrutinizing the repercussions of land use on landslide hazards and refrains from examining the impact of soil hydrological properties in the study area.

22.6 Recommendations

In future research endeavors, an exploration into rainfall thresholds instigating landslides could involve the integration of diverse rain satellite data sets like CMORPH, TRMM, CHIRPS, and others. This comparative analysis could be extended to various districts within Bali Province. Furthermore, a valuable avenue for future studies would be the implementation of varied methodologies to classify the levels of disaster impact at distinct geographical locations. It is recommended that forthcoming investigations consider integrating rain satellite products while taking into account environmental factors such as slope, soil moisture, and evapotranspiration. This holistic approach, coupled with a comprehensive review of disaster mitigation strategies, could be instrumental in refining early warning system maps for enhanced disaster preparedness and response.

22.7 Conclusion

The current research endeavors to apply and evaluate the efficacy of a double fusion approach in establishing diverse cumulative rainfall thresholds for landslide incidents across Badung Regency. The findings regarding the proficiency of the double fusion approach in setting rainfall thresholds in Badung Regency are summarized as follows:

1. The threshold line within the first quartile (Q1) exhibits commendable performance compared to both the second quartile (Q2) and third quartile (Q3).
2. The determined rainfall threshold for triggering landslides in Badung Regency occurs when the cumulative rainfall over 15 days preceding the event reaches 162.81 mm.

3. Comparative analysis of various fusion methods of SRDs in predicting rainfall-induced landslides underscores the superior performance of the second fusion over the initial fusion in Badung Regency. This is evidenced by an AUC value of 0.82 (82%) for a cumulative 15-day rainfall period.

This study marks the inaugural attempt to affirm the capability of double fusion SRDs (IMERG and GSMaP) in estimating rainfall-induced landslides on a regional scale, specifically at the regency level. The satisfactory performance of the double fusion is evident in determining cumulative rainfall thresholds for landslide occurrences in Badung Regency. The accessibility of near-real-time and high temporal-spatial resolution rainfall datasets emerges as a pivotal factor in accurately estimating landslides triggered by rainfall. The establishment of rainfall thresholds is anticipated to enhance community preparedness for landslide events. However, it is crucial to acknowledge that the threshold values triggering landslides are highly dependent on the unique characteristics and vulnerability levels of each area. Therefore, a thorough analysis of landslide occurrences across different vulnerability levels is imperative for a comprehensive understanding of the phenomenon.

Acknowledgements Thanks to the Regional Disaster Management Agency of Badung regency for furnishing the essential information and data for this study. Thanks to the Directorate of Research, Technology and Community Service of the Ministry of Education, Culture, Research and Technology for their financial support towards this research endeavor.

References

AghaKouchak A, Nasrollahi N, Habib E (2009) Accounting for uncertainties of the TRMM satellite estimates. Remote Sens 1:606–619. https://doi.org/10.3390/rs1030606

Agou VD, Pavlides A, Hristopulos DT (2022) Spatial modeling of precipitation based on data-driven warping of Gaussian processes. Entropy 24:1–21. https://doi.org/10.3390/e24030321

Allo ET (2010) Determaining rainfall thresholds for landslide initiation: a case study in wadaslintang watershed, wonosobo. Gadjah Mada University, Central Java Province

Alvioli M, Guzzetti F, Rossi M (2014) Scaling properties of rainfall induced landslides predicted by a physically based model. Geomorphol 213:38–47. https://doi.org/10.1016/j.geomorph.2013.12.039

Arrisaldi T, Wilopo W, Fathani TF (2021) Landslide susceptibility mapping and their rainfall thresholds model in Tinalah watershed, Kulon Progo district, Yogyakarta special region, Indonesia. J Appl Geol 6:112. https://doi.org/10.22146/jag.59185

Aryastana P, Liu C-Y, Jong-Dao Jou B, Cayanan E, Punay JP, Chen Y (2022) Assessment of satellite precipitation data sets for high variability and rapid evolution of typhoon precipitation events in the Philippines. Earth Sp Sci 9. https://doi.org/10.1029/2022EA002382

Aryastana P (2023) Grid satellite rainfall products potential application for developing I-D and E-D thresholds for landslide early alert system over Bali Island. 07. https://doi.org/10.30737/ukarst.v7i1.4318

Bengtsson L, Hagemann S, Hodges KI (2004) Can climate trends be calculated from reanalysis data? J Geophys Res Atmos 109:D11111. https://doi.org/10.1029/2004JD004536

Brunetti MT, Melillo M, Peruccacci S, Ciabatta L, Brocca L (2018) How far are we from the use of satellite rainfall products in landslide forecasting? Remote Sens Environ 210:65–75. https://doi.org/10.1016/j.rse.2018.03.016

Brunetti MT, Melillo M, Gariano SL, Ciabatta L, Brocca L, Amarnath G, Peruccacci S (2021) Satellite rainfall products outperform ground observations for landslide prediction in India. Hydrol Earth Syst Sci 25:3267–3279. https://doi.org/10.5194/hess-25-3267-2021

Chikalamo EE, Mavrouli OC, Ettema J, van Westen CJ, Muntohar AS, Mustofa A (2020) Satellite-derived rainfall thresholds for landslide early warning in Bogowonto catchment, central java, Indonesia. Int J Appl Earth Obs Geoinf 89:102093. https://doi.org/10.1016/j.jag.2020.102093

Department of Regional Development and Environment Organization of American States (1990) Disaster, planning and development: managing natural hazards to reduce loss

Dinku T, Ruiz F, Connor SJ, Ceccato P (2010) Validation and intercomparison of satellite rainfall estimates over Colombia. J Appl Meteorol Climatol 49:1004–1014. https://doi.org/10.1175/2009JAMC2260.1

Fatkhuroyan, Wati T, Sukmana A, Kurniawan R (2018) Validation of satellite daily rainfall estimates over Indonesia. Forum Geogr 32:170–180. https://doi.org/10.23917/forgeo.v32i2.6288

Fawcett T (2006) An introduction to ROC analysis. Pattern Recognit Lett 27:861–874. https://doi.org/10.1016/j.patrec.2005.10.010

Ferardi FD, Wilopo W, Fathani TF (2018) Rainfall thresholds for landslide prediction in Loano subdistrict, Purworejo district central java province. J ofApplied Geol 3:23–31

Gariano SL, Brunetti MT, Iovine G, Melillo M, Peruccacci S, Terranova O, Vennari C, Guzzetti F (2015) Calibration and validation of rainfall thresholds for shallow landslide forecasting in Sicily, southern Italy. Geomorphol 228:653–665. https://doi.org/10.1016/j.geomorph.2014.10.019

Gariano SL, Melillo M, Peruccacci S, Brunetti MT (2020) How much does the rainfall temporal resolution affect rainfall thresholds for landslide triggering? Nat Hazards 100:655–670. https://doi.org/10.1007/s11069-019-03830-x

Guo L, Jiang Z, Chen D, Le Treut H, Li L (2020) Projected precipitation changes over China for global warming levels at 1.5 °C and 2 °C in an ensemble of regional climate simulations: impact of bias correction methods. Clim Change 162:623–643. https://doi.org/10.1007/s10584-020-02841-z

Guzzetti F, Peruccacci S, Rossi M, Stark CP (2007) Rainfall thresholds for the initiation of landslides in central and southern Europe. Meteorol Atmos Phys 98:239–267. https://doi.org/10.1007/s00703-007-0262-7

He S, Wang J, Liu S (2020) Rainfall event-duration thresholds for landslide occurrences in China. Water (switzerland) 12. https://doi.org/10.3390/w12020494

Hidayat R, Sutanto SJ, Hidayah A, Ridwan B, Mulyana A (2019) Development of a landslide early warning system in Indonesia. Geosci 9:451. https://doi.org/10.3390/geosciences9100451

Hidayat R, Zahro AA (2020) Penentuan Ambang Curah Hujan untuk Memprediksi Kejadian Longsor. J Sumber Daya Air 16:1–10. https://doi.org/10.32679/jsda.v16i1.483

Huffman GJ, Bolvin DT, Braithwaite D, Hsu K-L, Joyce RJ, Kidd C, Nelkin EJ, Sorooshian S, Stocker EF, Tan J, Wolff DB, Xie P (2020) Integrated multi-satellite retrievals for the global precipitation measurement (GPM) mission (IMERG). pp 343–353

Iskandar I, Andika T, Wulandari W (2021) The model of nonstructural mitigation policy to the landslide prone residential areas in Lebong, Bengkulu. Yuridika 36:333. https://doi.org/10.20473/ydk.v36i2.22741

Katiraie-Boroujerdy P-S, Rahnamay Naeini M, Akbari Asanjan A, Chavoshian A, Hsu K, Sorooshian S (2020) Bias correction of satellite-based precipitation estimations using quantile mapping approach in different climate regions of Iran. Remote Sens 12:2102. https://doi.org/10.3390/rs12132102

Kim S, Parinussa RM, Liu YY, Johnson FM, Sharma A (2015) A framework for combining multiple soil moisture retrievals based on maximizing temporal correlation. Geophys Res Lett 42:6662–6670. https://doi.org/10.1002/2015GL064981

Kubota T, Hashizume H, Takahashi N, Shige S, Okamoto K, Ushio T, Aonashi K, Kachi M (2006) Global precipitation map using satelliteborne microwave radiometers by the GSMaP project: production and validation. In: 2006 IEEE international symposium on geoscience and remote sensing. IEEE, pp 2584–2587

Kubota T, Aonashi K, Ushio T, Shige S, Takayabu YN, Kachi M, Arai Y, Tashima T, Masaki T, Kawamoto N, Mega T, Yamamoto MK, Hamada A, Yamaji M, Liu G, Oki R (2020) Global satellite mapping of precipitation (GSMaP) products in the GPM Era. pp 355–373

Levizzani V, Cattani E (2019) Satellite remote sensing of precipitation and the terrestrial water cycle in a changing climate. Remote Sens 11. https://doi.org/10.3390/rs11192301

Liao Z, Hong Y, Wang J, Fukuoka H, Sassa K, Karnawati D, Fathani F (2010) Prototyping an experimental early warning system for rainfall-induced landslides in Indonesia using satellite remote sensing and geospatial datasets. Landslides 7:317–324. https://doi.org/10.1007/s10346-010-0219-7

Liu C-Y, Aryastana P, Liu G-R, Huang W-R (2020) Assessment of satellite precipitation product estimates over Bali Island. Atmos Res 244:105032. https://doi.org/10.1016/j.atmosres.2020.105032

Muntohar AS, Mavrouli O, Jetten VG, van Westen CJ, Hidayat R (2021) Development of landslide early warning system based on the satellite-derived rainfall threshold in Indonesia. In: Casagli N, Tofani V, Sassa K, Bobrowsky PT, Takara K (eds) Understanding and reducing landslide disaster risk. Springer, Cham, pp 227–235

Pratama GN, Suwarman R, Junnaedhi IDGA, Riawan E, Anugrah A (2017) Comparison landslide-triggering rainfall threshold using satellite data: TRMM and GPM in South Bandung area. IOP Conf Ser Earth Environ Sci 71. https://doi.org/10.1088/1755-1315/71/1/012003

Nauval F, Sinatra T, Awaludin A, Fatria D (2021) Performance evaluation of high-resolution satellite products in estimating rainfall condition over West Borneo. p 020009

Nikolopoulos EI, Destro E, Maggioni V, Marra F, Borga M (2017) Satellite rainfall estimates for debris flow prediction: an evaluation based on rainfall accumulation-duration thresholds. J Hydrometeorol 18:2207–2214. https://doi.org/10.1175/JHM-D-17-0052.1

Nomnafa FR, Krisnayanti DS, Ramang R, Udiana IM (2022) Penggunaan data Satelit TRMM terhadap Stasiun Curah Hujan di WS Noelmina. J Tek Pengair 13:1–11. https://doi.org/10.21776/ub.pengairan.2022.013.01.01

Park JY, Lee SR, Kim YT, Kang S, Lee DH (2020) A regional-scale landslide early warning system based on the sequential evaluation method: development and performance analysis. Appl Sci 10. https://doi.org/10.3390/APP10175788

Popescu ME, Trandafir AC (2014) Landslide risk assessment and mitigation. Bridg Eng Handbook, Second Ed Substruct Des 315–359. https://doi.org/10.1201/b15621

Pradhan AMS, Lee SR, Kim YT (2019) A shallow slide prediction model combining rainfall threshold warnings and shallow slide susceptibility in Busan, Korea. Landslides 16:647–659. https://doi.org/10.1007/s10346-018-1112-z

Pratiwi AM (2022) Regional statistics of Badung regency 2022. Central Bureau of Statistic

Rahmawati N, Rahayu K, Yuliasari ST (2021) Performance of daily satellite-based rainfall in groundwater basin of Merapi aquifer system, Yogyakarta. Theor Appl Climatol 146:173–190. https://doi.org/10.1007/s00704-021-03731-9

Raška P, Riezner J, Bfl M, Klimeš J (2023) Long-term landslide impacts and adaptive responses in rural communities: using historical cases to validate the cumulative causation approach. Int J Disaster Risk Reduct 93:103748. https://doi.org/10.1016/j.ijdrr.2023.103748

Rossi M, Luciani S, Valigi D, Kirschbaum D, Brunetti MT, Peruccacci S, Guzzetti F (2017) Statistical approaches for the definition of landslide rainfall thresholds and their uncertainty using rain gauge and satellite data. Geomorphol 285:16–27. https://doi.org/10.1016/j.geomorph.2017.02.001

Salahi A, Ashrafzadeh A, Vazifedoust M (2023) Remote sensing-based precipitation forecasting using cloud optical characteristics: threshold optimization and evaluation in Northern and Western Iran. Nat Hazards. https://doi.org/10.1007/s11069-023-06352-9

Sassa K, Konagai K, Tiwari B, Sassa S (2022) Progress in landslide research and technology

Segoni S, Piciullo L, Gariano SL (2018) A review of the recent literature on rainfall thresholds for landslide occurrence. Landslides 15:1483–1501. https://doi.org/10.1007/s10346-018-0966-4

Shi J, Yuan F, Shi C, Zhao C, Zhang L, Ren L, Zhu Y, Jiang S, Liu Y (2020) Statistical evaluation of the latest GPM-Era IMERG and GSMaP satellite precipitation products in the yellow river source region. Water 12:1006. https://doi.org/10.3390/w12041006

Sunilkumar K, Yatagai A, Masuda M (2019) Preliminary evaluation of GPM-IMERG rainfall estimates over three distinct climate zones with APHRODITE. Earth Sp Sci 6:1321–1335. https://doi.org/10.1029/2018EA000503

Tan ML, Duan Z (2017) Assessment of GPM and TRMM precipitation products over Singapore. Remote Sens 9. https://doi.org/10.3390/rs9070720

Vaittinada Ayar P, Vrac M, Mailhot A (2021) Ensemble bias correction of climate simulations: preserving internal variability. Sci Rep 11:3098. https://doi.org/10.1038/s41598-021-82715-1

Wang N, Lombardo L, Gariano SL, Cheng W, Liu C, Xiong J, Wang R (2021) Using satellite rainfall products to assess the triggering conditions for hydro-morphological processes in different geomorphological settings in China. Int J Appl Earth Obs Geoinf 102:102350. https://doi.org/10.1016/j.jag.2021.102350

Wei G, Lü H, Crow WT, Zhu Y, Wang J, Su J (2018) Comprehensive evaluation of GPM-IMERG, CMORPH, and TMPA precipitation products with gauged rainfall over mainland China. Adv Meteorol 2018. https://doi.org/10.1155/2018/3024190

Yatagai A, Kamiguchi K, Arakawa O, Hamada A, Yasutomi N, Kitoh A (2012) APHRODITE: constructing a long-term daily gridded precipitation dataset for Asia based on a dense network of rain gauges. Bull Am Meteorol Soc 93:1401–1415. https://doi.org/10.1175/BAMS-D-11-00122.1

Yu J, Li XF, Lewis E, Blenkinsop S, Fowler HJ (2020) UKGrsHP: a UK high-resolution gauge–radar–satellite merged hourly precipitation analysis dataset. Clim Dyn 54:2919–2940. https://doi.org/10.1007/s00382-020-05144-2

Yuda IWA, Prasetia R, As-Syakur AR, Osawa T, Nagai M (2020) An assessment of IMERG rainfall products over Bali at multiple time scale. E3S Web Conf 153:1–12. https://doi.org/10.1051/e3sconf/202015302001

Yuniawan RA, Rifa'i A, Faris F, Subiyantoro A, Satyaningsih R, Hidayah AN, Hidayat R, Mushthofa A, Ridwan BW, Priangga E, Muntohar AS, Jetten VG, van Westen CJ, den Bout BV, Sutanto SJ (2022) Revised rainfall threshold in the Indonesian landslide early warning system. Geosciences 12:129. https://doi.org/10.3390/geosciences12030129

Zhao C, Yao S, Ding Y, Zhao Q (2023) A gridded monthly precipitation merged rain gauge and satellite analysis dataset for the Tian Shan range between 1981 and 2019. J Appl Meteorol Climatol 62:691–708. https://doi.org/10.1175/jamc-d-21-0241.1

Chapter 23
Landslide Susceptibility and Risk Assessment in Hilly Regions of Bangladesh: A Geostatistical and Geospatial Modeling Approach for Sustainability

MD. Toufiq Hossain, Rowdra Dip Chackroborty, Labib Intisar, Sazzad Al Toufiq Shuvo, Abdullah Al Rakib, and Abdulla-Al Kafy

Abstract Landslides significantly threaten human life, infrastructure, and environmental balance. In the hilly regions of Bangladesh, including Sylhet and Rangamati, landslides are frequent, causing 727 deaths and 1017 injuries between 2000 and 2018. The northeastern section of Bangladesh is projected to receive over 500–600 mm of precipitation in 2023, breaking records over the past 122 years, according to the European Centre for Medium-Range Weather Forecasts (ECMWF). With an elevation range of 0 to 195 m above sea level and 18% of its total land area covered by water bodies, Rangamati is particularly vulnerable to landslides. Despite the devastating impact of landslides, susceptibility assessment and risk management strategies are lacking. This research aims to address this gap by developing a comprehensive framework for sustainable landslide risk mitigation using geostatistical and geospatial modeling techniques. Factors such as land use and land cover (LULC), elevation, slope, topographic wetness index (TWI), precipitation, lithology, soil type, normalized difference vegetation index (NDVI), and distance from roads are used to create a frequency ratio (FR) model and identify landslide susceptibility and risk zones. The resulting high-resolution landslide susceptibility map (LSM) and risk assessment models provide valuable insights for policymakers, land-use planners, and stakeholders involved in disaster risk reduction and sustainable development. By applying geostatistical and geospatial modeling techniques to assess landslide susceptibility, manage risk, and promote sustainability, this research enhances resilience to landslides and highlights the importance of proactive planning and informed decision-making in mitigating the impact of landslides for promoting sustainable development in hilly regions.

MD. T. Hossain (✉) · R. D. Chackroborty · L. Intisar · S. Al Toufiq Shuvo · A. Al Rakib · A.-A. Kafy
Department of Urban & Regional Planning, Rajshahi University of Engineering & Technology (RUET), Rajshahi 6204, Bangladesh
e-mail: toufiq.hossain2910@gmail.com

G. K. Panda et al. (eds.), *Landslide: Susceptibility, Risk Assessment and Sustainability*, Advances in Natural and Technological Hazards Research 52,
https://doi.org/10.1007/978-3-031-56591-5_23

Keywords Landslide susceptibility · Risk assessment · Geostatistical modeling · Geospatial modeling · NDVI · TWI · Sustainability · Bangladesh

23.1 Introduction

A landslide is a geological phenomenon encompassing various ground movements, such as rock falls, deep slope failures, and shallow debris flows (Rahman 2012). Unplanned development activities, overpopulation, settlement along hill slopes, and ineffective disaster risk reduction efforts are the anthropogenic contributors accompanying climate-change-induced increased torrential rainfall are the main reasons for the increase in landslide occurrence (Alam 2020). Over the past five decades, the United Nations (UN) has prioritized implementing disaster risk reduction initiatives. The United Nations officially designated 1990 to 1999 as the "International Decade for Natural Disaster Reduction" (Alexander 1993). Additionally, the United Nations' Sustainable Development Goals for 2015–2030 have established a precise aim to substantially reduce the number of fatalities and the overall impact on individuals affected by disasters worldwide (UN 2015). From 2004 to 2013, 115 of the 173 global landslides resulted in around 7098 fatalities and impacted 3 million people in Asia alone (Ahmed et al. 2014; IFRC 2014). Bangladesh has been a prime susceptible zone for landslides for the last decades. Most of Bangladesh's land is floodplain, while only 18% is hilly and mountainous (Islam and Uddin 2002). In Bangladesh, 17 persons perished in 1999, 13 perished in 2000, 91 perished in 2007, 54 perished in 2010, and 17 perished in 2011 (BWDB 2005). Over the last three decades, there have been around 200 fatalities and considerable economic and property damages. The unconsolidated sedimentary rocks in the hill tracts by the rivers and streams are the reasons for high landslide-susceptible areas (Rashid 1991; Brammer 1996).

In recent years, this landslide has been a significant concern in the Chittagong and Chittagong Hill Tracts regions of Bangladesh's southeast. Chittagong has experienced approximately 12 landslides over the past five decades (BWDB 2005). Kutupalong Rohingya Camp (KRC) in Cox's Bazar District (CBD), Bangladesh, saw roughly 257 landslides or slope collapses, resulting in five fatalities and the destruction of over 5000 shelters (Kamal et al. 2022). Sylhet is also a vulnerable location for landslides because of its topographic condition where about 10,000 households live in hazardous conditions due to deforestation and hillocks being chopped down. These families are not on a government list, and there are no overt efforts to help them get back on their feet. The number of individuals living in danger is unknown to the district administration (Debu 2022).

The most vulnerable locations to landslides in Sylhet are built-up areas and vegetation to built-up land use and land cover (LULC) change types, and primary economic activities like jhum cultivation and fishing are the worst affected by landslides, according to the characteristics, causes, and consequences of landslides in Rangamati District where the urbanized Bengali and Rohingya refugee communities are highly vulnerable to landslide (Abedin et al. 2020; Ahmed 2021). That's why an assessment of landslide vulnerability in hilly areas in Bangladesh is needed to understand the risk, improve preparedness, and reduce the impact of landslides (Edris and Alam 2020). Sylhet and Rangamati are essential from an environmental and economic standpoint. The selection of these areas is based on differences in the landslide situation and variables impacting natural hazards supporting the new range of research in geography.

The occurrence of any natural catastrophe may be traced mostly to the interaction of political and economic variables, insufficient adaptation to natural circumstances, and human activity. To have a comprehensive understanding of the constituents of a disaster, it is imperative to thoroughly examine both exceptional natural occurrences and societal development over time (Quarantelli 1998; Alexander 2000). So, the chapter aims to address this gap by developing a comprehensive framework for sustainable landslide risk mitigation using geostatistical and geospatial modeling techniques to assess landslide susceptibility, manage risk, and promote sustainability, this research enhances resilience to landslides on Sylhet and Rangamati.

The assessment methodology indeed includes numerous factors. In Bangladesh, slope angle aspect, LULC, elevation, geology, normalized difference vegetation index (NDVI), distance from the road, rainfall, and distance from stream have been given the highest importance in landslide susceptibility mapping (Chowdhury 2023). A multidisciplinary approach was used to research landslides in Bangladesh, focusing on the Chittagong region (Kafy et al. 2017). Also, the frequency ratio (FR) approach is used to assess landslide susceptibility because it possesses a success and prediction rate of 75% to 80% (Pratap and Vikram 2021). The FR value is one metric to assess the relationship between various causes and landslides. A moderate link with land sliding is shown by an FR value of less than 1, while a strong correlation is indicated by a value of more than 1 (Samanta et al. 2018). A more accurate map of vulnerable landslide sites is what the FR method's landslide susceptibility map (LSM) will provide (Mahdi et al. 2023). The comparative analysis between two locations (Sylhet and Rangamati) will add an immense dimension to the research that will help the policymakers build a comprehensive recommendation for capacity building against landslides. The mitigation of landslides in southeast Bangladesh is insufficient due to financial constraints (Rabby and Li 2019).

The main objective of this chapter is to analyze landslide susceptibility and risk assessment using geostatistical and geospatial modeling techniques to provide guidelines for policymakers and planners for the development of Sylhet and Rangamati. The chapter has provided specific guidance for each location. As the influencing factor for landslides in each location varies, the research will be beneficial for cost-efficient development. The order of upazilas and other parameters will be useful in determining the location and other aspects that will enhance landslide mitigation and protection. The overall research will provide an inventory of landslides and a comparison examination of two important locations with different topographies, environments, and climates. The research findings will provide qualitative assistance to improve the development process in Sylhet and Rangamati.

23.1.1 *Study Location*

The hill tracts of Bangladesh, including the Chittagong hilly regions and Sylhet hilly sides, are indeed considered landslide susceptible areas. Sylhet is situated in the northeastern region of Bangladesh at coordinates 24°32′0″ N and 91°52′0″ E on the northern side of the river Surma, encompassed by the Jaintia, Khasi, and Tripura hills. The physiography of Sylhet mainly consists of hill soils, which include many significant depressions referred to as "Haors" by the local population (Bangladesh Metrological Department 2009). A total of 27 wards in Sylhet are estimated to have a population density of 250.26 per square kilometer (Habibur et al. 2011). Its immense rainfall increase over the years from May to September is the prime factor for landslides (Bangladesh Metrological Department 2009). Rangamati is situated within the southeastern part delimited by India to the north and east, Bandarban district to the south, and Khagrachari and Chittagong districts to the west. It covers an expansive area of 6116.19 km^2, making it the largest district in the country in terms of geographical expansion with 10 administrative upazilas (Rangamati Hill District Council 2011). The research region encompasses 1145 km^2, specifically including landslide-vulnerable locations Rangamati Sadar, Kawkhali, and Kaptai, which individually cover 547 km^2, 339 km^2, and 259 km^2, respectively. This region is more significant and fragile in terms of biodiversity since 497 km^2 of it is covered in forest flora, whilst 218 km^2 of it is distinguished by riverine characteristics (BBS 2012). According to Rangamati district government, about 12,450 people experienced losses (UNPO 2017). Additionally, 1,500 dwellings were demolished, while 2,000 residences sustained partial damage due to landslides (UNDP 2017) (Fig. 23.1).

23.2 Methods and Materials

The research follows the FR method for analyzing the landslide susceptibility in Sylhet and Rangamati. These significant economic hubs of Bangladesh significantly influence the country's development. The previous year's landslide record and NASA landslide inventory data prepared the map. The workflow diagram represents the conditioning factors and procedures for the FR method to prepare LSM. The data sources required to conduct FR calculation are illustrated in (Table 23.1) for LSM (Fig. 23.2).

Elevation, topographic wetness index (TWI), LULC, slope, rainfall, lithology, NDVI, soil type, and distance from the road are the prime conditioning factors for the suitable method calculation. In landslide susceptibility research, slope, curvature, and aspect variables are commonly regarded as critical characteristics related to landslide conditions. These variables are typically derived using a digital elevation model (DEM) with a spatial resolution of 30 m. The research region was categorized into five distinct groups based on the significance of elevation and slope as primary determinants of landslide incidence. As the elevation height and slope angle drop,

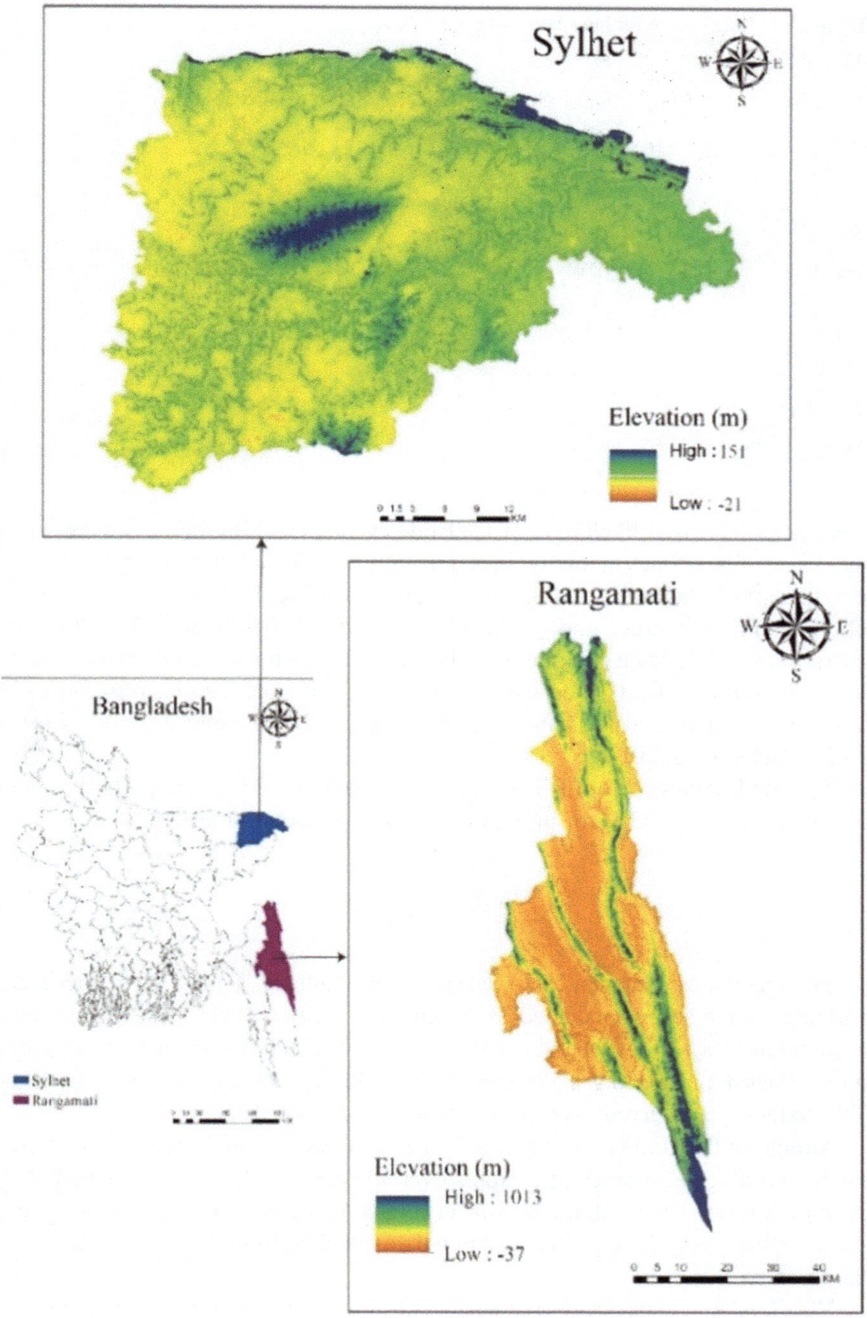

Fig. 23.1 Study area (Sylhet & Rangamati, Bangladesh)

Table 23.1 Data source for landslide influence factors

Factor	File type	Source
Elevation	DEM (30 m × 30 m)	USGS
TWI	DEM (30 m × 30 m)	USGS
LULC	Sentinel-2, Landsat 8 (10 m × 10 m)	ESRI
Slope	DEM (30 m × 30 m)	USGS
Rainfall	IDW	Bangladesh metrological Department
Lithology		USGS
NDVI	(30 m × 30 m)	USGS
Soil Type		BARC
Distance from Road		BARC

there is an increased likelihood of landslide events, resulting in a more significant incidence of landslides in low-lying and flat regions. Conversely, it is improbable for it to transpire at higher altitudes. Steep slopes accelerate the velocity of surface runoff, reducing the duration during which the soil can absorb the water. The influence of curvature is a significant component in the study of landslides. It has been categorized into three types: concave (negative), convex (positive), and flat (zero) surfaces. The amount of rainfall and sunshine received by terrain is influenced by conditioning factors (Jebur et al. 2014).

The conditioning factor for TWI was derived using DEM with a spatial resolution of 30 m, using Eq. (23.1), resulting in the following expression:

$$\mathrm{TWI} = \ln\left(\frac{A_s}{\tan\beta}\right) \tag{23.1}$$

The specific catchment area (m^2/m) and slope angle in degrees (β) are defined as variables in the study conducted by Regmi et al. (2010). TWI measures the water accumulation at various locations within a watershed and the inclination of water to move downslope due to gravitational forces (Moore et al. 1991). Typically, higher TWI values are observed in places prone to landslides.

Another influential element that affects landslides is rainfall data collected from the Meteorological Agency of Ethiopia over 30 years (1990–2021), especially from June to August. The rainfall map was generated using the IDW interpolation technique on the yearly average precipitation data obtained from the station sites within the research region.

NDVI is a popular and frequently utilized index. The vegetation index is commonly used in global climate and environmental change studies. Vegetation's sensitivity to the environment helps protect against natural disasters by influencing ecological balance and climate. Distance to Road: One factor that affects the likelihood of a landslip is the distance from a road. This factor refers to a road network

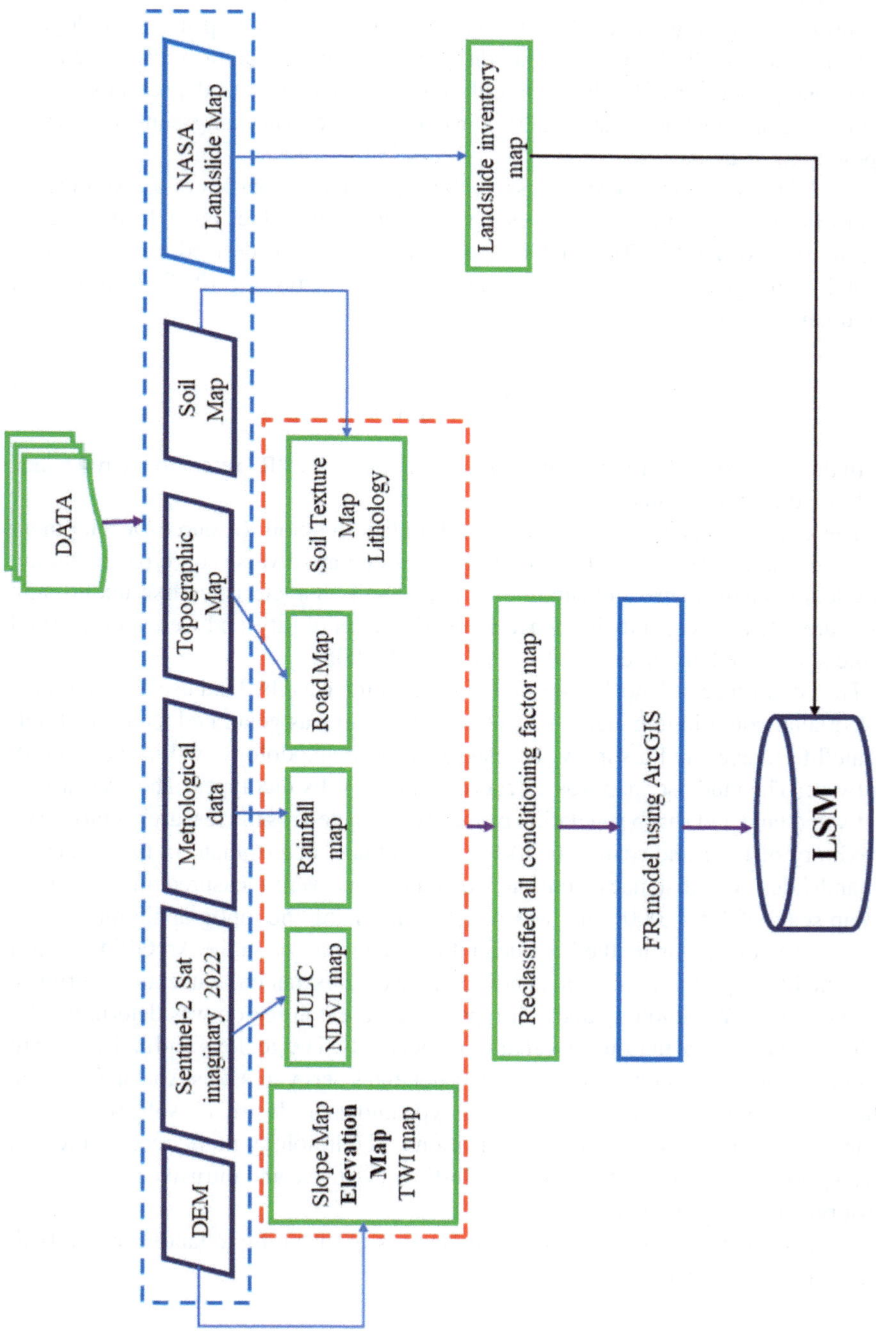

Fig. 23.2 Methodological framework

that serves as a dam or water barrier for a particular area. It has a reverse relationship with land sliding. Organizing mountain roadways needs construction projects such as cutting or digging slopes that will change and weaken the primary geological structure (Sofa et al. 2021). The road distance is a primary factor in planning landslide zone projections. Roads in the hilly region may be one of the reasons for the occurrence of landslides (Catena 2008); in due time, the road changes into a different topographic structure.

The NDVI is a metric used to assess the vegetation properties of a given region and its influence on landslides within a basin. The NDVI values range from -1 to $+1$, exhibiting variability. The NDVI conditioning factor was derived using Sentinel-2 satellite images with a spatial resolution of 30 m, using Eq. (23.2) as described (Pradhan et al. 2010).

$$\mathrm{NDVI} = \frac{IR - R}{IR + R}. \tag{23.2}$$

In this context, "IR" refers to the infrared bands, while "R" refers to the red bands within the electromagnetic spectrum.

The soil type data were acquired from Bangladesh's Soil Resource Development Institute (SRDI) (Haque 2006). The study areas contain diverse soil types including clay loam, sandy loam, and sandy clay loam. Soil characteristics like texture and moisture retention capacity influence landslide susceptibility, and were incorporated in the analysis (Rahman et al. 2018; Petley et al. 2005).

The occurrence of landslides is significantly influenced by land use, which serves as a crucial conditioning element. The land use map was generated using Sentinel-2 satellite images and a supervised classification methodology within the ArcGIS software. The land use map was categorized into six distinct types. The presence of cultivated land and shrub vegetation primarily characterizes the research region. The proximity of river and road networks is a significant determinant in the incidence of landslides. The distances from the Road and river were measured and shown on a map scale of 1:500,000. This was done by digitizing the topographic map of the research region and using the Euclidean distance technique in the ArcGIS tool. Soil type and lithology are significant conditioning elements that influence the occurrence of landslides. The lithology and soil type of the research region was determined by utilizing a geological map and a soil map, respectively. The geology and soil data were acquired from the Ethiopian Minister of Water and Energy in Addis Ababa, Ethiopia. The research region has a dominating soil type known as dystric nitisols, which falls under one of the eight soil type classifications. The lithology in the research region is categorized into four distinct classes. Tertiary extrusive and intrusive rocks are the most prevalent lithological types.

The LSM is prepared by summing the factors contributing to landslides for both Sylhet and Rangamati.

$$\text{LSM} = \sum \text{FR} * (\text{Elevation} + \text{TWI} + \text{LULC} + \text{Slope} + \text{Rainfall} \\ + \text{Lithology} + \text{NDVI} + \text{Soil Type} + \text{Distance from road}).$$

23.3 Results

A spatial mapping and statistical database were created for all nine influencing factors in landslide susceptibility, including LULC, elevation, slope, TWI, rainfall, lithology, soil type, NDVI, and distance from roads, along with their respective subcategories for both Sylhet and Rangamati district (Fig. 23.3).

23.3.1 Sylhet

Based on this spatial mapping of all nine factors, a statistical database was derived, which is illustrated in (Table 23.2 and Fig. 23.4).

LULC classification categorized the land into seven distinct classes: water, vegetation, flooded vegetation, cropland, buildup area, bare land, and range land. 'Crop Land' covers the majority of the land about 64.61%, and it has a low correlation to landslides with a FR of only 0.26. Surprisingly, even though the 'Buildup Area' covers a smaller land area of 13.85%, it exhibits a much higher FR value of 4.21, which is highly correlated with landslides. This is because the buildup of areas contributes to landslide susceptibility by clearing vegetation, disrupting natural drainage, altering soil properties, and increasing surface runoff (Quevedo et al. 2023). These changes weaken slope stability and enhance the risk of landslides, making them more likely in urbanized environments.

The study area displayed an elevation range spanning from −21 m to 151 m into five classes. Higher elevation may increase landslide hazards by intensifying slope steepness and terrain instability (Rabby et al. 2022). Conversely, lower altitudes, notably in flat or gently sloping regions, can foster landslide vulnerability as water accumulation raises pore pressure and diminishes soil stability. Notably, the class labeled '13.4 m to 47.8 m' showed a high FR value of 1.20, covering 62.56% of the land area.

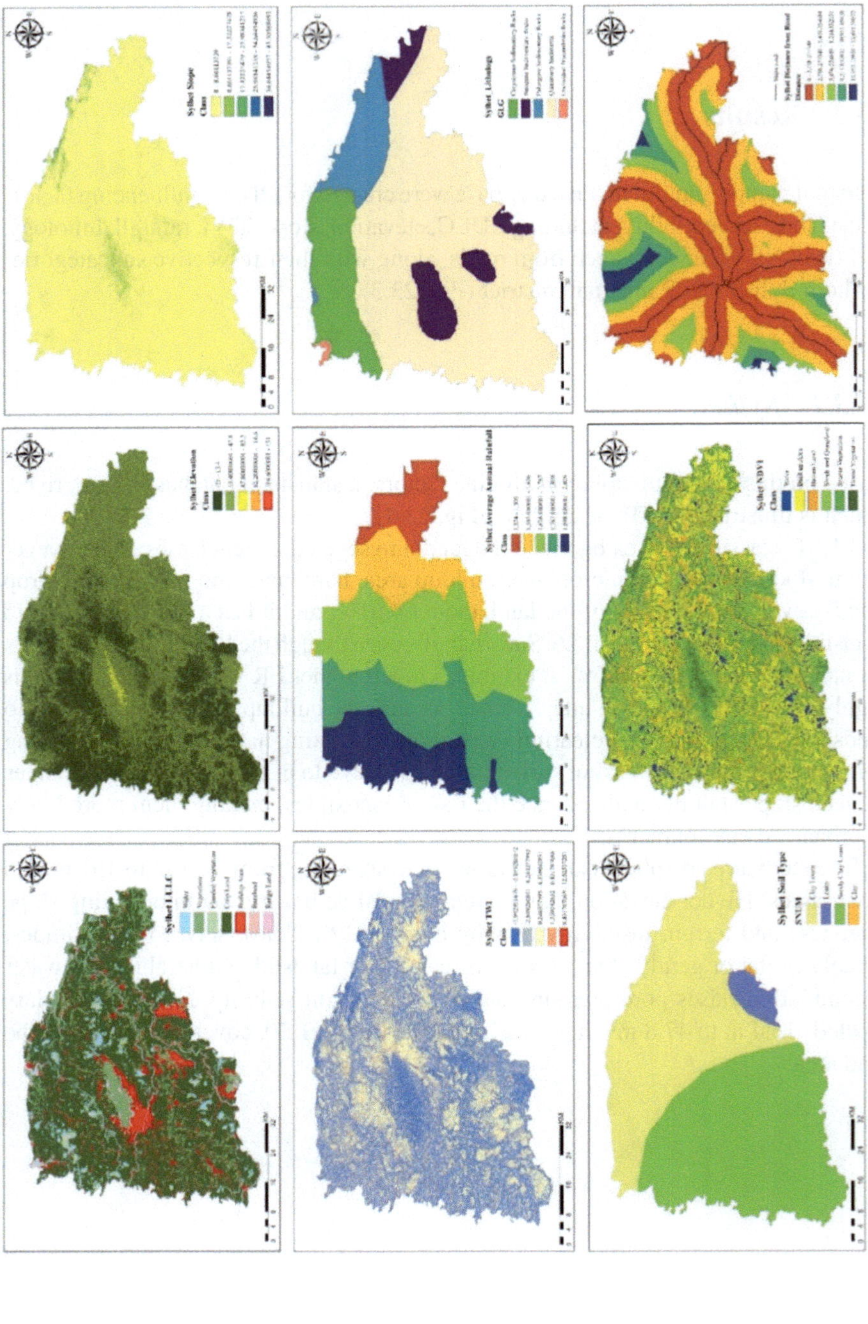

Fig. 23.3 Factors used for FR modeling (Sylhet)

Table 23.2 Conditioning factors used for Landslide susceptibility mapping through FR model (Sylhet)

Parameter	Class name or description	Histogram	% of histogram	Landslide number	% of Landslide number	FR
LULC	Water	2,252,687	6.59	2	16.67	2.53
	Vegetation	4,298,888	12.58	0	0.00	0.00
	Flooded vegetation	444,939	1.30	0	0.00	0.00
	Crop land	22,070,993	64.61	2	16.67	0.26
	Buildup area	4,729,986	13.85	7	58.33	4.21
	Bare land	39,226	0.11	0	0.00	0.00
	Range land	324,384	0.95	1	8.33	8.78
Elevation	−21–13.4	1,443,450	36.46	3	25.00	0.69
	13.4–47.8	2,476,386	62.56	9	75.00	1.20
	47.8–82.2	36,040	0.91	0	0.00	0.00
	82.2–116.6	2408	0.06	0	0.00	0.00
	116.6–151	342	0.01	0	0.00%	0.00
Slope (in degree)	0–8.66	3,821,341	98.34	11	91.67	0.93
	8.66–17.32	59,984	1.54	1	8.33	5.40
	17.32–25.91	4193	0.11	0	0.00	0.00
	25.91–34.64	358	0.01	0	0.00	0.00
	34.64–43.30	30	0.00	0	0.00	0.00
TWI	−7.93–3.84	1,552,684	39.36	4	33.33	0.85
	−3.84–0.24	1,347,342	34.16	4	33.33	0.98
	0.24–4.33	945,563	23.97	4	33.33	1.39
	4.33–8.43	88,942	2.25	0	0.00	0.00
	8.43–12.52	10,074	0.26	0	0.00	0.00
Rainfall	3749.34–4246.49	1,118,082	29.55	0	0.00	0.00
	4246.49–4743.64	1,456,208	38.49	4	33.33	0.87
	4743.64–5240.79	552,827	14.61	4	33.33	2.28
	5240.79–5737.94	380,005	10.04	1	8.33	0.83
	5737.94–6235.10	276,537	7.31	3	25.00	3.42
Lithology	Creyaceous sedimentary rocks	7543	14.22	7	58.33	4.10
	Neogene sedimentary rocks	4980	9.39	3	25.00	2.66
	Paleogene sedimentary rocks	5131	9.68	0	0.00	0.00
	Quaternary sediments	35,190	66.36	1	8.33	0.13
	Undivided Precambrian rocks	186	0.35	1	8.33	23.76

(continued)

Table 23.2 (continued)

Parameter	Class name or description	Histogram	% of histogram	Landslide number	% of Landslide number	FR
Soil type	Clay loam	24,713	46.94	5	41.67	0.89
	Loam	1682	3.19	0	0.00	0.00
	Sandy clay loam	26,211	49.78	7	58.33	1.17
	Clay	47	0.09	0	0.00	0.00
NDVI	Water	246,293	6.42	1	8.33	1.30
	Buildup area	907,229	23.66	4	33.33	1.41
	Barren land	689,540	17.98	2	16.67	0.93
	Shrub and grassland	1,166,047	30.41	5	41.67	1.37
	Sparse vegetation	747,499	19.49	0	0.00	0.00
	Dense vegetation	78,025	2.03	0	0.00	0.00
Distance from road	0–2738.27	15,757	41.98	12	100.00	2.38
	2738.27–5476.55	12,068	32.15	0	0.00	0.00
	5476.55–8214.83	6354	16.93	0	0.00	0.00
	8214.83–10,953.11	2524	6.72	0	0.00	0.00
	10,953.11–13,691.38	830	2.21	0	0.00	0.00

The susceptibility of a slope to landslides is impacted by its height and steepness. High, steep slopes are more prone to landslides due to increased gravitational forces, while gentle, low slopes have lower landslide risk because of better stability and reduced gravitational impact (Ramesh 2021). A spatial database on slope map (In Degree) was reclassified into five classes ranging from 0 to 43.30^0. The class labeled '0–8.66^0' stands out with a low FR ratio of 0.93, covering a significant 98.34% of the land area and being associated with 91.67% of landslide occurrences. On the other hand, the class '8.66^0–17.32^0' exhibits a high FR ratio of 5.40 but covers only 1.54% of the land, with 8.33% of landslide occurrences.

The TWI data is divided into five classes (−7.93 to −3.84, −3.84 to 0.24, 0.24 to 4.33, 4.33 to 8.43, 8.43 to 12.52), with the class '0.24 to 4.33' having a high FR value of 1.39 and all the landslide events occurred in the range of −7.93 to 4.33. According to (Różycka et al. 2017), low TWI values have the potential to affect slope stability, particularly in the top areas and on the steep head scarps of landslides, making them more susceptible to landslides. High TWI values also signify regions prone to water accumulation, escalating landslide risk. They often denote concentrated drainage paths, suggesting potential erosion and instability. In contrast, lower TWI values imply improved drainage, reducing landslide vulnerability.

Rainfall data is categorized into five classes spanning from 3749.34 mm to 6235.10 mm. The class '5737.94–6235.10' exhibits a high FR value of 3.42 as heavy rainfall in Bangladesh, particularly during the monsoon season, escalates the risk of landslides (Hossain 2020). Urban development on hillsides and settlement at their

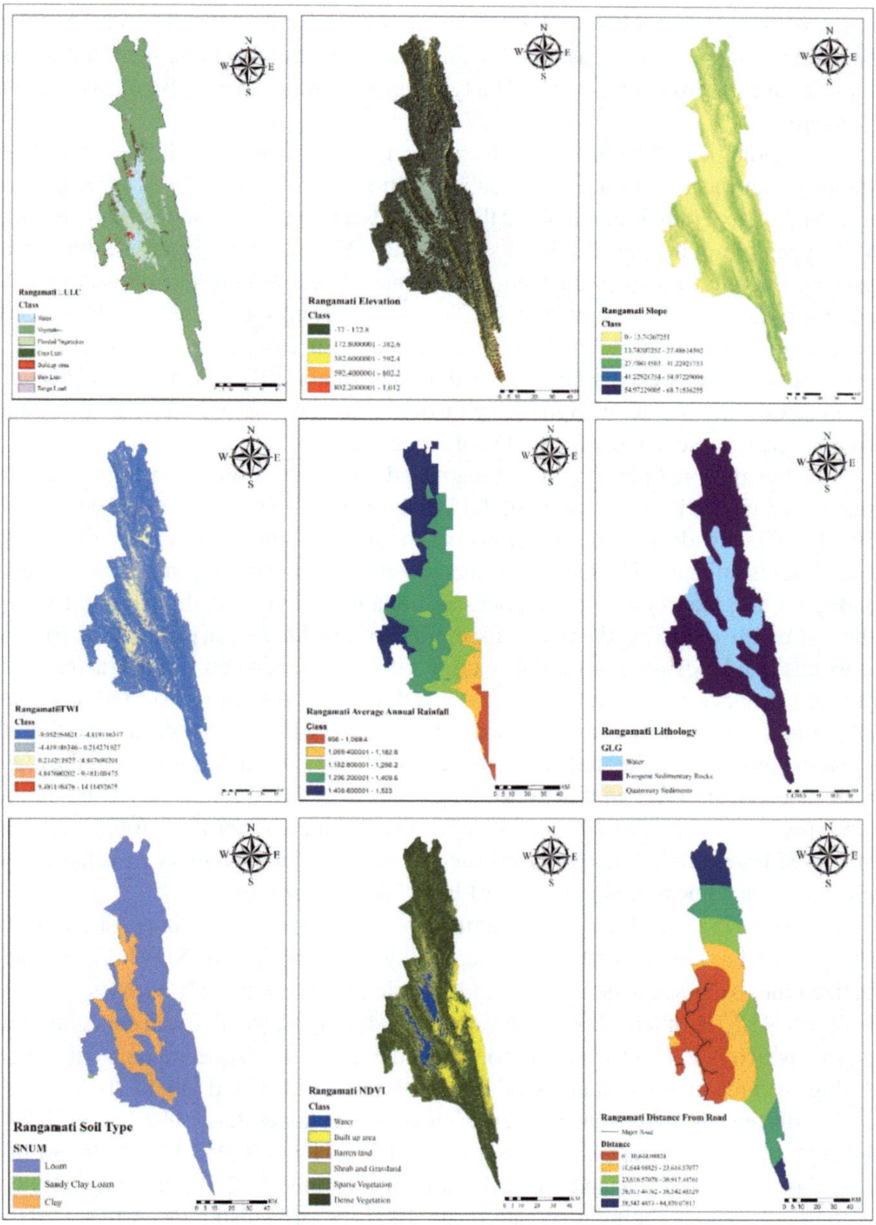

Fig. 23.4 Factors used for FR modeling (Rangamati)

bases worsens susceptibility. Rainfall impacts soil's pore water pressure, altering landslide danger, and emphasizing the importance of understanding rainfall patterns for effective hazard management. The type of rocks in the area is also classified into 5 categories.

The lithological composition of the basin plays a crucial role in influencing slope stability and the occurrence of landslides. Stronger rock formations offer greater resistance to external forces, making them less susceptible to landslides, while weaker rock types are more vulnerable to such events (Yalçın 2008). From a lithological perspective, areas with prevalent marl, limestone, shale, gypsum, and marly conglomerate formations are at a higher risk of landslides. Therefore, assessing the density of landslides within different geological structures is essential to understand their propensity for such events better (Abedini and Tulabi 2018). 'Undivided Precambrian Rocks' has a notably high FR value of 23.76, covering only 0.35% of the land area but being associated with 8.33% of landslides.

Soil type data includes clay loam, loam, and sandy clay loam. 'Sandy Clay Loam' stands out with a high FR value of 1.17 and covers 49.78% of the land area, with 58.33% of landslide occurrences. However, the susceptibility of different soil types to landslides is influenced by factors like composition, permeability, and shear strength, and sandy clay loam soils with a higher sand content exhibit better drainage and higher shear strength, rendering them less susceptible to landslides (Han et al. 2019a, b). The reason is that landslide susceptibility does not solely depend on soil characteristics; rainfall, distance from road factors, etc. influence it. As we see on the soil type map, most of the sandy clay loam portion is placed in the north-western and south-western zones. But at the same time, more rainfall happens in that zone, and the main road linkages are there in that zone. Clay loam soils, with a high clay content, are more prone to landslides due to poor drainage and high-water retention, leading to increased pore pressure and reduced shear strength. That's why this zone has a low FR but it has witnessed almost 42% of landslide occurrences.

The NDVI is a valuable tool for identifying landslide scars, tracking post-landslide changes, and assessing landslide susceptibility (Qu et al. 2020). NDVI data is categorized into 6 classes which 'Shrub and Grassland' has a relatively high FR value of 1.37 with 41.67% of landslide occurrences. Roads play a crucial role in concentrating runoff, which is evident from experience and existing landslide statistics observed during road reconstruction and widening projects (Abedini and Tulabi 2018).

The distance from the road is categorized into five classes (0–2738.27 m, 2738.27–5476.55 m, 5476.55–8214.83 m, 8214.83–10,953.11 m, 10,953.11–13,691.38 m). The class '0–2738.27' stands out with a high FR value of 2.38, covering 41.98% of the land area and being associated with 100.00% of landslide occurrences. This indicates a strong correlation between landslides and proximity to roads in this class as the proximity of roads to areas at risk of landslides, especially near rivers, heightens the landslide risk (Zhou et al. 2023).

23.3.2 Rangamati

Based on this spatial mapping of all nine factors, a statistical database was derived which is illustrated in (Table 23.3).

LULC is characterized by 7 distinct classes in which flooded vegetation, covering 3.08% of the area, exhibits a higher correlation of sporting an FR of 9.27. Flooded vegetation can make landslides more severe by reducing slope strength and root stability (Mirus et al. 2017). Slopes become more susceptible to collapse when floods sweep away vegetation. Because of the flooded vegetation, the soil is less cohesive and has higher pore pressures, which increases the risk of sliding. It also modifies the dynamics of water, impacting the stability of slopes. On the other hand, water plays a key role in triggering landslides by saturating and destabilizing soil as infiltration raises pore pressure, reducing soil strength, and causing slope failure (Abebe et al. 2021). Changes in land use like deforestation disrupt the natural water balance, increasing landslide risk. Steep terrain and poor land management compound this vulnerability, making water a critical factor in landslides. That's why, water occupies 6.17% of the total coverage area and is associated with a high FR of 2.32, which is highly correlated with landslide events.

Moving on to the elevation assessment, the Rangamati region displays an elevation range spanning from −37 to 1012 m. Notably, the class of −37 to 172.8 m, covering 74.12% of the area, is highly correlated with landslides, boasting an 85.71% correlation and an FR of 1.16. Other elevation classes exhibit lower or negligible correlations with landslide occurrences.

Slope, measured in degrees, is another crucial factor. The analysis categorizes slope data into five classes. The class spanning 13.7 to 27.4 degrees, covering 30.35% of the area, displays a strong correlation with landslides and a high FR of 3.30. In contrast, other slope classes show lower or no significant correlations with landslides.

TWI is divided into five classes and the class from −4.41 to 0.21 exhibits a higher FR of 2.43.

Rainfall data is also taken into account, and segmented into five distinct classes. The class within the range of 2488.14 to 2592.69 mm shows a high correlation with six landslide occurrences, resulting in a high FR of 3.11. Heavy rainfall is more susceptible to landslides but low rainfall can also indirectly cause landslides by drying up the soil, diminishing its cohesiveness, and increasing the susceptibility of slopes to collapse, particularly in steep places where constant rainfall is essential for soil stability (Ray and Lazzari 2020). Prolonged periods of low rainfall can cause soil desiccation and decreased moisture content, which raises the risk of landslides during periods of heavy rainfall.

The lithological composition is an influential factor, with Neogene Sedimentary Rocks covering 74.52% of the area and displaying a strong correlation with seven landslide occurrences, resulting in an FR of 1.34.

Soil type is another critical consideration, and the analysis reveals that Clay covers only 19.29% of the region, but is highly correlated with landslides at 85.71% landslide occurrence with an FR of 4.44. According to (Chen et al. 2010), clay content

Table 23.3 Conditioning factors used for landslide susceptibility mapping through FR model (Rangamati)

Parameter	Class name or description	Histogram	% of histogram	Landslide number	% of landslide number	FR
LULC	Water	4,787,643	6.17	1	14.29	2.32
	Vegetation	47,091,419	60.69	3	42.86	0.71
	Flooded vegetation	2,390,951	3.08	2	28.57	9.27
	Crop land	21,217,917	27.35	0	0.00	0.00
	Buildup area	894,206	1.15	1	14.29	0.19
	Bare land	83	0.00	0	0.00	0.00
	Range land	1,209,344	1.56	0	0.00	0.00
Elevation	−37–172.8	4,869,669	74.12	6	85.71	1.16
	172.8–382.6	1,302,929	19.83	1	14.29	0.72
	382.6–592.4	272,837	4.15	0	0.00	0.00
	592.4–802.2	111,246	1.69	0	0.00	0.00
	802.2–1012	13,692	0.21	0	0.00	0.00
Slope (in degree)	0–13.7	4,100,448	66.08	0	0.00	0.00
	13.7–27.4	1,883,125	30.35	7	100.00	3.30
	27.4–41.2	210,154	3.39	0	0.00	0.00
	41.2–54.9	10,953	0.18	0	0.00	0.00
	54.9–68.7	571	0.01	0	0.00	0.00
TWI	−9.05–4.41	4,496,591	63.80	3	42.86	0.67
	−4.41–0.21	1,240,896	17.61	3	42.86	2.43
	0.21–4.84	741,375	10.52	0	0.00	0.00
	4.84–9.48	560,589	7.95	1	14.29	1.80
	9.48–14.11	8663	0.12	0	0.00	0.00
Rainfall (in mm)	2383.60–2488.14	1,108,924	17.43	1	14.29	0.82
	2488.14–2592.69	1,753,364	27.56	6	85.71	3.11
	2592.69–2697.24	2,782,028	43.73	0	0.00	0.00
	2697.24–2801.79	600,009	9.43	0	0.00	0.00
	2801.79–2906.34	116,992	1.84	0	0.00	0.00
Lithology	Water	16,351	25.29	0	0.00	0.00
	Neogene sedimentary rocks	48,182	74.52	7	100.00	1.34
	Quaternary sediments	127	0.20	0	0.00	0.00
Soil type	Loam	50,299	80.61	1	14.29	0.18
	Sandy clay loam	63	0.10	0	0.00	0.00
	Clay	12,038	19.29	6	85.71	4.44
NDVI	Water	380,302	5.93	0	0.00	0.00

(continued)

Table 23.3 (continued)

Parameter	Class name or description	Histogram	% of histogram	Landslide number	% of landslide number	FR
	Buidup area	721,854	11.26	0	0.00	0.00
	Barren land	219,522	3.43	0	0.00	0.00
	Shrub and grassland	803,263	12.53	2	28.57	2.28
	Sparse vegetation	1,925,790	30.05	2	28.57	0.95
	Dense vegetation	2,358,592	36.80	3	42.86	1.16
Distance from road	0–10,644.08	19,602	30.69	6	85.71	2.79
	10,644.08–23,616.57	15,828	24.78	1	14.29	0.58
	23,616.57–38,917.44	13,171	20.62	0	0.00	0.00
	38,917.44–58,542.58	9116	14.27	0	0.00	0.00
	58,542.58–84,820.07	6160	9.64	0	0.00	0.00

significantly impacts soil mass failure, with moderate clay (5–10%) causing quick failures in short rainfall while low (2.5–5%) and high clay (>10%) soils need longer rain to fail. High clay reduces cohesion, aiding failure. Research suggests >2.5% clay is needed for landslides (Liu et al. 2021). Other soil type categories exhibit no or low correlations with landslide occurrences.

NDVI is categorized into six classes, a prominent factor causing landslides. According to (Asada and Minagawa 2023), dense vegetation, such as forests, can cause landslides by absorbing rainfall and blocking it, which results in runoff and soil saturation. Compared to forests, grasslands, and shrubs are more susceptible to landslides because of their weaker and less stable root systems than dense vegetation. That's why we can see that shrub and grassland demonstrate a relatively high correlation of 28.57% with two landslide occurrences and an FR of 2.28. Dense vegetation also exhibits high correlations with landslide events with an FR of 1.16.

Lastly, the distance from roads is examined and segmented into five classes where the class covering 0–10,644.08 m displays a strong correlation of 85.71% landslide events and an FR of 2.79 as areas closer to roads increase higher landslide occurrence (Zhou et al. 2023).

23.3.3 Comparison of Factors Between Sylhet and Rangamati

In both Sylhet and Rangamati regions, various factors contribute to landslide susceptibility, but the significance of these factors and their correlations differ between the two areas.

In Sylhet, 'Buildup Area' and 'Crop Land' significantly impact landslide susceptibility as urbanization disrupts natural drainage and soil properties, and also increase in deforestation increases susceptibility. Elevation influences susceptibility as steep slopes increase the chances of landslides. Heavy rainfall and certain geological formations increase susceptibility. Road proximity is a critical factor, with areas closer to roads showing a strong correlation with landslides. On the other hand, in Rangamati, features like 'Flooded Vegetation' and 'Water' exacerbate landslide susceptibility due to their effects on soil cohesion and slope stability. Slope, TWI, and heavy rainfall also contribute to susceptibility. 'Clay' soil type is a significant factor, with high clay content increasing cohesion and the risk of landslides. Dense vegetation and proximity to roads also play roles in landslide susceptibility. The key differences between the two regions lie in the factors influencing susceptibility. Sylhet's susceptibility is affected by land use changes and urbanization, while Rangamati experiences higher susceptibility due to flooded vegetation and water. Both areas share common factors like slope, TWI, and road proximity, highlighting their significance in landslide risk assessment.

23.3.4 FR Factors Ranking for Sylhet and Rangamati

Landslides are a common risk in terrain with slopes, resulting in fatalities in transportation routes, rural industrial sites, and urban areas (Froude and Petley 2018). Seismic activity, prolonged periods of heavy rain, creep fault occurrences, etc. are the prime reasons for landslides (Barbano et al. 2014). Risk assessment for landslides includes environmental, topographic, and social-based data sources. So, the factor-based analysis in the FR model provides immense details about the landslide susceptibility. Landslide databases of the highest caliber are necessary for assessing the danger of landslides where this landslide inventory is a fundamental tool for risk analysis and land planning, it represents a foundation of information and greatly aids the local authorities in making decisions (Froude and Petley 2018; Colombo et al. 2005). These NASA databases aid in the computation of the FR technique for Bangladesh's Sylhet and Rangamati during vulnerability assessments.

In Sylhet and Rangamati, the FR study details landslide vulnerability using nine parameters. In contrast to Sylhet, which is located in northeast Bangladesh, Rangamati symbolizes the southeast. The landslide causes at these two locations change depending on factors, as seen by the disparity in height and topographic conditions.

23.3.5 Comparing the Susceptibility Level of Landslide Between Sylhet and Rangamati

In comparing landslide susceptibility between Sylhet and Rangamati districts, it's evident that Sylhet has a greater number of villages and populations in the 'Moderate' zone compared to Rangamati (Fig. 23.5).

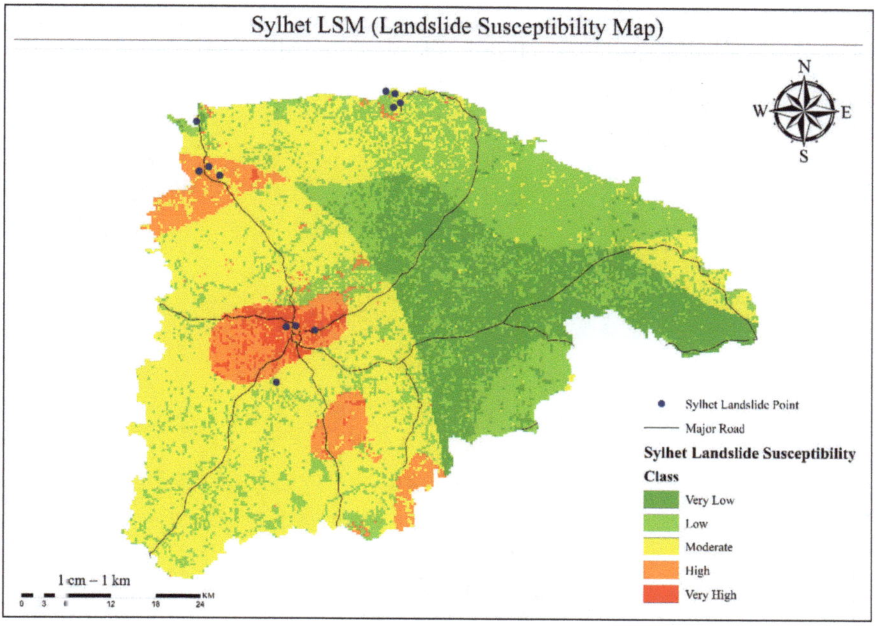

Fig. 23.5 Landslide susceptibility map (Sylhet)

Massive hill cutting in Sylhet has increased the risk of landslides (Islam et al. 2013). The problem is exacerbated by widespread hill damage brought on by housing firms breaking anti-hill cutting rules and misunderstanding the environmental effects. With its smaller land area and higher population density, Sylhet faces a more concentrated impact of land-use changes, deforestation, and construction activities that increase the risk of landslides. On the other hand, Rangamati has a higher percentage of villages with 'High' and 'Very High' ranges that are susceptible to landslides, and a higher number of people reside in these high-risk areas (Fig. 23.6).

Urbanization, forest conversion, uncontrolled hill cutting, and home-building on unstable slopes all increase the risk of landslides in hilly areas (Ahmed 2021). Especially in urbanized hill settlements, flash floods can be exacerbated by heavy monsoon rainfall. Furthermore, the topographical differences in these districts make it inherently more susceptible to landslides, especially during heavy rainfall. The statistics indicate the pronounced disparities in landslide susceptibility, highlighting the importance of considering geographical factors in disaster risk assessment and management.

The percentage of sensitive places varies according to the circumstances and FR factors in Sylhet and Rangamati. According to (Table 23.4), Rangamati has a high landslide vulnerability due to soil type, LULC, and TWI factors. The most vulnerable locations to landslides are those with vegetation to built-up LULC transition types, while the worst-affected industries include primary economic activities like fishing and jhum farming (Abedin et al. 2020). Meanwhile, lithology, LULC, and rainfall are

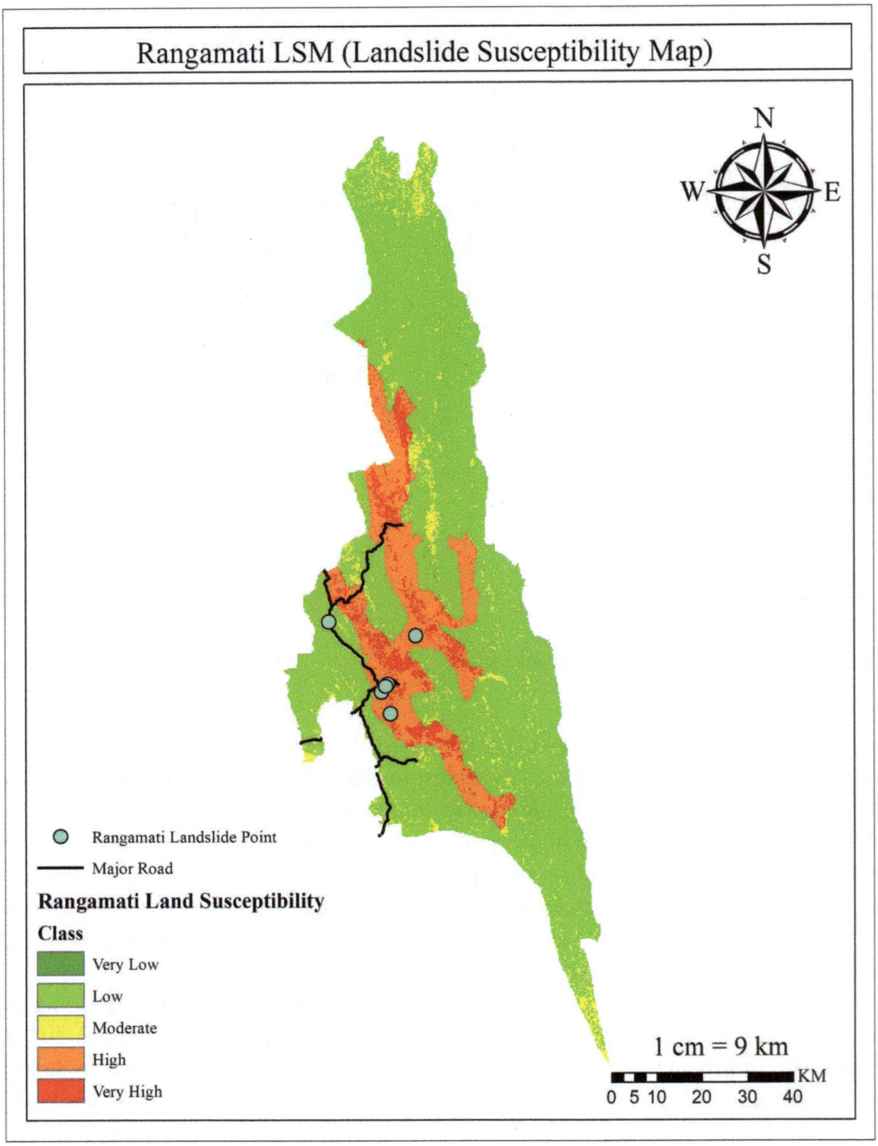

Fig. 23.6 Landslide susceptibility map (Rangamati)

major definers for the highly susceptible areas in Sylhet. The Surma River's discharge pattern is altering in tandem with the rainfall patterns in the northeastern region of Bangladesh which is an influential reason for disasters like floods, landslides, and so on (Akter et al. 2019). In Sylhet, the high landslide-prone zones are around 300 km^2 or nearly 10% of the total area; in Rangamati, the size is above 1000 km^2.

Table 23.4 FR ranking of factors between two regions

Parameter	Sylhet FR	Ranking	Rangamati FR	Ranking
LULC	15.77 (21.12%)	2	12.49 (31.05%)	1
Elevation	1.88 (2.52%)	9	1.88 (4.67%)	8
Slope (in degree)	6.33 (8.48%)	4	3.30 (8.2%)	7
TWI	3.21 (4.3%)	6	4.90 (12.18%)	2
Rainfall (mm)	7.40 (9.91%)	3	3.93 (9.77%)	5
Lithology	30.65 (41.04%)	1	1.34 (3.33%)	9
Soil type	2.06 (2.76%)	8	4.62 (11.48%)	3
NDVI	5.00 (6.70%)	5	4.40 (10.94%)	4
Distance from road	(3.19%)	7	3.37 (8.38%)	6

Table 23.5 Landslide susceptibility zone

Ranges	Sylhet villages	Rangamati villages	Sylhet population	Rangamati population	Sylhet area (km^2)	Rangamati area (km^2)
Very Low	473 (13.89%)	1 (0.07%)	495,349 (15.66%)	363 (0.07%)	588.44 (17.23%)	2.14 (0.04%)
Low	797 (23.4%)	1115 (73.6%)	820,129 (25.92%)	367,219 (71.24%)	999.21 (29.26%)	4363.79 (75.72%)
Moderate	1713 (50.29%)	81 (5.35%)	1,466,157 (46.35%)	27,620 (5.36%)	1505.10 (44.07%)	317.85 (5.52%)
High	311 (9.13%)	225 (14.85%)	285,510 (9.03%)	86,093 (16.7%)	257.77 (7.55%)	787.56 (13.67%)
Very High	112 (3.29%)	93 (6.14%)	96,348 (3.05%)	34,159 (6.63%)	64.48 (1.89%)	291.67 (5.06%)

Source Directorate General of Family Planning, MIS Unit (2014)

Roughly twenty villages in Sylhet and ten in Rangamati are located in the risk zones (Table 23.5).

To integrate policies and deliver services, the LSM offers comprehensive information on the people and locations that are vulnerable to landslides.

23.4 Discussion

This study provides important insights into landslide susceptibility and risk assessment in Bangladesh's Sylhet and Rangamati hilly regions using geostatistical and geospatial modeling techniques. The results highlight key differences in landslide causative factors between the two regions, with implications for targeted risk mitigation strategies.

In Sylhet, the FR analysis identified built-up areas, crop lands, lithology, rainfall, slope, and proximity to roads as key factors influencing landslide occurrence. Built-up areas exhibited the highest FR value (4.21), indicating significantly increased landslide susceptibility from urbanization, land use changes, and anthropogenic activities like deforestation and unplanned construction (Han et al. 2019a, b; Rahman et al. 2013). Rahman et al. (2013) noted that rapid and uncontrolled urban growth on slopes overlooking the Surma River basin escalates landslide risks in Sylhet City. Crop lands also contribute to landslides by replacing natural vegetation, disrupting drainage patterns, and increasing surface runoff (Quevedo et al. 2023). Specific lithological formations like cretaceous sedimentary rocks (FR 4.10) further increase susceptibility in Sylhet (Rabby and Li 2019). Areas of high rainfall, steep slopes, and proximity to roads likewise raise landslide hazards. These findings align with prior research on landslide conditioning factors in Bangladesh (Jebur et al. 2014; Kirschbaum et al. 2015).

In contrast, Rangamati's landslide susceptibility is strongly influenced by flooded vegetation, water, clay soils, slope, and road proximity. Flooded vegetation showed the highest FR (9.27), reflecting increased instability from soil saturation, loss of root cohesion, and altered drainage (Mirus et al. 2017). Water bodies also destabilize slopes, especially on steep terrain (FR 2.32) (Dahal and Hasegawa 2008). Heavy rainfall further elevates pore water pressure and reduces shear strength, making Rangamati's wet, vegetated slopes more prone to failure (Kirschbaum et al. 2015). Clay content is another critical factor, with clay soils exhibiting a high landslide correlation (85.71%) and an FR of 4.44. Louati et al. (2023) noted that clayey soils have higher moisture retention, lower permeability, and increased risk of sliding. Besides topographical and hydrological factors, proximity to roads remained influential in Rangamati like Sylhet.

The comparative FR analysis thus highlights the distinct landslide causation patterns based on the unique geographical settings of each region. While some common factors like roads exist, region-specific elements like lithology in Sylhet and flooded vegetation in Rangamati are major determinants of susceptibility. Integrating these differences is vital for targeted risk management (Reichenbach et al. 2014; Kirschbaum et al. 2015). Structural measures like retaining walls and improved drainage infrastructure in built-up areas can mitigate landslide impacts in Sylhet City (Rahman 2013). In Rangamati, restoring natural vegetation, avoiding developments on unstable clayey slopes, and regulating anthropogenic changes to hydrology are priority interventions (Mirus et al. 2017). Petley (2012) noted that susceptibility models should guide context-specific mitigation strategies.

The landslide susceptibility maps generated using the FR model also showcase the spatial patterns of landslide vulnerability across Sylhet and Rangamati. In Sylhet, the moderate risk zone covers the highest share of villages (50.29%) and population (46.35%), clustered around Sylhet City and other urbanized areas. In Rangamati, villages (73.60%) and people (71.24%) are predominantly concentrated in low landslide susceptibility areas, with the high risk zone accounting for 20.99% of villages and 22.46% of the population. These maps provide actionable information to local authorities for targeted disaster risk reduction planning based on vulnerability levels (Guzzetti et al. 2012).

Several limitations of this study provide avenues for future research. First, the analysis considered only past landslide locations. Using satellite data could strengthen it by integrating landslide triggering factors like rainfall thresholds (Kirschbaum et al. 2015; Petley 2012). Second, the FR model could include more landslide conditioning factors like geology and curvatures (Jebur et al. 2014). Lastly, different modeling techniques like logistic regression and machine learning can be used alongside FR for comparison and validation (Khosravi et al. 2018; Jebur et al. 2014). Nonetheless, this study delivers valuable insights into landslide susceptibility patterns in two major hilly regions of Bangladesh using robust geospatial analytics. The findings can guide evidence-based policies for landslide risk reduction and sustainable development.

23.5 Conclusion

This chapter delves into the complex realm of landslide susceptibility, with a focus on the various factors that influence this natural hazard, particularly in the Sylhet and Rangamati districts of Bangladesh. The analysis considers nine influential factors: LULC, elevation, slope, TWI, precipitation, lithology, soil type, NDVI, and distance from roads. By employing the FR model, the chapter provides valuable insights into the factors that shape landslide susceptibility in these regions.

The findings reveal that both regions are prone to landslides, but the underlying causes differ significantly. In Sylhet, susceptibility is tied to urbanization, land-use changes, and features such as 'Built-up Area' and 'Crop Land.' Conversely, Rangamati's susceptibility is largely due to elements like 'Flooded Vegetation' and 'Water,' which are influenced by unique topographical and environmental features, among other factors. Additionally, both regions' landslide susceptibility is influenced to varying degrees by elevation, slope, heavy rainfall, lithology, TWI, soil type, NDVI, and proximity to roads.

A comparative analysis between Sylhet and Rangamati underscores the importance of geographical and environmental factors in determining landslide susceptibility. Factors such as concentrated urbanization and deforestation in Sylhet, and unique topographical features in Rangamati, inherently increase their susceptibility to landslides. The Landslide Susceptibility Map (LSM) indicates that 18.73% of Rangamati's areas and 9.44% of Sylhet's areas are classified as highly to extremely susceptible to landslides.

This chapter plays a crucial role in identifying areas prone to landslides, enabling a swift determination of which regions are more susceptible to various factors. For instance, the Sylhet region is more vulnerable to changes in land use and urbanization, while Rangamati is more susceptible due to its topography. The comprehensive analysis provided in this chapter allows for a targeted approach to understanding the factors contributing to vulnerability in different regions. By pinpointing the specific challenges each region faces, we can devise effective strategies and take necessary measures to minimize landslide risks. This organized approach ensures a more effective and appropriate response to the specific vulnerabilities caused by different geographical and environmental conditions.

References

Abebe G, Getachew D, Ewunetu A (2021) Analysing land use/land cover changes and its dynamics using remote sensing and GIS in Gubalafito district, Northeastern Ethiopia. SN Appl Sci 4(1). https://doi.org/10.1007/s42452-021-04915-8

Abedin J, Rabby Y, Hasan I, Akter H (2020) An investigation of the characteristics, causes, and consequences of June 13, 2017, landslides in Rangamati District Bangladesh. Geoenvironmental Disasters 7:1–19. https://doi.org/10.1186/s40677-020-00161-z

Abedini M, Tulabi S (2018) Assessing LNRF, FR, and AHP models in landslide susceptibility mapping index: a comparative study of Nojian watershed in Lorestan province, Iran. Environ Earth Sci 77(11). Retrieved from: https://doi.org/10.1007/s12665-018-7524-1

Ahmed B (2021) The root causes of landslide vulnerability in Bangladesh. Landslides, 1–14. Retrieved from: https://doi.org/10.1007/s10346-020-01606-0

Ahmed B, Shahinoor M, Sonia R, Ferdous R, Huq F, Ara S (2014) Landslide inventory report of Chittagong Metropolitan Area, Bangladesh. Retrieved from: https://discovery.ucl.ac.uk/id/eprint/1467165/

Akter S, Howladar M, Ahmed Z, Chowdhury T (2019) The rainfall and discharge trends of Surma River area in North-eastern part of Bangladesh: an approach for understanding the impacts of climatic change. Environ Syst Res 8:1–12. https://doi.org/10.1186/s40068-019-0156-y

Alam E (2020) Landslide hazard knowledge, risk perception and preparedness in Southeast Bangladesh. Sustainability. Multidisciplinary Digital Publishing Institute. https://doi.org/10.3390/su12166305

Alexander D (2000) Confronting catastrophe: new perspectives on natural disasters. Terra Publishing, England

Alexander DE (1993) Natural disasters. CRC Press, Boca Raton

Asada H, Minagawa T (2023) Impact of vegetation differences on shallow landslides: a case study in ASO. Japan Water 15(18):3193. https://doi.org/10.3390/w15183193

Bangladesh Bureau of Statistics (2012) Population and housing census 2011: national report. Bangladesh bureau of statistics, Dhaka, Bangladesh. Retrieved from: https://bbs.gov.bd/WebTestApplication/userfiles/Image/Census2011/Chittagong/Rangamati/Rangamati%20at%20a%20glance.pdf

Bangladesh Meteorological Department (n.d.) Retrieved October 17, 2023, from https://live7.bmd.gov.bd/

Barbano M, Pappalardo G, Pirrotta C, Mineo S (2014) Landslide triggers along volcanic rock slopes in eastern Sicily (Italy). Nat Hazards 73:1587–1607. https://doi.org/10.1007/s11069-014-1160-1

Brammer H (1996) The geography of the soils of Bangladesh. University Press Limited, Dhaka

BWDB (2005) Flash flood events and mitigation management in Bangladesh. In: Workshop on managing flash floods and sustainable development in the Himalayas, 23–28 October 2005, Lhasa, Bangladesh Water Development Board (BWDB), Dhaka

Chen N, Zhou W, Yang CL, Hu G, Gao Y, Han D (2010) The processes and mechanism of failure and debris flow initiation for gravel soil with different clay content. Geomorphology 121(3–4):222–230. https://doi.org/10.1016/j.geomorph.2010.04.017

Chowdhury MS (2023) A review on landslide susceptibility mapping research in Bangladesh. Heliyon 9(7):e17972. Retrieved from: https://doi.org/10.1016/j.heliyon.2023.e17972

Colombo A, Lanteri L, Ramasco M, Troisi C (2005) Systematic GIS-based landslide inventory as the first step for effective landslide-hazard management. Landslides 2:291–301. https://doi.org/10.1007/S10346-005-0025-9

Dahal RK, Hasegawa S (2008) Representative rainfall thresholds for landslides in the Nepal Himalaya. Geomorphology 100(3–4):429–443. https://doi.org/10.1016/j.geomorph.2008.01.014

Debu D (2022) Thousands of Sylhet people are at risk of landslides. The Business Standard. Retrieved from: https://www.tbsnews.net/bangladesh/thousands-sylhet-people-risk-landslide-457870

Edris and Alam (2020) Landslide hazard knowledge, risk perception and preparedness in Southeast Bangladesh. Sustainability. https://doi.org/10.3390/SU12166305

Froude M, Petley D (2018) Global fatal landslide occurrence from 2004 to 2016. Nat Hazard. https://doi.org/10.5194/NHESS-18-2161-2018

Guzzetti F, Mondini AC, Cardinali M, Fiorucci F, Santangelo M, Chang KT (2012) Landslide inventory maps: new tools for an old problem. Earth-Sci Rev 112(1–2):42–66. https://doi.org/10.1016/j.earscirev.2012.02.001

Habibur Rahman M, Jannatul Fardusi M, Roy B, Raihan F (2011) Unplanned urbanization and hill cutting: a study on environmental change in Sylhet. BRAC Univ J VIII(2)

Han SS, Kim S, Lee S (2019) Landslide susceptibility mapping using frequency ratio, analytic hierarchy process, logistic regression, and artificial neural network methods at the Inje area, Korea. Environ Earth Sci 78(2). https://doi.org/10.1007/s12665-018-8035-x

Han Z, Su B, Li Y, Ma Y, Wang W, Chen G (2019) Comprehensive analysis of landslide stability and related countermeasures: a case study of the Lanmuxi landslide in China. Sci Rep 9(1). Retrieved from: https://doi.org/10.1038/s41598-019-48934-3

Haque MA (2006) The soils of Bangladesh. Springer,https://doi.org/10.1007/978-94-007-1128-0

Hossain ATMS (2020) Rainfall induced landslide hazards of Bangladesh: challenges, issues and sustainable development. Int J Sci Eng Res 11(7)

IFRC (2014) World disasters report 2014—focus on culture and risk. international federation of red cross and red crescent societies (IFRC), Geneva, Switzerland. Retrieved from: http://www.ifrc.org/en/publications-andreports/world-disasters-report/world-disasters-report-2013/

Islam M, Rahman MM, Mubin SH (2013) Hill cutting in Sylhet city: a case study. IOSR J Environ Sci 6(4):6–12

Islam MN, Uddin MN (2002) Country paper on hydrogeology, international workshop on arsenic issue in Bangladesh

Jebur MN, Pradhan B, Tehrany MS (2014) Optimization of landslide conditioning factors using very high-resolution airborne laser scanning (LiDAR) data at catchment scale. Remote Sens Environ 152:150–165. https://doi.org/10.1016/j.rse.2014.05.013

Kafy AA, Rahman MS, Ferdous L (2017) Exploring the association of land cover change and landslides in the Chittagong Hill Tracts (CHT): a remote sensing perspective. In: Proceedings of the international conference on disaster risk mitigation, Dhaka, Bangladesh

Kamal ASMM, Hossain F, Ahmed B, Sammonds P (2022) Analyzing the 27 July 2021 rainfall-induced catastrophic landslide event in the Kutupalong Rohingya Camp in Cox's Bazar, Bangladesh. Geoenvironmental Disasters 9(1). https://doi.org/10.1186/s40677-022-00219-0.

Khosravi K, Pham BT, Chapi K, Shirzadi A, Shahabi H, Revhaug I, Prakash I, Bui DT (2018) A comparative assessment of decision trees algorithms for flash flood susceptibility modeling at Haraz watershed, northern Iran. Sci Total Environ 627:744–755. https://doi.org/10.1016/j.scitotenv.2018.01.266

Kirschbaum D, Stanley T, Zhou Y (2015) Spatial and temporal analysis of a global landslide catalog. Geomorphology 249:4–15. https://doi.org/10.1016/j.geomorph.2015.03.016

Liu Y, Deng Z, Wang X (2021) The effects of rainfall, soil type, and slope on the processes and mechanisms of Rainfall-Induced shallow landslides. Appl Sci 11(24):11652. https://doi.org/10.3390/app112411652

Louati F, Trabelsi H, Alassaf Y, Jamei M, Taibi S (2023) Unsaturated permeability of clayey soil during desiccation: combined evaporation-shrinkage-cracking. Bull Eng Geol Environ 82(6). https://doi.org/10.1007/s10064-023-03202-7

Mahdi M, Bayes Ah, Singh RP, Manoj K, Yadav D, Silveira V, Paula da R, Fujitani T, Xing J (2023) Non-landslide sampling and ensemble learning techniques to improve landslide susceptibility mapping

Mirus BB, Smith JB, Baum RL (2017) Hydrologic Impacts of landslide disturbances: implications for remobilization and hazard persistence. Water Resour Res 53(10):8250–8265. https://doi.org/10.1002/2017wr020842

Moore ID, Grayson RB, Ladson AR (1991) Digital terrain modelling: a review of hydrological, geomorphological, and biological applications. Hydrol Process 5(1):3–30. https://doi.org/10.1002/hyp.3360050103

Petley D (2012) Global patterns of loss of life from landslides. Geology 40(10):927–930. https://doi.org/10.1130/g33217.1

Petley DN, Higuchi T, Petley DJ, Bulmer MH, Carey J (2005) Development of progressive landslide failure in cohesive materials. Geology 33(3):201–204. https://doi.org/10.1130/G21147.1

Pradhan B, Lee S, Buchroithner MF (2010) Remote sensing and GIS-based landslide susceptibility analysis and its cross-validation in three test areas using a frequency ratio model. Photogramm Fernerkund GeoInf 2010(1):17–32. https://doi.org/10.1127/1432-8364/2010/0037

Pratap R, Vikram G (2021) Landslide hazard, vulnerability, and risk assessment (HVRA), Mussoorie township, lesser Himalaya. Environment, Development and Sustainability, India. https://doi.org/10.1007/S10668-021-01449-2

Qu F, Qiu H, Sun H, Tang M (2020) Post-failure landslide change detection and analysis using optical satellite Sentinel-2 images. Landslides, 18(1):447–455. Retrieved from: https://doi.org/10.1007/s10346-020-01498-0

Quarantelli EL (1998) What Is a Disaster? Routledge, New York

Quevedo RP, Velástegui-Montoya A, Montalván-Burbano N, Morante-Carballo F, Korup O, Rennó CD (2023) Land use and land cover as a conditioning factor in landslide susceptibility: a literature review. Landslides 20(5):967–982. Retrieved from: https://doi.org/10.1007/s10346-022-02020-4

Rabby YW, Li Y (2019) Landslide inventory (2001–2017) of Chittagong hilly areas Bangladeshhttps://doi.org/10.20944/PREPRINTS201911.0269.V1

Rabby YW, Li Y, Abedin J, Sabrina S (2022) Impact of Land Use/Land Cover change on landslide susceptibility in Rangamati Municipality of Rangamati District, Bangladesh. ISPRS Int J Geo-Inf 11(2), 89. Retrieved from: https://doi.org/10.3390/ijgi11020089

Rabby YW, Li Y (2019) An integrated approach to map landslides in Chittagong Hilly Areas, Bangladesh, using Google Earth and field mapping. Landslides 16:633–645. https://doi.org/10.1007/s10346-018-1107-9

Rahman A, Zafor MA, Rahman MA (2013) Surface water quality and risk assessment in the vicinity of Sylhet City. Int J Water Resour Environ Eng 5(1):29–34. https://doi.org/10.5897/ijwree12.034

Rahman MS, Ahmed B, Di L (2017) Landslide initiation and runout susceptibility modeling in the context of hill cutting and rapid urbanization: a combined approach of weights of evidence and spatial multi-criteria. J Mt Sci 14:1919–1937. https://doi.org/10.1007/s11629-016-4220-z

Rahman T (2012) Landslide risk reduction of informal foothill settlements of Chittagong city through strategic design measure. Master's Thesis, BRAC University, Dhaka, Bangladesh, p 128

Ramesh G (2021) Slope and Landslide Stabilization: A Review. Indian J Struct Eng (IJSE) 1(2):13–16. Retrieved from: https://doi.org/10.35940/ijse.a1304.111221

Rangamati Hill District Council (2011) Overview of Rangamati. http://rhdcbd.org/old/about_rangamati.php

Rashid HE (1991) Geography of Bangladesh. University Press Limited, Dhaka

Ray RL, Lazzari M (2020) Landslides - investigation and monitoring. IntechOpen eBooks. https://doi.org/10.5772/intechopen.78130

Reichenbach P, Rossi M, Malamud BD, Mihir M, Guzzetti F (2018) A review of statistically-based landslide susceptibility models. Earth Sci Rev 180:60–91. https://doi.org/10.1016/j.earscirev.2018.03.001

Różycka M, Migoń P, Michniewicz A (2017) Topographic Wetness Index and Terrain Ruggedness Index in geomorphic characterisation of landslide terrains, on examples from the Sudetes, SW Poland. Zeitschrift Fur Geomorphologie, 61(2):61–80. Retrieved from: https://doi.org/10.1127/zfg_suppl/2016/0328

Samanta S, Pal DK, Palsamanta B (2018) Flood susceptibility analysis through remote sensing, GIS, and frequency ratio model. Appl Water Sci 8(2). Retrieved From: https://doi.org/10.1007/s13201-018-0710-1

Sifa SF, Mahmud T, Tarin MA, Haque DME (2019) Event-based landslide susceptibility mapping using weights of evidence (WoE0) and modified frequency ratio (MFR0) model: a case study of Rangamati district in Bangladesh. Geol Eco Landscapes. https://doi.org/10.1080/24749508.2019.1619222

UNDP (2017) The CHT was severely hit by four consecutive disasters in two weeks in June. Retrieved From: https://unpo.org/article/20199?id=20199

United Nations (UN) (2015) Transforming our world: the 2030 Agenda for Sustainable Development. Retrieved from: http://www.un.org/sustainabledevelopment/

UNPO (2017) Chittagong hill tracts: torrential rainstorms and wide-scale landslides leave thousands homeless. Retrieved from: https://unpo.org/error.php

Upazila Population Data (n.d.) Retrieved from http://www.dgfpmis.org/ss/admin/yearly_rep/population_mchthanaprocess.php

Yalçın A (2008) GIS-based landslide susceptibility mapping using analytical hierarchy process and bivariate statistics in Ardesen (Turkey): Comparisons of results and confirmations. CATENA 72(1):1–12. Retrieved from: https://doi.org/10.1016/j.catena.2007.01.003

Zhou M, Yuan M, Yang G, Mei G (2023) Risk analysis of road networks under the influence of landslides by considering landslide susceptibility and road vulnerability: a case study. Nat Hazard Res. https://doi.org/10.1016/j.nhres.2023.09.013

Chapter 24
Landslide Hazard and Risk Management Framework for Alaknanda Basin in the Indian Himalayan Region

Mitthan Lal Kansal and Sachchidanand Singh

Abstract Landslides in the Indian Himalayan Region (IHR) primarily result from extreme precipitation events, distinctive topography, and inadequate soil and water conservation practices. Managing such calamity is a daunting task. This study advocates an innovative rule-based approach for Landslide Risk Management (LRM). A distinct feature of this approach is a likelihood-consequence matrix (LCM) that formulates 25 combinations to segregate the basin into four risk zones: low, moderate, high, and extreme. Each combination aligns with a specific landslide likelihood and corresponding damage. Furthermore, integrating Geographic Information System (GIS) technology ensures effective spatial data management and visualization. The proposed rule-based model offers scalability, adaptability, and user-friendliness. The proposed methodology allows the stakeholders to take informed decisions for effective disaster response, evacuation plans, road closures, and long-term mitigation strategies. To illustrate the proposed methodology, the Alaknanda basin in the Indian Himalayan Region is considered a case study. The methodology adopted in this study seamlessly integrates the Likelihood of landslides depicted by Landslide Hazard Maps, with the anticipated Consequences, represented through comprehensive Soil Erosion Maps. It is observed that, 32.0% of the area falls under "Unlikely" likelihood, with 57.3% facing "Minor" consequences. "Low" risk covers nearly half the area (49.7%), while "Moderate," "High," and "Extreme" risks constitute 25.2%, 17.6%, and 7.5% respectively. Such an approach offers an expansive analysis, enriching the understanding of inherent vulnerabilities and potential impacts within the Alaknanda

M. L. Kansal (✉) · S. Singh
Water Resources Development and Management, Indian Institute of Technology, Roorkee, Roorkee, Uttarakhand, India
e-mail: mlk@wr.iitr.ac.in

S. Singh
e-mail: ssingh1@wr.iitr.ac.in

S. Singh
Western Himalayan Regional Center, National Institute of Hydrology Jammu, Jammu & Kashmir, India

G. K. Panda et al. (eds.), *Landslide: Susceptibility, Risk Assessment and Sustainability*, Advances in Natural and Technological Hazards Research 52,
https://doi.org/10.1007/978-3-031-56591-5_24

Basin. It is hoped that the proposed system will help planners and decision-makers in effective risk management on a scientific basis.

Keywords Landslide hazard · Risk management framework · Likelihood-consequence matrix · Geographic information system · Indian Himalayan Region

24.1 Introduction

The Indian Himalayan Region (IHR), with its distinctive topography, steep slopes, and extreme rainfall events, is a hotspot for frequent landslide incidents (Singh et al. 2023). A variety of factors, primarily fragile geology, unique topographic, and inadequate soil and water conservation practices, synergize to amplify the landslide risk in the region (Kansal and Singh 2022; Sikka et al. 2015). Consequently, the management of these landslides evolves as a challenging task, demanding an efficient, scientific, and systematic approach. The role of extreme precipitation events in triggering landslides in the IHR cannot be overstated. These events often result in rapid soil saturation, leading to an increase in pore water pressure and a reduction in the shear strength of soil and rocks. When combined with the region's steep slopes, this leads to a high susceptibility to landslides (Panwar et al. 2017a, b; Shukla et al. 2017; Singh and Kansal 2023a, b).

Numerous research studies conducted in the past have significantly contributed to our understanding of landslide risk, employing a variety of methods to assess both the probability of landslide occurrence and the potential severity of their consequences. These methods range from field-based assessments to sophisticated modeling techniques. Landslide Susceptibility Mapping, for instance, utilizes heuristic methods where experts evaluate factors such as slope, geology, and land use, alongside statistical analysis like logistic regression that uses historical data to identify correlations between various predisposing factors and landslide occurrences (Kinage et al. 2019; Oh et al. 2009; Regmi et al. 2014). In Landslide Hazard Assessment, deterministic models are employed to simulate slope stability under specific conditions using geotechnical data, while stochastic models introduce elements of randomness, often using probabilistic approaches to estimate landslide frequency and magnitude (Galve et al. 2015; Lee and Dan 2005; Park and Lee 2021). Monte Carlo Simulations are another valuable tool, using repeated random sampling to simulate a range of possible outcomes and provide a probability distribution of the risk (Wichmann and Becht 2003). Field Surveys and Monitoring, which involve instruments like inclinometers, extensometers, and ground-based LiDAR, provide crucial real-time data on ground movement and slope deformation (Metternicht et al. 2005; Sarkar et al. 2006; Zhang et al. 2016). Additionally, both Physical and Numerical Modeling are used to understand the physical processes involved in landslide initiation and movement, with physical models in laboratories and software tools simulating various scenarios of slope failure based on slope geometry, soil properties, and hydrological conditions (Reichenbach et al. 2018; Tran et al. 2018; Zhang et al. 2016). Together, these diverse and comprehensive techniques form the foundation of effective landslide risk estimation, addressing the multifaceted nature of this geohazard. However, there still exist gaps in synthesizing these various dimensions into an actionable and comprehensive framework for landslide risk management. This is where the integration of the Likelihood-Consequence Matrix (LCM) technique with Geospatial Technologies, such as Remote Sensing and GIS, becomes pivotal, enhancing the efficiency and comprehensiveness of landslide risk management. This synergistic approach combines the structured risk evaluation of LCM with the spatial analytical power of GIS and the extensive monitoring capacity of Remote Sensing to develop a more holistic, effective, and actionable strategy for addressing the complexities of landslide risk.

The LCM, with its 25 combinations of landslide likelihood and corresponding consequences, offers a nuanced and detailed approach to segregate the basin into four risk zones: low, moderate, high, and extreme. By amalgamating this model with GIS technology, the study advances beyond traditional landslide risk assessments to provide a more dynamic and adaptable tool for decision-making. The Alaknanda basin, characterized by its marked fluvial sedimentation, steep slopes, and intense rainfall conditions, serves as an ideal case study for this approach. The methodology adopted here bridges the existing research gaps by seamlessly integrating landslide hazard maps, which depict the likelihood of landslides, with comprehensive soil erosion maps, highlighting the consequences. This integrative approach not only enriches the understanding of the region's vulnerabilities but also offers a practical tool for planners and decision-makers. In essence, this study aims to contribute to the body of knowledge on landslide risk management in the IHR, providing a scientifically rigorous and practically applicable framework that can enhance the resilience of this ecologically and culturally significant region.

24.2 Landslide Likelihood and Consequences

The likelihood of landslides and their consequential effects in the IHR are an outcome of a complex interplay of geological, meteorological, and anthropogenic factors. An understanding of these intricate processes is paramount for developing efficient Landslide Risk Management (LRM) strategies.

24.2.1 Geological Factors

The geology of the IHR is primarily characterized by young, tectonically active, and complex geological formations. The region's diverse lithology, variable structural disposition, and weak geomechanical characteristics make it extremely susceptible to landslides (Kansal and Singh 2022). The geological properties such as slope gradient,

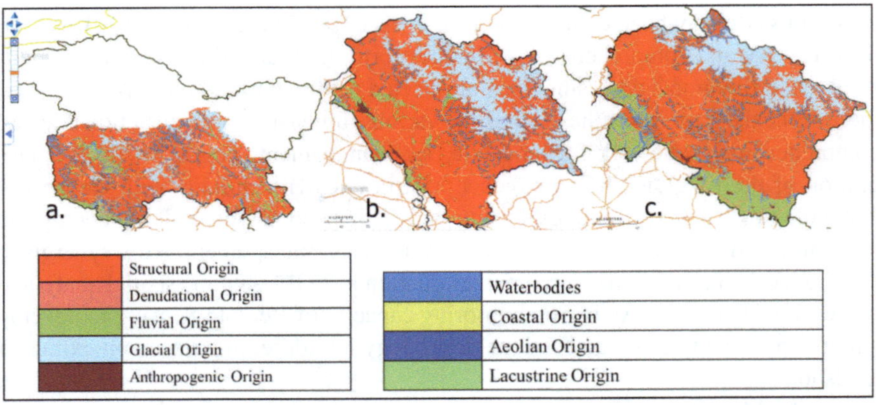

Fig. 24.1 The geomorphology of the important states in the Indian Himalayan Region **a** Jammu and Kashmir, **b** Himachal Pradesh and **c** Uttarakhand (*Source* https://bhuvan-app1.nrsc.gov.in/thematic/thematic/index.php)

slope aspect, lithology, and structure significantly affect the region's susceptibility to landslides (Sajwan and Sushil 2016). Frequent seismic activities due to the tectonic instability of the region further exacerbate the propensity of landslides. Earthquakes often act as a trigger for landslides, particularly in areas with a pre-existing slope instability (Sarkar et al. 2006). The geology of the IHR is a multifaceted tapestry woven from diverse origins and influences. Its structural origin lies in its tectonically active nature, shaping mountains and valleys, while denudational processes, such as weathering and erosion, further mould the landscape. Fluvial forces carve intricate patterns, forming valleys and floodplains, intertwined with the cold touch of glaciers that shape valleys and create moraines (Nardi and Rinaldi 2010). Anthropogenic influences, through activities like deforestation and construction, also imprint upon the landscape, altering its natural state. Waterbodies, whether forged by tectonic actions or glacial movements, sprinkle the region, contributing to its geomorphic diversity. Lacustrine origins emphasize the role of lakes in depositing sediments and reshaping surrounding areas. Embedded within these geomorphic alignments are the geological factors, with the IHR's young, dynamic formations, seismic activities, and varied slopes and lithologies intertwining to create a landscape both majestic and susceptible to challenges such as landslides (Arabameri et al. 2020; Chevuturi et al. 2015; Ocakoglu et al. 2002). The geomorphological map of important states in the IHR is shown in Fig. 24.1.

24.2.2 Meteorological Factors

Meteorological factors play a crucial role in triggering landslides in the IHR, with the landscape profoundly affected by the weather and climatic conditions. One of the

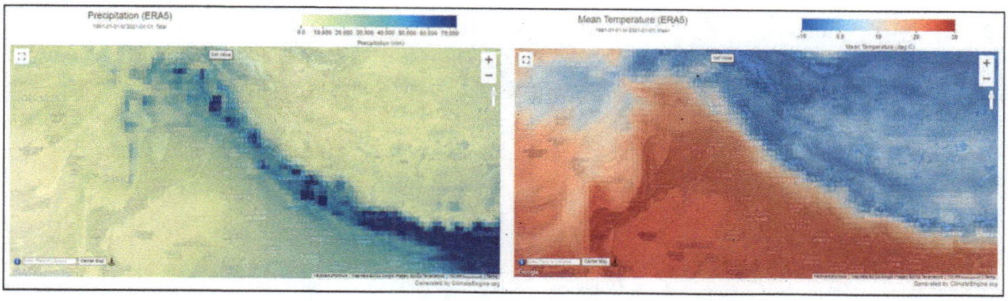

Fig. 24.2 The maps of total precipitation and mean temperature over Indian Himalayan Region in last 20 years (*Source* https://app.climateengine.org/climateEngine)

predominant meteorological contributors to landslides in the IHR is heavy precipitation, which encompasses both intense rainfall and snowfall. When the region experiences heavy rainfall, it leads to a rapid influx of water into the soil and rock layers on the slopes of the mountains. This sudden addition of water increases the pore water pressure within the geological materials, reducing their shear strength and making the slopes more susceptible to failure, leading to landslides (Singh et al. 2021). In addition to rainfall, the IHR also grapples with the consequences of snowmelt. As temperatures rise, snow and glaciers on the Himalayan peaks begin to melt, releasing vast amounts of water that flow down the slopes. This process, combined with rainwater, further exacerbates the saturation of soils, contributing to the instability of the slopes. Another meteorological aspect unique to this region is the occurrence of Glacial Lake Outburst Floods (GLOFs). GLOFs refer to the sudden release of a significant amount of water retained in a glacial lake. This could be due to factors like increasing temperatures, ice and rock avalanches into the lake, or seismic activities.

When such outbursts occur, they unleash massive volumes of water swiftly down the slopes, carrying with them debris and rocks, adding a tremendous force that can instigate landslides (Mir et al. 2018). The total rainfall and mean temperature over the span of 20 years is shown in Fig. 24.2.

24.2.3 Anthropogenic Factors

Anthropogenic activities in the IHR, including road and dam construction, rock quarrying, mining, riverbed encroachment, and uncontrolled tourism, profoundly exacerbate the vulnerability of the region to landslides (Fig. 24.3). Road construction and rock quarrying often involve extensive alteration and destabilization of slopes, disturbing the landscape's natural equilibrium. Dam construction modifies water loadings and flow patterns, further stressing the geological structures. Encroachments into riverbeds disrupt natural watercourses and exacerbate erosion, while unregulated tourism imposes additional pressures on the fragile ecosystems. Together, these human-induced activities significantly enhance landslide susceptibilities in the IHR, underscoring the urgent need for sustainable practices, meticulous planning,

Fig. 24.3 The compilation of anthropogenic activities in the IHR that lead to the landslides

and stringent regulation to safeguard the region's geomorphological integrity (Kansal and Singh 2021; Mishra 2015; Singh and Kansal 2022).

24.2.4 Consequences of Landslides

Landslides in the Indian Himalayan Region (IHR) unveil a cascade of multifaceted consequences that reverberate through the physical, economic, and societal dimensions of the region. Landslides manifest immediate and palpable physical impacts, prime among which is the tragic loss of human lives. The destructive prowess of landslides also extends to the built environment, unleashing damages upon properties, roads, bridges, and a myriad of infrastructural elements. Furthermore, inadequate soil and water conservation practices, coupled with significant anthropogenic activities, contribute to slope instability, thereby escalating landslide incidences (Fig. 24.4). Moreover, such demolitions compromise the structural integrity and functionality of critical infrastructures, thereby hampering mobility, accessibility, and the overall utility of the built environment (Abedini and Tulabi 2018; Martha et al. 2010). Economically, landslides impose a burden characterized by direct financial losses emanating from damages to properties and infrastructures. However, the economic repercussions extend beyond immediate losses, infiltrating the long-term economic vitality of the region. Landslides can disrupt economic activities, obstruct transport networks, and impede access to markets and resources, all of which can curtail economic productivity and propel communities into sustained financial hardships. On the societal front, landslides wield consequences that transcend material losses. Populations may be uprooted and displaced, ensuing in a labyrinth of challenges related to shelter, livelihoods, and basic services.

Fig. 24.4 The image showing recent landslides occurrences **a** Pithoragarh, Uttarakhand on Oct. 8, 2023, **b** Solan, Himachal Pradesh on Aug, 14, 2023, **c** Uttarkashi, Uttarakhand on Aug 15, 2023, **d** Shimla, Himachal Pradesh on Aug. 17, 2023

The abrupt disruption of essential services, such as water, electricity, and healthcare, further exacerbates the vulnerabilities of affected communities. Moreover, the tumultuous experiences associated with landslides can imbue communities with psychological stresses and traumas. Ecologically, landslides reshape landscapes and perturb ecosystems. They can induce the loss of forest cover and biodiversity, altering ecological compositions, and functionalities. Such ecological alterations can disrupt wildlife habitats, affect water resources, and contribute to broader environmental degradations (Rawat et al. 2012).

24.2.5 Relationship Between Likelihood and Consequences

The relationship between the likelihood of a landslide and its potential consequences is critical to understanding and managing landslide risk. High likelihood areas with significant potential consequences require urgent attention and robust mitigation strategies. Conversely, areas with low likelihood but high potential consequences cannot be ignored as they present a latent threat. The understanding of this relationship lays the foundation for the proposed rule-based approach for LRM. By incorporating both likelihood and consequence into a matrix, the model enables an objective, systematic, and structured assessment of landslide risk. This facilitates more informed and effective decision-making in terms of disaster response, mitigation strategies, and resource allocation. Understanding the likelihood and consequences of landslides is not just a scientific endeavor but also a crucial aspect of efficient disaster risk management. It ensures that our strategies and actions are informed, targeted, and effective, thereby reducing the potential impacts of landslides on life, property, and the environment in the Indian Himalayan Region.

24.3 Rule-Based Approach for LRM

Given the complex interplay of various factors contributing to the likelihood of landslides and their consequences in the Indian Himalayan Region (IHR), the establishment of a robust Landslide Risk Management (LRM) strategy is of utmost importance. This chapter outlines an innovative rule-based approach, centered around the Likelihood-Consequence Matrix (LCM), that facilitates systematic assessment and management of landslide risks.

24.3.1 Development of the Rule-Based Model for LRM Using the Likelihood-Consequence Matrix (LCM)

The development of the rule-based approach for LRM begins with the identification and quantification of landslide triggering factors. These factors include geological, meteorological, and anthropogenic factors that significantly influence the likelihood of landslides. The identification of these factors is accomplished by incorporating field observations, remote sensing data, and historical landslide records. Simultaneously, the potential consequences of landslides are evaluated. This is achieved by considering factors such as population density, infrastructure, economic activities, and other elements at risk in the event of a landslide.

The Likelihood-Consequence Matrix (LCM) (Table 24.1) is a crucial element of the rule-based model. This matrix formulates 25 unique combinations, each representing a specific landslide likelihood and corresponding potential consequence. The LCM essentially provides a platform to integrate the likelihood and consequence data, thereby enabling a holistic understanding of landslide risk. Each cell in the LCM is characterized by a unique combination of likelihood and consequence, which is then categorized into one of the four risk zones: low, moderate, high, and extreme. This categorization forms the basis for the segregation of the basin into different risk zones.

Based on the LCM following Rule-based approach are made:

Rule 1: Risk is "Low" when:
Likelihood is 'Rare', 'Unlikely', or 'Possible' and Consequences are 'Negligible'.
Likelihood is 'Rare' and Consequences are 'Minor' or 'Moderate'.

Rule 2: Risk is "Moderate" when:
Likelihood is 'Likely' or 'Certain' and Consequences are 'Negligible'.

Table 24.1 The Likelihood-Consequence Matrix

Risk			Likelihood				
Low, Moderate, High, Extreme			1	2	3	4	5
			Rare	Unlikely	Possible	Likely	Certain
Consequences	5	Catastrophic	5	10	15	20	25
	4	Major	4	8	12	16	20
	3	Moderate	3	6	9	12	15
	2	Minor	2	4	6	8	10
	1	Negligible	1	2	3	4	5

Likelihood is 'Unlikely' or 'Possible' and Consequences are 'Minor'.
Likelihood is 'Unlikely' and Consequences are 'Moderate'.
Likelihood is 'Rare' and Consequences are 'Major' or 'Catastrophic'.

Rule 3: Risk is "High" when:
Likelihood is 'Likely' or 'Certain' and Consequences are 'Minor'.
Likelihood is 'Possible' or "Likely" and Consequences are 'Moderate'.
Likelihood is 'Unlikely' or 'Possible' and Consequences are 'Major'.
Likelihood is 'Unlikely' and Consequences are 'Catastrophic'.

Rule 4: Risk is "Extreme" when:
Likelihood is 'Certain' and Consequences are 'Moderate', 'Major', or 'Catastrophic'.
Likelihood is 'Likely' and Consequences are 'Major', or 'Catastrophic'.
Likelihood is 'Possible' and Consequences are 'Catastrophic'.

24.3.2 Features and Benefits of the Rule-Based Approach

The rule-based approach for LRM offers numerous advantages.

Scalability: The rule-based model is designed to cater to different scales, ranging from local to regional, ensuring its relevance across varying geographical scales.

Adaptability: The model can be tailored to different contexts and adjusted as new data become available, making it dynamic and adaptive.

User-friendliness: The LCM provides a clear, easy-to-understand visual representation of landslide risk, making it a user-friendly tool for both technical experts and non-specialist stakeholders.

The proposed rule-based approach for LRM, through its incorporation of the LCM, brings together complex data on landslide likelihood and potential consequences into an accessible, adaptable, and scalable model. It promises a more systematic, structured, and comprehensive risk assessment process, ultimately leading to more informed and effective landslide risk management strategies in the Indian Himalayan Region. This model's innovative approach offers a significant step forward in managing and mitigating the profound impact of landslides in the region.

24.4 Segregation of Risk Zones

The segregation of risk zones, an integral part of the rule-based approach to Landslide Risk Management (LRM), provides a visual representation of the landslide susceptibility within the Indian Himalayan Region (IHR). The zoning process helps in quantifying the level of risk across the region, thereby informing effective risk mitigation and management strategies.

24.4.1 Criteria for Segregation

In this study, the regions are segregated into the four risk zones—low, moderate, high, and extreme—based on the combinations of landslide likelihood and potential consequences in terms of erosion. Each zone has specific criteria:

Low Risk Zone (Rule 1): These areas are characterized by a low likelihood of landslide occurrence and low potential consequences. They typically comprise stable geological formations and lack significant human or infrastructural elements at risk.

Moderate Risk Zone (Rule 2): These regions exhibit a moderate likelihood of landslides and/or moderate potential consequences. They could represent areas with less stable geology or those with some human or infrastructural elements at risk.

High Risk Zone (Rule 3): Areas falling within this zone show a high likelihood of landslide occurrence or high potential consequences. These are usually regions with unstable geological features or with significant human populations and infrastructure at risk.

Extreme Risk Zone (Rule 4): These areas are defined by a very high likelihood of landslide occurrence and severe potential consequences. These zones are often characterized by very unstable geology and a high density of vulnerable human and infrastructural elements.

24.4.2 Implications for Disaster Management

The segregation of risk zones has profound implications for disaster management. The zoning process not only identifies areas most susceptible to landslides, but it also helps prioritize areas for interventions based on their respective risk levels.

For Low and Moderate Risk Zones, regular monitoring of the region's stability should be performed. Here, the focus should be on maintaining the area's stability through appropriate land-use planning and promoting practices such as afforestation and proper drainage systems. High Risk Zone areas necessitate stringent monitoring and preemptive landslide mitigation measures. This may include slope stabilization, creating buffer zones, and restricting heavy construction. In addition, these regions require efficient early warning systems and evacuation plans. Very Risk Zones need urgent attention. Apart from stringent monitoring and mitigation measures, there may be instances where relocation of vulnerable communities is considered, given the extreme level of risk. These areas need robust disaster response strategies, including well-planned evacuation routes and emergency services. By segregating the region into distinct risk zones, the rule-based model for LRM provides a spatially explicit, organized, and systematic approach to manage landslide risk. Each zone has its unique requirements for risk management, making this approach a versatile tool for comprehensive disaster management planning in the IHR.

24.5 Integration of Geographic Information System (GIS)

Geographic Information Systems (GIS) provide an essential toolset for modern landslide risk management, allowing for the spatial analysis and visualization of complex environmental datasets. Integrating GIS into the rule-based approach amplifies its effectiveness, transforming raw data into meaningful, actionable insights. In the rule-based approach to Landslide Risk Management (LRM), Landslide Hazard Zonation is meticulously carried out using the Analytical Hierarchy Process (AHP) method, which involves structuring multiple attributes into a hierarchy, assessing them systematically to calculate and prioritize the likelihood of landslide occurrences (Sethia et al. 2018; Singh et al. 2023). This sophisticated method leverages multifaceted datasets, including geological, topographical, meteorological, and socio-economic factors, ensuring a comprehensive evaluation of landslide susceptibility within the area (Mishra 2015; Pareta et al. 2012). On the other hand, the consequential aspect of landslides, particularly soil erosion, is precisely assessed utilizing the Revised Universal Soil Loss Equation (RUSLE). This scientific methodology allows for the quantitative evaluation and mapping of erosion risks based on various critical factors such as rainfall erosivity, soil erodibility, slope length and steepness, cover management, and conservation practices (George et al. 2021; Gupta and Kumar 2017; Singh and Kansal 2021, 2023a, b). By employing RUSLE within the GIS framework, a detailed erosion map is generated, delineating the potential erosion consequences as a result of landslides. Subsequently, the GIS-based integration of these two distinct methodologies—AHP for Landslide Hazard Zonation and RUSLE for erosion assessment—culminates in a robust Likelihood-Consequence Matrix (LCM). This matrix, formed through a systematic combination of landslide likelihood and consequential erosion, facilitates the generation of a detailed, spatially explicit landslide risk map, enhancing the efficacy of Landslide Risk Management (LRM) strategies and actions within the study area.

24.6 Study Area: Alaknanda River Basin

The Alaknanda River Basin in the IHR (Fig. 24.5) spread across approximately 10,272 km^2, the basin nestles within varied elevations ranging from 600 m to 7800 m, and is latitudinally and longitudinally positioned between 30° 10′N to 31° 5′N and 78° 45′E to 80° 15′E, respectively (Panwar et al. 2017a, b; Singh and Kansal 2023a, b; Singh et al. 2022). The basin is marinated in a tropical monsoon climate, receiving an average annual rainfall of about 1600 mm, with a staggering 75% of this precipitation unfurling between June and September. This seasonal deluge, paired with the basin's dramatic topography characterized by sharp elevations and steep slopes, crafts a terrain that is incredibly susceptible to landslides and other natural calamities such as flash floods and cloudbursts.

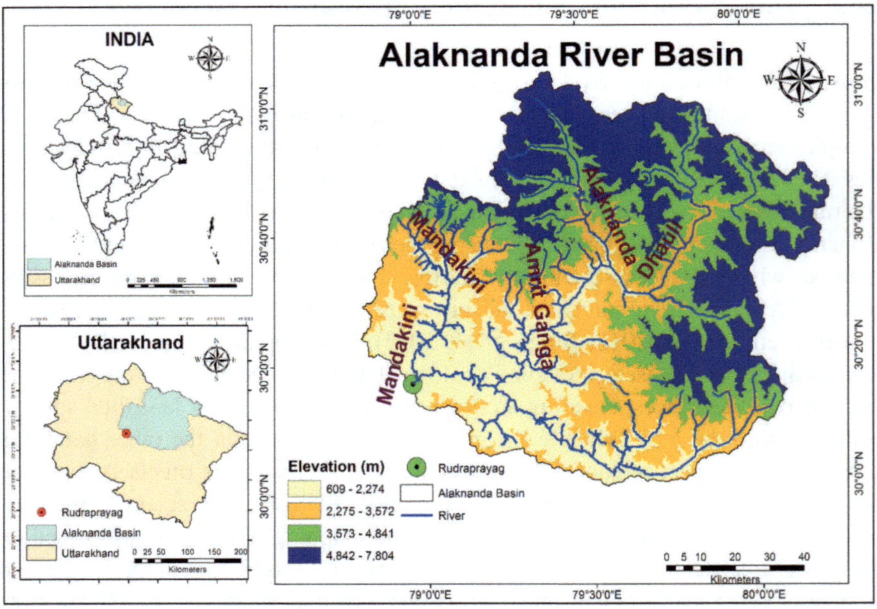

Fig. 24.5 The study area map of Alaknanda river basin

The escalating frequency of such natural disasters, augmented by anthropogenic pressures like hydropower projects and tourism, accentuates the basin's predisposition to significant fluvial sedimentation and landslides (Singh and Kansal 2023a, b). Such challenges necessitate the advent of rule-based approach for LRM effective landslide risk management.

24.7 Results and Discussion

24.7.1 Landslide Risk Map Based on the Rule-Based Model

In the Alaknanda Basin, a rigorous methodology is employed to assess and analyse landslide risks through the amalgamation of multiple datasets, resulting in the creation of a Likelihood-Consequence Matrix (LCM). A broad spectrum of parameters, such as elevation, slope, curvature, proximity to rivers, lineaments, geomorphology, landslide inventories, land use, soil characteristics, and rainfall, are meticulously integrated using AHP to evaluate the likelihood of landslides (Fig. 24.6) from landslide hazard zonation map (Singh et al. 2023). These variables, each contributing uniquely, encompass both intrinsic geological attributes and external influences, including climatic and anthropogenic factors.

For estimating the potential Consequences (Fig. 24.7) in terms of erosion due to landslides, the Revised Universal Soil Loss Equation (RUSLE) are employed. Factors constituting this include rainfall erosivity (R factor), soil erodibility (K factor), slope length and steepness (LS factor), cover-management (C factor), and support practices (P factor) (Singh and Kansal 2023a, b; Singh and Kansal 2021). These elements collectively provide a comprehensive perspective on the susceptibility of the region to erosion consequential to landslide occurrences. The synergy of these two extensive analyses within a GIS platform facilitates the formation of the LCM. Both the landslide likelihood and erosion consequence maps are refined and reclassified into five categorical risk levels, and subsequently interlinked through a matrix multiplication process. The final LCM Risk Map thus produced, embodies a harmonized representation of the region's vulnerability to landslides and the subsequent potential erosion impacts. This sophisticated approach provides a multifaceted, nuanced perspective on landslide risk in the Alaknanda Basin, acting as a pivotal tool for informed decision-making and the formulation of effective, strategic landslide risk management, and mitigation approaches.

Upon completion of the zoning process, a thorough analysis and interpretation of the results are undertaken. The outcomes show the spatial distribution of landslide risk across the Alaknanda Basin (Fig. 24.8), providing valuable insights into the area's most at risk. Notably, these areas often align with zones characterized by steep slopes, high levels of precipitation, and significant human activity.

The likelihood of landslides occurring in the area is categorized into five qualitative scales, as shown in Table 24.2, ranging from Rare to Certain. Most areas

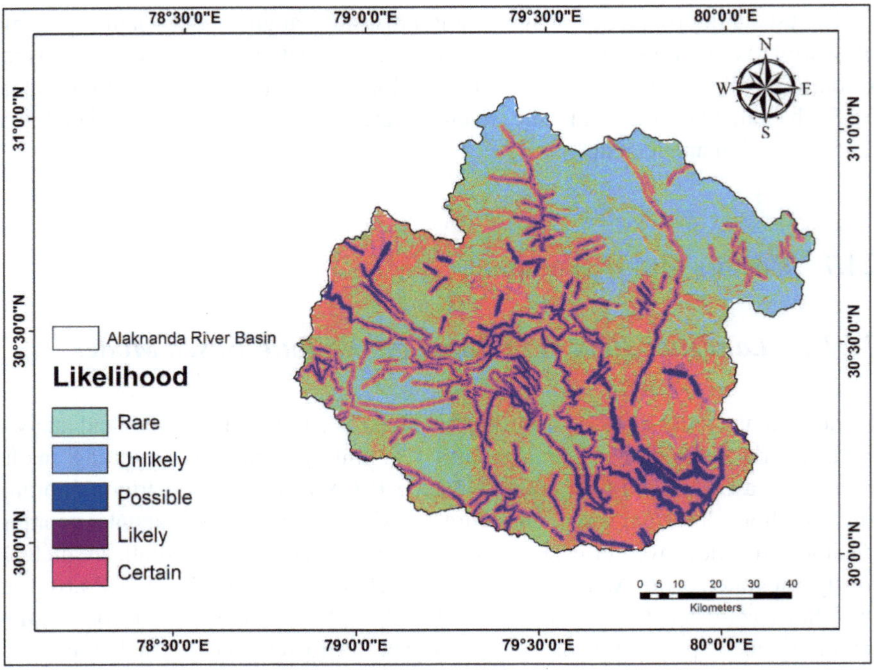

Fig. 24.6 The landslide likelihood map modified from Landslide hazard zonation map by Singh et al. (2023)

(32.0%) fall under the "Unlikely" category, followed by "Possible" at 28.0%. The consequences, or impact of the landslides, are also divided into five categories, from Minor to Catastrophic. A majority of the area (57.3%) is expected to experience "Minor" consequences in the event of a landslide. The risk is evaluated and categorized from Low to Extreme. Nearly half of the area (49.7%) is associated with "Low" risk levels, meaning that either the likelihood or the consequences (or both) are low in these areas. "Moderate" risk areas constitute 25.2%, while "High" and "Extreme" risk zones are less prevalent at 17.6% and 7.5% respectively. This classification enables decision-makers and planners to prioritize areas for risk mitigation strategies, focusing more on areas with higher combined risk levels ("High" and "Extreme") for effective resource allocation and planning. It provides a structured approach to visualize and understand the spatial distribution of landslide risks, supporting informed decision-making for disaster risk reduction. The ability of the model to assess landslide risks, classify the landscape into distinct risk zones, and present the information in a user-friendly format, underlines its potential for widespread application in the Indian Himalayan Region and other landslide-prone regions globally. The insights gained from this study hold significant implications for disaster management, providing a robust scientific basis for informed decision-making and strategic planning.

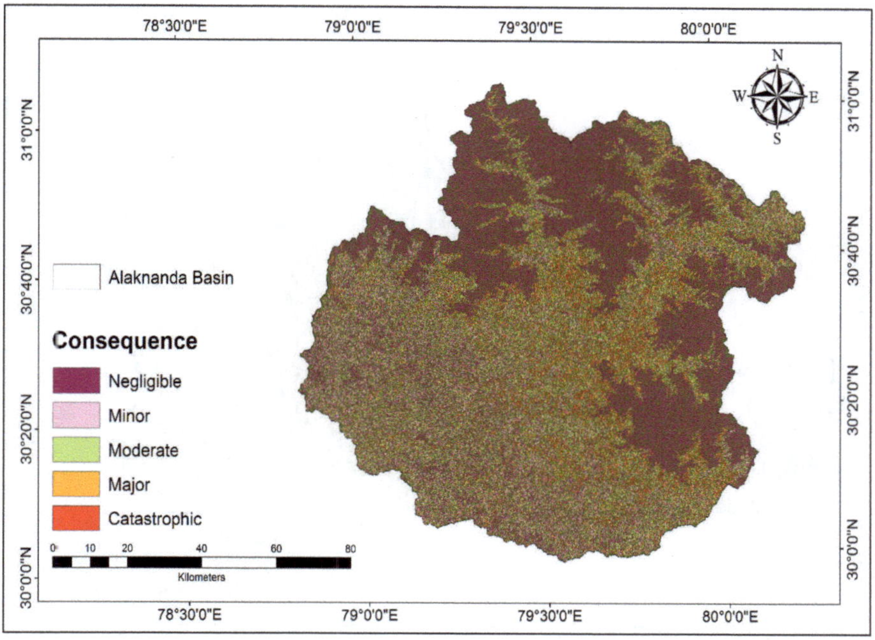

Fig. 24.7 The consequence map derived from soil erosion rate map modified from Singh and Kansal (2023a, b)

24.7.2 Discussion of Results

The findings of this study offer an updated perspective on landslide risk management, particularly within the Indian Himalayan Region. Recent advancements in the field increasingly emphasize integrating advanced geospatial technologies and data-driven models to enhance landslide prediction and risk management. For instance, research by Ma et al. (2022), Singh et al. (2023) underscores the importance of remote sensing data in identifying potential landslide areas, aligning with the application of GIS technology in the Likelihood-Consequence Matrix (LCM) framework of this study. Additionally, the work of Sun et al. (2023) on applying machine learning algorithms for landslide susceptibility mapping resonates with the rule-based model approach, highlighting a trend towards more sophisticated, data-centric methods. This study contributes to this evolving landscape by utilizing advanced technologies within a structured and practical Landslide Risk Management (LRM) framework. The LCM's ability to categorize risk into distinct zones based on quantifiable metrics addresses the need for more nuanced and actionable risk categorization methods in landslide-prone regions, a gap identified in earlier research like Aleotti and Chowdhury (1999), Guzzetti et al. (1995), Lee and Dan (2005), Ocakoglu et al. (2002).

Moreover, the integration of landslide likelihood with potential consequences into a cohesive model provides a comprehensive risk assessment tool, an aspect

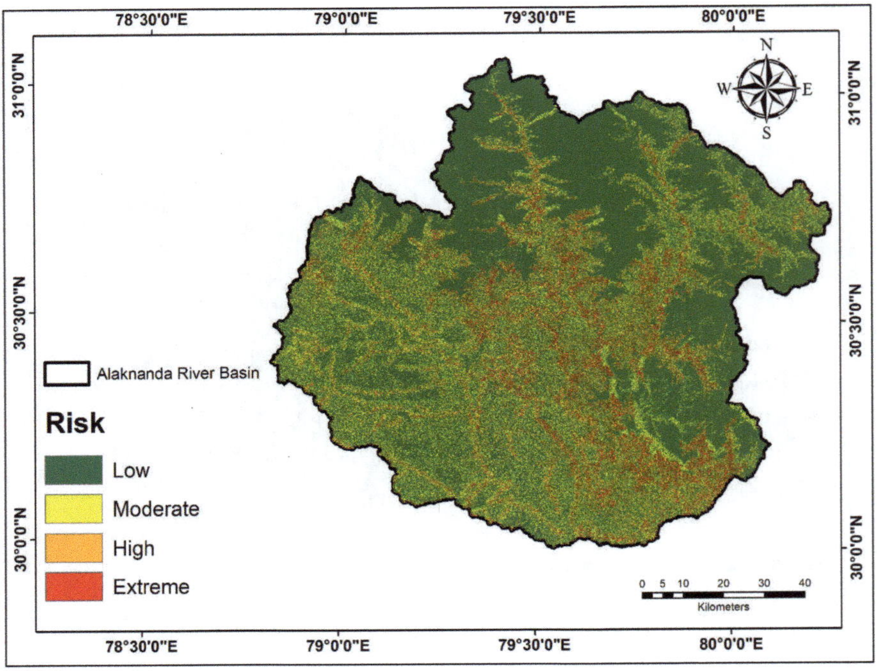

Fig. 24.8 The landslide risk map derived from the likelihood-consequences matrix

Table 24.2 The classification of Likelihood, Consequences and Risk

S. No.	Category	Qualitative scale	Value	Area (%)
1	Likelihood	Rare	1	16.7
2		Unlikely	2	32.0
3		Possible	3	28.0
4		Likely	4	15.1
5		Certain	5	8.3
1	Consequences	Minor	1	57.3
2		Negligible	2	17.1
3		Moderate	3	12.1
4		Major	4	8.3
5		Catastrophic	5	5.2
1	Risk	Low	1, 2 and 3	49.7
2		Moderate	4, 5 and 6	25.2
3		High	8, 9, 10 and 12	17.6
4		Extreme	15, 16, 20 and 25	7.5

previously underexplored. This integration offers a holistic understanding of the risk landscape, echoing recent studies, such as Pandey et al. (2022), Singh et al. (2023), which highlighted the necessity for comprehensive risk assessment models in disaster management. Yet, the limitations identified, such as the dependency on data quality and the need for a broader range of parameters, indicate directions for future research. These limitations highlight the importance of ongoing data collection and standardization and exploring methodologies that include socio-economic factors and climate change projections, as suggested by Kansal and Singh (2022), Kumar et al. (2018), Wei et al. (2018).

24.8 Further Recommendations for Landslide Management

24.8.1 Early Warning Systems

The development and implementation of advanced early warning systems is a crucial factor in effective landslide risk management. These systems can monitor environmental conditions and detect the precursors to landslides, such as increased ground movement or changes in hydrological conditions. They utilize a combination of remote sensing technologies, ground-based sensors, and meteorological data to provide real-time monitoring and alert notifications. This allows for timely evacuations and the implementation of other disaster response measures, significantly reducing the potential for loss of life and property. The success of these systems depends on the accuracy of the technology, the effectiveness of the communication channels used to disseminate warnings, and the preparedness and responsiveness of communities and relevant authorities. As such, regular maintenance, testing, and updates of the system and continued education and training for stakeholders are also essential (Kansal and Singh 2022; Lorenzo-Lacruz et al. 2019; Tullos et al. 2016).

24.8.2 Local Knowledge and Practices

Local knowledge is a reservoir of historical and experiential information that communities have cultivated over generations. This wealth of knowledge encompasses nuanced understandings of local terrains, weather patterns, and subtle indicators that precede landslides. Leveraging such knowledge allows for a more localized and context-specific approach to managing and mitigating landslide risks. When integrated with scientific methodologies and technological tools, local knowledge can enrich predictive models, hazard mapping, and early warning systems. Involving local communities in decision-making processes also fosters ownership and ensures that mitigation strategies are culturally sensitive and practically feasible (Arlikatti et al. 2018).

24.8.3 Improved Land Use Planning

Effective land-use planning is instrumental in mitigating the risks associated with landslides. It involves the strategic allocation and regulation of land uses, particularly in areas identified as susceptible to landslides. By imposing restrictions on the construction of buildings, roads, and other infrastructures in high-risk areas, the vulnerability of these structures to landslide hazards is reduced. Zoning laws and building codes, if meticulously crafted and enforced, can prevent the inception of potential risk factors, guiding development away from hazardous zones and ensuring that any development within vulnerable areas is resilient to landslides (Mehryar and Surminski 2021).

24.8.4 Soil and Water Conservation

Soil and water conservation techniques play a pivotal role in landslide management by enhancing the resilience of slopes and reducing the propensity for landslides. Strategies such as contour bunding and terracing modify the landscape to reduce the speed of water, encouraging infiltration and reducing surface runoff. Afforestation acts as a bio-engineering solution, where vegetation roots enhance soil cohesion, providing natural anchoring and reducing the risk of slope failure. Implementing proper drainage systems is also crucial in managing water flow, preventing the saturation and weakening of slopes. Collectively, these measures help in maintaining the structural integrity of the slopes, mitigating the adverse impacts of water, and promoting landscape stability (Chandrashekar et al. 2015; Tiwari and Kushwaha 2021).

24.8.5 Infrastructure Design

For effective landslide management, the design of infrastructure such as buildings and roads in landslide-prone areas should be meticulously curated, considering the intrinsic geological and hydrological attributes of the locale. The construction should be robust, envisaging the potential forces and movements instigated by landslides. Utilizing flexible design elements that can accommodate ground movements without succumbing to significant damages is imperative. This approach would mean employing engineering techniques and materials that augment the resilience of structures, ensuring they withstand the perturbations induced by landslides without catastrophic failures (Afriyanie et al. 2020).

24.8.6 Climate Change Adaptation

Recognizing the amplification of landslide risks due to climate change is essential. With climate change poised to intensify the frequency and magnitude of extreme weather events such as heavy rainfall, the strategies employed for landslide management must be adaptive and resilient to evolving climate scenarios. Enhancing the robustness of early warning systems to accurately predict and promptly relay information on imminent landslides under fluctuating climate patterns is crucial. Similarly, adaptive land-use planning that anticipates and integrates future climatic variabilities ensures that the landscape and infrastructures are predisposed to withstand exacerbated landslide occurrences (Kansal and Singh 2021).

24.8.7 Education and Awareness

Education and awareness form the bedrock of community resilience against landslides. Comprehensive campaigns aimed at elevating the community's understanding of landslide risks, precautionary measures, and appropriate responsive actions are indispensable. Through education, communities are empowered with the knowledge to identify potential landslide signs, adopt proactive risk mitigation measures, and execute effective response strategies during landslide events. These campaigns foster a culture of preparedness and vigilance, enabling communities to navigate the complexities of landslide risks with informed confidence and prudence (Wang et al. 2019).

24.8.8 Research and Development

To foster advancements in landslide management, continuous research and development (R&D) efforts are essential. Such initiatives could elucidate the complex mechanics of landslides, enhance predictive accuracies, and unveil innovative management strategies. The realm of R&D could embrace a spectrum of pursuits, ranging from the exploration of cutting-edge remote sensing technologies for precise landslide monitoring to the evolution of geotechnical investigation methodologies that unveil intricate subsurface realities. Furthermore, the development of sophisticated predictive models that amalgamate various parameters to forecast landslides with enhanced precision and reliability could be a pivotal R&D contribution (Ioris 2015).

24.8.9 Policy and Legislation

Robust policies and legislations serve as pillars in the architecture of landslide management. A tapestry of well-orchestrated laws and regulations, ranging from stringent building codes and regulations, meticulous land-use policies, to protective environmental regulations, can orchestrate a symphony of safety against the discord of landslide risks. Enforcing these legal frameworks with vigilance and integrity is imperative to ensure that activities in landslide-prone areas are conducted within the precincts of safety and sustainability (Krieger 2013; Wasson 2018).

24.8.10 Community-Based Disaster Management

Encouraging and empowering communities to be active participants in disaster management processes is vital. Community-based approaches could encompass the formulation and implementation of localized disaster management plans that resonate with the unique vulnerabilities and capacities of each community. Facilitating training programs for local emergency responders enhances the community's immediate response capabilities. Additionally, the instigation of community-based early warning systems can bolster real-time risk communication, enabling communities to undertake timely protective actions. Such community-centric strategies foster a sense of ownership, resilience, and collective efficacy in navigating the challenges posed by landslides (Otomofa et al. 2015).

24.9 Conclusions and Limitations

This study presents a detailed examination of the complex landslide challenges in the Indian Himalayan Region (IHR), a region highly prone to landslides due to a combination of extreme precipitation events, unique topography, and inadequate soil and water conservation practices. The physical, economic, and societal impacts of these landslides underscore the critical need for efficient disaster management strategies. Central to this study is the development and application of an innovative rule-based Landslide Risk Management (LRM) approach, anchored by a Likelihood-Consequence Matrix (LCM). This matrix effectively stratifies the IHR into four distinct risk zones—low, moderate, high, and extreme, correlating with varying levels of landslide risk and potential damage. The rule-based model, notable for its scalability, adaptability, and user-friendliness, is further enhanced by its integration with Geographic Information System (GIS) technology, facilitating efficient spatial data management and visualization. Applied to the Alaknanda Basin case study, the model demonstrates its utility in evaluating landslide risks and informing disaster response and mitigation strategies. Its potential for wider application across the IHR and other landslide-prone regions is significant, particularly as the model's effectiveness grows with the accumulation and integration of more data.

Nevertheless, it's important to acknowledge certain limitations inherent in this study. First, the reliance on available data means that the model's accuracy is contingent on the quality and comprehensiveness of the data inputs. In areas where data collection is sparse or non-standardized, the model's predictive power may be compromised. Additionally, while the LCM provides a robust framework for risk assessment, it may not fully capture the complex interplay of all geophysical and socio-economic factors that influence landslide risks. Future research could focus on expanding the model's parameters to include a broader range of factors in both likelihood and consequence assessments, thereby enhancing its precision and applicability. Another potential area for improvement is the refinement of GIS integration techniques, aiming for more nuanced spatial analyses and improved user interface design to facilitate easier interpretation and decision-making by stakeholders. Recognizing these limitations is crucial for guiding future research efforts and ensuring the continuous evolution and refinement of landslide risk management strategies.

References

Abedini M, Tulabi S (2018) Assessing LNRF, FR, and AHP models in landslide susceptibility mapping index: a comparative study of Nojian watershed in Lorestan province, Iran. Environ Earth Sci 77(11). https://doi.org/10.1007/s12665-018-7524-1

Afriyanie D, Julian MM, Riqqi A, Akbar R, Suroso DSA, Kustiwan I (2020) Re-framing urban green spaces planning for flood protection through socio-ecological resilience in Bandung City, Indonesia. Cities 101(10):102710. https://doi.org/10.1016/j.cities.2020.102710

Aleotti P, Chowdhury R (1999) Landslide hazard assessment: summary review and new perspectives. Bull Eng Geol Environ 58(1):21–44. https://doi.org/10.1007/s100640050066

Arabameri A, Tiefenbacher JP, Blaschke T, Pradhan B, Bui DT (2020) Morphometric analysis for soil erosion susceptibility mapping using novel gis-based ensemble model. Remote Sens 12(5). https://pubs.geoscienceworld.org/gsa/gsabulletin/article/63/9/923/4513/DYNAMIC-BASIS-OF-GEOMORPHOLOGY

Arlikatti S, Maghelal P, Agnimitra N, Chatterjee V (2018) Should I stay or should I go? Mitigation strategies for flash flooding in India. Int J Disaster Risk Reduct 27(August 2017):48–56. https://doi.org/10.1016/j.ijdrr.2017.09.019

Chandrashekar H, Lokesh KV, Sameena M, Roopa J, Ranganna G (2015) GIS-based morphometric analysis of two reservoir catchments of Arkavati River, Ramanagaram District, Karnataka. Aquat Procedia 4(Icwrcoe):1345–1353. https://doi.org/10.1016/j.aqpro.2015.02.175

Chevuturi A, Dimri AP, Das S, Kumar A, Niyogi D (2015) Numerical simulation of an intense precipitation event over Rudraprayag in the central Himalayas during 13–14 September 2012. J Earth Syst Sci 124(7):1545–1561. https://doi.org/10.1007/s12040-015-0622-5

Galve JP, Cevasco A, Brandolini P, Soldati M (2015) Assessment of shallow landslide risk mitigation measures based on land use planning through probabilistic modelling. Landslides 12(1):101–114. https://doi.org/10.1007/s10346-014-0478-9

George J, Kumar S, Hole RM (2021) Geospatial modelling of soil erosion and risk assessment in Indian Himalayan region—a study of Uttarakhand state. Environ Adv 4(August 2020):100039. https://doi.org/10.1016/j.envadv.2021.100039

Gupta S, Kumar S (2017) Simulating climate change impact on soil erosion using RUSLE model—a case study in a watershed of mid-Himalayan landscape. J Earth Syst Sci 126(3). https://doi.org/10.1007/s12040-017-0823-1

Guzzetti F, Carrara A, Cardinali M, Reichenbach P (1995) Stern, 1991 and one of the NAVZ (Northern Austral volcanic zone) volcanic events 3010 yr. Geomorphology 13(6):1995

Ioris AAR (2016) Agriculture, environment and development: international perspectives on water, land and politics. In: Agriculture, environment and development: international perspectives on water, land and politics. https://doi.org/10.1007/978-3-319-32255-1

Kansal ML, Singh S (2021) Flood management issues in hilly regions of Uttarakhand in India under changing climatic condition. In: Lowering risk by increasing resilience, issue 8. ICFM 8, pp 106–107. https://www.icfm.world/icmf8-book.html

Kansal ML, Singh S (2022) Flood management issues in hilly regions of Uttarakhand (India) under changing climatic conditions. Water 14(12):1879. https://doi.org/10.3390/w14121879

Kinage C, Kalgutkar A, Parab A, Mandora S, Sahu S (2019) Performance evaluation of different machine learning based algorithms for flood prediction and model for real time flood prediction. In: Proceedings—2019 5th international conference on computing, communication control and automation, ICCUBEA 2019, pp 2–8. https://doi.org/10.1109/ICCUBEA47591.2019.9128379

Krieger K (2013) The limits and variety of risk-based governance: the case of flood management in Germany and England. Regul Gov 7(2):236–257. https://doi.org/10.1111/rego.12009

Kumar A, Gupta AK, Bhambri R, Verma A, Tiwari SK, Asthana AKL (2018) Assessment and review of hydrometeorological aspects for cloudburst and flash flood events in the third pole region (Indian Himalaya). In: Polar science, vol 18, pp 5–20. https://doi.org/10.1016/j.polar.2018.08.004

Lee S, Dan NT (2005) Probabilistic landslide susceptibility mapping in the Lai Chau province of Vietnam: focus on the relationship between tectonic fractures and landslides. Environ Geol 48(6):778–787. https://doi.org/10.1007/s00254-005-0019-x

Lorenzo-Lacruz J, Garcia C, Morán-Tejeda E, Amengual A, Homar V, Maimó-Far A, Hermoso A, Ramis C, Romero R (2019) Hydro-meteorological reconstruction and geomorphological impact assessment of the October, 2018 catastrophic flash flood at Sant Llorenç, Mallorca (Spain). Nat Hazards Earth Syst Sci Discuss 2015(July):1–30. https://doi.org/10.5194/nhess-2019-226

Ma S, Shao X, Xu C (2022) Characterizing the distribution pattern and a physically based susceptibility assessment of shallow landslides triggered by the 2019 heavy rainfall event in Longchuan County, Guangdong Province, China. Remote Sens 14(17). https://doi.org/10.3390/rs14174257

Martha TR, Kerle N, Jetten V, van Westen CJ, Kumar KV (2010) Characterising spectral, spatial and morphometric properties of landslides for semi-automatic detection using object-oriented methods. Geomorphology 116(1–2):24–36. https://doi.org/10.1016/j.geomorph.2009.10.004

Mehryar S, Surminski S (2021) National laws for enhancing flood resilience in the context of climate change: potential and shortcomings. Climate Policy 21(2):133–151. https://doi.org/10.1080/14693062.2020.1808439

Metternicht G, Hurni L, Gogu R (2005) Remote sensing of landslides: an analysis of the potential contribution to geo-spatial systems for hazard assessment in mountainous environments. Remote Sens Environ 98(2–3):284–303. https://doi.org/10.1016/j.rse.2005.08.004

Mir RA, Jain SK, Lohani AK, Saraf AK (2018) Glacier recession and glacial lake outburst flood studies in Zanskar basin, western Himalaya. J Hydrol 564(May):376–396. https://doi.org/10.1016/j.jhydrol.2018.05.031

Mishra A (2015) Cloudburst and landslides in Uttarakhand: a nature's fury? Mausam 66(1):139–144. https://doi.org/10.54302/mausam.v66i1.374

Nardi L, Rinaldi M (2010) Modelling riverbank retreat by combining reach-scale hydraulic models with bank-scale erosion and stability analyses. River Flow 2010:1286–1291. http://vzb.baw.de/e-medien/river-flow-2010/PDF/B5/B5_02.pdf

Ocakoglu F, Gokceoglu C, Ercanoglu M (2002) Dynamics of a complex mass movement triggered by heavy rainfall: a case study from NW Turkey. Geomorphology 42(3–4):329–341. https://doi.org/10.1016/S0169-555X(01)00094-0

Oh HJ, Lee S, Chotikasathien W, Kim CH, Kwon JH (2009) Predictive landslide susceptibility mapping using spatial information in the Pechabun area of Thailand. Environ Geol 57(3):641–651. https://doi.org/10.1007/s00254-008-1342-9

Otomofa JO, Okafor BN, Obienusi EA (2015) Evaluation of the impacts of flooding on socio-economic activities in Oleh, Isoko South Local Government Area, Delta State. J Environ Earth Sci 5(18):155–171. http://citeseerx.ist.psu.edu/viewdoc/download?doi=10.1.1.1007.5309&rep=rep1&type=pdf

Pandey VK, Kumar R, Singh R, Kumar R, Rai SC, Singh RP, Tripathi AK, Soni VK, Ali SN, Tamang D, Latief SU (2022) Catastrophic ice-debris flow in the Rishiganga River, Chamoli, Uttarakhand (India). Geomat Nat Hazards Risk 13(1):289–309. https://doi.org/10.1080/19475705.2021.2023651

Panwar S, Agarwal V, Chakrapani GJ (2017a) Morphometric and sediment source characterization of the Alaknanda river basin, headwaters of river Ganga. Nat Hazards 87(3):1649–1671. https://doi.org/10.1007/s11069-017-2838-y

Panwar S, Gaur D, Chakrapani GJ (2017b) Total organic carbon transport by the Alaknanda River, Garhwal Himalayas, India. Arabian J Geosci 10(9). https://doi.org/10.1007/s12517-017-3003-3

Pareta K, Kumar J, Pareta U (2012) Landslide hazard zonation using quantitative methods in GIS. Int J Geospatial Eng Technol 1(1):1–9. www.globalskope.com

Park SJ, Lee D (2021) Predicting susceptibility to landslides under climate change impacts in metropolitan areas of South Korea using machine learning. Geomat Nat Hazards Risk 12(1):2462–2476. https://doi.org/10.1080/19475705.2021.1963328

Rawat PK, Pant CC, Tiwari PC, Pant PD, Sharma AK (2012) Spatial variability assessment of river line floods and flash floods in Himalaya: a case study using GIS. Disaster Prev Manag Int J 21(2):135–159. https://doi.org/10.1108/09653561211219955

Regmi NR, Giardino JR, McDonald EV, Vitek JD (2014) A comparison of logistic regression-based models of susceptibility to landslides in western Colorado, USA. Landslides 11(2):247–262. https://doi.org/10.1007/s10346-012-0380-2

Reichenbach P, Rossi M, Malamud BD, Mihir M, Guzzetti F (2018) A review of statistically-based landslide susceptibility models. In: Earth-science reviews, vol 180. Elsevier B.V., pp 60–91. https://doi.org/10.1016/j.earscirev.2018.03.001

Sajwan K, Sushil K (2016) A geological appraisal of slope instability in Upper Alaknanda Valley, Uttarakhand Himalaya, India. J Geol Geophys 5(5). https://doi.org/10.4172/2381-8719.1000258

Sarkar S, Kanungo DP, Patra AK (2006) Landslides in the Alaknanda valley of Garhwal Himalaya, India. Q J Eng Geol Hydrogeol 39(1):79–82. https://doi.org/10.1144/1470-9236/05-020

Sethia KK, Pandey P, Chattoraj S, Manickam S, Champati ray PK (2018) Mapping, modeling and simulation of snow avalanche in Alaknanda Valley, central Himalaya: hazard assessment. In: International geoscience and remote sensing symposium (IGARSS), 2018-July, pp 5150–5153. https://doi.org/10.1109/IGARSS.2018.8518516

Shukla S, Jain SK, Kansal ML, Chandniha SK (2017) Assessment of sedimentation in Pong and Bhakra reservoirs in Himachal Pradesh, India, using geospatial technique. Remote Sens Appl: Soc Environ 8(August):148–156. https://doi.org/10.1016/j.rsase.2017.08.008

Sikka DR, Ray K, Chakravarthy K, Bhan SC, Tyagi A (2015) Heavy rainfall in the Kedarnath valley of Uttarakhand during the advancing monsoon phase in June 2013. Curr Sci 109(2):353–361. https://doi.org/10.18520/cs/v109/i2/353-361

Singh S, Kansal ML (2021) Flash-flood induced Erosion-Prone area prioritization and potential soil loss estimation using RUSLE for Alaknanda River Basin. In: Book of abstracts of the second international conference on natural hazards and risks in a changing world, 2nd edn, issue October, pp 68–69. https://www.uni-potsdam.de/en/natriskchange/activities/second-international-natriskchange-conference-2021

Singh S, Kansal ML (2022) Cloudburst—a major disaster in the Indian Himalayan States. In: Kolathayar S, Pal I, Chian SC, Mondal A (eds) Civil engineering for disaster risk reduction. Springer Singapore, pp 115–126. https://doi.org/10.1007/978-981-16-5312-4_9

Singh S, Kansal ML (2023) A comparative study of morphometric, hydrologic, and semi-empirical methods for the prioritization of sub-watersheds against flash flood-induced landslides in a part of the Indian Himalayan Region. Environ Sci Pollut Res 01234567. https://doi.org/10.1007/s11356-023-30613-6

Singh S, Kansal ML (2023) Sub-basin prioritisation using RUSLE in a mountainous river basin of Uttarakhand (India). Environm Dev Sustain 0123456789:27. https://doi.org/10.1007/s10668-023-02989-5

Singh S, Dhote PR, Thakur PK, Chouksey A, Aggarwal SP (2021) Identification of flash-floods-prone river reaches in Beas river basin using GIS-based multi-criteria technique: validation using field and satellite observations. Nat Hazards 105(3):2431–2453. https://doi.org/10.1007/s11069-020-04406-w

Singh S, Dhal L, Yadav A, Kansal ML (2022) Impact of land use land cover on land surface temperature over Alaknanda river basin using geospatial techniques. In: Jha CS, Pandey A, Chowdary VM, Singh V (eds) Geospatial technologies for resources planning and management. Springer International Publishing, pp 709–726. https://doi.org/10.1007/978-3-030-98981-1_30

Singh S, Lal Kansal M, Tyagi A (2023) Landslide hazard zonation in the Alaknanda River Basin using innovative techniques. World Environ Water Resources Cong 2023:32–44. https://doi.org/10.1061/9780784484852.004

Sun D, Chen D, Zhang J, Mi C, Gu Q, Wen H (2023) Landslide susceptibility mapping based on interpretable machine learning from the perspective of geomorphological differentiation. Land 12(5). https://doi.org/10.3390/land12051018

Tiwari RN, Kushwaha VK (2021) Watershed prioritization based on morphometric parameters and PCA technique: a case study of Deonar River Sub Basin, Sidhi Area, Madhya Pradesh, India. J Geol Soc India 97(4):396–404. https://doi.org/10.1007/s12594-021-1697-z

Tran TV, Alvioli M, Lee G, An HU (2018) Three-dimensional, time-dependent modeling of rainfall-induced landslides over a digital landscape: a case study. Landslides 15(6):1071–1084. https://doi.org/10.1007/s10346-017-0931-7

Tullos D, Byron E, Galloway G, Obeysekera J, Prakash O, Sun YH (2016) Review of challenges of and practices for sustainable management of mountain flood hazards. Nat Hazards 83(3):1763–1797. https://doi.org/10.1007/s11069-016-2400-3

Wang C, Hou J, Miller D, Brown I, Jiang Y (2019) Flood risk management in sponge cities: the role of integrated simulation and 3D visualization. Int J Disaster Risk Reduct 101139. https://doi.org/10.1016/j.ijdrr.2019.101139

Wasson RJ (2018) Riverine flood hazard: disaster risk reduction in India. Proc Indian Natl Sci Acad 99(0):65–76. https://doi.org/10.16943/ptinsa/2018/49502

Wei SC, Li HC, Shih HJ, Liu KF (2018) Potential impact of climate change and extreme events on slope land hazard—a case study of Xindian watershed in Taiwan. Nat Hazards Earth Syst Sci 18(12):3283–3296. https://doi.org/10.5194/nhess-18-3283-2018

Wichmann V, Becht M (2003) Modelling of geomorphic processes in an alpine catchment. Rapport d'éude 1–14. https://doi.org/10.1201/9781420038101.ch12

Zhang K, Xue X, Hong Y, Gourley JJ, Lu N, Wan Z, Hong Z, Wooten R (2016) ICRESTRIGRS: a coupled modeling system for cascading flood-landslide disaster forecasting. Hydrol Earth Syst Sci 20(12):5035–5048. https://doi.org/10.5194/hess-20-5035-2016

Chapter 25
Integration Hybrid Multi Criteria Decision Making of GIS, AHP and TOPSIS in Evaluating Suitable Locations for CNG Fuel Stations: A Case of Bonab City, Iran

Mahdi Yousefi Nejad Attari, Babak Ejlaly, Fahime Lotfian Delouyi, and Mohadese Hoseingoli Poorasl

Abstract As the demand for vehicle usage continues to surge, the strategic placement of fuel stations within urban areas has become a topic of significant consideration. The location of a fuel station is a critical factor in ensuring accessibility and efficiency. Landslides stand out as one of the most destructive hazards in natural terrain areas, emphasizing the need for reliable susceptibility analysis methods to aid policymakers in facility placement. This chapter investigates the applicable criteria for the location of Compressed Natural Gas (CNG) fuel stations, using standards and expert opinions. The study employs Geographic Information System (GIS) as a quantitative technique for landslide susceptibility, integrating a comprehensive dataset. The research focuses on extracting potential alternatives through GIS and ranking them using the Technique for Order of Preference by Similarity to Ideal Solution (TOPSIS), Analytical Hierarchy Process (AHP), and a hybrid TOPSIS-AHP method. To assess the practicality of the framework, the model is implemented in Bonab City, East-Azerbaijan, Iran. Eight potential locations for CNG stations in Bonab are identified using GIS, and they are subsequently ranked through three multi-criteria decision-making methods. The results indicate that locations C and G, situated on Bonab-Tabriz road and Bonab-Maragheh road, respectively, are optimal for constructing a fuel station.

M. Y. N. Attari (✉) · M. H. Poorasl
Department of Industrial Engineering, Bonab Branch, Islamic Azad University, Bonab, Iran
e-mail: mahdi_108108@yahoo.com

B. Ejlaly
Department of Industrial Engineering, Malek Ashtar University of Technology, Tehran, Iran

F. L. Delouyi
Department of Mechanical Engineering, Faculty of Engineering, University of Zabol, 98613-35856 Zabol, Iran
e-mail: fahime.lotfian@uoz.ac.ir

G. K. Panda et al. (eds.), *Landslide: Susceptibility, Risk Assessment and Sustainability*, Advances in Natural and Technological Hazards Research 52,
https://doi.org/10.1007/978-3-031-56591-5_25

Keywords Landslide evaluation · Geographical information system · Multi-criteria decision analysis · Location of compressed natural gas station · Energy policy

25.1 Introduction

The problem of locating gas stations falls within a class of the issues referred to as semi-obnoxious/ambivalent facilities location problems. Natural gas production and consumption are substantial and growing contributors to climate change. When released into the atmosphere, it acts as a greenhouse gas, and when burned, it produces carbon dioxide. In comparison to other coal and biomass energies, gas can be burned more effectively to generate heat and power while producing less waste and impurities. However, emissions, venting, and unplanned recycling of gases along the supply chain can all contribute to a similar environmental impact. The challenge lies in determining optimal locations for facilities that have both positive and negative effects on the ecosystem, especially when receptors are sensitive to the adverse effects (Abdi et al. 2013). The emergence of problems and challenges due to the rapid population growth, especially in megalopolis areas, has compelled municipal managers to present optional strategies to address these issues (Amjad et al. 2021). Survival and profitability are significantly linked to the optimal selection of the business locations. A business's location is defined as the place where company operations are conducted (Ashkezari et al. 2018). Furthermore, achieving balanced development of elements and municipal service components is a consistent goal in municipal management. The network of fueling stations is among the municipal service network components crucial for fuel supply, playing a vital role in traffic, safety, and environmental considerations (Aragonés-Beltrán et al. 2014). Various multi-criteria decision methods (MCDM) can be found in the literature, including Analytic Hierarchy Process (AHP), Analytic Network Process (ANP), Technique for Order Preference by Similarity to Ideal Solution (TOPSIS), and Multi-Attribute Utility Technique (MAUT) (Chu 2002). In an MCDM problem, multiple alternative options are examined based on the constraints, preferences, and priorities of decision-makers (Doorga et al. 2019).

AHP proves to be an effective multi-criteria decision-making procedure, especially when dealing with alterity. It is particularly suitable when the evaluation criteria can be organized hierarchically. AHP empowers decision-makers to transform subjective assessments into objective measurements, providing a procedure to convert complex problems into hierarchical framework (Dong et al. 2018). Selecting the appropriate location for a facility within a city is of significant decisions for when implementing a comprehensive plan. This process necessitates the use of GIS as a powerful tool in for surveying, analyaimg, and evaluating various municipal issues (Elboshy et al. 2022). Over time, the application of GIS has extended to all aspects related to the land use (Hashemizadeh and Ju 2019).

Considering the irregular increase in fuel usage in current years, billion of dollars of national wealth has spent on imported fuels. Furthermore, Iran possesses abundant natural gas resources, ranking as the second-largest owner of gas reservation. Therefore, substituting CNG fuel emerges as one of the primary and leading solutions to address the challenge of high gasoline consumption.

CNG serves as a more environmentally friendly and cost-effective substitute for gasoline (petrol) and other automotive fuels. As of the end of 2019, nations leading in natural gas vehicles adoption included Iran (3.5 million), China (3.3 million), Pakistan (2.8 million), Argentina (2.5 million), Brazil (1.8 million), and India (1.8 million). While the energy efficiency of CNG engines is comparable to gasoline engines, it falls short when compared to modern diesel engines. The low injection pressure of gasoline/petrol cars converted to run on natural gas results in reduced delivered power (10–15%) while operating on natural gas.

On the other hand, vehicles designed specifically for CNG exhibit greater fuel efficiency owing to the their higher combustion efficiency (120–130). Moreover, expediting the establishment of necessary infrastructures for natural gas supply can accelerate progress toward achieving desired targets (Kaya et al. 2019). The decision to locate facilities such as CNG fuel stations is a crucial factor to design production system. Government organizations and international obligations regulate minimum distances between gas stations and critical sites like schools, hospitals, infrastructure networks, groundwater, and dams. Additionally, when determining the optimal location for such facilities, road network maps and infrastructures must be considered (Khan and Samadder 2015). Consideration of essential facility features and restrictions is essential in finding the optimal location for a new facility (Messaoudi et al. 2019).

This chapter presents a comprehensive study focused on the selection of a suitable CNG station in Bonab, a city in Iran. A detailed list of criteria is systematically extracted using GIS, facilitating the identification of potential locations. The study employs the TOPSIS, AHP, and a hybrid TOPSIS-AHP method to rank these potential CNG fuel stations. The main objectives of this chapter are as follows:

- To integrate Multi-Criteria Decision Making (MCDM) and GIS in the study area for the purpose of CNG station selection.
- To establish a list of comprehensive and novel criteria based on standards and expert opinions.
- To consider topographical conditions in identifying potential locations for CNG stations.
- To compare the results of TOPSIS, AHP, and the hybrid TOPSIS-AHP in the selection of CNG station locations.
- To assist decision-makers in the construction of new CNG stations in the designated area.

The remainder of this book chapter is organized as follows: Sect. 25.2, Related Works, introduces fundamental principles for mapping approaches and the Analytic Hierarchy Process (AHP), a multi-criteria decision-making procedure. Section 25.3, Methodology, outlines the problem and proposes a method for identifying optimal sites for CNG stations. The experimental approach involving Geographic Information System (GIS), AHP, and TOPSIS, along with the analytical results, is presented in Sect. 25.4, Preliminary Experiments and Findings. Section 25.5, Conclusions, scrutinizes the findings and suggests future directions.

25.2 Literature Review

25.2.1 The History of GIS

GIS has been proven to be a reliable tool for geographical analysis and maps digitization, contributing to the gradual enhancement of territory analysis efficiency and the elimination of inconsistencies. The inception of national GIS can be traced back to Canada, beginning in the late 1960s. Significant developments in GIS technology occurred during the 1970s and 1980s, enabling the utilization of "Geographical Information System" for analysis, design, and integration of statistical techniques, particularly in the study of governmental policies. While GIS technology has been entrenched in western countries such as Canada and the USA for over forty years, it remains a relatively novel technology in many third-world countries.

In Iran, the National Cartographic Center of Iran became the first center to adopt this technology in accordance with a parliamentary resolution in 1990. The initiation of GIS project activities commence in March 1992 in the Ministry of Industry, Mine and Trade. This system is now extensively utilized in its operations. Other institutes employing GIS technology include the Ministry of Roads and Urban Development, the Ministry of Agriculture, the International Institute of Earthquake Engineering and Seismology, and the Natural Resources and Watershed Management Organization. GIS is applied across various fields, serving as a decision support tool for territorial planning and the identification of environmentally feasible sites for various types of activities and installations. Its application can have a significant impact (Mensour et al. 2019).

Moreover, GIS also offers a uniform analysis framework for geographically related data (data with coordinates). Ozdemir and Sahin (2018) have presented GIS maps are made up of layers, each representing a distinct dataset. Points of importance, such as gas stations, schools, malls, parcels, streets, transportation systems, and polygons encompassing cities and communities, can all be found in these databases. A standard coordinates system is employed to register all layers.

25.2.2 Analytic Hierarchy Process

One of the multi-criteria decision-making methods is the Analytic Hierarchy Process (AHP). AHP is based on comparisons reflecting decision-makers' perceptions and interpretations of different choices (Salmani Moghadam et al. 2018). Utilizing a hierarchical framework and a pairwise comparison, AHP assists in selecting the optimal option based on numerous competing criteria. AHP approach has been widely applied across various domains. As the number of criteria increases, the complexity of the problem also rises. Effective evaluation of criteria becomes critical, and AHP has proven to be a valuable tool in addressing various decision-making challenges. AHP is among the most recommended approaches in the literature (Sari et al. 2020). Saraswat and Digalwat (2021) employed Fuzzy AHP approaches to prioritize sustainable energy alternatives. Consequently, AHP is employed in this study to evaluate potential locations for fuel plant installations.

25.2.3 Evaluation Criteria

Even though the exact determination of the various locating principles in the city is challenging due to the dynamic nature of urban issues, the following six criteria can be considered in line with land use goals:

1. Compatibility: The primary focus of urban planning is to locate numerous uses throughout the city and separate undesirable services from the others.
2. Comfort: Two essential components, distance, and time, are essential in determining locations. The most significant comfort components involve accessing necessary urban services and avoiding undesirable uses.
3. Efficiently: The pattern of urban land price is another critical parameter for determining land use locations.
4. Utility: In urban land use planning, utility and palatability refer to special effort to reserve natural sources, create open and pleasant space in road formation, and enhance the urban areas.
5. Health: Aiming to apply environmental and health regulations, this criterion seeks to reduce pollution from different uses and adhere to hygienic standards to ensure the health of the human.
6. Safety standards: This criterion is designed to protect urban and rural areas from potential hazards such as floods, earthquakes, and volcanoes.

Shahraki Shahdabadi et al. (2021) utilized the SAW, TOPSIS, and ELECTRE methodologies to address the optimal location for constructing a biomass power plant Tajbakhsh et al. (2018) developed a model for land allocation in the Bayg watershed, Iran, aiming to minimize surface runoff and sediment load while maximizing net income. Soha and Hartmann (2021) introduced a GIS-based methodology for site suitability and selection, considering public utility services such as power, gas, and water networks, as well as environmental, social, and techno-economic criteria. The authors applied a multi-criteria decision-making process to weigh these factors, identifying potential power-to-gas sites.

Tercan et al. (2020) proposed a GIS-based multi-criteria model for the site selection of offshore wind energy power plants, taking into account legal, political, and socio-economic characteristics. Yıldız et al. (2018) analyzed the thermal environment of Uzundere-Erzurum using satellite images from the Landsat Thematic Mapper sensor, focusing on the characterization of the thermal urban environment.

To best of our knowledge, previous works haven't considered the distinctive criteria presented in this study. To demonstrate the practically of the proposed framework, we conducted the analysis in Bonab city, East Azerbaijan province, Iran to determine the optimal location for CNG station. While many studies like that Sánchez-Lozano et al. (2013) on locating fuel stations have utilized mathematical and GIS methods, rarely have they incorporated a ranking of the proposed alternatives. In this research, we employed a hybrid TOPSIS-GIS-AHP method to rank the identified locations.

Table 25.1 provides a comprehensive overview of past research, examining aspects such as Research Purposes, Research Area, Method, and Results. By systematically extracting parameters from the literature review and subsequently comparing them with the current research (Table 25.2), we gain insights into the significance and indispensability of this study, highlighting its distinctiveness and advancement beyond prior research endeavors.

25.3 Methodology

25.3.1 Study Erea

Bonab is situated at 16° and 3 min and 15 s east longitude and 37° and 20 min and 30 s north latitude, with an elevation of 190 m above sea level. The city is approximately 835 km away from Tehran, the capital of Iran, and 114 km from Tabriz, the capital of East-Azerbaijan province. Furthermore, Bonab is approximately 8 km away from Urmia Lake.

This study is focused on identifying optimal locations for CNG fuel stations in Bonab and subsequently ranking these locations. The evaluation of alternatives is based on various criteria, including elevation, proximity to the gas network, distance to high-pressure electricity lines, closeness to water supply pipes, proximity to main roads, and accessibility to the fire station. The study employs Arc-GIS, AHP, TOPSIS, and a hybrid AHP-TOPSIS approach to comprehensively assess and prioritize the potential locations for CNG fuel stations in Bonab.

25.3.2 Statistical Society of Research

The study's statistical population encompasses 60 individuals, including experts from traffic transportation organizations, the governorship, municipality, and offices related to petroleum, gas, water, and electricity distribution, as well as professors in

Table 25.1 Literature review according to research purposes and research area and method

References	Research purposes	Research area	The method used	Results
Abdel-Basset et al. (2023)	Infrastructure for charging EVs (sustainable location)	Electric vehicle charging station (EVs)	A multi-criteria decision-making (MCDM)	Economic factor is the most important factor in selecting the optimal location
Asakereh et al. (2014)	Geographical mapping models	For solar energy farms	A fuzzy analytic hierarchy process (fuzzy AHP)	Results showed that 18.25% of the Shodirwan area is exploitable as solar farms
Chen et al. (2023)	Landslides collected	From the Sichuan Province of China (spatial distribution of landslides)	Geographical information system (GIS) platform	PGA, lithology, river density, and road density (more important factors)
Erdogmus et al. (2006)	Energy policies (cheap infrastructure)	In Turkey for further development	Using analytic network process (ANP)	Suitable fuel which can be used for residential
Ghorbani et al. (2019)	A two-step phase method to locate the CNG stations	In the two-way highways	Data envelopment analysis (DEA) model and a multi-objective flow refueling location	To aim of maximizing the traffic flow of the vehicles in the two way highways and reducing the cost of constructing fuel stations
Hosseini Dehshiri (2022)	Choosing the right location	For the sustainable development of offshore wind energy	Hybrid fuzzy-SWARA and fuzzy-WASPAS method	This is very high stability and can be used as an efficient method in future
Konstantinos et al. (2019)	The optimization of spatial planning	In order to identify the most suitable places for the installation of wind farms	Multi criteria decisionmaking methods (MCDM) and GIS	Easily overcome conflicting parameters
Mousaei and Hatefi (2014)	Select the premier fuel to develop	To develop in the value chain of natural gas (NG)	Decision support system (DSS)	Most effective and productive
Noorollahi et al. (2016)	Land suitability analysis	For solar farms exploitation	Using GIS and fuzzy analytic hierarchy process (FAHP)	Select the most excellent suitable provinces for exploiting solar PV plants

(continued)

Table 25.1 (continued)

References	Research purposes	Research area	The method used	Results
Ramya and Devadas (2019)	Suitable locations	For industrial development	Integration of GIS and decision making	A cost-effective tool for decision making in land-use planning and management
Ryu et al. (2023)	A web-based decision support system (DSS)	For hydrogen refueling station location	Decision support system (DSS)	Effectiveness of the DSS in solving optimization problems
Sarıkaya et al. (2022)	Site selection	Natural gas emergency response team centers (NGERTC)	Decision making methods (MCDM) and GIS	There are worth using in selecting suitable NGERTC locations
Strantzali et al. (2019)	A decision support approach	For evaluating liquefied natural gas supply options	A multicriteria analysis	Take advantage to assess future consortiums

Bonab city. The statistical sample, consisting of 51 participants for the first questionnaire aimed at ranking the criteria, was determined using the Cochran formula. Experts and specialists were identified through consultations with managers of relevant offices and organizations, and the distributed questionnaires were completed by the participants. Once the experts' opinions are collected, statistical analysis will be conducted, leading to the second questionnaire.

Given the specialized nature of the topic and the importance of meticulous attention to the paired comparisons questionnaire associated with the Analytic Hierarchy Process (AHP), insights from 10 highly proficient experts in the field of optimal CNG station locations were sought.

To ensure the narrative validity of the questionnaire, feedback from experts was utilized. Ten questionnaires, titled "experts' questionnaire," were distributed among professors of university, experts, and specialists at Bonab's municipality GIS section, and specialists at the petroleum products distribution company of East Azerbaijan province. After a thorough analysis of the responses, the experts affirmed the questionnaire's justification with over 90% agreement, offering valuable insights and suggestions.

For questionnaire reliability assessment, the Cronbach's alpha coefficient was computed using SPSS software, resulting in an alpha value of 0.93 for the 50 completed questionnaires. This exceeds the recommended threshold of 0.70, indicating that the questionnaire possesses acceptable reliability and validity. The computation was performed by experts in the products distribution offices of Bonab city.

Table 25.2 Parameters extracted from literature review

Parameters	Abdel-Basset et al. (2023)	Asakereh et al. (2014)	Chen et al. (2023)	Erdogmus et al. (2006)	Ghorbani et al. (2019)	Hosseini Dehshiri (2022)	Konstantinos et al. (2019)	Mousaei and Hatefi (2014)	Noorollahi et al. (2016)	Ramya and Devadas (2019)	Ryu et al. (2023)	Sarıkaya et al. (2022)	Strantzali et al. (2019)	Current chapter
MCDM	*	*		*		*	*	*	*	*		*	*	*
GIS		*	*				*		*	*		*		*
ANP	*			*				*						
TOPSIS							*	*		*				*
DSS							*				*			*
Landslides		*	*						*					
FAHP		*							*			*		
AHP							*			*			*	*
DEMATEL	*													
Energy policy	*			*	*	*		*				*	*	*
GIS—MCDM/ TOPSIS/AHP										*				*
Fuzzy-WASPAS						*								
Fuzzy-VIKOR						*								
ELECTRE III													*	
CNG stations					*									*
DEA					*									
Sustainable	*	*				*								*
Uncertainty	*	*				*						*		

(continued)

Table 25.2 (continued)

Parameters	Abdel-Basset et al. (2023)	Asakereh et al. (2014)	Chen et al. (2023)	Erdogmus et al. (2006)	Ghorbani et al. (2019)	Hosseini Dehshiri (2022)	Konstantinos et al. (2019)	Mousaei and Hatefi (2014)	Noorollahi et al. (2016)	Ramya and Devadas (2019)	Ryu et al. (2023)	Sarıkaya et al. (2022)	Strantzali et al. (2019)	Current chapter
Geographical detector model			*											
Multi-objective	*			*	*	*			*	*	*	*		*
Multi-criteria locating	*				*					*	*	*		*
Multicriteria analysis	*	*		*		*	*	*	*	*	*	*	*	*

25.3.3 Criteria in Fuel Supply Stations' Locating

There has been an increase in demand for natural resource systems during the last decades. Natural gas energy is one of the essential natural resources. In the context of this research, a comprehensive examination of the literature, industry standards, and criteria put forth by the petroleum products distribution office of the country, coupled with consultations with experts in the field, led to the identification of 12 crucial factors influencing the location of fuel stations. These factors were subsequently categorized into compatible and incompatible criteria. A detailed breakdown of each of these components, including their access radius and safety privacy considerations, is provided in Table 25.3.

Table 25.3 Various criteria sources, access radius, and safety privacy

Criteria sources	Calculation method	Safety privacy	Access radius	Criteria
Gas office map	Spatial layer in GIS	50 m	500 m	High-pressure gas line 250 psi
Electricity office map	Spatial layer in GIS	50 m	500 m	High-pressure electricity line
Water office map	Spatial layer in GIS	–	500 m	Water network
Municipality map	Spatial layer in GIS	–	500 m	Main road network
Municipality map	Spatial layer in GIS	–	Conventional and proper access	Fire station centers
Municipality map	Spatial layer in GIS	–	Conventional and proper access	Green space
Municipality map	Spatial layer in GIS	–	Conventional and proper access	Barren lands
Municipality map	Spatial layer in GIS	1000 m	–	Residential areas
Municipality map	Spatial layer in GIS	500 m	–	Health centers
Municipality map	Spatial layer in GIS	1000 m	–	Historical places
Municipality map	Spatial layer in GIS	500 m	–	Business trade centers
Municipality map	Spatial layer in GIS	250 m	–	Rivers

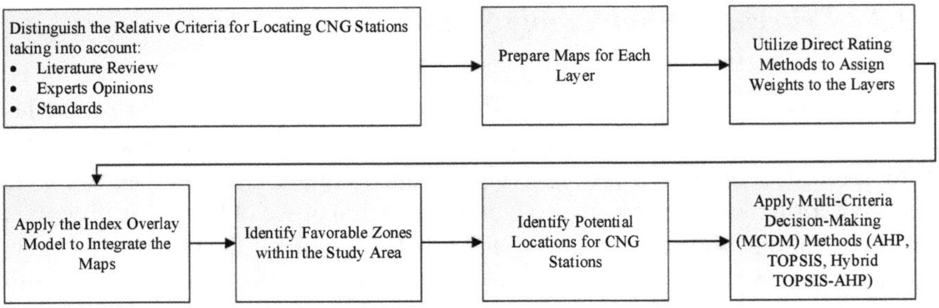

Fig. 25.1 Overview of information analyzing steps flowchart in CNG location

25.3.4 Preparing the Information for Analyzing

Spatial and descriptive data were initially input into ArcGIS 10.3 software, leveraging its capabilities for subsequent data analysis. The array of software employed in this research includes:

- Excel, utilized for implementing TOPSIS method.
- AutoCAD, employed for digitization purposes.
- ArcGIS, utilized for the analysis of plans and specification of alternatives.
- Expert Choice, chosen for implementing the AHP model.

Figure 25.1 provides a comprehensive visualization, elucidating each step of the analysis process. By following these steps, the model aims to systematically identify and prioritize suitable locations for CNG stations based on a set of defined criteria.

25.3.5 Information Studying Phase

This stage of the research has a great significance. The outcomes obtained through the direct rating method undergo a normalization process. Information from twelve layers, created using the AutoCAD program, is transferred to ArcGIS for comprehensive data evaluation aiming to identify potential CNG stations within the study area. After ArcGIS software is employed to overlay these layers within a geographical information system, the potential options are ranked utilizing the AHP, TOPSIS, and a hybrid TOPSIS-AHP methodology.

25.4 Results and Analysis

As depicted in Fig. 25.1, this research has been investigated as a case study on Bonab city. We have considered various CNG station locations within both Bonab city and other areas within the province. To facilitate this study, we have extrapolated the distances between existing stations in various cities and applied them to our primary focus, Bonab city (Fig. 25.2).

25.4.1 *Applying the AHP Method*

AHP is a decision-making technique widely applied in various fields for handling complex problems involving multiple criteria and alternatives. Developed by Thomas L. Saaty in the 1970s, AHP provides a systematic approach to decision analysis, allowing decision-makers to compare, rank, and evaluate alternatives based on rational comparisons. The process begins with the clear identification of the objective, the criteria, and the potential candidates.

First of all, the decision problem must be structured hierarchically, with the objective at the top, criteria in the middle layers, and potential alternatives at the bottom.

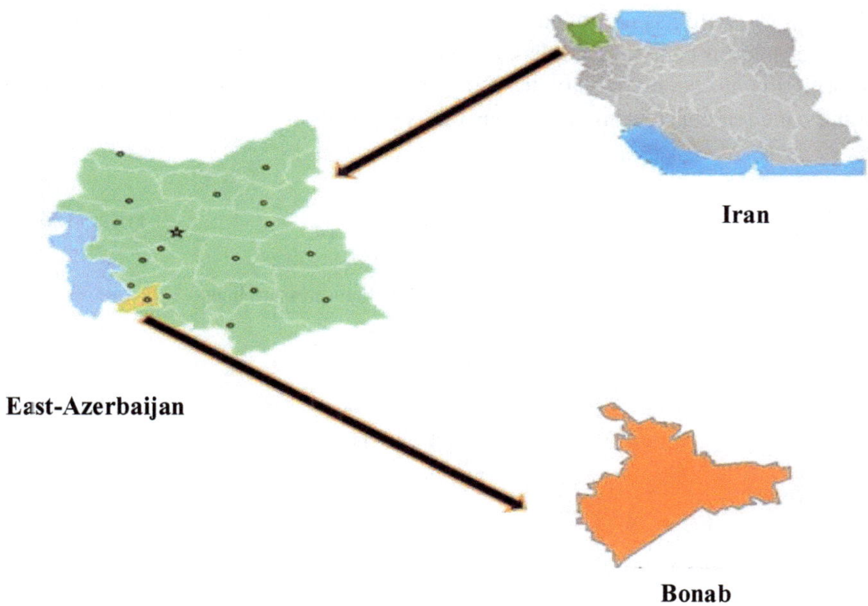

Fig. 25.2 A basic frame about the location of Bonab city and the East-Azerbaijan province

Using a numerical scale, decision-makers are asked to perform pairwise comparisons of items within each level of the hierarchy (within criteria and within alternatives for each criterion).

Then Consistency checks are applied to ensure the reliability of the comparisons. The pairwise comparison results are synthesized to derive the weights of criteria, representing their relative importance in achieving the decision objective. Then, preferences are synthesized to generate a preference matrix, which is then used to compute the overall ranking of each alternative.

25.4.2 Land Use Maps of Bonab City

In adherence to the safety standards outlined by the oil products distribution office, paramount importance is placed on ensuring the safety of residents in the vicinity of petrol stations. This commitment aims to mitigate potential environmental issues arising from gas leakage or explosions. The specific considerations and privacies essential for safeguarding the well-being of residents are detailed in Table 25.4.

Figure 25.3 presents the land use map of Bonab city, encompassing crucial privacy considerations. This includes the privacy map of the high-pressure gas network in Bonab (A), the privacy map of high-pressure electricity lines (B), privacy considerations for water pipes (C), residential areas (D), the privacy map of health centers (E), the privacy map of trade centers (F), and the privacy map of historical places (G). The GIS software has identified limitations in specific regions of the city, implying that fuel stations cannot be implemented in areas subject to these seven criteria. This information has been crucially derived from GIS-generated plans, underscoring the significance of considering these constraints in the establishment of fuel stations.

Table 25.4 Scoring important and necessary to establish CNG fuel stations

Row	Criteria	Privacy (m)
1	Gas high-pressure lines	50
2	Electricity high-pressure lines	50
3	Water pipes	50
4	Residential areas	1000
5	Health centers	500
6	Trade centers	500
7	Historical places	1000

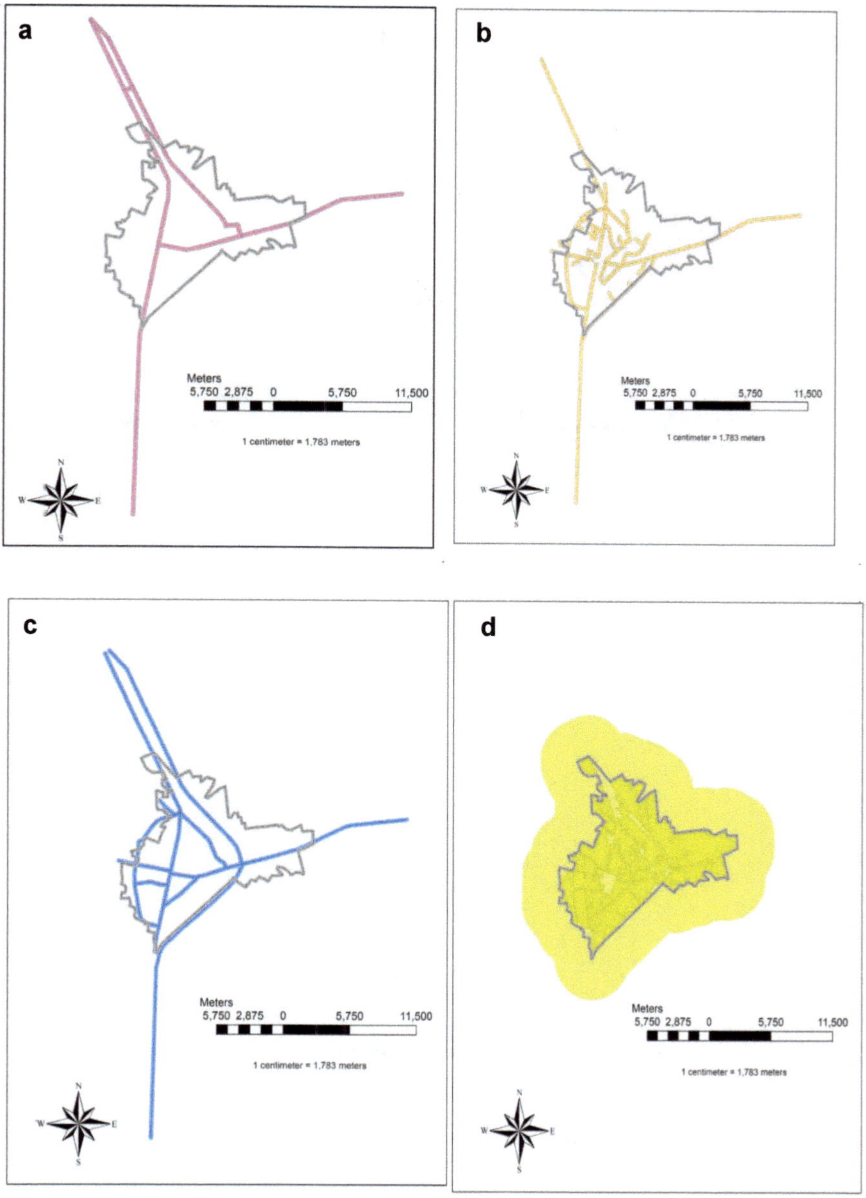

Fig. 25.3 Land use maps of Bonab city; **a** privacy map of the 250 psi gas network in Bonab, **b** privacy map of high-pressure electricity lines in Bonab, **c** privacy map of water pipes in Bonab, **d** privacy map of residential areas in Bonab, e privacy map of health centers in Bonab, **f** privacy map of trade centers in Bonab, **g** privacy map of historical places in Bonab

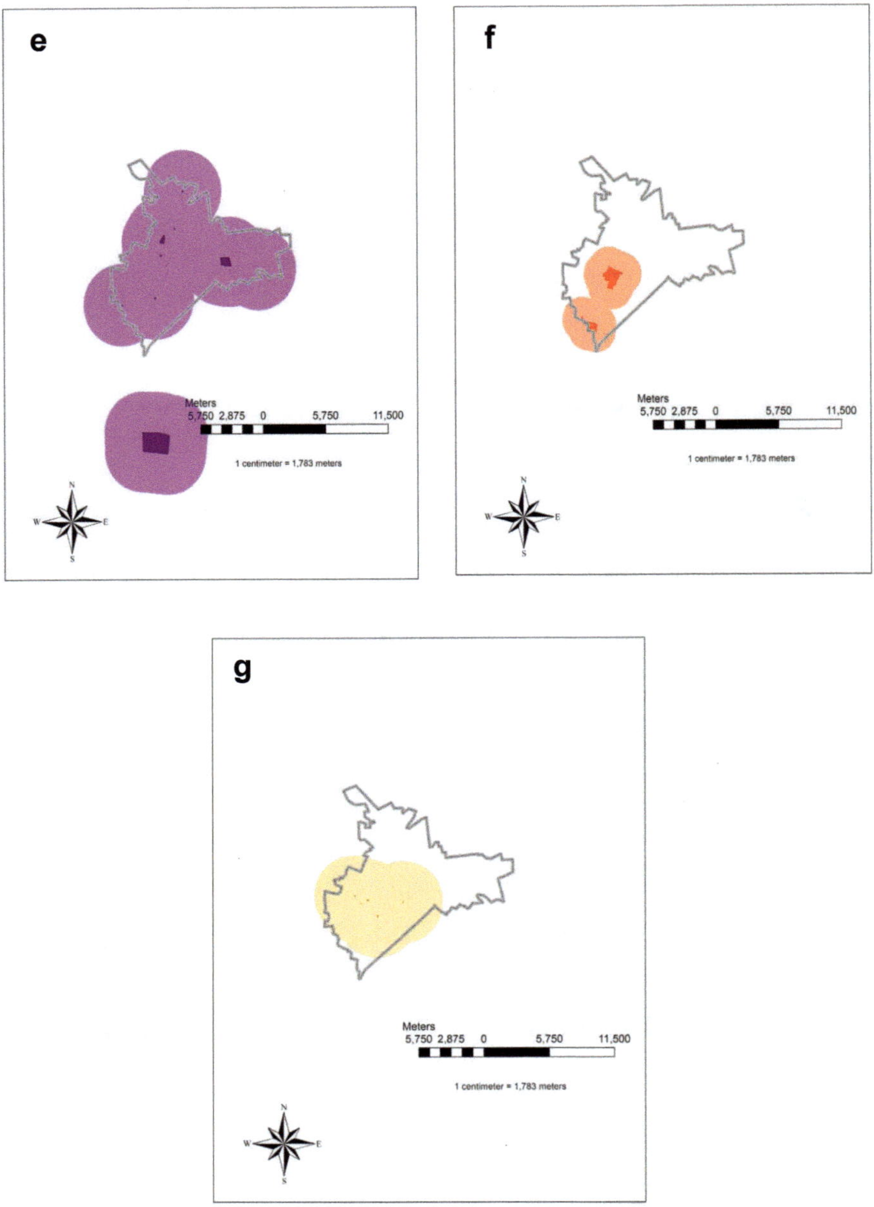

Fig. 25.3 (continued)

25.4.3 Locating the Fuel Stations

The methodology employed for the spatial determination of CNG fuel stations in Bonab city within this study is the Index Overlay model in the integration of maps. Utilizing the direct rating method, experts were asked to assign a score to each criterion, reflecting the relative importance of each criterion. The prioritization of criteria is detailed in Table 25.5 and is as follows.

The outcomes of layer combination with assigned weights and the resulting determined ranks are visually presented in Fig. 25.4.

Table 25.5 Ranking of effective criteria in locating the CNG stations based on direct rating method

Priorities	Criteria
1	Proximity to 250 psi gas network
2	Distance from residential areas
3	Proximity to high-pressure electricity lines
4	Distance from historical buildings
5	Proximity to fire stations
6	Proximity to the main roads
7	Proximity to water supply
8	Proximity to the green spaces
9	Distance from trade centers
10	Distance from rivers
11	Distance from health centers
12	Proximity to barren lands

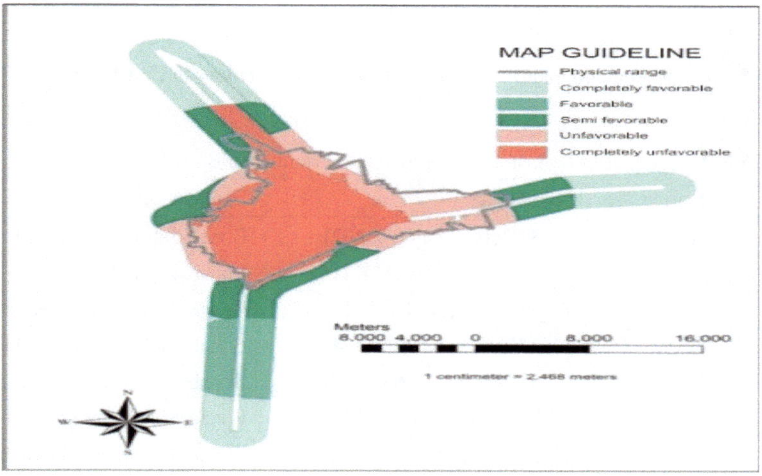

Fig. 25.4 Adequate map layers combination in locating of Bonab city CNG stations

After applying the Index Overlay method for the combination of layers and eliminating unsuitable locations for the establishment of CNG stations, including areas with privacy concerns and other inappropriate uses identified in earlier maps, the precise locations of the eight proposed CNG fuel stations in Bonab are visually depicted in Fig. 25.5.

The precise locations for establishing CNG stations in Bonab city, along with their corresponding geographical coordinates, are delineated in Table 25.6.

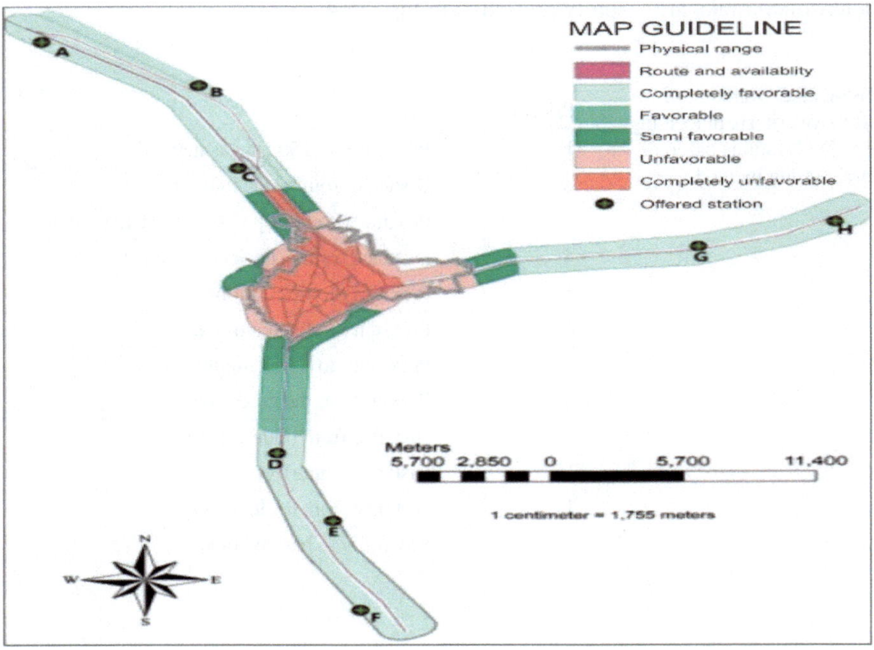

Fig. 25.5 Obtained potential locations for establishing CNG stations in Bonab city

Table 25.6 Geographic coordinates of proposed CNG stations in Bonab City

Row	Station	Coordinates	Explanation
1	A	(N: 37/36052, N: 46/05402)	Bonab-Tabriz 10 km road
2	B	(N: 37/238724, N: 46/04216)	Bonab-Tabriz 6 km road
3	C	(N: 37/44178, N: 45/95611)	Bonab-Tabriz 3 km road
4	D	(N: 37/30437, N: 46/04864)	Bonab-Malekan 2 km road
5	E	(N: 37/28991, N: 46/04797)	Bonab-Malekan 5 km road
6	F	(N: 37/27126, N: 46/05158)	Bonab-Malekan 9 km road
7	G	(N: 37/34783, N: 46/10526)	Bonab-Maragheh 3 km road
8	H	(N: 37/35278, N: 46/15715)	Bonab-Maragheh 6 km road

25.4.4 TOPSIS Ranking

This ranking methodology employs entropy techniques to consider variables from the expert's perspective, especially in bipolar scenarios. The decision matrix, constructed from the expert poll and its compilation, is formulated using the geometric average. In the case of positive variables, zero signifies the lowest value, while ten representing the highest value. Conversely, for negative variables, zero indicates maximum importance, while ten denotes the minimum value. The outcomes of alternative ranking using direct rating method and TOPSIS method are detailed in Table 25.8.

25.4.5 AHP Ranking

Following the identification of proposed alternatives for establishing CNG fuel stations, a questionnaire was developed to rank these stations based on AHP method. This questionnaire was distributed among ten experts and specialists from the Bonab gas, water, and electricity office, as well as the municipality and the oil products distribution, considering the unique expertise required for such stations. Subsequently, the data collected from these pairwise comparisons were imported into the Expert Choice software to calculate the rankings of potential locations. The symbols employed in the Expert Choice software for introducing the criteria are presented in Table 25.7.

Figure 25.6 depicts the pairwise comparison of distinguished criteria. In Fig. 25.7, the prioritization of criteria is reported, with C_1, signifying proximity to a 250-psi gas station network with a weight of 0.237, emerging as the most important factor. Additionally, C_{10}, representing distance from residential areas with a weight of 0.183, and C_2, denoting proximity to high-pressure electricity lines with a weight of 0.157,

Table 25.7 Criteria and symbols for stabilizations of CNG fuel stations

Row	Criteria	Symbol
1	Proximity to 250 psi gas network	C_1
2	Proximity to high-pressure electricity lines	C_2
3	Proximity to water supply	C_3
4	Proximity to the main roads	C_4
5	Proximity to fire stations	C_5
6	Proximity to the green spaces	C_6
7	Proximity to barren lands	C_7
8	Distance from health centers	C_8
9	Distance from trade centers	C_9
10	Distance from residential areas	C_{10}
11	Distance from historical buildings	C_{11}
12	Distance from rivers	C_{12}

follow in priority. The incompatibility rate of 0.09 underscores the consistency of comparisons, as it is below the threshold of 0.1, affirming the validity of respondents' credentials.

In summary, the analysis reveals that C and G stations, each assigned weights of 0.145, hold the majority shares, establishing them as the most favorable locations for establishing the CNG fuel station in Bonab city. The prioritized arrangement of proposed stations based on the AHP results is presented in Table 25.8.

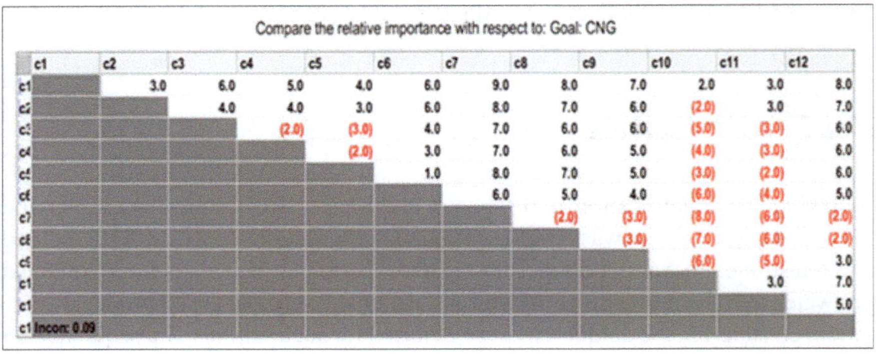

Fig. 25.6 Pairwise comparison of criteria

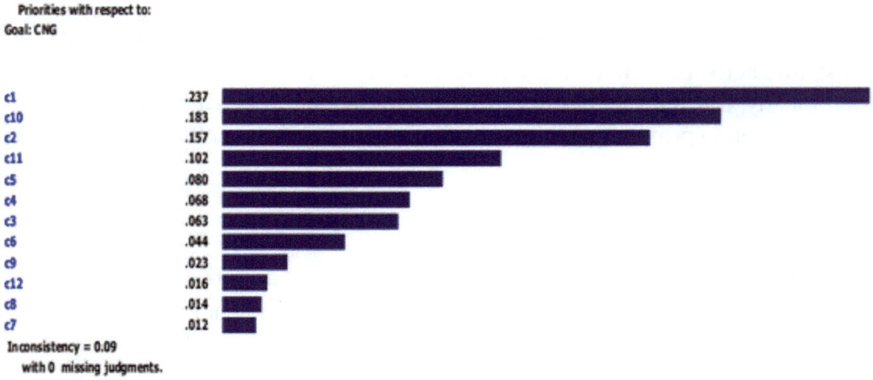

Fig. 25.7 Weights of criteria

Table 25.8 Ranking of alternatives

Station	A	B	C	D	E	F	G	H
AHP-ranking	3	6	1	8	7	5	2	4
TOPSIS-ranking	3	5	1	6	7	8	2	4
AHP-TOPSIS-ranking	3	5	1	8	6	7	2	4

25.4.6 TOPSIS-AHP Ranking

In this section, the weights derived from the AHP method are utilized as inputs for the TOPSIS method. The resulting AHP-TOPSIS ranking is presented in the final row of Table 25.8.

25.5 Discussion

Bonab City holds a strategic position in the western part of the country, serving as a vital transit hub. Its proximity to central provinces and accessibility to trans-regional markets make it a focal point for attracting and generating passenger trips. The establishment of a new Compressed Natural Gas (CNG) fuel station is crucial for enhancing passenger welfare. In pursuit of this objective, eight potential locations for CNG stations in Bonab were identified using GIS. Following a comprehensive evaluation based on selected criteria, eight locations met the qualifications and homogeneity criteria.

Subsequently, the ranking of these alternatives was calculated, employing three models to assess their viability. Both direct rating method and AHP revealed that proximity to a 250-psi gas station network holds utmost importance. The results consistently identified locations C, G, and A as optimal choices across the three proposed methods, demonstrating the robustness and reliability of the findings.

To further validate the results, experts were consulted, taking into account the strategic aspects of the locations. Considering factors such as impact on economic growth, accessibility to factories and industrial zones, and proximity to the transition road connecting western provinces to central provinces of Iran, stations A, B, and C emerged as top-ranked choices. Stations D, E, and F, situated along the Bonab-Malekan road connecting western provinces to the Tehran-Tabriz highway, secured the second position. G and H stations, located on the Bonab-Maragheh road, claimed the third spot, with a lower importance attributed to this route compared to Bonab-Tabriz and Bonab-Malekan roads in terms of traffic volume, strategic location, and economic growth impact.

In conclusion, alternative location C is unequivocally affirmed as the most preferred site for constructing a new CNG station in Bonab City, based on a comprehensive evaluation of both technical criteria and strategic considerations.

25.5.1 Conclusion and Future Studies

The primary purpose of this study is the best locating of Bonab CNG fuel stations and to rank those stations. Initially, through reference to existing standards and consultations with experts from Bonab city in East-Azerbaijan province's oil products distribution firms, a comprehensive set of criteria was identified. These criteria include proximity to 250 psi gas network, proximity to high-pressure electricity lines, proximity to water supply pipes, proximity to main roads, proximity to fire station in this city, distance from health centers, distance from residential areas, distance from business and trade centers, distance from historical monuments and distance from rivers. These criteria are recognized as essential factors influencing the location of CNG stations.

The collected data underwent thorough analysis within the ARC GIS software. Subsequently, utilizing the Index Overlay plan, the study classified locations into five main sections: entirely favorable, favorable, semi-favorable, unfavorable, and completely unfavorable. From this classification, the precise positioning of eight candidate stations within the favorable section was determined. To rank these potential locations, three distinct methodologies—TOPSIS, AHP, and a hybrid TOPSIS-AHP approach—were applied.

In implementing the TOPSIS method, a direct rating method was employed to determine the criteria weights. A questionnaire was designed based on key criteria, which was then distributed among experts specializing in the distribution of gas, electricity, water, and oil products. The TOPSIS method was subsequently employed to ascertain the ranking of these potential locations. For the AHP ranking, a separate questionnaire was devised to facilitate pairwise comparisons of alternatives grounded in the identified criteria. Finally, the hybrid TOPSIS-AHP method was deployed to derive a comprehensive ranking of the alternatives. The findings consistently identified locations C, G, and A as optimal choices across all three proposed methods.

Looking ahead, future research in this domain should explore additional techniques and factors. While the Index Overlay model was employed in this study, investigating alternative scientific methods such as mathematical modeling or FRLM could be pursued. Given that the study's criteria were selected from a myriad of potential factors influencing location decisions, there is an opportunity to expand the model by incorporating additional considerations, such as lands, topography, slope, or the ease of passage. Furthermore, the methodology applied in this study could be adapted for locating various other facilities and urban services, offering a versatile framework for broader applications.

References

Abdel-Basset M, Gamal A, Hezam IM, Sallam KM (2023) Sustainability assessment of optimal location of electric vehicle charge stations: a conceptual framework for green energy into smart cities. Environ Dev Sustain 68:37–48

Abdi A, Alavi MPN, Mobasheri H (2013) Allocation of CNG stations in urban street networks based on GIS approach and prioritization with AHP and topsis methods (case study: Rasht City). Res J Appl Sci Eng Technol 5(12):3262–3272

Amjad F, Agyekum EB, Shah LA, Abbas A (2021) Site location and allocation decision for onshore wind farms, using spatial multi-criteria analysis and density-based clustering. A techno-economic-environmental assessment, Ghana. Sustain Energy Technol Assess 47:101503

Aragonés-Beltrán P, Chaparro-González F, Pastor-Ferrando JP, Pla-Rubio A (2014) An AHP (analytic hierarchy process)/ANP (analytic network process)-based multi-criteria decision approach for the selection of solar-thermal power plant investment projects. Energy 66:222–238

Asakereh A, Omid M, Alimardani R, Sarmadian F (2014) Developing a GIS-based fuzzy AHP model for selecting solar energy sites in Shodirwan region in Iran. Int J Adv Sci Technol 68:37–48

Ashkezari AD, Hosseinzadeh N, Chebli A, Albadi M (2018) Development of an enterprise geographic information system (GIS) integrated with smart grid. Sustain Energy Grids Netw 14 25–34

Chen Z, Song D, Du Y, Dong L (2023) Investigation on the spatial distribution of landslides in Sichuan Province, southwest China. Geomat Nat Haz Risk 14(1):2232085

Chu TC (2002) Facility location selection using fuzzy TOPSIS under group decisions. Int J Uncertain Fuzziness Knowl-Based Syst 10(06):687–701

Doorga JR, Rughooputh SD, Boojhawon R (2019) Multi-criteria GIS-based modelling technique for identifying potential solar farm sites: a case study in Mauritius. Renew Energy 133:1201–1219

Dong XM, Li Y, Pan YL, Huang YJ, Cheng XD (2018) Study on urban fire station planning based on fire risk assessment and GIS technology. Procedia Eng 211:124–130

Elboshy B, Alwetaishi M, Aly RM, Zalhaf AS (2022) A suitability mapping for the PV solar farms in Egypt based on GIS-AHP to optimize multi-criteria feasibility. Ain Shams Eng J 13(3):101618

Erdogmus S, Aras H, Koc E (2006) Evaluation of alternative fuels for residential heating in Turkey using analytic network process (ANP) with group decision-making. Renew Sustain Energy Rev 10:269–279

Ghorbani R, Ghousi R, Makui A (2019) Location of compressed natural gas stations using multi-objective flow refueling location model in the two-way highways: a case study in Iran. J Ind Syst Eng 2(12):95–112

Hashemizadeh A, Ju Y (2019) Project portfolio selection for construction contractors by MCDM–GIS approach. Int J Environ Sci Technol 16(12):8283–8296

Hosseini Dehshiri SS (2022) New hybrid multi criteria decision making method for offshore windfarm site location in Persian Gulf, Iran. Ocean Eng 256:111498

Kaya I, Colak M, Terzi F (2019) A comprehensive review of fuzzy multi criteria decision making methodologies for energy policy making. Energ Strat Rev 24:207–228

Khan D, Samadder SR (2015) A simplified multi-criteria evaluation model for landfill site ranking and selection based on AHP and GIS. J Environ Eng Landsc Manag 23(4):267–278

Konstantinos I, Georgios T, Garyfalos A (2019) A decision support system methodology for selecting wind farm installation locations using AHP and TOPSIS: case study in Eastern Macedonia and Thrace region, Greece. Energy Policy 132:232–246

Mensour ON, El Ghazzani B, Hlimi B, Ihlal A (2019) A geographical information system-based multi-criteria method for the evaluation of solar farms locations: a case study in Souss-Massa area, southern Morocco. Energy 182:900–919

Messaoudi D, Settou N, Negrou B, Settou B (2019) GIS based multi-criteria decision making for solar hydrogen production sites selection in Algeria. Int J Hydrogen Energy 44(60):31808–31831

Mousaei A, Hatefi MA (2014) A decision support system (DSS) to select the premier fuel to develop in the value chain of natural gas. Iranian J Oil Gas Sci Technol 4(3):60–76

Noorollahi E, Fadai D, Akbarpour Shirazi M, Ghodsipour SH (2016) Land suitability analysis for solar farms exploitation using GIS and fuzzy analytic hierarchy process (FAHP)—a case study of Iran. Energies 9:1–24

Ozdemir S, Sahin G (2018) Multi-criteria decision-making in the location selection for a solar PV power plant using AHP. Measurement 129:218–226

Ramya S, Devadas V (2019) Integration of GIS, AHP and TOPSIS in evaluating suitable locations for industrial development: a case of Tehri Garhwal district, Uttarakhand, India. J Clean Prod 238:117872

Ryu H, Lee D, Shin J, Song M, Lee S, Kim H, Kim B (2023) A web-based decision support system (DSS) for hydrogen refueling station location and supply chain optimization. Int J Hydrog Energy 48:36223–36239

Salmani Moghadam M, Khodabandehlo H, Yousefi F, Samadi M (2018) Optimal location of natural gas refueling stations using geographic information system and AHP model (case study: district 4 of Tehran). Geogr Human Relationsh 1(2):417–438

Sánchez-Lozano JM, Teruel-Solano J, Soto-Elvira PL, García-Cascales MS (2013) Geographical information systems (GIS) and multi-criteria decision making (MCDM) methods for the evaluation of solar farms locations: case study in south-eastern Spain. Renew Sustain Energy Rev 24:544–556

Saraswat SK, Digalwar AK (2021) Evaluation of energy alternatives for sustainable development of energy sector in India: an integrated Shannon's entropy fuzzy multi-criteria decision approach. Renew Energy 171:58–74

Sarıkaya MS, Yanalak M, Karaman H (2022) Site selection of natural gas emergency response team centers in Istanbul metropolitan area based on GIS and FAHP. Int J Geo-Inf 11:571

Sari F, Kandemir İ, Ceylan DA, Gül A (2020) Using AHP and PROMETHEE multi-criteria decision-making methods to define suitable apiary locations. J Apic Res 59(4):546–557

Shahraki Shahdabadi R, Maleki A, Haghighat S, Ghalandari M (2021) Using multi-criteria decision-making methods to select the best location for the construction of a biomass power plant in Iran. J Therm Anal Calorim 145(4):2105–2122

Soha T, Hartmann B (2021) Complex power-to-gas plant site selection by multi-criteria decision-making and GIS. Energy Convers Manag: X:100168

Strantzali E, Aravossis K, Livanos GA, Nikoloudis C (2019) A decision support approach for evaluating liquefied natural gas supply options: Implementation on Greek case study. J Clean Prod 222:414–423

Tajbakhsh SM, Memarian H, Kheyrkhah A (2018) A GIS-based integrative approach for land use optimization in a semi-arid watershed. Glob J Environ Sci Manag 4(1):31–46

Tercan E, Tapkın S, Latinopoulos D, Dereli MA, Tsiropoulos A, Ak MF (2020) A GIS-based multi-criteria model for offshore wind energy power plants site selection in both sides of the Aegean Sea. Environ Monit Assess 192(10):1–20

Yıldız ND, Avdan U, Yılmaz S, Matzarakis A (2018) Thermal map assessment under climate and land use changes; a case study for Uzundere Basin. Environ Sci Pollut Res 25(1):940–951

Chapter 26
Geospatial Modeling of Potential Landslide Hazard Estimation for Better Management in the Bandarban District of Bangladesh

Md. Sofi Ullah

Abstract This study presents a geospatial assessment of potential landslide hazards in the Bandarban District of the Chittagong Hill Tracts (CHT) in Bangladesh. Landslides are a severe threat to mountainous regions, causing loss of life and significant economic damage worldwide. The study employs a Geospatial weighted overlaying technique, assigning values on a scale of 1–5 and 100 for factors influencing landslides. These factors encompass historical landslide occurrences, land use patterns, rainfall, elevation, slope characteristics, soil types, geological features, distances to rivers, roads, stream orders, and socio-economic variables like household density, population density, income levels, and education. These values are determined in consultation with local communities and domain experts. The geospatial model categorizes the Bandarban District into five distinct levels of landslide hazards, ranging from "Very High Hazard" to "Very Low Hazard." "Very Low Hazard" areas, constituting 13.39% of the total hazard area, pose the least risk but still require basic preparedness measures and educational initiatives. "Low Hazard" areas, covering 36.04% of the hazard area, necessitate lower mitigation priority but ongoing awareness and preparedness efforts. The "Moderate Hazard" areas, making up 39.39% of the hazard area, require a multi-faceted approach for risk reduction, including land-use regulations, reforestation, and community-based disaster risk reduction programs. "High Hazard" areas, though smaller at 11.00% of the hazard area, demand immediate attention with actions such as engineering solutions, land use planning, and proactive disaster preparedness. The "Very High Hazard" area, representing only 0.18% of the hazard area, requires the most urgent focus for mitigation measures.

Keywords Geospatial modeling · Remote sensing · Landslide assessment · Geographic Information Systems (GIS) · Landslide susceptibility

Md. S. Ullah (✉)
Department of Geography and Environment, University of Dhaka, Dhaka 1000, Bangladesh
e-mail: sofiullah.geoenv@du.ac.bd

G. K. Panda et al. (eds.), *Landslide: Susceptibility, Risk Assessment and Sustainability*, Advances in Natural and Technological Hazards Research 52,
https://doi.org/10.1007/978-3-031-56591-5_26

26.1 Introduction

Landslide hazard assessment emerges as a critical imperative, especially in rugged terrains like Bandarban District. The region's complex dynamics, influenced by intricate topography, monsoonal rainfall, and rapid urbanization, demand a nuanced understanding. The vulnerability of Bandarban is amplified by its diverse demographic composition, including indigenous communities (Das et al. 2018a, b) deeply connected to natural resources, underscoring the urgency of implementing measures for sustainable development.

Recent inquiries bring attention to the increasing peril of landslide catastrophes in the Chittagong Hill Tracts (CHT) of Bangladesh, resulting in noteworthy human and infrastructural detriments (Rabby and Li 2019). This elevation in jeopardy is ascribed to sundry elements, including heightened instances of intense precipitation, augmented population density, swift urbanization, topographical alterations through hill cutting, deforestation, and a deficiency in cultural acumen (Alam 2020; Ahmed 2017; Khan et al. 2020a, b). The repercussions of landslides on elevated communities, notably in the CHT, have been heightened, causing a considerable loss of lives between 2000 and 2018, further compounded by the influx of Rohingya refugees in 2017, involving substantial deforestation for habitation (Ahmed et al. 2020a, b; UNHCR 2020, Zaman et al. 2020).

Approximately 3 million individuals, presently encounter substantial landslide perils in the CHT (UNOCHA 2017). In the context of disaster risk reduction (DRR), two predominant perspectives are identified: one that accentuates natural hazards as the instigators of disasters and another spotlighting the intricate interplay of socioeconomic and sociocultural phenomena (Kelman 2020; Wisner et al. 2004).

While conventional landslide investigations primarily concentrate on mapping hazards (Chen et al 2020a, b) and quantifying risks, this exploration posits that landslide disasters frequently have their roots in anthropogenic elements such as environmental deterioration, unsustainable progress, cultural hindrances, and governance predicaments (Guzzetti et al. 2020; van Westen et al. 2006; Glade 2003). Vulnerability, delineated as the human facet of risk, encompasses diverse components amplifying susceptibility to hazard repercussions, encompassing geographical, communal, financial, ecological, governmental, societal, cultural, and institutional elements (Birkmann et al. 2013). Resilience, representing a community's capacity (Das et al 2018a, b) to endure, adapt to, and rebound from landslide repercussions sustainably, is paramount in this context (UNDRR 2017).

The numerous studies underscore the significance of comprehending and tackling geological, climatic, and anthropogenic factors influencing landslide susceptibility in the hilly terrains of Bangladesh (BBS 2009a, b; Brammer 2012; Cui et al. 2021; Hasnat et al. 2018; Khan 1991; Mannan 2002). Furthermore, it illuminates the vulnerability of mountain inhabitants in Bangladesh, attributing their susceptibility to aspects such as impoverishment, land deprivation, inaccessibility to natural assets, lack of education, and impacts of climate change (Hassan et al. 2005; Huddleston and Ataman 2003; UNEP-WCMC 2002). Strategies for managing and mitigating landslide risks must account for these elements for efficacious results.

Bangladesh demonstrated remarkable achievements in meeting the Millennium Development Goals (MDGs), notably in reducing poverty, improving primary education, promoting gender equality, and lowering child and maternal mortality rates (UNDP 2015). Following the conclusion of the MDGs in 2015, Bangladesh enthusiastically embraced the Sustainable Development Goals (SDGs) with 17 goals and 169 targets set for accomplishment by 2030 (UNDP 2015).

Among the Sustainable Development Goals (SDGs), one of the most important objectives is addressing climate-related disasters such as cyclones, floods, riverbank erosion, landslides, etc. The present study specifically focuses on landslide hazards because increased precipitation is likely to affect landslides in Bangladesh. Studies indicate that the trend is on the rise, accompanied by increasing fatalities due to landslides, often caused by heavy rainfall (Guo et al. 2019) in the hilly regions of Bangladesh (Alam 2020). Therefore, to meet SDG by 2030, urgent action will be taken to combat climate change and its impacts, aligning with SDG 13: Climate Action (UN 2020).

Landslide hazard assessment is a critical imperative, particularly in regions characterized by rugged terrain (Glade et al. 2005). Bandarban District in Bangladesh presents a unique and compelling scenario due to its intricate topography and heightened susceptibility to landslides, further exacerbated by monsoonal rainfall (Hossain et al. 2018). Understanding the complex dynamics of this region is essential for effective landslide risk assessment and mitigation.

Bandarban's vulnerability is amplified by rapid urbanization, which has the potential to significantly escalate landslide risks by altering land use and intensifying human activities in areas prone to landslides (Shahid et al. 2018).

The study of landslide hazards has a rich history, with various researchers contributing to our understanding. Dai and Lee (2002) conducted extensive research on landslide characteristics and slope instability, particularly using GIS techniques, which are invaluable for understanding landslide hazards. Van Den Eeckhaut et al. (2007) have contributed to new approaches for landslide susceptibility mapping in various soil conditions, which can be highly relevant in regions like Bandarban. Carrara et al. (1991) utilized GIS techniques (Huang et al. 2019a, b; Fekete and Sziebig 2017) and statistical models (Chimidi et al. 2017; Khan et al. 2017) for evaluating landslide hazard, offering valuable insights for similar regions.

Wieczorek et al. (2002) have studied landslides and surficial geology in the Seattle area, providing insights into landslide processes. Crozier and Glade (2005) offer a comprehensive overview of landslide hazard and risk concepts, enriching the understanding of these critical aspects. Petley (2012) has examined global patterns of loss of life from landslides, shedding light on the worldwide impact of these events. Guzzetti et al. (2012) have explored landslide inventory mapping, a valuable tool in landslide hazard assessment. Varnes (1978) has classified different types of slope movements and processes, contributing to the understanding of landslide mechanisms.

Sassa and Wang (2019a, b) discuss landslides in relation to environmental change, which is particularly relevant in the context of changing climate patterns. Crozier

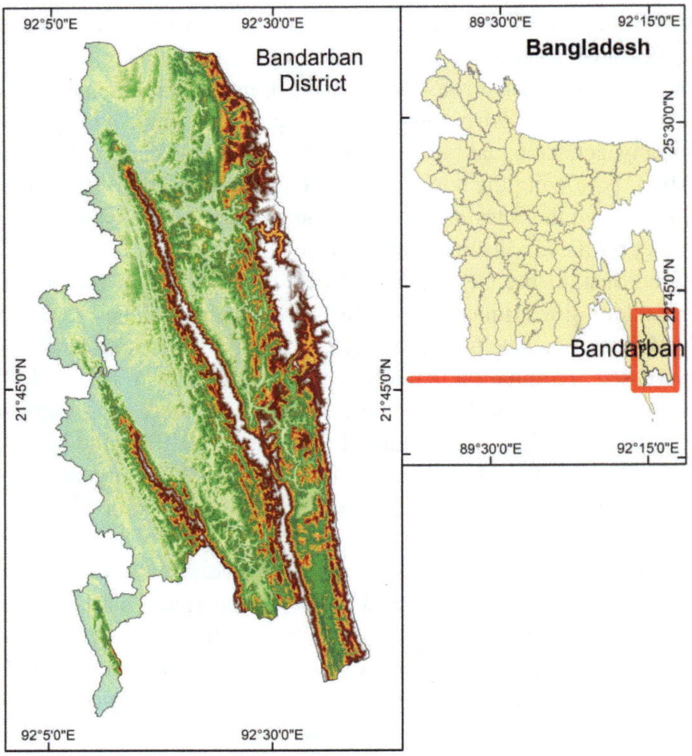

Fig. 26.1 Study area—Bandarban district

and Glade (2005) have extensively addressed landslide hazard and risk, providing essential insights into these critical aspects of geohazards.

In this context, geospatial modeling plays a pivotal role (Johnson et al. 2021) in estimating potential landslide hazards for effective management in the Bandarban District of Bangladesh (Fig. 26.1). These models contribute to proactive landslide risk assessment, allowing for better management strategies and sustainable development in the region.

26.1.1 Justification and Rationale of the Study

Recent inquiries highlight the susceptibility of Bandarban District to landslides and their impacts on livelihoods, which are exacerbated by various characteristics. Factors such as increasing rainfall, human interventions, and climate change contribute to the heightened risk of landslides. The study aligns with Sustainable Development Goal

(SDG) 13: Climate Action. Recognizing environmental factors is crucial for developing effective mitigation and adaptation strategies. Studies of this nature require attention to meet the SDGs by 2030 and overcome associated challenges.

Rapid urbanization and human intervention in the Bandarban District are identified as significant factors escalating landslide risks (Fig. 26.1). Understanding the interplay between human intervention and landslide susceptibility is crucial for devising effective mitigation strategies.

The study's rationale is grounded in the recurring, severe threats faced by Bandarban district, resulting in tragic loss of lives and valuable resources. Historical analysis, supported by previous research (Smith et al. 2019; Jones et al. 2020), reveals a troubling pattern of approximately 600 recorded landslide incidents, emphasizing the persistent and escalating nature of this issue over time (Gupta 2018).

Despite various landslide models proposed in prior studies (Brown et al. 2017; Patel et al. 2018), there is a notable deficiency in adequately addressing the critical aspect of landslide hazard estimation. This identified research gap, highlighted in recent literature (Johnson et al. 2021), underscores the urgent need for a comprehensive geospatial modeling approach to assess potential landslide hazards in Bandarban. By addressing this crucial gap and building upon the insights of previous researchers, the study aims to make a significant contribution to the better management of landslide risks in the district, ultimately safeguarding lives and invaluable resources for future generations.

26.1.2 Aims and Objectives of the Study

The study aims to identify potential landslide hazard areas and explore better management options to secure the livelihood of the region. The detailed objectives are as follows:

1. To map landslide hazards in the Bandarban District.
2. To estimate and quantify landslide hazards in the Bandarban District.
3. To identify better management options for securing livelihoods in the Bandarban District.

26.1.3 Landslide Risk Assessment- Techniques and Factors

Various techniques are employed for the comprehensive assessment of landslide hazard and risk. Geospatial methods (Johnson et al. 2021), such as remote sensing and Geographic Information System (GIS), provide valuable insights into terrain morphology, precipitation patterns, and geological attributes. Topographic and geological analyses involve assessing factors like slope, aspect, elevation, rock types, soil properties, and fault lines. Community-based approaches, including community engagement and the integration of indigenous knowledge (Fekete et al. 2020a, b), play a vital role, aligning with principles of resilience and enhancing the effectiveness of scientific modeling. Machine learning techniques (Lu et al. 2020), through comparative assessments of models, offer improved accuracy in predicting landslides by capturing complex contributing factors. Real-time data techniques, such as the utilization of rainfall data and the integration of remote sensing and GIS, contribute to timely responses and the development of advanced early warning systems. Together, these techniques form a holistic approach to landslide hazard and risk assessment, enabling proactive measures to mitigate risks and protect lives and infrastructure.

Geospatial techniques play a pivotal role in landslide modeling, offering insights into hazards and aiding mitigation strategies (Biswas and Ranjan 2021). Tien Bui et al. (2016) highlight the transformative impact of geospatial technology in assessing landslides, considering factors such as terrain morphology, precipitation patterns, land cover, and geological attributes. Remote sensing, emphasized by Shahid et al. (2018), is crucial for monitoring land use changes in Bandarban, providing insights into how rapid urbanization and land use alterations heighten landslide risks over time. The integration of GIS (Lee and Pradhan 2019) with geospatial techniques (Johnson et al. 2021), demonstrated by Carrara et al. (1991), is instrumental in landslide modeling.

The assessment of landslide hazards relies significantly on analyzing topographic and geological factors, a point underscored by various studies, including Chimidi et al. (2017). Arefi et al. (2015) stress the importance of topographic characteristics such as slope, aspect, and elevation in evaluating landslide susceptibility, particularly in the mountainous terrain of Bandarban. Geological surveys, highlighted by Anderson and Holcombe (2003), Van Den Eeckhaut et al. (2007), Zhou et al. (2018a, b), Crozier et al. (2010), and Huang et al. (2019a, b), provide essential insights into geological attributes affecting landslides, including rock types, soil properties, and fault lines, crucial for hazard assessment.

Community-based approaches are instrumental in mitigating landslide risks in regions like Bandarban, where indigenous communities have profound connections to their environment (Sultana et al. 2019; Das et al. 2018a, b). These community-based approaches not only empower local residents but also complement scientific modeling efforts, aligning with principles advocated by Cardona et al. (2012a, b, c). Involving local communities and tapping into their indigenous knowledge enhances overall resilience and contributes to more effective and context-specific landslide risk reduction strategies.

Machine learning has emerged as a promising tool for landslide prediction, significantly improving accuracy and effectiveness in hazard assessment (Ahmed et al. 2020a, b; Khan et al. 2020a, b). Comparative assessments of various machine learning models provide valuable insights into their effectiveness in capturing complex factors contributing to landslides. Researchers gain a more comprehensive understanding of landslide susceptibility (Chen et al. 2020a, b), enabling proactive measures to mitigate risks and protect lives and infrastructure.

Real-time rainfall data plays a pivotal role in assessing landslide hazards, enabling timely responses to potential disasters in regions like Bandarban (Rahman et al. 2018; Tariqul et al. 2020). The temporal relationship between rainfall events and landslides provides valuable insights into triggers, emphasizing the importance of real-time rainfall data in understanding and predicting landslides. Utilizing remote sensing and GIS technologies effectively monitors and assesses rainfall-induced landslides (Li et al. 2016), significantly improving the accuracy of prediction and early warning systems.

26.2 Causes of Landslides

Landslides, as identified by Varnes, can be attributed to a complex interplay of geological, topographical, climatic, and human-related factors. Geological conditions are foundational in landslide initiation, with bedrock type, soil composition, fault lines, and geological structures significantly influencing susceptibility to landslides (Varnes 1978). These inherent geological factors can weaken or destabilize slopes, increasing the risk of mass movements.

The topographical configuration of an area plays a pivotal role in landslide occurrence (Fig. 26.2). Steep slopes, particularly when subjected to heavy rainfall or seismic activity, are more prone to landslides (Varnes 1978). The combination of steep terrain and external triggers can create conditions conducive to slope failures.

Rainfall and water are critical elements in landslide causation. Excessive rainfall can saturate the soil, increasing its weight and reducing cohesion, thereby rendering it prone to sliding (Varnes 1978). The influence of precipitation, especially during intense or prolonged periods, cannot be understated in landslide susceptibility.

The presence or absence of vegetation also factors into landslide susceptibility. Vegetation stabilizes slopes by binding soil with its roots. Deforestation and

Police and emergency services observe the aftermath of the landslide in Bandarban on July 23, 2017. Source: Dhaka Tribune, 23 Jul 2017	Jul 03, 2018 A landslide in Bandarban following heavy rains has left three of a family killed in the hill tract district's Lama Upazila while another death was reported in a landslide near the district town. Source: Bangla Tribune, Jul 03, 2018

Fig. 26.2 The topographical configuration and landslides in Bandarban District

land clearing activities can significantly heighten landslide risk by removing this stabilizing factor (Varnes 1978).

Human activities contribute significantly to landslide hazards (Fig. 26.2). Construction, mining, excavation, and poor land use planning can alter the landscape and soil composition, exacerbating the potential for landslides (Varnes 1978). Inappropriate human-induced changes to the natural environment can increase the vulnerability of an area.

Additionally, seismic activity, particularly earthquakes, can induce landslides by shaking the ground and destabilizing slopes (Varnes 1978). Volcanic eruptions also pose a landslide risk due to the rapid movement of volcanic materials down slopes. Changes in climate patterns, including prolonged periods of rainfall or rapid snowmelt, can influence the occurrence of landslides, highlighting the relationship between climatic factors and slope stability (Varnes 1978).

Moreover, landslides may have underlying causes, such as the gradual weathering and erosion of rock over time, which weakens slopes until they become susceptible to sliding (Varnes 1978). These underlying causes may develop over extended periods, eventually leading to landslide events.

26.3 Landslides Classification

Landslides, diverse geological phenomena, demand nuanced study for effective risk assessment. This classification details rotational, translational, block slides, rockfalls, toppling landslides, earth flows, creep, and lateral slides. Debris flows and avalanches, marked by rapid downslope movements, underscore the dynamic and destructive nature of these events, guiding strategies for risk mitigation and landscape protection. The above types can be grouped as Movement Slides and Debris Slides.

Movement Type Landslides: Landslides encompass a diverse range of geological phenomena, each characterized by distinct mechanisms and features. To effectively assess landslide risks and devise appropriate mitigation strategies, it is crucial to differentiate between various types of landslides. Rotational landslides (Fig. 26.3 A) involve material movement in a concave, rotating manner along a curved slip surface, often occurring in cohesive soils or rock (Hungr et al. 2014). Translational landslides (Fig. 26.3 B) exhibit more planar movement, with material sliding along a distinct surface, common in both cohesive and non-cohesive soils (Hungr et al. 2014). Block slides (Fig. 26.3 C) entail the sudden descent of large, coherent rock masses along steep slopes, posing substantial hazards (Hungr et al. 2014). Rockfalls (Fig. 26.3 D) are characterized by the rapid free-fall or bouncing of individual rocks down steep slopes, triggered by factors like weathering, freeze–thaw cycles, or seismic activity (Hungr et al. 2014). Toppling landslides (Fig. 26.3 E) involve the forward rotation of rock or soil blocks around a pivot point at the base of a slope, commonly occurring on steep cliffs or rocky outcrops (Hungr et al. 2014). Earth flows (Fig. 26.3 H) involve the downslope movement of fine-grained, water-saturated materials, typically occurring in cohesive soils and influenced by factors like rainfall (Hungr et al. 2014). Creep (Fig. 26.3 I) is a slow, continuous downslope movement of soil and rock, barely perceptible over short periods but leading to significant landscape deformation over time driven by the force of gravity on gentle slopes (Cruden and Varnes 1996). Lateral slides (Fig. 26.3J) refer to the lateral movement of material along a nearly horizontal plane, often occurring in areas with weak bedding planes or structural weaknesses (Hungr et al. 2014).

Debris Flow and Landslides: Debris flows (Fig. 26.3F) represent a distinctive type of landslide characterized by fast-moving mixtures of water-saturated rock, soil, and debris, often resembling flowing mud. Triggered by events such as heavy rainfall, snowmelt, or volcanic activity, debris flows pose significant threats due to their rapid downhill movement (Hungr et al. 2014). Another hazardous type is the debris avalanche (Fig. 26.3G), involving the rapid downslope movement of a mixture of rock soil, vegetation, and other debris. Particularly destructive and dangerous, debris avalanches are prevalent in mountainous regions (Hungr et al. 2014). Both debris flow and debris avalanche events underscore the dynamic and often destructive nature of mass wasting events, emphasizing the importance of understanding and mitigating their associated risks, especially in vulnerable landscapes.

Each type of landslide presents its own set of challenges for assessment and mitigation. Factors such as slope characteristics, geological conditions, and triggering events determine the specific attributes and hazards associated with each type. Consequently, a comprehensive understanding of these landslide categories is crucial for effectively safeguarding communities and infrastructure in regions prone to landslides.

26.4 Materials and Method

26.4.1 Materials

Both primary and secondary data were utilized to analyze the issue. Secondary data sources included a 1:25,000 scale topographic map of SoB (Survey of Bangladesh), soil texture data from SRDI (Soil Resource Development Institute), geological compactness data from GSB (Geological Survey of Bangladesh), rainfall variability data from the Bangladesh Meteorological Department (BMD), and a land-use map created using Landsat satellite data. Additionally, Digital Elevation Model (DEM) data was collected from the USGS GloVis and considered in the analysis. Furthermore, household location and demographic data were gathered from HKI and various government sources, while roads and other spatial features were collected from the Roads and Highways Department (R&HD) and SoB. Finally, the primary dataset of the landslide inventory created by Ullah in 2018 (Ullah 2021) was used in this analysis.

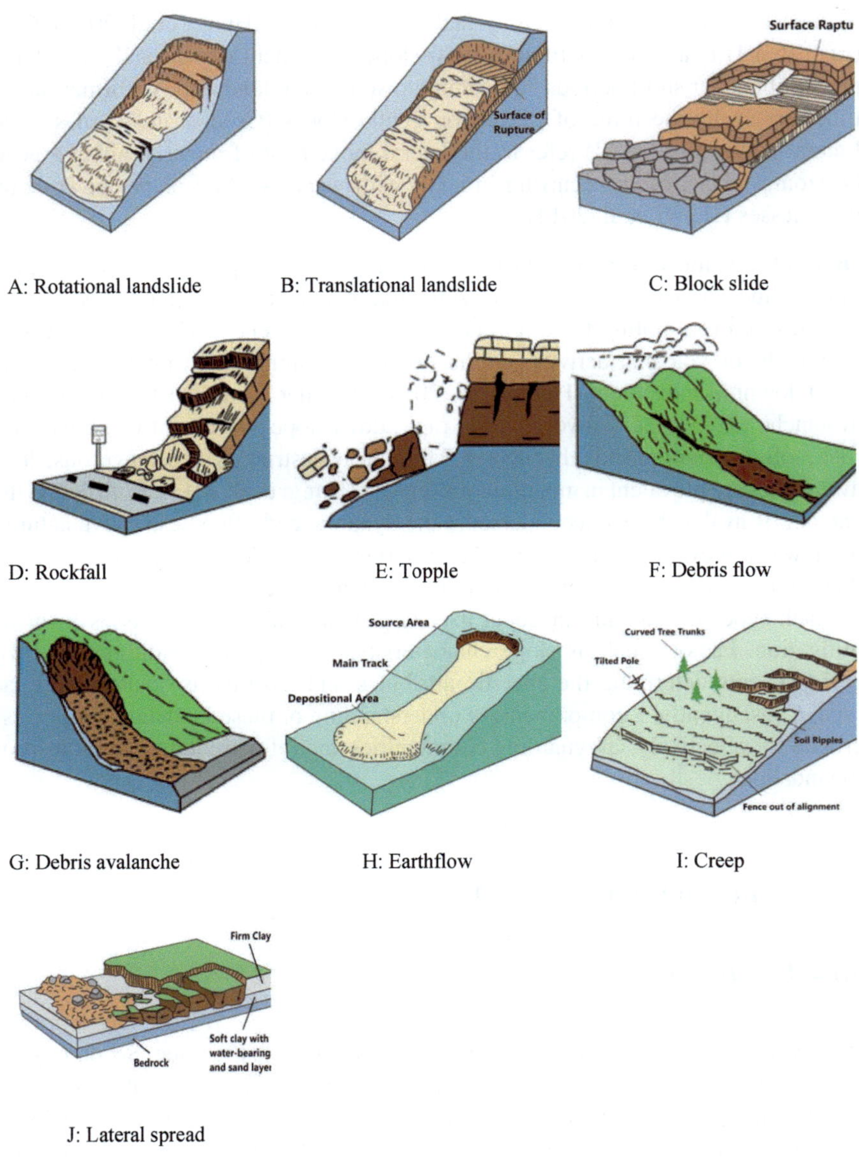

Fig. 26.3 Classification and types of landslide. *Source* Hungr et al. (2014), Varnes (1978)

26.4.2 Method

Statistical significance was tested for the relationship between the landslide inventory and factors including slope, elevation, distance to road features, distance to main streams, soil types, and geology. Similarly, the correlation between vegetation coverage and landslides, as well as settlement density and landslides, was examined to build the model. All variables were assigned weights on a scale of 1–5, indicating their level of influence (Fig. 26.4). The considered variables are described below:

Rainfall: Data on the average total rainfall over 50 years during the monsoon season were collected from the Bangladesh Meteorological Department (BMD). The data were analyzed and classified to understand the spatial variability of rainfall and to establish its relationship with historical landslides (Fig. 26.5A).

Geology: Geology forms the foundation for understanding soil compactness. Geological data were collected from the Bangladesh Geological Survey (GOB) and analyzed to investigate and estimate landslide hazards in Bandarban District (Fig. 26.5B).

Soil types: Soil type data for Bandarban District were collected from the Bangladesh Resource Development Institute (SRDI), investigated, and utilized to model and estimate landslide hazards in Bandarban District (Fig. 26.5C).

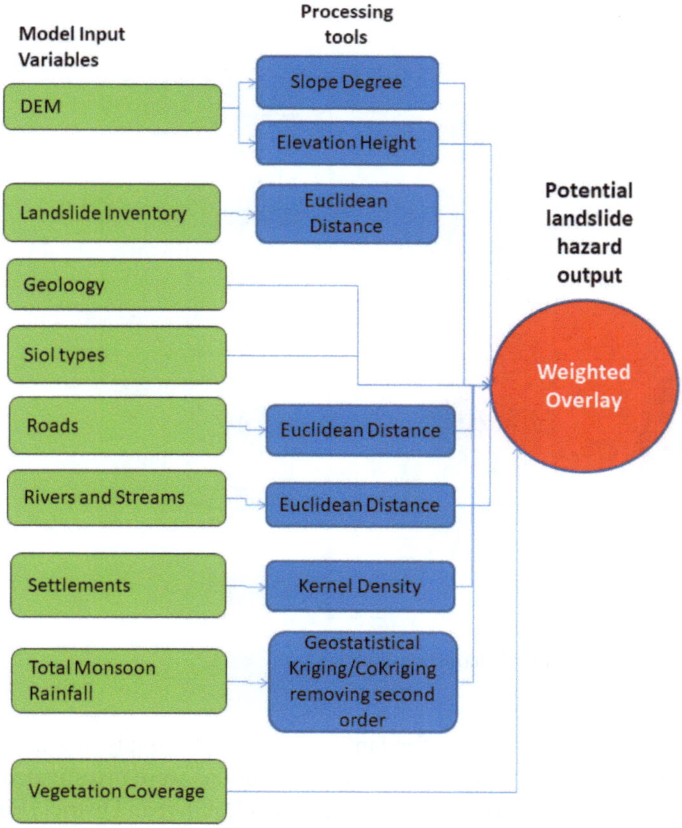

Fig. 26.4 Model of the potential landslide hazard estimate

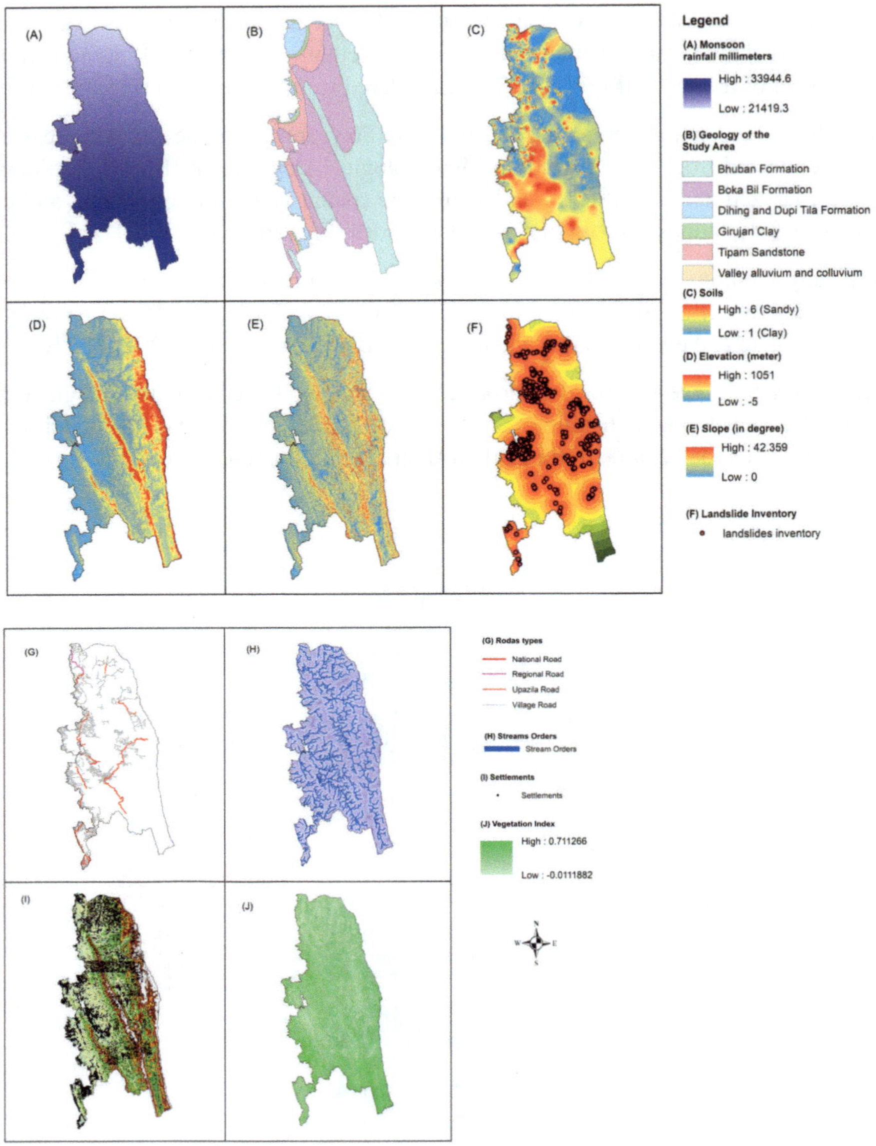

Fig. 26.5 Variables of landslide hazards estimation

DEM: The DEM elevation is a fundamental dataset for analyzing and modeling landslides. Consequently, historical landslide elevations were investigated using the SRTM 30-m DEM to estimate landslide hazards in Bandarban District (Fig. 26.5D).

Slope: DEM slope is another important variable for modeling landslide hazards. In the Bandarban data simulation, it is shown that approximately 75% of landslides occurred with slopes of 10 degrees or less (Fig. 26.5E). For this reason, landslide hazard estimation models consider this portion of the data very significant.

Landslide Inventory: The historical landslide inventory is the primary dataset for modeling landslide hazards. In 2018, a landslide inventory was created by Ullah (2021), which was collected and used as the main dataset to investigate the conditions of other data for use in the model to assess landslide hazards in Bandarban District (Fig. 26.5F).

Road types: Road construction is one of the main human interventions on the earth's surface that can contribute to landslides. In Bandarban District, various types of roads exist, including access roads, foot trails, national roads, regional roads, upazila roads, village roads, and district roads. These roadside areas are particularly vulnerable to landslides. Most landslides occur in proximity to national roads, regional roads, Upazila roads, and district roads (Fig. 26.5G). Landslide hazard estimation modeling has rationally considered these issues.

River/Streams: The human tendency is to settle near rivers and streams for water accessibility. Therefore, the Euclidean distance to the rivers and streams was calculated, and the data were input into the model to assess landslide hazards (Fig. 26.5H).

Settlement: Settlers in the Bandarban District are involved in jhum cultivation. Therefore, in the vicinity of settlements, the probability of landslide hazards is high. For this reason, settlement data were used with a similar algorithm for Euclidean distance and input into the model to calculate landslide hazards (Fig. 26.5I).

Vegetation Cover: Vegetation coverage is very important for protecting against landslides. Therefore, in the model for estimating landslide hazards, Bandarban District's vegetation coverage and Normalized Difference Vegetation Index (NDVI) values were taken into consideration (Fig. 26.5J).

26.5 Results and Discussion

26.5.1 Potential Landslide Hazard in the Bandarban District

The description of Bandarban District's Potential Landslide Hazard is presented on a scale of 1–5, with categories ranging from "Very Low Hazard" to "Very High Hazard." The preceding sections provide an overview of the potential landslide hazard conditions in the district (Table 26.1 and Fig. 26.6).

Table 26.1 Potential landslide hazards in the Bandarban District

Category of the landslide hazard	Area (sq. km)	Percept of landslide hazard
Very low hazard	604.16	13.39
Low hazard	1625.78	36.04
Moderate hazard	1777.18	39.39
High hazard	496.16	11.00
Very high hazard	8.32	0.18

Source Quantification of Landslide Hazard from Geospatial Model Output, 2023

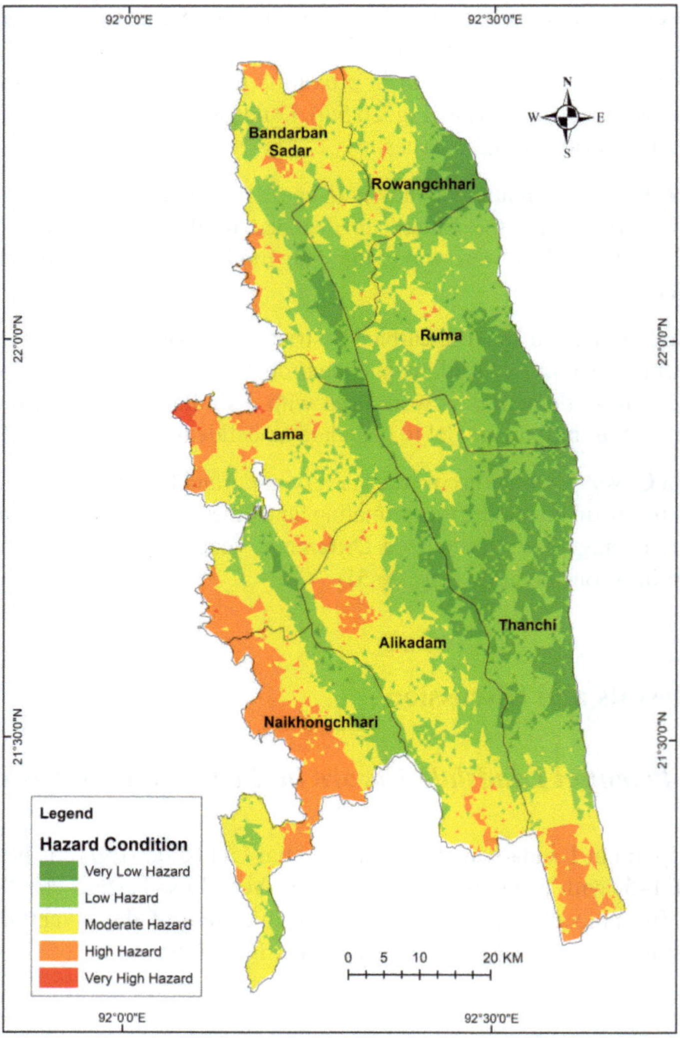

Fig. 26.6 Potential landslide hazard in the Bandarban District. *Source* Geospatial model result based on spatial data, 2023

26.5.1.1 Very Low Hazard

The largest and least hazardous category, "Very Low Hazard," encompasses 604.16 square kilometers, constituting 13.39% of the total hazard area, suggesting that it poses the least risk (Table 26.1 and Fig. 26.6). The largest category, "Very Low Hazard," can receive the least priority for mitigation efforts, as they pose the least risk. However, even in these areas, basic preparedness measures and education should be maintained to create a culture of safety and resilience.

26.5.1.2 Low Hazard

The "Low Hazard" areas, encompassing 36.04% of the total hazard area (1625.78 km^2), present a unique challenge in landslide hazard management (Table 26.1 and Fig. 26.6). While "Low Hazard" areas require lower priority for mitigation efforts due to their lower risk levels, it is essential to maintain awareness and preparedness in these regions. This ensures that communities are well-informed and can respond effectively in case of any unforeseen landslide events.

26.5.1.3 Moderate Hazard

The "Moderate Hazard" areas, constituting 39.39% of the total hazard area, represent a significant portion of the landscape with a moderate level of landslide risk (Table 26.1 and Fig. 26.6). While these areas do not require immediate, high-priority mitigation efforts like the "High Hazard" and "Very High Hazard" zones, they are not to be overlooked in landslide hazard management.

Mitigating risk in "Moderate Hazard" areas involves a multi-faceted approach. This may include the implementation of land-use regulations to control construction in vulnerable zones, reforestation initiatives to enhance slope stability, and community-based disaster risk reduction programs (Fekete et al. 2020a, b) to prepare residents for potential landslide events. These measures are crucial for reducing the overall risk landscape, enhancing community resilience, and contributing to a safer environment.

26.5.1.4 High Hazard

The "High Hazard" areas, covering approximately 11.00% of the total hazard area, present a significant level of risk in landslide hazard management (Table 26.1 and Fig. 26.6). Although they are smaller in size compared to the "Moderate Hazard" areas, the higher level of risk makes them a critical focus for immediate mitigation efforts.

Addressing the risk in "High Hazard" areas necessitates prompt and targeted actions. This may involve implementing engineering solutions, such as slope stabilization measures and erosion control, to reduce the potential impact of landslides. Improved land use planning, including zoning regulations that restrict construction in highly vulnerable areas, is essential to enhance safety.

Furthermore, proactive disaster preparedness and emergency response planning are vital to safeguard communities in "High Hazard" zones. This may include the development of early warning systems, evacuation plans, and community education initiatives to raise awareness and ensure that residents are well-prepared to respond in case of a landslide event.

26.5.1.5 Very High Hazard

The "Very High Hazard" area, as indicated in Table 26.1, represents the most critical and extreme level of landslide hazard within the Bandarban District. This area covers 8.32 square kilometers, constituting only 0.18% of the total hazard area, but it signifies a severe risk that demands immediate attention and mitigation measures (Fig. 26.6).

26.5.2 Potential Landslide Hazard in Upazilas of Bandarban District

In the Table 26.2 and Figs. 26.7 and 26.8, the hazard levels are categorized into five distinct classifications, each represented as a percentage: Very Low Hazard, Low Hazard, Moderate Hazard, High Hazard, and Very High Hazard. This information is invaluable for evaluating, managing, and prioritizing areas for risk mitigation and preparedness measures within the region. However, it's worth noting that data for the "Very High Hazard" category is missing for most Upazilas, suggesting the need for comprehensive hazard assessments (Table 26.2).

26.5.2.1 Potential Landslide Hazard in Alikadam Upazila

Alikadam Upazila displays a diverse range of hazard levels (Figs. 26.7 and 26.8). It is primarily characterized by Moderate Hazard, constituting 54.40% of the area, indicating a substantial degree of vulnerability. Low Hazard follows closely at 33.01%, highlighting the presence of certain risks. High Hazard and Very Low Hazard account for 6.74% and 5.85%, respectively. Unfortunately, no data is available for the Very High Hazard category in this Upazila (Table 26.2).

Table 26.2 Upazila-wise potential landslide hazard in the Bandarban District (in percent)

Potential Landslide hazard	Upazila Name						
	Alikadam	Bandarban	Lama	Naikhongchhari	Rowangchhari	Ruma	Thanchi
Very low hazard	5.85	7.30	4.31	0.52	14.43	22.47	29.25
Low hazard	33.01	27.11	22.00	22.46	41.31	58.67	43.26
Moderate hazard	54.40	54.70	54.65	40.99	43.35	18.54	19.38
High hazard	6.74	10.89	17.86	36.03	0.91	0.32	8.11
Very high hazard			1.19				

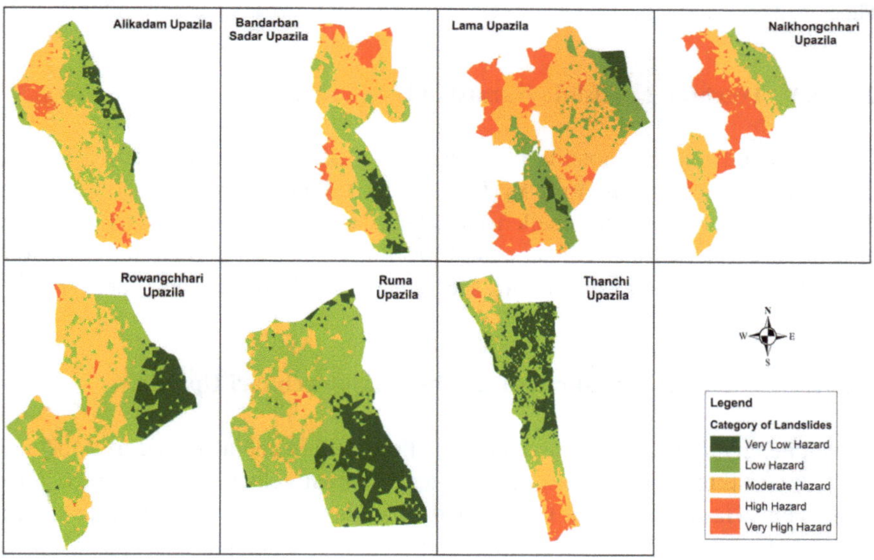

Fig. 26.7 Upazila-wise potential landslide hazard in the Bandarban District

26.5.2.2 Potential Landslide Hazard in Bandarban Upazila

Bandarban Upazila exhibits a broader range of hazard levels. Most notably, it registers a Moderate Hazard level of 54.70%, indicating significant vulnerability (Figs. 26.7 and 26.8). Low Hazard and High Hazard levels, at 27.11% and 10.89% respectively, suggest a moderate risk environment. Additionally, the presence of Very High Hazard at 1.19% implies the need for heightened preparedness measures, while Very Low Hazard stands at 7.30% (Table 26.2).

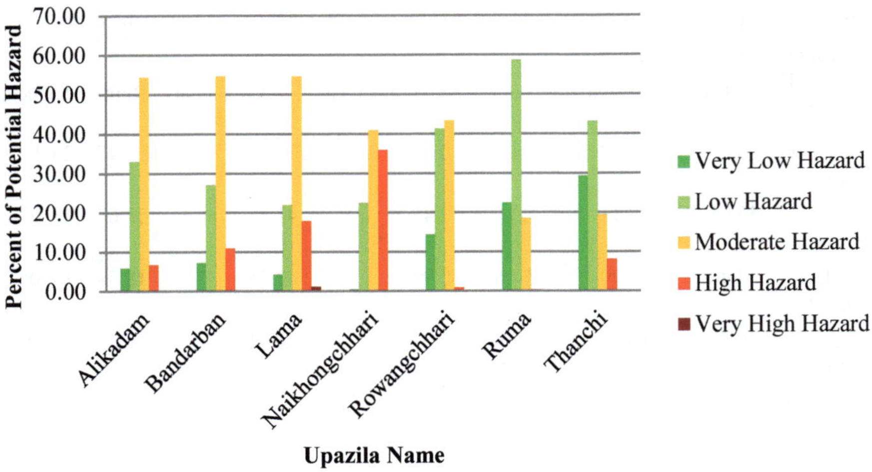

Fig. 26.8 Potential landslide hazard (in percent) in different Upazila of Bandarban District

26.5.2.3 Potential Landslide Hazard in Lama Upazila

Lama Upazila shows a notable prevalence of Moderate Hazard at 54.65%, indicating significant exposure to risks (Figs. 26.7 and 26.8). Low Hazard levels are at 22.00%, pointing to ongoing challenges that need attention. High Hazard levels stand at 17.86%, signifying areas of concern. Unfortunately, data for Very High Hazard is unavailable, while Very Low Hazard accounts for 4.31% of the area (Table 26.2).

26.5.2.4 Potential Landslide Hazard in Naikhongchhari Upazila

In Naikhongchhari Upazila, the majority of the area is characterized by Moderate Hazard, constituting 40.99%, suggesting a substantial vulnerability (Figs. 26.7 and 26.8). High Hazard levels at 36.03% and Low Hazard at 22.46% indicate significant risk factors. Very Low Hazard is at a mere 0.52%, emphasizing the need for preparedness. Regrettably, data for Very High Hazard is missing (Table 26.2).

26.5.2.5 Potential Landslide Hazard in Rowangchhari Upazila

Rowangchhari Upazila primarily falls under the categories of Low Hazard at 41.31% and Moderate Hazard at 43.35% (Figs. 26.7 and 26.8). It presents a different hazard profile compared to the other Upazilas, with Very Low Hazard at 14.43%. High Hazard is minimal at 0.91%. However, data for Very High Hazard is absent (Table 26.2).

26.5.2.6 Potential Landslide Hazard in Ruma Upazila

Ruma Upazila demonstrates a predominant Low Hazard level at 58.67%, indicating ongoing risk factors (Figs. 26.7 and 26.8). Very Low Hazard stands at 22.47%, suggesting areas with lower vulnerability. Moderate Hazard accounts for 18.54%, while High Hazard is minimal at 0.32%. Unfortunately, data for Very High Hazard is not available in this Upazila (Table 26.2).

26.5.2.7 Potential Landslide Hazard in Thanchi Upazila

Thanchi Upazila presents a mixed profile of hazard levels, with Low Hazard at 43.26%, Moderate Hazard at 19.38%, and High Hazard at 8.11% (Figs. 26.7 and 26.8). Notably, Very Low Hazard encompasses 29.25% of the area, pointing to areas with lower vulnerability. As with other Upazilas, data for Very High Hazard is absent (Table 26.2).

26.5.2.8 Potential Landslide Hazard and Management Issues

The analysis of "High Hazard" areas in various Upazilas reveals the need for prioritized risk mitigation. Naikhongchhari Upazila stands out with a substantial High Hazard level of 36.03% (Table 26.2), signifying an urgent requirement for interventions aimed at reducing the heightened risks in this region. Lama Upazila also presents a noteworthy High Hazard level at 17.86%, emphasizing the importance of allocating resources for risk reduction measures (Table 26.2). Bandarban Upazila, with a High Hazard level of 10.89%, requires vigilant preparedness and mitigation efforts to minimize potential risks (Figs. 26.7 and 26.8). While other Upazilas exhibit lower percentages of High Hazard, the data underscores the necessity to prioritize Naikhongchhari, Lama, and Bandarban Upazilas due to their significantly higher vulnerability to high-risk factors.

In evaluating the "Moderate Hazard" category across the Upazilas, it becomes evident that specific areas warrant prioritized attention and focused mitigation efforts. Notably, Alikadam, Bandarban, and Lama Upazilas exhibit substantial Moderate Hazard levels ranging from 54.40% to 54.70% (Table 26.2). These regions necessitate immediate action and resource allocation to reduce vulnerabilities and enhance preparedness. Thanchi and Rowangchhari Upazilas also require targeted efforts to address their specific risk factors, with Moderate Hazard levels at 19.38% and 43.35%, respectively (Figs. 26.7 and 26.8). Although Ruma Upazila has a slightly lower Moderate Hazard percentage at 18.54%, it remains essential to consider its vulnerabilities. Prioritizing these Upazilas contributes to a more resilient and secure region, given their higher degree of risk in the "Moderate Hazard" category, necessitating proactive measures to mitigate these risks effectively.

26.6 Summary, Conclusions and Recommendation

26.6.1 Summary and Conclusions

The geospatial model results for potential landslide hazard in the Bandarban District reveal a varied landscape of risk, ranging from "Very Low Hazard" to "Very High Hazard." This comprehensive analysis provides valuable insights into the distribution of landslide vulnerability, guiding policymakers, and stakeholders in formulating effective risk mitigation and preparedness strategies.

In summary, "Very Low Hazard" areas cover the largest portion, constituting 13.39% of the total hazard area. These areas, while posing the least risk, should not be neglected. Basic preparedness measures and educational initiatives must be maintained to foster a culture of safety and resilience. "Low Hazard" areas, covering 36.04% of the hazard area, require lower mitigation priority but should retain awareness and preparedness efforts to ensure communities can respond effectively to unforeseen landslide events.

"Moderate Hazard" areas, encompassing 39.39% of the hazard area, present a significant landscape with moderate landslide risk. A multi-faceted approach is needed to mitigate this risk, including land-use regulations, reforestation initiatives, and community-based disaster risk reduction programs.

"High Hazard" areas, although smaller in size at 11.00% of the hazard area, demand critical focus due to their higher risk. Addressing this risk requires immediate actions such as engineering solutions, land use planning, and proactive disaster preparedness.

The "Very High Hazard" area, covering just 0.18% of the hazard area, represents the most extreme level of landslide risk. Immediate attention and mitigation measures are imperative for this highly vulnerable region.

26.6.2 Recommendation

To effectively manage landslide hazards in both the Bandarban District and the hilly areas of Chittagong city, a comprehensive and tailored approach is recommended. The allocation of resources and prioritization of mitigation efforts should be based on risk levels, with a primary focus on "Very High Hazard" and "High Hazard" areas, followed by attention to "Moderate Hazard" regions. Public awareness and education programs must be maintained and strengthened across all hazard levels, fostering a culture of safety and resilience within the communities.

In vulnerable zones, especially in "Moderate Hazard" and "High Hazard" areas, the implementation of stringent land-use regulations is imperative to control construction and restrict development in highly vulnerable regions, thereby minimizing the potential impact of landslides. Concurrently, reforestation initiatives in "Moderate

Hazard" areas, along with the integration of vegetation and bio-engineering techniques, prove highly effective in stabilizing slopes and preventing erosion. The use of jute geo-textiles and the cultivation of vetiver grass are particularly recommended as bio-engineering measures to enhance slope stability and reduce the risk of landslides, contributing to long-term resilience.

Simultaneously, the integration of advanced technologies, such as the Finite Element Method for numerical modeling, should be embraced to ensure precise risk assessment and adaptive planning. This technological approach enhances the accuracy of landslide predictions and allows for more effective planning and response strategies. Furthermore, integrating a message chatbox to share hazard information with people in hilly areas can reduce landslide risks.

The establishment of a warning system based on rainfall data is essential for empowering communities to proactively respond to potential landslide risks. Early warnings provide critical time for evacuation and preparedness. Community involvement through awareness programs is equally vital for fostering resilience and preparedness at the grassroots level. Through these collective efforts, the comprehensive framework aims to reduce the potential impact of landslides and foster greater community resilience in the face of this escalating threat.

Due to the absence of a national database on historical landslides, creating a community-based historical landslide inventory and maintaining data for landslide modeling is lagging. The delay is attributed to the lack of data maintenance for such kinds of events, which is essential for modeling landslide hazards.

In the near future, updating historical landslide data and modeling the geospatial correlation between landslides and socio-economic variables can address one of the gaps. Another important area of research is geospatial real-time landslide modeling, which is crucial for proactive warning systems.

References

Ahmed S (2017) Urbanization and its impact on landslide vulnerability in the Chittagong hill tracts. Bangladesh Geogr Rev 29(1):45–58

Ahmed M, Verma AK, Mubeen R, Malik AS (2020a) Landslide susceptibility mapping using an integrated machine learning model. Geocarto Int 35(8):932–950

Ahmed R, Imran M, Islam M (2020) Landslide susceptibility assessment in Rangamati hill district of Bangladesh using machine learning algorithms. Geocarto Int 1–23

Alam E (2020) Landslide hazard knowledge, risk perception and preparedness in southeast Bangladesh. Sustainability 12(16):6305. https://doi.org/10.3390/su12166305

Anderson MG, Holcombe EA (2003) Slope stability and instability: introduction. In: Geomorphology of desert environments, pp 165–188. Springer

Arefi AM, Haghizadeh A, Alimardani R, Yavari A (2015) Landslide susceptibility mapping based on the weight of evidence model: a case study from the Azadbar region, Iran. Geomorphology 243:150–164

Bangladesh Bureau of Statistics (BBS) (2009a) Compendium of environment statistics of Bangladesh. Statistics Division, Ministry of Planning, Government of Bangladesh (GoB), Dhaka

Bangladesh Bureau of Statistics (BBS) (2009b) Statistical Yearbook of Bangladesh. Statistics Division, Ministry of Planning, Government of Bangladesh (GoB), Dhaka

Birkmann J, Cardona OD, Carreño ML, Barbat AH, Pelling M, Schneiderbauer S, ... Welle T (2013) Framing vulnerability, risk and societal responses: the MOVE framework. Natural Hazards 67(2):193–211

Biswas S, Ranjan R (2021) Geospatial techniques play a pivotal role in landslide modeling, offering insights into hazards and aiding mitigation strategies

Brammer H (2012) The physical geography of Bangladesh. The University Press Limited, Dhaka, Bangladesh

Brown R, White L, Green S (2017) A review of landslide models in the context of Bandarban district (Unpublished report)

Cardona OD, Barbat AH, Cerda A (2012a) A tool for the evaluation of city resilience: the case of earthquakes in Bogotá—Colombia. Habitat Int 36(1):111–122

Cardona OD, Van Aalst MK, Birkmann J, Fordham M, McGregor G, Perez R (2012b) Determinants of risk: exposure and vulnerability. In: Managing the risks of extreme events and disasters to advance climate change adaptation, pp 65–108. Cambridge University Press

Cardona OD, van Aalst MK, Birkmann J, Fordham M, McGregor G, Perez R, Pulwarty RS et al (2012c) Determinants of risk: exposure and vulnerability. In: Field CB, Barros V, Stocker TF, Qin D, Dokken DJ, Ebi KL, Mastrandrea MD et al (eds) Managing the risks of extreme events and disasters to advance climate change adaptation. A Special Report of Working Groups I and II of the Intergovernmental Panel on Climate Change, pp 65–108. Cambridge University Press, Cambridge, United Kingdom, and New York

Carrara A, Cardinali M, Guzzetti F, Reichenbach P (1991) Use of GIS technology in the prediction and monitoring of landslide hazard. Nat Hazards 3(1):77–90

Chen W, Pourghasemi HR, Naghibi SA, Koohikamali S, Zhang K (2020a) Landslide susceptibility mapping using J48 decision tree with AdaBoost, bagging, and rotation forest ensembles in the Guangchang area (China). Geocarto Int 35(11):1181–1210

Chen Y, Zhang X, Wei X, Liu S (2020) Landslide susceptibility mapping in a hilly area using remote sensing and GIS: a case study in Xingzi County, China. Geocarto Int 1–18

Chimidi G, Raghuvanshi TK, Suryabhagavan KV (2017) Landslide hazard evaluation and zonation in and around Gimbi town, western Ethiopia—a GIS-based statistical approach. Appl Geomat 9:219–236. FDOI. https://doi.org/10.1007/s12518-017-0195-x

Crozier MJ, Glade T (2005) Landslide hazard and risk: issues, concepts, and approach. Geomorphology 66(1–4):1–20

Crozier MJ, Preston NJ, Fullerton LR (2010) Landslide hazard and risk: issues, concepts and approach. In: Landslide hazards and risk, pp 3–27

Cruden DM, Varnes DJ (1996) Landslide types and processes. In: Landslides: investigation and mitigation (Transportation Research Board Special Report 247), pp 36–75. National Academy Press

Cui Y, Bao P, Xu C, Ma S, Zheng J, Fu G (2021) Landslides triggered by the Mw 6.6 Hokkaido, Japan earthquake on 6 September 2018: an updated inventory and retrospective hazard assessment. Earth Sci Inf 14(2):247–258. https://doi.org/10.1007/s12145-020-00544-8

Dai J, Lee CF (2002) Landslide characteristics and slope instability analysis using GIS, Lantau Island, Hong Kong. Geomorphology 42(3–4):213–228

Das I, Jana S, Saha S (2018a) Indigenous knowledge and practices for landslide vulnerability reduction in the Darjeeling Himalaya, India. Nat Hazards 94(3):1427–1445

Das S, Sarma JN, Goswami DC (2018b) Community-based landslide risk reduction: a case study from the hilly terrain of Karbi Anglong district. Assam, India

Fekete A, Sziebig G (2017) A geographic information system (GIS) based tool for flood risk mapping and disaster management. ISPRS Int J Geo Inf 6(7):209

Fekete A, Halász S, Hermann T (2020a) Indigenous knowledge and modern technology: spatial participatory mapping for disaster preparedness among the Ngäbe-Buglé in Panama. Geoforum 114:237–249

Fekete A, Moore D, Fiedrich F (2020b) Disaster risk reduction in Bangladesh: the role of indigenous and traditional knowledge in community-based approaches. Int J Disaster Risk Reduction 45:101501

Glade T (2003) Landslide occurrence as a response to land use change: a review of evidence from New Zealand. CATENA 51(3–4):297–314

Glade T, Anderson M, Crozier M (2005) Landslide hazard and risk: Issues, concepts, and approach. In: Landslide hazard and risk, pp 1–40. John Wiley & Sons

Guo C, Zhang H, Chen W, Lai CG, Yang L, Niu X (2019) Rainfall-triggered landslide susceptibility assessment using a coupled model with GIS and machine learning algorithm: a case study in a low-hilly area, China. Landslides 16(7):1365–1380

Gupta S (2018) Landslides in Bandarban: an ongoing issue. Environ Disaster Rev 4(2):87–101

Guzzetti F, Gariano SL, Peruccacci S, Brunetti MT, Marchesini I, Rossi M, Melillo M (2020) Geographical landslide early warning systems. Earth-Sci Rev 200:102973

Guzzetti F, Mondini AC, Cardinali M, Fiorucci F, Santangelo M, Chang KT (2012) Landslide inventory maps: new tools for an old problem. Earth Sci Rev 112(1–2):42–66

Hasnat GNT, Kabir MA, Hossain MA (2018) Major environmental issues and problems of South Asia, particularly Bangladesh. In: Hussain CM (ed) Handbook of environmental materials management. Springer International Publishing AG. https://doi.org/10.1007/978-3-319-58538-3_7-1

Hassan R, Scholes R, Ash N (eds) (2005) Ecosystems and human well-being: current state and trends. Volume 1: Findings of the condition and trends. In: Working Group of the Millennium Ecosystem Assessment. Island Press, Washington

Hossain MS, Tarekul Islam M, Rahman MS (2018) Landslide susceptibility mapping using GIS and remote sensing: a case study of Bandarban, Bangladesh. Geomat Nat Haz Risk 9(1):1529–1546

Huang RQ, Huang JH, Lin W, Chen W (2019a) Mechanisms and thresholds of large-scale landslides in catastrophic sedimentary basins: the Longjia gully landslide, China. Eng Geol 250:29–40

Huang W, Chen Y, Yi S, Lin Y, Li T, Du Y (2019b) GIS-based landslide susceptibility assessment: a case study in the Zhaotong region, China. Geocarto Int 34(2):202–218

Huddleston B, Ataman E (2003) Towards a GIS-based analysis of mountain environments and populations, Environment and Natural Resources Working Paper No. 10. Food and Agriculture Organization of the United Nations, Rome

Hungr O, Leroueil S, Picarelli L (2014) The Varnes classification of landslide types, an update. Landslides 11(2):167–194

Johnson E, Lee A, Clark D (2021) Identifying research gaps in landslide hazard estimation: a literature review. Int J Geospat Res 8(2):104–122

Jones M, Williams T, Harris L (2020) Patterns of landslide incidents in Bandarban district (Unpublished research paper)

Kelman I (2020) Disaster by choice: how our actions turn natural hazards into catastrophes. Oxford University Press, UK

Khan FH (1991) Geology of Bangladesh. University Press Limited, Dhaka

Khan IH, Lee S, Oh HJ, Choi JH (2017) Landslide susceptibility assessment using GIS-based statistical models and remote sensing data in tropical monsoonal hilly area: a case study of the Nilgiris District, India. Remote Sens 9(6):543

Khan MI, Pradhan B, Lee S (2020a) Landslide susceptibility assessment and prediction: a comparative evaluation of support vector machine, random forest, and extreme learning machine models. Remote Sens 12(3):512

Khan MRH, Ahmed B, Islam MS (2020b) Landslide susceptibility assessment in Bandarban District of Bangladesh using DEM-based geospatial techniques. Geoenviron Disasters 7(1):1–16

Lee S, Pradhan B (2019) Landslide susceptibility assessment using a GIS-based approach: a case study of the Gyeongju area, Korea. Geocarto Int 34(6):657–677

Li L, Zhang Y, Ding L, Yan E (2016) Monitoring rainfall-induced landslides using remote sensing data: a case study in the Three Gorges area. Landslides 13(6):1259–1272

Lu X, Damm B, Trani L, Wieczorek GF (2020) Applications of machine learning methods in landslide susceptibility assessment: a review. Comput Geosci 139:104477
Mannan A (2002) Stratigraphic evolution and geochemistry of the Neogene Burma Group, Burma Basin, Sleet, Bangladesh. (Doctoral dissertation, Faculty of Science, University of Oulu)
Patel K, Miller J, Garcia M (2018) Landslide models for hazard estimation: a review and analysis. Geohazards J 6(3):149–166
Petley DN (2012) Global patterns of loss of life from landslides. Geology 40(10):927–930
Rabby YW, Li Y (2019) An integrated approach to map landslides in Chittagong Hilly Areas, Bangladesh, using Google Earth and field mapping. Landslides 16(3):633–645
Rahman MS, Dhar A, Akhter SH (2018) Landslide hazard assessment and mapping using GIS and remote sensing: a study in the Cox's Bazar area, southeast Bangladesh. Geoenviron Disasters 5(1):20
Sassa K, Wang F (2019) Landslides in relation to environmental change. In: Handbook of engineering geomorphology, pp 673–692. Springer
Sassa K, Wang F (2019) Landslides in relation to environmental change: relevance in the context of changing climate patterns
Shahid S, Behrawan H, Yusoff MK (2018) Impact of rapid urbanization on landslide susceptibility in Bandarban, Bangladesh. Nat Hazards 92(1):213–229
Smith A, Johnson P, Davis R (2019) Landslide incidents in Bandarban district: a historical analysis (Unpublished report)
Sultana MS, Nahar S, Stojanovic J (2019) Indigenous knowledge and community-based approaches for landslide risk reduction in the Chittagong Hill tracts, Bangladesh. Geoenviron Disasters 6(1):7
Tariqul IM, Rahman MS, Haque A (2020) Landslide hazard zonation using multi-criteria evaluation and GIS: a case study of Bandarban district in Bangladesh. Geocarto Int 1–23
Tien Bui D, Hoang ND, Revhaug I, Dick ØB (2016) Landslide susceptibility assessment in the Hoa Binh province of Vietnam: a comparison of the Levenberg-Marquardt and quasi-newton optimization algorithms in training adaptive neuro-fuzzy inference system. Geomorphology 259:105–118
Ullah MS (2021) Geospatial modeling of landslide vulnerability and simulating spatial correlation with associated factors in Bandarban District. The Dhaka Univ J Earth Environ Scie 8(2):51–66. Faculty of Earth and Environmental Sciences, University of Dhaka
UNDP (2015) Sustainable development goals. https://www.undp.org/sustainable-development-goals
UNDRR (2017) UNISDR terminology on disaster risk reduction—2009. The United Nations Office for Disaster Risk Reduction (UNDRR). https://www.undrr.org/terminology. Accessed 22 Aug 2020
UNEP-WCMC (2002) Mountain watch: environmental change and sustainable development in mountains. United Nations Environmental Programme (UNEP), Nairobi
UNHCR (2020) Rohingya refugee influx and its impact on deforestation in the Chittagong hill tracts. UNHCR Rep Environ Impact 7:1–20
United Nations (2020) Sustainable development goal 13: climate action. https://bangladesh.un.org/en/sdgs/13. Accessed 15 Dec 2023
UNOCHA (2017, June 20) Bangladesh: report on landslides. United Nations Resident Coordinators Office, Bangladesh. United Nations Office for the Coordination of Humanitarian Affairs (UNOCHA). https://www.humanitarianresponse.info/en/operations/bangladesh/document/bangladesh-report-landslides-20-june-2017. Accessed 21 Aug 2020
Van Den Eeckhaut M, Vanwalleghem T, Poesen J, Govers G, Verstraeten G, Demaree G (2007) New trends in studying soil erosion in European landscapes. The Example of Belgium. Geomorphology 91(1–2):47–56
Van Den Eeckhaut M, Poesen J, Govers G (2007) New approaches for landslide susceptibility mapping: a case study in Bandarban, Bangladesh. Geomorphology 90(1–2):145–161

van Westen CJ, Van Asch TW, Soeters R (2006) Landslide hazard and risk zonation—why is it still so difficult? Bull Eng Geol Env 65(2):167–184

Varnes DJ (1978) Slope movement types and processes. In: Landslides: analysis and control, pp 11–33. National Academy Press

Wieczorek GF, Jibson RW, Schulz WH (2002) Landslides and surficial geology in the Seattle area: site investigation techniques, slope stability analysis, and the inventory of landslides. In: Engineering geology in Washington: Volume II, Washington Division of Geology and Earth Resources Information Circular, vol 88, pp 133–148

Wisner B, Blaikie P, Cannon T, Davis I (2004) At risk: natural hazards, people's vulnerability and disasters. Routledge, Oxon

Zaman S, Sammonds P, Ahmed B, Rahman T (2020) Disaster risk reduction in conflict contexts: lessons learned from the lived experiences of Rohingya refugees in Cox's Bazar, Bangladesh. In: J Disaster Risk Reduction 50:101694

Zhou CH, Yin KL, Zhou TY, Lu SX (2018a) Rockfall susceptibility assessment and its verification using field investigation, remote sensing, and GIS in the Three Gorges area, China. Landslides 15(11):2235–2251

Zhou W, Zhang M, Wang D, Zhao T, Li G, Huang D (2018b) Landslide susceptibility mapping in Mizhi County (China) using integrated machine learning models. Geomorphology 305:59–75

Chapter 27
Correlating Circular Failure Charts, Limit Equilibrium and Finite Element Method Based Safety Factor for Circular Failure Type Landslides

Harsh Varshney, Tariq Siddique, Atif Ahamad, and Wali Akhtar

Abstract Landslides in the form of circular failures often occur in homogeneous geomaterials like soil, debris and highly weathered or fragile rocks. In slope engineering practices, Factor of Safety (FoS) is commonly determined using Circular Failure Charts (CFCs), Limit Equilibrium Methods (LEM) and Finite Element Methods (FEM). The present study is an attempt to assess the disparities and interrelationships in FoS using CFCs, LEM and FEM. FoS calculations were performed manually and using an automated Windows-based open-source tool FS Calculator 1.0. Pertinent geomechanical (cohesion, friction, unit weight) and geometrical properties (slope angle and height) of fifty-one slopes from nine case records were considered in the present analysis. A linear correlation occurs in FoS using CFC and LEM i.e., $\text{FoS}(\text{CFC}_{\text{FS}}) = 2.7592 \times \text{FoS}(\text{LEM}) - 2.2082$, with the coefficient of determination of 0.99. However, the linear relationship between SRF and FoS (LEM) is $\text{FoS}(\text{CFC}_{\text{FS}}) = 0.889 \times \text{SRF}(\text{FEM}) + 0.0438$ having coefficient of determination of 0.63 only. It connotes that FoS using CFCs and LEM is relatively well in comparison to FoS from CFCs and FEM. It is also evident that CFCs provide relatively more conservative FoS than FEM and LEM. Among all three proxies, FEM relies on a wider set of geomechanical parameters and evaluates stability under static and dynamic conditions with capability of applying wider range of remedial measures. Since, most of the FEM tools are commercial, open-source CFCs based FS Calculator may be used and proposed relationships may be utilized as correction factor.

Keywords Landslides · Circular failure charts (CFCs) · Limit equilibrium methods (LEM) · Finite element methods (FEM) · Factor of safety (FoS)

H. Varshney · T. Siddique (✉) · A. Ahamad · W. Akhtar
Department of Geology, Aligarh Muslim University, Aligarh 202002, India
e-mail: tariqsiddiqueiitr@gmail.com

G. K. Panda et al. (eds.), *Landslide: Susceptibility, Risk Assessment and Sustainability*, Advances in Natural and Technological Hazards Research 52,
https://doi.org/10.1007/978-3-031-56591-5_27

27.1 Introduction

Landslide is a significant problem in hilly tectonically active regions such as the Himalayas. Landslide events pose a serious threat to life and infrastructure and escalate major issues of geoenvironmental imbalance. Landslide is a gravity-driven movement of Slope-forming material involving downward and outward motion, including sliding, falling, flowing, or any combination of these processes (Cruden and Varnes 1996; Siddque and Pradhan 2018). Ever rising human interventions are one of the prime causes to escalate frequency of landslides (Wan et al. 2010; Rahman et al. 2011; Ansari et al. 2021; Siddique et al. 2022). Owing to the haste in which the road networks are being aligned, the innate geological and geotechnical factors are ignored during the planning and designing phases. Geological, geotechnical and geomechanical properties of slope-forming material significantly influence the stability of slopes. Several dynamic factors, including earthquakes, precipitation, cloudburst events can trigger frequent landslides (Petley et al. 2007; Kumar et al. 2017; Pradhan et al. 2018; Koley et al. 2022; Khali et al. 2023). From monsoon to dry season, a significant fluctuation in groundwater can be witnessed, which causes rampant landslides during and post monsoon (Dahal et al. 2009; Panda et al. 2023). To avert disasters, it is vital to conduct geotechnical assessments of vulnerable sections. For such geotechnical appraisal, a thorough understanding of geological and geotechnical conditions is required to select the most appropriate and viable method. In heterogeneous rock mass, failures dominantly occur along weak planes like bedding, joints, foliation and faults (Rao 2020; Akhtar et al. 2023). These weak planes are the zones of least strength and offer sliding of rock mass. In case of homogeneous mass like soil slopes, debris slopes and shattered rocks, a well-defined structural failure pattern no longer exists and slope fails along the zone of least resistance forming an arched or circular surface (Hoek and Bray 1981; Wyllie 2018). Generally, these failures initiate along tensional cracks at the crown portion of slopes. These cracks allow the percolation of water into the slope forming material, which cause reduction in shear strength of the material. Different approaches are being used to assess the stability for circular failure type landslides using deterministic and sensitivity analysis. Factor of Safety (FoS) is an index of prime importance for such geotechnical evaluation of slopes and widely assessed by LEM, CFCs and FEM. LEM is also known as method of slices in which the slope is divided into number of slices or columns of equal width and forces acting on these slices are evaluated and integrated to determine resistive and driving forces (Hoek and Bray 1981; Pantelidis and Griffiths 2013; Wyllie 2018; Ahmad et al. 2023). LEM based CFCs is a semi-empirical conventional technique to determine FoS values of slopes. The major limitation of LEM is the assumption of failure surface, which may pose slight subjectivity in outcomes. LEM and CFCs are not reliable tools for slopes having complex internal deformations (Hoek and Bray 1981; Siddque and Pradhan 2018). To cope with these assumptions of failure surface and complex geotechnical behavior of slopes, FEM can be utilized, which is a robust technique and provide static and dynamic slope stability analyses (Singh et al. 2013; Siddque and Pradhan 2018; Pradhan et al. 2019; Komadja et al. 2021).

Slope stability analyses using FEM is a time-consuming process and require extensive geotechnical and geomechanical input data to run simulations. Moreover, all widely used FEM based tools are commercial. Contrary, the use of CFCs is simple, quick and provides appreciably reliable results. In this study, the outcomes of a recently developed Windows-based tool freeware tool 'FS Calculator 1.0' by Aalam et al. (2023) have been compared with the advanced FEM. Utilizing CFCs, LEM and FEM, the present study also attempted to evaluate the degree of relationship in FoS obtained by different proxies.

27.2 Study Area

In the present study, nine case studies, including fifty-one debris or soil slopes from various parts of the world, were collected from the literature and examined (Table 27.1). Six studies are from the Himalayan region and two from Indonesia and one from Tamil Nadu. Most of these studies were conducted on debris slopes and highly fractured and jointed metamorphic rocks. For the present analyses, the required geotechnical data was taken from these geotechnical case records. A brief geological background of each section has been discussed below.

Table 27.1 Case studies taken into consideration

Authors and year	Study area	General geology of study area
Chakraborty and Anbalagan (2003)	Uttarkashi-Bhatwari road section of the Uttarkashi Gangotri route	Garhwal Lesser and Higher Himalayas
Anbalagan and Parida (2013)	Harmony Village Karnaprayag Gwaldam Road, Chamoli	Garhwal Lesser Himalayas
Anbalagan et al. (2008)	Harmony Village Karnaprayag Gwaldam Road, Chamoli	Garhwal Lesser Himalayas
Dudeja et al. (2017)	Yamuna bridge to Judo village	Southern Lesser Himalayan division of Yamuna Valley
Kannan et al. (2011)	Bodi-Bodimettu Ghat section, Theni district Tamilnadu	Charnokites and Migmatite
Singh and Kumar (2021)	Chamba District of Himachal Pradesh NH 154A	Chamba Formation, Himalayas
Luden et al. (2021)	Bolango Ulu Dam, Bone Bolango Regency Gorontalo Province, Indonesia	Bone Diorite unit
Dinata et al. (2021)	Tanju dam, Dompu, west Nusa Tenggara	Volcanic rocks of Quaternary age and alluvial deposit
Siddque and Pradhan (2018)	NH-58, Rishikesh and Devprayag, Uttarakhand	Kumaun Lesser Himalayas

According to Chakraborty and Anbalagan (2008) Uttarkashi-Bhatwari road section lies in Garhwal Lesser Himalaya and Higher Himalaya. The considered section is dissected by several thrust faults including Main Central Thrust (MCT), Vaikrita Thrust (VT), Krol Thrust (KT) and Ramgarh Thrust (RT). Munsiari Formation and Nagthat-Berinag Formation are two major lithounits found in the study area, which are the part of Almora Group and Jaunsar Group respectively. Many irregularly oriented faults are also marked in the area. Rocks found in the area are highly folded having a general trend of NW-SE.

The Harmony village is a part of Chamoli district of Uttarakhand state, India that lies in the Survey of India toposheet number 53 J/8. The topography of the slide area is rugged and undulating and geologically falls in the Garhwal Lesser Himalaya. As per Anbalagan et al. (2008), Anbalagan and Parida (2013) the area has medium to low grade metamorphic rocks, found as nappes/klippe overlying metamorphosed to unmetamorphosed sedimentary rocks. The age of metamorphic rocks is estimated around to be Proterozoic.

Dudeja et al. (2017) conducted geotechnical study along Yamunotri pilgrimage route that lies in Tehri Garhwal district of Uttarakhand. Geologically the area lie in the Southern Lesser Himalayas and rocks belongs to Jaunsur Group, Mussoorie Group. Lithologically, it consists of phyllite and phyllitic slates with inter-bedded bands of dirty brown fine-grained quartzites. Slates and phyllites are thinly foliated, folded and friable.

According to Kannan et al. (2011) Bodi-Bodimettu Ghat section lies in the Western Ghats. The East flowing Puliuttu River is the main stream in the region. Geologically, the area is comprised of high-grade metamorphic rocks, including quartzite, charnockite, pyroxene granulite and migmatite. These metamorphic rocks belong to Khondalite group, Charnokite group and Migmetite complex. High degree of weathering has been profoundly observed in the area and marked by the presence of joints and well-developed foliations.

Singh and Kumar (2021) studied the slopes roughly parallel to Ravi River and lies in Chamba district of Himachal Pradesh. The rocks of the study area belong to Late Proterozoic to Triassic age. The region mainly comprised of metamorphic rocks of Bhalai Formation, Chamba Formation, Manjir Formation, Batal Formation, Katarigali Formation, and Mandi Granites, whereas slate, quartzite and gneiss are locally present along the selected stretch along NH-154A.

The geotechnical study by Luden et al. (2021) is concerned about inlet and outlet sections of diversion tunnel of Bolango Ulu Dam of Gorontalo Province, Indonesia. The study area mainly comprises of diorite and alluvial deposits of Late Miocene age. Normal faults and shear joints are main geological features.

According to Dinata et al. (2021), the study area is located at the drainage tunnel of Tanju Dam of West Nusa Tenggara. The drainage tunnel is situated at a ridge extending 2.2 km in NE-SW direction. The area is composed of volcanic rock products and alluvial deposits. The volcanic rocks belongs to Quaternary age, which majorly comprises breccias lava and tuffs.

Siddque and Pradhan (2018) performed stability and sensitivity analysis on road cut debris slopes along NH-58 from Rishikesh to Devprayag, Uttarakhand. Geologically, it is a part of Kumaun Lesser Himalaya, which mainly comprises rocks of Sirmur Group, Jaunsar Group, Mussorie Group, Damatha Group and Ramgarh Group. The road section is dissected by several faults and thrust, including thrust near Shivpuri town. Recurrent nature of landslide can be witnessed along such disturbed zones.

27.3 Materials and Methods

The term landslide encapsulates different types of land movement, including falls, creep, spreads, flows and other intricate movements. However, landslide mainly refers to the sliding motion of slope-forming material. "Mass wastage" is a more accurate term to describe all kinds of slope movements. Slope stability analysis is the process through which the stability of a slope can be estimated. This analysis compares the shear stresses with the shear strength of the slope-forming mass along the zone of least resistance (Dudeja et al. 2017). Numerous parameters, such as geotechnical properties of geomaterials, pore pressure distribution, mineralogical and chemical properties are taken into consideration. Surface conditions such as tension cracks and drainage are considered to evaluate the stability of the slope. Usually, circular failure occurs when the particle size in the slope-forming material is very small compared to the slope size and there is a lack of interlocking between the particles (Hoek and Bray 1981). Debris or soil slopes, highly shattered and altered rock may offer circular failures. According to Hoek and Bray (1981), circular failures occur in homogeneous mass in which mechanical properties do not vary greatly with different directions of loading. In such cases, CFCs can be a convenient and helpful tool to evaluate the stability of the slope. CFCs were generated by Hewlett-Packard 9100 B Calculator with graph plotting facilities (Hoek and Bray 1981). These charts consider Mohr-Coulomb failure criteria.

The Factor of Safety (FoS) is the ratio of resisting and driving forces. FoS is one of the most frequently applied methods of slope design for civil engineering, which can be employed in various geological conditions (Wyllie 2018). If the FoS is less than 1, the slope is precarious and susceptible to failure. Whereas, if FoS is from 1 to 1.2, the slope is almost stable but vulnerable to failure under dynamic triggers. Greater than 1.3 is a reasonable threshold for the stability of slopes. Shear Strength Reduction (SSR) approach is most widely applied for computing the FoS. SRF is an equivalent index to FoS (Gover and Hammah 2013; Hammah et al. 2005; Sharma et al. 2017). SSR involves the reduction of shear strength parameters of slope-forming material till it fails (Kainthola et al. 2012; Siddque and Pradhan 2018). The basic principle of SSR in finite element analysis is to reduce cohesion and friction until shear failure occur (Matsui and San 1992). Many researchers applied SSR technique in stability analysis of slopes using FEM tools (Gover and Hammah 2013; Singh et al. 2013;

Gupta et al. 2016; Pradhan et al. 2019; Solanki et al. 2019; Komadja et al. 2021; Roul et al. 2021; Das et al. 2023; Thakur et al. 2023).

The Limit Equilibrium Method (LEM) is the method to analyze the stability of a potentially failing mass by comparing driving forces to the resisting forces along the plane of failure, which allows for determining whether the slope is in a state of equilibrium. LEM is a widely employed concept to compute the stability of debris or soil slope (Siddque and Pradhan 2018). LEM utilizes the Mohr-Coulomb expression to calculate the shear strength of the surface of failure. In this method, the potential slide mass is divided into 'n' number of small slices. Limit state conditions are assumed and equilibrium is analyzed by applying different techniques. Many workers have contributed to the development of LEM-based methods. Fellenius proposed the first method of LEM, referred as Ordinary or the Swedish method in 1936. Later, Janbu (1954, 1955, 1973), Bishop (1955) simplified; Bishop (1955) rigorous; Lowe and Karafiath (1960), Morgenstern-Price (1965), Spencer (1967, 1973); Corps of Engineer (1982) were introduced. Chugh (1986) devised a method of general limit equilibrium as an extension of Spencer and Morgenstern-Price method, which considers force and momentum equilibrium simultaneously (Sharma 2002). Detailed note on disparities, failure surface, inter-slice forces and momentum equilibrium considerations in different LEM are highlighted in Table 27.2.

FoS using CFCs is a chart-based semi-empirical technique. It relies on the limit equilibrium approach that offers a quick assessment of FoS (Luden et al. 2021; Wyllie 2018). Circular failure charts were devised by Hoek and Bray (1981) and are being widely applied for quick deterministic analysis. Recently, Aalam et al. (2023) introduced a sensitivity analysis function in CFCs. This method requires a minimal amount of laboratory testing and field data, making it a cost-effective and time-saving efficient approach. CFCs are a series of five charts primarily based on different groundwater conditions. Other parameters are cohesion (c), friction angle (φ), slope height (H) and unit weight (ϒ), to calculate a dimensionless ratio, which is drawn on a suitable chart according to the existing groundwater conditions. A Windows-based computer tool 'FS Calculator 1.0' has been employed to determine the FoS. This software uses the CFCs to compute FoS. It is open-source software that can compute FoS and perform sensitivity analysis in a few seconds. The tool has been developed in Python programming language with the Tkinter library. Performing sensitivity analysis manually is a tedious task and can have significant errors in the results. To resolve this limitation, 'FS Calculator 1.0' offers single and multi-parameter sensitivity analysis functions by implementing one-at-a-time (OAT) approach (Aalam et al. 2023). According to the developers, the average percent error in computing the FoS using the software is 0.37%, making it a reliable tool for FoS determination. The weightage of individual parameters influencing the stability of the slope can also be determined by performing sensitivity analysis.

FEM is one of the widely used practical approaches based on numerical modeling that can be used to evaluate the stability of slopes. According to Krahn (2003), the lack of a fundamental physical stress-strain relationship in LEM is one of the major limitations associated with LEM, due to which these methods are unable to compute a realistic stress distribution. Whereas, FEM has an advantage over conventional limit

Table 27.2 Various LE methods (Aryal 2006; Sharma 2002)

Limit equilibrium methods	Developers or proposed by	Consideration of					
		Force equilibrium		Momentumum equilibrium	Failure surface		Inter-slice normal (E) and shear (T) forces
		x	y		Circular	Non-circular	
The ordinary method	Fellenius (1936)	No	No	Yes	Yes	No	None
Bishop's simplified	Bishop (1955)	Yes	No	Yes	Yes	Yes	E only
Janbu's simplified	Janbu (1954a, 1954b, 1973)	Yes	Yes	No	Yes	Yes	E only
Lowe and Karafiath's	Lowe and Karafiath (1960)	Yes	Yes	No	No	Yes	Both
Corps of Engineers	U.S. Army Corps of Engineers (1982)	Yes	Yes	No	No	Yes	Both
Spencer's	Spencer (1967, 1973)	Yes	Yes	Yes	Yes	Yes	Both
Bishop's rigorous	Bishop (1955)	Yes	Yes	Yes	Yes	No	Both
Janbu's generalized	Janbu, (1954a, 1954b, 1973)	Yes	Yes	No	Yes	Yes	Both
Sarma's	Sarma (1973)	Yes	Yes	Yes	Yes	Yes	Both
Morgenstern-price	Morgenstern and Price (1965)	Yes	Yes	Yes	Yes	Yes	Both

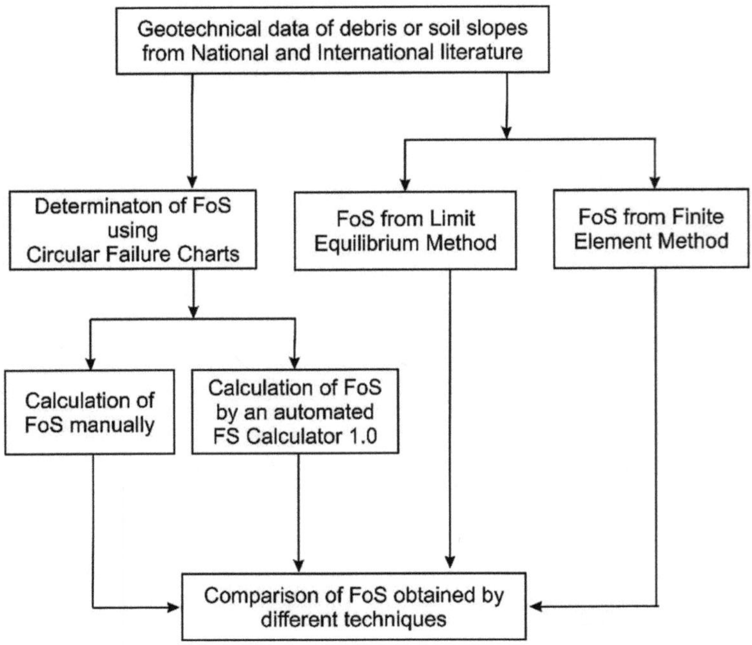

Fig. 27.1 Flowchart depicting the adopted methodology

equilibrium techniques due to its ability to compute displacements and deformation (Wyllie 2018). The primary benefit of using FEM over LEM is that the critical sliding surface is determined automatically without any prior assumption of the slide surface. Several attempts have been made to compare the outcomes of LEM and FEM (Ayob et al. 2019; Liu et al. 2015; Potgieter and Jacobsz 2019; Rawat and Gupta 2016) and it was found that translational as well as rotational equilibrium are self-satisfied by FEM. In this study, the outcomes of the above-discussed techniques are compared and brief methodology has been illustrated using flowchart (Fig. 27.1).

27.4 Results

The stability of slopes is governed by the geotechnical and geometrical parameters of the slope. FoS calculations were performed manually and using FS Calculator (Table 27.3 and Fig. 27.2). The geotechnical parameters of the slope mass have a more significant influence than the geometrical parameters. In slope engineering practices, shear strength parameters including, cohesion and friction are of prime importance. Shear strength is major resistive force in the slope movement. Cohesion is the ability of soil fragments to hold together. It has positive relation linear relationship with FoS. Soil or debris with substantial clay and silt fraction has greater cohesion.

Friction angle also plays an important role in governing the stability of the slope. The presence of water from discharge or groundwater reduces the shear strength of the soil mass. As higher friction angle resists the movement of the slope. However, the unit weight shows an inverse relation with the FoS. Apart from geotechnical parameters, geometrical parameter controls driving forces. Increasing slope angle and height contributes to the greater normal component of effective stresses acting on slope-forming material, thereby escalating driving forces. Using the above geotechnical and geometrical parameters, semi-empirical chart-based conventional technique is employed herein. The values of FoS obtained using the 'FS Calculator 1.0' are almost identical to FoS by performing manual calculations, reflecting that the automated tool provides correct results in fractions of seconds.

Among all considered case studies, twelve slopes, which include the determination of FoS by LEM were selected to compare FoS by LEM and FS Calculator 1.0 (Fig. 27.3). There exists a linear relationship $\text{FoS}(\text{CFC}_{\text{FS}}) = 2.7592 \times \text{FoS}(\text{LEM}) - 2.2032$ with a coefficient of regression 0.99, indicating that both corroborate well (Fig. 3a). However, SRF obtained from FEM follows the trend but with a relatively lesser coefficient of determination and slope. The relationship between FoS (CFC_{FS}) and SRF(FEM) is given by $\text{FoS}(\text{CFC}) = 0.889 \times \text{SRF}(\text{FEM}) + 0.0438$ with a coefficient of regression 0.63. It can be inferred that the results of CFCs correlate relatively better with LEM than FEM.

The FoS obtained by LEM, FEM and CFCs show a notable agreement with each other regarding trend only (Fig. 27.4). CFCs tend to yield a more conservative estimate of a Factor of Safety than FEM and LEM. In terms of consideration of parameters, FEM is the most exhaustive.

27.5 Discussion

The CFCs based 'FS Calculator 1.0' generates reliable results and validated with manual calculations. It also marginalized the calculation errors, which may rise due to tedious chart-based solutions. The sensitivity analysis feature provided by the 'FS Calculator 1.0' can be utilized to analyze the effect of different parameters on stability very quickly which can be very useful to design a stable slope.

The relationship between FoS obtained by using 'FS Calculator 1.0' and LEM gives coefficient of regression 0.99, which indicate a very strong relationship between the variables. As the Circular Failure Charts are also based on LEM, there is a strong relationship between these two. This relationship can be used to determine FoS by LEM, in case of unavailability of commercial LEM software like SLIDE (Rocscience). The coefficient of regression of the relationship between 'FS Calculator 1.0' and FEM is 0.63, which also shows a good relationship, but comparatively of lesser degree. The comparison of trend of CFCs and FEM based FoS highlights that CFCs provide relatively conservative and lower bound solutions. Therefore, it is recommended that these limitations of limit equilibrium approaches must be revisited for rectification and may be considered as future scope in CFCs. Meanwhile, proposed

Table 27.3 FoS determined by manual calculation and using FS calculator 1.0

Authors	Slope no./ slope ID	Cohesion (C in KN/m^2)	Friction angle (Φ in degrees)	Unit weight (γ in KN/m^3)	Height (H in meters)	Slope angle (θ in degrees)	Chart number	FoS	
								Manually	Using FS calculator 1.0
Chakraborty and Anbalagan (2008)	14	88.26	31.00	16.67	35.0	40	1, 2, 3	2.1, 1.9, 1.8	2.1, 1.9, 1.8
	15-A	25.52	41.00	16.67	18.0	50	1, 2, 3	1.7, 1.6, 1.4	1.7, 1.6, 1.4
Anbalagan and Parida (2013)	Section 1 AA	13.63	29.00	16.18	183.0	40	1, 2, 3	0.8, 0.6, 0.5	0.7, 0.6, 0.5
	Section 2	13.04	33.00	16.18	176.0	40	1, 2, 3	0.8, 0.8, 0.5	0.7, 0.7, 0.6
	Section 3	12.16	36.00	16.18	210.0	42	1, 2, 3	0.9, 0.6, 0.6	0.7, 0.6, 0.5
Anbalagan et al. (2008)	Section A-A	17.72	34.00	16.18	165.0	40	1, 2, 3	0.9, 0.8, 0.7	0.9, 0.8, 0.7
	Section BB	16.96	34.00	16.18	170.0	40	1, 2, 3	1, 0.8, 0.7	0.9, 0.8, 0.7
	Section CC	15.81	36.00	16.18	170.0	40	1, 2, 3	0.9, 0.8, 0.6	0.9, 0.8, 0.7
Dudeja et al. (2017)	Kyarkali AB	17.55	39.50	15.98	162.0	40	1, 2, 3	1, 0.9, 0.7	1, 0.9, 0.8
	Kyarkali CD	19.22	38.23	15.98	168.0	40	1, 2, 3	1.1, 0.9, 0.7	1, 0.9, 0.8
	Niste A AB	20.01	30.79	16.18	67.0	41	1, 2, 3	1, 0.9, 0.8	1, 0.9, 0.8
	Kandriyana AB	18.83	36.35	15.49	77.0	45	1, 2, 3	1, 1, 0.8	1, 1, 0.8
	Niste B AB	19.42	41.02	17.85	153.0	42	1, 2, 3	1.1, 0.9, 0.7	1, 0.9, 0.8
	Niste B CD	18.24	43.94	17.85	146.0	38	1, 2, 3	1.3, 1.1, 0.9	1.2, 1.1, 0.9
Kannan et al. (2011)	S1	20.00	34.00	15.88	4.25	75	1, 2, 3, 4, 5	1.8, 1.7, 1.7, 1.6, 1.6	1.8, 1.8, 1.7, 1.6, 1.6

(continued)

Table 27.3 (continued)

Authors	Slope no./slope ID	Cohesion (C in KN/m^2)	Friction angle (Φ in degrees)	Unit weight (γ in KN/m^3)	Height (H in meters)	Slope angle (θ in degrees)	Chart number	FoS	
								Manually	Using FS calculator 1.0
	S2	30.00	38.00	15.86	8.25	70	1, 2, 3, 4, 5	1.8, 1.7, 1.7, 1.6, 1.5	1.8, 1.8, 1.7, 1.6, 1.5
	S3	27.50	34.00	12.42	8.25	80	1, 2, 3, 4, 5	1.5, 1.5, 1.5, 1.4, 1.4	1.5, 1.5, 1.5, 1.4, 1.4
	S4	25.00	34.00	13.84	7.75	70	1, 2, 3, 4, 5	1.7, 1.7, 1.6, 1.5, 1.4	1.7, 1.7, 1.6, 1.5, 1.4
	S5	24.38	38.00	16.67	10.5	65	1, 2, 3, 4, 5	1.5, 1.4, 1.4, 1.3, 1.2	1.5, 1.4, 1.4, 1.3, 1.2
	S6	23.75	36.00	16.22	12.25	65	1, 2, 3, 4, 5	1.3, 1.3, 1.2, 1.2, 1	1.3, 1.3, 1.2, 1.2, 1.1
	S7	33.75	36.00	16.49	28.0	65	1, 2, 3, 4, 5	1, 1, 1, 0.9, 0.8	1.1, 1, 1, 0.9, 0.8
	S8	22.50	39.00	18.17	16.0	68	1, 2, 3, 4, 5	1.1, 1.1, 1, 0.9, 0.8	1.1, 1.1, 1, 0.9, 0.8
	S9	32.92	38.00	18.17	35.0	64	1, 2, 3, 4, 5	0.9, 0.9, 0.9, 0.8, 0.7	1, 1, 0.9, 0.8, 0.7
	S10	20.42	38.00	18.17	17.0	78	1, 2, 3, 4, 5	0.8, 0.7, 0.7, 0.7, 0.5	0.8, 0.8, 0.8, 0.7, 0.5
	S11	23.33	38.00	18.17	15.5	69	1, 2, 3, 4, 5	1.1, 1.1, 1, 0.9, 0.8	1.1, 1.1, 1, 0.9, 0.8
	S12	10.62	42.00	18.17	12.0	68	1, 2, 3, 4, 5	1, 0.9, 0.9, 0.8, 0.6	1, 0.9, 0.9, 0.8, 0.6

(continued)

Table 27.3 (continued)

Authors	Slope no./slope ID	Cohesion (C in KN/m^2)	Friction angle (Φ in degrees)	Unit weight (γ in KN/m^3)	Height (H in meters)	Slope angle (θ in degrees)	Chart number	FoS	
								Manually	Using FS calculator 1.0
Singh and Kumar (2021)	L1S1	16.00	30.00	18.00	8.0	70	1, 2, 3	1, 1, 1	1, 1, 1
	L2S2	20.00	30.00	18.50	19.0	60	1, 2, 3	0.9, 0.9, 0.8	0.9, 0.9, 0.8
	L4S3	10.78	32.00	18.00	14.5	50	1, 2, 3	1, 1, 0.9	1, 1, 0.9
	L7S4	22.00	30.00	18.00	12.0	80	1, 2, 3	0.8, 0.8, 0.8	0.8, 0.8, 0.8
	L8S5	20.00	35.00	18.00	12.0	80	1, 2, 3	0.9, 0.8, 0.8	0.9, 0.8, 0.8
	L11S6	10.00	31.00	18.94	11.0	50	1, 2, 3	1.1, 1, 0.9	1.1, 1, 0.9
	L15S7	11.76	35.00	18.87	178.0	40	1, 2, 3	1, 0.7, 0.6	0.7, 0.6, 0.5
	L17S8	40.00	26.00	18.00	20.0	80	1, 2, 3	0.8, 0.8, 0.8	0.8, 0.8, 0.8
	L20S9	14.00	35.00	18.88	20.0	50	1, 2, 3	1.1, 1, 0.9	1.1, 1, 0.9
	L21S10	15.00	30.00	18.18	12.2	50	1, 2, 3	1.2, 1.1, 1	1.2, 1.1, 1
	L22S11	20.00	32.00	18.88	11.0	50	1, 2, 3	1.4, 1.4, 1.3	1.4, 1.4, 1.3
	L23S12	22.00	31.00	18.98	15.0	60	1, 2, 3	1.1, 1.1, 1	1.1, 1.1, 1
	L28S13	13.00	31.00	18.00	29.8	50	1, 2, 3	0.9, 0.8, 0.7	0.9, 0.8, 0.7
Luden et al. (2021)	IL 1	20.20	20.93	17.71	12.5	50	2	1.0	1.0
	OL 1	22.16	22.45	19.03	10.0	45	2	1.3	1.3
Dinata et al. (2021)	Inlet	350.0	40.00	24.77	12.0	45	1	7.8	7.5
	Outlet	400.0	45.00	25.60	12.0	45	1	8.8	8.6
Siddque and Pradhan (2018)	L1	24.05	38.40	19.20	42.0	58	1	1.0	0.9
	L2	21.04	33.20	20.40	19.0	56	1	1.1	1.0

(continued)

Table 27.3 (continued)

Authors	Slope no / slope ID	Cohesion (C in KN/m^2)	Friction angle (Φ in degrees)	Unit weight (γ in KN/m^3)	Height (H in meters)	Slope angle (θ in degrees)	Chart number	FoS	
								Manually	Using FS calculator 1.0
	L3	26.42	34.80	24.30	20.0	69	1	0.8	0.8
	L4	24.30	38.80	19.70	40.0	65	1	0.7	0.8
	L5	21.22	38.90	22.80	20.0	68	1	0.9	0.9
	L6	26.25	37.90	24.00	26.0	65	1	0.9	0.9
	L7	28.97	38.10	21.30	45.0	61	1	0.9	0.9
	L8	25.28	37.00	18.90	18.0	67	1	1.1	1.4

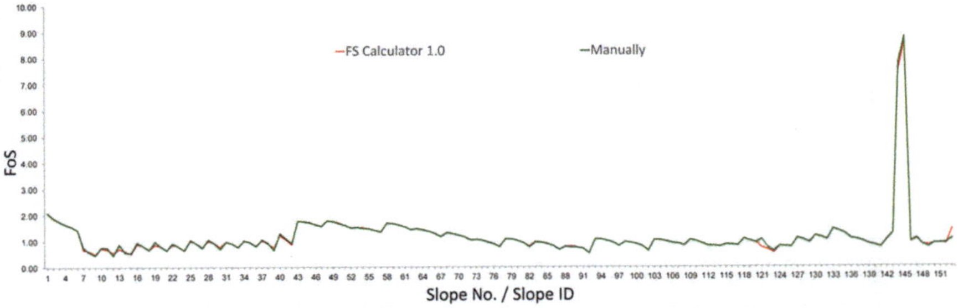

Fig. 27.2 FoS obtained by manual calculations and using FS calculator 1.0

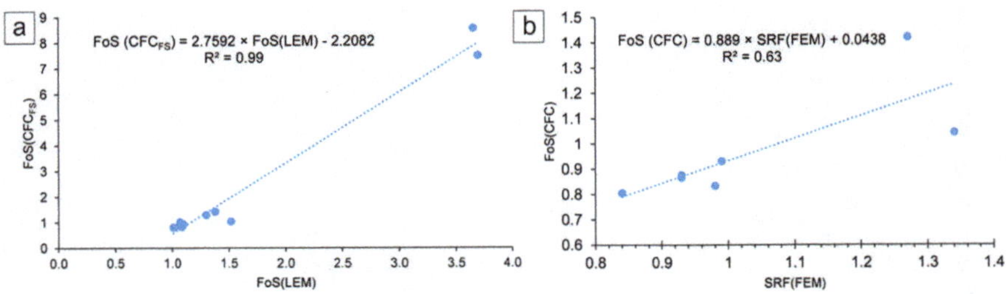

Fig. 27.3 Bivariate graphs depicting the linear relationship among FoS obtained by different approaches: **a** LEM and FS calculator 1.0; **b** CFC and FEM

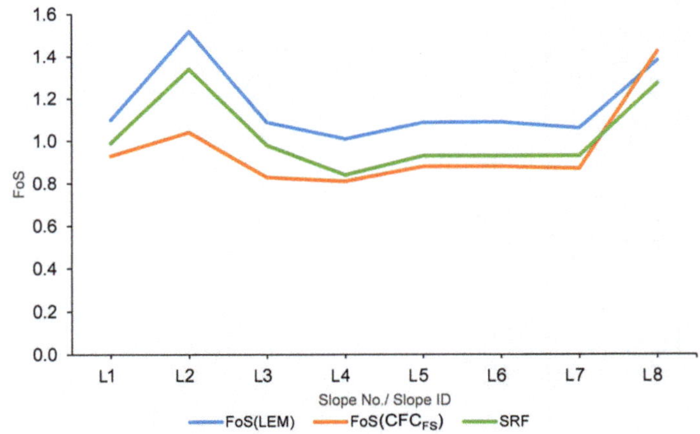

Fig. 27.4 Graphical representation of FoS by LE, FE and CFCs methods

correlations and likewise relationships may be utilized as correction factor to the FoS obtained by different limit equilibrium approaches.

Earlier comparative assessment studies highlighting safety factor from LEM and FEM by by several researchers including Hammah et al. (2004), Koca and Koca

(2020), Adolphe et al. (2022), Mebrahtu et al. (2022) concluded that both shows good agreement with each other and FEM yields slightly smaller FoS values than LEM. The present analysis also confirms that FoS by FEM is marginally smaller than LEM.

27.6 Conclusions

Deterministic analysis using FoS values is an integral part of slope engineering practices. CFCs, LEM and FEM are widely available techniques, but some disparities exist among these methods. FoS obtained by an automated tool FS Calculator 1.0 correlates well with manual calculations. Apart from deterministic analysis, FS Calculator 1.0 is an effective tool for sensitivity analysis. This tool can be very beneficial to geotechnical engineers, as it provides a quick evaluation of the stability of slopes in terms of FoS and also capable in determining the most influencing geomechanical or geometrical parameter. The comparative analysis provided a significant insight to assess the outcomes and geotechnical judgment of slopes. FoS by LEM and CFCs correlates relatively better ($R^2 = 0.99$) than LEM and FEM ($R^2 = 0.63$). Such conclusions were derived from earlier literature also and conforms the relationships between applied proxies. Among all three techniques, FEM considers more geomechanical inputs, making it the most reliable approach in deterministic analysis Proposed correlations may be utilized as a correction factor in FoS obtained by CFCs and LEM. Since, LEM provide higher FoS values than FEM, it is emphasized that the results of limit equilibrium may be corrected using proposed relationship between LEM and FEM. This research significantly enhances the comprehension of FoS using different techniques, with particular relevance to the Himalayan region. The major limitation of this study is that the limited dataset was used for correlation among different proxies. Furthermore, there are only five charts of discrete nature for groundwater conditions, whereas intermittent situations are also possible in the real field conditions. So, there is a scope of modification in CFCs by adding some other parameters to enhance the correlation between FoS by using CFCs and FEM.

Acknowledgements The authors express their deepest gratitude to the Chairperson, Department of Geology, Aligarh Muslim University, Aligarh, for providing the laboratory facilities.

References

Aalam U, Siddique T, Pradhan SP (2023) FS Calculator: a windows-based application for factor of safety assessment and sensitivity analyses by circular failure charts. Earth Sci Inf 16:1917–1928

Adolphe K, Noel WN, Harmel OO, Gampio MU, Goncalves AO, Lacaba RG, Florent B (2022) Comparative study of slope soils analysis by using limit equilibrium and finite element methods. Int J Geosci 13(12):1089–1102

Ahmad F, Samui P, Mishra SS (2023) Probabilistic analysis of slope using bishop method of slices with the help of subset simulation subsequently aided with hybrid machine learning paradigm. Indian Geotech J 1–21. https://doi.org/10.1007/s40098-023-00796-3

Akhtar W, Siddique T, Haris PM, Ahamad A, Mondal MEA (2023) An integrated geotechnical study using Q-slope method and factor of safety appraisal along NH-5 from Solan to Shimla, India. J Earth Syst Sci 132(3):106

Anbalagan R, Parida S (2013) Geoenvironmental problems due to harmony landslide in Garhwal Himalaya, Uttarakhand, India. Int J Emerg Technol Adv Eng 3(3):553–559

Anbalagan R, Kohli A, Chakraborty D (2008) Geotechnical evaluation of Harmony landslide on Karnaprayag-Gwaldam road, Uttarakhand Himalaya. Curr Sci 94(12):1613–1619

Ansari T, Kainthola A, Singh KH, Singh TN, Sazid M (2021) Geotechnical and micro-structural characteristics of phyllite derived soil; implications for slope stability, Lesser Himalaya, Uttarakhand, India. CATENA 196:104906

Aryal KP (2006) Slope stability evaluations by limit equilibrium and finite element methods. PhD Thesis. Norwegian University of Science and Technology, Trondheim, Torgarden, Norway, 146p

Ayob M, Kasa A, Sulaiman MS, Miniandi ND, Yusoff AH (2019) Slope stability evaluations using limit equilibrium and finite element methods. Int J Adv Sci Technol 28(18):27–43

Bishop AW (1955) The use of the slip circle in the stability analysis of slopes. Geotechnique 5(1):7–17

Chakraborty D, Anbalagan R (2008) Landslide hazard evaluation of road cut slopes along Uttarkashi-Bhatwari road, Uttaranchal Himalaya. J Geol Soc India 71(1):115–124

Chugh AK (1986) Variable interslice force inclination in slope stability analysis. Soils Found 26(1):115–121

Corps of Engineers (1982) Slope stability manual EM-1110-2-1902. Department of the Army, Washington, DC, Office of the Chief of Engineers

Cruden DM, Varnes DJ (1996) Landslides: investigation and mitigation. Chapter 3-Landslide types and processes. Transp Res Board Spec Report 36–75

Dahal RK, Hasegawa S, Yamanaka M, Dhakal S, Bhandary NP, Yatabe R (2009) Comparative analysis of contributing parameters for rainfall-triggered landslides in the Lesser Himalaya of Nepal. Environ Geol 58:567–586

Das S, Pandit K, Sarkar S, Kanungo DP (2023) Stability and hazard assessment of the progressive zero landslide in the Kalimpong region of Darjeeling Himalaya, India. Geotech Geol Eng 1–17. https://doi.org/10.1007/s10706-023-02641-z

Dinata EU, Indrawan IGB, Idrus A (2021) Slope stability analysis of the drainage tunnel portal in Tanju Dam, Dompu District, West Nusa Tenggara. E3S Web Conf 325:01016

Dudeja D, Bhatt SP, Biyani AK (2017) Stability assessment of slide zones in Lesser Himalayan part of Yamunotri pilgrimage route, Uttarakhand, India. Environ Earth Sci 76:54

Fellenius W (1936) Calculation of the stability of earth dams. Proc Second Congr Large Dams 4:445–463

Gover S, Hammah R (2013) A comparison of finite elements (SSR) and limit-equilibrium slope stability analysis by case study: geotechnical engineering. Civ Eng 2013(3):31–34

Gupta V, Bhasin RK, Kaynia AM, Kumar V, Saini AS, Tandon RS, Pabst T (2016) Finite element analysis of failed slope by shear strength reduction technique: a case study for Surabhi Resort Landslide, Mussoorie township, Garhwal Himalaya. Geomat Nat Haz Risk 7(5):1677–1690

Hammah RE, Curran JH, Yacoub TE, Corkum B (2004) Stability analysis of rock slopes using the finite element method. In: Proceedings of the ISRM regional symposium EUROCK 2004 & 53rd Geomechanics Colloquium. Schubert

Hammah RE, Yacoub TE, Corkum B, Curran JH (2005) A comparison of finite element slope stability analysis with conventional limit-equilibrium investigation. In: Proceedings of the 58th Canadian geotechnical and 6th joint IN-CNC and CGS groundwater specialty conferences

Hoek E, Bray JW (1981) Rock slope engineering. Revised 3rd edition. The Institute of Mining And Metallurgy, London, pp 226–256

Janbu N (1954) Stability analysis of slopes with dimensionless parameters. PhD thesis Harvard University Soil Mechanics Series, No. 46, Harvard University, Cambridge, Massachusetts

Janbu N (1955) Application of composite slip surfaces for stability analysis. Eur Conf Stab Earth Slopes 1955(3):43–49

Janbu N (1973) Slope stability computations. In: Hirschfeld RC, Poulos SJ (eds) Textbook. Embankment: Dam Engineering. John Wiley and Sons Inc, New York, pp 47–86

Kainthola A, Singh PK, Wasnik AB, Sazid M, Singh TN (2012) Finite element analysis of road cut slopes using Hoek and Brown failure criterion. Int J Earth Sci Eng 5(5):1100–1109

Kannan M, Saranathan E, Anbalagan R (2011) Macro landslide hazard zonation mapping-case study from Bodi-Bodimettu Ghats Section, Theni District, Tamil Nadu-India. J Indian Soc Rem Sens 39:485–496

Khali H, Rawat KS, Chatterjee RS (2023) Dynamic association of slope movements in the Uttarakhand Himalaya: a critical review on the landslide susceptibility assessment. Geomat Nat Haz Risk 14(1):2273214

Koca KT, Koca MY (2020) Comparative analyses of finite element and limit-equilibrium methods for heavily fractured rock slopes. J Earth Syst Sci 129:49

Koley B, Nath A, Saraswati S, Chatterjee U, Bandyopadhyay K, Bhatta B, Ray BC (2022) Assessment of spatial distribution of rain-induced and earthquake-triggered landslides using geospatial techniques along North Sikkim Road Corridor in Sikkim Himalayas, India. GeoJournal 1–39. https://doi.org/10.1007/s10708-022-10585-9

Komadja GC, Pradhan SP, Oluwasegun AD, Roul AR, Stanislas TT, Laïbi RA, Adebayo B, Onwualu AP (2021) Geotechnical and geological investigation of slope stability of a section of road cut debris-slopes along NH-7, Uttarakhand, India. Results Eng 10:100227

Krahn J (2003) The limits of limit equilibrium analyses. Can Geotech J 40(3):643–660

Kumar A, Asthana AKL, Priyanka RS, Jayangondaperumal R, Gupta AK, Bhakuni SS (2017) Assessment of landslide hazards induced by extreme rainfall event in Jammu and Kashmir Himalaya, northwest India. Geomorphology 284:72–87

Liu SY, Shao LT, Li HJ (2015) Slope stability analysis using the limit equilibrium method and two finite element methods. Comput Geotech 63:291–298

Lowe J (1960) Stability of earth dams upon drawdown. In: Proceedings of 1st Pan American Conference on soil mechanics and foundation engineering, Mexico

Luden AS, Indrawan IGB, Karnawati D (2021) Slope stability analyses by circular failure chart and limit equilibrium methods: the inlet and outlet of diversion tunnel of Bolango Ulu Dam, Indonesia. E3S Web Conf 325:01015

Matsui T, San KC (1992) Finite element slope stability analysis by shear strength reduction technique. Soils Found 32(1):59–70

Mebrahtu TK, Heinze T, Wohnlich S, Alber M (2022) Slope stability analysis of deep-seated landslides using limit equilibrium and finite element methods in Debre Sina area, Ethiopia. Bull Eng Geol Env 81(10):403

Morgenstern NU, Price VE (1965) The analysis of the stability of general slip surfaces. Geotechnique 15(1):79–93

Panda SD, Kumar S, Pradhan SP, Singh J, Kralia A, Thakur M (2023) Effect of groundwater table fluctuation on slope instability: a comprehensive 3D simulation approach for Kotropi landslide, India. Landslides 20(3):663–682

Pantelidis L, Griffiths DV (2013) Stability of earth slopes. Part I: two-dimensional analysis in closed-form. Int J Numer Anal Meth Geomech 37(13):969–1986

Petley DN, Hearn GJ, Hart A, Rosser NJ, Dunning SA, Oven K, Mitchell WA (2007) Trends in landslide occurrence in Nepal. Nat Hazards 43:23–44

Potgieter JT, Jacobsz SW (2019) Comparing the factors of safety from finite element and limit equilibrium analyses in lateral support design. J South Afr Inst Civil Eng 61(4):29–41

Pradhan SP, Panda SD, Roul AR, Thakur M (2019) Insights into the recent Kotropi landslide of August 2017, India: a geological investigation and slope stability analysis. Landslides 16:1529–1537

Pradhan SP, Vishal V, Singh TN (2018) Finite element modelling of landslide prone slopes around Rudraprayag and Agastyamuni in Uttarakhand Himalayan terrain. Nat Hazards 94:181–200

Rahman AU, Khan AN, Collins AE, Qazi F (2011) Causes and extent of environmental impacts of landslide hazard in the Himalayan region: a case study of Murree, Pakistan. Nat Hazards 57:413–434

Rao SK (2020) Characterization, modelling and engineering of rocks and rockmasses. Indian Geotech J 50:1–95

Rawat S, Gupta AK (2016) Analysis of a nailed soil slope using limit equilibrium and finite element methods. Int J Geosynthet Ground Eng 2:1–23

Roul AR, Pradhan SP, Mohanty DP (2021) Investigation to slope instability along railway cut slopes in Eastern Ghats mountain range, India: a comparative study based on slope mass rating, finite element modelling and probabilistic methods. J Earth Syst Sci 130:1–25

Sarma SK (1973) Stability analysis of embankments and slopes. Geotechnique 23(3):423–433

Sharma LK, Umrao RK, Singh R, Ahmad M, Singh TN (2017) Stability investigation of hill cut soil slopes along National highway 222 at Malshej Ghat, Maharashtra. J Geol Soc India 89:165–174

Sharma S (2002) Slope stability concepts. In: Abramson LW, Lee TS, Sharma S, Boyce GM (eds) Slope stability and stabilization methods (2nd ed). John Wiley and Sons Inc, pp 329–461

Siddique T, Haris PM, Pradhan SP (2022) Unraveling the geological and meteorological interplay during the 2021 Chamoli disaster, India. Nat Hazards Res 2(2):75–83

Siddque T, Pradhan SP (2018) Stability and sensitivity analysis of Himalayan road cut debris slopes: an investigation along NH-58, India. Nat Hazards 93:577–600

Singh K, Kumar V (2021) Slope stability analysis of landslide zones in the part of Himalaya, Chamba, Himachal Pradesh, India. Environ Earth Sci 80:332

Singh PK, Wasnik AB, Kainthola A, Sazid M, Singh TN (2013) The stability of road cut cliff face along SH-121: a case study. Nat Hazards 68:497–507

Solanki A, Gupta V, Bhakuni SS, Ram P, Joshi M (2019) Geological and geotechnical characterisation of the Khotila landslide in the Dharchula region, NE Kumaun Himalaya. J Earth Syst Sci 128:1–14

Spencer E (1967) A method of analysis of the stability of embankments assuming parallel interslice forces. Geotechnique 17(1):11–26

Spencer E (1973) Thrust line criterion in embankment stability analysis. Geotechnique 23(1):85–100

Thakur M, Kumar N, Dhiman RK, Malik JN (2023) Geological and geotechnical investigations of the Sataun landslide along the Active Sirmauri Tal Fault, Sataun, Northwestern Himalaya, India. Landslides 20(5):1045–1063

Wan S, Lei T, Chou T (2010) A novel data mining technique of analysis and classification for landslide problems. Nat Hazards 52:211–230

Wyllie DC (2018) Rock slope engineering: civil applications, 5th edn. CRC Press, 621p

MIX
Papier aus verantwortungsvollen Quellen
Paper from responsible sources
FSC® C105338

If you have any concerns about our products,
you can contact us on
ProductSafety@springernature.com

In case Publisher is established outside the EU,
the EU authorized representative is:
Springer Nature Customer Service Center GmbH
Europaplatz 3, 69115 Heidelberg, Germany

Printed by Libri Plureos GmbH
in Hamburg, Germany